BIOLOGY
FOR A
CHANGING
WORLD

Michèle Shuster
New Mexico State University

Janet Vigna
Grand Valley State University

Gunjan Sinha

Matthew Tontonoz

W. H. Freeman and Company • New York

Publisher Kate Ahr Parker

Senior Acquisitions Editor Marc Mazzoni

Developmental Editors Andrea Gawrylewski and Susan Weisberg

Associate Director of Marketing Debbie Clare

Managing Editor for First Editions Elaine Palucki, Ph.D.

Senior Media Editor Patrick Shriner

Supplements Editor Amanda Dunning

Assistant Editor Anna Bristow

Managing Editor Philip McCaffrey

Project Editors Leigh Renhard and Dana Kasowitz

Art Director Diana Blume

Text Designers Matthew Ball and Diana Blume

Senior Illustration Coordinator Bill Page

Artwork Precision Graphics

Photo Editors Christine Buese and Ted Szczepanski

Photo Researcher Elyse Rieder

Production Manager Ellen Cash

Composition MPS Limited, a Macmillan Company

Printing and Binding Quad Graphics–Versailles

Library of Congress Control Number: 2011930178

ISBN-13: 978-1-4292-9702-8
ISBN-10: 1-4292-9702-6

Printed in the United States of America

Second printing

W. H. Freeman and Company
41 Madison Avenue
New York, NY 10010
Houndmills, Basingstoke RG21 6XS, England
www.whfreeman.com

To our teachers and students: You are our inspiration

About the Authors

Michèle Shuster, Ph.D., is an assistant professor in the biology department at New Mexico State University in Las Cruces, New Mexico. She focuses on the scholarship of teaching and learning, studying introductory biology, microbiology, and cancer biology classes at the undergraduate level, as well as working on several K-12 science education programs. Michèle is an active participant in programs that provide mentoring in scientific teaching to postdoctoral fellows, preparing the next generation of undergraduate educators. She is the recipient of numerous teaching awards, including a Donald C. Roush Excellence in Teaching Award at NMSU. Michèle received her Ph.D. from the Sackler School of Graduate Biomedical Sciences at Tufts University School of Medicine, where she studied meiotic chromosome segregation in yeast.

Janet Vigna, Ph.D., is an associate professor in the biology department at Grand Valley State University in Allendale, Michigan. As a member of the Integrated Science Program, she teaches courses in genetics and science education for preservice teachers, and is active in a variety of K-12 science education programs. She has been teaching university-level biology for 14 years, with a special focus on effectively teaching biology to nonmajors. Her current research focuses on the environmental effects of the biological pesticide *Bacillus thuringiensis israelensis* on natural frog communities. She received her Ph.D. in microbiology from the University of Iowa.

Gunjan Sinha is a freelance science journalist who writes regularly for *Scientific American, Science,* and *Nature Medicine.* Her article on the biochemistry of love, "You Dirty Vole," was published in *The Best American Science Writing 2003.* She holds a graduate degree in molecular genetics from the University of Glasgow, Scotland, and currently lives in Berlin, Germany.

Matthew Tontonoz has been a developmental editor for textbooks in introductory biology, cell biology, evolution, and environmental science. He received his B.A. in biology from Wesleyan University, where he did research on the neurobiology of birdsong, and his M.A. in the history and sociology of science from the University of Pennsylvania, where he studied the history of the behavioral and life sciences. His writing has appeared in *Science as Culture.* He lives in Brooklyn, New York.

About the Publishers

ALL OF US INVOLVED in science education understand the importance of scientific literacy. How do we get the attention of a nonscientist? And if we can get it, how do we keep it—not only for the duration of the course, or the chapter in a textbook, but beyond?

How do we convey in our courses and our textbooks not just what we know but also how science is done? These are the challenges we hope to address with our new series of textbooks specifically for the nonscientist.

With this series, W. H. Freeman and *Scientific American* join forces not just to engage nonscientists but also to equip them with critical life tools.

DISTINGUISHED by a discerning editorial vision and a long-standing commitment to superior quality, W. H. Freeman works closely with top researchers and educators to develop superior teaching and learning materials in the sciences. We know that a dedicated instructor and the right textbook have the power to change the world—one student at a time.

SCIENTIFIC AMERICAN

COMMITTED to bringing first-hand developments in modern science to its audience, *Scientific American* has long been the world's leading source for science and technology information, featuring more articles by Nobel laureates than any other consumer magazine. The oldest continuously published magazine in the United States, *Scientific American* has been independently ranked among the top 10 U.S. consumer media outlets as "Most Credible" and "Most Objective."*

*Erdos & Morgan 2008-2009 Opinion Leader Survey

From the Authors

The development of this book has taken us all on an extremely long and winding road, on which we have met fascinating people and had incredible experiences. The authors would like to thank Elizabeth Widdicombe, Kate Parker, and the folks at W. H. Freeman and Company and *Scientific American* for supporting this vision for biology education. They recognized our diverse strengths and brought us together to make this vision a reality. We have learned so much from one another on this challenging and rewarding professional journey, and none of us has likely worked so hard and so passionately on a project as we all have on this one.

We would like to thank all of the people who were interviewed and generously contributed information for these chapters. Their stories are central to the impact that this book will have on the students we teach. They are authentic examples of biology in a changing world, and they bring this book to life.

A special thank you is required for our Senior Acquisitions Editor, Marc Mazzoni, for his unwavering encouragement and ability to bring stable direction and support to the project. Developmental Editors Andrea Gawrylewski and Susan Weisberg and Assistant Editor Anna Bristow have spent many hours in the pages of this book, editing the details, managing our chaos, and smoothing our rough edges. We thank them for their dedication, patience, experience, and expertise. Thanks go to Patrick Shriner and Amanda Dunning for their tireless work on our media and supplements program. And we must thank Elaine Palucki, who has been with us from the very beginning, bringing enthusiasm and a fresh voice to our discussions. Elaine has recruited an outstanding pool of reviewers for this project, to whom we owe a debt of gratitude.

Many thanks to the production team, Leigh Renhard, Dana Kasowitz, Philip McCaffrey, Nancy Brooks, Matthew Ball, Diana Blume, Bill Page, Christine Buese, Ted Szczepanski, Elyse Rieder, Ellen Cash, and all the people behind the scenes at W. H. Freeman for translating our ideas into a beautiful, cohesive product. We would like to thank Rachel Rogge and Jan Troutt at Precision Graphics for their outstanding work on the Infographics. We appreciate their patience with the many edits and quick timelines throughout the project. They do amazing work.

We'd like to thank Debbie Clare for her enthusiasm and hard work in promoting this book in the biology education community. We thank the enthusiastic group of salespeople who connect with biology educators across the country and do a wonderful job representing this book.

The authors would like to thank our families and friends who have been close to us during this process. They have been our consultants, served as sounding boards about challenges, celebrated our successes, shared our passions, and supported the extended time and energy we often diverted away from them to this project. We are grateful for their patience and unending support.

And finally, a sincere thank you to our many teachers, mentors, and students over the years who have shaped our views of biology and the world, and how best to teach about one in the context of the other. You are our inspiration.

Brief Contents

Contents

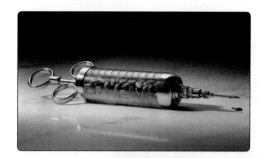

5. Energy Flow and Photosynthesis 81

Mighty Microbes Can scientists make algae into the next global fuel source? 82

6. Dietary Energy and Cellular Respiration 101

Supersize Me? Changing our culture of eating 102

7. DNA Structure and Replication 121

Biologically Unique How DNA helped free an innocent man 122

Milestones in Biology The Model Makers 137
Watson, Crick, and the structure of DNA 138

8. Genes to Proteins 143

Medicine from Milk Scientists genetically modify animals to make medicine 144

9. Cell Division and Mitosis   169

10. Mutations and Cancer   187

11. Single-Gene Inheritance and Meiosis   203

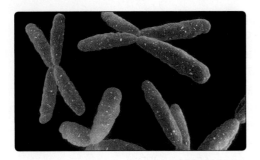

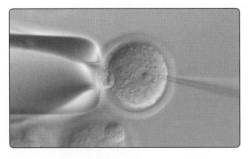

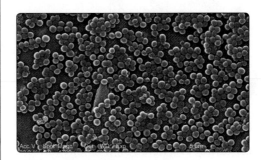

23. Ecosystem Ecology 449

The Heat Is On From migrating maples to shrinking sea ice, signs of a warming planet 450

24. Sustainability 471

Eco-Metropolis Designing the city of the future 472

25. Overview of Physiology 493

Man versus Mountain Physiology explains a 1996 disaster on Everest 494

30. Plant Physiology 607

Q & A: Plants Plants have evolved a unique set of solutions to nature's challenges 608

Biology for a Changing World at a Glance

Biology for a Changing World was written by a team of two full-time college biology instructors and two science writers, with extensive input from nonmajors biology instructors across the United States and Canada. The authors identified newsworthy stories to convey key concepts, then gathered feedback from instructors to ensure that these stories are relevant, useful, and (most important) interesting to students.

Key Features

- Engaging stories carry students through each chapter, demonstrating how biology relates to their daily lives.

- Magazine-style design balances words and images while providing students with the learning tools they need.

> "This format does exactly what I would like to do—it takes a real-life example as an application of the material and uses the example in the form of an unfolding story to *both* teach the material to the student and at the same time demonstrate why and how the material is important to society and the student today."
> —Chris Haynes, Shelton State Community College

Chapter 8 Genes to Proteins

Medicine from milk

Scientists genetically modify animals to make medicine

In a Massachusetts barn nestled among willow and oak trees, rows of juglike machines drone in a constant hum. Goats, dozens of them, are being milked. But this is no ordinary dairy operation. This farm is among several worldwide practicing the art of "pharming"—using genetically modified animals to churn out therapeutic drugs.

The first drug produced from such **transgenic** animals is already available, manufactured by GTC Biotherapeutics, a firm based in Framingham, Massachusetts. The drug consists of a human protein called antithrombin that was extracted from transgenic goats' milk. Antithrombin is most commonly used to treat patients who either inherit or acquire a deficiency of the antithrombin protein, which puts them at risk of developing dangerous blood clots.

For decades, scientists had extracted antithrombin from human blood donations. But blood contains only small amounts of antithrombin, and the supply depends on the number of blood donors. Transgenic goats, however, can produce massive amounts of the drug in a relatively short period of time. Moreover, relying on a herd of goats instead of human volunteers ensures a consistent supply. And because the animals live in a controlled environment, there is less risk of transmitting infections such as HIV and hepatitis to healthy people through contaminated donor blood.

Because of all these advantages, some people are predicting that transgenic animals may one day replace human donors as the source for therapeutic agents extracted from blood. "This is very exciting, it is novel and has great potential for where we can go with this new technology," Bernadette Dunham, director of the FDA's Center for Veterinary Medicine, told

"This is very exciting, it is novel and has great potential for where we can go with this new technology." —Bernadette Dunham

TRANSGENIC
Refers to an organism that carries one or more genes from a different species.

PROTEIN
A macromolecule made up of repeating subunits known as amino acids, which determine the shape and function of a protein. Proteins play many critical roles in living organisms.

AMINO ACIDS
The building blocks of proteins. There are 20 different amino acids.

the *Washington Post* in February 2009, when the company's drug for antithrombin deficiency was approved for market.

Antithrombin: From Gene to Protein
Antithrombin is a **protein**. Recall from Chapter 2 that proteins are one of the four main macromolecules that make up cells. Proteins have myriad functions in the body: they allow our muscles to contract, give our hair and skin its texture, and facilitate the thousands of chemical reactions that occur in our cells. In fact, proteins play a huge role in all basic cellular functions. Proteins can perform such a variety of different tasks because they come in many shapes and sizes.

All proteins are made of the same building blocks called **amino acids**. There are 20 different amino acids in all. All amino acids have the same basic core structure, but each also has a unique chemical side group that distinguishes the amino acids from one another. Amino acids bond together to form linear chains. The human antithrombin protein is a chain of 432 amino acids. Many human proteins are in this size range, but chain lengths vary from just a few to thousands of amino acids. The longest human protein, titin, is a single chain of 34,350 amino acids.

The sequence of amino acids in any given chain makes each chain unique, and also determines how that chain ultimately folds into a

SERGEITAIHAKNIGHT/DREAMSTIME.COM

What Is DNA and Where Is It Found?

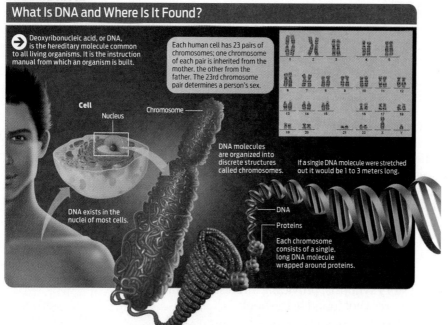

Deoxyribonucleic acid, or DNA, is the hereditary molecule common to all living organisms. It is the instruction manual from which an organism is built.

Each human cell has 23 pairs of chromosomes; one chromosome of each pair is inherited from the mother, the other from the father. The 23rd chromosome pair determines a person's sex.

Cell

Nucleus

Chromosome

DNA molecules are organized into discrete structures called chromosomes.

If a single DNA molecule were stretched out it would be 1 to 3 meters long.

DNA exists in the nuclei of most cells.

DNA

Proteins

Each chromosome consists of a single, long DNA molecule wrapped around proteins.

"The graphics are head and shoulders above anything in other texts that target a nonscience audience."
—Mark Bucheim, University of Tulsa

- **Infographics** *Scientific American*-style illustrations teach core biological concepts by combining easy-to-follow images with straightforward explanations. Each graphic provides a complete picture of a fundamental scientific principle.

- End-of-chapter pedagogy provides a point-by-point review of each chapter's key concepts:

  - **Test Your Knowledge** self-tests are aligned with each chapter's key concepts
  - **Know It** questions assess general comprehension
  - **Use It** questions assess whether students can apply what they've learned

"This is a great way to reinforce the 'science of the story.' The Know It and Use It segments. reinforce scientific information and allow the student to apply concepts to everyday situations."
—Pamela Anderson Cole, Shelton State Community College

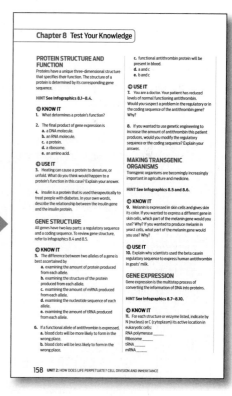

chromosome in the nucleus where it can be used again in transcription.

to the ribosome, it to the growing chain (Infographic

he human genome y thousands of dif- ins, each one is er from a starting s. In the same way abet can spell hun- ls, the basic set of reds of thousands

TRANSFER RNA (tRNA)
A type of RNA that helps ribosomes assemble chains of amino acids during translation.

ANTICODON
The part of a tRNA molecule that binds to a complementary mRNA codon.

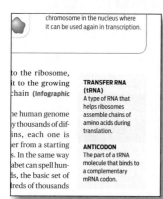

- Running Glossary provides immediate, concise definitions for key terms.

Chapter 8 Test Your Knowledge

PROTEIN STRUCTURE AND FUNCTION
Proteins have a unique three-dimensional structure that specifies their function. The structure of a protein is determined by its corresponding gene sequence.

HINT See Infographics 8.1–8.4.

KNOW IT
1. What determines a protein's function?

2. The final product of gene expression is
   a. a DNA molecule.
   b. an RNA molecule.
   c. a protein.
   d. a ribosome.
   e. an amino acid.

USE IT
3. Heating can cause a protein to denature, or unfold. What do you think would happen to a protein's function in this case? Explain your answer.

4. Insulin is a protein that is used therapeutically to treat people with diabetes. In your own words, describe the relationship between the insulin gene and the insulin protein.

GENE STRUCTURE
All genes have two key parts: a regulatory sequence and a coding sequence. To review gene structure, refer to Infographics 8.4 and 8.5.

KNOW IT
5. The difference between two alleles of a gene is best ascertained by
   a. examining the amount of protein produced from each allele.
   b. examining the structure of the protein produced from each allele.
   c. examining the amount of mRNA produced from each allele.
   d. examining the nucleotide sequence of each allele.
   e. examining the amount of tRNA produced from each allele.

6. If a functional allele of antithrombin is expressed,
   a. blood clots will be more likely to form in the wrong place.
   b. blood clots will be less likely to form in the wrong place.

c. functional antithrombin protein will be present in blood.
   d. a and c
   e. b and c

USE IT
7. You are a doctor. Your patient has reduced levels of normal functioning antithrombin. Would you suspect a problem in the regulatory or in the coding sequence of the antithrombin gene?

8. If you wanted to use genetic engineering to increase the amount of antithrombin this patient produces, would you modify the regulatory sequence or the coding sequence? Explain your answer.

MAKING TRANSGENIC ORGANISMS
Transgenic organisms are becoming increasingly important in agriculture and medicine.

HINT See Infographics 8.5 and 8.6.

KNOW IT
9. Melanin is expressed in skin cells and gives skin its color. If you wanted to express a different gene in skin cells, which part of the melanin gene would you use? Why? If you wanted to produce melanin in yeast cells, what part of the melanin gene would you use? Why?

USE IT
10. Explain why scientists used the beta casein regulatory sequence to express human antithrombin in goats' milk.

GENE EXPRESSION
Gene expression is the multistep process of converting the information of DNA into proteins.

HINT See Infographics 8.7–8.10.

KNOW IT
11. For each structure or enzyme listed, indicate by N (nucleus) or C (cytoplasm) its active location in eukaryotic cells:
RNA polymerase ____
Ribosome ____
tRNA ____
mRNA ____

Media and Supplements

Biology for a Changing World is supported by a robust set of study and teaching resources and products. These support materials have been written by a team of experienced nonmajors educators and are tied together by peer-reviewed Learning Objectives for each chapter. These objectives allow instructors to identify the core concepts that most challenge their students and enable them to target student needs earlier and more effectively. In addition, they provide instructors with a way to demonstrate that their students have mastered specific chapter goals.

Our program is outlined below, please ask your sales representative to see our supplement sampler for more clarification.

- **Instructor Resources**
  - **Story Abstracts** The abstracts offer a brief story synopsis, providing interesting details relevant to the chapter and to the online resources not found in the book.
  - **Active Learning Activities** Our activities aim to enhance the student's natural curiosity and to inspire critical thinking about the topics. These will also provide alternative examples to the stories in the text.
  - **Clicker Questions** Designed to be used by students working in teams as well as in large lectures.
  - **Optimized Figure JPEGS and PowerPoints** Infographics are optimized and split apart to be used for projection in large lecture halls.
  - **Stepped Art Sequences and Animations** Every piece of art in the text is interactive in some way, either through an art sequence or an animation.
  - **Lecture PowerPoints**–Prebuilt lectures to help with the transition to a new textbook.
  - **Test Questions/Quizzes** All assessment is organized into the textbook's "Know it" and "Use it" categories.

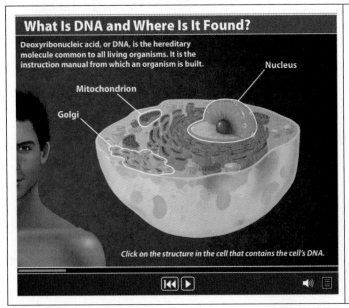
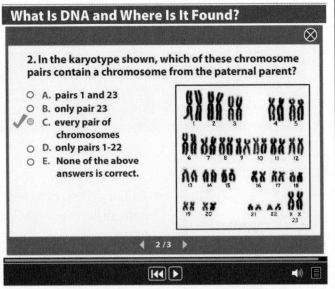

Interactive Infographic Tutorial

- **Instructor Products**
  - **Test Bank/Computerized Test Bank** More than 100 questions per chapter presented in a sortable, searchable platform.
  - **Interactive e-book** Priced lower than the printed textbook and featuring seamlessly integrated interactive resources and study tools.
  - **BioPortal** A learning space to help instructors administer their courses by combining our fully customizable e-book, instructor resources, student resources, news feeds, and homework management tools.
  - **Course Management System e-packs** Available for Blackboard, WebCT, and other course management platforms.
  - **Faculty Lounge** The only publisher-provided Web site linking the nonmajors biology community, where instructors can share lecture ideas, videos, animations, and other resources.
  - **Instructors Resource DVD**

- **Student Resources**
  - **Key Term Flashcards** Students can drill and learn the most important terms in each chapter using interactive flashcards.
  - **Lecture Companion Art** The Infographics for each chapter are available as PDF files that students can download and print before lectures.
  - **Quizzing with Feedback** Response-specific feedback helps explain concepts and correct student misunderstandings.
  - **Interactive Infographics** All Infographics in the text include an animated interactive tutorial or an infographic activity.
  - **LearningCurve** A new learning tool that evaluates what students know and don't know and provides them with a personalized study plan to guide their study of each chapter.

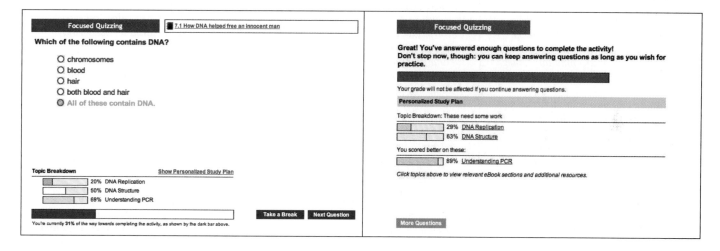

- **Student Products**
  - **Interactive e-book** Priced lower than the printed textbook and featuring seamlessly integrated interactive resources and study tools.
  - **BioPortal** A learning space combining our fully customizable e-book, student resources, news feeds, and homework management tools.
  - **Printed Study Guide** Covers all the topics in each chapter, breaking down each infographic by offering students clear learning objectives and providing questions to test their critical thinking.
  - **Free Book Companion Web Site** Featuring most student resources in an online format.

Acknowledgments

We would like to thank the many reviewers who have helped with the development of this text.

Stephanie Aamodt, *Louisiana State University-Shreveport*

Marilyn Abbott, *Lindenwood University*

Julie Adams, *Ohio Northern University*

Tadesse A. Addisu, *Northern Virginia Community College-Annandale*

Adjoa Ahedor, *Rose State College*

Ann Aguanno, *Marymount Manhattan College*

Zulfiqar Ahmad, *East Tennessee State University*

Mark Ainsworth, *Seattle Central Community College*

Carol Allen, *Montgomery College*

Pamela Anderson Cole, *Shelton State Community College*

Ken Andrews, *East Central University*

Corrie Andries, *Central New Mexico Community College*

Josephine Arogyasami, *Southern Virginia University*

Joseph A. Arruda, *Pittsburg State University*

Tami Asplin, *North Dakota State University*

Kim Atwood, *Cumberland University*

Felicitas Avendano, *Grandview University*

James Backer, *Daytona State College*

David Bailey, *Delta College*

Andy Baldwin, *Mesa Community College*

Mary Ball, *Carson-Newman College*

Verona Barr, *Heartland Community College*

Tina Beams Jones, *Shelton State Community College*

Lynne Berdainer, *Gainesville State College*

Christine Bezotte, *Elmira College*

Bill Rogers, *Ball State University*

Curtis Blankenspoor, *Calvin College*

Lisa Boggs, *Southwestern Oklahoma State University*

Cheryl Boice, *Lake City Community College*

Larry Boots, *University of Montavello*

Barbara Boss, *Keiser University*

Brenda Bourns, *Seattle University*

Bradley Bowden, *University of Connecticut*

Mark Boyland, *Union University*

Dean Bratis, *Villanova University*

Mimi Bres, *Prince George's Community College*

Randy Brewton, *University of Tennesee-Knoxville*

Marguerite (Peggy) Brickman, *University of Georgia*

Clay Britton, *Methodist University*

Gregory Brown, *McGill University*

Carole Browne, *Wake Forest University*

Sara Browning, *Palm Beach Atlantic University*

Joseph Bruseo, *Holyoke Community College*

Mark Buchheim, *University of Tulsa*

Anne Bunnell, *East Carolina University*

Jamie Burchill, *Troy University*

Greg Butcher, *Centenary College of Louisiana*

David Byres, *Florida State College at Jacksonville-South Campus*

Carolee Caffrey, *Hofstra University*

Jane Caldwell, *Washington & Jefferson College*

Jamie Campbell, *Truckee Meadows Community College*

Shillington Cara, *Eastern Michigan University*

Michael Carr, *Oakton Community College*

Dale Casamata, *University of North Florida*

Deborah A. Cato, *Wheaton College*

Jeannie Chapman, *University of South Carolina-Upstate*

Xiaomei Cheng, *Mount St. Mary's College*

Steven D. Christenson, *Brigham Young University-Idaho*

Kimberly Cline-Brown, *University of Northern Iowa*

Yvonne Cole, *Lindenwood University*

Claudia Cooperman, *Philadelphia University*

Erica Corbett, *Southeastern Oklahoma State University*

David Corey, *Midlands Technical College-Beltline Campus*

Cathy Cornett, *University of Wisconsin-Platteville*

Frank Coro, *Miami-Dade College-InterAmerican Campus*

Angela Costanzo, *Hawaii Pacific University*

Richard Cowart, *University of Dubuque*

Jan Crook-Hill, *North Georgia College and State University*

Peter Cumbie, *Winthrop University*

Kathleen L. Curran, *Wesley College*

Jennifer Cymbola, *Grand Valley State University*

Gregory Dahlem, *Northern Kentucky University*

Don Dailey, *Austin Peay State University*

Michael S. Dann, *Pennsylvania State University*

Farahad Dastoor, *University of Maine*

Cara L. Davies, *Ohio Northern University*

Renne Dawson, *University of Utah*

Nishantha de Silva, *Lock Haven University*

Jodi Denuyl, *Grand Valley State University*

Elizabeth DeStasio, *Lawrence University*

Chris Dobson, *Grand Valley State University*

Therese Dudek, *Kiswaukee College*

Denise Due-Goodwin, *Vanderbilt University*

Jacquelyn Duke, *Baylor University*

Michael Edgehouse, *Cabrillo College*

Susan S. Epperson, *University of Colorado at Colorado Springs*

Paul Farnsworth, *University of New Mexico*

Steven Fields, *Winthrop University*

Lynn Firestone, *Brigham Young University-Idaho*

Teresa Fischer, *Indian River Community College*

Carey Fox, *Brookdale Community College*

Karen Francl, *Radford University*

Barbara S. Frank, *Idaho State University*

Diane Fritz, *Northern Kentucky University*

Richard Gardner, *Southern Virginia University*

Shelley Garrett, *Guilford Technical Community College*

Phil Gibson, *University of Oklahoma*

Julie L. Glenn, *Gainesville State College-Oconee Campus*

Inna Goldenberg, *Oakton Community College*

Stephen Gomez, *Central New Mexico Community College*

Brad Goodbar, *College of the Sequoias*

Kate Goodrich, *Widener University*

Sherri Graves, *Sacramento City College*

Madoka Gray-Mitsumune, *Concordia University*

Bradley Griggs, *Piedmont Technical College*

Cheryl Hackworth, *West Valley College*

Janelle Hare, *Morehead State University*

Katherine Harris, *Hartnell College*

Joe Harsh, *Butler University*

Roberta Hayes, *St. Johns College of Liberal Arts and Sciences*

Chris Haynes, *Shelton State Community College*

Steve Heard, *University of New Brunswick*

Jason Heaton, *Samford University*

Susan Hengeveld, *Indiana University*

Kelly Hogan, *University of North Carolina-Chapel Hill*

Andrew Holmgren, *Heartland Community College*

Ann Marie Hoskinson, *Minnesota State University-Mankato*

Tim Hoving, *Grand Rapids Community College*

Tonya Huff, *Riverside Community College*

Evelyn Jackson, *University of Mississippi*

Laurie Johnson, *Bay College*

Tanganika K. Johnson, *Southern University and A&M College*

David Jones, *Dixie State College*
Jackie Jordan, *Clayton State University*
Marian Kaehler, *Luther College*
John Kell, *Radford University*
Michael Kennedy, *Missouri Southern State University*
Janine Kido, *Mt. San Antonio College*
Kerry Kilburn, *Old Dominion University*
Dennis J. Kitz, *Southern Illinois University-Edwardsville*
Cindy Klevickis, *James Madison University*
Jeannifer Kneafsey, *Tulsa Community College*
Brenda Knotts, *Eastern Illinois University*
Olga Kopp, *Utah Valley College*
Ari Krakowski, *Laney College*
Dan Krane, *Wright State*
Wendy A. Kuntz, *Kapiolani Community College*
Holly Kupfer, *Central Piedmont Community College*
Dale Lambert, *Tarrant County College*
Kirkwood Land, *University of the Pacific*
Elaine Larsen, *Skidmore College*
Mary Lehman, *Longwood University*
Beth Leuck, *Centenary College of Louisiana*
Robert Levine, *McGill University*
Patrick Lewis, *Sam Houston State University*
Tammy Liles, *Bluegrass Community College*
Susanne Lindgren, *California State University-Sacramento*
Matthew Linton, *University of Utah*
Cynthia Littlejohn, *University of Southern Mississippi*
Madelyn Logan, *North Shore Community College*
Ann S. Lumsden, *Florida State University*
Will Mackin, *Elon University*
Paul H. Marshall, *Northern Essex Community College*
Mary Martin, *Northern Michigan University*
Ron Mason, *Mt. San Jacinto College-Menifee*
Helen Mastrobuoni, *County College of Morris*
Amie Mazzoni, *Fresno City College*
Rob McCandless, *Methodist*
Brett McMillan, *McDaniel College*
Malinda McMurry, *Morehead State University*
Michael McVay, *Green River Community College*
Scott Medler, *State University of New York-Buffalo*
Judith Megaw, *Indian River State College*
Diane L. Melroy, *University of North Carolina-Wilmington*
Paige Mettler-Cherry, *Lindenwood University*
Jim Mickle, *North Carolina State University*
Hugh Miller, *East Tennessee State University*
Scott Moody, *Ohio University*
John Moore, *Taylor University*
Lia Muller, *San Diego Mesa College*
Ann Murkowski, *North Seattle Community College*
Shawn Nordell, *University of St. Louis*
Peter Oelkers, *University of Michigan-Dearborn*

Margaret Oliver, *Carthage College*
Joanna Padolina, *Virginia Commonwealth University*
Karen Pasko, *Emmanuel College*
Forrest E. Payne, *University of Arkansas at Little Rock*
Joseph Peabody, *Brigham Young University-Independent Study*
Linda Peters, *Holyoke Community College*
Stephanie Toering Peters, *Wartburg College*
William Pietraface, *State University of New York-Oneonta*
Joel Piperberg, *Millersville University*
Gregory J. Podgorski, *Utah State University*
Jeff Podos, *University of Massachusetts-Amherst*
Therese Poole, *Georgia State University*
Michelle Priest, *Irvine Valley College*
Kenneth Pruitt, *University of Texas at Brownsville*
Dianne Purves, *Crafton Hills College*
Scott Quinton, *Johnson County Community College*
Logan Randolph, *Polk State College*
Nick Reeves, *Mt. San Jacinto College-Menifee*
Kim Regier, *University of Colorado-Denver*
Nancy Rice, *Western Kentucky University*
Stanley Rice, *Southeastern Oklahoma State University*
Brendan Rickards, *Gloucester County College*
Jennifer Robbins, *Xavier University*
Laurel Roberts, *University of Pittsburg*
Peggy Rolfsen, *Cincinnati State Technical and Community College*
Amy Rollins, *Clayton State University*
Deanne Roquet, *Lake Superior College*
Karen Rose, *Shelton State Community College*
Barbara Salvo, *Carthage College*
Ken Saville, *Albion College*
Michael Sawey, *Texas Christian University*
Karen Schaffer, *Northwest Missouri State University*
Daniel Scheirer, *Northeastern University*
Bronwyn Scott, *Bellevue College*
David Serrano, *Broward College-Central Campus*
Marilyn Shopper, *Johnson County Community College*
Laurie Shornicle, *University of Missouri-St. Louis*
Brad Shuster, *New Mexico State University*
Tamara Sluss, *Kentucky State University*
Patricia Smith, *Valencia Community College-East Campus*
Sharon Smth, *Florida State College at Jacksonville-Deerwood Center*
Adrienne Smyth, *Worcester State University*
James Sniezek, *Montgomery College*
Andrea Solis, *Mount St. Mary's University*
Anna Bess Sorin, *University of Memphis*
Carol St. Angelo, *Hofstra University*
Wendy Stankovich, *University of Wisconsin-Platteville*
Rob Stark, *California State*

University-Bakersfield
Amanda Starnes, *Emory University*
Alicia Steinhardt, *Hartnell College*
Bethany Stone, *University of Missouri*
Christine Stracey, *Westminster College*
Sheila Strawn, *University of Oklahoma*
Steve Taber, *Saginaw Valley State University*
John R. Taylor, *Southern Utah University*
Sonia Taylor, *Lake City Community College*
Don Terpening, *State University of New York-Ulster*
Pamela Thineson, *Century College*
Janice Thomas, *Montclair State University*
Paula Thompson, *Florida State College at Jacksonville-North Campus*
Heather Throop, *New Mexico State University*
Sanjay Tiwary, *Hinds Community College-Raymond Campus*
Jeff Travis, *State University of New York-Albany*
Eileen Underwood, *Bowling Green University*
Craig Van Boskirk, *Florida State College at Jacksonville-Deerwood Center*
Bina Vanmali, *University Missouri-Columbia*
José Vázquez, *New York University*
R. Steve Wagner, *Central Washington University*
Rebekah Waikel, *Eastern Kentucky University*
Timothy Wakefield, *John Brown University*
Helen Walter, *Mills College*
Paul Wanda, *Southern Illinois University-Edwardsville*
Katherine Warpeha, *University of Illinois at Chicago*
Arthur C. Washington, *Florida Agricultural and Mechanical University*
Amanda Waterstrat, *Eastern Kentucky University*
Kathy Webb, *Bucks Country Community College*
Karen Wellner, *Arizona State University*
Mike Wenzel, *California State University-Sacramento*
Brad Wetherbee, *University of Rhode Island*
Alicia Whatley, *Troy University*
Robert S. Whyte, *California University of Pennsylvania*
Tara Williams-Hart, *Lousiana State University-Shreveport*
Christina Wills, *Rockhurst University*
Carol Wymer, *Morehead State University*
Lan Xu, *South Dakota State University*
Rick Zechman, *California State University-Fresno*
Michelle Zjhra, *Georgia Southern University*
Elena Zoubina, *Bridgewater College*
Jeff Zuiderveen, *Columbus State University*

Java Report

Java Report

Making sense of the latest buzz in health-related news

In 1981, a study in the *New England Journal of Medicine* made headlines when it reported that drinking two cups of coffee a day doubled a person's risk of getting pancreatic cancer; five or more cups a day supposedly tripled the risk. "Study Links Coffee Use to Pancreas Cancer," trumpeted the *New York Times*. "Is there cancer in the cup?" asked *Time* magazine. The lead author of the study, Dr. Brian MacMahon of the Harvard School of Public Health, appeared on the *Today* show to warn of the dangers of coffee. "I will tell you that I myself have stopped drinking coffee," said MacMahon, who had previously drunk three cups a day.

Just five years later, MacMahon's research group was back in the news reporting in the same journal that a second study had found *no* link between coffee and pancreatic cancer. Subsequent studies, by other authors, also failed to reproduce the original findings.

A sometime health villain, coffee's reputation seems to be on the rise. Recent studies have suggested that, far from causing disease, the beverage may actually help *prevent* a number of conditions–everything from Parkinson disease and diabetes to cancer and tooth decay. A 2010 CBS News headline announced, "Java Junkies Less Likely to Get Tumors," and a blog proclaimed, "Morning Joe Fights Prostate Cancer." The September 2010 issue of *Prevention* magazine ran an article titled "Four Ways Coffee Cures."

Not everyone is buying the coffee cure, however. Public health officials are increasingly alarmed by our love affair with–some might say, addiction to–caffeine. Emergency rooms are reporting more caffeine-related admissions, and poison control centers are receiving more calls related to caffeine "overdoses." In response, the state of California is even considering forcing manufacturers to put warning

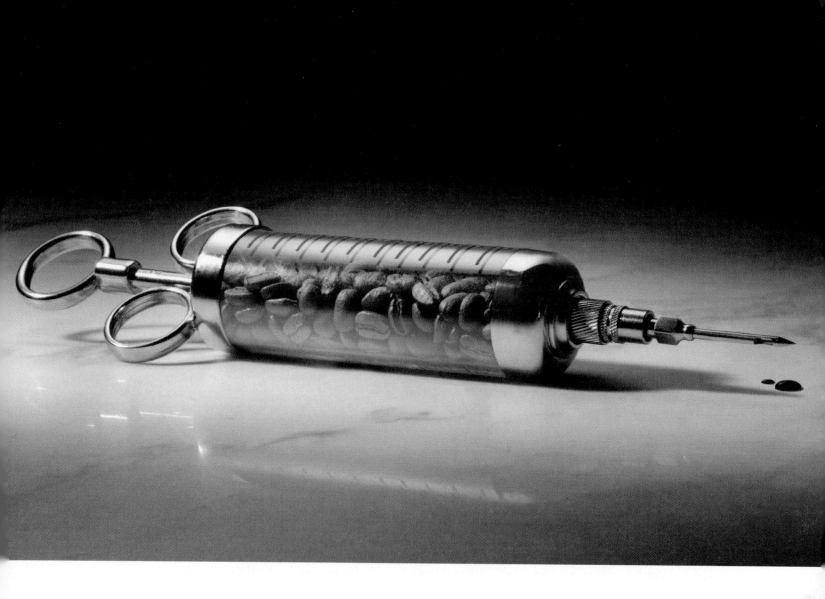

labels on energy drinks. Nevertheless, caffeine's "energizing" effect is advertised on nearly every street corner, where, increasingly, you're also likely to find a coffee shop; as of 2010, there were 222 Starbucks within a five-mile radius of a Manhattan zip code according to Foodio54.com; nationally, the average within the same radius is 10.

Conflicting messages like these are all too common in the news. From the latest cancer therapies to the ecological effects of global warming, a steady but often contradictory stream of scientific information vies for our increasingly Twitter-size attention spans.

Why the mixed messages? Are researchers making mistakes? Are journalists getting their facts wrong? While both of these possibilities may be true at times, the bigger problem is widespread confusion over the nature of science and the meaning of scientific evidence.

"Consumers are flooded with a firehose of health information every day from various media sources," says Gary Schwitzer, publisher of the consumer watchdog blog HealthNewsReview. org and former director of health journalism at the University of Minnesota. "It can be—and often is—an ugly picture: a bazaar of disinformation." Too often, he says, the results of studies are reported in incomplete or misleading ways.

Consider the grande cup of coffee or the Red Bull you may have had with breakfast this morning. Why might consuming coffee or caffeine be

> **Consumers are flooded with a firehose of health information every day.**
> —Gary Schwitzer

The national average number of Starbucks within a five-mile radius of a single zip code is 10.

associated with such dramatically different results? The risks or benefits of a caffeinated beverage may depend on the amount a person drinks—one cup versus a whole pot. Or maybe it matters *who* is drinking the beverage. The *New England Journal of Medicine* study, for example, looked at hospitalized patients only. Would the same results have been seen in people who weren't already sick? Sometimes, to properly evaluate a scientific claim, we need to look more closely at how the science was done (Infographic 1.1).

Science Is a Process

Science is less a body of established facts than a way of knowing—a method of seeking answers to questions on the basis of observation and experiment. Scientists draw conclusions from the best evidence they have at any one time, but the process is not always easy or straightforward. Conclusions based on today's

evidence may be modified in the future as other scientists ask different—and sometimes better—questions. Moreover, with improved technology, researchers may uncover better data; new information can cast old conclusions in a new light. Science is a never-ending process.

Let's say you want to investigate the "energizing" effects of coffee scientifically—how might you go about it? A logical place to start would be your own personal experience. You may notice that you feel more awake when you drink coffee. It seems to help you concentrate as you pull an all-nighter to finish a paper. Such informal, personal observations are called **anecdotal evidence.** It's a type of evidence that may be interesting but is often unreliable, since it wasn't based on systematic study. You could perhaps poll your classmates to find out if they experience coffee in the same way.

> ### Science is less a body of established facts than a way of knowing.

Conflicting Conclusions

→ A variety of studies published in peer-reviewed scientific journals report different conclusions about the risks and benefits of coffee. In order for the public to understand and use these outcomes to its advantage, a closer look at the scientific process and the factors that surround coffee drinking is necessary.

Scientific studies report that drinking coffee...

- May cause pancreatic cancer
- Is linked to infertility and low infant birth weight
- Lowers the risk of Parkinson disease
- Does not cause pancreatic cancer
- Reduces risk of ovarian cancer

So, is it really the coffee?
Or other factors associated with drinking coffee?

- Chemicals naturally present in coffee, including caffeine
- The climate and soil in which different coffee plants are grown (which in turn influences the chemicals in coffee)
- How the beans are roasted and processed
- How much coffee a person drinks
- The gender, age, and general health of a coffee drinker
- Other social factors, such as whether coffee is consumed with a meal or with a cigarette, or with other foods and beverages that may interact in some way with coffee
- Other unknown factors that just happen to correlate with coffee drinking

TESTABLE
A hypothesis is testable if it can be supported or rejected by carefully designed experiments or nonexperimental studies.

FALSIFIABLE
Describes a hypothesis that can be ruled out by data that show that the hypothesis does not explain the observation.

EXPERIMENT
A carefully designed test, the results of which will either support or rule out a hypothesis.

Nevertheless, this anecdotal evidence might lead you to formulate a question: Does coffee improve mental performance? To get a sense of what information currently exists on the subject, you could read relevant coffee studies that have already been conducted, available in online databases of journal articles or in university libraries. Generally, you can trust the information in scientific journals because it has been subject to **peer review,** meaning that independent and unbiased experts have critiqued the soundness of a study before it was published. The aim of peer review is to weed out sloppy research, as well as overstated claims, and thus to ensure the integrity of the journal and its

scientific findings. To further reduce the chance of bias, authors must declare any possible conflicts of interest and name all funding sources (for example, pharmaceutical or biotechnology companies). With this information, reviewers and readers can view the study with a more critical eye.

Based on what you learn from reading journal articles, you could formulate a **hypothesis** to explain how coffee improves mental performance. A hypothesis is a narrowly focused statement that is **testable** and **falsifiable,** that is, it can be proved wrong. A hypothesis represents one possible answer to the question under investigation. One hypothesis to explain coffee's effects, for example, is that drinking coffee improves memory. Another might be: high levels of caffeine increase concentration. Not all explanations will be *scientific* hypotheses, though. Statements of opinion, and hypotheses that use supernatural or mystical explanations that cannot be tested or refuted, fall outside the realm of scientific explanation. (Some call such explanations "pseudoscience"; astrology is a good example.)

With a clear scientific hypothesis in hand– "coffee improves memory"–the next step is to test it, generating evidence for or against the idea. If a hypothesis is shown to be false–"coffee does not improve memory"–it can be rejected and removed from the list of possible answers to the original question. On the other hand, if data support the hypothesis, then it will be accepted, at least until further testing and data show otherwise. Because it is impossible to test whether a hypothesis is true in every possible situation, a hypothesis can never be proved true once and for all. The best we can do is support the hypothesis with an exhaustive amount of evidence **(Infographic 1.2).**

There are multiple ways to test a hypothesis. One is to design a controlled **experiment** in which you measure the effects of coffee drinking on a group of subjects. In 2002, Lee Ryan, a psychologist at the University of Arizona, decided to do just that. Ryan noticed that memory is often optimal early in the morning in adults over age 65 but tends to decline as the day goes on. She also noticed that many adults report feeling more alert after drinking caffeinated coffee. She therefore hypothesized that drinking coffee might prevent this decline in memory, and devised an experiment to test her hypothesis.

First she collected a group of participants–40 men and women over age 65, who were active, healthy, and who reported consuming some form of caffeine daily. She then randomly divided these people into two groups: one that would get caffeinated coffee, and one that would receive decaf. The caffeine group is known as the **experimental group,** since caffeine is what's being tested in the experiment. The decaf group is known as the **control group**–it serves as the basis of comparison. Both groups were given memory tests at 8 A.M. and again

at 4 P.M. on two nonconsecutive days. The experimental group received a 12-ounce cup of regular coffee containing approximately 220–270 mg of caffeine 30 minutes before each test. The control group received a **placebo:** a 12-ounce cup of decaffeinated coffee containing no more than 5 to 10 mg of caffeine per serving.

By administering a placebo, Ryan could ensure that any change observed in the experimental group was a result of consuming caffeine and not just any hot beverage. Moreover, participants did not know whether they were drinking regular or decaf, so a **placebo effect** was also ruled out. In addition, all participants were forbidden to eat or drink any other caffeine-containing foods or drinks–like chocolate, soda, or coffee–for at least four hours before

EXPERIMENTAL GROUP
The group in an experiment that experiences the experimental intervention or manipulation.

CONTROL GROUP
The group in an experiment that experiences no experimental intervention or manipulation.

PLACEBO
A fake treatment given to control groups to mimic the experience of the experimental groups.

INFOGRAPHIC 1.2

Science Is a Process: Narrowing Down the Possibilities

Multiple scientists doing multiple experiments narrow down the pool of possible hypotheses. Those that are rigorously tested and supported by other experiments emerge with greatest confidence.

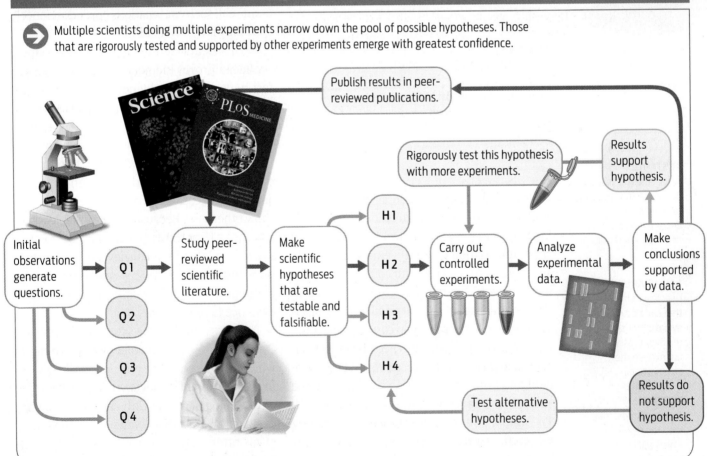

Publish results in peer-reviewed publications.

Initial observations generate questions.

Q1 · Q2 · Q3 · Q4

Study peer-reviewed scientific literature.

Make scientific hypotheses that are testable and falsifiable.

H1 · H2 · H3 · H4

Rigorously test this hypothesis with more experiments.

Carry out controlled experiments.

Analyze experimental data.

Make conclusions supported by data.

Results support hypothesis.

Results do not support hypothesis.

Test alternative hypotheses.

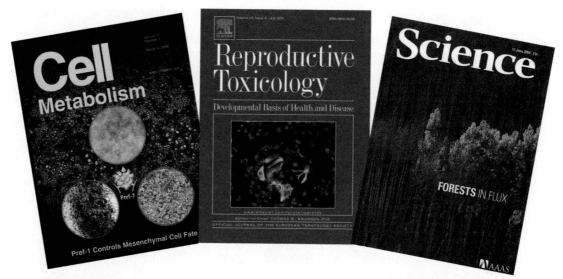

The studies in scientific journals are reviewed by experts before publication to ensure accuracy.

PLACEBO EFFECT
The effect observed when members of a control group display a measurable response to a placebo because they think that they are receiving a "real" treatment.

INDEPENDENT VARIABLE
The variable, or factor, being deliberately changed in the experimental group.

DEPENDENT VARIABLE
The measured result of an experiment, analyzed in both the experimental and control groups

SAMPLE SIZE
The number of experimental subjects or the number of times an experiment is repeated. In human studies, sample size is the number of subjects.

each test. Thus, the control group was identical to the experimental group in every way except for the consumption of caffeine.

In this experiment, caffeine consumption was the **independent variable**—the factor that is being changed in a deliberate way. The tests of memory are the **dependent variable**—the outcome that may "depend" on caffeine consumption.

Ryan found that people who drank decaffeinated coffee did worse on tests of memory function in the afternoon compared to the morning. By contrast, the experimental group who drank caffeinated coffee performed equally well on morning and afternoon memory tests. The results, which were reported in the journal *Psychological Science*, support the hypothesis that caffeine, delivered in the form of coffee, improves memory—at least in certain people (Infographic 1.3).

Because other factors might, in theory, explain the link between coffee and mental performance (perhaps coffee drinkers are more active, and their physical activity rather than their coffee consumption explains their mental performance), it's too soon to see these results as proof of coffee's memory-boosting powers. To win our confidence, the experiment must be repeated by other scientists and, if possible, the methodology refined.

Size Matters

Consider the size of Ryan's experiment—40 people, tested on two different days. That's not a very big study. Could the results have simply been due to chance? What if the 20 people who drank caffeinated coffee just happened to have better memory?

One thing that can strengthen our confidence in the results of a scientific study is **sample size**. Sample size is the number of individuals participating in a study, or the number of times an experiment or set of observations is

Anatomy of an Experiment

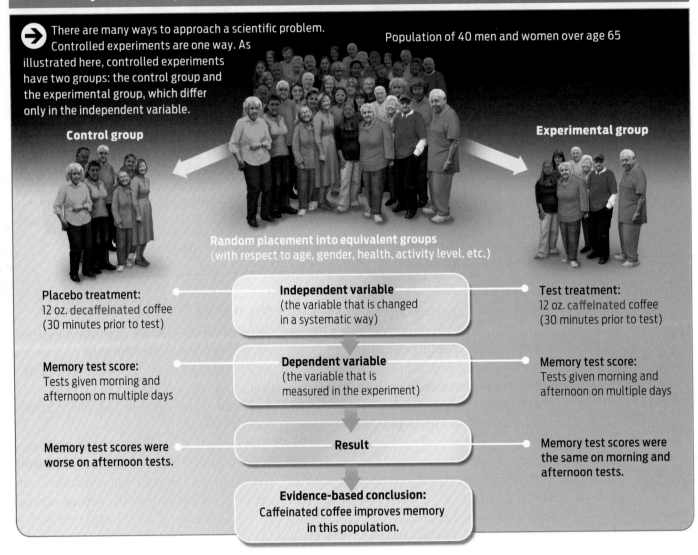

There are many ways to approach a scientific problem. Controlled experiments are one way. As illustrated here, controlled experiments have two groups: the control group and the experimental group, which differ only in the independent variable.

Population of 40 men and women over age 65

Control group

Experimental group

Random placement into equivalent groups
(with respect to age, gender, health, activity level, etc.)

Placebo treatment:
12 oz. decaffeinated coffee (30 minutes prior to test)

Independent variable
(the variable that is changed in a systematic way)

Test treatment:
12 oz. caffeinated coffee (30 minutes prior to test)

Memory test score:
Tests given morning and afternoon on multiple days

Dependent variable
(the variable that is measured in the experiment)

Memory test score:
Tests given morning and afternoon on multiple days

Memory test scores were worse on afternoon tests.

Result

Memory test scores were the same on morning and afternoon tests.

Evidence-based conclusion:
Caffeinated coffee improves memory in this population.

---

repeated. The larger the sample size, the more likely the results will have **statistical significance**—that is, they will not be due to random chance (Infographic 1.4).

News reports are full of statistics. On any given day, you might hear that 75% of the American public opposes a piece of legislation. Or that 15% of a group of people taking a medication experienced a certain unpleasant side effect—like nausea or suicidal thoughts—compared to, say, 8% of people taking a placebo. Are these differences significant or important? Whenever you hear such numbers being cited, it's important to keep in mind the total sample size. In

the case of the side effects, was this a group of 20 patients (15% of 20 patients is 3 people), or was it 2,000? Only with a large enough sample size can we be confident that the results of a given study are statistically significant and represent something more than chance. Moreover, it's important to consider the population being studied. For example, do the people reporting their views on a piece of legislation represent a broad cross section of the public, or are most of them watchers of the same television network, whose views lie at one extreme? Likewise, in Ryan's study, are the 65-year-old self-described "morning people"

STATISTICAL SIGNIFICANCE
A measure of confidence that the results obtained are "real," rather than due to random chance.

Sample Size Matters

 The more data collected in an experiment, the more you can trust the conclusions.

Data from only eight participants:

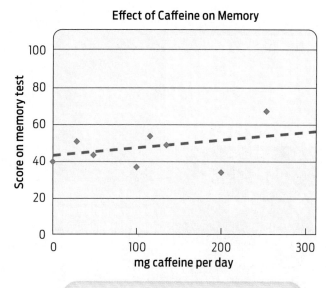

Conclusions drawn from these data might suggest that caffeine has only a slight positive influence on memory, a 15% average increase, but could easily be inconclusive, because of the small sample size.

Data from dozens of participants:

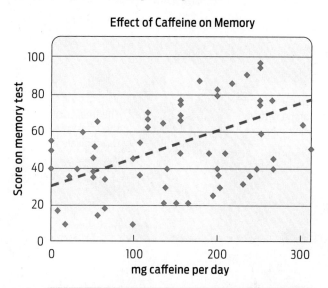

These data show a more convincing positive effect of caffeine on memory, a 45% average increase, because it is supported by more data. A statistical analysis would show that this positive influence is significant — in other words, it is not due to chance.

---

who regularly consume coffee representative of the wider population?

If you search for "caffeine and memory" on PubMed.gov (a database of medical research papers), you'll see that the memory-enhancing properties of caffeine is a well-researched topic. Many studies have been conducted, at least some of which tend to support Ryan's results. Generally, the more experiments that support a hypothesis, the more confident we can be that it is true. A hypothesis that continues to hold up after many years of rigorous testing may eventually be considered a **scientific theory.** Note that the word "theory" in science means something very different from its colloquial meaning. In everyday life we may

In science, a theory is the best explanation we have for an observed phenomenon.

say something is "just a theory," meaning it isn't proved. But in science, a theory is an explanation that is supported by a large body of evidence compiled over time by numerous researchers, and which remains the best explanation we have for an observed phenomenon **(Infographic 1.5).**

This Is Your Brain on Caffeine

Caffeine is a stimulant. It is in the same class of psychoactive drugs as cocaine, amphetamines, and heroin (although less potent than these, and acting through different chemical pathways). Caffeine boosts not just memory and mental activity but physical activity as well. One study, in 2004, found that 33% of 193 track and field

SCIENTIFIC THEORY
A hypothesis that is supported by many years of rigorous testing and thousands of experiments.

Everyday Theory vs. Scientific Theory

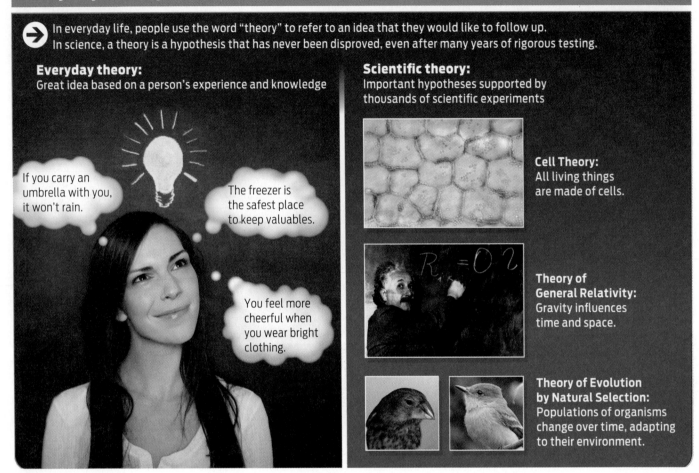

→ In everyday life, people use the word "theory" to refer to an idea that they would like to follow up.
In science, a theory is a hypothesis that has never been disproved, even after many years of rigorous testing.

Everyday theory:
Great idea based on a person's experience and knowledge

If you carry an umbrella with you, it won't rain.

The freezer is the safest place to keep valuables.

You feel more cheerful when you wear bright clothing.

Scientific theory:
Important hypotheses supported by thousands of scientific experiments

Cell Theory:
All living things are made of cells.

$R_{1k} = O 2$

Theory of General Relativity:
Gravity influences time and space.

Theory of Evolution by Natural Selection:
Populations of organisms change over time, adapting to their environment.

athletes and 60% of 287 cyclists said they consumed caffeine to enhance their performance. Recognizing caffeine's reputation as a performance-enhancing drug, the International Olympic Committee prohibited athletes from using it until 2004 (when it decided to allow it, presumably because it had become too common a substance to regulate).

While the exact mechanisms are not fully understood, scientists think that caffeine exerts its energizing effect by counteracting the actions of a chemical in the brain called adenosine. Adenosine is the body's natural sleeping pill–its concentration increases in the brain while you are awake and by the end of the day promotes drowsiness. Caf-

Some researchers contend that coffee's mind-boosting effects are an indirect result of the cycle of dependency.

feine blocks the effect of adenosine in the brain and keeps us from falling asleep.

Though our understanding of the chemistry is relatively new, humans have enjoyed coffee's kick for more than a thousand years. It's said that an Ethiopian goatherd found his goats acting unusually frisky one afternoon after munching the leaves of a small bush. Chewing a few of the shrub's berries himself, he got a caffeine buzz, and the rest was history. Today, caffeine is the most wildly used stimulant on the planet (Table 1.1).

In fact, consumption of caffeinated beverages has skyrocketed in the past 25 years; for example, young people now drink far more soda than milk. A 2009 study in the journal *Pediatrics*

INFOGRAPHIC 1.6

Caffeine Side Effects

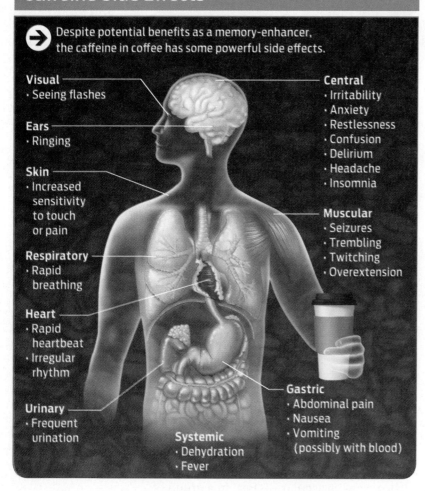

→ Despite potential benefits as a memory-enhancer, the caffeine in coffee has some powerful side effects.

Visual
· Seeing flashes

Ears
· Ringing

Skin
· Increased sensitivity to touch or pain

Respiratory
· Rapid breathing

Heart
· Rapid heartbeat
· Irregular rhythm

Urinary
· Frequent urination

Systemic
· Dehydration
· Fever

Central
· Irritability
· Anxiety
· Restlessness
· Confusion
· Delirium
· Headache
· Insomnia

Muscular
· Seizures
· Trembling
· Twitching
· Overextension

Gastric
· Abdominal pain
· Nausea
· Vomiting (possibly with blood)

found that teenagers consume up to 1,458 mg of caffeine a day–nearly five times the recommended maximum adult dose of 300 mg. Caffeine can cause anxiety, jitters, heart palpitations, trouble sleeping, dehydration, and more serious symptoms–especially in people who are sensitive to it. In 2007, two high school students in Colorado Springs, Colorado, were hospitalized with stomach pain, nausea, and vomiting after drinking one 8-ounce can of Spike Shooter, a potent beverage that packs a walloping 300 mg of caffeine–the equivalent of almost four Red Bulls (Infographic 1.6).

For regular coffee drinkers who crave their morning buzz, such symptoms are unlikely to convince them to kick the habit. This may be because, like many other psychoactive substances, caffeine is addictive. Those who drink a significant amount of coffee every day may notice that they don't feel quite right if they skip a day; they may be cranky or get a headache. These are symptoms of withdrawal. In fact, some researchers contend that coffee's mind-boosting effects are an indirect result of the cycle of dependency. Improvement in mood or performance following a cup of coffee, they say, may simply represent relief from withdrawal symptoms rather than any specific beneficial property of coffee.

To test this dependency hypothesis, scientists could conduct an experiment. They could compare the effects of drinking coffee in two groups: one group of regular coffee drinkers who had abstained from coffee for a short period, and another group of non–coffee drinkers. Does coffee give both groups a boost, or only the regular coffee drinkers looking for their fix?

In fact, this very experiment was done in 2010 by a group of researchers at the University of Bristol in England. Their study, published in the journal *Neuropsychopharmacology,* looked at caffeine's effect on alertness. Researchers gave caffeine or a placebo to 379 participants and asked them to take a test that rated their level of alertness. The study found that caffeine did not boost alertness in non–coffee drinkers compared to those drinking a placebo (although it did boost their level of anxiety and headache). Heavy coffee drinkers, on the other hand, experienced a steep drop in alertness when given the placebo.

"What this study does is provide very strong evidence for the idea that we don't gain a benefit in alertness from consuming caffeine," the study author, Peter Rogers, said. "Although we feel alert, that's just caffeine bringing us back to our normal state of alertness." Of course, this doesn't really explain why people get hooked on coffee in the first place.

Finding Patterns

Performing controlled laboratory experiments like those discussed above is one way that scientists try to answer questions. Another approach is to make careful observations or comparisons

of phenomena that exist in nature. This is the approach taken by scientists who study **epidemiology**–the incidence of disease in populations–or some other area, like the movement of stars or the nature of prehistoric life, that cannot be directly manipulated.

For example, if an epidemiologist wanted to learn about the relationship between cigarette smoking and lung cancer, he could compare the rates of lung cancer in smokers and nonsmokers, but he could not actually perform an experiment in which he made people smoke cigarettes and waited to see whether or not they got cancer. Such an experiment would be highly unethical.

Although epidemiological studies do not provide the immediate gratification of a laboratory experiment, they do have certain advantages. For one thing, they can be relatively inexpensive to conduct, since often the only procedure involved is a participant questionnaire. And you can study factors that are considered harmful, such as excess alcohol or smoking, that you would be unable to test experimentally. Finally, epidemiological studies have the power of numbers and time. The Framingham Heart Study, for example, is a famous epidemiological study that has tracked rates of cardiovascular disease in a group of people and their descendants in Framingham, Massachusetts, in order to identify common risk factors. Begun in 1948, the study has been going on for decades and has provided mountains of data for researchers in many fields, from cardiology to neuroscience.

Most of the health studies featured in the news are epidemiological studies. Consider a study on coffee and Parkinson disease published in the *Journal of the American Medical Association (JAMA)* in 2000. Researchers examined the relationship between coffee drinking and the incidence of Parkinson disease, a condition that afflicts more than 1 million people in the United States, including men and women of all ethnic groups. There is no known cure, only palliative treatments to help lessen symptoms, which include trembling limbs and difficulty coordinating speech and movement.

EPIDEMIOLOGY
The study of patterns of disease in populations, including risk factors.

TABLE 1.1

How Much Caffeine Is in Our Beverages?

The FDA Recommends No More than 65 mg of Caffeine in 12 oz.

BEVERAGE	SERVING SIZE	QUANTITY OF CAFFEINE
Coffee	8 oz	95 mg and up
Red Bull	8.3 oz (1 can)	76 mg
Rockstar	8 oz (half can)	80 mg
Amp	8.4 oz (1 can)	74 mg
Coke Classic	12 oz (1 can)	35 mg
Mountain Dew	12 oz (1 can)	54 mg
Barq's Root Beer	12 oz (1 can)	23 mg
Sprite	12 oz (1 can)	0 mg

Source: Mayo Clinic

For more than 30 years, researchers at the Veterans Affairs Medical Center in Honolulu followed more than 8,000 Japanese-American men, gathering all sorts of information about them: their age, diet, health, smoking habits, and other characteristics. Of these men, 102 developed Parkinson disease. What did these 102 men have in common? Epidemiologists found that most of them did not drink caffeinated beverages—no coffee, soda, or caffeinated tea.

By contrast, coffee drinkers had a lower incidence of Parkinson disease. In fact, those who drank the most coffee were the least likely to get the disease. Men who drank more than two 12-ounce cups of coffee each day had one-fifth the risk of getting the disease compared to non–coffee drinkers.

So does coffee prevent Parkinson disease? The occurrence and progression of many diseases are affected by a complex range of factors, including age, sex, diet, genetics, and exposure to bacteria and environmental chemicals, as well as lifestyle factors like drinking, smoking, and exercise. Although the study discussed here suggests a link–or **correlation**–between caffeine and lower incidence of Parkinson disease, it does not necessarily show that caffeine prevents the disease. In other words, correlation is not causation. Perhaps the people who like to drink coffee have different brain chemistry, and it's this different brain chemistry that explains the differing incidence of Parkinson disease among coffee drinkers (**Infographic 1.7**).

Indeed, other studies have found that cigarette smoking also correlates with a lower risk of Parkinson disease. Both coffee drinking and smoking could be considered types of thrill seeking, behavior observed in people who enjoy the "high" they get from stimulants such as caffeine or nicotine. The lower risk of Parkinson disease among coffee drinkers might therefore result from thrill-seeking brain chemistry that also happens to resist disease–rather than being caused by either smoking or drinking coffee per se.

Moreover, the study followed Japanese-American men. Would the same relationship of caffeine and Parkinson disease be seen in other ethnic groups or in women? Several

CORRELATION
A consistent relationship between two variables.

Correlation Does Not Equal Causation

→ While the data shown below show a convincing **correlation** between reduced caffeine intake and an increased risk of Parkinson disease, it is impossible to state that less coffee **causes** Parkinson disease. Other factors that were not tested or controlled for could be causing the reduced risk.

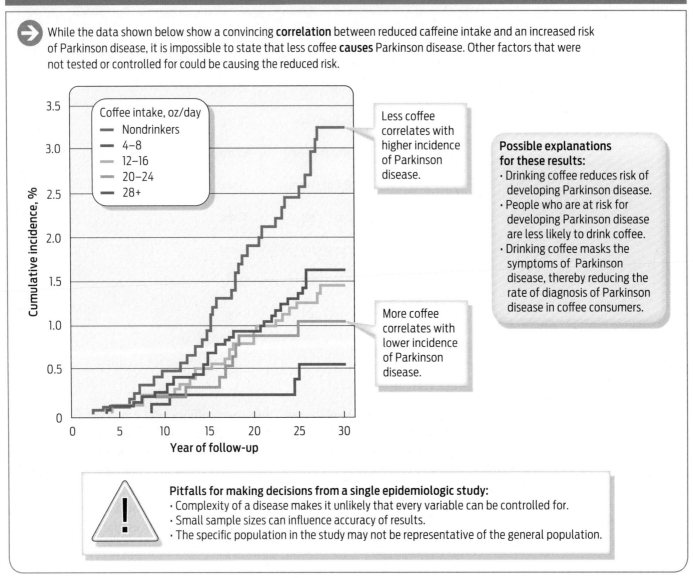

Coffee intake, oz/day
- Nondrinkers
- 4–8
- 12–16
- 20–24
- 28+

Less coffee correlates with higher incidence of Parkinson disease.

More coffee correlates with lower incidence of Parkinson disease.

Possible explanations for these results:
- Drinking coffee reduces risk of developing Parkinson disease.
- People who are at risk for developing Parkinson disease are less likely to drink coffee.
- Drinking coffee masks the symptoms of Parkinson disease, thereby reducing the rate of diagnosis of Parkinson disease in coffee consumers.

Pitfalls for making decisions from a single epidemiologic study:
- Complexity of a disease makes it unlikely that every variable can be controlled for.
- Small sample sizes can influence accuracy of results.
- The specific population in the study may not be representative of the general population.

SOURCE: ROSS ET AL., JAMA 2000; 283:2671–2679

other epidemiological studies have found a correlation between caffeine consumption and a lower incidence of Parkinson disease in men of other ethnicities. But in women the results have been inconclusive. All in all, there's still no direct evidence that caffeine actually prevents the disease in either men or women.

"While our study found a strong correlation between coffee drinkers and low rates of Parkinson's disease," said the study's lead author, G. Webster Ross in a press release issued by the U.S. Department of Veterans Affairs, "we have not identified the exact cause of this effect. I'd like to see these findings used as a basis to help

other scientists unravel the mechanisms that underlie Parkinson's onset."

To get a clearer picture of caffeine's role in Parkinson disease, researchers could conduct a type of experiment known as a **randomized clinical trial,** in which the effects of coffee are measured directly under controlled conditions. One could divide a population into two groups, put one group on coffee and the other on decaf, and then follow both groups for a number of years to see which one had the higher incidence of disease. The problem with such a study is that it is often very expensive to conduct, and it can be difficult to get

RANDOMIZED CLINICAL TRIAL
A controlled medical experiment in which subjects are randomly chosen to receive either an experimental treatment or a standard treatment (or placebo).

INFOGRAPHIC 1.8

From the Lab to the Media: Lost in Translation

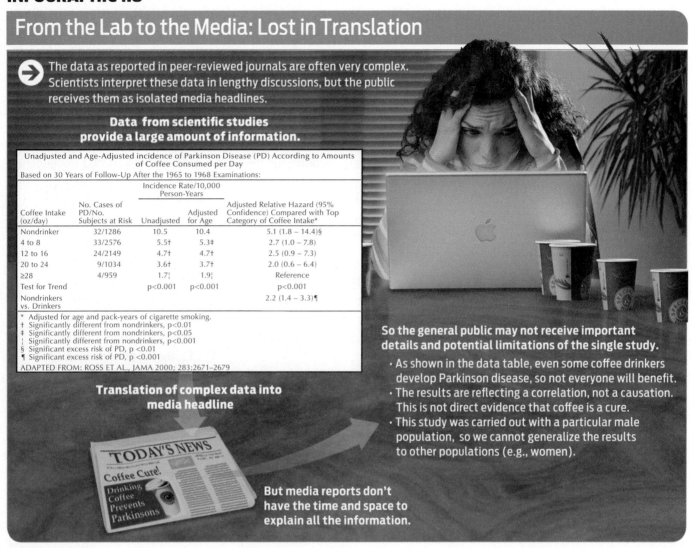

→ The data as reported in peer-reviewed journals are often very complex. Scientists interpret these data in lengthy discussions, but the public receives them as isolated media headlines.

Data from scientific studies provide a large amount of information.

Unadjusted and Age-Adjusted incidence of Parkinson Disease (PD) According to Amounts of Coffee Consumed per Day
Based on 30 Years of Follow-Up After the 1965 to 1968 Examinations:

Coffee Intake (oz/day)	No. Cases of PD/No. Subjects at Risk	Incidence Rate/10,000 Person-Years		Adjusted Relative Hazard (95% Confidence) Compared with Top Category of Coffee Intake*
		Unadjusted	Adjusted for Age	
Nondrinker	32/1286	10.5	10.4	5.1 (1.8 – 14.4)§
4 to 8	33/2576	5.5†	5.3‡	2.7 (1.0 – 7.8)
12 to 16	24/2149	4.7†	4.7†	2.5 (0.9 – 7.3)
20 to 24	9/1034	3.6†	3.7†	2.0 (0.6 – 6.4)
≥28	4/959	1.7¦	1.9¦	Reference
Test for Trend		$p<0.001$	$p<0.001$	$p<0.001$
Nondrinkers vs. Drinkers				2.2 (1.4 – 3.3)¶

* Adjusted for age and pack-years of cigarette smoking.
† Significantly different from nondrinkers, $p<0.01$
‡ Significantly different from nondrinkers, $p<0.05$
¦ Significantly different from nondrinkers, $p<0.001$
§ Significant excess risk of PD, $p <0.01$
¶ Significant excess risk of PD, $p <0.001$
ADAPTED FROM: ROSS ET AL., JAMA 2000; 283:2671–2679

Translation of complex data into media headline

TODAY'S NEWS
Coffee Cure!
Drinking Coffee Prevents Parkinsons

But media reports don't have the time and space to explain all the information.

So the general public may not receive important details and potential limitations of the single study.

· As shown in the data table, even some coffee drinkers develop Parkinson disease, so not everyone will benefit.
· The results are reflecting a correlation, not a causation. This is not direct evidence that coffee is a cure.
· This study was carried out with a particular male population, so we cannot generalize the results to other populations (e.g., women).

people to stick to the regimen for the length of the study. (And such studies are unethical if the experimental treatment is likely to cause harm.)

Getting Beyond the Buzz

While a lower risk of Parkinson disease represents a potential boon to coffee drinkers, the news for caffeine addicts isn't all good. Over the years, epidemiological studies have linked caffeine consumption to *higher* rates of various diseases, including osteoporosis, fibrocystic breast disease, and bladder cancer. As with the link to Parkinson disease, however, such correlations do not necessarily prove that caffeine causes any of these diseases.

Nevertheless, such studies are often quite influential and newsworthy–like the supposed link between coffee and pancreatic cancer that made headlines in 1981. That study was based on a single epidemiological study, which was later discounted by further research.

Journalists face unique challenges in covering health news, says Gary Schwitzer of HealthNewsReview.org: "They must cover complex topics, do it quickly, creatively, accurately, completely and with balance–and then be sure they don't 'dumb it down' too much for a general news audience. . . . If they can't do it right, they must realize the *harm* they can do by reporting inaccurately, incompletely, and in an imbalanced way" (Infographic 1.8).

Journalists and scientists aren't the only ones who bear the responsibility of determining what information is trustworthy. As consumers and citizens, we can become more knowledgeable about how science is done and which studies deserve to influence our behavior. Whether it's the latest media report linking cell phones to brain tumors or vaccines to autism, the only way to really judge the value of a study is to sift through the evidence ourselves. Of course, to do that, we might first need a cup of coffee. ∎

▶ Summary

∎ Science is an ongoing process in which scientists conduct carefully designed studies to answer questions or test hypotheses.

∎ Scientific hypotheses are tested in controlled experiments or in nonexperimental studies, the results of which can support or rule out a hypothesis.

∎ Scientific hypotheses can be supported by experimental data but cannot be proved absolutely, as future experiments or technologies may provide new findings.

∎ The strength of the conclusions of a scientific study depends on, among other factors, the type of study carried out and the sample size.

∎ Every experiment should have a control—a group that is identical in every way to the experimental group except for one factor: the independent variable.

∎ The independent variable in an experiment is the one being deliberately changed in the experimental group (e.g., coffee intake). The dependent variable is the measured result of the experiment (e.g., effect of coffee on memory).

∎ Often a control group takes a placebo, a fake treatment that mimics the experience of the experimental group.

∎ In epidemiological studies, a relationship between an independent variable (such as caffeine intake) and a dependent variable (such as development of Parkinson disease) does not necessarily mean one caused the other; in other words, correlation does not equal causation.

∎ A randomized clinical trial is one in which test subjects are randomly chosen to receive either a standard treatment (or placebo) or an experimental treatment (e.g., caffeine).

∎ Scientists rely on peer-reviewed scientific reports to learn about new advances in the field. Peer review helps to ensure that the scientific results are valid as well as accurately and fairly presented.

∎ Most of the general public relies on media reports for their scientific information. Media reports are not always completely accurate in how they portray the conclusions of the scientific studies.

∎ Scientific theories are different from everyday theories. A scientific theory has withstood the test of time and extensive testing and is supported by a significant body of evidence.

PROCESS OF SCIENCE

Science is a method of seeking answers to questions on the basis of observation and experiment.

HINT See Infographics 1.1. and 1.2.

⊗ KNOW IT

1. When scientists carry out an experiment, they are testing a
   a. theory.
   b. question.
   c. hypothesis.
   d. control.
   e. variable.

2. Of the following, which is the earliest step in the scientific process?
   a. generate a hypothesis
   b. analyze data
   c. conduct an experiment
   d. draw a conclusion
   e. ask a question about an observation

⊗ USE IT

3. When a scientist reads a scientific article in a scientific or medical journal, he or she is confident that the report has been peer reviewed. What does this mean? Why is peer review important?

DESIGNING EXPERIMENTS

Many considerations go into the design and implementation of a scientific experiment.

HINT See Infographics 1.3–1.4.

⊗ KNOW IT

4. In a controlled experiment, which group receives the placebo?
   a. the experimental group
   b. the control group
   c. the scientist group
   d. the independent group
   e. all groups

5. In the studies of coffee and memory discussed, the independent variable was _____ and the dependent variable was _____.
   a. caffeinated coffee; decaffeinated coffee
   b. memory; caffeinated coffee
   c. caffeine; memory
   d. memory; caffeine
   e. decaffeinated coffee; caffeinated coffee

⊗ USE IT

6. You are working on an experiment to test the effect of a specific drug on reducing the risk of breast cancer in postmenopausal women. Describe your control and experimental groups with respect to age, gender, and breast cancer status.

7. Design a randomized clinical trial to test the effects of caffeinated coffee on brain activity. Design your study so that the results will be as broadly applicable as possible.

EVALUATING EVIDENCE

Many factors can influence the strength of a scientific claim.

HINT See Infographics 1.4–1.8.

⊗ KNOW IT

8. From what you have read in this chapter, would you say a 21-year-old Caucasian female can count on caffeinated coffee to reduce her risk of Parkinson disease?
   a. yes, because the results of a peer-reviewed study showed that drinking caffeinated beverages reduced the risk of Parkinson disease
   b. no, because subjects in that peer-reviewed study were Japanese-American males; it cannot be inferred that the same results would hold for Caucasian females
   c. no; she would have to restrict her consumption of coffee to decaffeinated coffee to reduce her risk of Parkinson disease
   d. yes; coffee is known to reverse the symptoms of Parkinson disease
   e. There is no data on the relationship between drinking caffeinated beverages and Parkinson disease because it would be unethical to conduct such an epidemiological study.

9. In which type of study would you have the most confidence?
   a. a randomized clinical trial with 10,000 subjects
   b. a randomized clinical trial with 5,000 subjects
   c. an epidemiological study with 15,000 subjects
   d. an endorsement of a product by a movie star
   e. a report on a study presented by a new organization

10. Your friend's mother has always been a coffee addict. She recently received a diagnosis of Parkinson disease. Does her experience negate the results of the *JAMA* study described in this chapter? Why or why not?

11. Depending on the television station that you watch, you may have seen advertisements that show beautiful people with clear skin who claim that a specific skin care product is "scientifically proven" to reduce acne. The product reportedly gave these people their glowing, clear skin.
   a. Is their testimony itself strong enough evidence for you to act on? Why or why not?
   b. What kind of scientific evidence would convince you to spend money on this product? Explain your answer.

SCIENCE AND ETHICS

12. You know that scientific reports are subject to peer review before being published in scientific journals. Do you think that scientists should also review media reports about their studies and work to correct any misleading statements? Why or why not? Who is ultimately responsible for what is reported in the popular press?

13. Your grandmother has told you about the changes she is making to her diet because of stories she has read in the news. Make a checklist of things she should consider before changing her behavior.

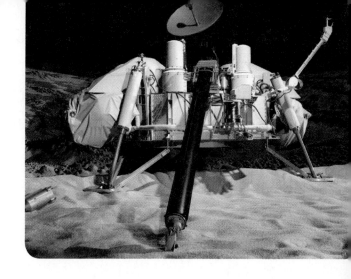

What Is Life?

What Is Life?

Evidence from space heats up an age-old debate

With a flash of fiery light, a shooting star streaks across the night sky. Thirteen thousand years later, on December 27, 1984, geologist Roberta Score picks up that shooting star and holds it in her mittens. It is a grapefruit-size rock, weighing just over 4 pounds, whose dark gray-green color stands out sharply against the brilliant white of the frozen Antarctic ice cap.

Score is one of a team of six researchers with ANSMET, the Antarctic Search for Meteorites program, who for six weeks every year crisscross the mile-thick ice in snowmobiles, searching for booty from space.

Score knew the rock was special as soon as she saw it. Coated in a layer of molten glass, or fusion crust, it had the telltale sign of having blazed through the atmosphere, but was otherwise unique in color and texture. "Yowza-Yowza," wrote the team in their field report. The first meteorite to be catalogued that year, it was named ALH84001, after Allan Hills, the patch of ice where it was discovered.

Each year, tens of thousands of such meteorites, often called shooting stars, fall to earth. Most are commonplace chunks of interstellar debris left over from the dawn of the solar system. But this one was special. At 4.5 billion years old, it is by far the oldest of only a handful of meteorites known to have come from Mars. NASA scientists believe the rock was kicked off the surface of Mars and jettisoned into space when a comet or meteorite struck that planet some 16 million years ago. It then floated in space until nudged again, this time toward earth.

In 1996, just 12 years after its discovery, the rock was catapulted into international fame when a team of NASA researchers claimed to have found evidence of Martian life inside it. Presenting their findings in the journal *Science*, lead author David McKay, a planetary scientist at NASA's Johnson Space Center,

The surface of Mars.

described what he said was convincing evidence of "primitive life on early Mars" found within the ancient rock.

> **"If this discovery is confirmed, it will surely be one of the most stunning insights into our universe that science has ever uncovered."** —Bill Clinton

The report sent shock waves through the press: "Life on Mars: Official," proclaimed the UK *Daily Mirror*. "We're Not Alone," echoed the Montreal *Gazette*. "E.T., phone Mars," requested the *Boston Globe*. President Bill Clinton held a press conference to mark the occasion, declaring, "Today, rock 84001 speaks to us across all those billions of years and millions of miles. It speaks of the possibility of life. If this discovery is confirmed, it will surely be one of the most stunning insights into our universe that science has ever uncovered."

And yet what began with excitement and fanfare quickly took a decidedly sour turn when other researchers stepped up to cast doubt on the evidence. The microscopic findings in the meteorite could have been produced without life, skeptics argued. NASA scientists had overblown the significance of their findings, critics said.

The *Viking Lander 1* spacecraft.

The Martian meteorite ALH84001.

Two years after NASA's historic announcement, biologist Andrew Knoll of Harvard University told *Science,* "You would have a hard time finding even a small number of people who are enthused by the idea of life being recorded in this meteorite."

But what would definitive evidence of Martian life look like? Would we even recognize it? These are not just idle questions mulled over by imaginative *Star Trek* fans. They go to the heart of a fundamental debate in biology, one that has been raging since Aristotle: What is life?

The Search for Alien Life

NASA's search for life on Mars began in 1964, when the *Mariner 4* spacecraft photographed the planet during a deep-space flyby, providing us with the first up-close pictures of the red planet. The photographs revealed a dry, rocky landscape, more reminiscent of our lifeless moon than the lush, blue marble we call home.

But looks can be deceiving, so NASA followed up its *Mariner* missions with *Viking Lander 1*— the first spacecraft to land on the Martian surface, touching down on July 20, 1976.

Equipped with mechanical arms that could grab and test Martian soil, *Viking Lander 1* was designed to look for signs of life. NASA scientists hypothesized that if life were present in the soil, then they should be able to measure its chemical signature. Was anything emitting

HOMEOSTASIS
The maintenance of a relatively constant internal environment.

ENERGY
The ability to do work. Living organisms obtain energy from food, which they either make using the energy of sunlight or consume from the environment.

INFOGRAPHIC 2.1

Five Functional Traits of Life

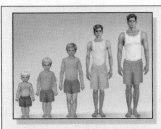

Growth:
For unicellular (one-cell) organisms, this is an increase in cell size prior to reproduction. For multicellular organisms, growth refers to an increase in an organism's size, as the number of cells making up the organism increases.

Reproduction:
The process of producing new organisms. Offspring are similar, but not necessarily identical, to their parents in general structure, function, and properties.

Homeostasis:
Organisms maintain a stable internal environment, even when the external environment changes.

Sense and Respond to Stimuli:
Organisms respond to stimuli in many ways. For example, they may move toward a food source or move away from a threatening predator.

Obtain and Use Energy:
All living organisms require an input of energy to power their activities. Organisms obtain energy from food (which they either produce themselves or consume from the environment). Chemical reactions convert that energy into usable forms. The sum total of all these reactions is metabolism.

METABOLISM
All the chemical reactions taking place in the cells of a living organism that allow it to obtain and use energy.

to be breaking down the nutrients and producing carbon dioxide gas. More intriguing, when the experiment was repeated after the soil was heated to a very high temperature (a temperature that would kill most life), no carbon dioxide was measured. The researchers interpreted this experiment as evidence for Martian life.

But the results were far from definitive. Subsequent analyses revealed that the Martian soil could have produced carbon dioxide through strictly abiotic (that is, nonliving) means by a chemical reaction similar to combustion. Whether or not living organisms were responsible for the carbon dioxide remained unclear.

In looking for a specific chemical reaction, the NASA scientists were employing a definition of life based on what living things *do*. Biologists generally agree that–on earth at least–all living things have in common five functional traits, traits that rocks and sand will never have and robots don't yet have. Specifically, living things (1) grow and (2) reproduce: they increase in size and produce offspring that are similar but not necessarily identical to their parents. Living things also (3) maintain a relatively stable internal environment in the face of changing external circumstances–producing heat when they're cold, for example–a phenomenon known as **homeostasis.** To maintain homeostasis, (4) living things sense and respond to their environment, as when a plant grows toward sunlight. And to carry out these and other life-defining activities, (5) all living organisms obtain and use **energy,** the power to do work. Energy comes from sunlight or food, which living things break down through a series of chemical reactions, the sum total of which is called **metabolism** (Infographic 2.1).

In looking for carbon dioxide, the NASA scientists were looking for evidence of chemical metabolism. But the inconclusive results of the experiment demonstrate why it is risky to rely on any one functional trait as the defining feature of life: it's always possible to come up with an exception to the rule.

For example, the ability to reproduce would seem to be a fundamental principle of life–and it

carbon dioxide, for example, as many organisms on earth do? Researchers put Martian soil in a sterile container filled with nutrients and waited to see what would happen.

Initially, the results seemed promising: something in the Martian soil did indeed seem

is. Yet this definition alone would exclude some entities that are clearly alive, such as mules, which are sterile and thus cannot reproduce. Similarly, if the sole definition is that living things consume energy and grow, we could claim that fire is alive, and yet that doesn't seem right.

Carol Cleland is a philosopher at the University of Colorado and a member of NASA's Astrobiology Institute who has spent a lot of time thinking about the problem of defining life. At a NASA-sponsored conference on astrobiology held in 2006 she said, "There's a serious problem with trying to answer the question 'What is life?' and designing a search for life based upon definitions. Yet this is something that's been commonly done." It's what NASA's *Viking* mission did, for example. The problem with this strategy, she explained, is that a functional "definition" of life will match only our current beliefs about life; it leaves no room for life that functions or behaves differently from the way it does on earth.

Another approach, which NASA scientists have also used to look for life on Mars, is to search for the distinctive chemical building blocks of life. Regardless of how it functions, at its most basic level all life is a chemical concoction, a chemical soup. We can therefore analyze life, in part, by analyzing that soup's ingredients.

ELEMENT
A chemically pure substance that cannot be chemically broken down; each element is made up of and defined by a single type of atom.

MATTER
Anything that takes up space and has mass.

ATOM
The smallest unit of an element that cannot be chemically broken down into smaller units.

INFOGRAPHIC 2.2

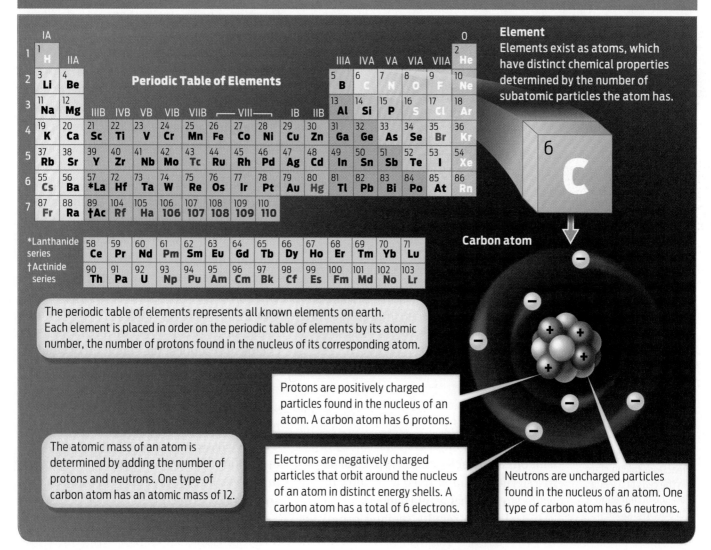

All Matter on Earth Is Made of Elements

Periodic Table of Elements

The periodic table of elements represents all known elements on earth. Each element is placed in order on the periodic table of elements by its atomic number, the number of protons found in the nucleus of its corresponding atom.

Element
Elements exist as atoms, which have distinct chemical properties determined by the number of subatomic particles the atom has.

Carbon atom

Protons are positively charged particles found in the nucleus of an atom. A carbon atom has 6 protons.

The atomic mass of an atom is determined by adding the number of protons and neutrons. One type of carbon atom has an atomic mass of 12.

Electrons are negatively charged particles that orbit around the nucleus of an atom in distinct energy shells. A carbon atom has a total of 6 electrons.

Neutrons are uncharged particles found in the nucleus of an atom. One type of carbon atom has 6 neutrons.

INFOGRAPHIC 2.3

Carbon Is a Versatile Component of Life's Molecules

→ Molecules are chains of atoms linked by covalent bonds. The element carbon is a key component of the molecules of living organisms because it can form multiple covalent bonds.

Carbon can form multiple covalent bonds:

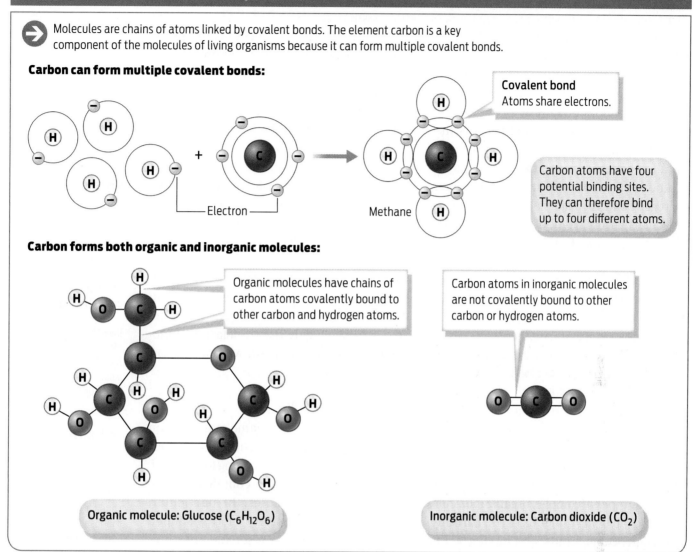

Covalent bond
Atoms share electrons.

Electron

Methane

Carbon atoms have four potential binding sites. They can therefore bind up to four different atoms.

Carbon forms both organic and inorganic molecules:

Organic molecules have chains of carbon atoms covalently bound to other carbon and hydrogen atoms.

Carbon atoms in inorganic molecules are not covalently bound to other carbon or hydrogen atoms.

Organic molecule: Glucose ($C_6H_{12}O_6$)

Inorganic molecule: Carbon dioxide (CO_2)

PROTON
A positively charged subatomic particle found in the nucleus of an atom.

ELECTRON
A negatively charged subatomic particle with negligible mass.

NEUTRON
An electrically uncharged subatomic particle found in the nucleus of an atom.

NUCLEUS
The dense core of an atom.

Life's Recipe

So far, all life we know of–from amoeba to leaf to zebra–uses the same basic chemical recipe: a stew of carbon-based ingredients floating in a broth of water. Carbon is one of approximately 100 different **elements** found on earth. Elements are substances that cannot be broken down by chemical means into smaller substances. They are considered the fundamental components of anything that takes up space or has mass–the **matter** in the universe.

The smallest unit of an element that still retains the property of an element is an **atom.** What gives each atom its identity is the specific number of positively charged **protons,** negatively charged **electrons,** and neutral **neutrons** each atom has.

The element carbon, for example, is made up of atoms with six protons, six electrons, and six neutrons. The relatively heavy protons and neutrons are packed into the atom's dense core, or **nucleus,** while the tiny electrons orbit around it (Infographic 2.2).

Carbon is the fourth most common element in the universe and the second most common element in your body. In fact, just six elements make up the bulk of you: oxygen (65%), carbon (18.5%), hydrogen (9.5%), nitrogen (3.3%), and phosphorus and sulfur (2%).

Carbon has unique properties that make it an ideal backbone for life. Most important, it easily forms long chains and rings. You can think of carbon atoms as having four attachment, or bonding, sites. By sharing electrons with its neighbors, carbon can form **covalent bonds** with two, three, or four other atoms, giving the element enormous versatility.

When atoms are linked by covalent bonds, they form **molecules.** Living things are made up of so-called **organic molecules,** which have a backbone of carbon with at least one carbon-hydrogen bond. An example of a simple organic molecule is glucose, a type of sugar. Its molecular formula is $C_6H_{12}O_6$. This means that each molecule of glucose has 6 carbon atoms, 12 hydrogen atoms, and 6 oxygen atoms. Glucose is a ring-shaped molecule, with the carbon atoms forming the backbone of the ring. Carbon dioxide (CO_2), however, is an **inorganic molecule**—it does not have a carbon-carbon backbone and a carbon-hydrogen bond (**Infographic 2.3**).

When astrobiologists (and science fiction writers) talk about "carbon-based life forms," they are talking about our chemical makeup of organic molecules. The particular organic molecules that NASA scientists hoped to find in Martian soil during the *Viking* mission were any of the four types of complex organic molecules that make up living things on earth: **carbohydrates, proteins, lipids,** and **nucleic acids.** Every molecule forming the structure of your body can be classified as one of these organic molecules. Your skin, for example, is composed of the proteins collagen and elastin, and the padding in your soft spots is composed of lipids, also known as fats.

Carbohydrates, proteins, and nucleic acids can be quite large and are therefore considered **macromolecules.** Macromolecules share a similar organization in that they are composed of subunits called **monomers** linked together in a chain. When two or more monomers join together they form a **polymer.** Carbohydrates, for example, are polymers made up of linked monomers called **monosaccharides;** similarly, proteins are made up of subunits called **amino acids** that are bonded together; and

nucleic acids are polymers composed of **nucleotides** that form long chains (see **Up Close: Molecules of Life**).

Despite careful efforts, NASA's *Viking* probe failed to find any of these life-defining organic molecules in Martian soil. At the time, NASA's conclusion was that the Martian surface is self-sterilizing, meaning that no living organisms could survive in the harsh conditions. The combination of intense solar radiation, the extreme dryness of the soil, and a soil chemistry resembling combustion all make the Martian surface a particularly inhospitable place. Not to mention the fact that it's extremely cold: $-120°C$ ($-184°F$) in the pre-dawn winter.

More recently, some researchers have argued that any organic molecules present on Mars would be quickly broken down and destroyed by the highly reactive Martian atmosphere. This could explain why none were detected by *Viking*.

Traces of Ancient Life

Viking's failure to find organic molecules on Mars made the 1996 discoveries in ALH84001 all the more surprising. According to NASA scientists, ALH84001 clearly contains carbon-based organic molecules. In particular, scientists found a variety of ring-shaped organic molecules that resemble ones produced when living things burn or decay. While the presence of such organic molecules does not in itself prove the presence of life—they can be produced without life—NASA scientists argued that their location within the meteorite, near other potential markers of life, strengthened the case for life on Mars.

Where did these organic molecules come from if *Viking* did not detect them in Martian soil but they are clearly present in ALH84001? The meteorite definitely came from Mars. Scientists know this because the trapped gases in the rock perfectly match the profile of gases recorded by *Viking Lander 1*. Scientists believe that ALH84001 is a piece of volcanic rock that was churned up from deep within the Martian surface. Since ALH84001 likely came from sub-

COVALENT BOND
A strong chemical bond resulting from the sharing of a pair of electrons between two atoms.

MOLECULE
Atoms linked by covalent bonds.

ORGANIC MOLECULE
A molecule with a carbon-based backbone and at least one C–H bond.

INORGANIC MOLECULE
A molecule that lacks a carbon-based backbone and C–H bonds.

CARBOHYDRATE
An organic molecule made up of one or more sugars. A one-sugar carbohydrate is called a monosaccharide; a carbohydrate with multiple linked sugars is called a polysaccharide.

PROTEIN
An organic molecule made up of linked amino acid subunits.

LIPIDS
Organic molecules that generally repel water.

NUCLEIC ACIDS
Organic molecules made up of linked nucleotide subunits; DNA and RNA are examples of nucleic acids.

terranean Mars, could there be life beneath the surface of the red planet?

It's a distinct possibility. NASA hopes one day to be able to answer this question definitively by drilling deep into the Martian crust and hauling Martian soil back to earth for analysis–but that mission is a long way off. In the shorter term, NASA plans to send a rover to Mars in 2011 to explore Martian soil more closely than was possible with *Viking*. Known as the *Mars Science Laboratory*, or *Curiosity*, the rover will be able to perform a variety of extremely sensitive chemical tests on the soil, including ones designed to detect minute quantities of amino acids, carbohydrates, lipids, and nucleic acids.

Given the problems of defining life–knowing what to look for–what does philosopher Cleland think of NASA's plan to search for life using organic molecules as an indicator? "I think it is a good idea," she says, noting that they're what life on earth is made of. But she

> **"We shouldn't lock ourselves into a definition that might blind us to the presence of unfamiliar forms of life."**
> –Carol Cleland

warns that we shouldn't turn the detection of organic molecules into an absolute requirement for life, or make it our definition. "Because our experience of life is limited to a single example–familiar earth life," she explains, "we shouldn't lock ourselves into a definition that might blind us to the presence of unfamiliar forms of life should we be so fortunate to encounter them."

Martian Bacteria?

Besides organic molecules, the really tantalizing find in meteorite ALH84001 was the presence of what looked like the fossilized remains of microscopic organisms. A widely publicized photo that has since become famous shows what looks to be a wormlike creature inching its way through a sample of the meteorite. Other pictures of the meteorite show jelly bean-shaped structures resembling bacteria.

The tiny fossilized "beans" found in ALH84001 resemble a type of bacteria on earth

MACROMOLECULES
Large organic molecules that make up living organisms; they include carbohydrates, proteins, and nucleic acids.

MONOMER
One chemical subunit of a polymer.

POLYMER
A molecule made up of individual subunits, called monomers, linked together in a chain.

MONOSACCHARIDE
The building block, or monomer, of a carbohydrate.

AMINO ACID
The building block, or monomer, of a protein.

NUCLEOTIDE
The building block, or monomer, of a nucleic acid.

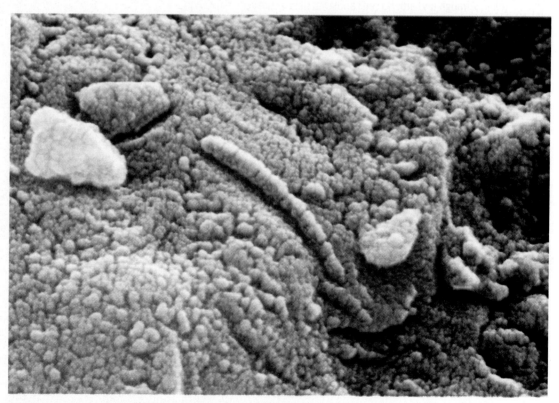

Microscopic fossil-like structures found in the ALH84001 meteorite.

a. Carbohydrates Are Made of Monosaccharides

Carbohydrates are made up of repeating subunits known as monosaccharides, or simple sugars. Carbohydrates act as energy-storing molecules in many organisms. Other carbohydrates provide structural support for cells.

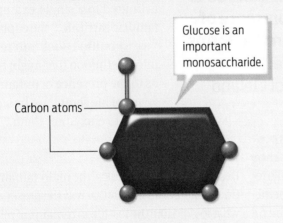

Glucose is an important monosaccharide.

Carbon atoms

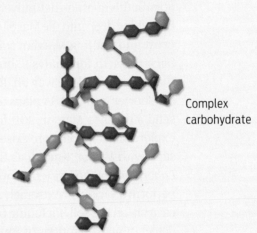

Complex carbohydrate

Monosaccharides
The backbone of carbon atoms in monosaccharides is most often arranged in a ring.

Complex carbohydrates
Monosaccharides like glucose can be bonded together in straight or branching chains called complex carbohydrates.

b. Proteins Are Made of Amino Acids

Proteins are polymers of different small repeating units called amino acids joined together by peptide bonds. Proteins carry out many functions in cells. They help speed up the rate of chemical reactions. They also move things through and around cells and even help entire cells move.

Amino Acid
There are 20 different amino acids found in proteins. Each amino acid shares a common "core" structure (shown in green).

Amino group

Carboxyl group

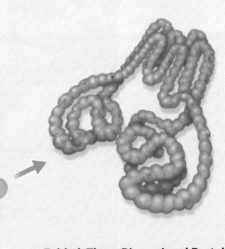

Linear Strand of Amino Acids
Different amino acids have different "side chains" (highlighted in different colors).

met   leu   gly   val   leu   ala   ser   gln   pro   trp

Folded, Three-Dimensional Protein
Proteins do not function properly until they fold up into their unique three-dimensional shape.

c. Lipids Are Hydrophobic Molecules

There are different types of lipids, each with a distinct structure and function. Lipids are not made up of repeating subunits or building blocks, but they are all hydrophobic molecules, meaning they don't mix with water.

Saturated

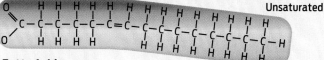
Unsaturated

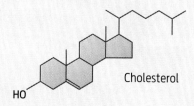
Cholesterol

Fatty Acids
Fatty acids contain long chains of carbon atoms bonded to one another and to hydrogen atoms.

Sterols
Sterols have four connected carbon rings. Cholesterol is a sterol that's an important component of cell membranes. Other sterols may be hormones or color-inducing pigments.

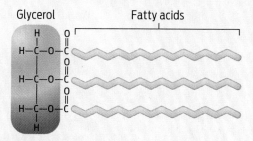
Glycerol   Fatty acids

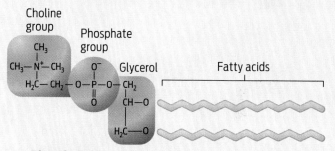
Choline group   Phosphate group   Glycerol   Fatty acids

Triglycerides
Triglycerides, also known as fat, have three fatty acid chains attached to a glycerol molecule. Fats store large amounts of energy and also provide padding and thermal insulation.

Phospholipids
Phospholipids have two fatty acid chains and a phosphate group attached to a glycerol molecule. Phospholipids are an important component of cell membranes.

d. Nucleic Acids Are Made of Nucleotides

Nucleic acids are polymers of repeating subunits known as nucleotides. There are two types of nucleic acids, DNA and RNA, each of which is made up of slightly different types of nucleotides. DNA and RNA are critical for the storage, transmission, and execution of genetic instructions.

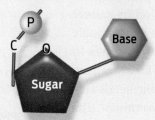

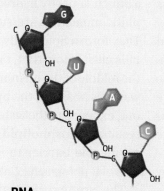

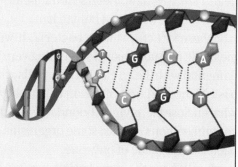

Nucleotide
Nucleotides share a common "core" structure, including a phosphate group and a sugar, which varies slightly between DNA and RNA. Each of the five different nucleotides differs by virtue of the individual base.

RNA
RNA molecules consist of only one linear chain of bonded nucleotides.

DNA
A DNA molecule consists of two chains of bonded nucleotides twisted into a helical shape.

A Layer Rich in Phospholipids Defines Cell Boundaries

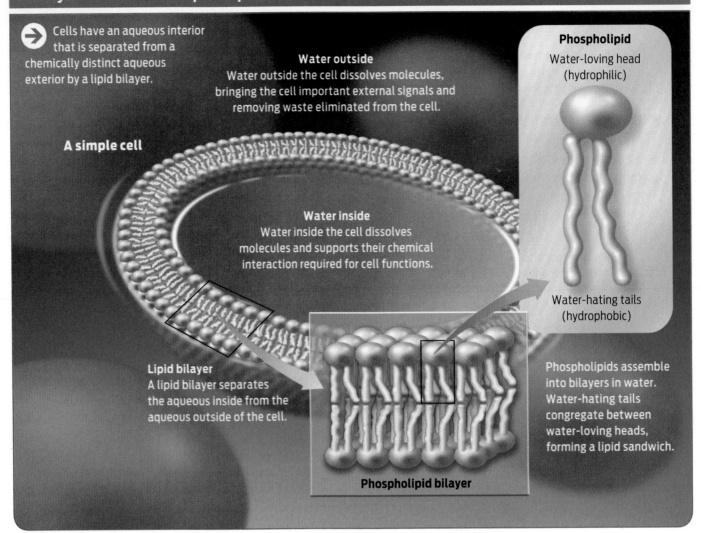

Cells have an aqueous interior that is separated from a chemically distinct aqueous exterior by a lipid bilayer.

Water outside
Water outside the cell dissolves molecules, bringing the cell important external signals and removing waste eliminated from the cell.

A simple cell

Water inside
Water inside the cell dissolves molecules and supports their chemical interaction required for cell functions.

Phospholipid
Water-loving head (hydrophilic)

Water-hating tails (hydrophobic)

Lipid bilayer
A lipid bilayer separates the aqueous inside from the aqueous outside of the cell.

Phospholipids assemble into bilayers in water. Water-hating tails congregate between water-loving heads, forming a lipid sandwich.

Phospholipid bilayer

known as nanobacteria–"nano" for their exceptionally tiny size. The largest of the fossilized beans are 100 nanometers in diameter–less than 1/100 the width of a human hair.

If these structures *were* bacteria, it would mean life on Mars shared something very fundamental with life on earth: cells. **Cells** are the basic structural unit of life on earth; they are what enclose life, giving it boundaries. Humans contain trillions of cells; some organisms, like bacteria, are made of only one.

All cells have the same basic structure: they are water-filled sacs bounded by a membrane rich in lipids. The membrane is essentially a sandwich of lipids. In particular, the lipid membrane is made of a type of lipid called a **phospholipid.** Each phospholipid has one **hydrophobic** ("water-fearing") end that

repels water and a **hydrophilic** ("water-loving") end that attracts it. What happens when a bunch of partly hydrophobic, partly hydrophilic molecules are surrounded by water? They form a lipid sandwich: the hydrophobic tails cluster together, burying themselves in the middle of the membrane, as far away from water as possible; the hydrophilic heads face out, exposed to the watery environment. The resulting **phospholipid bilayer** forms a semipermeable barrier to substances on either side of it **(Infographic 2.4).**

The original team of NASA researchers argued that at least some of these oval lumps could be the remains of bacteria-like organisms. But other scientists were skeptical, arguing that the lumps were far too small to house the necessary components of living cells. More likely,

CELL
The basic structural unit of living organisms.

PHOSPHOLIPID
A type of lipid that forms the cell membrane.

PHOSPHOLIPID BILAYER
A double layer of lipid molecules that characterizes all biological membranes.

they said, the structures were formed from non-living chemical processes that just happened to form oval shapes.

Chris McKay, an astrophysicist with NASA's Ames Research Center who is generally sympathetic to the quest to find life on Mars, is skeptical of the famous "nano-worm." There is "no evidence that these shapes had anything to do with biology," says McKay.

Though the case for cellular life in ALH84001 has been weakened, it has not been completely ruled out. According to many researchers, the strongest evidence for life in the meteorite is the presence of so-called magnetite grains—tiny magnetic particles composed of iron that are found alongside the bacteria-like beans. On earth, similar magnetite particles are used by certain bacteria as a kind of navigation device, like a magnet in a compass. In the bacterial compass, the magnetite grains are arranged end to end, like beads on a string.

> "Liquid water is the key requirement in the search for life."
> —Chris McKay

"The magnetite grains remain intriguing enough that I am sure that this will be one of the first things investigated on a Mars sample return," says Chris McKay. "If we find the magnetite grains aligned in 'string-of-pearls' fashion this would be good evidence of a biological origin." It other words, it would mean Martian bacteria, and therefore Martian cells.

Follow the Water

In their search for extraterrestrial life, astrobiologists often say, "Follow the water." Water is viewed as a proxy for life because it is so crucial to life on earth. Water makes up 75% to 85% of a cell's weight. All of life's chemical reactions take place in water, and many living things can survive only a few days without it.

A simple Mickey Mouse-shaped molecule consisting of one oxygen atom bound to two hydrogens, water comes pretty close to being a miracle substance. It is a universal **solvent,** capable of dissolving just about any substance—even gold. Water transports all of life's dissolved molecules, or **solutes,** from place to

place—whether through a cell, a body, or an ecosystem. Life, in essence, is a water-based **solution.**

But water is more than just a stage on which the chemical reactions in question take place. It is a principal actor. Many biological molecules, like proteins and DNA, have the necessary shapes they do only because of the surrounding water that they interact with.

What makes water such a good solvent? Because the electrons in a water molecule are shared unequally between the oxygen and hydrogen atoms, water is considered a **polar molecule.** With a partial negative and a partial positive charge on either end, water is an excellent solvent for other polar molecules with partial charges and substances like salt that contain **ionic bonds.** Ionic bonds are strong bonds formed between oppositely charged **ions.** By surrounding each charged ion, water dissolves the bond between them (**Infographic 2.5**).

When astrobiologists speak about the importance of water for life, they make an important qualification: *liquid* water. Frozen water is found throughout the universe; there are abundant quantities on Mars, for example. But only on earth does water exist primarily in its liquid form at ambient temperature and pressure.

"Liquid water is the key requirement in the search for life," says astrophysicist Chris McKay. "The other worlds of the solar system have enough light, enough carbon, and enough of the other key elements for life. Water in the liquid form is rare."

Why is water liquid at room temperature? Essentially, it's because water molecules are "sticky." Each water molecule has a partial charge on each end and can therefore form electrostatic attractions, known as **hydrogen bonds,** with one another and with other molecules. These hydrogen bonds act as a kind of glue holding water molecules together and keeping them liquid at room temperature. You can see water's stickiness wherever you look:

a drop of water clinging to a leaf despite the downward pull of gravity, for example, or an insect able to land on the surface of a pond (Infographic 2.6).

Compared to other molecules its size, water also has a large liquid range—freezing at 0°C (32°F) and boiling at 100°C (212°F). That's because water molecules can absorb a lot of energy before they get hot and vaporize (that is, turn into a gas)—again because of their tenacious hydrogen bonds. Because of these bonds, water gets hot more slowly than do other liquids and also holds onto heat longer. And water's liquid range can be extended even further: add salt to water and you can lower the freezing point to −46°C (−50°F); increase the pressure and you can bump up the boiling point to over 343°C (650°F). It's because there is so much salt in seawater that most oceans don't freeze in winter.

Finally, unlike most substances on earth, water has the unusual property of being less dense as a solid than as a liquid: ice floats. And because it does, fish can live beneath frozen lakes in winter and not turn into ice cubes—which is good for both the fish and us.

Given its amazing properties, scientists want to find out whether liquid water exists on Mars. From the *Viking* missions, scientists know that frozen water exists in the form of large ice caps on the surface of Mars and also as a layer of permafrost just beneath the surface. In 2008, NASA's *Phoenix Lander* provided further evidence of frozen water in Martian soil. But so far no liquid water has been found. Scientists suspect that the Martian atmosphere is so thin and so cold that any liquid water would rapidly evaporate or freeze.

Though liquid water is not present on the surface of Mars today, many scientists suspect that liquid water—lots of it—once covered the planet. Clues to this ancient water can be seen all over the Martian surface, which in many places is carved out like sections of the Grand Canyon. The *Phoenix Lander* also found telltale signs of liquid water's past on the surface of Mars in the form of salt deposits like those you can see when seawater evaporates.

Water Is a Good Solvent Because It Is Polar

→ Water is a polar molecule because electrons are not shared equally between the oxygen and hydrogen atoms. Electrons are pulled closer to the oxygen atom than to the hydrogen atoms, creating a slightly negative oxygen atom and slightly positive hydrogen atoms. The partial charges on each water molecule can interact with charged ions or other molecules, allowing water to "coat" or dissolve the hydrophilic solutes.

Table salt is made up of oppositely charged ions, which are attracted to one another and form ionic bonds.

Water is a polar molecule because one end of it is more positive and the other is more negative.

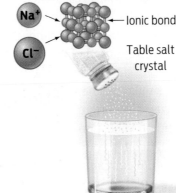

Na⁺
Cl⁻
Ionic bond
Table salt crystal

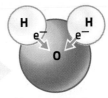

More positively charged hydrogen side

More negatively charged oxygen side

Charged substances dissolve in water:

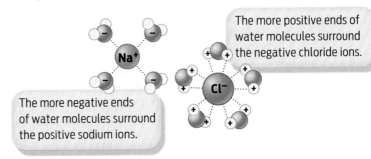

The more positive ends of water molecules surround the negative chloride ions.

The more negative ends of water molecules surround the positive sodium ions.

Additional support for the presence of ancient water comes from meteorite ALH84001. Crevices of the meteorite are filled with carbon-rich globules that resemble those produced by bacteria on earth. Scientists believe these globules could have formed only if liquid water had once percolated through the meteorite, carrying CO_2 from the Martian atmosphere into the rock.

Where all this water went, no one knows. But some scientists suspect that liquid water

pH
A measure of the concentration of H⁺ in a solution.

ACID
A substance that increases the hydrogen ion concentration of solutions, making them more acidic.

BASE
A substance that reduces the hydrogen ion concentration of solutions, making them more basic.

Water Is "Sticky" Because It Forms Hydrogen Bonds

→ When many water molecules are near one another, the partially positive hydrogen atoms of some molecules are attracted to the partially negative oxygen atoms of neighboring water molecules. These attractions are hydrogen bonds, weak electrical attractions.

Hydrogen Bonds — two polar molecules are attracted to each other

Partial negative charge

δ^-

O

H

H

δ^+

δ^+

Partial positive charge

Hydrogen bond between opposite partial charges

Cohesion
Hydrogen bonding between water molecules is strong enough to defy gravity, allowing water to flow up stems of even the tallest plants. This cohesive property supports life, for example by providing surface tension on lakes for insects to land on.

Adhesion
The partial charges on water molecules allow them to readily bind to many surfaces, making them wet. Leaves can collect water for the organisms that live on them.

may still exist beneath the surface of the planet, and may even bubble to the surface periodically, as is suggested by photographs of apparent water flows taken in 2004 and 2005 by NASA's *Mars Global Surveyor* satellite. The possibility that water existed on Mars in the past, and may still exist today beneath the surface, raises the question of whether a belowground habitat exists that is conducive to life.

If there is water within Mars, would it have the properties of earth water? Depending on what's dissolved in it, water can have a large range of characteristics—from caustic drain cleaner and ammonia to tart lemon juice and cavity-causing soda. The different chemical properties of water-based solutions reflect their **pH,** the concentration of hydrogen ions (H^+) in a solution, which is defined as ranging from 0 to 14. Here's the background of measurement by pH: water molecules (H_2O) can split briefly into separate hydrogen (H^+) and hydroxide (OH^-) ions. In pure water, the number of separated H^+ ions is by definition exactly equal to the number of separated OH^- ions, and the pH is therefore 7, or neutral. Acidic solutions, or **acids,** have a higher concentration of hydrogen ions (H^+) and a pH closer to 0. When acids are added to water, they increase the concentration of hydrogen ions and make the solution more acidic. Basic solutions, or **bases,** on the other hand, have a lower concentration of H^+ ions and a pH closer to 14. Bases remove H^+ ions from a solution, thereby increasing the proportion of OH^- ions.

Strong acids and bases are highly reactive with other substances, which makes them destructive to the molecules in a cell. Also, many biochemical reactions take place only at a certain pH. Living things are thus extremely sensitive to changes in pH, and most function best when their pH stays within a specific range. The pH of human blood ranges from about 7.35 to 7.4. If that pH were to fall even slightly, to 7, our biochemistry would malfunction and we would die.

The *Viking* experiments determined that the pH of Martian soil is roughly 7.2. The *Phoenix Lander* recently calculated it at 7.7—mild enough to grow asparagus, as the mission's chief chemist put it. (Infographic 2.7)

"Weird Life"

So far, NASA's search for life on Mars has hewn very closely to our understanding of life on earth, where living things seem to share certain chemical and structural properties, like carbon-based molecules and cells. Nevertheless, there are a few exceptions, or boundary cases, that seem to bend the rules of life on earth. **Viruses** are an example. Viruses reproduce and pass their genetic information on to new viruses, but they are not made of cells at all. Instead, they are infectious particles consisting of a protein shell that encloses genetic information. Viruses reproduce by infecting a host cell and hijacking its cellular machinery to make copies of itself. Other noncellular, self-reproducing entities include **prions,** infectious proteins that are responsible for mad cow disease and related human and animal illnesses. Whether or not viruses and prions are truly alive is hotly debated among scientists.

If viruses and prions bend the rules, then might not Martian life as well? In 2008, the National Academy of Sciences issued a "weird life" report suggesting that NASA not be so narrowly focused on water and organic molecules

A binocular microscopic view of carbonate globules in ALH84001.

Solutions Have a Characteristic pH

→ The pH of a solution is a measure of the concentration of hydrogen ions (H⁺) in it. Solutions with a low concentration of H⁺ ions have a basic pH (greater than pH 7). Solutions with a high concentration of H⁺ ions have an acidic pH (a pH of less than 7). Both acids and bases can be damaging because they are highly reactive with other substances. A neutral solution has a pH of 7.

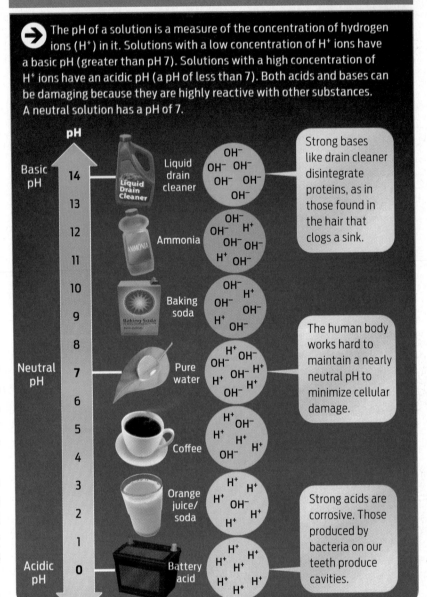

Basic pH — 14, 13, 12, 11, 10, 9, 8
Liquid drain cleaner
Ammonia
Baking soda

Neutral pH — 7, 6
Pure water

5, 4 — Coffee

3, 2, 1 — Orange juice/soda

Acidic pH — 0 — Battery acid

Strong bases like drain cleaner disintegrate proteins, as in those found in the hair that clogs a sink.

The human body works hard to maintain a nearly neutral pH to minimize cellular damage.

Strong acids are corrosive. Those produced by bacteria on our teeth produce cavities.

in its search for life on other planets. True, water may be crucial to life on earth, but that doesn't mean that other solvents—ammonia or methane, for example—could not support life elsewhere, the report noted. The report also urged the space agency to avoid being "fixated on carbon," even though carbon forms the scaffold of life on earth. Other elements, like silicon, for example, could in theory provide a functional scaffold for life on other planets.

VIRUS
An infectious agent made up of a protein shell that encloses genetic information.

PRION
A protein-only infectious agent.

Our first panoramic view of the surface of Mars.

"I spend my time and energy in the search for evidence of life on Mars." —Chris McKay

Recent discoveries by biologists who study microorganisms are also challenging our notions of what life looks like and where it can survive. Microscopic organisms have been found growing just about anywhere, from radioactive waste and boiling geysers to sunless deep-sea vents and Arctic springs made almost entirely of salt. Such extreme-loving organisms reveal that life is nothing if not adaptive. Could similarly adaptive organisms have once inhabited Mars? Might they still? At least some astrobiologists are cautiously optimistic.

"I spend my time and energy in the search for evidence of life on Mars," says Chris McKay. "Obviously, this is because I think there must have been life there and we have a good chance of finding evidence of it." ■

⊙ Summary

■ "Life" is difficult to define in universal terms because we have only a single example of it to consider: life on earth.

■ On earth, living organisms share a number of functional characteristics: they grow and reproduce, maintain homeostasis, sense and respond to their environment, and rely on energy to carry out their functions.

■ All matter is composed of elements, of which there are about 100 on earth. Each element has a unique atomic structure, with a particular number of protons, neutrons, and electrons.

■ When atoms share pairs of electrons, they form covalent bonds, making molecules.

■ On earth, living organisms are made up of organic molecules, those containing a backbone of the element carbon.

■ Four types of carbon-based organic molecules make up living things: proteins, carbohydrates, nucleic acids, and lipids.

■ Living organisms on earth are made of cells, which contain water and are surrounded by a membrane of lipids; cells are the smallest unit of life.

■ Water is a polar molecule, with a partial positive and a partial negative charge.

■ Water has many properties that make it a crucial component of life on earth: it is a good solvent, it is "sticky," it regulates heat well, and it floats when frozen.

■ Substances, like salt, that easily dissolve in water are considered hydrophilic; substances, like lipids, that do not dissolve in water are hydrophobic.

■ The concentration of H^+ ions in a solution determines its pH. Most chemical reactions in cells take place at a nearly neutral pH.

■ If life is found on other planets, it may or may not use the chemical framework used by life on earth.

PROPERTIES OF LIFE

Life on earth is marked by certain functional, structural, and chemical properties.

HINT See Infographics 2.1–2.4 and Up Close: Molecules of Life.

⊜ KNOW IT

1. Which of the following is *not* a generally recognized characteristic of most (if not all) living organisms?
   a. the ability to reproduce
   b. the ability to maintain homeostasis
   c. the ability to obtain energy directly from sunlight
   d. the ability to sense and respond to the environment
   e. the ability to grow

2. What is homeostasis? Why it is important to living organisms?

3. The basic building blocks of life are
   a. DNA molecules.
   b. cells.
   c. proteins.
   d. phospholipids.
   e. inorganic molecules

4. What subatomic particles are located in the nucleus of an atom?
   a. protons
   b. neutrons
   c. electrons
   d. protons, neutrons, and electrons
   e. protons and neutrons

5. When an atom loses an electron, what happens?
   a. It becomes positively charged.
   b. It becomes negatively charged.
   c. It becomes neutral.
   d. Nothing happens.
   e. atoms cannot lose an electron because atoms have a defined number of electrons

6. What does it mean to say a macromolecule is a polymer? Give an example.

7. A collection of amino acids could be used to build a
   a. protein.
   b. complex carbohydrate.
   c. triglyceride.
   d. nucleic acid.
   e. cell

⊜ USE IT

8. How would you assess whether or not a possibly living organism from another planet were truly alive?

9. Which of the characteristics of living organisms (if any) allow you to distinguish between living and formerly living (that is, dead) organisms? Explain your answer.

10. What are the arguments for and against viruses being considered living organisms?

11. If, in a mound of dirt, you had evidence that carbon dioxide was being consumed and converted to glucose, what could you conclude about the presence of a living organism? Explain your answer.

12. How does a sterol, such as cholesterol, differ from a triglyceride? Structurally, what do triglycerides and phospholipids have in common?

WATER: THE SOLVENT OF LIFE

Water has many properties that make it a suitable medium for living things and their chemical reactions.

HINT See Infographics 2.5–2.7.

⊜ KNOW IT

13. Is olive oil hydrophobic or hydrophilic? What about salt? Explain your answer.

14. Two water molecules can bond to each other by _____ bonding; this is an example of _____.
   a. hydrogen; adhesion
   b. covalent; adhesion
   c. non-covalent; cohesion
   d. covalent; cohesion
   e. hydrogen; cohesion

15. Coffee, tea, or any water-based beverage with sugar in it is an example of a(n) _____ solution.
   a. What is the solvent in such a beverage?
   b. What is the solute in such a beverage?

16. As an acidic compound dissolves in water, the pH of the water _____.
   a. becomes higher
   b. remains neutral
   c. becomes lower
   d. doesn't change
   e. becomes basic

17. The bond between the oxygen atom and a hydrogen atom in a water molecule is a(n) _____ bond.
   a. covalent
   b. hydrogen
   c. ionic
   d. hydrophobic
   e. noncovalent

18. How do ionic bonds compare to hydrogen bonds? What are the similarities and differences?

⊙ USE IT

19. Why do olive oil and aqueous vinegar tend to separate in salad dressing? Will added salt dissolve in the oil or the vinegar? Explain your answer.

20. Why do deserts cool off more at night than do seaside towns?

21. Which of the following would be most likely to dissolve in olive oil?
   a. a polar molecule
   b. a nonpolar molecule
   c. a hydrophilic molecule
   d. a and c
   e. b and c

SCIENCE AND ETHICS

22. One approach to finding out if there is life on Mars is to bring Martian dirt samples to earth for analysis. What are possible considerations for science and society if a Martian life form is released on earth? If an earth life form is introduced onto Mars?

Wonder Drug

Wonder Drug

How a chance discovery in a London laboratory revolutionized medicine

On a September morning in 1928, biologist Alexander Fleming returned to his laboratory at St. Mary's Hospital in London after a short summer vacation. As usual, the place was a mess–his bench piled high with the petri dishes on which he was growing bacteria. On this day, as Fleming sorted through the plates, he noticed that one was growing a patch of fluffy white mold. It had been contaminated, likely by a rogue mold spore that had drifted in from a neighboring laboratory.

Fleming was about to toss the plate in the sink when he noticed something unusual: wherever mold was growing, there was a zone around the mold where the bacteria did not seem to grow. Curious, he looked under a microscope and saw that the bacterial cells near the mold had burst, or lysed. Something in the mold was killing the bacteria.

Experiments confirmed that the mold was capable of killing many kinds of bacteria, including *Streptococcus, Staphylococcus,* and *Pneumococcus.* Fleming published his results in 1929 in the *British Journal of Experimental Pathology.* He named the antibacterial substance "penicillin," after the fungus producing it, *Penicillium notatum.* It was the birth of the first **antibiotic**.

Fleming was not the first to notice the bacteria-killing property of *Penicillium,* but he was the first to study it scientifically and publish the results. In fact, Fleming had been looking for bacteria-killing substances for a number of years, ever since he had served as a medical officer in World War I and witnessed soldiers dying from bacteria-caused infections. He had already discovered one such antimicrobial agent–the chemical lysozyme–which he detected in his own tears and nasal mucus, so he knew what bacteria-killing signs to look for.

If you've ever seen a piece of moldy bread or rotting fruit, then you've met the *Penicillium* fungus. It doesn't look very impressive, but the

ANTIBIOTIC
A chemical that can slow or stop the growth of bacteria; many antibiotics are produced by living organisms.

Fleming in his lab.

chemical it produces ushered in a whole new age of medicine. For the first time, doctors had a way to treat such deadly illnesses as bacterial pneumonia, syphilis, and meningitis. As physician Lewis Thomas, former president of Memorial Sloan-Kettering Cancer Center in New York City, wrote in his 1992 memoir *Fragile Species,* "We could hardly believe our eyes on seeing that bacteria could be killed off without at the same time killing the patient. It was not just amazement, it was a revolution" (**Infographic 3.1**).

CELL THEORY
The concept that all living organisms are made of cells and that cells are formed by the reproduction of existing cells.

Bug Bullet

What makes antibiotics special is not just their ability to kill bacteria. After all, cyanide kills bacteria just fine. The important thing about antibiotics is that they exert their destructive effects on bacteria without (typically) harming their human or animal host, even if taken internally.

> **"We could hardly believe our eyes on seeing that bacteria could be killed off without at the same time killing the patient. It was not just amazement, it was a revolution."** –Lewis Thomas

Although Fleming didn't know it at the time, penicillin and other antibiotics preferentially kill bacteria because they target what is unique about bacterial cells. According to the **cell theory,** all living things are made of cells, and

INFOGRAPHIC 3.1

How Penicillin Was Discovered

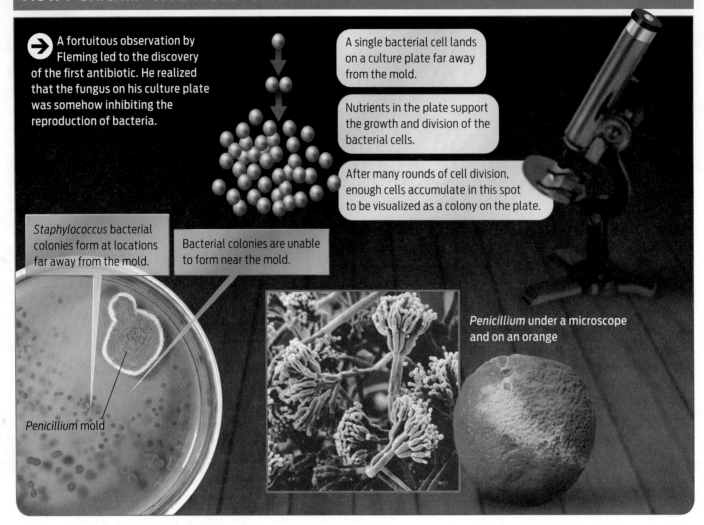

→ A fortuitous observation by Fleming led to the discovery of the first antibiotic. He realized that the fungus on his culture plate was somehow inhibiting the reproduction of bacteria.

A single bacterial cell lands on a culture plate far away from the mold.

Nutrients in the plate support the growth and division of the bacterial cells.

After many rounds of cell division, enough cells accumulate in this spot to be visualized as a colony on the plate.

Staphylococcus bacterial colonies form at locations far away from the mold.

Bacterial colonies are unable to form near the mold.

Penicillium mold

Penicillium under a microscope and on an orange

Figure 2 from Alexander Fleming's 1929 paper, showing the response of different bacteria to penicillin.

every new cell comes from the division of a pre-existing one. But not all cells are alike. Cells come in many shapes and sizes and perform various functions, depending on where they are found **(Infographic 3.2)**. Moreover, they fall into two fundamentally different categories: **prokaryotic** or **eukaryotic.** Prokaryotic cells are relatively small and lack internal membrane-bound compartments, called **organelles.** Eukaryotic cells, by contrast, are much larger and contain many such organelles. Penicillin and other antibiotics target structures that are unique to prokaryotic cells.

To understand why antibiotics affect prokaryotic and eukaryotic cells differently, it helps

PROKARYOTIC CELLS
Cells that lack internal membrane-bound organelles.

EUKARYOTIC CELLS
Cells that contain membrane-bound organelles, including a central nucleus.

ORGANELLES
The membrane-bound compartments of eukaryotic cells that carry out specific functions.

Cell Theory: All Living Things Are Made of Cells

 All living organisms are composed of cells. These cells arise from the reproduction of existing cells. Different cells have different structures and functions.

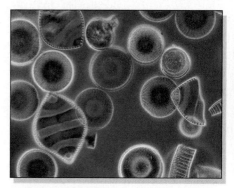

Diatoms: single-cell eukaryotes

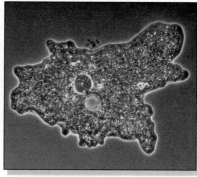

Amoeba (a protozoan): a single-cell eukaryote

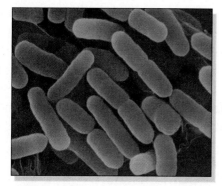

Bacteria: single-cell prokaryotes

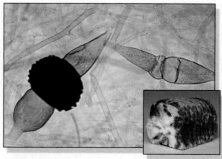

Molds (fungi): single and multicellular eukaryotic cells

Elodea (an aquatic plant): a multicellular eukaryote

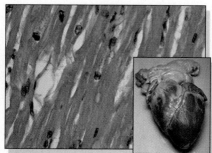

Humans (these are heart cells): multicellular eukaryotes

CELL MEMBRANE
A phospholipid bilayer with embedded proteins that forms the boundary of all cells.

CYTOPLASM
The gelatinous, aqueous interior of all cells.

RIBOSOME
A complex of RNA and protein that carries out protein synthesis in all cells.

NUCLEUS
The organelle in eukaryotic cells that contains the genetic material.

to understand first what the two cell types have in common. All cells, both prokaryotic and eukaryotic, are surrounded by a **cell membrane.** This flexible yet sturdy structure forms a boundary between the external environment and the cell's watery **cytoplasm** and literally holds the cell together. Partly hydrophobic, partly hydrophilic molecules known as phospholipids make up the bulk of the cell membrane, and proteins embedded in the membrane perform particular functions, such as transporting nutrients in and wastes out. The cell membrane forms a semipermeable barrier to substances on either side of it (**Infographic 3.3**).

In addition to a flexible cell membrane, both prokaryotic and eukaryotic cells have two

other elements in common: **ribosomes,** which synthesize the proteins that are crucial to cell function; and DNA, the molecule of heredity.

Beyond these three features, however–cell membrane, ribosomes, and DNA–the two cell types are structurally quite different. In a prokaryotic cell, for instance, the DNA floats freely within the cell's cytoplasm, while in a eukaryotic cell it is housed within a central command center called the **nucleus.** The nucleus is one of many organelles found within eukaryotic cells, but not in their simpler prokaryotic cousins (**Infographic 3.4**).

Penicillin kills bacteria because of one important difference between prokaryotic and eukaryotic cells. Unlike human and other animal cells, most bacteria are surrounded by a **cell wall.** This rigid structure is what allows bacteria to survive in watery environments—say, your intestines or a pond.

Water has a tendency to move across cell membranes from lower to higher solute concentration, a process called **osmosis.** In a low-solute environment, water will tend to rush into the solute-rich cytoplasm of a cell, causing it to swell. This swelling is potentially fatal to bacteria. Without a cell wall, bacterial cells would fill up with water and burst. Their sturdy cell wall, however, counteracts this osmotic pressure, keeping too much water from rushing in. (Eukaryotic cells are protected from osmotic pressure by the cholesterol in their cell membrane.)

What makes the bacterial cell wall rigid is the molecule **peptidoglycan,** a polymer made of sugars and amino acids that link to form a chainlike sheath around the cell. Different bacterial walls can have different structures, but all have peptidoglycan, which is found only in bacteria. By interfering with the synthesis of

CELL WALL
A rigid structure enclosing the cell membrane of some cells that helps the cell maintain its shape.

INFOGRAPHIC 3.3

Membranes: All Cells Have Them

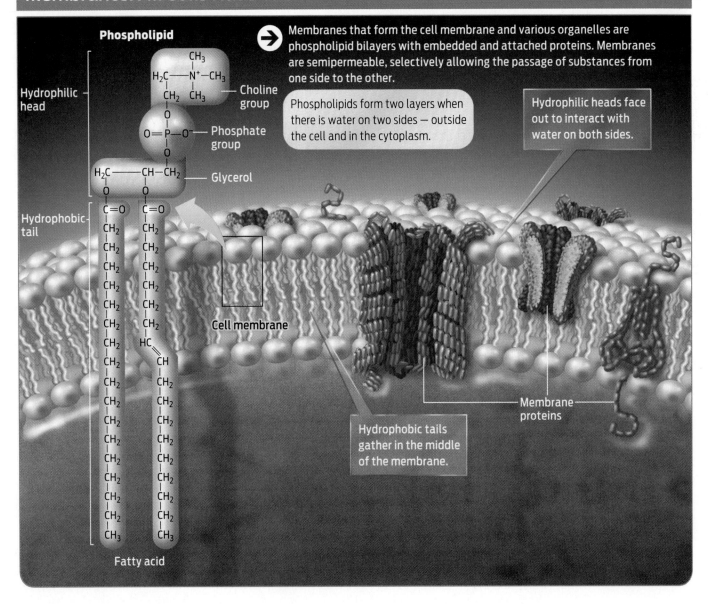

Membranes that form the cell membrane and various organelles are phospholipid bilayers with embedded and attached proteins. Membranes are semipermeable, selectively allowing the passage of substances from one side to the other.

Phospholipid

Hydrophilic head

CH₃
H₂C—N⁺—CH₃
CH₂ CH₃ — Choline group
O
O=P—O⁻ — Phosphate group
O
H₂C——CH—CH₂ — Glycerol
O O
C=O C=O
CH₂ CH₂
CH₂ CH₂
CH₂ CH₂
CH₂ CH₂
CH₂ CH₂
CH₂ CH₂
CH₂ HC
CH₂ CH
CH₂ CH₂
CH₂ CH₂
CH₂ CH₂
CH₂ CH₂
CH₂ CH₂
CH₃ CH₃

Hydrophobic tail

Fatty acid

Cell membrane

Phospholipids form two layers when there is water on two sides — outside the cell and in the cytoplasm.

Hydrophilic heads face out to interact with water on both sides.

Hydrophobic tails gather in the middle of the membrane.

Membrane proteins

Prokaryotic and Eukaryotic Cells Have Different Structures

→ While all cells have a cell membrane, cytoplasm, ribosomes and DNA, there are specific structural differences between prokaryotic and eukaryotic cells. Eukaryotic cells contain a variety of membrane-enclosed organelles while prokaryotic cells do not.

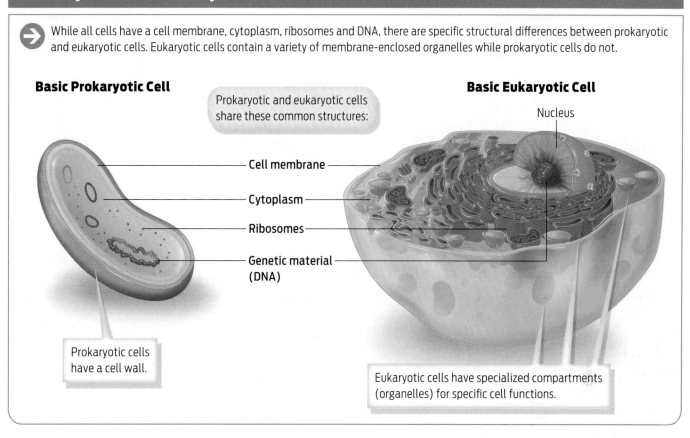

Basic Prokaryotic Cell

Prokaryotic and eukaryotic cells share these common structures:

Cell membrane

Cytoplasm

Ribosomes

Genetic material (DNA)

Prokaryotic cells have a cell wall.

Basic Eukaryotic Cell

Nucleus

Eukaryotic cells have specialized compartments (organelles) for specific cell functions.

OSMOSIS
The diffusion of water across a semipermeable membrane from an area of lower solute concentration to an area of higher solute concentration.

PEPTIDOGLYCAN
A macromolecule that forms all bacterial cell walls and provides rigidity to the cell wall.

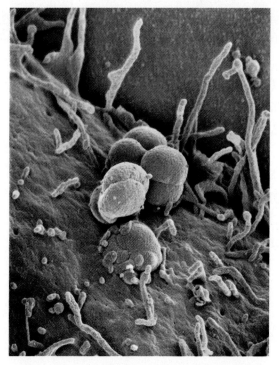

Scanning electron micrograph of the bacteria that cause gonorrhea.

peptidoglycan, penicillin weakens the cell wall, which is then no longer able to counteract osmotic water pressure. Eventually, the cell bursts (Infographic 3.5).

Bacteria are not the only organisms with a cell wall (plant cells and fungi have them, too), but they are the only ones that have a cell wall made of peptidoglycan–which is why penicillin is such a selective bacteria killer.

Ironically, despite its remarkable killing powers, penicillin was not immediately recognized as a medical breakthrough when it was first discovered. In fact, Fleming didn't think his mold had much of a future in medicine. At the time, the idea that an antiseptic agent could kill bacteria without at the same time harming the patient was unheard of, so Fleming never considered that penicillin might be taken internally. Nor was he a chemist, so he lacked the expertise to isolate and purify the active ingredient from the mold. While he found that his

Some Antibiotics Target Bacterial Cell Walls

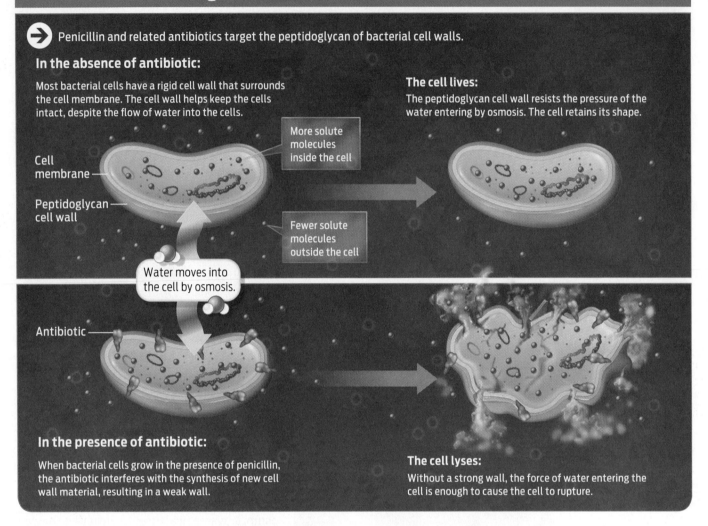

➡ Penicillin and related antibiotics target the peptidoglycan of bacterial cell walls.

In the absence of antibiotic:

Most bacterial cells have a rigid cell wall that surrounds the cell membrane. The cell wall helps keep the cells intact, despite the flow of water into the cells.

Cell membrane

Peptidoglycan cell wall

More solute molecules inside the cell

Fewer solute molecules outside the cell

Water moves into the cell by osmosis.

The cell lives:

The peptidoglycan cell wall resists the pressure of the water entering by osmosis. The cell retains its shape.

Antibiotic

In the presence of antibiotic:

When bacterial cells grow in the presence of penicillin, the antibiotic interferes with the synthesis of new cell wall material, resulting in a weak wall.

The cell lyses:

Without a strong wall, the force of water entering the cell is enough to cause the cell to rupture.

---

mold juice made a "reasonably good" topical antiseptic, he noted in a 1940 paper that "the trouble of making it seemed not worth while," and largely gave up working on it.

Ten years would pass before anyone reconsidered Fleming's mold. By then, history had intervened and given new urgency to the search for antibacterial medicines.

From Fungus to Pharmaceutical

On September 1, 1939, Germany invaded Poland, plunging the world into war for the second time in a generation. With the horrors of World War I still seared into memory, many feared the

With few other antibacterial medicines available, penicillin suddenly became the focus of research during World War II.

death toll that would result from the hostilities. Millions of soldiers and civilians had died in World War I, many not as a result of direct combat injuries but from infections resulting from surgeries meant to treat those injuries. With few other antibacterial medicines available, penicillin suddenly became the focus of research during World War II.

In 1938, Ernst Chain, a German-Jewish biochemist, was working in the pathology department at Oxford University, having fled Germany for England in 1933 when the Nazis came to power. Both Chain and his supervisor, Howard Florey, were interested in the biochemistry of antibacterial substances. Chain stumbled across

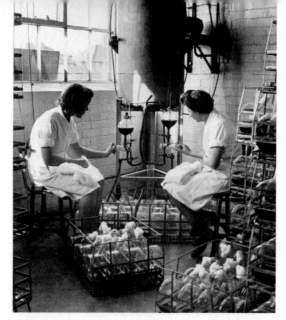

Manufacturing penicillin in 1943: culture flasks are filled with the nutrient solution in which penicillin mold is grown.

Fleming's 1929 paper on penicillin and set about trying to isolate and concentrate the active ingredient from the mold, which he succeeded in doing by 1940. Chain's breakthrough allowed Florey's group to begin testing the drug's clinical efficacy. They injected the purified chemical into bacteria-infected mice and found that the mice were quickly rid of their infection. Human trials followed next, in 1941, with the same remarkable result.

As encouraging as these results were, there was one nagging problem: it took up to 2,000 liters of mold fluid to obtain enough pure penicillin to treat one person. The Oxford doctors used almost their whole supply of the drug treating their first patient, a policeman ravaged by a staphylococcal infection. The team stepped up their purification efforts—even culturing the mold in patients' bedpans and re-purifying the drug from patients' urine—but there was no way they could keep up with demand.

The turning point came in 1941, when Oxford scientists approached the U.S. government and asked for help in growing penicillin on a large scale. The method they devised took advantage of something the United States had in abundance: corn. Using a by-product of large-scale corn processing as a culture medium in which to grow the fungus, the scientists were able to produce penicillin in much greater quantities.

At first, all the penicillin harvested from U.S. production plants came from Fleming's original strain of *Penicillium notatum*. But researchers continued to look for more potent strains to improve yields. In 1943, they got lucky: researcher Mary Hunt discovered one such strain growing on a ripe cantaloupe in a Peoria, Illinois, supermarket. This new strain, called *Penicillium chrysogenum,* produced more than 200 times the amount of penicillin as the origi-

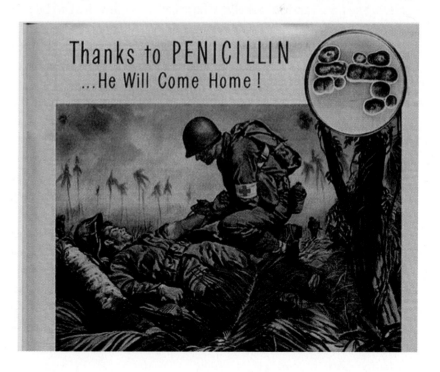

Thanks to PENICILLIN
...He Will Come Home!

"For the first time in human history, most people felt that infectious disease was ceasing to be a threat."

nal strain. With it, production of the drug soared. By the time the Allies invaded France on June 6, 1944–D-day–they had enough penicillin to treat every soldier that needed it. By the following year, penicillin was widely available to the general public.

"Penicillin seemed to justify a carefree attitude to infection," says medical historian Robert Bud, principal curator of the Science Museum in London. "In western countries, for the first time in human history, most people felt that infectious disease was ceasing to be a threat, and sexually infectious disease had already been conquered. For many it seemed cure would be easier than prevention."

Yet, as effective as penicillin was, it was effective only against certain types of bacteria; against others, it was powerless.

Stockpiling the Antibiotic Arsenal

As Fleming knew, most of the bacterial world falls into one of two categories, **Gram-positive** and **Gram-negative;** these names reflect the way bacterial cell walls trap a dye known as Gram stain (after its discoverer, the Danish scientist Hans Christian Gram). Fleming found that while penicillin easily killed Gram-positive bacteria like *Staphylococcus* and *Streptococcus*, it had little effect on Gram-negative bacteria like *E. coli* and *Salmonella*, whose cell walls have an extra layer of lipids surrounding them. This extra lipid layer prevents penicillin from reaching the peptidoglycan beneath it.

The discovery that penicillin was effective only on Gram-positive bacteria led other researchers in the 1940s to look for other antibiotics that could kill Gram-negative bacteria. The first such broad-spectrum antibiotic was streptomycin, discovered in 1943 by Albert Schatz and Selman Waksman at Rutgers University. In addition to killing Gram-negative bacteria, streptomycin was the first effective treatment for the deadly bacterial disease tuberculosis.

Like other antibiotics in the class known as aminoglycosides, streptomycin works by interfering with protein synthesis on bacterial ribosomes. Ribosomes are the molecular machines that assemble a cell's proteins. While both eukaryotic and prokaryotic cells have ribosomes, their ribosomes are different sizes and have different structures. Because streptomycin targets features specific to bacterial ribosomes, it doesn't harm the human who is taking it (**Infographic 3.6**).

Antibiotics can also target bacteria by inhibiting a bacterium's ability to make a critical vitamin or to copy its DNA before dividing. When this happens, the bacterium dies instead of reproducing.

"Penicillin seemed to justify a carefree attitude to infection. . . . For many it seemed cure would be easier than prevention." –Robert Bud

---

Why can broad-spectrum antibiotics, like streptomycin or gentamicin, kill Gram-negative bacteria when penicillin cannot? It's because these drugs have a chemical structure that allows them to pass more easily through the outer lipid layer of the Gram-negative bacterial cell wall. Although natural penicillin cannot pass this layer, many modern synthetic varieties of penicillin, known collectively as beta-lactams, can.

Crossing Enemy Lines

For any drug to be effective, it has to reach its designated target. In the case of many antibiotics, that means getting inside the cell to do their work. How do antibiotics penetrate a cell's outer defenses?

In all cells, the cell membrane acts as a barrier to transport, allowing only certain substances to pass through it.

With its densely packed collection of hydrophobic phospholipid tails, the cell membrane prevents many large molecules, like glucose, and hydrophilic substances, like sodium ions, from wandering across the cell membrane. In fact, the only things that do cross the membrane easily are small, uncharged molecules like oxygen (O_2), which can travel relatively easily across by a process known as **simple diffusion.**

Simple diffusion takes advantage of the natural tendency of dissolved substances to spread

GRAM-POSITIVE
Refers to bacteria with a cell wall that includes a thick layer of peptidoglycan that retains the Gram stain.

GRAM-NEGATIVE
Refers to bacteria with a cell wall that includes a thin layer of peptidoglycan surrounded by an outer lipid membrane that does not retain the Gram stain.

SIMPLE DIFFUSION
The movement of small, hydrophobic molecules across a membrane from an area of higher concentration to an area of lower concentration; simple diffusion does not require energy.

INFOGRAPHIC 3.6

Some Antibiotics Inhibit Prokaryotic Ribosomes

Ribosomes are responsible for the synthesis of proteins in both prokaryotic and eukaryotic cells, but their structure is slightly different in the two types of cells. Antibiotics that interfere with prokaryotic ribosomes leave eukaryotic ribosomes unaffected.

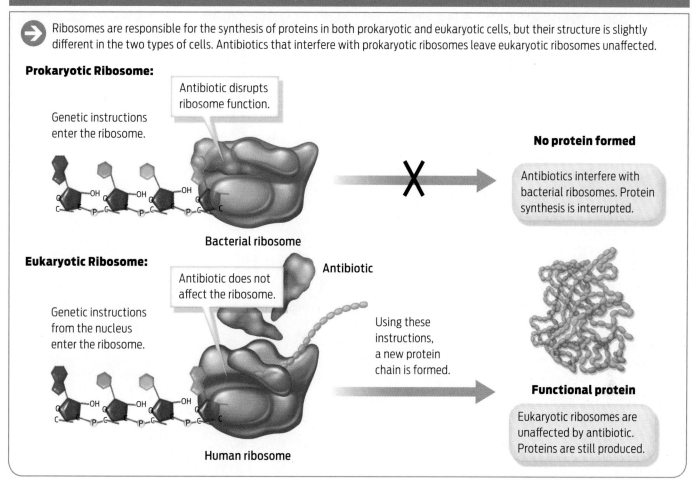

Prokaryotic Ribosome:

Antibiotic disrupts ribosome function.

Genetic instructions enter the ribosome.

Bacterial ribosome

No protein formed

Antibiotics interfere with bacterial ribosomes. Protein synthesis is interrupted.

Eukaryotic Ribosome:

Antibiotic does not affect the ribosome.

Antibiotic

Genetic instructions from the nucleus enter the ribosome.

Using these instructions, a new protein chain is formed.

Human ribosome

Functional protein

Eukaryotic ribosomes are unaffected by antibiotic. Proteins are still produced.

TRANSPORT PROTEINS
Proteins involved in the movement of molecules across the cell membrane.

FACILITATED DIFFUSION
The process by which large or hydrophilic solutes move across a membrane from an area of higher concentration to an area of lower concentration with the help of transport proteins.

out from an area of higher concentration to one of lower concentration–think of food coloring diffusing in a glass of water. Because the substance is moving from the side of the membrane with a higher concentration to the side with a lower concentration, no energy is required to move substances across the membrane. Take oxygen, for example. The concentration of oxygen molecules, which are small and uncharged, is often higher outside the cell and lower inside. This concentration difference, or gradient, allows oxygen to diffuse easily into the cell–a good thing, because the cell needs oxygen in order to survive.

But the cell also needs some large or hydrophilic molecules in order to survive–one of them is glucose, the cell's energy source. To move such molecules across the membrane the cell makes use of **transport proteins.** Transport proteins sit in the membrane bilayer with

one of their ends outside the cell and the other inside. By acting as a kind of channel, carrier, or pump, transport proteins provide a passageway for those large or hydrophilic molecules to cross the membrane. They are also very specific: a protein that transports glucose will not transport calcium ions, for example. The cells of your body contain hundreds of types of transport proteins.

Some antibiotics are small hydrophobic molecules that can cross the cell membrane directly by simple diffusion–tetracycline, for example. Others, including penicillin and streptomycin, require the assistance of transport proteins. Transport proteins can move substances either up or down a concentration gradient. When a substance uses a transport protein to move down a concentration gradient, the process is called **facilitated diffusion.** Like simple diffusion, facilitated diffusion requires no energy

Molecules Move across the Cell Membrane

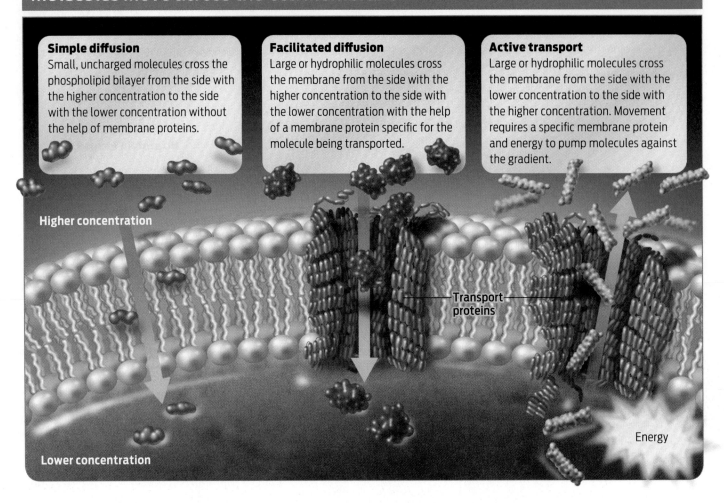

Simple diffusion
Small, uncharged molecules cross the phospholipid bilayer from the side with the higher concentration to the side with the lower concentration without the help of membrane proteins.

Facilitated diffusion
Large or hydrophilic molecules cross the membrane from the side with the higher concentration to the side with the lower concentration with the help of a membrane protein specific for the molecule being transported.

Active transport
Large or hydrophilic molecules cross the membrane from the side with the lower concentration to the side with the higher concentration. Movement requires a specific membrane protein and energy to pump molecules against the gradient.

Higher concentration

Transport proteins

Lower concentration

Energy

---

since the substance is moving from a higher to a lower concentration. Facilitated diffusion is the way many antibiotics pass through bacterial cell membranes.

Just because an antibiotic makes it inside a bacterial cell, however, doesn't mean it will stay there. Some bacteria have transport proteins that can actively pump the antibiotic back out of the cell. This bacterial counteroffensive measure is an example of **active transport,** in which proteins pump a substance uphill from an area of lower concentration to an area of higher concentration, a process that requires energy. In this case, active transport keeps the antibiotic concentration in the bacterial cell low, but the cell must expend energy to keep pumping the antibiotic out (Infographic 3.7).

Pumping antibiotics out of the bacterial cell is one way bacteria can resist the destructive power of an antibiotic. Other ways include chemically breaking down the antibiotic with enzymes. Why would bacteria have such built-in mechanisms for counteracting or resisting drugs? Remember that penicillin was originally isolated from a living organism, a fungus. Streptomycin was originally isolated from microorganisms living in soil. Microorganisms have evolved chemical defenses as a way to protect themselves from other organisms. In turn, these organisms have evolved countermeasures that give them resistance. Humans thus find themselves embroiled in a battle originally waged solely between microorganisms. We have "amplified a local warfare among microbes in a few grams of soil into a global planetary war between Man and Microbe," writes Alexander Tomasz, a microbiologist at the Rockefeller University, in the book *Fighting Infection in the 21st Century.* In the early 1980s Tomasz helped discover how penicillin works, and is now an expert on antibiotic resistance.

ACTIVE TRANSPORT
The energy-requiring process by which solutes are pumped from an area of lower concentration to an area of higher concentration with the help of transport proteins.

Eukaryotic Cells Have Organelles

→ Humans and other animals, as well as plants, fungi, and protists, are eukaryotes—they are made up of eukaryotic cells, containing many organelles. Some organelles are found in all eukaryotic cells; other organelles are found in only a subset of eukaryotes.

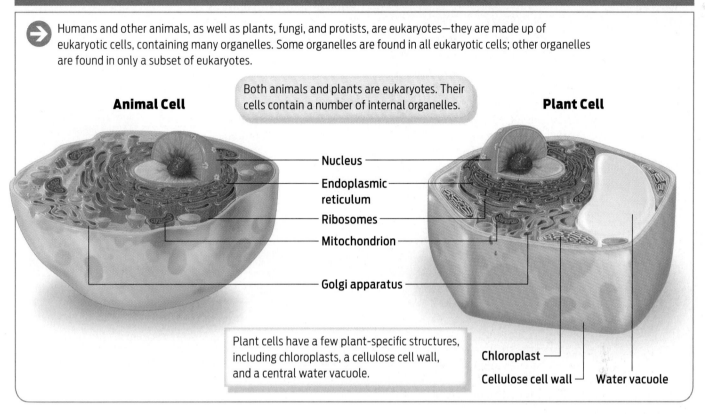

Animal Cell

Both animals and plants are eukaryotes. Their cells contain a number of internal organelles.

Plant Cell

Nucleus

Endoplasmic reticulum

Ribosomes

Mitochondrion

Golgi apparatus

Plant cells have a few plant-specific structures, including chloroplasts, a cellulose cell wall, and a central water vacuole.

Chloroplast

Cellulose cell wall — Water vacuole

NUCLEAR ENVELOPE
The double membrane surrounding the nucleus of a eukaryotic cell.

MITOCHONDRIA
Membrane-bound organelles responsible for important energy-conversion reactions in eukaryotes.

ENDOPLASMIC RETICULUM
A membrane-enclosed series of passages in eukaryotic cells in which proteins and lipids are synthesized.

Your Inner Bacterium

Antibiotics kill bacteria but leave humans unharmed because their cells have different structures. Of all the ways that prokaryotic and eukaryotic cells differ, the most obvious is the complexity of eukaryotic cells compared to their smaller prokaryotic cousins. In particular, eukaryotic cells—both animal and plant cells—are characterized by the presence of multiple, distinct membrane-bound organelles (Infographic 3.8).

You can think of a eukaryotic cell as a miniature factory with an efficient division of labor. Each organelle is separated from the cell's cytoplasm by a membrane similar to the cell's outer membrane, and each performs a distinct function.

The nucleus is the defining organelle of eukaryotic cells (from the Greek *eu,* meaning "good" or "true" and *karyon,* meaning "nut"

Antibiotics kill bacteria but leave humans unharmed because their cells have different structures.

or "kernel"). It is surrounded by the **nuclear envelope,** a double membrane made of two lipid bilayers. The nucleus encloses the cell's DNA and acts as a kind of control center. Important reactions for interpreting the genetic instructions contained in DNA take place in the nucleus.

Other organelles in a eukaryotic cell perform other specialized tasks. **Mitochondria** are the cell's power plants—they help extract energy from food and convert that energy into a useful form. Humans who inherit or develop defects in their mitochondria usually die—an indication of just how important these organelles are (see **Up Close: Eukaryotic Organelles**).

Much like the plumbing system of a building, the **endoplasmic reticulum (ER)** is a vast network of membrane-covered "pipes" that serve as a transport system throughout the cell. With the help of a protein "packaging plant" known

Nucleus

The nucleus is the defining organelle of eukaryotic cells. The nucleus is separated from the cytoplasm by a double membrane (two phospholipid bilayers), known as the nuclear envelope. The nuclear envelope controls the passage of molecules between the nucleus and cytoplasm. The nucleus contains the DNA, the stored genetic instructions of each cell. In addition, important reactions for interpreting the genetic instructions occur in the nucleus.

DNA
(genetic material)

Nuclear envelope

Endoplasmic Reticulum

The endoplasmic reticulum (ER) is an extensive, membranous intracellular "plumbing" system that is critical for the production of new proteins. The "rough ER" has a rough appearance because it is studded with ribosomes that are making proteins. The rough ER is contiguous with the "smooth ER," the site of lipid production.

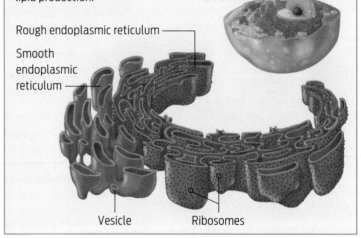

Rough endoplasmic reticulum

Smooth endoplasmic reticulum

Vesicle   Ribosomes

Golgi

The Golgi is a series of flattened membrane compartments, whose purpose is to process and package proteins produced in the rough endoplasmic reticulum. The processed molecules are packaged into membrane vesicles, then targeted and transported to their final destinations.

2. As the proteins make their way through the Golgi, they are processed.

3. Proteins are then packaged into transport vesicles, which deliver the proteins to their final destination.

1. Transport vesicle delivers proteins from the rough endoplasmic reticulum to the Golgi.

Transport vesicle

The Nucleus, Endoplasmic Reticulum and Golgi Work Together to Produce and Transport Proteins

Nucleus

2. Proteins are made in the ER and packaged into vesicles for transport to the Golgi.

3. Proteins receive final modifications in the Golgi. They are packaged into vesicles for transport to the site of protein function.

Cell membrane

1. The nucleus provides instructions for protein production.

Secreted from cell

Various locations within cell

Endoplasmic reticulum   Golgi

Mitochondria

Mitochondria are found in almost all eukaryotes, including plants. Mitochondria have two membranes surrounding them. The inner one is highly folded. Mitochondria carry out critical steps in the extraction of energy from food, and the conversion of that "trapped" energy to a useful form. They are the cell's "power plants."

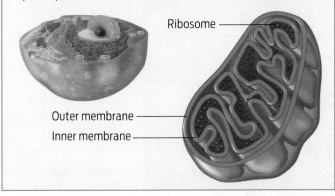

Ribosome

Outer membrane
Inner membrane

Chloroplast

Chloroplasts are organelles found in algae and in the green parts of plants. Chloroplasts have two membranes surrounding them, as well as an internal system of stacked membrane discs. Chloroplasts are the sites of photosynthesis, the reactions that plants use to capture the energy of sunlight in a usable form.

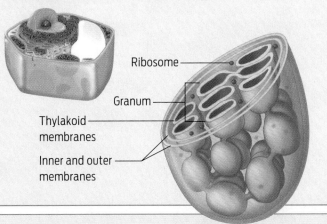

Ribosome

Granum

Thylakoid
membranes

Inner and outer
membranes

Lysosome

Lysosomes are the cell's "recycling centers." Full of digestive enzymes, lysosomes break down worn out cell parts or molecules so they can be used to build new cellular structures.

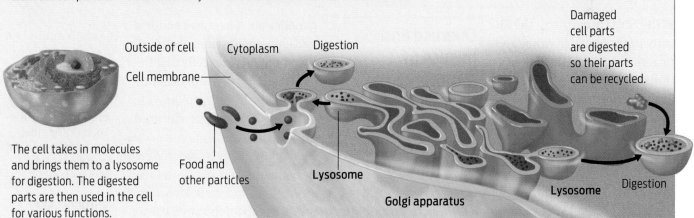

Outside of cell   Cytoplasm   Digestion

Cell membrane

Damaged cell parts are digested so their parts can be recycled.

The cell takes in molecules and brings them to a lysosome for digestion. The digested parts are then used in the cell for various functions.

Food and other particles

Lysosome

Golgi apparatus

Lysosome   Digestion

Cytoskeleton

The cytoskeleton is a meshwork of protein fibers that carry out a variety of functions, including cell support, cell movement and movement of structures within cells. Each type of cytoskeletal fiber has a specific structure and function.

Cell membrane

Microfilament   Intermediate filaments   Microtubule

A supermarket pharmacy manager retrieves a bottle of antibiotics from the shelf.

as the **Golgi apparatus**, the ER transports newly synthesized proteins to specific destinations, such as the cell membrane, other organelles, and even extracellular destinations like the bloodstream.

Other eukaryotic organelles include the **chloroplast,** responsible for photosynthesis in plants, and **lysosomes,** the cell's recycling centers, which digest and recycle molecules. In addition to these membrane-bound structures, a vast network of protein fibers called the **cytoskeleton** allows cells to move and maintain their shape, much the same way that your skeleton does.

Prokaryotic cells carry out similar functions of energy conversion and protein transport, but they don't contain these processes within separate organelles; everything occurs in the cytoplasm.

How did eukaryotic cells develop their factory-like compartments? That question has long intrigued biologists. One fascinating hypothesis was proposed in the 1960s by biologist Lynn Margulis, who argued that eukaryotic or-

Overuse and misuse of antibiotics have led to an epidemic of antibiotic-resistance.

ganelles such as mitochondria and chloroplasts were once free-living prokaryotic cells that become incorporated–engulfed–by other free-living prokaryotic cells in a process dubbed **endosymbiosis.**

Although many considered endosymbiosis a crazy idea at first, quite a bit of evidence now supports it. Mitochondria and chloroplasts are about the same size as bacteria, and to reproduce they divide in a manner similar to prokaryotic cells. Both mitochondria and chloroplasts have circular strands of DNA, just like prokaryotic cells. They also contain ribosomes that are similar in structure to prokaryotic ribosomes–so similar, in fact, that some antibiotics that target prokaryotic ribosomes can affect the ribosomes in eukaryotic mitochondria, which accounts for both the toxicity and the side effects of these antibiotics.

Winning the Battle, Losing the War
To those who first benefited from its healing powers, penicillin seemed a wonder drug, a magic bullet. A once-lethal bacterial infection

GOLGI APPARATUS
An organelle made up of stacked membrane-enclosed discs that packages proteins and prepares them for transport.

CHLOROPLAST
An organelle in plant and algal cells that is the site of photosynthesis.

LYSOSOME
An organelle in eukaryotic cells filled with enzymes that can degrade worn-out cellular structures.

CYTOSKELETON
A network of protein fibers in eukaryotic cells that provides structure and facilitates cell movement.

Drug-resistant strains of *Neisseria gonorrhoeae* are increasing in many countries.

demic of such antibiotic-resistance, which the Centers for Disease Control and Prevention calls "one of the world's most pressing public health problems."

Fleming himself warned against this very danger. In his own research, he found that whenever too little penicillin was used or when it was used for too little time, populations of bacteria emerged that were resistant to the antibiotic. In a 1945 interview in the *New York Times*, Fleming warned that improper use of penicillin could lead to the survival and reproduction of virulent strains of bacteria that are resistant to the drug. He was right. In 1945, when penicillin was first introduced to the public, virtually all strains of *Staphylococcus aureus* were sensitive to it. Today, more than 90% of *Staphyloccocus aureus* strains are resistant to the antibiotic that once conquered this common microbe. (For more on antibiotic-resistant bacteria, see Chapter 14.)

> **Today, more than 90% of *Staphylococcus aureus* strains are resistant to the antibiotic that once conquered this common microbe.**

Because of the alarming growth in antibiotic-resistant superbugs, drug companies and researchers are trying to develop new antibiotics. One strategy they employ is to tweak the chemical structure of existing antibiotics just enough that a bacterium cannot disable it. Another approach is to look for antibiotics that target other bacterial weaknesses.

But all these efforts would be nothing without the man who gave a moldy petri dish a second glance nearly a century ago. That famous dish now sits in the museum at St. Mary's Hospital in London. For his pioneering research, Alexander Fleming—along with Oxford researchers Howard Florey and Ernst Chain—was awarded a Nobel prize in 1945. ∎

could now be cleared in a matter of days with a course of antibiotic. Today, some of the most commonly prescribed drugs are antibiotics.

Antibiotics are so common, in fact, that many people routinely take them when they catch a cold or the flu. But antibiotics are powerless against these ills. That's because viruses, not bacteria, cause colds and flu. Since viruses are not made of cells—and according to the cell theory are not even considered to be alive—they can't be killed with an antibiotic.

But that doesn't stop people from trying. In 2010, the American College of Physicians estimated that of the more than 133 million courses of antibiotics prescribed in the United States each year, as many as 50% are prescribed for colds and other viral infections. What's more, many patients who are prescribed antibiotics for bacterial infections use them improperly. Taking only part of a prescribed dose, for example, can spare some harmful bacteria living in the body, and those bacteria that survive are often heartier and more resistant to the antibiotic than the ones that were killed. Such overuse and misuse of antibiotics have led to an epi-

ENDOSYMBIOSIS
The theory that free-living prokaryotic cells engulfed other free-living prokaryotic cells billions of years ago, forming eukaryotic organelles such as mitochondria and chloroplasts.

▶ Summary

■ Antibiotics are chemicals, originally produced by living organisms, that selectively target and kill bacteria.

■ According to the cell theory, all living organisms are made of cells. New cells are formed when an existing cell reproduces.

■ There are two types of cells, distinguished by their structure: prokaryotic and eukaryotic.

■ Prokaryotic cells lack membrane-bound organelles; eukaryotic cells have a variety of membrane-bound organelles.

■ All cells are enclosed by a cell membrane made up of phospholipids and proteins. The cell membrane controls passage of molecules between the exterior and the cytoplasm of the cell.

■ Small hydrophobic molecules can cross cell membranes by the process of simple diffusion.

■ Large or hydrophilic molecules need to be transported across the membrane with the help of membrane proteins.

■ Facilitated diffusion is transport down a concentration gradient; it does not require energy. Active transport is transport up a concentration gradient; it requires energy.

■ Bacteria are surrounded by a cell wall containing peptidoglycan, a molecule not found in eukaryotes. Some antibiotics, like penicillin, work by preventing peptidoglycan synthesis.

■ All cells have ribosomes, complexes of RNA and proteins that synthesize new proteins.

■ Despite their common function, the structure of prokaryotic and eukaryotic ribosomes differs. Some antibiotics, like streptomycin, work by interfering with prokaryotic ribosomes.

■ Eukaryotic cells contain a number of specialized organelles including a nucleus, endoplasmic reticulum, Golgi apparatus, mitochondria, chloroplasts, and other organelles, each of which carries out a distinct function.

■ Eukaryotic cells likely evolved as a result of endosymbiosis, the engulfing of one single-cell prokaryote by another.

■ Increased and sometimes inappropriate use of antibiotics has lead to the emergence of antibiotic-resistant bacteria. Infections caused by these bacteria are very hard to treat.

CELLULAR BASIS OF LIFE

Cells are the fundamental unit of life. All living things are made of cells, and all cells come from the division of pre-existing cells.

HINT **See Infographics 3.2 and 3.4.**

⊘ KNOW IT

1. Describe the cell theory.

2. Which of the following statements best explains why bacteria are considered living organisms?
    a. They can cause disease.
    b. They are made up of biological macromolecules.
    c. They move around.
    d. They are made of cells.
    e. They contain organelles.

3. What are the two main types of cells found in organisms?

⊘ USE IT

4. Consider the distinction between living and nonliving things.
    a. If all living things are made of cells, should a virus be considered alive? What about the infectious agent—a prion consisting of a single protein—responsible for mad cow disease?
    b. Following from your answer to part a, are all disease-causing agents—pathogens—alive?

5. According to the cell theory, all living organisms are made of cells. More specifically what do all living organisms have in common? For example, do all living organisms carry genetic instructions? Do their cells all have a nucleus? What other features do they have in common?

MEMBRANES AND TRANSPORT

All cells are surrounded by a membrane that contains the cell's contents and acts as a semipermeable barrier to substances on either side of it. Many substances move across the membrane with the help of proteins.

HINT **See Infographics 3.3 and 3.7.**

⊘ KNOW IT

6. The two major components of cell membranes are
    a. phospholipids and DNA.
    b. DNA and proteins.
    c. peptidoglycan and phospholipids.
    d. peptidoglycan and proteins.
    e. phospholipids and proteins.

7. If a solute is moving through a phospholipid bilayer from an area of higher concentration to an area of lower concentration without the assistance of a protein, then the manner of transport must be
    a. active transport.
    b. facilitated diffusion.
    c. simple diffusion.
    d. any of the above, depending on the solute
    e. Solutes cannot cross phospholipid bilayers.

8. Consider the movement of molecules across the cell membrane.
    a. What do simple diffusion and facilitated diffusion have in common?
    b. What do active transport and facilitated diffusion have in common?

⊘ USE IT

9. Why does facilitated diffusion require membrane transport proteins while simple diffusion does not?

10. Sugars are large, hydrophilic molecules that are important energy sources for cells. How can they enter cells from an environment with a very high concentration of sugar?
    a. by simple diffusion
    b. by osmosis
    c. by facilitated diffusion
    d. by active transport
    e. by using ribosomes

11. Many foods—for example, bacon and salt cod—are preserved with high concentrations of salt. How can high concentrations of salt inhibit the growth of bacteria? (Think about the high solute concentration of the salty food, relative to the solute concentration in the bacterial cells. Now think about what will happen to the water in the bacterial cells under these conditions. What do you think will happen to the cells as a result?)

PROKARYOTIC VS. EUKARYOTIC CELLS

Prokaryotic cells and eukaryotic cells have different structures. Antibiotics are effective because of these differences.

HINT **See Infographics 3.4–3.6 and 3.8.**

➲ KNOW IT

12. Penicillin interferes with the synthesis of
   a. bacterial cell membranes.
   b. peptidoglycan.
   c. the nuclear envelope.
   d. membrane proteins.
   e. ribosomes.

13. Bacteria have _____ cells, defined by the _____.
   a. prokaryotic; presence of a cell wall
   b. eukaryotic; presence of organelles
   c. eukaryotic; absence of a cell wall
   d. prokaryotic; absence of organelles
   e. eukaryotic; absence of organelles

14. Which of the following is associated with eukaryotic cells but not with prokaryotic cells?
   a. cell membrane
   b. cell wall
   c. DNA
   d. ribosome
   e. nucleus

15. Briefly describe the structure and function of each eukaryotic organelle listed:
   a. mitochondrion
   b. nucleus
   c. endoplasmic reticulum
   d. chloroplast

➲ USE IT

16. If you treated a bacterial infection with two different antibiotics, one that stopped bacterial reproduction and one (penicillin, for example) that inhibited the production of new peptidoglycan, would this use of penicillin be effective? Explain your answer.

17. If bacterial cells were placed in a nutrient-containing solution that had the same solute concentration as the cytoplasm, and which also contained penicillin, would the cells burst? Explain your answer.

18. Fungi are eukaryotic organisms. Why is it more challenging to develop treatments for fungal infections (for example, yeast infections, athlete's foot, and certain nail infections) than for bacterial infections?

19. Some inherited syndromes, for example Tay-Sachs disease and MERFF (myoclonic epilepsy with ragged red fibers), interfere with the function of specific organelles. MERFF disrupts mitochondrial function. From what you know about mitochondria, why do you think the muscles and the nervous system are the predominant tissues affected in MERFF? (Think about the activity of these tissues compared to, say, skin.)

SCIENCE AND ETHICS

20. Many patients attempt to pressure their physician to prescribe antibiotics for colds. If you were a doctor, would you prescribe an antibiotic for a cold? How would you explain your decision to your patient?

Powerfoods

Powerfoods

Foods fit to fight chronic disease

Food giant Nestlé is perhaps best known for its chocolate, cereals, and other sundry foods–products that taste good but aren't always good for your health. But the company sells much more than just snack foods. At its mountainside research laboratory near Lausanne, Switzerland, for more than 10 years company scientists have been developing products with a nutritional bang. In September 2010, Nestlé established a separate division, Nestlé Health Science, to develop products that exclusively target diseases and also to stake a claim in a rapidly growing market. The company's goal: to keep chronic diseases at bay with food containing therapeutic ingredients.

Some of these products are already on the market. In 2006, for example, the company introduced to the United States and Canada a snack drink called Boost Glucose Control that doesn't cause dangerous spikes in blood-sugar levels, which is a concern for people with **diabetes.** Another Nestlé product, a low-fat milk containing less saturated fat than others on the market, may cut cholesterol levels; and a yogurt fortified with calcium may help stave off the bone-thinning disease known as **osteoporosis.** These so-called functional foods, also called nutraceuticals–a play on the word "pharmaceuticals"–represent the next wave in food science.

Food manufacturers have long tinkered with their food products by pumping in "healthful" extras, from vitamins to oat bran. However, very few of these foods with added nutrients have been tested in clinical trials to prove that they actually improve health. But tough food labeling laws in the European Union and the threat of more stringent regulation in the United States are forcing large food companies such as Nestlé and others to change. With their nutraceuticals, Nestlé aims "to provide health benefits whose value has been demonstrated and justified by science," says Nina Backes, media spokesperson at Nestlé.

But do they work? And is eating such "manufactured" food healthier than eating freshly prepared food?

DIABETES
A disease characterized by abnormally high blood-sugar levels.

OSTEOPOROSIS
A disease characterized by thinning bones.

Food Is Nutrition

Walk into any grocery store and you're sure to find orange juice with added calcium, eggs containing higher levels of omega-3 fatty acids, and cereal packed with extra oat bran. Popular trends in food products include adding DHA, a type of fat believed by some to promote healthy brain and eye development in children, and probiotics for digestive health. Obviously, food provides the body with something necessary. But what exactly does food provide? And why do food manufacturers feel they need to alter or "improve" it?

All food is a source of two essential things required for life: **nutrients** and **energy.** Nutrients are all the chemical components our bodies need to live, grow, and repair themselves. Energy is what powers our activities—everything from thinking to running. While both nutrients and energy are crucial components of food, we discuss them separately. This chapter focuses on the nutrient component of food–food as chemical building blocks; Chapters 5 and 6 consider energy–the fuel component of food–in more detail.

When experts talk about eating a nutritious diet, they mean one that provides all the nutrients our bodies need in appropriate amounts. Nutrients that the body requires in large amounts are called **macronutrients.** The macronutrients found in our diet include proteins, carbohydrates, and fats–three of the four organic molecules discussed in Chapter 2. Nucleic acids are also provided in food, but in much smaller amounts, so while they are still considered nutrients, they are not *macro*nutrients. Because most foods contain mixtures of these macronutrients, most of us who eat a

NUTRIENTS
Components in food that the body needs to grow, develop, and repair itself.

ENERGY
The ability to do work, including building complex molecules.

MACRONUTRIENTS
Nutrients, including proteins, carbohydrates, and fats, that organisms must ingest in large amounts to maintain health.

varied diet that includes vegetables, oils, grains, meat, and dairy products can easily obtain all the nutrients our bodies need. A cheeseburger, for example, contains all three types of molecule; but the proportion of each one varies in different foods. Animal products typically contain more protein per gram relative to carbohydrate; most plant products contain more carbohydrate relative to protein (Infographic 4.1).

Macronutrients are an essential component of our diet because they provide our cells with crucial building blocks. When we eat, our digestive systems break down large molecules into smaller subunits. Proteins are broken down into amino acids; carbohydrates into simple sugars; fats into fatty acids and glycerol; and nucleic acids into individual nucleotides.

The body then uses these subunits as building blocks to make new cellular structures. Amino acids, for example, are ultimately assembled into proteins that have many different functions in the body, such as moving substances within and between cells or serving as communication molecules. Other food subunits, such as simple sugars, are used to build cell-surface markers and energy-storage molecules. And fats provide the building blocks to form cell membranes. Breaking down food to build up our bodies means that, quite literally, we are what we eat (Infographic 4.2).

Eating a balanced diet is therefore important to supply the building blocks to produce healthy, functioning cells. To a certain extent, our bodies can compensate for a deficiency in one or another nutrient by synthesizing it from other chemical components. For example, if a particular amino acid is in short supply, cells may be able to make it from another amino acid that is in excess. But there are some nutrients our bodies can't manufacture from scratch, and which must be obtained pre-assembled from our diet. These are known as **essential nutrients.** For example, from starting materials in food, our cells can synthesize 12 of the 20 amino acids it needs to make proteins. The other eight must be obtained pre-assembled from our diet. Because our body can't manufacture them,

The Nestlé research center near Lausanne, Switzerland.

these eight amino acids are called **essential amino acids.** Animal sources such as meat, eggs, fish, and dairy products are the richest sources of essential amino acids.

Why Manufacture Food?

If food naturally contains nutrients, why would anyone want to eat foods with nutrients added? The answer is complex, and has its roots in our changing relationship to cooking, work, and leisure. Processed, ready-to-eat foods were initially marketed as a convenience and became very popular during the 1950s and 1960s, when more women were joining the workforce and had less time to prepare and cook food.

Frozen foods, for example, allowed people more time for work or for leisure. But freezing food had a downside. Freezing generally degrades taste. And so an entire industry populated by food manufacturers and food scientists grew up around the need to create food that tasted good, was convenient, and that could be

ESSENTIAL NUTRIENT
A substance that cannot be synthesized by the body and must be obtained pre-assembled from the diet, including certain amino acids and fatty acids, vitamins, and minerals.

ESSENTIAL AMINO ACIDS
Eight amino acids the human body cannot synthesize and must obtain from food.

INFOGRAPHIC 4.1

Food Is a Source of Macronutrients

→ The most important dietary macronutrients are carbohydrates, proteins, and fats. While most foods contain all of them, one or two macronutrients predominate in each food type. A well-balanced diet is one that includes variety of foods to ensure that the body gets enough of each macronutrient to grow and remain healthy.

Carbohydrates	Proteins	Fats
Fruits and Veggies	Meats	Dairy
Grains	Dairy	Meats
Legumes	Legumes	Oils

frozen or stored on shelves for long periods of time. To achieve these goals, frozen meals were pumped with extra salt and fat to compensate for poor taste. To give a product a longer shelf life, food scientists devised preservatives and other additives, such as partially hydrogenated vegetable oils–oils that are chemically altered to remain solid at room temperature–to help keep baked and fried foods crispy. These chemically altered foods were typically less

Today, the average American over 15 years old spends a mere 34 minutes a day on food preparation and cleanup.

nutritious than freshly prepared foods because processing degrades nutrients. Because they suited Americans' lifestyles, we bought them–along the way affording huge profits to the companies that produced them.

Today, the average American over 15 years old spends a mere 34 minutes a day on food preparation and cleanup, according to the 2008 American Time Use Survey published by the U.S. Bureau of Labor Statistics; that's less than

Macronutrients Build and Maintain Cells

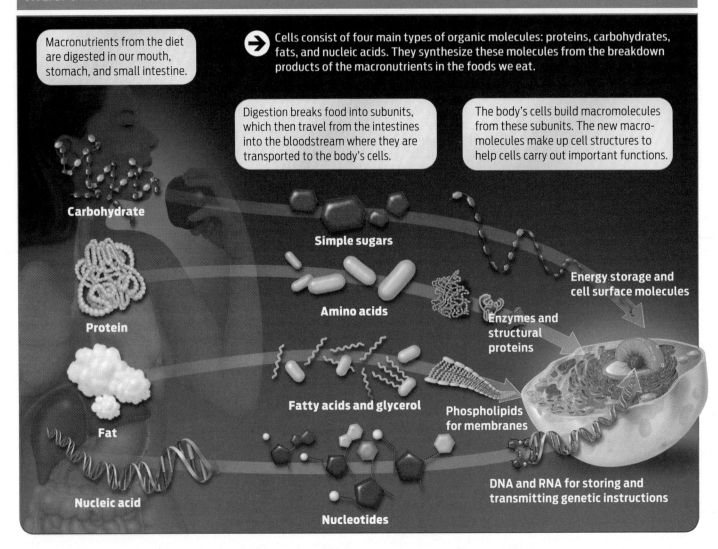

Macronutrients from the diet are digested in our mouth, stomach, and small intestine.

→ Cells consist of four main types of organic molecules: proteins, carbohydrates, fats, and nucleic acids. They synthesize these molecules from the breakdown products of the macronutrients in the foods we eat.

Digestion breaks food into subunits, which then travel from the intestines into the bloodstream where they are transported to the body's cells.

The body's cells build macromolecules from these subunits. The new macro- molecules make up cell structures to help cells carry out important functions.

Carbohydrate

Protein

Fat

Nucleic acid

Simple sugars

Amino acids

Fatty acids and glycerol

Nucleotides

Enzymes and structural proteins

Phospholipids for membranes

Energy storage and cell surface molecules

DNA and RNA for storing and transmitting genetic instructions

half the time that we spent cooking and cleaning during the 1960s. And consumption of processed foods is at historic levels.

Paradoxically, Americans have become more health-conscious than ever. We have access to the latest health news, much of which focuses on reporting the health benefits of this or that nutrient. In response to consumer demand, the food industry several years ago began trying to make its processed food more healthful by putting nutrients in and taking food additives—such as the hydrogenated oil known as trans fat that scientists have found to harm health—out.

Oat bran, for example, a natural ingredient found in oatmeal, has been shown to lower cholesterol, reduce blood pressure, and consequently cut the risk of heart disease. So food manufacturers have added oat bran to many types of foods, including bread and pasta. Likewise, omega-3 fatty acids—a type of essential fatty acid that is naturally found in such fatty fish as mackerel, sardines, and salmon—may promote cardiovascular health and carry other health benefits. Consequently, many companies have started adding omega-3 fatty acids to products such as margarine, cereals, and eggs. And many foods, from breakfast cereal to orange

Chemical food processing degrades nutrients.

juice to milk, are regularly fortified with vitamins and minerals.

What's in a Label?

The problem for consumers has been that loose government regulation of such products has made it difficult to distinguish marketing hype from the real deal. The Food and Drug

Oat bran and omega-3 fatty acids have been shown to carry health benefits.

Administration (FDA) requires that all prepared foods and dietary supplements have labels detailing nutrient content and ingredients. Manufacturers are allowed to make health claims on their labels as long as the claim isn't misleading or false, and as long as it complies with published guidelines (if there are any for that particular claim).

Currently, the FDA recognizes 17 specific health claims, each linking a food or dietary ingredient to a disease or a health-related condition. For example, if a company wants to promote a calcium-containing food as helpful in preventing osteoporosis, the FDA allows this specific claim language: "Adequate calcium as part of a healthful diet, along with physical activity, may reduce the risk of osteoporosis in later life." Companies can also promote a food by stating what it does in the body, so the following claims are acceptable: "Calcium builds strong bones" and "Fiber maintains bowel regularity."

However, a company can make a claim for a food product that is not one of the officially recognized 17 simply by adding a disclaimer, for instance, "The FDA has not evaluated this claim." And, more significantly, false or misleading claims brought to the FDA's attention by advocacy groups or consumers result in nothing more than a letter of reprimand asking the company to remove or reword the claim or face a mandatory recall of the product.

To be fair, clinical studies to assess whether a product benefits health are laborious and expensive to conduct. And there is no requirement to conduct such research. Therefore, there is very little evidence—such as published clinical studies—to support the claims made for most food products fortified with added nutrients. However, there may be evidence in the scientific literature that supports the health benefits of specific ingredients.

Given the promise of large profits, companies often push the limits of what is allowed—or what is accurate—on their product labels. Consequently, American market shelves are awash with products labeled with all sorts of health claims in often confusing and sometimes misleading language.

In Europe, however, the landscape is quite different. Since 2007, food manufacturers that want to market products with health claims must apply to the European Food Safety Authority for approval before introducing the product to the market. Scientific evidence documenting the claim must be submitted with the application.

Tighter regulation in Europe is one reason that large food manufacturers such as Nestlé have been pumping more money into research to support product claims. And because functional foods are by their very nature intended to provide health benefits beyond basic nutrition, companies that produce them are more eager to get government approval to be able to market them with specific health claims.

Digestion, Enzymes, and Metabolism

Nestlé's Boost Glucose Control drink, for example, contains fiber, a type of carbohydrate known to slow down digestion. Most functional foods, in fact, are built on the way our bodies naturally digest and use specific types of food molecules. Digestion is the process of breaking down the huge food molecules into smaller pieces so that our bodies can use them. It is a series of chemical reactions that take place throughout the digestive system. In our mouths, stomachs, and small intestines, chemical reactions break the bonds that hold food molecules together.

For most of us, carbohydrates constitute the largest portion of our diet. Bread, pasta, and rice are rich in carbohydrates. When we eat a plate of pasta, for example, our digestive system breaks down the carbohydrates into smaller sugar molecules. These sugar molecules are absorbed from the small intestine into the bloodstream, in which they are transported to cells in the rest of the body for use in building cell structures and carrying out cell functions.

To break down any macromolecule into its constituent parts, however, the chemical reactions that take place during digestion require the help of chemical facilitators called **enzymes,** which are specialized proteins that speed up the rate of a chemical reaction.

Enzymes work by lowering the amount of energy required to nudge a chemical reaction into motion. Enzymes substantially reduce **activation energy,** and so the reaction occurs more easily.

Enzymes bind to molecules called **substrates.** The part of the enzyme that binds to substrates is called its **active site.** Each enzyme is made so that its active site fits only one particular substrate molecule or group of highly similar substrate molecules. For example, each type of digestive enzyme in our mouth, stomach, and intestine binds to one specific type of organic molecule present in food.

Because of enzymes our bodies are able to break apart the chemical bonds in food molecules to release their component building blocks. Reactions that break molecules into smaller units—those that digest food, for instance—are called **catabolic reactions.** Reactions that build organic molecules from their simpler building blocks—such as those that build new muscle—are known as **anabolic reactions.** All the chemical reactions that take place inside our bodies are collectively called **metabolism (Infographic 4.3).**

Digesting Carbs

In some diseases, normal metabolism goes awry. People with diabetes, for example, have trouble controlling sugar levels in their blood. Healthy people have ways of regulating sugar, the end product of carbohydrate digestion. Cells in the pancreas, a small organ located near the stomach, secrete a hormone called **insulin** in response to high blood-sugar levels. Insulin is a protein—it is a chain of amino acids produced by the pancreas. When insulin binds to most cells in the body, it enables them to absorb sugar from the blood. People with type 1 diabetes, however, cannot make insulin; in people with type 2 diabetes, the receptors on their cells respond poorly to insulin, triggering the pancreas to release more insulin, and eventually the pancreas can "burn out," leading to insulin deficiency and elevated blood sugar. Over time, high blood sugar can lead to serious impairments,

Nestlé's Boost Glucose Control drink contains fiber, a type of carbohydrate known to slow down digestion.

ENZYME
A protein that speeds up the rate of a chemical reaction.

ACTIVATION ENERGY
The energy required for a chemical reaction to proceed. Enzymes accelerate reactions by reducing their activation energy.

SUBSTRATE
A compound or molecule that an enzyme binds to and on which it acts.

ACTIVE SITE
The part of the enzyme that binds to substrates.

CATABOLIC REACTION
Any chemical reaction that breaks down complex molecules into simpler molecules.

Enzymes Facilitate Chemical Reactions

→ Cells require enzymes to break down and build up macromolecules. Enzymes are proteins that speed up chemical reactions by reducing the amount of activation energy required to set them in motion.

a. Catabolic Reaction: Bonds are broken

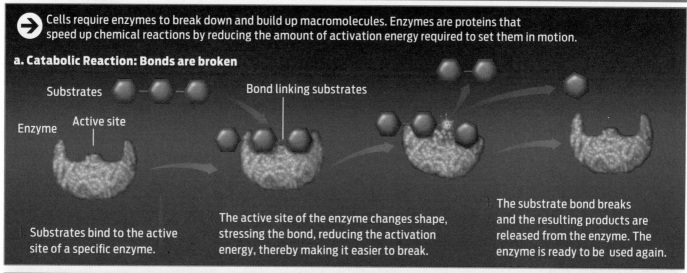

Substrates

Bond linking substrates

Enzyme   Active site

Substrates bind to the active site of a specific enzyme.

The active site of the enzyme changes shape, stressing the bond, reducing the activation energy, thereby making it easier to break.

The substrate bond breaks and the resulting products are released from the enzyme. The enzyme is ready to be used again.

b. Anabolic Reaction: Bonds are created

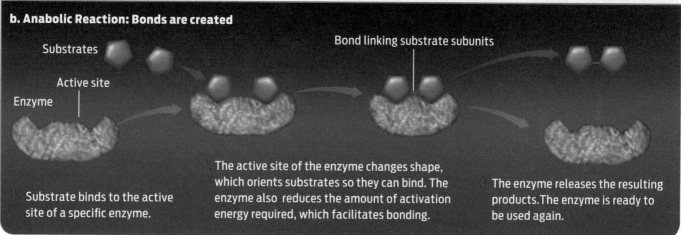

Substrates

Bond linking substrate subunits

Active site

Enzyme

Substrate binds to the active site of a specific enzyme.

The active site of the enzyme changes shape, which orients substrates so they can bind. The enzyme also reduces the amount of activation energy required, which facilitates bonding.

The enzyme releases the resulting products. The enzyme is ready to be used again.

ANABOLIC REACTION
Any chemical reaction that combines simple molecules to build more-complex molecules.

METABOLISM
All biochemical reactions occurring in an organism, including reactions that break down food molecules and reactions that build new cell structures.

INSULIN
A hormone secreted by the pancreas that regulates blood sugar.

including cardiovascular disease, kidney failure, and blindness.

To combat high blood sugar, many diabetics inject themselves with insulin. But as insulin causes sugar in the blood to rush into cells, blood sugar can plummet quickly. Low blood sugar is equally dangerous: it can cause sweating, shakiness, hunger, dizziness, and nausea. To stave off these highs and lows, diabetics are advised to regulate the amount of sugar in their diet.

Because the body breaks down most carbohydrates into sugars, carbohydrates present the most trouble for diabetics—patients must keep track of the amount of carbs in their diets. Not all carbohydrates are the same, however, nor do all types cause spikes in blood sugar.

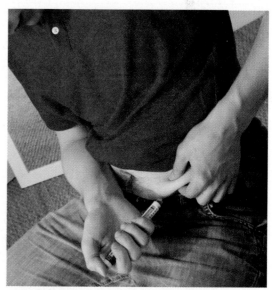

To combat high blood sugar, many diabetics inject themselves with insulin.

Most of the carbohydrates we eat are **complex carbohydrates**–large molecules with branch-like extensions found in plant and meat products. Because they are made of many smaller sugar molecules bound together, complex carbohydrates are also called **polysaccharides.** **Starch** is a complex carbohydrate found in plant products such as rice and potatoes; **glycogen** is a complex carbohydrate found in chicken and steak.

Enzymes in our digestive tract break complex carbohydrates into their component subunits. The smallest carbohydrate subunit is called a **simple sugar**, or **monosaccharide**. The most common simple sugar released from food is glucose, one of two sugars found in table sugar (Infographic 4.4).

But there are some carbohydrates that humans can't digest; one of them is fiber. **Fiber** is a type of indigestible complex carbohydrate found in fruits and vegetables. Humans lack the necessary enzyme to break down fiber, so most fiber passes undigested through the digestive system and out in feces. Although not technically a nutrient because it is not absorbed by the body, fiber is an important part of a healthful diet–it can lower cholesterol and decrease our risk of various cancers. It also

COMPLEX CARBOHYDRATE (POLYSACCHARIDE)
A carbohydrate made of many simple sugars linked together, that is, a polymer of monosaccharides; examples are starch and glycogen.

STARCH
A complex plant carbohydrate made of linked chains of glucose molecules; a source of stored energy.

INFOGRAPHIC 4.4

Complex Carbohydrates Are Broken Down into Simple Sugars

Meats, vegetables, fruits, and grains are rich sources of complex carbohydrates, also called polysaccharides. Digestion breaks down complex carbohydrates into their monosaccharide subunits, also called simple sugars. Not all complex carbohydrates are digestible by humans.

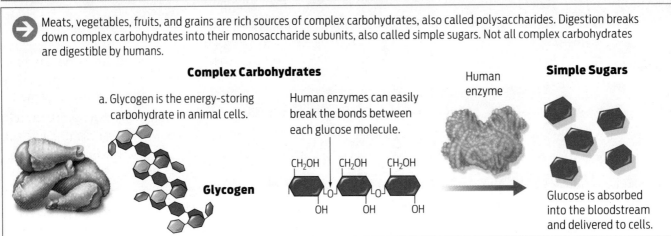

Complex Carbohydrates

Simple Sugars

a. Glycogen is the energy-storing carbohydrate in animal cells.

Glycogen

Human enzymes can easily break the bonds between each glucose molecule.

Human enzyme

Glucose is absorbed into the bloodstream and delivered to cells.

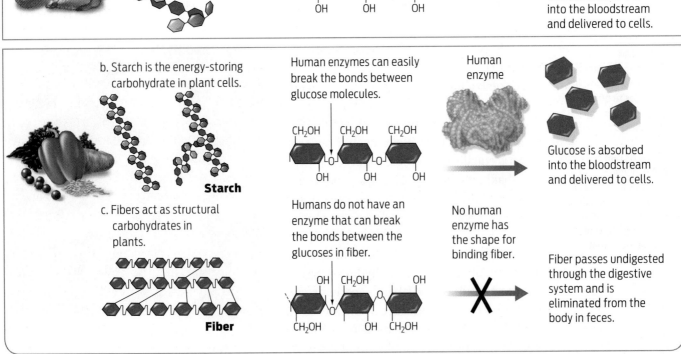

b. Starch is the energy-storing carbohydrate in plant cells.

Starch

Human enzymes can easily break the bonds between glucose molecules.

Human enzyme

Glucose is absorbed into the bloodstream and delivered to cells.

c. Fibers act as structural carbohydrates in plants.

Fiber

Humans do not have an enzyme that can break the bonds between the glucoses in fiber.

No human enzyme has the shape for binding fiber.

Fiber passes undigested through the digestive system and is eliminated from the body in feces.

helps regulate digestion. One of the primary therapeutic ingredients in Boost Glucose Control is fiber.

Because diabetics have trouble balancing their blood-sugar levels, doctors recommend that they avoid foods made of simple sugars—candy, for example—and instead eat complex carbohydrates such as potatoes and oats (in addition to other foods required for a healthful diet). That's because complex carbohydrates are digested more slowly than simple sugars, which means that blood-sugar levels rise more slowly as well. But as anyone who has ever dieted knows, sticking to a restrictive diet is hard. And it's hard even for diabetics. What's more, diabetics suffer the same consequences of dieting that normal people do: deprivation can lead to bingeing on foods that they should avoid.

To stave off dangerous fluctuations in blood sugar, Nestlé's Boost Glucose Control drink is an alternative to other snacks. It is essentially a mix of protein, digestible complex carbohydrates, and fiber. Combining the two types of carbohydrates makes the digestible carbohydrates less accessible to enzymes, thus slowing the release of sugar even more. Protein also takes longer to digest than does carbohydrate. Slowing down digestion reduces the risk of surges in blood sugar that are dangerous for diabetics (**Infographic 4.5**).

Does it work? "Some food additives, like dietary fiber, do have solid evidence behind them," says Jeya Henry, professor of human nutrition at Oxford Brookes University, in Oxford, England, who also heads the university's Functional Foods Centre. In addition to helping regulate blood sugar, fiber appears to cut the risk of many other diseases, too, including heart disease and diseases of the gastrointestinal tract.

In 2009, Henry co-authored a paper that showed that beta-glucan—a type of fiber normally found in oats and barley—when added in certain doses to a type of flat bread significantly slowed down rises in blood-sugar levels relative to blood-sugar levels in people who ate flatbread without beta-glucan. This is important, he says, because some studies have shown that processing or cooking foods with beta-glucan can degrade it. The study showed that mild cooking does not necessarily affect beta-glucan's effectiveness.

Of course, anyone can always eat foods that naturally contain high amounts of fiber—whole-grain breads and lentils, for example—but such foods aren't palatable to everyone, says Henry. In fact, the average American eats only about 15 grams of fiber a day; the recommended amount is 20 to 40 grams, depending on sex, age, and other factors.

Nestlé has shown in one clinical study that their Boost Glucose Control drink does not cause spikes in blood sugar (whereas most snack foods with a high simple-sugar content cause abnormally high blood-sugar levels in diabetics). However, the company is primarily relying on evidence already published in the scientific literature that has shown that each individual active ingredient—the specific types of protein, carbohydrates, and fiber—in its drink can help diabetics control swings in blood-sugar levels.

> "Some food additives, like dietary fiber, do have solid evidence behind them."
> —Jeya Henry

Fighting Chronic Disease

Developing foods with health benefits isn't Nestlé's only goal. Nestlé is, after all, a company, and companies are in the business of earning money. For Nestlé, foods marketed to diabetics presented a lucrative opportunity—according to the World Health Organization (WHO), about 171 million people around the world suffer from diabetes. And their numbers are likely to more than double by 2030.

Many chronic diseases, however, can be either prevented or slowed down with diet and lifestyle changes—which is why they make a good target for food manufacturers. Take osteoporosis, for example. In America alone about 10 million people already have osteoporosis and

GLYCOGEN
A complex animal carbohydrate made of linked chains of glucose molecules; a source of stored energy.

SIMPLE SUGAR (MONOSACCHARIDE)
A carbohydrate made up of a single sugar subunit; an example is glucose.

FIBER
A complex plant carbohydrate that is not digestible by humans.

Fiber Helps Regulate Blood Sugar in Type 2 Diabetics

→ To maintain healthy blood sugar levels, the body must produce and respond to the hormone insulin, which stimulates cells to take up sugar. People with diabetes either cannot make insulin (Type 1 diabetes) or do not respond well to insulin (Type 2 diabetes). In the absence of insulin activity, blood sugar levels remain high. Eating fiber can help regulate blood sugar levels.

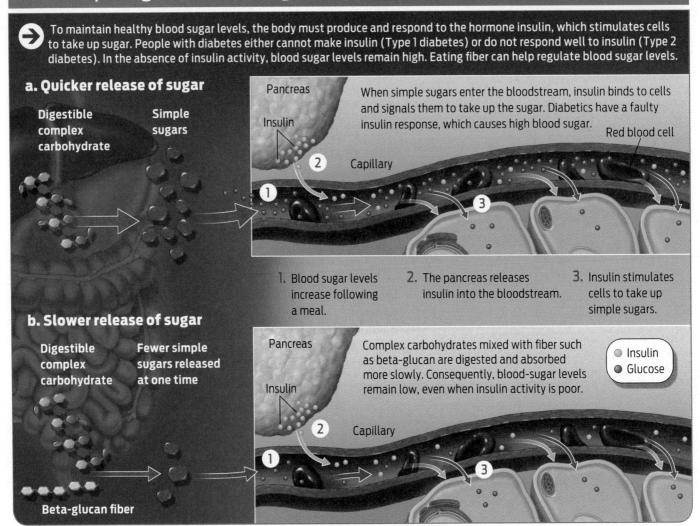

a. Quicker release of sugar

Digestible complex carbohydrate → Simple sugars

Pancreas
Insulin

When simple sugars enter the bloodstream, insulin binds to cells and signals them to take up the sugar. Diabetics have a faulty insulin response, which causes high blood sugar.

Red blood cell

Capillary

1. Blood sugar levels increase following a meal.
2. The pancreas releases insulin into the bloodstream.
3. Insulin stimulates cells to take up simple sugars.

b. Slower release of sugar

Digestible complex carbohydrate → Fewer simple sugars released at one time

Pancreas
Insulin

Complex carbohydrates mixed with fiber such as beta-glucan are digested and absorbed more slowly. Consequently, blood-sugar levels remain low, even when insulin activity is poor.

○ Insulin
● Glucose

Capillary

Beta-glucan fiber

---

another 40 million suffer from low bone density, a condition that can lead to osteoporosis.

Studies have shown that a diet rich in certain nutrients and exercise can either slow the disease or prevent it altogether. So Nestlé scientists have come up with something that they claim may help fight osteoporosis: a powdered drink mix that contains nutrients necessary to build bone.

Building and Rebuilding Bone

Although they seem fixed, our bones are in constant flux. They build up and break down in cycles over a lifetime. By the time we are 20 to 30 years old, our bones are the most dense they will ever be. After that, bone mass deteriorates—in women the process accelerates after menopause. Over a lifetime, women lose about 35% of dense surface bone and 50% of spongy interior bone; men lose 20% of their surface bone and about 30% of their interior bone.

While bone gain and loss is a natural phenomenon, nutrition and exercise influence how dense our bones eventually become and can slow down bone loss. The **mineral** calcium, for example, is especially important for bone health because bone is primarily made up of calcium. Around 99% of the body's calcium is locked up in our bones. Calcium is what makes our bones rigid and strong. The mineral phosphorus is the second most important constituent of bone. About 85% of our body's store of

MINERAL
An inorganic chemical element required by organisms for normal growth, reproduction, and tissue maintenance; examples are calcium, iron, potassium, and zinc.

COFACTOR
An inorganic substance, such as a metal ion, required to activate an enzyme.

COENZYME
A small organic molecule, such as a vitamin, required to activate an enzyme.

VITAMIN
An organic molecule required in small amounts for normal growth, reproduction, and tissue maintenance.

phosphorus is found in the skeleton. Besides keeping bones strong, calcium and phosphorus carry out other functions in the body as well. When our diets lack calcium or phosphorus, the body breaks down bone to release these stored minerals. Over time, robbing bone to compensate for dietary mineral deficiencies can contribute to bone thinning and weakening.

Environmental factors also affect bone density. Weight-bearing exercise, for example, has been shown to increase bone density. Muscular contractions during weight-bearing exercise stress bone tissue, stimulating the body to deposit calcium and phosphorus into bone (Infographic 4.6).

To deposit minerals into bone, our cells rely on specific enzymes that speed up bone-building reactions. As discussed, enzymes bind to and act on substrates, and this interaction is specific. Enzymes don't do their jobs alone, however. Most require accessory, or "helper," chemicals to function. These accessory chemicals are called **cofactors.** Cofactors include inorganic metals such as zinc, copper, and iron. Cofactors can also be organic molecules, in which case they are called **coenzymes.** Most **vitamins**, among them vitamin C, are important coenzymes. Without cofactors and coenzymes that bind to enzymes and enable them to bind substrates, cell metabolism would grind to a halt.

INFOGRAPHIC 4.6

Diet and Exercise Keep Bones Dense

→ Diet and lifestyle contribute to bone density changes over time. Bones are primarily made of the minerals calcium and phosphorus, which a healthy diet provides.

Foods like milk and broccoli provide us with the minerals calcium and phosphorus.

Normal bone matrix

Osteoporosis

Dietary deficiencies of vitamin cofactors as well as the minerals calcium and phosphorus can cause bone to thin.

Muscle contractions during weight-bearing exercise stress bone, stimulating bone cells to deposit calcium and phosphorus into bone tissue, provided these minerals are present in the diet.

Since our bodies only need them in small amounts, minerals and vitamins are examples of **micronutrients** (in contrast to the macronutrients, which we need in larger amounts). But just because our bodies require only small amounts of them doesn't mean micronutrients aren't important components of our diet. In fact, micronutrient deficiency can have serious consequences for health. Iron deficiency causes the blood disease known as anemia, for example, and lack of vitamin C causes a tissue-deteriorating disease known as scurvy. (To prevent the effects of scurvy while on long sea voyages, British sailors used to eat limes and other citrus fruit, which are high in vitamin C; hence the nickname "limeys.")

Food producers routinely add to foods some micronutrients that are hard to obtain from natural sources. Iodine, for example, is added to table salt (in "iodized salt") to prevent goiter, an abnormal thickening of the neck caused by an enlarged thyroid gland due to a lack of dietary iodine.

Bone health also relies on a host of micronutrients. In addition to the minerals calcium and phosphorus, bone requires the metals zinc and magnesium and the vitamins C, D, and K. These micronutrients are important cofactors that assist the enzymes that build bone. Zinc, for example, activates an enzyme that helps deposit phosphorus from our diets into our bones. Vitamin C helps build collagen, the protein that forms the scaffold of bone onto which calcium and phosphorus are added (Infographic 4.7).

To support bone health, Nestlé markets a powdered drink mix that contains the minerals calcium and phosphorus in addition to magnesium, zinc, and vitamins C, D, and K—essentially all the micronutrients necessary to build bone tissue. The company markets the mix to people with osteoporosis and also adds it to products, such as yogurt, that are intended for those who may need a nutritional boost, like children in developing countries who don't get enough of these nutrients in their diet.

Although Nestlé hasn't conducted clinical studies on the bone-preserving drink, the company is relying on indirect evidence to support its health claims. Many studies show that these specific vitamins and minerals help maintain bone density.

While the indirect evidence is strong, not everyone is convinced that such evidence can prove that a food product can benefit health. "Direct proof of activity and benefit should be required for any claim," says José Ordovas, director of the nutrition and genomics laboratory at Tufts University.

"Nutraceuticals, probiotics, and so on are a good concept," he adds, "but they have to be targeted to people who will really benefit from them. Some companies are funneling a good amount of research and development money and resources into designing experiments and interventions to support their claims. But the costs of such studies are much higher than the potential financial benefits, which prevents intervention studies that are large enough and long enough to prove the real benefits of any product."

Other experts are downright dismissive. "A sensible approach would be deep skepticism," says Marion Nestle, a professor of food science and nutrition at New York University (and no relation to the founder of the Nestlé company). "The purpose of nutraceuticals is marketing, not health."

The European Food Safety Authority (EFSA) has been dealing with thousands of applications for health claims on food labels and turning them down in droves, Nestle adds. "EFSA doesn't think the science justifies the claims. I don't either."

Follow the Pyramid

Most public health experts advise that, rather than buy functional-food products, people eat a healthful and varied diet to ensure they get adequate levels of micronutrients. One recent study published in the *American Journal of Clinical Nutrition*, for example, concluded that, aside

> **Many studies show that these specific vitamins and minerals help maintain bone density.**

MICRONUTRIENTS
Nutrients, including vitamins and minerals, that organisms must ingest in small amounts to maintain health.

Enzymes Require Micronutrient Cofactors

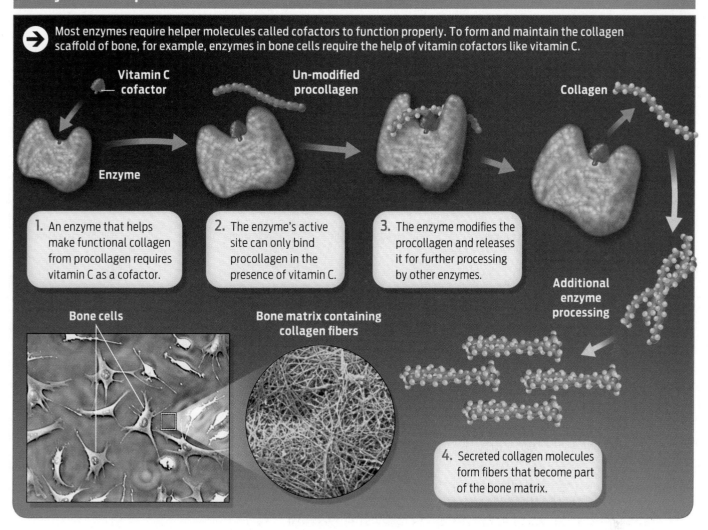

→ Most enzymes require helper molecules called cofactors to function properly. To form and maintain the collagen scaffold of bone, for example, enzymes in bone cells require the help of vitamin cofactors like vitamin C.

Vitamin C cofactor

Un-modified procollagen

Collagen

Enzyme

1. An enzyme that helps make functional collagen from procollagen requires vitamin C as a cofactor.

2. The enzyme's active site can only bind procollagen in the presence of vitamin C.

3. The enzyme modifies the procollagen and releases it for further processing by other enzymes.

Additional enzyme processing

Bone cells

Bone matrix containing collagen fibers

4. Secreted collagen molecules form fibers that become part of the bone matrix.

from vitamin D, "the other micronutrient needs for optimizing bone health can be easily met by a healthy diet that is high in fruits and vegetables." Five servings a day of fruits and vegetables should do it (Table 4.1).

Moreover, researchers are finding that supplements or the nutrients in fortified foods may not be equivalent to the nutrients obtained directly from whole foods. Studies that have followed large groups of people who take vitamin E or beta-carotene supplements to stave off cancer, for example, have found no benefit from these supplements. Meanwhile, epidemiology studies show that people who eat lots of fruits and vegetables and diets rich in vitamin E have a lower risk of developing certain cancers. The reason for the discrepancy isn't clear. Some speculate that vitamin supplementation to stave off disease might work better in people who are deficient in vitamins—most studies haven't separated healthy participants from deficient ones. Others argue that vitamins may confer health benefits only in combination with other as yet unidentified plant compounds. Or, as Ordovas points out, functional foods may simply work better in some people than in others. The research defining such populations, however, has yet to be done. Whatever the reason, it is

TABLE 4.1

A Sample of Micronutrients in Your Diet

 A diet rich in vegetables and dairy products meets our micronutrient needs. Otherwise, nutritional supplements can help fill in dietary gaps.

Minerals: Inorganic elements not synthesized by the body.

MINERAL	FUNCTION	FOOD SOURCES	PROBLEMS OF DEFICIENCY	PROBLEMS OF EXCESS
Calcium	Bone and teeth formation, blood clotting	Dairy products, green vegetables, legumes	Osteoporosis, stunted growth	Kidney stones
Iron	Components of hemoglobin in red blood cells; carries oxygen throughout the body	Green vegetables, beef, liver	Anemia, fatigue, dizziness, headaches, poor concentration	Constipation, risk of type 2 diabetes
Potassium	Electrolyte balance, muscle contraction, nerve function	Fruits, vegetables, meat	Muscle weakness, neurological disturbances	Muscle weakness, heart failure
Sodium	Electrolyte balance, muscle contraction, nerve function	Salt, bread, milk, meat	Muscle cramps, reduced appetite, neurological disturbances	High blood pressure

Water-Soluble Vitamins: Organic molecules not synthesized by the body. Excess vitamin is excreted in urine and so does not harm health.

VITAMIN	FUNCTION	FOOD SOURCES	PROBLEMS OF DEFICIENCY	PROBLEMS OF EXCESS
B$_1$ (thiamine)	Cofactor for enzymes involved in energy metabolism and nerve function	Leafy vegetables, whole grains, meat	Heart failure, depression	None
Folate	Cofactor for enzymes involved in DNA synthesis and cell production	Dark green vegetables, nuts, legumes, whole grains	Neural tube defects, anemia	None
B$_{12}$	Cofactor for enzymes involved in the breakdown of fatty acids and amino acids and nerve cell maintenance	Meat, milk, eggs	Anemia, neurological disturbances	None
C	Cofactor for enzymes involved in collagen synthesis; improves iron absorption and immunity	Citrus fruits	Scurvy, poor wound healing	None

Fat-Soluble Vitamins: Organic molecules not synthesized by the body (except vitamin D). Excess vitamin is stored in fat cells and can harm health.

VITAMIN	FUNCTION	FOOD SOURCES	PROBLEMS OF DEFICIENCY	PROBLEMS OF EXCESS
A (retinol)	Component of eye pigment, supports skin, bone, and tooth growth, supports immunity and reproduction	Fruits and vegetables, liver, egg yolk	Skin problems, blindness	Headaches, intestinal pain, bone pain
D	Calcium absorption, bone growth	Fish, dairy products, eggs	Bone deformities	Kidney damage
E	Antioxidant, supports cell membrane integrity	Green leafy vegetables, legumes, nuts, whole grains	Neural tube defects, anemia, digestive-health problems	Fatigue, headaches, blurred vision, diarrhea
K	Supports synthesis of blood clotting factors	Green leafy vegetables, cabbage, liver	Abnormal blood clotting, bruising	Liver damage, anemia

Food Pyramids: Guides to a Balanced Diet

 While different food pyramids may vary in the recommended quantity of certain food groups, all recommend that grains, fruits, and vegetables occupy the largest proportion of a person's diet and that refined sugars and sweets occupy the smallest. Some pyramids, like this one from the Harvard School of Public Health, include healthy oils, alcohols, supplement and exercise recommendations.

USE SPARINGLY:
RED MEAT & BUTTER
REFINED GRAINS: WHITE RICE, BREAD & PASTA
POTATOES
SUGARY DRINKS & SWEETS
SALT

OPTIONAL:
ALCOHOL IN MODERATION
(Not for everyone)

DAIRY (1–2 servings a day) OR
VITAMIN D/CALCIUM SUPPLEMENTS

HEALTHY FATS/OILS:
OLIVE, CANOLA, SOY, CORN,
SUNFLOWER, PEANUT
& OTHER VEGETABLE OILS;
TRANS-FREE MARGARINE

DAILY MULTIVITAMIN
PLUS EXTRA VITAMIN D
(For most people)

NUTS, SEEDS, BEANS & TOFU   FISH, POULTRY & EGGS

WHOLE GRAINS:
BROWN RICE,
WHOLE WHEAT PASTA,
OATS, ETC.

VEGETABLES & FRUITS   HEALTHY FATS/OILS   WHOLE GRAINS

DAILY EXERCISE & WEIGHT CONTROL

**The Healthy Eating Pyramid by the
Harvard School of Public Health**

clear that no supplement or food additive can replace the health benefits provided by food in its most natural and least processed form–at least not yet.

To help and persuade consumers to eat more fresh foods, public health experts and the U.S. Department of Agriculture have over the past 20 years been devising and updating food pyramids that show the relative importance of each food group and how much from each group we should be eating. While there are many food pyramids, each with slightly different recommendations, all emphasize whole grains as a staple food and recommend eating refined grains and sweets only sparingly **(Infographic 4.8)**.

Nestlé scientists don't disagree. "It is always better to obtain your nutrients by eating a balanced diet in reasonable quantities," says Brian Fern, a research scientist at Nestlé. But

Micronutrient deficiency can have serious health consequences, including osteoporosis (left) and goiter (right).

people don't always have the time to eat right, he adds.

As an example, Fern points to vitamin D. Public health experts are increasingly ringing alarm bells over studies that show Americans are not getting enough of the vitamin, which the body synthesizes when exposed to sunlight. For most people in America with fair to olive skin, 15 minutes a day of strong sunshine is sufficient. But because most people today work in offices, many don't spend enough time outdoors. And in the most northern states, the sun isn't strong enough in the winter to produce vitamin D from the small amount of skin that is exposed.

For people who don't spend much time in the sun, taking a vitamin D supplement might be a good idea, says Fern. In America, milk is fortified with vitamin D. But, for various reasons, not everyone drinks milk. Oily fish such as sardines and mackerel also contain vitamin D, but many Americans eat relatively little fish.

> ## "The public shouldn't think that consuming these foods can compensate for unhealthy dietary or lifestyle habits."
> —José Ordovas

And while nutritionists such as Jeya Henry and José Ordovas agree that some supplements or functional foods can help consumers struggling to conform to diets or meet standard nutrient recommendations, they advise against putting too much faith in substitutes. "Too often the public trusts and consumes products based on promises and unfounded claims," says Ordovas. ∎

▶ Summary

■ Food is a source of both nutrients and energy. Nutrients are all the chemicals required to build and maintain cells and tissues. Cell functions are powered by energy.

■ Nutrients required in large amounts are known as macronutrients; nutrients required in smaller amounts are known as micronutrients. Both are essential for good health.

■ Macronutrients include proteins, carbohydrates, and fats; these are among the organic molecules that make up our cells.

■ Digestion breaks down organic molecules into smaller subunits, which are then used by cells to build cell structures and carry out cell functions.

■ Enzymes are proteins that accelerate the rate of chemical reactions.

■ Enzymes speed up reactions by binding specifically to substrates and reducing the activation energy necessary for a reaction to occur. Enzymes mediate both bond-breaking (catabolic) and bond-building (anabolic) reactions.

■ Complex carbohydrates are a major component of a healthy diet. Foods such as bread, pasta, and rice are rich in complex carbohydrates.

■ With the help of enzymes, the body breaks down complex carbohydrates such as starch and glycogen into individual molecules of the sugar glucose. The complex carbohydrate fiber is not digestible by humans.

■ People with diabetes have trouble regulating blood-sugar levels and are advised to eat complex carbohydrates and fiber rather than simple sugars.

■ Many enzymes require small "helper" chemicals known as cofactors to function. Cofactors are required to maintain healthy bones and other body parts.

■ Micronutrients such as minerals and vitamins, found abundantly in fruits and vegetables, are important cofactors.

NUTRITION AND METABOLISM

Food is a source of nutrients that our bodies need in order to build the macromolecules of cells and function properly. Metabolic reactions break down and build macromolecules.

HINT See Infographics 4.1, 4.2, 4.5, 4.6, 4.8.

➔ KNOW IT

1. Digesting starch releases which of the following subunits?
   a. amino acids
   b. fatty acids
   c. monosaccharides
   d. fiber
   e. proteins

2. The digestion of starch is a type of metabolic reaction known as a(n) _____ reaction.

3. Which of the following foods is the most protein-rich?
   a. olive oil
   b. chocolate
   c. baby back ribs
   d. whole-wheat bread
   e. orange juice

4. Which subunit is needed in order to build DNA?
   a. amino acids
   b. monosaccharides
   c. starch
   d. fatty acids
   e. nucleotides

5. Eating which of the following will cause blood-sugar levels to spike most rapidly?
   a. protein
   b. starch
   c. fiber
   d. fats
   e. starch and fiber combined

6. Which of the following disorders is characterized by poor regulation of blood-sugar levels?
   a. obesity
   b. osteoporosis
   c. diabetes
   d. scurvy
   e. all of the above

➔ USE IT

7. Which of the following molecules can be built directly from protein subunits?
   a. insulin
   b. cell membrane phospholipids
   c. DNA
   d. glycogen
   e. beta-glucan

8. What effect would a meal heavy in starch and fiber have on blood-sugar levels compared to a meal heavy in starch alone?

9. Starch and fiber are both complex carbohydrates made of joined glucose subunits. From the way our bodies digest both types of carbohydrates, would you say a diet rich in fiber would produce higher or lower blood-sugar levels? Explain your answer.

10. Cell membranes are made of phospholipids. What are the dietary source(s) of phospholipids?

11. For a person with type 1 diabetes, will insulin be more effective if taken between meals or at meal times? Explain your answer.

12. Nestlé's bone density–boosting drink, described in this chapter, contains, among other ingredients, calcium and vitamins C and D. Which of these three nutrients—calcium, vitamin C, and vitamin D—does each of the following foods contain? Does any food contain all three?
   a. organic milk, straight from the cow, with no supplementation
   b. milk fortified with vitamin D
   c. oranges
   d. orange juice fortified with calcium
   e. sardines

ENZYME FUNCTION

Enzymes speed up the rate of chemical reactions in the body. Many enzymes require "helper" substances to function.

HINT See Infographics 4.3, 4.4, and 4.7.

➔ KNOW IT

13. The substrate of an enzyme is
   a. an organic accessory molecule.
   b. what is released at the end of an enzyme-speeded reaction.
   c. the shape of the enzyme.
   d. one of the amino acids that make up the enzyme.
   e. what the enzyme acts on.

14. Compare and contrast enzyme cofactors and coenzymes.

15. Enzymes speed up chemical reactions by
    a. increasing the activation energy.
    b. decreasing the activation energy.
    c. breaking bonds.
    d. forming bonds.
    e. giving off energy

16. Why is vitamin C important for bone health?
    a. It is one of the minerals that hardens bone.
    b. It helps the skin produce vitamin D.
    c. It is a coenzyme for the enzymes in bone cells that help build bone.
    d. It blocks the enzymes involved in the breakdown of bone.
    e. It is needed for calcium to be absorbed from food.

⊙ USE IT

17. If the shape of an enzyme's active site were to change, what would happen to the reaction that enzyme usually speeds up?

18. Most of the enzymes in our bodies work best at body temperature. Given this fact, why might a high fever be dangerous to the body?

19. What is osteoporosis? If a woman is at risk for developing osteoporosis, what measures can she take to reduce that risk? For each such factor, explain how it acts to reduce the risk.

20. What is the relationship between sunlight and bone health?

SCIENCE AND ETHICS

21. What are the advantages and disadvantages of eating a "nutritionally engineered" product versus relying on a balanced diet to maintain health? Would your answer change if you were making the same list of pros and cons for a person with type 1 diabetes? What about for a woman whose bones were beginning to thin?

22. Consider the potential benefits of a cereal bar containing beta-glucan marketed specifically to diabetics.
    a. This cereal bar has not yet been tested in people with diabetes. Do you think that there is enough evidence to market this nutraceutical food product to this group of people? Explain your answer.
    b. How would you design an experiment that would test the effectiveness of this cereal bar for diabetics? Explain the rationale for your experimental design and the data that you would collect in order to support or reject marketing claims for this product.

Mighty Microbes

Mighty Microbes

Can scientists make algae into the next global fuel source?

As an engineer working for the Navy SEALS in 1978, Jim Sears took a nighttime scuba dive off the coast of Panama City, Florida, one of many he took to do underwater research. The dive started out routinely, but then, suddenly, glowing phosphorescent algae appeared as if out of nowhere. When Sears put his hands out in front of him, sparkling streamers of microbes came off his fingertips. "It was magical," he recalls.

Sears is an inventor with many and varied inventions to his credit. When working for the Navy in the 1970s and 1980s, he built an underwater speech descrambler and a portable mine detector, among other things. Later he moved on to more creative ventures, including a "hump-o-meter" that could tell farmers when their animals were in heat or mating.

But the seeds of his real claim to fame weren't sown until 2004, when Sears was working on agricultural electronics. That's when he began to turn his attention toward what he felt was the world's biggest problem: dwindling fossil fuel reserves. After he did some thinking and a little research, the tiny, glowing organisms that had wowed him during his dive more than a decade earlier came to mind, in part because of a Web site he stumbled across that discussed the unique properties of algae. He realized suddenly that they might be able to help.

Algae are perhaps best known for the green, red, or bluish hue they give to the surfaces of ponds and lakes, but they have other unique characteristics, too. They were among the first eukaryotic life forms to appear on our planet, and in many ways we can thank them for our very existence, because they fill the atmosphere with the oxygen that supports the majority of life on earth.

Algae also have the amazing ability to convert the energy of sunlight into forms of energy usable by other organisms. This makes them prodigious fuel-producers. Algae convert the

> **Algae also have the amazing ability to convert the energy of sunlight directly into forms of energy usable by other organisms.**

Algal fuel plants (like the one illustrated here) could be the wave of the future.

energy of sunlight and produce not just their own food but also an oil that is very similar to common vegetable oil. It accumulates inside the microbes' tiny cells, and once extracted, it can be processed to make biodiesel, gasoline, or jet fuel. "The more I looked into them, the more amazing they were," Sears says.

And that's good news, because America is desperate for new fuels. After all, Americans burn through 378 million gallons of gasoline a day, enough to fill about 540 Olympic-size swimming pools. And despite the fact that our demand will likely increase over the course of the next 25 years, the sources of our precious gasoline–petroleum reserves buried deep underground– are finite, take millions of years to replenish, and largely lie outside U.S. borders (Table 5.1).

Scientists and politicians are now turning toward alternatives such as biofuels, that is, fuels made from biological material. In an effort to end our addiction to oil, in 2007 President George W. Bush signed the Energy Independence and Security Act, which requires the United States to produce 36 billion gallons of renewable fuels by 2022, of which 21 billion gallons must be advanced biofuels (and not corn-based ethanol) (Infographic 5.1).

Bioluminescent algae glow along the shore.

U.S. Energy Consumption

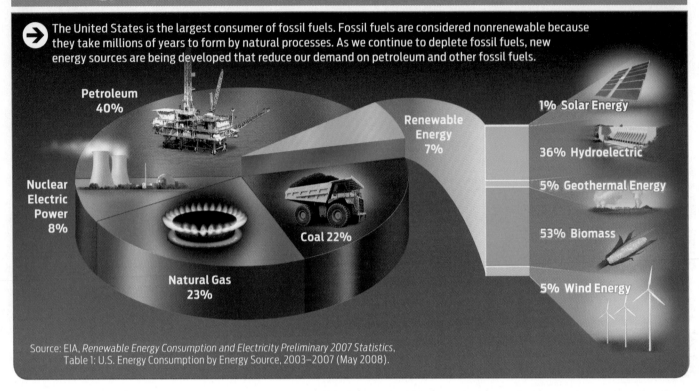

→ The United States is the largest consumer of fossil fuels. Fossil fuels are considered nonrenewable because they take millions of years to form by natural processes. As we continue to deplete fossil fuels, new energy sources are being developed that reduce our demand on petroleum and other fossil fuels.

Petroleum 40%

Nuclear Electric Power 8%

Coal 22%

Natural Gas 23%

Renewable Energy 7%

1% Solar Energy

36% Hydroelectric

5% Geothermal Energy

53% Biomass

5% Wind Energy

Source: EIA, *Renewable Energy Consumption and Electricity Preliminary 2007 Statistics,* Table 1: U.S. Energy Consumption by Energy Source, 2003–2007 (May 2008).

Convinced by the promise of algae biofuels, in 2006, Sears founded Solix, one of the first biotechnology companies working to mass-produce biodiesel from algae. Though he is no longer involved with Solix, the company is still going strong. In 2009, it began commercially producing its algae-based fuel, with the goal of making the equivalent of 3,000 gallons of oil per acre of cultivated algae. Other companies are getting on the algae bandwagon—as of

---

As of 2010, there were more than 150 companies dedicated to making fuel from algae.

2010, there were more than 150 companies dedicated to making fuel from algae. In January 2009, Continental Airlines flew its first plane powered in part by jet fuel made from algae, and in September of that year, a modified Toyota Prius dubbed Algaeus drove 3,750 miles across the country powered by a fuel mix of algal and conventional gasoline, plus batteries. Algae, many say, are the future.

Energy Basics

Energy isn't just needed to fly planes and drive cars, of course. **Energy**—defined as the capacity to do work—is critical to all life on earth. Energy powers every activity we perform, from the more obvious ones like breathing, thinking, and running to less obvious activities like building the molecules that make up our bodies. Without a source of energy, all life on

ENERGY
The capacity to do work. Cellular work includes processes such as building complex molecules and moving substances in and out of the cell.

A traffic jam in Los Angeles.

TABLE 5.1

Largest Fossil Fuel Reserves by Country as of 2005*

COAL		NATURAL GAS		OIL	
United States	119,327	Russian Federation	43,038	Saudi Arabia	36,038
Russian Federation	68,699	Iran	24,066	Iran	18,754
India	60,843	Qatar	23,205	Iraq	15,686
China	58,900	Saudi Arabia	6,210	Kuwait	13,845
Australia	39,033	United Arab Emirates	5,432	United Arab Emirates	13,340
South Africa	32,500	United States	4,908	Venezuela	10,875
Kazakhstan	19,810	Nigeria	4,707	Russian Federation	10,153
Ukraine	16,809	Algeria	4,122	Kazakhstan	5,404
Poland	9,333	Venezuela	3,884	Libya	5,337
Colombia	4,280	Iraq	2,853	Nigeria	4,894
Brazil	3,371	Kazakhstan	2,700	United States	3,996
Reserves in other countries	29,705		36,724		25,455
Global reserves	**462,612**		**161,848**		**163,777**

*In million tonnes of oil equivalent. One tonne equals approximately 2,205 lbs.
Source: EarthTrends and BP plc, 2006.

earth would grind to a halt, like a cell phone with a dying battery.

Organisms can't simply create energy when they need it, however–as we'll see, energy cannot be created–they must obtain it from an outside source. Humans and other animals obtain the energy they need by eating food. We've already seen that our digestive systems break down food to obtain nutrients (Chapter 4). As these molecules are further broken down, the energy stored in the molecules is made available to do work. The bonds that hold molecular subunits together represent a form of stored **chemical energy**; breaking these bonds releases that stored energy, making it available to power cell functions.

Algae could be the world's next energy source because these tiny organisms are very efficient energy converters–the oil they produce is rich in chemical energy. All they need in order to make this oil is sunlight, carbon dioxide, water, and two key nutrients: nitrogen and phosphorus. Give them these tidbits, and algae grow rapidly–

CHEMICAL ENERGY
Potential energy stored in the bonds of biological molecules.

The Toyota Algaeus, powered by algae, gasoline, and batteries.

Algae Capture Energy in Their Molecules

➔ Algae can use sunlight, carbon dioxide, and nutrients to produce a high volume of oil readily used to produce biofuel, in addition to carbohydrates and proteins that can be useful as additional energy sources.

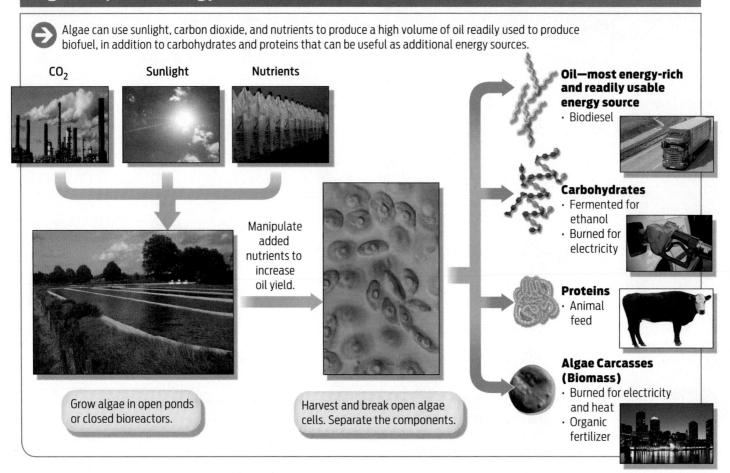

CO_2

Sunlight

Nutrients

Grow algae in open ponds or closed bioreactors.

Manipulate added nutrients to increase oil yield.

Harvest and break open algae cells. Separate the components.

Oil—most energy-rich and readily usable energy source
· Biodiesel

Carbohydrates
· Fermented for ethanol
· Burned for electricity

Proteins
· Animal feed

Algae Carcasses (Biomass)
· Burned for electricity and heat
· Organic fertilizer

some strains double their volume in 12 hours–all the while accumulating organic molecules inside their tiny cells. These molecules include oils that can be used to make fuel and sugars that can be converted into biofuels like ethanol and butanol. The latter can be mixed with gasoline to power hybrid cars **(Infographic 5.2)**.

Sears wasn't the first to consider algae's fuel potential. In 1978–the same year Sears took his fateful night dive–the U.S. Department of Energy started its Aquatic Species Program, with the goal of exploring algae's fuel possibilities. But when oil fell to $20 a barrel in 1996, the government abandoned the program, assuming that oil made from algae would always be too expensive. Now, with oil prices much more volatile and overall creeping higher and higher, biofuel from algae has become an attractive option again.

How do algae stack up against other biofuels? Compared to corn-based ethanol and fuels made from plants like soybeans and oil palm, algae take the prize in terms of how much fuel they can produce for the amount of space they take up. According to the U.S. Department of Energy, if American soybean farmers had converted all of their crops in 2007–on about 67 million acres of land–into biofuel, they would have provided the country with only enough on-road diesel to meet 6% of the nation's needs **(Table 5.2)**.

On the other hand, if farmers had grown algae on this same amount of land in open ponds or containers, they would have produced enough fuel to supply all our country's diesel needs. And because algae can be cultivated on land that is unsuitable for food crops, production of algal biofuel doesn't compete with or take resources away from food production (which is a common criticism of corn-derived ethanol). Extracting fuel from algae is also less energy-intensive than for other biofuel sources. To make ethanol, for

TABLE 5.2

How Green Are Biofuels?

 Biofuels are getting a bad rap as stories of rising food prices and shortages fill the news. But the environmental, energy, and land-use impacts of the crops used to make the fuels vary dramatically. Current fuel sources—corn, soybeans, and canola—are more environmentally damaging than alternatives that are under development.

CROP	Fuel source USED TO PRODUCE	GREENHOUSE GAS EMISSIONS* (KG OF CO_2 CREATED PER MEGAJOULE OF ENERGY PRODUCED)	Use of resources during growing, harvesting, and refining of fuel				% OF EXISTING U.S. CROP LAND NEEDED TO PRODUCE ENOUGH FUEL TO MEET HALF OF U.S. DEMAND	PROS AND CONS
			WATER	FERTILIZER	PESTICIDE	ENERGY		
Corn	Ethanol	81–85	High	High	High	High	157–262	Technology ready and relatively cheap; reduces food supply
Sugar cane	Ethanol	4–12	High	High	Medium	Medium	46–57	Technology ready; limited as to where sugar cane will grow
Switch grass	Ethanol	–24	Medium-low	Low	Low	Low	60–108	Won't compete with food crops; technology not ready
Wood residue	Ethanol, biodiesel	N/A	Medium	Low	Low	Low	150–250	Uses timber waste and other debris; technology not fully ready
Soy-beans	Biodiesel	49	High	Low-medium	Medium	Medium-low	180–240	Technology ready; reduces food supply
Rape-seed, canola	Biodiesel	37	High	Medium	Medium	Medium-low	30	Technology ready; reduces food supply
Algae	Biodiesel	–183	Medium	Low	Low	High	1–2	Potential for huge production levels; technology not ready

*Emissions produced during the growing harvesting, refining, and burning of fuel; gasoline is 94, diesel is 83. A megajoule is 1 million joules.

Source: Groom et al., *Conservation Biology*, 2008.

INFOGRAPHIC 5.3

Energy Is Conserved

→ Energy in the universe is neither created nor destroyed, but is converted from one form to another. Stored potential energy, for example, can be converted to kinetic energy, as the cyclist below illustrates.

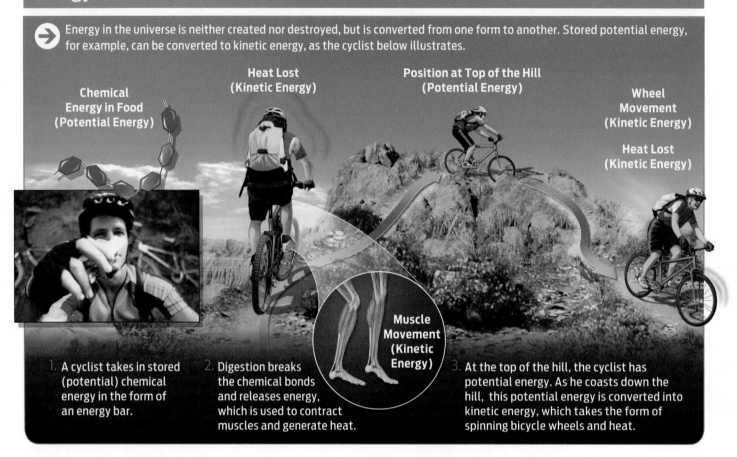

Chemical
Energy in Food
(Potential Energy)

Heat Lost
(Kinetic Energy)

Position at Top of the Hill
(Potential Energy)

Wheel
Movement
(Kinetic Energy)

Heat Lost
(Kinetic Energy)

Muscle
Movement
(Kinetic
Energy)

1. A cyclist takes in stored (potential) chemical energy in the form of an energy bar.

2. Digestion breaks the chemical bonds and releases energy, which is used to contract muscles and generate heat.

3. At the top of the hill, the cyclist has potential energy. As he coasts down the hill, this potential energy is converted into kinetic energy, which takes the form of spinning bicycle wheels and heat.

---

instance, farmers typically start with corn, which requires more energy to grow than algae do. Then they have to ferment the corn into ethanol, and finally they must harvest the ethanol. It's a multistep process that requires precious farmland, clean water, and energy inputs. In addition, energy has to be transferred, first from sunlight into corn, then from corn into ethanol. Algae are far simpler life forms, and therefore more efficient: using sunlight and carbon dioxide, they produce fuel automatically. And they can be grown on poor-quality lands that could not be used for other crops.

With so much talk about dwindling energy reserves, it's tempting to think that energy is something that we simply use up over time. But energy cannot be created or destroyed. When energy is transferred from food or fuel to the organisms or products that use it, that energy is not destroyed—it merely changes form. This is a principle known as the "conservation of energy."

Consider a cyclist who eats a cereal bar before an uphill ride. The bar contains **potential energy** in the chemical bonds that hold the molecules of that bar together. When the cyclist eats and digests the bar, digestion breaks those chemical bonds, and the stored potential energy is released. As the cyclist climbs the hill, his body converts this potential energy into the **kinetic energy** of muscle contraction and **heat**. At the top of the hill, he relies once again on potential energy to get him downhill. His relatively higher position means that he has "positional" potential energy.

As the cyclist coasts down the hill, the friction of the wheels on the ground converts his positional potential energy into the kinetic energy of moving wheels and heat. From start to finish, from cereal bar to spinning wheels, energy is converted from one form into another. With every conversion, though, some energy is lost as heat (Infographic 5.3).

POTENTIAL ENERGY
Stored energy.

KINETIC ENERGY
The energy of motion or movement.

HEAT
The kinetic energy generated by random movements of molecules or atoms.

Energy Transformation Is Not Efficient

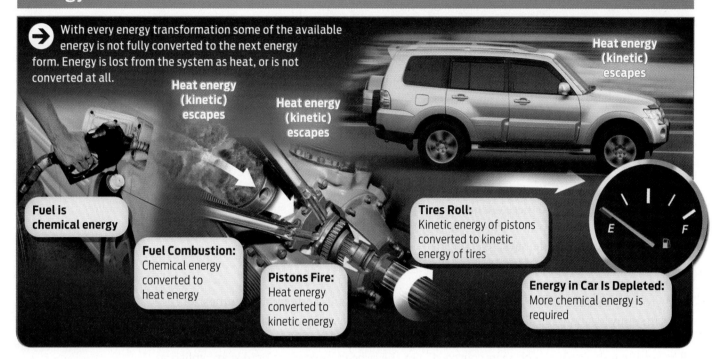

→ With every energy transformation some of the available energy is not fully converted to the next energy form. Energy is lost from the system as heat, or is not converted at all.

Heat energy (kinetic) escapes

Heat energy (kinetic) escapes

Heat energy (kinetic) escapes

Fuel is chemical energy

Fuel Combustion: Chemical energy converted to heat energy

Pistons Fire: Heat energy converted to kinetic energy

Tires Roll: Kinetic energy of pistons converted to kinetic energy of tires

Energy in Car Is Depleted: More chemical energy is required

E   F

Biofuel from algae has a similar life story: algae use the energy from the sun to produce oil molecules called triacylglycerols that contain energy-storing bonds. When these bonds are broken during combustion, they release large amounts of energy that can be used to power machines. The chemical energy in the triacylglycerols is converted to heat energy that warms gas molecules. The expansion of the heated gas molecules (kinetic energy) pushes the pistons in a car's engine, which cause the wheels to move (kinetic energy).

The conversion of energy from one form to another isn't 100% efficient, though. With each energy transformation, a bit of energy is lost to the environment as heat. This is why our bodies heat up when we exercise and car engines are warm after being driven. In the case of a car engine, the generation of heat serves a purpose (to power the pistons), but heat loss to the outside of the car is inefficient. This inefficiency is the reason we need to keep supplying energy to any system. We eat three meals a day to replenish the energy our bodies have lost as heat and converted into the chemical energy of cells and the kinetic energy of movement. Similarly, cars need a new tank of fuel after they have burned

through the last one as they convert the chemical energy of the fuel into the heat and kinetic energy of motor movement. Just how good your car is at converting the chemical energy of gas into the kinetic energy of car speed determines your mpg–your miles per gallon. If an engine doesn't combust efficiently, some of the fuel molecules will undergo chemical reactions and be converted to other molecules–like pollutants–rather than power the pistons as heat. If the pistons can't use the heat efficiently, the heat will leave the car without powering the wheels. At each step of energy transformation, energy is lost from the car system and into the environment, and then we're back to the fuel pump one more time (Infographic 5.4).

Energy's Beginnings

Tom Allnutt calls himself an "algae guy." While a student at Virginia Tech in the 1970s, he decided on a whim to take a class in phycology–the study of algae–and immediately he was hooked. Algae fascinate him in part because they can survive pretty much anywhere, from the scalding thermal vents in Yellowstone National Park to the dry bitter cold of Antarctica, where Allnutt later spent 3 years as a scientist diving in freezing lakes and probing rock fissures for signs of the tiny organisms.

Now Allnutt has turned his attention to algae biofuels. He is the Senior Vice President of Research & Development at Phycal, a biotechnology company based in Ohio. In 2010, Phycal received $24 million in federal funding to build a 40-acre pilot facility to grow algae in shallow ponds, extract their oils, and test the oil for its viability as a commercial fuel. The plant will be located in Hawaii, where there is almost constant daytime sunlight –and therefore plenty of energy for the algae to convert into fuel. "We're taking the easy stuff first," he says, referring to the company's decision to locate their pilot plant in a sunny spot. It's a common choice: many algae-growing companies have positioned themselves in sunny locations like California.

Why is steady sun so important? For almost all living things on earth, the ultimate source of energy is the sun. Sunlight plays a role in the energy carried in a cereal bar or a barrel of oil. The sun functions like a giant thermonuclear reactor: it converts matter into the energy of sunlight. Obviously, humans and other animals can't use the power of sunlight directly, since they have no means to capture it and convert it into a usable form. But organisms like plants and algae are able to capture the energy of sunlight and convert it into a form of chemical energy that can be used to sustain the growth of that organism, can be eaten by other organisms, or can fuel energy-requiring reactions. The process by which plants and algae convert the energy of sunlight into the chemical energy of energy-rich molecules is called **photosynthesis.**

Photosynthesis is critical to life on earth because it is the primary mechanism that makes energy available to almost all living organisms. Photosynthesis is the specialty of **autotrophs**—organisms such as plants, algae, and certain bacteria that can use the energy of sunlight to build organic molecules. Their name means, literally, "self-feeders"–and as you might

> **For almost all living things on earth, the ultimate source of energy is the sun.**

expect, autotrophs make their own food (Infographic 5.5).

Capturing Energy: Photosynthesis

How exactly do autotrophs–which include not just algae, but also plants like wheat, corn, rice, soybeans, and photosynthetic bacteria–use sunlight to create molecules? These molecules can be used for growth, making up the body of the maturing organism, or as energy sources that power cellular reactions. The process of photosynthesis can be summarized in the following equation:

$$\text{Sunlight} + \text{Water} + \text{Carbon dioxide} \rightarrow \text{Oxygen} + \text{Glucose}$$

That is, plants and other photosynthesizers use the energy of sunlight to make the molecule glucose using carbon dioxide as a source of carbon. In the process, water molecules are split and oxygen is given off as a by-product.

Carbon dioxide consumption is yet another benefit associated with algae and essentially all photosynthetic organisms, since they use CO_2 as carbon source. Carbon dioxide is not just the gas that plants and algae take in during photosynthesis, it is also the gas that is released by burning fossil fuels. As we'll see in Chapter 23, fossil fuels such as coal, petroleum, and natural gas are the compressed remains of once-living photosynthetic organisms that have formed over millions of years; burning these fuels releases this stored carbon dioxide into the atmosphere. Carbon dioxide is also a greenhouse gas that is accumulating in the atmosphere and is in part responsible for increasing temperature levels around the globe. By pulling carbon dioxide out of the atmosphere, algae and photosynthetic organisms such as plants help mitigate the effects of climate change. Algae's need for carbon also presents a challenge, however: in order to grow large numbers of these microbes, companies need to provide their algae with more carbon dioxide than is readily available in the atmosphere. In other words, they have to have

PHOTOSYNTHESIS
The process by which plants and other autotrophs use the energy of sunlight to make energy-rich molecules using carbon dioxide and water.

AUTOTROPHS
Organisms such as plants, algae, and certain bacteria that capture the energy of sunlight by photosynthesis.

another source, which can be costly. Sears's company, Solix, has set up its first biofuel production plant next to a beer manufacturer that produces carbon dioxide as a by-product of brewing. The company simply siphons off this carbon dioxide and feeds it to its algae, thus helping them grow.

Another boon of using algae as a source of biofuel is that, because algae remove CO_2 from the atmosphere as they grow, they release little net CO_2 when they are burned–they are basically just returning to the atmosphere the same carbon they just removed. In addition, compared to traditional diesel, studies show that biodiesel releases less carbon monoxide and particulate matter. Algae-based biofuels aren't

The power plant project of Arizona Public Service includes housing algae in tubes where they multiply as part of the process of producing biodiesel.

INFOGRAPHIC 5.5

Autotrophs Convert Light Energy into Chemical Energy

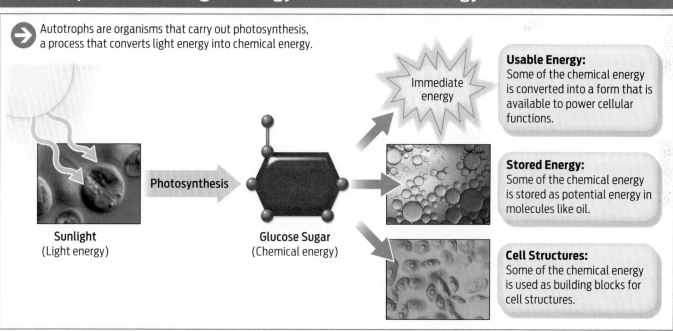

→ Autotrophs are organisms that carry out photosynthesis, a process that converts light energy into chemical energy.

Sunlight
(Light energy)

Photosynthesis

Glucose Sugar
(Chemical energy)

Immediate energy

Usable Energy:
Some of the chemical energy is converted into a form that is available to power cellular functions.

Stored Energy:
Some of the chemical energy is stored as potential energy in molecules like oil.

Cell Structures:
Some of the chemical energy is used as building blocks for cell structures.

There are three basic types of autotrophs:

Plants

Algae

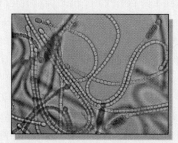

Some bacteria (e.g., cyanobacteria)

Photosynthesis Captures Sunlight to Make Food

Photosynthesis is the process by which plants and other autotrophs use the energy of sunlight to make food. In plants and algae, photosynthesis occurs in an organelle called the chloroplast, found in cells that make up the green parts of the plant. Photosynthesis has two main steps.

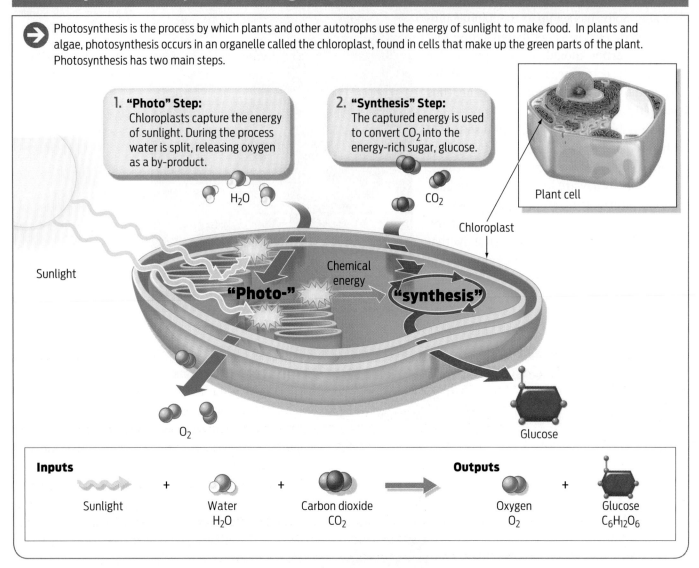

1. **"Photo" Step:**
Chloroplasts capture the energy of sunlight. During the process water is split, releasing oxygen as a by-product.

2. **"Synthesis" Step:**
The captured energy is used to convert CO_2 into the energy-rich sugar, glucose.

H_2O

CO_2

Plant cell

Chloroplast

Sunlight

"Photo-"

Chemical energy

"synthesis"

O_2

Glucose

Inputs				Outputs		
Sunlight	+	Water H_2O	Carbon dioxide CO_2	Oxygen O_2	+	Glucose $C_6H_{12}O_6$

only renewable, then, they are also more environmentally friendly when they are burned.

Photosynthetic algae (and plants) complete photosynthesis in two parts: a "photo" part and a "synthesis" part. During the "photo" part, light energy is captured in chemical form. During the "synthesis" part, this chemical energy is used to generate glucose molecules using the carbon atoms of carbon dioxide. The entire process occurs in an organelle present in leaves and algae called the **chloroplast (Infographic 5.6).**

Through photosynthesis, algae can convert up to 6% of the sun's radiation into new cell mass, according to the U.S. Department of Energy–most crop plants convert only 1-2% of

the sun's radiation into new cell mass. While glucose is the major product of photosynthesis, other smaller sugars are produced during the "synthesis" reactions. Glucose and these other sugars provide the building materials for a variety of metabolic reactions in the cell–for example, the assembly of amino acids for protein synthesis, and the synthesis of the oils that make up biofuels.

Mass-producing algae, however, is not as simple as just putting microbes in a pond with some carbon dioxide, sitting back, and watching them grow. One problem is that algae need to be stirred frequently to incorporate carbon dioxide, a process that requires a lot of extra energy. Aurora Biofuels, an algae company

CHLOROPLAST
The organelle in plant and algae cells where photosynthesis occurs.

Solazyme founders Jonathan Wolfson (left) and Harrison Dillion examine green algae.

Photosynthetic algae can be expensive to maintain because they are either grown in large transparent tanks called photobioreactors, which are costly, or they are grown in open ponds, which require lots of fresh water and constant monitoring for contamination. Wolfson and Harrison looked for a solution. They knew that there were rare strains of algae that do not rely on sunlight, instead surviving on sugar. If they used these algae to make biofuel, could they save money? Algae that could be grown in closed, nonphotosynthetic vats and fed cane sugar could conceivably be much cheaper. With support from their investors, Wolfson and Harrison changed their business model and gave the new idea a whirl. It was a good decision: they now believe that they can make oil using these nonphotosynthetic algae that cost only $60 to $80 a barrel, in part because they can save many overhead costs. Many other algae companies still use photosynthetic algae, and they have developed cost-effective ways to make them work, but for Solazyme, the rare sugar-eating algae are a better choice.

Organisms like ones that Solazyme uses—organisms that can't photosynthesize and must eat other organisms or molecules produced by other organisms to obtain energy—are called **heterotrophs** ("other-feeders"). It's a group that includes Solazyme's sugar-feeding algae, plus all animals, fungi, and most bacteria. When humans and other heterotrophs eat plants—or eat animals that have eaten plants—specialized processes release chemical energy stored in the molecules of the plant (or animal) body. Organisms can then either use this energy to grow (in which case the energy will become stored in the chemical bonds making up their bodies), or to move and power other chemical reactions in cells, like making oil and other organic molecules. The result: energy is converted from one form into another and flows from organism to organism—and eventually, from organism to fuel, to machine.

From Sun to Fuel

In order for photosynthetic organisms to convert the sun's energy, they must be able to capture it from the sun. But how can something as

based in Alameda, California, gets around this difficulty by using the carbon dioxide that is pumped into the ponds to drive circulation, rather than using a paddle wheel. This saves about four-fifths of the energy normally required. Another problem for algae growers is that in the summer, sun intensity is so high that it actually oversaturates the microbes, limiting their growth. Algae evolved to live below the surface of the ocean, where light is limited, so they don't handle lots of sun very well.

To overcome this hurdle, in 2007 researchers at the University of California at Berkeley engineered algae with smaller antennae, the body parts algae use to absorb and regulate sunlight, similar to the way radio antennas absorb electromagnetic signals. These smaller antennae help them avoid oversaturation and increase growth.

A Cheaper Strategy?

There are more than 100,000 species of algae, and it may be more economical to make biofuel out of certain species. Harrison Dillon and Jonathan Wolfson, the founders of Solazyme, a San Francisco-based algae biofuels company, know this well. In 2005, while attempting to develop biofuel from photosynthetic algae, they came to a shocking realization: the oil they were producing was going to cost approximately $1,000 a gallon.

HETEROTROPHS
Organisms, such as humans and other animals, that obtain energy by eating organic molecules that were made by other organisms.

The Energy in Sunlight Travels in Waves

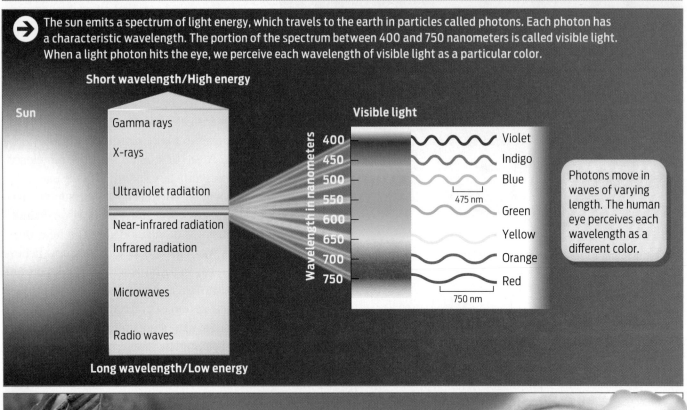

➡️ The sun emits a spectrum of light energy, which travels to the earth in particles called photons. Each photon has a characteristic wavelength. The portion of the spectrum between 400 and 750 nanometers is called visible light. When a light photon hits the eye, we perceive each wavelength of visible light as a particular color.

Short wavelength/High energy

Sun

Gamma rays

X-rays

Ultraviolet radiation

Near-infrared radiation

Infrared radiation

Microwaves

Radio waves

Long wavelength/Low energy

Visible light

Wavelength in nanometers

400 — Violet
450 — Indigo
500 — Blue
550 — 475 nm
600 — Green
650 — Yellow
700 — Orange
750 — Red

750 nm

Photons move in waves of varying length. The human eye perceives each wavelength as a different color.

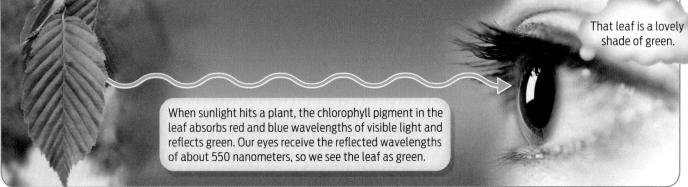

That leaf is a lovely shade of green.

When sunlight hits a plant, the chlorophyll pigment in the leaf absorbs red and blue wavelengths of visible light and reflects green. Our eyes receive the reflected wavelengths of about 550 nanometers, so we see the leaf as green.

intangible as sunlight carry energy? If you've ever walked barefoot across a sandy beach on a hot summer day, you know that sunlight is a potent source of heat energy. You may also have a sense that certain colors absorb or reflect sunlight better than others—on a sunny day, wearing a white shirt keeps you cooler than a black one.

These properties reflect the nature of **light energy,** which is part of the electromagnetic spectrum. Light energy exists in discrete particles called **photons.** Each photon carries a certain amount of energy, determined by its wavelength. Photons of different wavelengths contain different amounts of energy, and some of these wavelengths, when viewed by the human eye and interpreted by the human brain, appear to us as different colors (Infographic 5.7).

Take green plants. When sunlight hits a plant, its leaves absorb red and blue wavelengths and reflect green wavelengths—which is why plants appear green to our eyes. The molecule that absorbs and reflects light, called **chlorophyll,** is a crucial player in plant photosynthesis. It is

LIGHT ENERGY
The energy of the electromagnetic spectrum of radiation.

PHOTONS
Packets of light energy, each with a specific wavelength and quantity of energy.

chlorophyll that actually captures the energy of sunlight. During the "photo" reaction, chlorophyll molecules absorb energy from the red and blue wavelengths of sunlight. In addition to chlorophyll, algae contain other pigment molecules that absorb and reflect certain wavelengths of light. These other pigments give red and blue (and even golden) algae their distinctive colors. But chlorophyll is the main pigment involved in photosynthesis. When red and blue photons of sunlight hit chlorophyll, the electrons in its atoms become excited. These excited electrons are used to generate an energy-carrying molecule known as **adenosine triphosphate (ATP)**, which is used in the "synthesis" part of photosynthesis to make sugar. (We'll talk more about ATP, the cell's "energy currency," in Chapter 6.)

Plants use the sugar they make as food. They can link simple sugars together to make more complex sugar molecules such as starches and to form plant products such as wheat grains. They can also use the chemical energy stored in the bonds of the sugar molecules to power cellular reactions, and they can use the actual molecules as building materials in the anabolic reactions that make proteins, nucleic acids, and lipids—all the organic molecules critical to building a plant body.

Algae are vital because, using the energy of sunlight, they produce a tremendous amount of lipid molecules: oil. In some algae, oil constitutes up to half of their dry weight. Some algae species secrete the oil in order to adhere to other cells, whereas others accumulate the oil inside their bodies in order to control buoyancy. Still others use the oil as an efficient way to store energy. For reasons that are not yet entirely clear, oil production is highest when algae are grown under stressful conditions—for instance, when they are deprived of nitrogen, an essential nutrient—but the algae do not multiply as quickly under these conditions, so most companies do not try to grow algae this way.

Biofuel and Beyond

In the end, photosynthesis accomplishes two main things. First, it converts light energy from the sun into chemical energy that can be used as food and fuel by plants, animals, and humans. Second, it captures carbon dioxide gas from the air and incorporates those carbon atoms into sugar, a process called **carbon fixation.** By converting inorganic gaseous carbon into an organic form that can be eaten by animals or used by plants to grow and increase their biomass, carbon fixation is ultimately the way carbon enters the global energy chain (**Infographic 5.8**)

Carbon fixation is Jim Sears's favorite topic these days. Having left Solix in 2007, Sears is now the Chief Technology Officer of A2BE Carbon Capture, a company based in Boulder, Colorado, that is looking for ways to reduce carbon dioxide levels in the atmosphere. If carbon dioxide is necessary for photosynthesis, why would anyone want to reduce carbon dioxide levels? As we'll discuss more in Chapter 23, carbon dioxide is a greenhouse gas, contributing to the greenhouse effect and therefore to global warming. Plants, algae, and other photosynthetic organisms all help to temper the effects of global warming by pulling carbon dioxide out of the atmosphere and fixing it into organic sugars. If scientists could find a way to enhance this natural process, it would be a boon to the planet.

Believe it or not, Sears says, the healthy soil in your backyard is actually photosynthetic. In soil, tiny bacteria called cyanobacteria thrive, and they perform photosynthesis.

These cyanobacteria are good not only for the soil, but for the entire planet. Like algae, they perform photosynthesis, which means that in addition to absorbing sunlight, they capture carbon from the atmosphere. They then convert the carbon into forms that provide energy and nutrition to colonies of other microorganisms buried deep within the soil. One square meter of healthy, undisturbed soil can remove 30 grams of atmospheric carbon per year.

The problem is that approximately 2 billion out of the earth's 13 billion total hectares of landmass have been damaged by human activity—construction and fires are among the biggest culprits. According to Sears, it can take anywhere from 30 to 3,000 years for soil

CHLOROPHYLL
The pigment present in the green parts of plants that absorbs photons of light energy during the light reactions of photosynthesis.

ADENOSINE TRIPHOSPHATE (ATP)
The molecule that cells use to power energy-requiring functions.

CARBON FIXATION
The conversion of inorganic carbon (for example, CO_2) into organic forms (for example, sugars).

Photosynthesis: A Closer Look

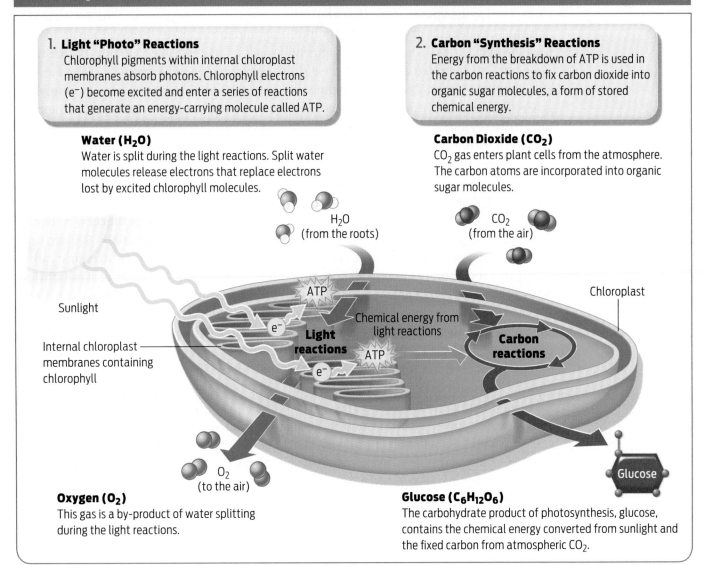

1. Light "Photo" Reactions
Chlorophyll pigments within internal chloroplast membranes absorb photons. Chlorophyll electrons (e^-) become excited and enter a series of reactions that generate an energy-carrying molecule called ATP.

2. Carbon "Synthesis" Reactions
Energy from the breakdown of ATP is used in the carbon reactions to fix carbon dioxide into organic sugar molecules, a form of stored chemical energy.

Water (H_2O)
Water is split during the light reactions. Split water molecules release electrons that replace electrons lost by excited chlorophyll molecules.

Carbon Dioxide (CO_2)
CO_2 gas enters plant cells from the atmosphere. The carbon atoms are incorporated into organic sugar molecules.

H_2O
(from the roots)

CO_2
(from the air)

Sunlight

Internal chloroplast membranes containing chlorophyll

e^-

Light reactions

e^-

ATP

Chemical energy from light reactions

ATP

Carbon reactions

Chloroplast

Glucose

O_2
(to the air)

Oxygen (O_2)
This gas is a by-product of water splitting during the light reactions.

Glucose ($C_6H_{12}O_6$)
The carbohydrate product of photosynthesis, glucose, contains the chemical energy converted from sunlight and the fixed carbon from atmospheric CO_2.

microorganisms to regenerate after being destroyed—and in the meantime, the damaged soil is unable to remove carbon dioxide from the atmosphere.

Sears, however, has a solution. His new company takes small samples of microorganisms from healthy soil, grows them in a contained facility, and then transplants them to damaged soil, where they spread out and thrive. He estimates that if 1 billion hectares of land

> **"Algae truly are the foundations of our entire planet."**
> **—Jim Sears**

were restored in this way, one-seventh of the world's greenhouse gas problem would be solved because of the vast amounts of carbon dioxide that would be pulled out of the atmosphere by the photosynthetic cyanobacteria in the regenerated soil.

When you think about it, it's amazing that organisms like algae and cyanobacteria that seem so simple could be so vital to life on earth. But they are. Not only did they provide the

Jim Sears examines a restored soil sample.

planet with the first breaths of atmospheric oxygen millions of years ago, but soon they could become the world's most important fuel source as well as a potential solution to climate change. All this from single-cell organisms that have just one major claim to fame: they can convert the energy of sunlight into energy-rich organic molecules. "Algae truly are the foundations of our entire planet," says Sears. ■

▶ Summary

■ All living organisms require energy to live and grow. The ultimate source of energy on earth is the sun.

■ Photosynthesis is a series of chemical reactions that captures the energy of sunlight and converts it into chemical energy in the form of sugar and other energy-rich molecules. This energy is used by all living organisms to fuel cellular processes.

■ Photosynthesis can be divided into two main parts: a "photo" part, during which the pigment chlorophyll captures light energy, and a "synthesis" part, during which captured energy is used to fix carbon dioxide into glucose.

■ Photosynthetic organisms are known as autotrophs; they include plants, algae, and some bacteria. Animals do not photosynthesize; they are known as heterotrophs.

■ Energy is neither created nor destroyed, but is converted from one form into another, a principle known as the conservation of energy.

■ Kinetic energy is the energy of motion and includes heat and light energy. Potential energy is stored energy and includes chemical energy.

■ Energy flows from the sun, is captured and transferred through living organisms, and then flows back into the environment as heat.

■ Energy conversions are inefficient. Some energy is lost as heat with every conversion of energy.

■ The energy-rich molecules produced by some photosynthetic algae include oils that can be used as an energy source to power automobiles and aircraft. These alternative fuels show great promise in terms of sustainable consumption.

PHOTOSYNTHESIS
Photosynthesis is the process by which the energy of sunlight is captured and stored as chemical energy.

HINT **See Infographics 5.2, 5.5–5.8.**

➡ KNOW IT
1. The energy of sunlight exists in the form of
   a. glucose.
   b. photons.
   c. gamma rays.
   d. ions.
   e. particles.

2. Which photon wavelength contains the greatest amount of energy?
   a. violet
   b. red
   c. green
   d. yellow
   e. blue

3. Why does algae appear green?

4. Glucose is a product of photosynthesis. Where do the carbons in glucose come from?
   a. starch
   b. cow manure
   c. carbon dioxide
   d. water
   e. soil

5. Compare and contrast the ways photosynthetic algae and animals obtain energy.

6. Mark each of the following as an INPUT (I) or an OUTPUT (O) of photosynthesis.
   Oxygen _____
   Carbon dioxide _____
   Photons _____
   Glucose _____
   Water _____

➡ USE IT
7. Global warming is linked to elevated atmospheric carbon dioxide levels. How might this affect photosynthesis? If global warming should cause ocean levels to rise, in turn causing forests to be immersed in water, how would photosynthesis be affected?

8. Why are energy-rich lipids from algae more "useful" as a fuel than energy-rich sugars and other carbohydrates produced by photosynthetic organisms like corn and wheat?

ENERGY FLOW
Energy is initially captured by autotrophs and flows through other organisms and machines. As energy flows, some of it is lost as heat.

HINT **See Infographics 5.2, 5.3, and 5.4.**

➡ KNOW IT
9. The energy in an energy bar is _____ energy. The energy of a cyclist pedaling is _____ energy.
   a. light; chemical
   b. potential; chemical
   c. chemical; kinetic
   d. potential; potential
   e. kinetic; potential

10. Kinetic energy is best described as
    a. stored energy.
    b. light energy.
    c. the energy of movement.
    d. heat energy
    e. any of the above, depending on the situation.

➡ USE IT
11. If you wanted to get the most possible energy from photosynthetic algae, should you eat algae directly or feed algae to a cow, and then eat a burger made from that cow? Explain your answer.

ALGAE AND BIOFUELS
Algae that produce large amounts of lipids are being developed as new and sustainable fuel sources.

HINT **See Infographics 5.1, 5.2, and 5.5.**

➡ KNOW IT
12. Photosynthetic algae are
    a. eukaryotic autotrophs.
    b. prokaryotic autotrophs.
    c. eukaryotic heterotrophs.
    d. prokaryotic heterotrophs.

13. Which of the following is/are necessary for biofuel production by algae?
- **a.** sunlight
- **b.** sugar
- **c.** CO_2
- **d.** soil
- **e.** all of the above
- **f.** a & b
- **g.** a & c

➲ USE IT

14. Many types of algae can divert the sugars they make by photosynthesis into pathways to make biodiesel, a fuel that essentially consists of lipids. Biodiesel is a promising replacement for fossil fuels. Describe the energy transfers required to make biodiesel and explain why biodiesel might be a more promising fuel than lipids extracted from animals.

15. What do you think are some of the advantages and disadvantages of growing algae in enclosed tubes or bags compared to growing them in open vats? Be sure to explain your reasoning.

SCIENCE AND ETHICS

16. Many biofuels require arable land as part of their production process. Discuss competing needs for arable lands in the context of human needs for food and fuel, and how algae may alleviate this tension.

Supersize Me?

Supersize me?

Changing our culture of eating

For years Paul Rozin, a professor of psychology at the University of Pennsylvania, was baffled by this question: How are the French able to eat rich cheeses, butter-laden sauces, fatty meat, and still stay slimmer than Americans?

As of 2008, a whopping 72% of American men and 64% of American women were overweight or obese. Compare these figures to the corresponding ones in France: about 56% of men and 40% of women.

Americans aren't the only ones tipping the scales in greater numbers. People around the world are getting heavier. As of 2005, approximately 1.6 billion adults over the age of 15 were overweight and at least 400 million were obese, according to the World Health Organization (WHO). By 2020, WHO expects those numbers to double, a figure that will amount to almost half of today's global population. Our increasing girth alone wouldn't be a problem were it not for

> **How are the French able to eat rich cheeses, butter-laden sauces, fatty meat, and still stay slimmer than Americans?**

what comes with it: obesity has ushered in increased rates of heart disease, diabetes, and other related illnesses. In fact, by 2020 chronic illnesses resulting from obesity will likely kill more people than infectious diseases.

Why are people getting heavier? Biologists argue that humans are predisposed to gain weight. Throughout human evolution, famine was the norm, and people had to work hard to grow or hunt and gather their food. Our bodies have adapted by storing extra food as fat for times when food is scarce.

How heavy is too heavy? Even with a few extra pounds here and there, most people still fall within a healthy weight range. Only when our total body fat passes a certain point do the scales tip toward unhealthy. That point depends on a number of factors, including gender, body type, and frame size. To get a rough estimate, some health care professionals rely on the **body mass index (BMI).** The BMI estimates body fat from

BODY MASS INDEX (BMI)
An estimate of body fat based on height and weight.

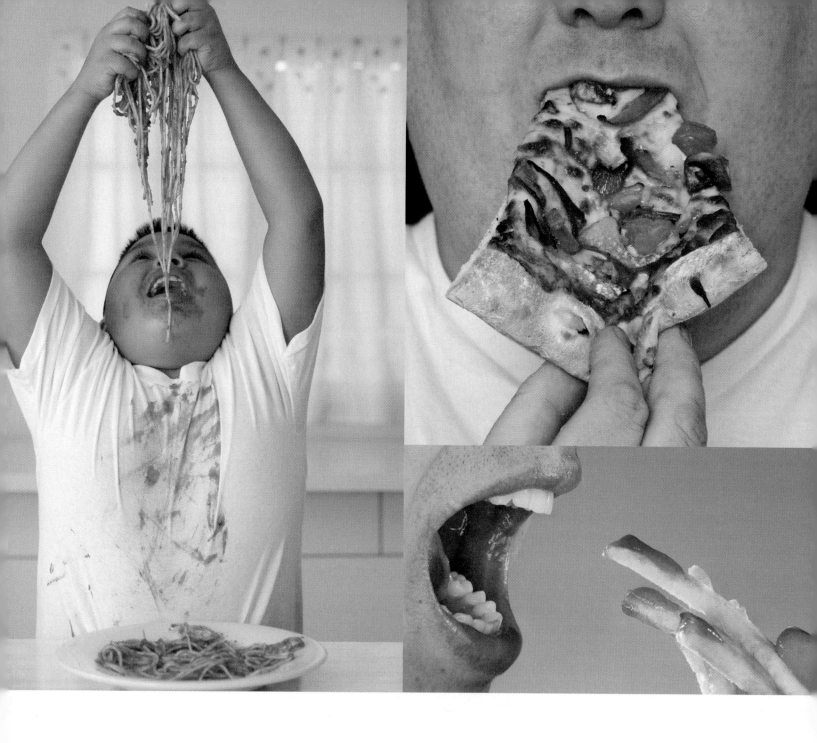

the indirect measures of height and weight. People with a BMI between 25 and 29.9 are considered overweight; people with a BMI of 30 and above are considered **obese**. BMI can be misleading, however. Athletes and people with more muscle mass will sometimes register as overweight or obese when in fact they are perfectly healthy **(Infographic 6.1)**.

But sociologists peg another weight-gain culprit: eating behavior. Culture, they argue, is as much to blame for the obesity epidemic as biology. Some societies have remained relatively thin, they point out, despite similar biology. But when these societies adopt American eating habits, which include fast-food, snacks, and soda, they tend to put on pounds, too, which is why Rozin wanted to study the eating habits of the French.

Rozin has found that culture affects not only *how* we eat but also *how much*. In studies conducted in America and in France, Rozin found that portion sizes in America are often bigger than those in France. In other words, one reason that a greater percentage of Americans are heavier than the French is simply because in America, more food is spooned onto our plates. Eating more food than our bodies need means more food-energy stored as fat.

"Culture is underrated as a contributing factor to unhealthy eating," says Rozin. "The idea of what a proper meal is and your own habits are largely instituted by the culture in which you live."

Clearly there's little we can do about our biology. But culture is another matter. There is a growing movement in America to reign in what has been the cultural norm of unhealthful eating. For example, public health experts have been lobbying the government to pass legislation that would improve people's access to more-heathful foods—fresh fruits and vegetables, for example, that are nutritious and also low in Calories—to counteract our larger serving sizes. Many local governments have already banned restaurants from using an unhealthful type of fat to fry foods. And Congress has passed legislation that limits what types of food can be sold in schools so that children aren't filling themselves up with Calorie-dense food with little nutritional value. Many are pressing the U.S. government to do even more to change the way Americans eat (Infographic 6.2).

"Culture is underrated as a contributing factor to unhealthy eating." —Paul Rozin

"Unhealthy food choices have become the default food choice," says Kelly Brownell, director of the Rudd Center for Food Policy and Obesity at Yale University. "The question is what can we do about it?"

What's in a Meal?

A few years ago, Rozin and colleagues at the Centre National de la Recherche Scientifique in France set out to see just how much more Americans eat than their French counterparts. They compared portion sizes at eleven restaurants in Philadelphia and in Paris. Their results, published in the journal *Psychological Science*, didn't surprise them. The average portion size in the Paris restaurants weighed 277 grams. By contrast, the average portion size in the Philadelphia restaurants weighed 346 grams—25% more! Even restaurant chains like McDonald's served smaller portions of certain foods: in

INFOGRAPHIC 6.1
Body Mass Index (BMI)

A BMI chart provides an indirect measure of body fat, based on the ratio of body height to weight. Because it does not take muscle mass or frame size into account, BMI is only an estimate of body fat. Some people may register as overweight even though they are a healthy weight.

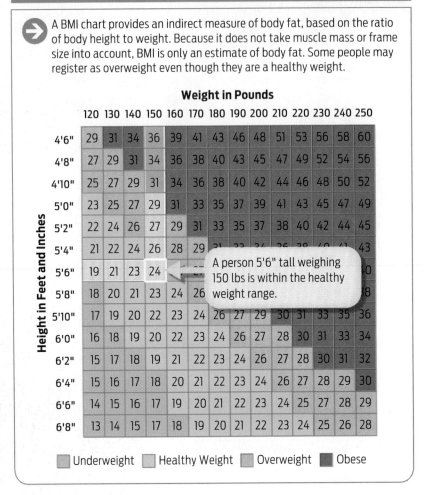

INFOGRAPHIC 6.2

Obesity Is Influenced by Biology and Culture

+

=

Biological History
Famine was common. Our bodies have evolved to hoard energy in the form of body fat to get them through times when food was scarce.

Cultural Influence
An abundance of high-fat, processed food is increasingly common in developed countries.

Modern Obesity
Today people consume many more Calories than during any other time in history because food is abundant. Our bodies store extra Calories as fat, as they have been evolutionarily programmed to do.

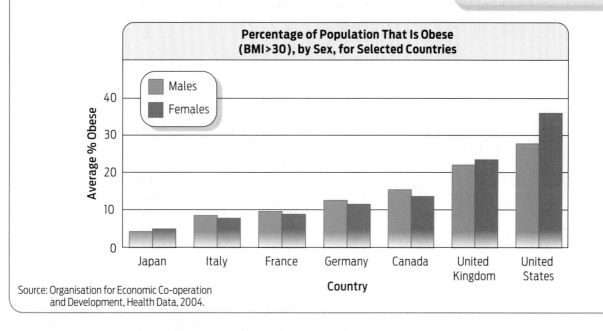

Percentage of Population That Is Obese (BMI>30), by Sex, for Selected Countries

Legend: Males, Females

Y-axis: Average % Obese (0, 10, 20, 30, 40)

X-axis (Country): Japan, Italy, France, Germany, Canada, United Kingdom, United States

Source: Organisation for Economic Co-operation and Development, Health Data, 2004.

Paul Rozin, University of Pennsylvania psychology professor, is examining the differences between American and French eating culture.

Paris, medium fries weighed 90 grams, large fries 135 grams; in Philadelphia, medium fries weighed 155 grams and large fries 200 grams.

Once they had surveyed the restaurant scene, Rozin's team went further. They compared the sizes of packaged food in American and in French supermarkets, and they found the same trend: the portions of the majority of food items they tested—from ice cream to chewing gum to yogurt—were smaller in France. Even portion sizes for ingredients in the most commonly used French cookbook were smaller

than those in the celebrated American favorite, *Joy of Cooking* (Infographic 6.3).

How does portion size contribute to weight gain? The answer lies in the energy content of the food we eat. We've already seen, in Chapter 4, that all food consists of mixtures of carbohydrates, fats, proteins, and nucleic acids, and that the relative proportion of each macronutrient varies in different types of food. Meat, for example, contains more protein per unit weight than do potatoes; potatoes have more carbohydrates than meat does. To nourish our bodies, we must eat a balanced diet that includes appropriate amounts of all macronutrients. But food is not only a source of nutrition; it is also a source of chemical energy that powers our activities. Food is fuel.

For a number of the items McDonald's serves, portion sizes are larger in Philadelphia than in Paris.

INFOGRAPHIC 6.3

Americans Eat Large Portions

→ Researchers compared portion sizes in restaurants in Philadelphia to those in Paris. In all but one restaurant, U.S. portions were larger at least half the time. The average portion size in Paris was 277 grams; the average size in Philadelphia was 346 grams. Philadelphians eat an average of 25% more food than Parisians at every meal.

For sampled menu items, U.S. restaurants consistently serve larger portion sizes.

Numbers >1 reflect larger portion sizes in the U.S., compared to France

Table 1. Restaurant portion sizes

Restaurant in Paris	Restaurant in Philadelphia	No. of items sampled/ No. larger in U.S.	Mean size ratio (U.S./France)
McDonald's	McDonald's	6/4	1.28
Hard Rock Cafe	Hard Rock Cafe	2/0	0.92
Pizza Hut	Pizza Hut	2/2	1.32
Häagen-Dazs	Häagen-Dazs	2/2	1.42
French: local bistro	French: local bistro	1/1	1.17
Quick	Burger King	5/4	1.36
Local Chinese	Local Chinese	6/4	1.72
Italian: Bistro Romain	Olive Garden	3/2	1.02
Crêpes: local	Crêpes: local	4/2	1.04
Ice cream: local	Ice cream: local	2/2	1.24
Pizza: local	Pizza: local	2/2	1.32

ROZIN, P ET AL. PSYCHOLOGICAL SCIENCE. 203, 14:450–454.

Food Powers Cellular Work

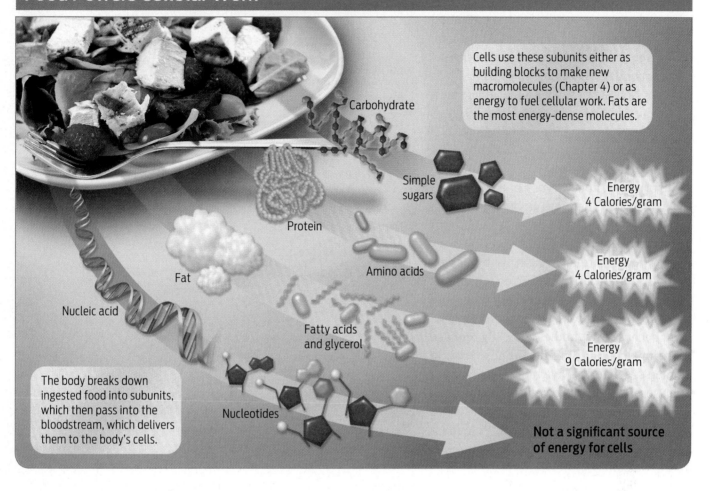

Cells use these subunits either as building blocks to make new macromolecules (Chapter 4) or as energy to fuel cellular work. Fats are the most energy-dense molecules.

Carbohydrate

Simple sugars

Protein

Fat

Amino acids

Nucleic acid

Fatty acids and glycerol

Energy 4 Calories/gram

Energy 4 Calories/gram

Energy 9 Calories/gram

The body breaks down ingested food into subunits, which then pass into the bloodstream, which delivers them to the body's cells.

Nucleotides

Not a significant source of energy for cells

---

calorie
The amount of energy required to raise the temperature of 1 gram of water by 1° Celsius.

CALORIE
1,000 calories or 1 kilocalorie (kcal); the capital "C" in Calorie indicates "kilocalorie." The Calorie is the common unit of energy used in food nutrition labels.

Scientists measure energy in units called calories. A **calorie** (in lower case) is the amount of energy required to raise the temperature of 1 gram of water by 1° Celsius. In essence, a calorie is a measurement of energy—the capacity to perform a certain amount of work. On most food labels, the amount of energy stored is listed in kilocalories, which are also referred to as kcals or as Calories (the capital "C" indicates that kilocalories, not calories, are meant). One **Calorie** is equal to 1,000 calories, or 1 kcal.

Of all the organic molecules, fats are the most energy dense: each gram of fat stores approximately 9 Calories in its chemical bonds. Proteins and carbohydrates are about half as energy dense: each gram stores about 4 Calories. Clearly, a 200-gram serving of fatty bacon contains many more Calories than does a 200-gram serving of asparagus (**Infographic 6.4**).

All our activities—everything from thinking and digesting to sleeping and running—require energy. So all bodies expend some Calories each day just to stay alive. A person's daily energy needs largely depend on gender, age, body type, and activity levels. A sedentary college-age average-size male, for example, would need to ingest anywhere between 2,200 and 2,400 Calories per day to power his activities and maintain his weight, whereas a football player would need more than 3,200 Calories a day to power and maintain his. Exercise or other physical activities require additional energy beyond the basic life-sustaining energy needs of the body. Consequently, athletes, or those who exercise a great deal, generally need to eat more to fuel their activities than do their less-active peers.

Exactly how much more should an athlete eat? Consider the college football player, who must consume 800 to 1,000 Calories more than his sedentary roommate. An average cheeseburger contains anywhere between 400 and 600 Calories depending on its size, so the football player would need about two extra cheeseburgers per day. That's probably less than you thought. Now suppose that same athlete ate a cheeseburger off-season. It would take 1.5 hours of slow swimming, or 2.5 hours of walking, or 3 hours of cycling at 5.5 miles per hour, to use up that extra energy (Table 6.1).

Surprisingly, not everyone burns energy at the same rate. There are people who seem to be able to eat to their heart's content and hardly gain an ounce. And there are those who seem to gain weight just by looking at food. Genetics plays a large role in how much food each one of us actually needs, but there are other factors, too—gender, for one. Men, because their bodies naturally produce more muscle-building hormones than do women's, generally have more muscle mass and therefore need to eat more than women do. Since muscle cells burn more calories than do fat cells, the ratio of muscle mass compared to fat content in our bodies is another factor.

Putting on Pounds

Our bodies are fairly efficient at extracting energy from food, but we humans eat not only for sustenance but also for pleasure—which is where problems can arise. Many of us eat more food than our bodies need. We also have a natural preference for fatty and sugary foods because such foods are energy dense. For our ancestors, it was likely important to load up on those foods to store energy for times when food was scarce. Today, this ancient taste preference has become a vice that snack food companies have become very good at exploiting. For the large majority of us, when we eat Calories beyond what our bodies require, the extra energy is stored in one of two places: as **glycogen** in muscle and liver cells, or as **triglycerides** in fat cells.

Glycogen is the energy-storing carbohydrate found in animal cells. You can think of glycogen as a short-term storage system. When we require

TABLE 6.1

Calories In, Calories Out

CALORIES IN SELECT FOODS

FOOD	CALORIES
8 oz. unsweetened green tea	2
1 large slice whole wheat bread	79
½ cup cooked white rice	102
12 oz. nonfat milk	120
12 oz. cola	140
1 glazed doughnut	200
1 slice thick-crust cheese pizza	256
1 Starbucks grande mocha frappucino with whipped cream	380
1 McDonald's Big Mac	540
1 Burger King Whopper	670

CALORIES BURNED DURING SELECT ACTIVITIES*

ACTIVITY	CALORIES/HOUR
Sleeping	55
Sitting	85
Standing	100
Office work	140
Golf (walking)	240
Gardening (planting)	250
Walking (3 mph)	280+
Tennis	350+
Biking (moderate)	450+
Jogging (5 mph)	500+
Swimming (active)	500+
Hiking	500+
Power walking	600+
Cycling (stationary)	650
Squash	650+
Running	700+

*Approximate number of Calories burned per hour by a 150-pound woman.

Glycogen and Fat Store Excess Calories

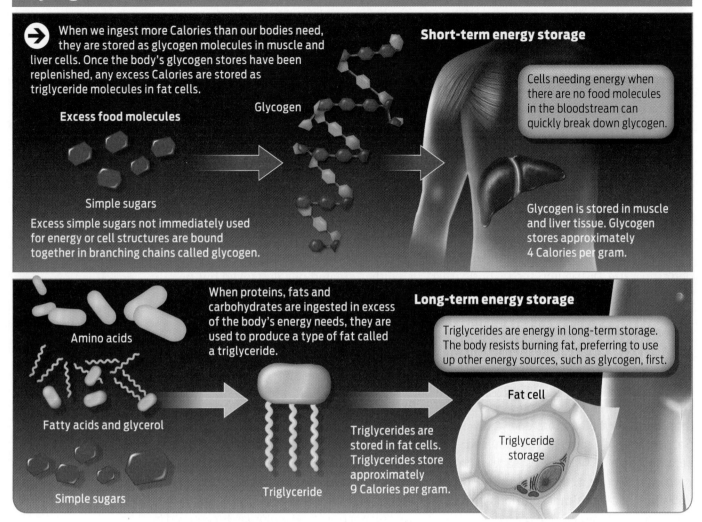

→ When we ingest more Calories than our bodies need, they are stored as glycogen molecules in muscle and liver cells. Once the body's glycogen stores have been replenished, any excess Calories are stored as triglyceride molecules in fat cells.

Short-term energy storage

Excess food molecules

Glycogen

Cells needing energy when there are no food molecules in the bloodstream can quickly break down glycogen.

Simple sugars

Excess simple sugars not immediately used for energy or cell structures are bound together in branching chains called glycogen.

Glycogen is stored in muscle and liver tissue. Glycogen stores approximately 4 Calories per gram.

Amino acids

When proteins, fats and carbohydrates are ingested in excess of the body's energy needs, they are used to produce a type of fat called a triglyceride.

Long-term energy storage

Triglycerides are energy in long-term storage. The body resists burning fat, preferring to use up other energy sources, such as glycogen, first.

Fat cell

Triglyceride storage

Fatty acids and glycerol

Triglycerides are stored in fat cells. Triglycerides store approximately 9 Calories per gram.

Simple sugars

Triglyceride

GLYCOGEN
A complex animal carbohydrate, made up of linked chains of glucose molecules, that stores energy for short-term use.

TRIGLYCERIDE
A type of lipid found in fat cells that stores excess energy for long-term use.

short bursts of energy—as in a sprint, for example—the body breaks down glycogen to obtain energy. However, because a gram of glycogen stores only half as many Calories as a gram of fat (about 4 Calories per gram versus 9), our bodies would have to carry around twice as much glycogen to store the same amount of Calories. So our bodies store most excess Calories as fat, which actually allows us to carry around less weight overall. The downside, however, is that it takes sustained activity to burn fat. The body burns fats only after it has already used up food molecules in the bloodstream and in stored glycogen.

For our ancestors who lived during times of frequent famine, this system of storing Calories as fat would have come in handy. Their bodies

could burn fat for energy to carry them through times of food scarcity. Today, people in most of the developed world have plenty of food. But they are largely sedentary and eat more Calories than they need—which is why they've started to pack on, and keep on, the pounds (Infographic 6.5).

Because each type of energy-rich organic molecule that we ingest—whether protein, carbohydrate, or fat—stores a different amount of energy, it's not only how much we eat but also what we eat that contributes to weight gain. We are more likely to gain weight from a portion of ice cream than an equivalent portion of broccoli, for example, because ice cream contains more fat—and therefore more Calories—than broccoli.

Some scientists have used this fact to argue that there must be some factor other than small portion size that explains why the French have remained relatively thin. The French may eat less, but they are also world renowned for their penchant for buttery, creamy sauces, dense desserts, and fatty meats. If the French load up on fat, which has more Calories, how do they manage to keep the weight off? Research such as Rozin's has shown that even though the French may eat more fat-laden or fat-heavy foods than Americans, it still has to do with portion size: they still consume fewer total Calories.

How do the French manage to eat only small portions? In France and many other European countries, small is the cultural norm. The French don't super size. Distributors such as Costco that sell bulk items don't exist in France, at least not yet. Research by several groups suggest that a person presented with a bigger package of, say, M&M candies will take more from it than when presented with a smaller package. A 2007 study led by Jennifer Fischer at Baylor College of Medicine, for example, found that preschool-age children consumed 33% more energy when the portion size of the meal was doubled. This behavior combined with meals made with a high ratio of energy-dense ingredients—oil, butter, and sugar—is a significant contributor to childhood obesity, these authors concluded.

Or it may have to do with the *way* the French eat; not only do they eat smaller portions at each meal, they don't snack, they don't run for second helpings, and they don't skip meals. Mireille Guiliano in her book *French Women Don't Get Fat* described how she gained 20 pounds during her 5-month stay in America. She snacked, she drank a lot of soda, and she ate standing up, she wrote. She found that she had forgotten how to enjoy the taste of food she was used to in France and compensated by eating larger portions. Part of her diet plan when she returned to France, she wrote, was quitting in-between-meals snacking and reacquainting herself with the French culture of eating—part of which involves eating only a few bites of any dish, just enough to satisfy the taste buds, and then pushing the plate away.

The French also spend more time eating. In his study, Rozin compared the average time people spent eating at McDonald's in Paris and in Philadelphia. In Paris, the average time of the meal was 22 minutes; in Philadelphia, only 14. While scientists don't know for sure how longer meals help people eat less, they speculate that taking it slow may help people enjoy their food more and recognize when they are full.

However it is that the French are able to curb their appetites, one thing is clear: they weigh less because they either eat less or are more active and burn more energy. The only way to gain weight is by taking in more Calories than we expend. In other words, our waistlines obey the principle of conservation of energy: energy is neither created nor destroyed but merely converted from one form into another. If more food energy is taken in than is used to power cellular reactions and physical movement, the excess (minus what is released as heat with every energy conversion) is stored as fat.

Extracting Energy from Food

Getting energy from food seems simple enough: we eat food and we have energy. To provide us with fuel, food goes through a series of complex biochemical reactions that convert the chemical energy stored in food into a form of fuel we can use. Energy from food is ultimately captured in

Research shows that large portion sizes are behind weight gain.

ATP: The Energy Currency of Cells

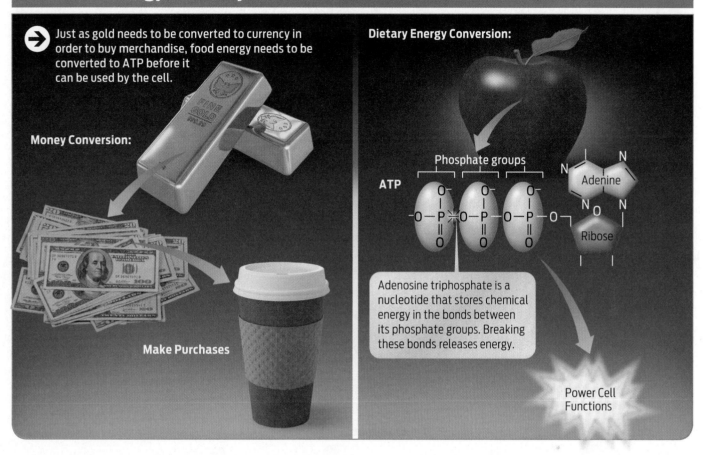

→ Just as gold needs to be converted to currency in order to buy merchandise, food energy needs to be converted to ATP before it can be used by the cell.

Money Conversion:

Make Purchases

Dietary Energy Conversion:

Phosphate groups

ATP

Adenine

Ribose

Adenosine triphosphate is a nucleotide that stores chemical energy in the bonds between its phosphate groups. Breaking these bonds releases energy.

Power Cell Functions

ADENOSINE TRIPHOSPHATE (ATP)

The molecule that cells use to power energy-requiring functions; the cell's energy "currency."

AEROBIC RESPIRATION

A series of reactions that occurs in the presence of oxygen and converts energy stored in food into ATP.

a molecule called **adenosine triphosphate (ATP)** that our cells use to carry out energy-requiring functions.

You can think of food as a bar of gold: it has a great deal of value, but if you carried that gold bar to your local convenience store, you wouldn't be able to buy even a cup of coffee with it. You would first have to convert your gold bar into bills and coins. ATP is the energetic equivalent of bills and coins; it's currency that your body can actually spend **(Infographic 6.6)**.

To make ATP, our bodies first break down food molecules into their smaller subunits through the process of digestion. Once released from food, such subunits as fatty acids, glycerol, amino acids, and sugars leave the small intestine and enter the bloodstream, which transports them to the body's cells. Inside the cells, enzymes break apart the bonds holding these subunits together. The energy stored in those bonds is then captured and converted into the molecular bonds that make up ATP. When ATP bonds are broken, energy is released, allowing cells to "spend" their ATP currency and carry out normal cellular functions.

The primary process that all eukaryotic organisms, including plants, use to convert energy stored in food molecules to form ATP is called **aerobic respiration.** Of the subunits released from food, sugar–in the form of glucose–is the most common source of energy for all organisms, from bacteria to humans. The aerobic respiration of glucose can be summarized by this equation:

Glucose + Oxygen ⟶
Carbon dioxide + Water
+ Energy (+ heat)

That is, the bonds holding the glucose molecule together are broken. Oxygen from the air we

Aerobic Respiration Transfers Food Energy to ATP

During aerobic respiration, our cells use the oxygen we inhale to help extract energy from food. Cells convert the energy stored in food molecules into the bonds of adenosine triphosphate (ATP), the cell's energy currency.

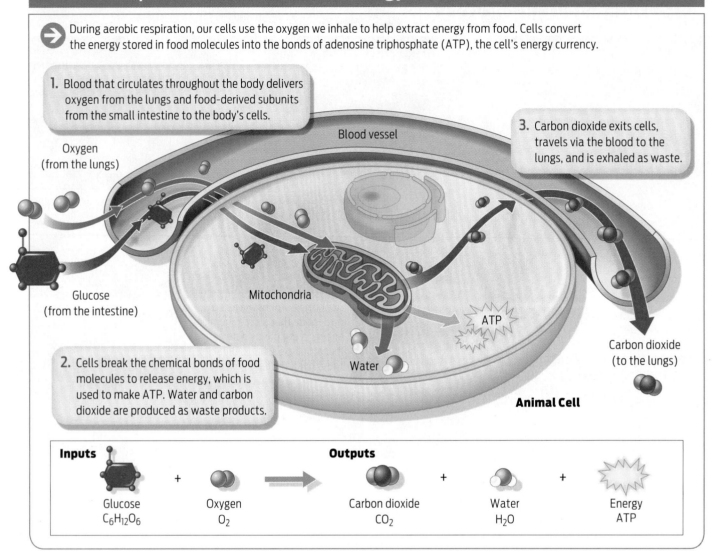

1. Blood that circulates throughout the body delivers oxygen from the lungs and food-derived subunits from the small intestine to the body's cells.

Oxygen
(from the lungs)

Blood vessel

3. Carbon dioxide exits cells, travels via the blood to the lungs, and is exhaled as waste.

Glucose
(from the intestine)

Mitochondria

ATP

Water

Carbon dioxide
(to the lungs)

2. Cells break the chemical bonds of food molecules to release energy, which is used to make ATP. Water and carbon dioxide are produced as waste products.

Animal Cell

Inputs			Outputs			
Glucose $C_6H_{12}O_6$	+	Oxygen O_2	→ Carbon dioxide CO_2	+	Water H_2O	+ Energy ATP

breathe is consumed in the process. When the bonds of glucose are broken, the energy released is used to form ATP and heat. Water and carbon dioxide are given off as waste products of the process (**Infographic 6.7**).

Aerobic respiration is a multistep process that takes place in different parts of the cell. The initial steps of this process, known as **glycolysis,** take place in the cell's cytoplasm. Glycolysis is a series of chemical reactions that splits glucose into two smaller molecules. The products of glycolysis then enter the cell's mitochondria, where the last two steps of aerobic respiration occur.

During the second step of aerobic respiration, the **citric acid cycle,** a series of reactions strips electrons from the bonds between car-

bon atoms. The process releases CO_2, which is ultimately exhaled from an organism's lungs.

The electrons stripped from the carbon bonds are carried to the inner membranes of the mitochondria, where they go through the last step of aerobic respiration: the **electron transport chain.** Electrons stripped from the bonds in glucose contain a lot of potential energy. During electron transport, these energetic electrons are passed like hot potatoes down a chain of molecules, mostly proteins, in the inner mitochondrial membrane. Eventually the electrons pass to oxygen molecules, which accept the electrons and combine with hydrogen atoms to produce water. As electrons pass down the chain, they supply the energy needed

GLYCOLYSIS
A series of reactions that breaks down sugar into smaller units; glycolysis takes place in the cytoplasm and is the first step of both aerobic respiration and fermentation.

CITRIC ACID CYCLE
A set of reactions that takes place in mitochondria and helps extract energy (in the form of high-energy electrons) from food; the second step of aerobic respiration.

Aerobic Respiration: A Closer Look

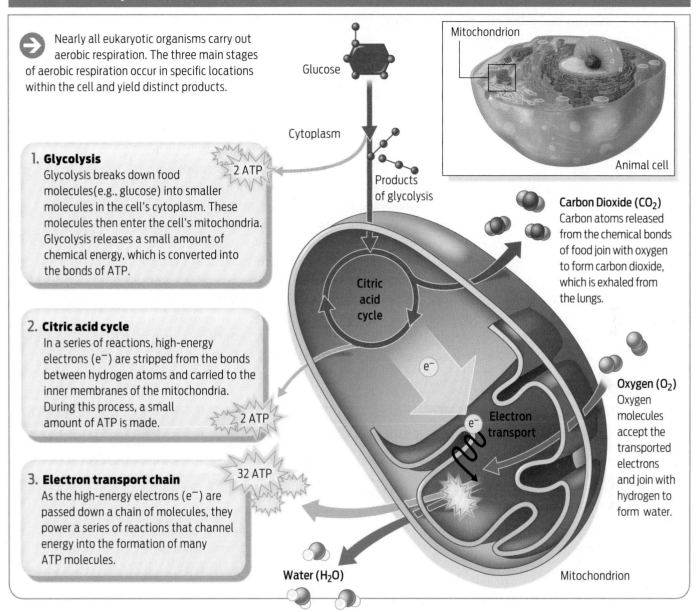

Nearly all eukaryotic organisms carry out aerobic respiration. The three main stages of aerobic respiration occur in specific locations within the cell and yield distinct products.

Mitochondrion

Animal cell

Glucose

Cytoplasm

Products of glycolysis

1. Glycolysis
Glycolysis breaks down food molecules(e.g., glucose) into smaller molecules in the cell's cytoplasm. These molecules then enter the cell's mitochondria. Glycolysis releases a small amount of chemical energy, which is converted into the bonds of ATP.

2 ATP

2. Citric acid cycle
In a series of reactions, high-energy electrons (e^-) are stripped from the bonds between hydrogen atoms and carried to the inner membranes of the mitochondria. During this process, a small amount of ATP is made.

2 ATP

3. Electron transport chain
As the high-energy electrons (e^-) are passed down a chain of molecules, they power a series of reactions that channel energy into the formation of many ATP molecules.

32 ATP

Citric acid cycle

e^-

e^- Electron transport

Carbon Dioxide (CO_2)
Carbon atoms released from the chemical bonds of food join with oxygen to form carbon dioxide, which is exhaled from the lungs.

Oxygen (O_2)
Oxygen molecules accept the transported electrons and join with hydrogen to form water.

Water (H_2O)

Mitochondrion

ELECTRON TRANSPORT CHAIN
A process that takes place in mitochondria and produces the bulk of ATP during aerobic respiration; the third step of aerobic respiration.

FERMENTATION
A series of chemical reactions that takes place in the absence of oxygen and converts some of the energy stored in food into ATP. Fermentation produces far less ATP than does aerobic respiration.

to form ATP. This electron transport chain produces the bulk of ATP (Infographic 6.8).

We've focused on glucose, but cells can also burn fats and amino acids for fuel during aerobic respiration. Because fats generally have more carbon-hydrogen bonds than do sugars and amino acids, they have more electrons to be stripped in the citric acid cycle. More electrons stripped means that more ATP molecules are produced during electron transport (which also explains why a gram of fat contains more Calories than a gram of sugar or protein).

Feel the Burn
Aerobic respiration requires a continual source of oxygen ("aerobic" means "in the presence of oxygen"). If the rate at which cells consume oxygen exceeds the rate at which they take it in when we breathe, aerobic respiration comes to a halt; the electron transport chain has no oxygen to which it can deliver electrons. While glycolysis still occurs in the absence of oxygen, its products are shunted into a different process, called **fermentation,** which takes place in the cell's cytoplasm (as opposed to the

mitochondria). Instead of carbon dioxide, fermentation in humans and other animals produces a waste product called lactic acid.

Because fermentation bypasses both the citric acid cycle and the electron transport chain, much less ATP is produced–only about 2 molecules of ATP from each molecule of glucose compared to 36 ATP produced by aerobic respiration (Infographic 6.9).

In humans, fermentation takes place primarily during bursts of energy-intensive tasks, such as sprinting or power weight-lifting. It is, in essence, a back-up plan for times when oxygen isn't available. (The panting you experience on a treadmill is your body's way of trying to obtain more oxygen.) But for many organisms, like certain fungi and bacteria, fermentation is the main way of obtaining energy. In some of these organisms, fermentation produces alcohol rather than lactic acid as a by-product. Humans

take advantage of these fermentation reactions when they make alcoholic beverages. Brewer's yeast, for example, is a fungus that ferments sugar in the absence of oxygen, producing alcohol as a result. Humans use brewer's yeast to make beer and wine.

Since fermentation does not break glucose down as completely as does aerobic respiration, there is still quite a bit of carbohydrate energy left in such beverages as beer and wine, about 7 Calories per gram–which explains why most weight-loss diets eliminate alcohol.

Even during aerobic respiration, however, our bodies don't convert every Calorie in food into ATP. The chemical processes aren't 100% efficient, so some energy is always released as heat, which keeps your body warm.

It's important to remember that aerobic respiration does not create energy–it only extracts it from food. All the food we eat–

INFOGRAPHIC 6.9

Fermentation Occurs When Oxygen Is Scarce

Glycolysis occurs whether or not oxygen is present. In the absence of oxygen, fermentation reactions follow glycolysis. Fermentation occurs in the cytoplasm and converts the products of glycolysis into lactic acid (or alcohol in some organisms). The only ATP produced is the small amount produced during glycolysis.

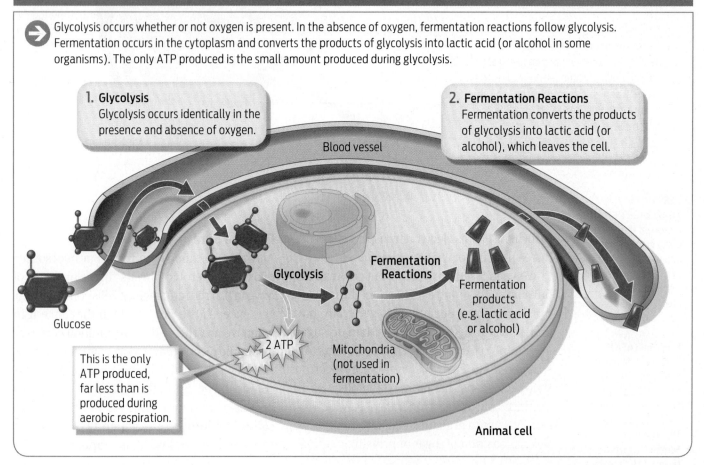

1. **Glycolysis**
Glycolysis occurs identically in the presence and absence of oxygen.

2. **Fermentation Reactions**
Fermentation converts the products of glycolysis into lactic acid (or alcohol), which leaves the cell.

Blood vessel

Glycolysis

Fermentation Reactions

Fermentation products (e.g. lactic acid or alcohol)

Glucose

2 ATP

This is the only ATP produced, far less than is produced during aerobic respiration.

Mitochondria (not used in fermentation)

Animal cell

In humans, fermentation takes place primarily during energy-intensive tasks.

whether burger, chicken leg, or Caesar salad–originally gets its energy from the sun, by way of photosynthesis. Photosynthesizers such as plants and algae capture the energy of sunlight and convert it into chemical energy stored in sugar. We then eat this sugar (or eat animals that have eaten this sugar), and that stored energy becomes available to us. Plants benefit from the relationship, too: plants use our carbon dioxide waste as raw material for making sugar during photosynthesis. In this way, photosynthesis and respiration form a continual cycle, with the outputs of one process serving as the inputs of the other (Infographic 6.10).

The Culture of Eating

Some scientists are interested in understanding how humans metabolize food in order to find a means of blocking some Calorie-dense molecules from being absorbed by the body. Food

INFOGRAPHIC 6.10

Photosynthesis and Aerobic Respiration Form a Cycle

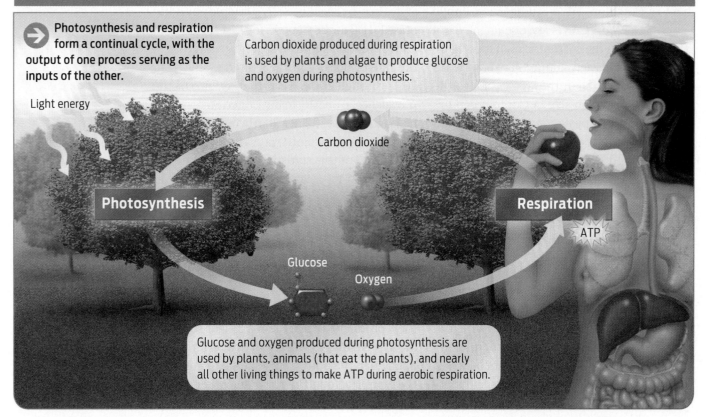

Photosynthesis and respiration form a continual cycle, with the output of one process serving as the inputs of the other.

Carbon dioxide produced during respiration is used by plants and algae to produce glucose and oxygen during photosynthesis.

Light energy

Carbon dioxide

Photosynthesis

Respiration

ATP

Glucose

Oxygen

Glucose and oxygen produced during photosynthesis are used by plants, animals (that eat the plants), and nearly all other living things to make ATP during aerobic respiration.

industry scientists, for example, have developed diet drugs that keep the body from absorbing some of the fat molecules in food. Food manufacturers would surely increase their markets if they could develop foods that pass through the intestines without being absorbed. (You may have heard of potato chips made with olestra, a fat substitute that is not absorbed by the intestines, and so passes through the body as waste, with famously unpleasant side effects.)

Of course, many food manufacturers specialize in low-fat and fat-free foods, which are less Calorie dense. But these efforts at tackling the problem of obesity offer temporary solutions at best. Studies have shown that 90% of dieters who avoid specific foods to lose weight over the short term regain most of their lost weight over time. That's because most people find it difficult to stick to the food restrictions prescribed by diets and tend to revert to their former eating habits.

A more permanent change in eating culture, on the other hand, offers a more sustainable fix because it emphasizes a change in eating habits, not the type of food itself. In fact, Americans didn't always consume large portion sizes. Lisa Young and Marion Nestle, both of the Department of Nutrition, Food Studies, and Public Health at New York University, documented in a 2002 study that the current sizes of fries, hamburgers, and soda at restaurants are now two to five times larger than they were before the 1970s, when portion size began to creep up. Single servings of pasta, muffins, steaks, and bagels today now exceed the government-recommended serving size by 480%, 333%, 224%, and 195%, respectively. And cookies exceed the standard by a factor of 8, according to Nestle's research. In America, bigger is better.

The French are eating more, too, and the results are evident: the number of obese French people has grown from 8.6% in 1997 to 13.1% in 2006, according to a 2008 study published by Marie-Aline Charles at the Institut National de la Santé et la Recherche Médicale in Villejuif, France. As American music, movies, and clothing have become pervasive in other countries, so, too, have our eating habits. More and more French people are eating large amounts of nutri-

Senator Kel Seliger displays a box of candy in the Texas Senate during debate on a bill that would ban trans fats in restaurants—he opposed the bill.

tionally poor, energy-dense foods. And as jobs increasingly place people in front of computers, they have become less physically active. More French people are eating on the go, eating fast food, and spending less time enjoying formal meals. Much to the dismay of public health experts, French eating culture is tipping toward unhealthful.

But not all obesity news is bad. Recent studies suggest that the rate of obesity in the United States may be leveling off. A 2010 study by Katherine Flegal at the Centers for Disease Control and Prevention found no significant increase in the rate of obesity in the United States from 2003 through 2008. Nevertheless, at 30% the prevalence of obesity in the United States is still high and remains higher than in most European countries (and double the prevalence in France).

Because of this statistic, there is a growing trend in the United States to legislate changes in the foods people eat. Some cities, for example, have banned the use of **trans fat,** a type of hydrogenated vegetable fat that studies have shown contributes to heart disease. Commercially prepared foods such as cookies, French fries, doughnuts, and margarine often contain trans fat, which food manufacturers add to their products to give them a longer shelf life or a pleasing texture. Hydrogenated fat behaves in the body much like **saturated fat**, the type of fat found in butter and other animal products. Studies have shown that eating large

TRANS FAT
A type of vegetable fat which has been hydrogenated, that is, hydrogen atoms have been added, making it solid at room temperature.

SATURATED FAT
An animal fat, such as butter; saturated fats are solid at room temperature.

amounts of saturated fat can clog arteries. By contrast, **unsaturated fats**, which come from plants and are liquid at room temperature, are considered more healthful (although they are still high in Calories).

"Actions by governments are the only way conditions will change enough to have a major public health impact," says Kelly Brownell. For America's obesity woes, Brownell blames the food industry for their relentless marketing of unhealthful foods, agricultural and trade policies that promote unhealthful diets, and economic policies that make unheathful foods cheaper than healthful ones.

Not everyone agrees, however, that it is the government's job to restrict our food choices. Americans equate choice in food with democracy, argues Paul Rozin. "We could also over-respond to what many perceive as an obesity epidemic and that could be dangerous. It would restrict individual freedom."

Besides, while such government legislation would restrict *what* we eat, it wouldn't really affect *how* we eat. Rozin hopes people change their behavior voluntarily. For example, people could fit more exercise into their daily routines, climbing stairs instead of using an elevator or parking the car farther away from the entrance at the mall. To combat large portion sizes in restaurants, Rozin advocates ordering less, sharing dishes, or as Mireille Guiliano recommends, simply not eating everything that is on our plates.

The U.S. eating culture can change, Rozin says, but not overnight. Look at cigarette smoking: "It took 50 years to get cigarette smoking to decline and they [cigarettes] are much more harmful to health." ■

● Summary

■ The macronutrients in our food (proteins, carbohydrates, and fats) are sources of dietary energy.

■ Fats are the most energy-rich organic molecules in our diet. Fats contain twice as many Calories per gram as carbohydrates and proteins.

■ When we consume more Calories than we use, our bodies store the excess energy in the bonds of glycogen and body fat. Fats store more energy than does glycogen.

■ Cells carry out chemical reactions that break down food to obtain usable energy in the form of ATP.

■ In the presence of oxygen, aerobic respiration produces large amounts of ATP from the energy stored in food.

■ Aerobic respiration occurs in three stages: (1) glycolysis, (2) the citric acid cycle, and (3) electron transport. The first stage occurs in the cytoplasm, the latter two in mitochondria. Electron transport produces the bulk of ATP.

■ In the absence of oxygen, fermentation follows glycolysis and produces lactic acid in animals (or, in some organisms, alcohol). Fermentation produces far less ATP than does aerobic respiration.

■ Exercise helps burn stored Calories. A combination of eating fewer Calories and exercising more will result in weight loss, although hereditary factors play a large role in determining a person's weight.

■ During exercise, glycogen is used first. Stored fats are tapped only when glycogen stores have been depleted, as might occur during long periods of exercise.

■ Photosynthesis and respiration form a cycle: the carbon dioxide given off by animals, plants, and all organisms that perform aerobic respiration is used by photosynthesizers to make glucose and oxygen during photosynthesis.

FOOD IS ENERGY

Each type of organic molecule found in food stores a different amount of energy. Both what we eat and how much we eat contribute to weight gain.

HINT **See Infographics 6.1–6.5.**

KNOW IT

1. Which type of organic molecule stores the most energy per gram?

    a. proteins
    b. starch
    c. nucleic acid
    d. fats (triglycerides)
    e. glycogen

2. A moderately active 21-year-old female has a choice of eating a 2,500-Calorie meal primarily of protein or a 2,500-Calorie meal primarily of sugar. What would be the result, in terms of energy, of choosing one over the other?

    a. Nothing; she would burn all these Calories, given her age, gender, and activity level.
    b. She would store the excess Calories as protein, regardless which meal she ate.
    c. She would store the excess Calories as protein if she ate the protein meal, and as glycogen if she ate the sugar meal.
    d. In either case, once her glycogen stores are replenished, she will store the excess Calories as fat.
    e. Regardless of the number of Calories, she will get more energy from the sugar meal.

3. A 5′6″ female weighs 167 pounds. Use Infographic 6.1 to determine her BMI. Would she be considered underweight, normal weight, overweight, or obese?

4. If you exercise for an extended period of time, you will use energy first from _____, then from _____.

    a. fats; glycogen
    b. proteins; fats
    c. glycogen; proteins
    d. fats; proteins
    e. glycogen; fats

5. Preparing foods with trans fats specifically increases the risk of which of the following conditions?

    a. cardiovascular disease
    b. obesity
    c. excessive weight gain
    d. arthritis
    e. colon cancer

USE IT

6. If you frequently crave French fries (that is, starchy potatoes fried in fat), how could you modify your lifestyle to eat fries without gaining weight? Explain your answer.

7. Consider a well-trained 130-pound female marathon runner. She has just loaded up on a carbohydrate meal and has the maximum amount of stored glycogen (6.8 grams of glycogen per pound of body weight).

    a. How many grams of glycogen is she storing?
    b. How many Calories does she have stored as glycogen?
    c. If this same number of Calories were stored as fat, how much would it weigh?
    d. Suppose she decides to go for a run at a pace of 9 miles per hour (she will be running 6.5-minute miles). Given her weight, she will burn 885 Calories per hour at this pace. How long will it take her to deplete her glycogen stores? How many miles can she run before her glycogen supplies run out? Will she be able to complete a 26.2-mile marathon?
    e. Once her glycogen supplies run out, what has to happen if she wants to keep running?

8. If the French eat meals with a higher fat content, why are the French on average not fatter than Americans?

AEROBIC RESPIRATION AND FERMENTATION

These reactions convert energy stored in food into usable forms.

HINT **See Infographics 6.6–6.10.**

KNOW IT

9. Which step is not correctly matched with its cellular location?

    a. glycolysis—cytoplasm
    b. citric acid cycle—mitochondria
    c. fermentation—mitochondria
    d. electron transport—mitochondria
    e. none of the above—they are all correctly matched

10. Compared to aerobic respiration, fermentation produces _____ ATP.
  a. much more
  b. the same amount of
  c. a little less
  d. much less
  e. no

11. We obtain carbohydrates by eating them. How do plants obtain their carbohydrates?
  a. by eating them
  b. by cellular respiration
  c. by fermentation
  d. from the soil
  e. by photosynthesis

12. In the presence of oxygen we use _____ to fuel ATP production. What process do plants use to fuel ATP production from food?
  a. aerobic respiration; photosynthesis
  b. aerobic respiration; aerobic respiration
  c. fermentation; aerobic respiration
  d. fermentation; photosynthesis
  e. glycolysis; photosynthesis

➔ USE IT

13. Draw a carbon atom that is part of a CO_2 molecule that you just exhaled. Using a written description or a diagram, trace what happens to that carbon atom as it is absorbed by the leaf of a spinach plant and then what happens to the carbon atom when you eat that leaf in a salad.

14. Given 1 gram of each, which of the following would yield the greatest amount of ATP by aerobic respiration?
  a. fat
  b. protein
  c. carbohydrate
  d. nucleic acid
  e. alcohol

15. Explain how the presence or absence of oxygen affects ATP production. (The terms *aerobic respiration* and *fermentation* should appear in your answer.)

16. If you ingest carbon in the form of sugar, how is that carbon released from your body?
  a. as sugar
  b. as fat
  c. as CO_2
  d. as protein
  e. in urine

SCIENCE AND ETHICS

17. Why do you think that longer meal times translate into fewer Calories consumed?

18. Do you think the government has a responsibility to regulate the information provided on nutrition labels? Explain your answer.

19. If the government were to issue tax incentives to reduce obesity in the United States, which of the following do you think would be most effective? Explain your choice.
  a. taxing foods high in fat
  b. giving tax breaks for people who join gyms or health clubs
  c. giving rebates for purchasing fresh fruits and vegetables
  d. paying enhanced salaries for teachers in elementary and middle schools to provide education about diet and nutrition

Biologically Unique

Biologically unique

How DNA helped free an innocent man

Roy Brown thought the police were just checking up on him when an officer knocked on his door one day in May 1991. Brown, a self-professed hard drinker who earned a living selling magazine subscriptions, had only a week before been released after serving an 8-month prison term. His crime: threatening to kill the director of the Cayuga County Department of Social Services in upstate New York. A caseworker had deemed Brown unfit to care for his 7-year-old daughter. Furious, Brown had threatened to kill the director and other workers. But he had served his time. What could the officer want from him now?

Three days earlier, police had found the battered body of a woman lying in the grass about 300 feet from the farmhouse where she lived. Someone had burned the place to the ground. The body was identified as that of Sabina Kulakowski, a social worker at the Cayuga County Department of Social Services. The crime was horrific. The murderer had beaten the 49-year-old Kulakowski, bit her several times, dragged her outside, and then stabbed and strangled her

to death. It was obvious that Kulakowski had struggled; her body was covered with defensive wounds.

Although Kulakowski was not involved in Brown's case, officers arrested Brown that day on suspicion of murder. Eight months later, a jury found Brown guilty of homicide and sentenced him to prison for 25 years to life. The prosecution argued that Brown's motive was revenge against the Department of Social Services. But what really nailed the case was testimony from an expert who stated that bite marks on the victim's body matched Brown's teeth.

Brown, however, maintained his innocence. "I never knew Ms. Kulakowski, and I had nothing to do with that woman's death . . . I am truly innocent," he told the court and onlookers after the verdict had been announced.

Even from jail Brown never stopped trying to prove his innocence. He repeatedly petitioned, in vain, for a retrial. Then something unexpected happened. Brown uncovered additional evidence that strongly suggested he was not the perpetrator. The evidence was so compelling, in

fact, that in late 2004, after Brown had spent 12 years in prison, his lawyers decided to contact the Innocence Project–a nonprofit organization founded in 1992 by Peter Neufeld and Barry Scheck of the Benjamin N. Cardozo School of Law in New York City. Their mission: to use DNA evidence to free people wrongly convicted of crimes.

When the jury convicted Brown in 1992, analyzing crime scene evidence for traces of DNA wasn't established practice yet, so DNA was rarely used as evidence in criminal cases. But about a decade later, using DNA evidence in court cases became standard practice as science increasingly showed that it was an extremely accurate way to match crime scene evidence to perpetrators.

DNA as Evidence

How can scientists use DNA to identify a person? The answer lies in the chemical makeup of this molecule, often referred to as the "blueprint of life." **Deoxyribonucleic acid,** or **DNA,** is the hereditary molecule that is common to all life forms–from plants to bacteria to fungi–and that is passed from parents to offspring. DNA serves as the instruction manual from which we are built; it's the reason why you look like your parents, an aunt, or perhaps even a grandparent.

Where can you find DNA? The molecule exists inside the nucleus of almost every cell in our body in the form of **chromosomes,** strands of DNA wound around proteins. Humans have 23 pairs of chromosomes in the vast majority of their cells; we inherit one chromosome of each pair from our mother and the other from our father. Since there is DNA in most of our cells, scientists can collect such evidence as blood, semen, saliva, or hair from a crime scene and extract DNA from it to identify a perpetrator (**Infographic 7.1**).

DNA testing has helped the Innocence Project free more than 240 people from prison since 1992, including more than a dozen who served time on death row. The technology has not only given these people their lives back, but has also thrown a spotlight on our flawed criminal justice system. Why were people wrongly convicted and placed on death row? Innocence Project lawyers have found the usual suspects: dishonest witnesses, unscrupulous police officers, apathetic or overburdened lawyers, mistakes in eyewitness identification. But perhaps even more important, DNA technology has helped find ways to improve the system.

"DNA is only one example of how advances in science have made the criminal justice system more reliable," says Neufeld. "But what we really hope to do now is use DNA as the gold standard of reliability to weed out junk science."

Unreliable Evidence

Indeed, "junk science" convicted Roy Brown. The only physical evidence linking Brown to the case was his teeth. A dentist hired by the prosecution testified that the bite marks on Kulakowski's body matched Brown's teeth. But the bite marks came from someone with six upper teeth–Brown had only four, the defense pointed out. The prosecution's witness further argued that Brown could have twisted Kulakowski's skin while biting her and filled in the gaps.

INFOGRAPHIC 7.1

What Is DNA and Where Is It Found?

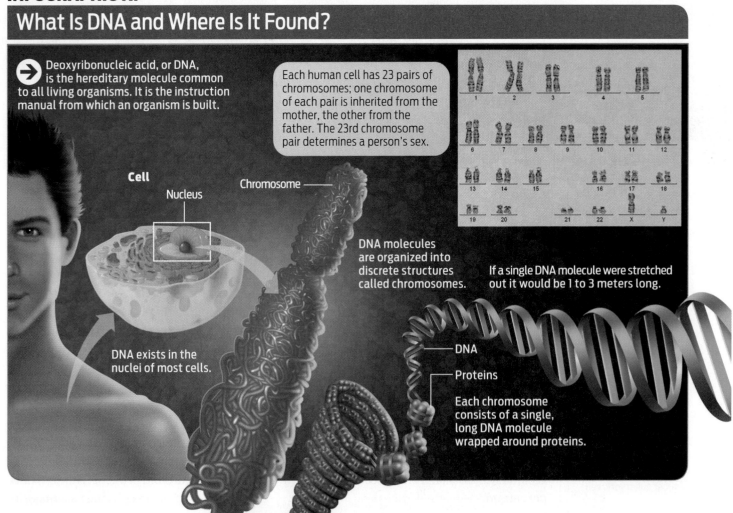

→ Deoxyribonucleic acid, or DNA, is the hereditary molecule common to all living organisms. It is the instruction manual from which an organism is built.

Each human cell has 23 pairs of chromosomes; one chromosome of each pair is inherited from the mother, the other from the father. The 23rd chromosome pair determines a person's sex.

Cell
Nucleus
Chromosome

DNA exists in the nuclei of most cells.

DNA molecules are organized into discrete structures called chromosomes.

If a single DNA molecule were stretched out it would be 1 to 3 meters long.

DNA
Proteins

Each chromosome consists of a single, long DNA molecule wrapped around proteins.

Peter Neufeld and Barry Scheck, founders of the Innocence Project.

Some evidence commonly presented in criminal cases can be unreliable, including bite-mark analysis. In fact, studies show error

"What we really hope to do now is use DNA as the gold standard of reliability to weed out junk science." —Peter Neufeld

NUCLEOTIDES
The building blocks of DNA. Each nucleotide consists of a sugar, a phosphate, and a base. The sequence of nucleotides (As, Cs, Gs, Ts) along a DNA strand is unique to each person.

DOUBLE HELIX
The spiral structure formed by two strands of DNA nucleotides bound together.

rates—the rate at which experts have falsely identified bite marks as belonging to a particular person—as high as 91%. Hair analysis can be equally troublesome. In dozens of cases, Innocence Project lawyers found that forensic scientists had testified that hairs from crime scenes matched the accused, explains Neufeld. But when scientists subsequently tested the DNA inside the follicle cells from those hairs, the DNA didn't match.

The problem is that hair analyses, performed under a microscope, can reveal only certain characteristics; it can distinguish whether hair is human or animal, show a person's ancestry (because of ethnic differences in hair texture), or whether the hair has been dyed, cut in a cer-

tain way, pulled out, and where on the body it came from. Hair samples can exclude a suspect, but not positively identify one.

By contrast, each person's DNA is unique. To understand how DNA varies from person to person, consider its structure. DNA is made up of two strands of molecules, which are called **nucleotides,** linked together in long chains. Each nucleotide has three parts: a sugar, a phosphate, and a base. The phosphate group of one nucleotide binds to the sugar group of the next nucleotide to form a long strand of nucleotides. Then, the two strands of linked nucleotides are bound together and twisted around each other to form a spiral-shaped **double helix.** The sugars and phosphates form the outside "backbone" of the helix and the bases form the internal "rungs," like steps on a twisting ladder. The bases are held together by hydrogen bonds **(Infographic 7.2).**

The nucleotide rungs, made up of bases, are most useful in DNA profiling. There are four different possible nucleotide bases: adenine (A),

DNA Is Made of Two Strands of Nucleotides

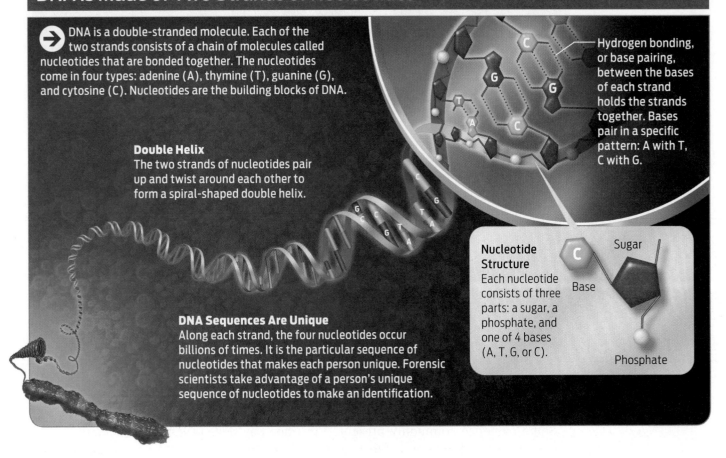

→ DNA is a double-stranded molecule. Each of the two strands consists of a chain of molecules called nucleotides that are bonded together. The nucleotides come in four types: adenine (A), thymine (T), guanine (G), and cytosine (C). Nucleotides are the building blocks of DNA.

Double Helix
The two strands of nucleotides pair up and twist around each other to form a spiral-shaped double helix.

DNA Sequences Are Unique
Along each strand, the four nucleotides occur billions of times. It is the particular sequence of nucleotides that makes each person unique. Forensic scientists take advantage of a person's unique sequence of nucleotides to make an identification.

Hydrogen bonding, or base pairing, between the bases of each strand holds the strands together. Bases pair in a specific pattern: A with T, C with G.

Nucleotide Structure
Each nucleotide consists of three parts: a sugar, a phosphate, and one of 4 bases (A, T, G, or C).

Sugar
Base
Phosphate

---

thymine (T), guanine (G), and cytosine (C). In DNA, these four nucleotide bases are repeated over and over, billions of times, in different combinations. To identify perpetrators, forensic scientists examine the specific sequence of nucleotide bases along one strand of a person's DNA–the precise order of As, Ts, Gs, and Cs. With the exception of identical twins, no two people share exactly the same order of DNA nucleotides.

Brown Gets a Break

At the time Brown was convicted, DNA evidence was scarcely used in law enforcement. Crime scene evidence was, however, routinely stored, which was lucky for Brown. In 2003, Brown filed a Freedom of Information Act request to obtain copies of all documents relating to his case.

To identify perpetrators, forensic scientists examine the specific sequence of nucleotide bases along one strand of a person's DNA.

The additional evidence included four affidavits collected by the Cayuga County Sheriff's Department the day after the murder–documents that neither Brown nor his lawyers had ever seen. In the affidavits, four people described the suspicious behavior of another man: Barry Bench. Bench was the brother of Kulakowski's former boyfriend. The Bench family owned the farmhouse in which Kulakowski had been living.

The affidavits stated that on the day of Kulakowski's murder, Bench argued with his then girlfriend, Tamara Heisner, went to a local bar, and only returned home between 1:30 and 1:45

Kary Mullis receiving a Nobel Prize (in 1993) for developing the polymerase chain reaction.

A.M.–the same time the victim's neighbors alerted the fire department that the farmhouse was ablaze.

The statements further noted that Bench, who came home highly intoxicated, had left the bar at approximately 12:30 A.M. That left 60 to 75 minutes unaccounted for until he arrived home–although he lived only a mile from the bar. When Bench came home, he immediately went inside to "wash up," according to Heisner.

Brown realized that Bench would have had to drive by the farmhouse to get home from the bar and thought it strange that Bench would not have noticed the raging fire on his own property. While not conclusive, this new evidence was enough to prompt Brown's lawyers to contact the Innocence Project for help.

Meanwhile, Brown decided to write Bench a letter detailing what he had found and urge him to confess. He warned him of his intent to obtain a DNA test on evidence from the murder. "Judges can be fooled and juries make mistakes," he wrote, "[but] when it comes to DNA testing

there's no mistakes. DNA is God's creation and God makes no mistakes."

Five days after Brown mailed his letter, Bench threw himself in front of an Amtrak passenger train and died instantly.

Soon after, the Innocence Project team took on Brown's case and filed a motion to have Kulakowski's nightshirt tested at a New York State crime lab for DNA. The nightshirt was not only bloodstained, it was also stained with saliva. Since both saliva and blood contain cells that carry DNA, scientists could chemically extract the DNA from the cells to create a **DNA profile** of the perpetrator.

Making More DNA

In theory, creating a DNA profile is simple enough, but there is a huge practical hurdle: having enough crime scene DNA to analyze. Although all body fluids and materials contain cells that house our DNA, the amount left at crime scenes is very small. Without some way to increase the amount of DNA in a saliva stain,

Roy Brown at his home in upstate New York.

that can vastly increase the amount of DNA in a sample.

In devising PCR, Mullis took advantage of a cell's natural ability to copy DNA, a process called **DNA replication.** DNA replication happens throughout our lives whenever cells reproduce. It is a remarkably accurate process that happens at mind-boggling speeds—about 1,000 nucleotides per second. On a human scale, that's like a car speeding down the highway at 300 miles per hour, weaving in and out of traffic without hitting any other cars.

To understand how DNA replication works, it's important to note that the two strands of nucleotides in a DNA helix do not pair up randomly, but rather in a consistent pattern: A always pairs with T, and G always pairs with C. Because of this patterned pairing, the two strands are said to be **complementary,** meaning that they fit together like pieces of a puzzle. During DNA replication, each comple-

for example, DNA would be useless as evidence.

Chemist Kary Mullis solved that problem in the mid-1980s when he developed a groundbreaking technique called the **polymerase chain reaction (PCR)**—a chemical reaction

POLYMERASE CHAIN REACTION (PCR)
A laboratory technique used to replicate, and thus amplify, a specific DNA segment.

DNA REPLICATION
The natural process by which cells make an identical copy of a DNA molecule.

COMPLEMENTARY
Two strands of DNA are said to be complementary in that A always pairs with T, and G always pairs with C.

INFOGRAPHIC 7.3

DNA Structure Provides a Mechanism for DNA Replication

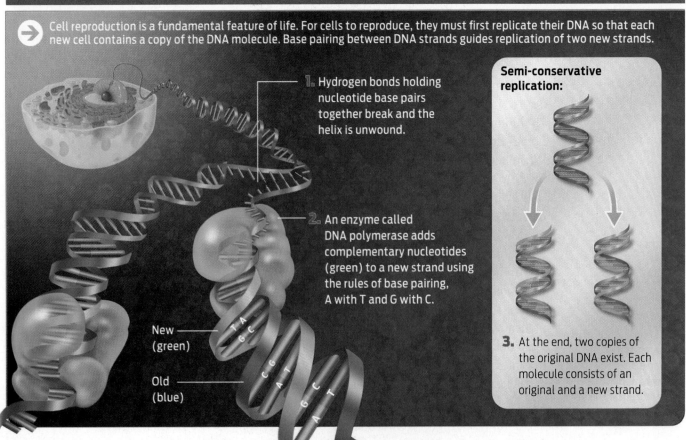

Cell reproduction is a fundamental feature of life. For cells to reproduce, they must first replicate their DNA so that each new cell contains a copy of the DNA molecule. Base pairing between DNA strands guides replication of two new strands.

1. Hydrogen bonds holding nucleotide base pairs together break and the helix is unwound.

2. An enzyme called DNA polymerase adds complementary nucleotides (green) to a new strand using the rules of base pairing, A with T and G with C.

New (green)

Old (blue)

Semi-conservative replication:

3. At the end, two copies of the original DNA exist. Each molecule consists of an original and a new strand.

The Polymerase Chain Reaction Amplifies Small Amounts of DNA

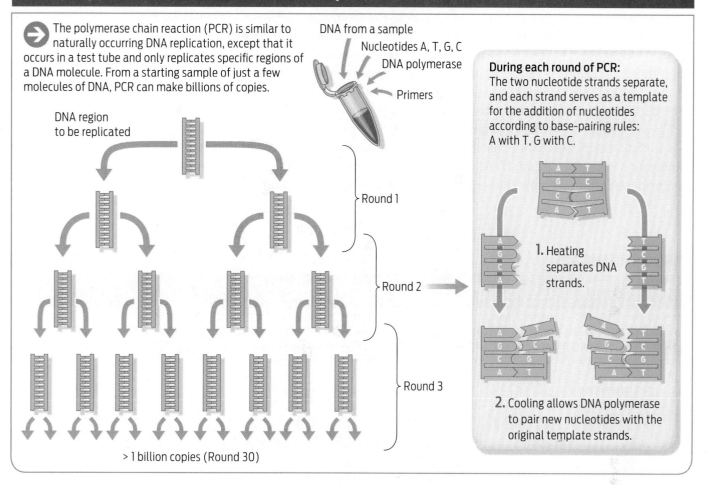

→ The polymerase chain reaction (PCR) is similar to naturally occurring DNA replication, except that it occurs in a test tube and only replicates specific regions of a DNA molecule. From a starting sample of just a few molecules of DNA, PCR can make billions of copies.

DNA from a sample
Nucleotides A, T, G, C
DNA polymerase
Primers

DNA region to be replicated

Round 1

Round 2

Round 3

> 1 billion copies (Round 30)

During each round of PCR:
The two nucleotide strands separate, and each strand serves as a template for the addition of nucleotides according to base-pairing rules: A with T, G with C.

1. Heating separates DNA strands.

2. Cooling allows DNA polymerase to pair new nucleotides with the original template strands.

DNA POLYMERASE
An enzyme that "reads" the sequence of a DNA strand and helps to add complementary nucleotides to form a new strand during DNA replication.

SEMI-CONSERVATIVE
DNA replication is said to be semi-conservative because each newly made DNA molecule has one original and one new strand of DNA.

mentary strand of DNA serves as a template for the creation of a new complementary strand. First, the helix unwinds and the two strands "unzip." Then, an enzyme known as **DNA polymerase** pairs new nucleotides to each individual DNA strand, matching A with T and C with G. The end result is two molecules of DNA. Because each replicated DNA molecule is made up of one original and one new strand, DNA replication is said to be **semi-conservative (Infographic 7.3).**

PCR is similar to DNA replication that occurs naturally, except that it takes place in a test tube. To a small sample of DNA, scientists add nucleotides, the DNA polymerase enzyme, and primers—short segments of DNA that act as

From a starting sample of just a few molecules of DNA, PCR can make billions of copies.

guideposts and flag the section to which DNA polymerase should bind to begin replication. The DNA is first heated to separate the strands, and then cooled to allow new nucleotides to be added. From a starting sample of just a few DNA molecules, PCR can make billions of copies of a specific region of the DNA in less than a few hours **(Infographic 7.4).**

DNA from the Crime Scene

A New York State crime lab used PCR to amplify DNA from various items of evidence collected by Cayuga County law enforcement officials during their original investigation of Kulakowski's homicide. The evidence included remnants of cotton swabs used to sample bite marks on the victim; the saliva- and bloodstained night-shirt; fingernail clippings; and vaginal swabs from the victim.

To extract DNA from any forensic sample, scientists typically use chemicals to separate cells from other material, such as fabric. The specific type of chemical used depends on the starting material. Other types of laboratory machinery, such as a centrifuge–a machine that spins samples at high speeds to separate materials–in combination with other chemicals help to further extract DNA from a sample. DNA extraction is usually the most painstaking step of the process because it can be difficult to obtain enough cells in a forensic sample to yield enough DNA for PCR. Also, improperly stored samples can degrade too much to be useful. Samples can also become contaminated with foreign DNA from improper handling, which could render results useless.

In Brown's case, the laboratory's first report on the stained nightshirt was disappointing. Technicians hadn't been able to obtain any DNA from the bite-mark swab. The lab's second

INFOGRAPHIC 7.5

DNA Profiling Uses Short Tandem Repeats

DNA profiling takes advantage of the fact that no two people have the same exact nucleotide sequence. The specific regions of DNA that forensic scientists analyze are those that contain short tandem repeats (STRs). STRs are short stretches of repeated DNA sequences. People differ in the number of copies of an STR sequence found along their chromosomes.

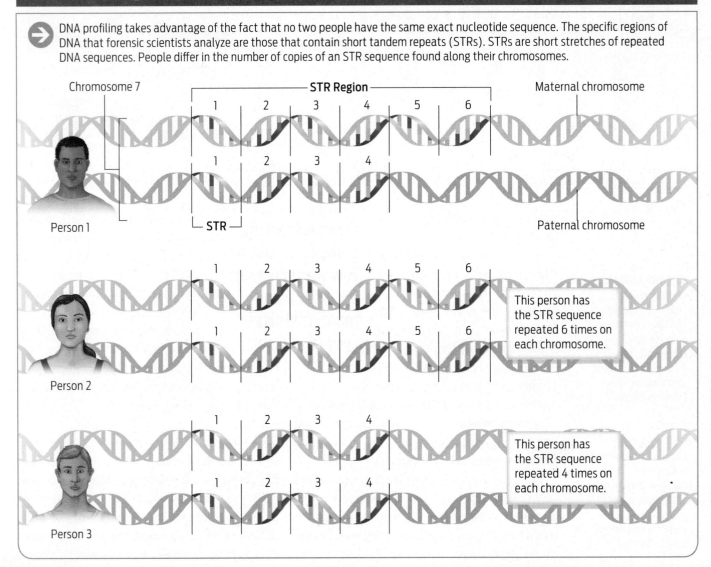

Each human cell contains vast amounts of DNA— each cell carries on the order of 3 billion nucleotide base pairs in its genome.

report was more conclusive: seven separate stained pieces of the victim's nightshirt contained DNA.

Moreover, the report went on to state that six of the pieces yielded mixtures of DNA, containing DNA from the victim and from another person who was male.

DNA Profiling: How It Works

Once cells from crime scene evidence are recovered, the next step is to analyze the DNA contained within them. Human cells contain vast amounts of DNA—there are on the order of 3 billion nucleotide base pairs within the 23 chromosomes that compose the human **genome.** What's more, because each cell carries two sets of chromosomes—one set inherited from each parent—there are approximately 6 billion nucleotide base pairs per cell. Figuring out the sequence of every nucleotide in the genome would be extremely time consuming and expensive. So instead, forensic scientists use a short cut—they employ PCR to amplify specific segments of DNA, and analyze just these segments. These segments are known as short tandem repeats.

Short tandem repeats (STRs) are blocks of repeated DNA sequences found at points along our chromosomes. They are a bit like nonsense words in our DNA: the sequence AGCT repeated over and over again, for example. While all of us have STRs in the same places along our chromosomes, the exact length of each STR varies from person to person. At a single STR site, one person may carry the AGCT sequence repeated six times while another person might carry the AGCT sequence repeated four times. Also, since we inherit two copies of every chromosome, every person has two copies of each STR, and

they can be of two different lengths—four repeats of AGCT on one chromosome and six repeats of AGCT on the other chromosome, for example. It is these differences in STR lengths that forensic scientists use to distinguish between individuals (**Infographic 7.5**).

In addition to comparing lengths of AGCT repeats, DNA profilers use STRs because these DNA sequences differ from other types of DNA found in our genome. Scientists divide DNA sequences into two general categories: those that contain instructions for making proteins, called **coding regions,** and those that do not. STRs are found in the so-called **noncoding regions** of our DNA. While coding sequences are extremely similar from person to person (99% identical, to be precise), noncoding sequences vary much more between individuals. This variation in noncoding sequences provides a kind of genetic fingerprint, which can be used to identify someone uniquely.

Because they do not code for proteins, scientists had long thought that noncoding regions served no purpose and dubbed them "junk DNA." They now know that these regions, which make up about 98% of the human genome, do have important functions. For example, some of the noncoding DNA plays a regulatory role, controlling how and when the coding regions are used. And of course the presence in our genome of these noncoding regions is what makes DNA profiling possible. (We'll have more to say about coding and noncoding regions of DNA in Chapter 8, when we discuss gene expression and protein synthesis.)

To create a DNA profile, scientists first employ PCR to increase the amount of DNA at multiple STR regions. Second, they use a method called **gel electrophoresis** to separate the replicated STRs according to their length. Shorter STRs—those with fewer numbers of repeats—are smaller and travel farther in the gel; longer ones do not travel as far. When visualized with fluorescence, the separated segments of DNA create a specific pattern of bands that is unique to each person. It is this unique pattern that is called a DNA profile. Scientists can then compare band patterns or profiles among people

GENOME
One complete set of genetic instructions encoded in the DNA of an organism.

SHORT TANDEM REPEATS (STRs)
Sections of a chromosome in which DNA sequences are repeated.

CODING REGIONS
Sequences of DNA that serve as instructions for making proteins.

NONCODING REGIONS
DNA sequences that do not hold instructions to make proteins.

GEL ELECTROPHORESIS
A laboratory technique that separates fragments of DNA by size.

to match DNA from a crime scene to DNA from a suspect. Such a DNA profile has other applications, too–paternity or ancestry testing, for example (Infographic 7.6).

DNA Profiling and the Law

Since 1994, the federal government has been collecting DNA profiles of offenders in the National DNA Index System (NDIS), a database that contains more than 4.8 million profiles from criminals convicted of specific crimes in all 50 states. Each profile consists of a banding pattern that represents 15 specific STR regions scattered throughout our genomes. Forensic scientists typically describe the likelihood that any two unrelated people will have the same number of repeated sequences at all 15 regions as 1 in some number of quintillions (billions of billions) (Infographic 7.7).

"Judges can be fooled and juries make mistakes,
[but] when it comes to DNA testing there's no mistakes."
—Roy Brown

So far, the database of DNA profiles has been helpful in more than 116,000 cases. More significantly, DNA evidence is helping to change the criminal justice system for the better. That there have been so many people determined to be innocent suggests that many more may have been wrongly convicted but lack the evidence to support their cases. In the majority of criminal cases, there is no DNA evidence.

Recognizing the flaws in our criminal justice system, Innocence Project lawyers are working with several states to change the way law enforcement operates. For example, studies have shown that witnesses more accurately identify perpetrators if they are shown suspects one at a time instead of in a group line-up. They're also helping to force changes in the way interrogations are conducted and videotaped to reduce the possibility of forced confessions. In

INFOGRAPHIC 7.6

Creating a DNA Profile

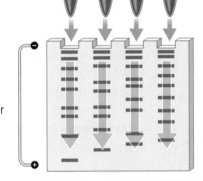

1. Collect cells and extract DNA from crime scene evidence.

White blood cell   Cheek cell in saliva   Skin cell

STR 1   STR 2   STR 3

2. PCR amplify multiple STR regions.

STRs from saliva sample

STRs from Suspect:
A   B   C

3. Separate STRs using gel electrophoresis.

PCR products are inserted into a gel. An electrical current applied to the gel causes polar DNA to migrate through it. Shorter fragments travel further while longer fragments remain near the top.

Different STRs
If a person has two different STR lengths, there will be two bands, the two green bands for Suspects A and C, for example.

Identical STRs
If a person has two identical STR lengths, they will be represented by a single band, the thick green band for Suspect B, for example.

4. The gel (right) shows the results of amplifying three different STR regions (green, red and blue bands below) in a crime scene sample and three suspects. As PCR products travel through the gel they create a specific pattern of bands based on the number of repeats present.

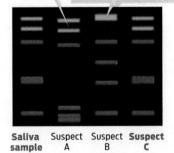

Saliva sample   Suspect A   Suspect B   Suspect C

Conclusion
While some individuals match at some STRs, only suspect C perfectly matches the crime scene saliva sample.

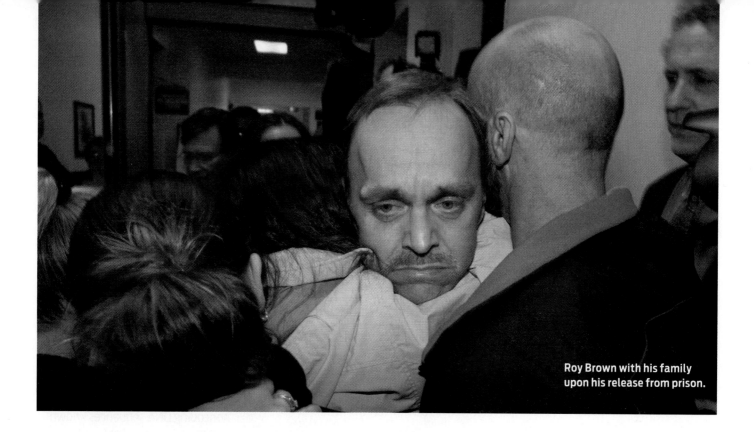

Roy Brown with his family upon his release from prison.

INFOGRAPHIC 7.7

DNA Profiling Uses Many Different STRs

To create a DNA profile, scientists analyze 15 different STRs (yellow boxes) scattered among our chromosomes. Sharing the same number of repeats at any particular STR is relatively common — typically 5% to 20% of people share the same pattern at any one STR site. But it is the combined pattern of STR repeats at multiple sites that is unique to a person; the more STRs tested, the more discriminating the test becomes.

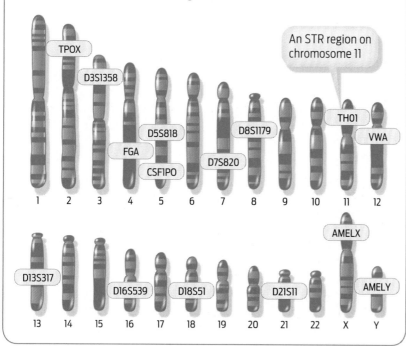

An STR region on chromosome 11

TPOX
D3S1358
D5S818
FGA
CSF1PO
D7S820
D8S1179
TH01
VWA

1 2 3 4 5 6 7 8 9 10 11 12

D13S317
D16S539
D18S51
D21S11
AMELX
AMELY

13 14 15 16 17 18 19 20 21 22 X Y

addition, they are lobbying for legislation to ensure that evidence from crimes is properly collected and maintained, and also to ensure that anyone convicted of a crime can gain access to DNA testing.

"The key is that DNA really gives us an opportunity to start making the other institutions in the system more scientific and reliable as well," says Neufeld.

Vindication

The DNA that the New York State crime lab extracted from the victim's nightshirt contained a mixture of DNA from the victim and another person who was male. Analysis showed that this male DNA, however, did not match Roy Brown's. DNA evidence excluded him as Kulakowski's murderer.

Additional testing eventually linked that DNA evidence to Barry Bench. After Bench's suicide, of course, he couldn't provide DNA directly. So lawyers pursued the next best option: a DNA sample voluntarily donated by Bench's biological daughter, Katherine Eckstadt. Because we all receive one set of chromosomes from our mother, and one set from our father, half of Katherine Eckstadt's DNA would have come from her father, and therefore would show great similarity to his. The test yielded dramatic results—a 99.99% probability that the man who deposited his saliva on Sabina Kulakowski's nightshirt was Eckstadt's father, Bench.

To clinch the case, Cayuga County prosecutors eventually agreed to have Bench's body exhumed for DNA tests—which matched the DNA from the saliva stains.

"We've had a lot of crazy cases," says Nina Morrison, the Innocence Project attorney who handled Brown's case, "but this is really up there with the best of them . . . the client solving his own case . . . it's insane." Brown was cleared of all charges and is now putting his life back together. ■

▶ Summary

■ DNA is the hereditary molecule of all living organisms. DNA contains instructions for building an organism.

■ DNA sequences determine the genetic uniqueness and relatedness of individuals.

■ The DNA in a eukaryotic cell is packaged into chromosomes located in the nucleus.

■ Humans have 23 pairs of chromosomes in their cells—one chromosome of each pair inherited from the mother, the other from the father.

■ DNA is a double-stranded molecule, which forms a spiral structure known as a double helix.

■ Each strand of DNA is made of nucleotides bonded together in a linear sequence.

■ There are four distinct nucleotides: adenine (A), thymine (T), guanine (G), and cytosine (C).

■ The two linear strands of a DNA molecule are bound together by complementary pairing of A with T and G with C.

■ Complementary pairing of DNA strands guides DNA replication, a fundamental part of cell reproduction.

■ PCR enables scientists to vastly increase the number of copies of specific DNA sequences.

■ Forensic scientists use noncoding DNA sequences known as STRs to create a DNA profile.

■ STRs are blocks of repeated sequences of DNA. People differ in the number of times the sequences are repeated along their chromosomes.

■ A DNA profile is more accurate and reliable than many other forms of evidence.

DNA STRUCTURE AND REPLICATION

Two strands of DNA nucleotides bound together form a spiral structure called a double helix. DNA strands are complementary: A always pairs with T; G always pairs with C. The nucleotide sequence of one strand dictates the nucleotide sequence of the other.

HINT See Infographics 7.1–7.4.

➔ KNOW IT

1. Which of the following is not a nucleotide found in DNA?
   a. A, adenine
   b. T, thymine
   c. C, cytosine
   d. G, guanine
   e. U, uracil

2. If the sequence of one strand of DNA is AGTCTAGC, what is the sequence of the complementary strand?
   a. AGTCTAGC
   b. CGATCTGA
   c. TCAGATCG
   d. GTCGACGC
   e. GCTAGACT

3. In addition to the base, what are the other components of a nucleotide?
   a. sugar and polymerase
   b. phosphate and sugar
   c. phosphate and polymerase
   d. phosphate and helix
   e. helix, sugar

➔ USE IT

4. Complete the statements below, and then number them to indicate the order of these two major steps necessary to copy a DNA sequence during PCR.

   The enzyme _____ "reads" each template strand and adds complementary nucleotides to make a new strand.

   The two original strands of the DNA molecule are separated by means of _____.

5. Given this segment of a double-stranded DNA molecule, draw the two major steps involved in DNA replication.

   ATCGGCTAGCTACGGCTATT TACGGCATAT
   TAGCCGATCGATGCCGATAAATGCCGTATA

6. A series of statements is presented below. Mark each statement as true (T) or false (F).
   a. G pairs with T. _____
   b. Genetic information is passed on to the next generation in the form of DNA molecules. _____
   c. All DNA sequences encode information to produce proteins. _____
   d. Each person carries the same number of STR repeats on their maternal and paternal chromosomes. _____
   e. DNA evidence can be obtained from saliva left in a bite mark. _____

7. Explain why the statements that you marked as true in Question 6 are in fact true.

8. Rewrite the statements that you marked as false in Question 6 to make them true.

GENETIC VARIABILITY AND DNA PROFILING

DNA sequences determine genetic uniqueness and relatedness of people. Forensic scientists take advantage of this uniqueness by amplifying segments of DNA called short tandem repeats (STRs) to create a DNA profile.

HINT See Infographics 7.5 and 7.6.

➔ KNOW IT

9. Which STR will have migrated farthest through an electrophoresis gel?
   a. (GAAG) repeated twice
   b. (GAAG) repeated three times
   c. (AGCT) repeated five times
   d. (GAAG) repeated seven times
   e. (AGCT) repeated seven times

10. An individual's STR may vary from the same STR of another individual by virtue of
    a. the order of nucleotides.
    b. which specific bases are present.
    c. the specific chromosomal location of the STR in each individual.
    d. the number of times a particular sequence is repeated.
    e. the number of coding regions

11. Which of the following represents genetic variation between individuals?
    a. whether or not G pairs with C or T
    b. the presence of both coding and noncoding sequences in their genomes

c. the number of chromosomes in the nucleus

d. the sequence of nucleotides along the length of each chromosome

e. the number of chromosomes received from each parent

12. A person has an STR with the sequence GACCT repeated six times on one chromosome and eight times on the other chromosome. If this STR were amplified by PCR, and the PCR products run on a gel, which lane shows the corresponding banding pattern (see gel at top right)? Note that the marker lane (M) has fragments starting at 10 nucleotides (at the bottom) and increasing in 10-nucleotide increments.

13. The gel at middle right shows the DNA profile of STRs from four sources: blood from crime scene evidence (E), suspect A, suspect B, and the victim (V). An eyewitness identified suspect A as fleeing the apartment building where the crime occurred. Suspect B was picked up at a local convenience store after using bloodstained money.

a. From the DNA profiles shown, can you draw any conclusions about where the crime scene DNA came from?

b. Can you draw any conclusions about relationships among the people profiled? Explain your reasoning.

CHROMOSOMES, CRIME SCENES, AND PATERNITY TESTING

We can use DNA profiling to establish paternity. Each person inherits 23 chromosomes from his or her mother and 23 chromosomes from his or her father. Females inherit one X chromosome from their mother, and a second X chromosome from their father. Males inherit one X chromosome from their mother and a Y chromosome from their father. It is the presence of the Y chromosome that determines maleness.

HINT See Infographics 7.1, 7.6, and 7.7.

⊜ KNOW IT

14. The _____ chromosomes in a human cell from inside the cheek are found in the _____ .

a. 46; cytoplasm

b. 23; nucleus

c. 24; cytoplasm

d. 46; nucleus

e. 22; nucleus

15. Each chromosome contains

a. DNA only.

b. proteins only.

c. DNA and proteins.

d. the same number of genes and STRs.

e. the entire genome of a cell.

⊜ USE IT

16. Look at Infographic 7.7. From the STRs used in forensic investigations, which STRs would be particularly useful in determining whether crime scene evidence was left by a female or a male?

17. Explain your response to Question 16, stating the number of STR copies you would expect to see if the perpetrator was female and if the perpetrator was male.

18. The gel at bottom right shows a DNA profile using five STRs. The lane labeled W is a mother and the lane labeled C is her child. The lanes labeled M1 and M2 are two men, either of whom, according to the mother, could be the father of the child.

a. Circle the STR bands that the child (C) inherited from its mother (W).

b. Use the DNA profiles to determine which man is the father of the child.

SCIENCE AND ETHICS

19. Scientists used DNA from Barry Bench's daughter to pinpoint Bench as a possible suspect, as his DNA was not on file anywhere. Similarly, in cases of disasters, DNA evidence is sometimes required in order to identify victims. If a victim doesn't have a DNA profile on file, identity must be reconstructed by comparing the victim's DNA profile to relatives'. These situations illustrate that a DNA profile database has the potential to be useful in cases in which DNA-based identification is required. However, such a database is controversial. What arguments can you make for and against "banking" people's DNA profiles in a database? If such a database existed, what restrictions would you place on it?

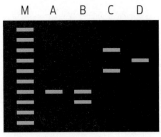

Gel for Question 12

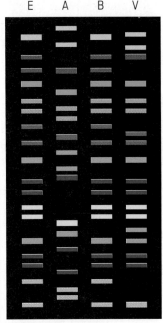

Gel for Question 13

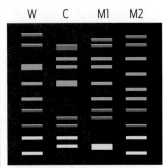

Gel for Question 18

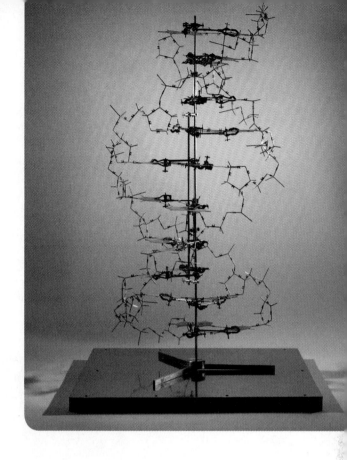

The Model Makers

➔ What You Will Be Learning

Two scientists piece together clues—some from other researchers—to solve the DNA puzzle.

The Model Makers

Watson, Crick, and the structure of DNA

In 1953, with inflated egos, James Watson and Francis Crick announced to a crowd in their favorite pub in Cambridge, England, that they had found the secret of life. Given the nature of their discovery, they had every right to boast: they had finally revealed the structure of DNA. They had succeeded where other scientists had failed, and they were proud. And although they likely didn't realize it at the time, theirs was a discovery that would usher in an explosion of research on DNA and genetics and push forward the study of anthropology, evolution, and medicine.

But Watson and Crick's success wasn't merely the result of a marriage between two great minds. Scientific breakthroughs rarely result from single scientists working in a vacuum. Rather, breakthroughs happen after many incremental discoveries made by different scientists over years. And so it was with DNA. In addition to their own insight, Watson and Crick built on the discoveries of others. They also had luck on their side.

> **It was a discovery that would usher in an explosion of research on DNA and genetics.**

James Watson was an American scientist who in 1951 accepted a research position at Cambridge University in England, where he met Francis Crick. Crick was a Ph.D. student at the time, studying protein structure with a technique called x-ray crystallography.

Given their varied backgrounds, the men didn't appear obvious collaborators. Watson was a prodigy. Twelve years younger than Crick, he had earned his Ph.D at 22. Crick, by contrast, was a late bloomer. He was 38 years old by the time he had his Ph.D. But what they did share was intellectual curiosity. Both had changed their research focus several times. By their own admission, both were more interested in solving current hot topics in science—like the structure of DNA—than pursuing the more obscure science that each had trained to do.

Although DNA was first observed in cell nuclei in the late 1860s, it took almost a century before scientists realized its importance. For a long time, the prevailing belief was that proteins carried the genetic information. But by the late 1940s it was well accepted that DNA, not protein, was the genetic material. Scientists still, however, didn't understand the structure of DNA, and solving that structure became a quest for many scientists of the day.

When Crick and Watson came to the problem, they already knew, from the work of other scientists, that DNA was made up of nucleotides containing a sugar, a phosphate group, and one of four bases: adenine, thymine, cytosine, or guanine. But how did the elements fit together? To answer this question, they took inspiration from the scientist Linus Pauling. Pauling had been studying the structure of proteins and had built a molecular model showing that some proteins exist as a single-stranded, twisting helix. He backed up his model with lab experiments to prove his structure was correct. If an eminent scientist like Pauling could model a structure without first conducting laboratory experiments, Watson and Crick thought they might be able to do the same with DNA.

Using wire and metal, Watson and Crick began building scale models of DNA on the basis of existing evidence about the chemical structure of nucleotides. They initially built a three-helix model with the phosphate groups on the inside and the bases radiating outward. But experts who analyzed the structure deemed it chemically unstable.

Then came a crucial finding. In 1951, Watson attended a lecture by a young scientist named Rosalind Franklin. In her laboratory at King's College, London, she had been creating x-ray diffraction pictures of DNA. X-ray diffraction analyzes the way x-rays bounce off a sample of material to determine characteristics of the sample such as physical structure and chemical composition. Franklin had observed that increasing the humidity of a DNA sample could cause it to elongate. She speculated that if the phosphate part of the DNA formed the *outside* of the molecule, water available from an increase in humidity would readily interact with it. This form of DNA would more closely resemble that found in the aqueous cell environment.

The DNA Puzzle

→ It was known that DNA was made of nucleotides that included a deoxyribose sugar, a phophate group, and one of four nitrogenous bases. But no one had yet figured out how the nucleotides fit together to produce a DNA molecule.

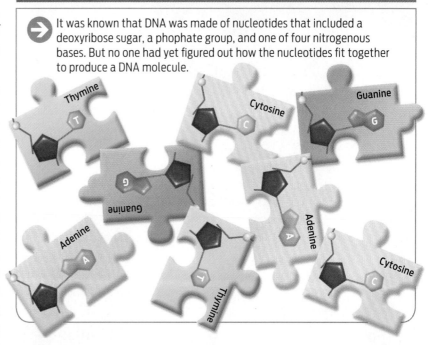

Rosalind Franklin and the Shape of DNA

→ Franklin's 1951 x-ray diffraction studies of DNA showed that the structure was likely helical, involving two strands that run in opposite directions, and that the phosphate groups were on the outside of the molecule. X-ray diffraction involves shooting x-rays at a crystallized version of a molecule and recording on film how the x-rays scatter when they bounce off its surface. The image shown here is a view from the end of a DNA molecule looking down its center.

Rosalind Franklin
July 25, 1920–April 16, 1958

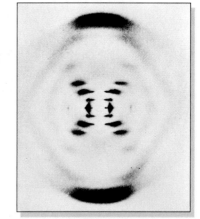

X-ray diffraction image of DNA
The signature "X" in the middle of this picture suggests a double-stranded helical structure. The symmetry of the image suggests that the molecule is uniform in width along its length.

Erwin Chargaff's Work Provided a Clue to Base Pairing

→ Erwin Chargaff studied the nitrogenous bases of DNA. He found that no matter which DNA molecule he analyzed, it always contained equal amounts of adenine and thymine bases and equal amounts of cytosine and guanine bases. These data suggested that adenine must pair with thymine and that cytosine must pair with guanine. This base pairing further supported the idea of a double-stranded DNA structure.

Erwin Chargaff
August 11, 1905–June 20, 2002

Within any DNA molecule:
% adenine = % thymine
% cytosine = % guanine

Chargaff's Rule of Base Pairing

Thymine   Adenine

Cytosine   Guanine

Franklin's contribution didn't end there. She also discovered other important facts about the structure of DNA. Working with a graduate student, Raymond Gosling, she found that her x-ray diffractions confirmed that the elongated form of DNA had all the characteristics of a twisting helix. Maurice Wilkins, who was Franklin's peer, was also studying DNA

The final double-helix model so perfectly fit the experimental data that the scientific community almost immediately accepted it.

structure at the time. In 1953, Wilkins saw Franklin's best unpublished x-ray picture of DNA and showed it to Watson without Franklin's knowledge. "The instant I saw the picture my mouth fell open," Watson recalled in his memoir of the discovery, *The Double Helix*, published in 1968. The sneak preview "gave several of the vital helical parameters."

With that clue in hand, Watson and Crick then took a crucial conceptual step and suggested that the molecule was made of two chains of nucleotides. Each formed a helix, as Franklin had found, but they spiraled in oppo-

site directions. To construct the model, Crick also built on a discovery made a few months earlier. In 1952, Erwin Chargaff had found that DNA contained equal amounts of adenine and thymine and equal amounts of guanine and cytosine. This information helped Watson and Crick deduce how the bases were paired: adenine with thymine and cytosine with guanine.

The final double-helix model so perfectly fit the experimental data that the scientific community almost immediately accepted it. Watson, Crick, and Wilkins published their paper on the structure of DNA in the prominent journal *Nature* in the same issue alongside Franklin's. In 1962, Watson, Crick, and Wilkins shared the Nobel prize in physiology or medicine. But what about Franklin? Franklin had died of cancer in 1958, at the age of 37. Nobel prizes aren't awarded posthumously.

Controversy over whether Franklin has been adequately recognized continues. Although Watson and Crick acknowledged her contribution to their research in their article in *Nature*, the extent to which her input helped them build their DNA model was revealed only much later in Watson's

The Structure Is Finally Known: The DNA Double Helix

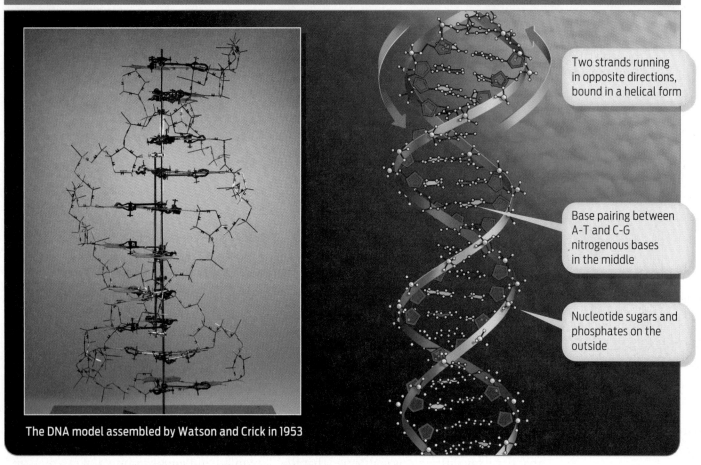

The DNA model assembled by Watson and Crick in 1953

Two strands running in opposite directions, bound in a helical form

Base pairing between A-T and C-G nitrogenous bases in the middle

Nucleotide sugars and phosphates on the outside

1968 book, published 10 years after Franklin's death. For example, at the time *Nature* published the papers on DNA structure, Franklin's paper was perceived as mere supporting evidence. But it was her data that helped Watson, Crick, and Wilkins clinch the structure. Some historians argue that sexist attitudes prevented her from receiving the acclaim she deserved before she died. At the time, female scientists in the biomedical sciences were few and frequently confronted by negative attitudes from their male peers. "I'm afraid we always used to adopt–let's say, a patronizing attitude towards her," Crick publicly commented after Watson's book was published. He added that if Franklin had lived, "It would have been impossible to give the prize to Maurice and not to her" because "she did the key experimental work."

Although it was quite normal for colleagues to share data, some have even argued that Wilkins took Franklin's critical x-ray diffraction photos without her knowledge or consent and showed them to Watson out of jealousy or disdain.

"It would have been impossible to give the prize to Maurice and not to [Franklin because] she did the key experimental work." –Francis Crick

Despite controversy, Franklin's contribution to the discovery has never been completely ignored, and she is now recognized as having been a top-notch scientist: her notebooks show that without her thorough scientific research and original ideas, we would have had to wait much longer for what is still considered to be one of the most important discoveries in biology. ∎

Medicine from Milk

Medicine from milk

Scientists genetically modify animals to make medicine

In a Massachusetts barn nestled among willow and oak trees, rows of juglike machines drone in a constant hum. Goats, dozens of them, are being milked. But this is no ordinary dairy operation. This farm is among several worldwide practicing the art of "pharming"– using genetically modified animals to churn out therapeutic drugs.

The first drug produced from such **transgenic** animals is already available, manufactured by GTC Biotherapeutics, a firm based in Framingham, Massachusetts. The drug consists of a human protein called antithrombin that was extracted from transgenic goats' milk. Antithrombin is most commonly used to treat patients who either inherit or acquire a deficiency of the antithrombin protein, which puts them at risk of developing dangerous blood clots.

For decades, scientists had extracted antithrombin from human blood donations. But blood contains only small amounts of antithrombin, and the supply depends on the number of blood donors. Transgenic goats, however, can produce massive amounts of the drug in a relatively short period of time. Moreover, relying on a herd of goats instead of human volunteers ensures a consistent supply. And because the animals live in a controlled envi-

---

"This is very exciting, it is novel and has great potential for where we can go with this new technology." —Bernadette Dunham

---

ronment, there is less risk of transmitting infections such as HIV and hepatitis to healthy people through contaminated donor blood.

Because of all these advantages, some people are predicting that transgenic animals may one day replace human donors as the source for therapeutic agents extracted from blood. "This is very exciting, it is novel and has great potential for where we can go with this new technology," Bernadette Dunham, director of the FDA's Center for Veterinary Medicine, told

TRANSGENIC
Refers to an organism that carries one or more genes from a different species.

PROTEIN
A macromolecule made up of repeating subunits known as amino acids, which determine the shape and function of a protein. Proteins play many critical roles in living organisms.

AMINO ACIDS
The building blocks of proteins. There are 20 different amino acids.

the *Washington Post* in February 2009, when the company's drug for antithrombin deficiency was approved for market.

Antithrombin: From Gene to Protein

Antithrombin is a **protein.** Recall from Chapter 2 that proteins are one of the four main macromolecules that make up cells. Proteins have myriad functions in the body: they allow our muscles to contract, give our hair and skin its texture, and facilitate the thousands of chemical reactions that occur in our cells. In fact, proteins play a huge role in all basic cellular functions. Proteins can perform such a variety of different tasks because they come in many shapes and sizes.

All proteins are made of the same building blocks called **amino acids.** There are 20 different amino acids in all. All amino acids have the same basic core structure, but each also has a unique chemical side group that distinguishes the amino acids from one another. Amino acids bond together to form linear chains. The human antithrombin protein is a chain of 432 amino acids. Many human proteins are in this size range, but chain lengths vary from just a few to thousands of amino acids. The longest human protein, titin, is a single chain of 34,350 amino acids.

The sequence of amino acids in any given chain makes each chain unique, and also determines how that chain ultimately folds into a

Machines milk rows of goats.

three-dimensional structure–the protein itself. A protein's three-dimensional structure is important because it determines a protein's function. Some proteins, such as the antibody molecules that are a critical part of our defenses, or the hemoglobin that carries oxygen in our red blood cells, are made up of more than one folded amino acid chain. A protein's final overall shape–which is dictated by the placement of amino acid side groups– determines its specific function **(Infographic 8.1)**. Antithrombin folds into a compact globular structure. This structure is maintained through numerous chemical interactions between amino acid side groups. The position of every amino acid in the chain is important, contributing to the protein's overall shape and therefore its optimal function.

INFOGRAPHIC 8.1

Amino Acid Sequence Determines Protein Shape and Function

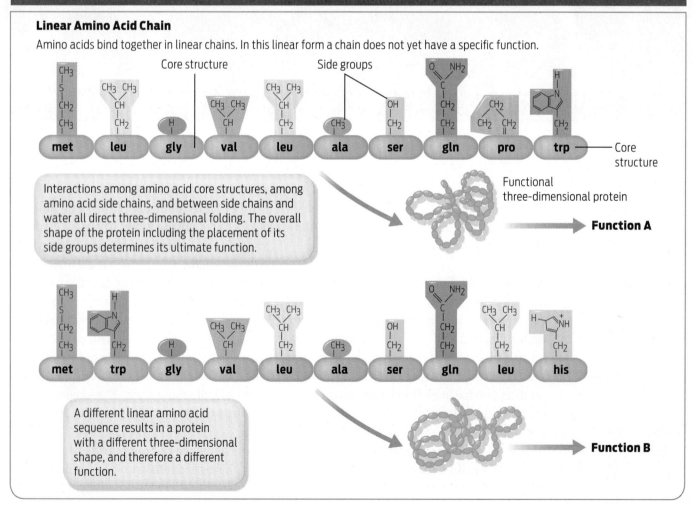

Linear Amino Acid Chain

Amino acids bind together in linear chains. In this linear form a chain does not yet have a specific function.

Core structure

Side groups

met leu gly val leu ala ser gln pro trp — Core structure

Interactions among amino acid core structures, among amino acid side chains, and between side chains and water all direct three-dimensional folding. The overall shape of the protein including the placement of its side groups determines its ultimate function.

Functional three-dimensional protein

Function A

met trp gly val leu ala ser gln leu his

A different linear amino acid sequence results in a protein with a different three-dimensional shape, and therefore a different function.

Function B

Chromosomes Include Gene Sequences That Code for Proteins

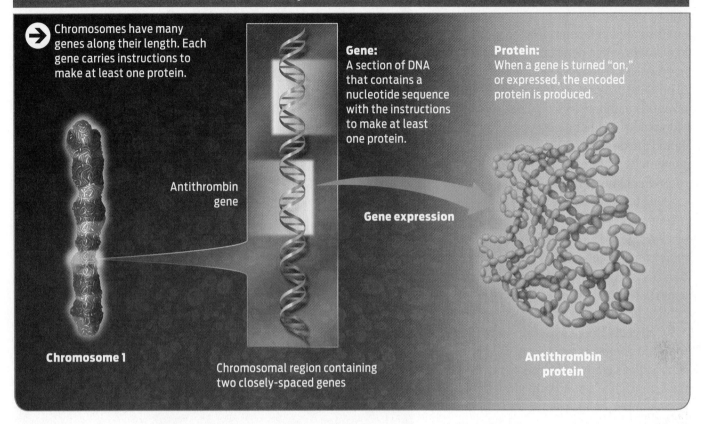

Chromosomes have many genes along their length. Each gene carries instructions to make at least one protein.

Antithrombin gene

Gene:
A section of DNA that contains a nucleotide sequence with the instructions to make at least one protein.

Protein:
When a gene is turned "on," or expressed, the encoded protein is produced.

Gene expression

Chromosome 1

Chromosomal region containing two closely-spaced genes

Antithrombin protein

GENE
A sequence of DNA that contains the information to make at least one protein.

GENE EXPRESSION
The process of using DNA instructions to make proteins.

GENOTYPE
The genetic make-up of an organism.

PHENOTYPE
The physical attributes of an organism, including both observable and internal or non-observable traits.

Because proteins play such important roles in the body, most drugs are either chemicals that interact with specific proteins, or, like antithrombin, are themselves protein molecules.

Where do proteins come from? Just as cells make DNA out of building blocks that we, in part, obtain from food, cells also make proteins using amino acid building blocks from our diet. But DNA and protein are not equals when it comes to their function in cells. Rather, the relationship between the two is hierarchical, with one directing the production of the other.

The instructions to make proteins are encoded in our DNA–our genes. A **gene** is a segment of DNA that contains instructions for making at least one protein. Genes are particular nucleotide sequences organized along the length of chromosomes, with each chromosome carrying a unique set of genes. The process of synthesizing a protein from the information encoded in a gene is called **gene expression (Infographic 8.2)**. When we say a

gene inside a cell is "expressed," we mean that the cell is making the protein encoded by that gene. In other words, our genes are the master instruction manual of our bodies; they dictate what proteins are made, when, and how many. Another way of saying this is that genes provide our **genotype,** but it is the proteins specified by those genes that, to a large extent, determine our physical traits, or **phenotype.**

The antithrombin gene, for example, sits on chromosome 1 and holds instructions to make a chain of 432 amino acids that folds into the antithrombin protein. When cells express the antithrombin gene, it means they produce antithrombin protein.

In the body, antithrombin protein prevents blood from clotting. The protein plays a regulatory role by inactivating enzymes that promote blood clotting. In this way, antithrombin prevents blood from clotting in the wrong place and causing a stroke or a heart attack.

Some people, however, can become antithrombin deficient because of heart, liver, or kidney disease, or cancer. Others can inherit antithrombin deficiency from a parent. Inherited antithrombin deficiency isn't rare—about 1 in every 5,000 people is born with the inability to produce this protein. People with too little or no antithrombin carry a high risk of developing blood clots inside blood vessels, a condition called thrombosis, and consequently they sometimes require antithrombin transfusions to prevent these vessel-blocking clots from forming (Infographic 8.3).

Remember, each human cell has two copies of every gene in our genome. When people inherit antithrombin deficiency, it doesn't mean they don't have the antithrombin gene. Rather, it means that both their copies of the antithrombin gene are defective. This can happen because, as for all genes, there exist different versions of the antithrombin gene. Alternative versions of genes differ slightly in their sequence of nucleotides, much like words with different spellings (for example, color, colour; theater, theatre). Different versions of a gene with such alternative nucleotide "spellings" are called **alleles.** Sometimes, the allele of a gene encodes a protein with an abnormal shape. If a protein's shape is too contorted, it may not be able to do its job and the protein is nonfunctional. Having a nonfunctional protein is as harmful as not having one at all. If a person's two alleles of the

ALLELES
Alternative versions of the same gene that have different nucleotide sequences.

INFOGRAPHIC 8.3

Antithrombin Deficiency Can Cause Blood Clots

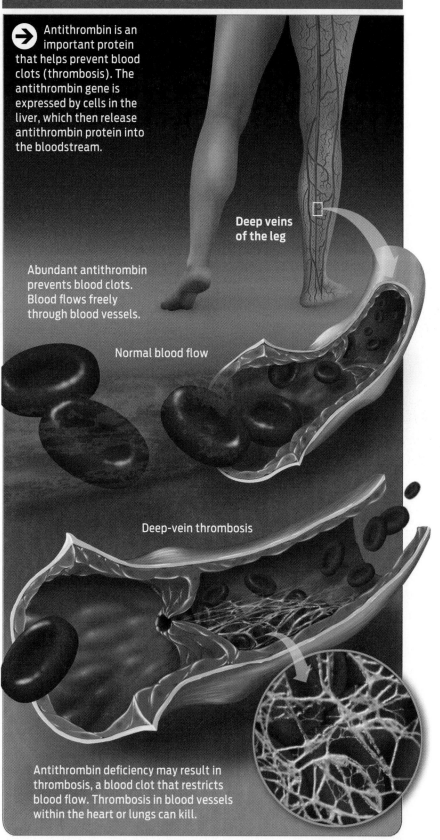

→ Antithrombin is an important protein that helps prevent blood clots (thrombosis). The antithrombin gene is expressed by cells in the liver, which then release antithrombin protein into the bloodstream.

Deep veins of the leg

Abundant antithrombin prevents blood clots. Blood flows freely through blood vessels.

Normal blood flow

Deep-vein thrombosis

Antithrombin deficiency may result in thrombosis, a blood clot that restricts blood flow. Thrombosis in blood vessels within the heart or lungs can kill.

Genes to Proteins: Different Alleles Influence Phenotype

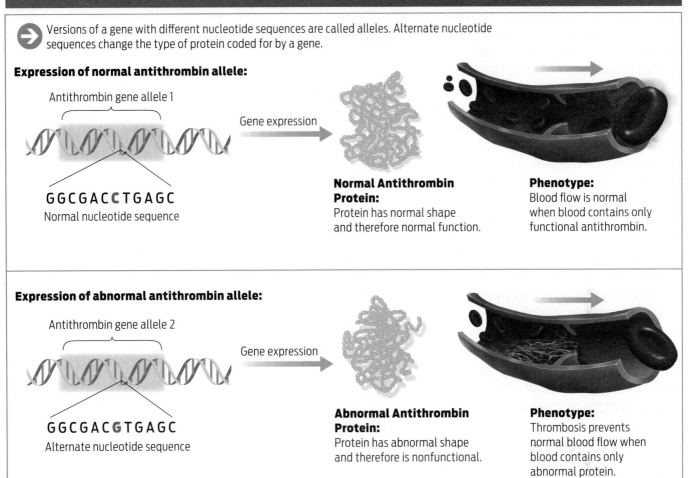

Versions of a gene with different nucleotide sequences are called alleles. Alternate nucleotide sequences change the type of protein coded for by a gene.

Expression of normal antithrombin allele:

Antithrombin gene allele 1

Gene expression

GGCGAC**C**TGAGC
Normal nucleotide sequence

Normal Antithrombin Protein:
Protein has normal shape and therefore normal function.

Phenotype:
Blood flow is normal when blood contains only functional antithrombin.

Expression of abnormal antithrombin allele:

Antithrombin gene allele 2

Gene expression

GGCGAC**G**TGAGC
Alternate nucleotide sequence

Abnormal Antithrombin Protein:
Protein has abnormal shape and therefore is nonfunctional.

Phenotype:
Thrombosis prevents normal blood flow when blood contains only abnormal protein.

---

antithrombin gene both code for defective proteins, antithrombin deficiency results **(Infographic 8.4).**

People with inherited antithrombin deficiencies usually take medication to thin their blood and prevent clots. At times when the risk of clots is high–during surgery, for example–they receive antithrombin treatment.

But it takes 50,000 blood donors to produce 1 kilogram of antithrombin. A single transgenic goat can produce the same amount in her milk in just one year, according to GTC Biotherapeutics.

Making Transgenic Goats

More than 20 years ago, when Harry Meade was working as a research scientist at a company

called Biogen, it occurred to him that producing drugs in a mammal's milk might be more efficient than existing methods of large-scale protein production. The mammary gland is a natural protein factory, he reasoned. To nourish their young, all mammals produce proteins and secrete them into their milk.

Meade had been experimenting with getting genes from various organisms into hamster cells, which could be grown in large numbers in the laboratory, allowing scientists to purify significant amounts of protein from the cells. This method was effective, and in fact is still used today to express and harvest proteins of interest. But wouldn't it be more efficient, Meade wondered, to transfer a gene into a large mammal, such as a goat, so that the gene is expressed by the mammary gland? That way, when the goat lactates, the protein will collect in the goat's milk. This method would produce much greater quantities of a protein than could be produced by hamster cells in stainless steel vats.

To work on this project, Meade co-founded GTC Biotherapeutics. At his new company he helped devise a technique to create transgenic animals, the first of which was a transgenic goat. The basic idea is simple: isolate the gene of interest from a human chromosome and then insert it into the genome of a goat embryo. But in order to make sure the human gene is expressed properly in goat mammary glands, Meade and his colleagues had to create a hybrid gene that was part human and part goat.

The technique that Meade used takes advantage of the fact that every gene has two parts: a **regulatory sequence** and a **coding sequence.** Regulatory sequences are like on/off switches for genes; they determine when, where, and how much protein a gene makes. Coding sequences determine the identity of a protein–they specify the amino acid makeup **(Infographic 8.5).**

Meade realized that if he could attach the coding sequence of the human antithrombin gene to the regulatory sequence of a goat gene that is expressed only in the animal's mammary cells, he could get the goat's mammary cells to make the human protein. In other words, he could dupe the goat's mammary glands into making the human antithrombin protein and secreting it as part of the goat's milk.

About 10 years ago, Meade and his team successfully attached the coding sequence of the human antithrombin gene to the regulatory sequence of a goat gene expressed only by the mammary glands (the beta casein gene). Using the regulatory sequence of the beta casein gene ensures that the gene of interest, antithrombin, is expressed only in the mammary cells, and not in any other tissues. The antithrombin protein will be found solely in the goat's milk, and nowhere else. We'll look at the details of how genes are expressed later. For now, it's important to note that scientists took advantage of a natural phenomenon: since beta casein protein

INFOGRAPHIC 8.5

The Two Parts of a Gene

→ Genes are organized into two parts. Regulatory sequences determine when and how much protein a gene makes. Coding sequences determine the amino acid sequence of the encoded protein, which determines its shape and function.

Gene

Regulatory sequence: Controls the timing, location, and amount of gene expression.

Coding sequence: Determines the sequence of amino acids in the protein.

is normally found only in goat's milk, the antithrombin protein would also be found only in the animal's milk. With the exception of the mammary glands, no other tissue would express the antithrombin gene, so as not to disrupt the tissue's normal function and harm the animal.

With the coding sequence of the human antithrombin gene fused to the regulatory sequence of the goat beta casein gene, the researchers could begin the process of putting the transgene inside a goat. Using a long, thin needle, a GTC scientist injected the gene construct into a fertilized single-cell goat embryo. He then implanted this transgenic embryo into a surrogate mother. As the embryo grew and the cells divided, the inserted gene replicated and was passed on to every cell in the developing goat **(Infographic 8.6).**

This is the technique the company used to create the first transgenic goats; today, GTC uses newer and more efficient methods to get gene constructs into animals.

Animals aren't the only organisms that have been genetically modified by humans. Much of the corn you eat today is transgenic–it contains genes from a soil bacterium. There are strains of transgenic soybeans, transgenic tomatoes, and transgenic insects. Trans-

REGULATORY SEQUENCE
The part of a gene that determines the timing, amount, and location of protein produced.

CODING SEQUENCE
The part of a gene that specifies the amino acid sequence of a protein. Coding sequences determine the identity, shape, and function of proteins.

Making a Transgenic Goat

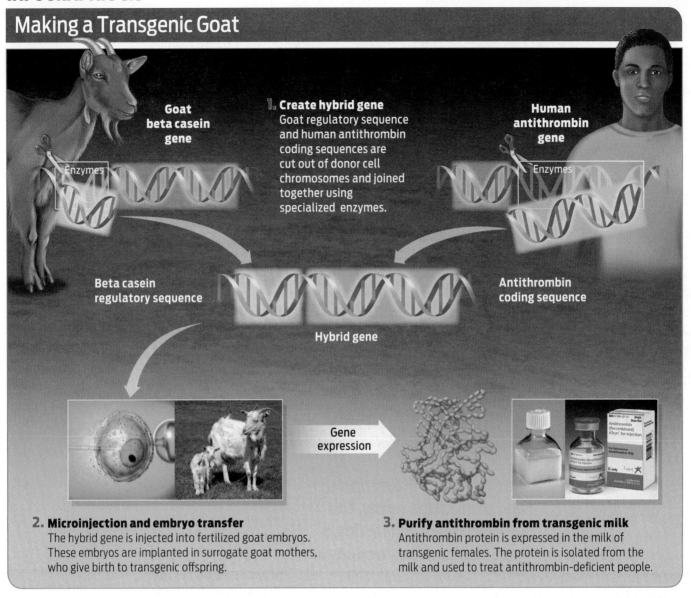

Goat beta casein gene

Enzymes

1. Create hybrid gene
Goat regulatory sequence and human antithrombin coding sequences are cut out of donor cell chromosomes and joined together using specialized enzymes.

Human antithrombin gene

Enzymes

Beta casein regulatory sequence

Antithrombin coding sequence

Hybrid gene

Gene expression

2. Microinjection and embryo transfer
The hybrid gene is injected into fertilized goat embryos. These embryos are implanted in surrogate goat mothers, who give birth to transgenic offspring.

3. Purify antithrombin from transgenic milk
Antithrombin protein is expressed in the milk of transgenic females. The protein is isolated from the milk and used to treat antithrombin-deficient people.

GENETICALLY MODIFIED ORGANISM (GMO)
An organism that has been genetically altered by humans.

GENE THERAPY
A type of treatment that aims to cure disease by replacing defective genes with functional ones.

genic organisms are also called **genetically modified organisms (GMOs).** Transgenic crops such as corn and soybean usually contain genes for natural pesticides, which help the plants fight pests and reduce the amount of pesticide a farmer must use. Transgenic animals serve many purposes; sometimes they are used in research to study a gene's function, other times they can be used for a specific commercial purpose, such as producing medicines or other marketable products. Spider silk, for example, is a very strong, resilient fiber that can be produced in transgenic animals or in plants that carry spider genes.

Such gene-swapping technology also has an important application in medicine: in **gene therapy** scientists attempt to replace a person's defective gene with a healthy one, an approach that can already treat, and in some cases cure, debilitating diseases such as severe combined immunodeficiency syndrome (SCID)—a disorder in which babies are born with deficient immune systems. Researchers hope that gene therapy might one day help treat several types of disorders caused by defective genes, such as cystic fibrosis, Huntington disease, and hemophilia.

Despite the many actual and potential benefits of genetic engineering, mixing and match-

Goats raised for medicine are kept in controlled environments.

ing genes inspires debate among scientists, environmentalists, and the general public.

Many people also object to human meddling with the biology of organisms that have evolved naturally. There are also other concerns, such as what might happen to a natural population of organisms if their genetically modified cousins were to escape into the environment and mate with the natural population; the consequences are unpredictable.

Although the idea of genetically engineering animals may be disquieting to some, humans have been tampering with the natural evolution of farm animals for centuries by selectively breeding them to have desirable traits. Moreover, from an animal-rights point of view, transgenic goats are treated no differently from goats farmed for their milk and meat.

That said, the prospect of being able to genetically modify–even clone whole organisms–for

human purposes raises legitimate questions about how to conduct genetic engineering safely and humanely. For example, many people who find nothing ethically troubling about using gene therapy to treat human diseases would nonetheless find the prospect of cloning humans abhorrent. In this case, however, goats are being modified to save human lives–a much less controversial use of the technology.

Making Proteins, or How Genes Are Expressed

Transgenic or not, all organisms make proteins from genes in the same way. So far, we've been discussing gene expression in abstract terms: a gene provides instructions to make proteins. But what are those instructions? How is the antithrombin protein actually made by goat cells?

In order to get from a gene to a protein, cells carry out two major steps: **transcription** and **translation.** Briefly, transcription is the process of using DNA to make a **messenger RNA (mRNA)** copy of the gene. Translation is the process of using this mRNA copy as a set of instructions to assemble amino acids into a protein (**Infographic 8.7**).

Why two separate steps? As the names *transcription* and *translation* imply, the process is like converting a text into another language. In this case, the text to be translated is a valuable, one-of-a-kind document: DNA. Just as you would be forbidden to borrow a rare manuscript from the library, and would instead have to rewrite the characters in it onto another sheet, the cell cannot take DNA out of its library–the nucleus. It must first make a copy–the mRNA. The cell can then take this mRNA copy into the cytoplasm, where it is translated into a protein.

Let's take a closer look at both steps. Transcription begins when an enzyme called **RNA polymerase** binds to the regulatory sequence of DNA just ahead of a gene's coding sequence. At that site, cellular machinery unwinds the

> **As the names *transcription* and *translation* imply, the process is like converting a text into another language.**

TRANSCRIPTION
The first stage of gene expression, during which cells produce molecules of messenger RNA (mRNA) from the instructions encoded within genes.

TRANSLATION
The second stage of gene expression. Translation "reads" mRNA sequences and assembles the corresponding amino acids to make a protein.

MESSENGER RNA (mRNA)
The RNA copy of an original DNA sequence made during transcription.

RNA POLYMERASE
The enzyme that accomplishes transcription. RNA polymerase copies a strand of DNA into a complementary strand of mRNA.

Gene Expression: An Overview

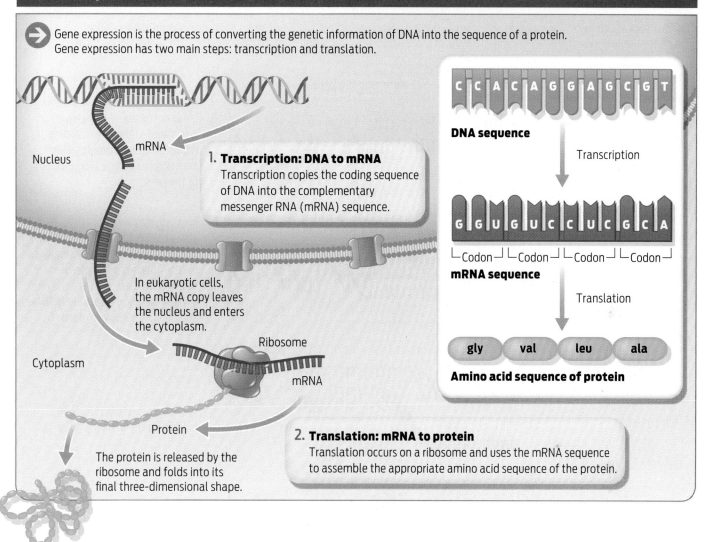

→ Gene expression is the process of converting the genetic information of DNA into the sequence of a protein. Gene expression has two main steps: transcription and translation.

mRNA

Nucleus

1. Transcription: DNA to mRNA
Transcription copies the coding sequence of DNA into the complementary messenger RNA (mRNA) sequence.

In eukaryotic cells, the mRNA copy leaves the nucleus and enters the cytoplasm.

Ribosome

Cytoplasm

mRNA

Protein

The protein is released by the ribosome and folds into its final three-dimensional shape.

2. Translation: mRNA to protein
Translation occurs on a ribosome and uses the mRNA sequence to assemble the appropriate amino acid sequence of the protein.

C C A C A G G A G C G T

DNA sequence

Transcription

G G U G U C C U C G C A

└Codon┘ └Codon┘ └Codon┘ └Codon┘

mRNA sequence

Translation

gly   val   leu   ala

Amino acid sequence of protein

RIBOSOME
The cellular machinery that assembles proteins during the process of translation.

CODON
A sequence of three mRNA nucleotides that specifies a particular amino acid.

DNA double helix and RNA polymerase begins moving along one DNA strand. As it moves, the RNA polymerase "reads" the DNA sequence and synthesizes a complementary mRNA strand according to the rules of base pairing. The same rules of base pairing we discussed in the context of DNA apply here, with one difference: RNA nucleotides are made with the base uracil (U) instead of thymine (T). So the complementary base pairs are C with G and A with U (Infographic 8.8).

As its moniker states, messenger RNA serves to relay information. Once the mRNA copy is made, it leaves the nucleus and attaches to a complex cellular machinery called the **ribosome.** This is the start of translation.

During translation, the ribosome "reads" the mRNA message and assembles a chain of amino acids. The ribosome acts like a factory in which mRNA serves as the instruction manual that specifies which amino acids should be joined together to form chains. Amino acids are specified by groups of three nucleotides called **codons.** Each codon is like a word: its letters name a particular amino acid (for example, the codon GGU specifies the amino acid glycine).

Although ribosomes are protein-assembling factories, they don't house all the parts needed to make proteins. Rather, they rely on a delivery system to bring the appropriate amino acids to the assembly site. The delivery system is

Transcription: A Closer Look

In eukaryotic cells, transcription occurs in the nucleus and copies a DNA sequence into a corresponding mRNA sequence. RNA polymerase is the key enzyme involved. In prokaryotic cells, transcription occurs in the cytoplasm, where DNA is located.

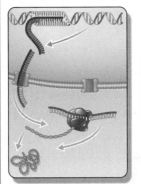

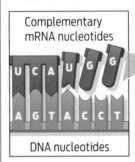

Complementary mRNA nucleotides

DNA nucleotides

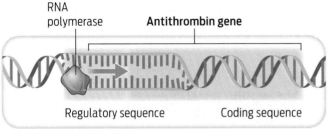

RNA polymerase

Antithrombin gene

Regulatory sequence

Coding sequence

1. RNA polymerase (pink circle) binds to the regulatory sequence just ahead of the gene's coding region. The DNA strands unwind, exposing the coding sequence of the gene.

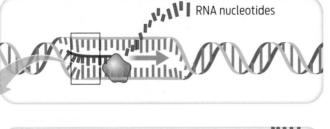

RNA nucleotides

2. RNA polymerase moves along the DNA strand. As it moves, it "reads" the DNA coding sequence and synthesizes a complementary mRNA strand according to the rules of base pairing, except that in RNA, adenine (A) pairs with uracil (U).

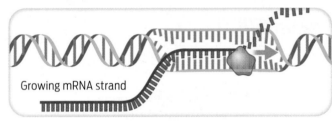

Growing mRNA strand

3. As the mRNA strand is formed, it detaches from the DNA sequence. The DNA reforms its double-stranded helix.

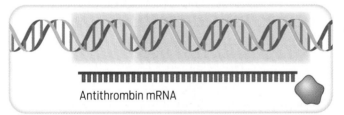

Antithrombin mRNA

4. Once the mRNA molecule is complete, it leaves the nucleus. The gene remains part of the chromosome in the nucleus where it can be used again in transcription.

---

another type of RNA called **transfer RNA (tRNA),** which physically transports amino acids to the ribosome. Each tRNA is structured like an adaptor: one end binds to an amino acid, the other end binds to mRNA. The part that binds mRNA is called the **anticodon** because it base-pairs in a complementary fashion with an mRNA codon. When the amino acid-toting tRNA finds its codon match, it releases the

The genetic code is universal, meaning that it is the same in all living organisms.

amino acid to the ribosome, which adds it to the growing amino acid chain (**Infographic 8.9**).

Although the human genome codes for many thousands of different proteins, each one is pieced together from a starting set of a mere 20 amino acids. In the same way that the 26 letters in the alphabet can spell hundreds of thousands of words, the basic set of amino acids can make hundreds of thousands

TRANSFER RNA (tRNA)
A type of RNA that helps ribosomes assemble chains of amino acids during translation.

ANTICODON
The part of a tRNA molecule that binds to a complementary mRNA codon.

Translation: A Closer Look

In the cytoplasm, the ribosome reads the mRNA sequence and "translates" it into a chain of amino acids to make a protein.

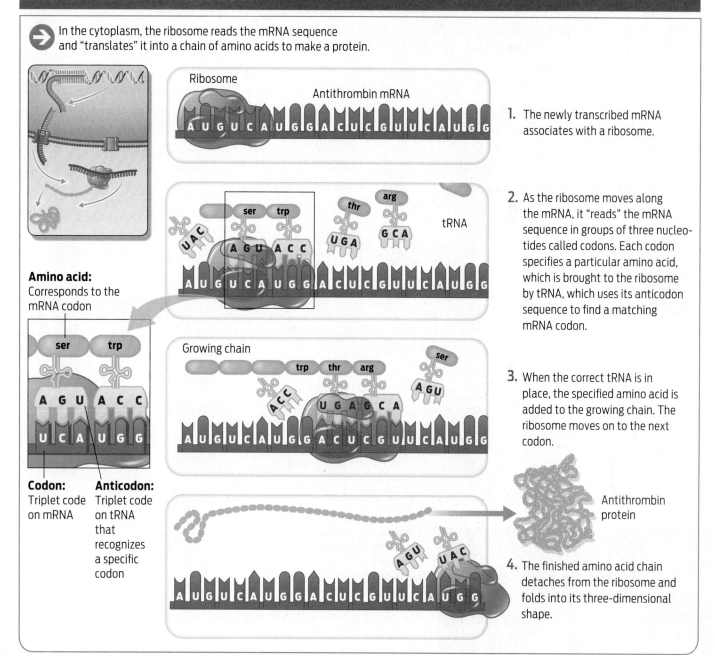

Ribosome

Antithrombin mRNA

A U G U C A U G G A C U C G U U C A U G G

1. The newly transcribed mRNA associates with a ribosome.

ser trp

thr arg

UAC

AGU ACC

UGA

GCA

tRNA

A U G U C A U G G A C U C G U U C A U G G

Amino acid: Corresponds to the mRNA codon

2. As the ribosome moves along the mRNA, it "reads" the mRNA sequence in groups of three nucleotides called codons. Each codon specifies a particular amino acid, which is brought to the ribosome by tRNA, which uses its anticodon sequence to find a matching mRNA codon.

ser trp

A G U A C C

U C A U G G

Codon: Triplet code on mRNA

Anticodon: Triplet code on tRNA that recognizes a specific codon

Growing chain

trp thr arg

ser

ACC

UGAGCA

AGU

A U G U C A U G G A C U C G U U C A U G G

3. When the correct tRNA is in place, the specified amino acid is added to the growing chain. The ribosome moves on to the next codon.

Antithrombin protein

AGU

UAC

A U G U C A U G G A C U C G U U C A U G G

4. The finished amino acid chain detaches from the ribosome and folds into its three-dimensional shape.

GENETIC CODE
The particular amino acids specified by particular mRNA codons.

of proteins. The rules by which mRNA codons specify amino acids are known as the **genetic code.**

The genetic code is universal, meaning that it is the same in all living organisms. It is because the code is universal that the mammary cells of a goat carrying the human gene for antithrombin are able to express the gene and produce antithrombin protein in its milk (**Infographic 8.10**).

The Advantages of GMOs and "Pharming"

One of the primary advantages of using transgenic animals to churn out protein drugs is that scientists can produce more complex proteins in a mammal's milk than they can from cell culture–the traditional way that scientists have produced many protein drugs. Because the mammary gland is a natural protein factory, mammalian milk already contains dozens of

The Genetic Code Is Universal

 Codons are groups of three-nucleotide sequences within chains of mRNA. Codons specify particular amino acids according to the universal genetic code. Since the genetic code is universal, the same gene will be transcribed and translated into the same protein in all cells and organisms.

Second letter

First letter	U	C	A	G	Third letter
U	UUU, UUC Phenylalanine (Phe); UUA, UUG Leucine (Leu)	UCU, UCC, UCA, UCG Serine (Ser)	UAU, UAC Tyrosine (Tyr); UAA Stop codon, UAG Stop codon	UGU, UGC Cysteine (Cys); UGA Stop codon; UGG Tryptophan (Trp)	U C A G
C	CUU, CUC, CUA, CUG Leucine (Leu)	CCU, CCC, CCA, CCG Proline (Pro)	CAU, CAC Histidine (His); CAA, CAG Glutamine (Gln)	CGU, CGC, CGA, CGG Arginine (Arg)	U C A G
A	AUU, AUC, AUA Isoleucine (Iso); AUG Methionine (Met); start codon	ACU, ACC, ACA, ACG Threonine (Thr)	AAU, AAC Asparagine (Asn); AAA, AAG Lysine (Lys)	AGU, AGC Serine (Ser); AGA, AGG Arginine (Arg)	U C A G
G	GUU, GUC, GUA, GUG Valine (Val)	GCU, GCC, GCA, GCG Alanine (Ala)	GAU, GAC Aspartic acid (Asp); GAA, GAG Glutamic acid (Glu)	GGU, GGC, GGA, GGG Glycine (Gly)	U C A G

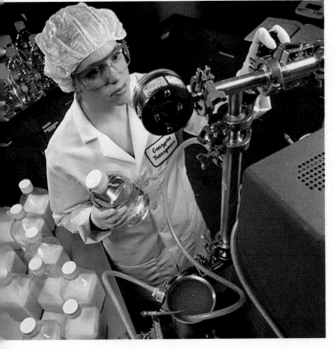

It is easier to scale up medicine derived from milk than to rely on blood donations.

complex proteins that are expressed at high levels.

By contrast, with hamster cells in culture, "You are asking the cell to produce a product in a way and in an environment completely different from what it would naturally produce," according to Thomas Newberry, former vice president for communications at GTC. And when scientists do succeed in getting such cells to produce complex proteins, they are expensive to make in large volumes; consequently the drugs are very expensive.

As an example, Newberry points to another human protein drug, one called factor VIIa. This is a protein that helps blood clot. Some people with the genetic disease hemophilia are born with clotting factor alleles that either encode nonfunctional clotting factor protein or no clotting factor protein at all. Consequently, if a hemophiliac cuts himself, he must be quickly transfused with clotting factor–otherwise he may bleed to death. Doctors usually give patients factor VIIa protein to restore normal coagulation and prevent excessive bleeding.

Companies that sell the protein drug produce it via cell culture. But it is extremely expensive: one milligram of factor VIIa can cost up to $1,000. Fortunately, hemophiliacs don't need much of the clotting factor to stop bleeds. But because of the drug's expense, they are given the drug only in an emergency. Over time, however, such rescue therapy, while minimizing initial cost, can be detrimental to patients' health: joints and other areas where bleeding typically occurs can become stiff and difficult

---

"I won't be surprised if people start to think that reinjecting blood products into other people is 'barbaric.'" —Thomas Newberry

---

to move. Transgenic animals can produce human clotting factors in their milk in large volumes for about one tenth of the amount of money it would take to produce the same proteins using cell culture. In fact, GTC Biotherapeutics is working on establishing transgenic animals to produce two other human clotting proteins: factor VIII and factor IX. In the FDA press release announcing the approval of GTC's antithrombin deficiency medication, Dunham was quoted: "I am pleased that this approval makes possible another source of an important human medication."

"We have the potential to build an abundant and controlled supply for any plasma protein," says Newberry, who predicts that protein drugs extracted from human blood may become a thing of the past. "In the future, I won't be surprised if people start to think that reinjecting blood products into other people is 'barbaric.'" ∎

▶ Summary

∎ Genes provide instructions to make proteins. The process of using the information in genes to make proteins is known as gene expression.

∎ Proteins are folded chains of amino acids that make up cell structures and help cells to function properly.

∎ Amino acid sequences determine the shape and function of a protein.

∎ Many drugs act on proteins in the body, or are themselves proteins. ∎

∎ A change in the DNA sequence of a gene can change the corresponding amino acid sequence, and therefore the function, of a protein.

∎ Different versions of the same gene, those with different nucleotide sequences, are called alleles.

∎ Every gene has two parts: a coding sequence and a regulatory sequence. The coding sequence determines the identity of a protein and the regulatory sequence determines where, when, and how much of the protein is produced.

∎ Gene expression occurs in two stages, transcription and translation, which take place in separate compartments in eukaryotic cells.

∎ Transcription is the first step of gene expression and copies the information stored in DNA into mRNA. Transcription occurs in the nucleus.

∎ Translation is the second step of gene expression and uses the information stored in mRNA to assemble a protein. Translation occurs in the cytoplasm.

∎ Proteins are assembled by ribosomes with the help of tRNA.

∎ The genetic code is the set of rules by which DNA sequences are translated into protein sequences; the code is shared by all living organisms.

∎ Through genetic engineering, genes from one species of organism can be inserted into the genome of another species of organism to make a transgenic organism.

PROTEIN STRUCTURE AND FUNCTION

Proteins have a unique three-dimensional structure that specifies their function. The structure of a protein is determined by its corresponding gene sequence.

HINT See Infographics 8.1–8.4.

➔ KNOW IT

1. What determines a protein's function?

2. The final product of gene expression is
   a. a DNA molecule.
   b. an RNA molecule.
   c. a protein.
   d. a ribosome.
   e. an amino acid.

➔ USE IT

3. Heating can cause a protein to denature, or unfold. What do you think would happen to a protein's function in this case? Explain your answer.

4. Insulin is a protein that is used therapeutically to treat people with diabetes. In your own words, describe the relationship between the insulin gene and the insulin protein.

GENE STRUCTURE

All genes have two key parts: a regulatory sequence and a coding sequence. To review gene structure, refer to Infographics 8.4 and 8.5.

➔ KNOW IT

5. The difference between two alleles of a gene is best ascertained by
   a. examining the amount of protein produced from each allele.
   b. examining the structure of the protein produced from each allele.
   c. examining the amount of mRNA produced from each allele.
   d. examining the nucleotide sequence of each allele.
   e. examining the amount of tRNA produced from each allele.

6. If a functional allele of antithrombin is expressed,
   a. blood clots will be more likely to form in the wrong place.
   b. blood clots will be less likely to form in the wrong place.

   c. functional antithrombin protein will be present in blood.
   d. a and c
   e. b and c

➔ USE IT

7. You are a doctor. Your patient has reduced levels of normal functioning antithrombin. Would you suspect a problem in the regulatory or in the coding sequence of the antithrombin gene? Why?

8. If you wanted to use genetic engineering to increase the amount of antithrombin this patient produces, would you modify the regulatory sequence or the coding sequence? Explain your answer.

MAKING TRANSGENIC ORGANISMS

Transgenic organisms are becomingly increasingly important in agriculture and medicine.

HINT See Infographics 8.5 and 8.6.

➔ KNOW IT

9. Melanin is expressed in skin cells and gives skin its color. If you wanted to express a different gene in skin cells, which part of the melanin gene would you use? Why? If you wanted to produce melanin in yeast cells, what part of the melanin gene would you use? Why?

➔ USE IT

10. Explain why scientists used the beta casein regulatory sequence to express human antithrombin in goats' milk.

GENE EXPRESSION

Gene expression is the multistep process of converting the information of DNA into proteins.

HINT See Infographics 8.7–8.10.

➔ KNOW IT

11. For each structure or enzyme listed, indicate by N (nucleus) or C (cytoplasm) its active location in eukaryotic cells:
RNA polymerase _____
Ribosome _____
tRNA _____
mRNA _____

12. The sequence of a strand of DNA of a gene is AGATACGAAACA.

    a. Write the sequence of the complementary strand of DNA.

    b. Write the sequence of the mRNA that is complementary to the original DNA strand.

    c. Refer to the genetic code in Infographic 8.10 to translate this short stretch of RNA. How many amino acids does it encode? What are they? (Remember that translation always begins at a start codon.)

⊖ USE IT

13. A change in DNA sequence can affect gene expression and protein function. What would be the impact of each of the following changes? How, specifically, would each affect protein or mRNA structure, function, and levels?

    a. a change that prevents RNA polymerase from binding to a gene's regulatory region

    b. a change in the coding sequence that changes the amino acid sequence of the protein

    c. a change in the regulatory region that allows transcription to occur at much higher levels

    d. a combination of the changes in b and c.

14. The 18th codon in the coding sequence of a gene's mRNA is CCA.

    a. What amino acid is encoded by this codon?

    b. What amino acid would be encoded if the codon changed to CCG?

    c. What amino acid would be encoded if the codon changed to CUA?

SCIENCE AND ETHICS

15. Some people with diabetes would die without insulin because their bodies can no longer produce this protein. Historically, scientists purified insulin from the pancreas of pigs. Human insulin is now produced by inserting an artificial gene construct into bacteria. What are the ethical pros and cons of each type of insulin?

Sequence Sprint

➡ What You Will Be Learning

The Human Genome Project was a massive undertaking that continues to spur new technology and discoveries in scientific research and medicine.

Sequence Sprint

Venter and Collins race to decode the human genome

It started out as a bold fantasy: the entire sequence of human DNA spelled out for scientists to examine at will. Knowledge of the human genome would be an indispensable medical tool. Scientists could, for example, scan the genome to hunt for genes that confer susceptibility to disease, which might lead to better treatments. It could enable diagnostic tests that could help predict the risk of developing a genetically based disease. Fields other than medicine would benefit, too. Comparing the human genome to the genomes of other organisms, for example, might shed light on our own evolution. The possible benefits to science were endless.

But when an international group of scientists met in the early 1980s and first floated the idea of sequencing the human genome, they faced skepticism. Some scientists found the idea absurd, especially given its then-estimated $3 billion price tag. Others thought the potential benefits were illusory because the technology to sequence genes was rudimentary. Some simply deemed the task impossible.

Over the years, however, the idea gained both scientific and political support. In 1988, Congress funded both the National Institutes of Health (NIH) and the U.S. Department of Energy to explore the novel concept. By 1990, the collaborative effort to sequence the entire string of more than 3 billion As, Gs, Cs, and Ts that make up a human genome–the Human Genome Project (HGP)–was officially under way.

When the project was officially launched in 1990, the NIH appointed James Watson, the co-discoverer of the structure of DNA, to head and coordinate it. The ambitious and mammoth undertaking involved more than 20 institutions spread around the globe, in China, France, Germany, Japan, the United Kingdom, and other countries, as well as in the United States.

Initially, the researchers set about sequencing every nucleotide on every chromosome that makes up the entire genome. Automated sequencers then available could sequence only

> **Some scientists found the idea absurd, especially given its then-estimated $3 billion price tag.**

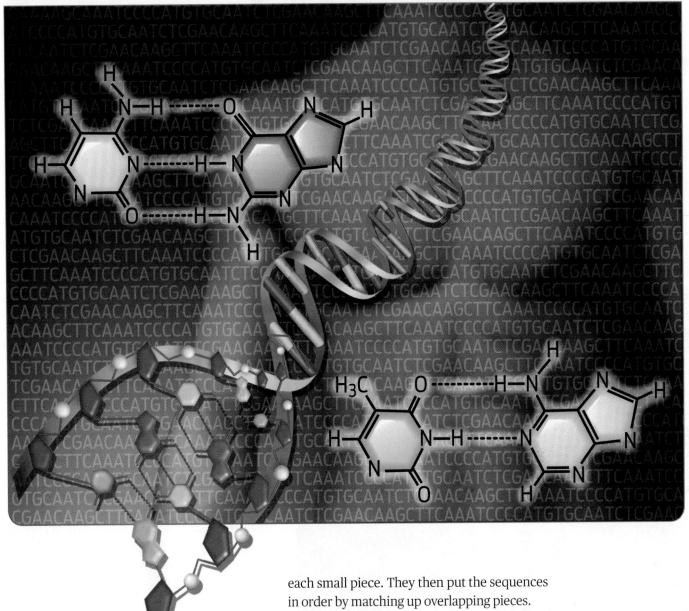

each small piece. They then put the sequences in order by matching up overlapping pieces.

To understand sequencing a bit better, imagine photocopying this sentence 10 times. Now imagine taking all the paper copies and tearing them at different points. Depending on where you tear, each copy will contain pieces with different parts of the same word. One copy might contain only the "ima" part of the word "imagine," while another might contain the whole word. You could use the piece containing the entire word to piece together two pieces with "ima" and "gine." Using this approach, scientists estimated it would take 15 years to finish the entire sequence of 3 billion bases.

about 500 nucleotides at a time, so researchers developed a way to manage large DNA sequences in batches, an approach that became known as "hierarchical shotgun sequencing." They first chemically broke human chromosomes into large segments and created a map indicating which piece belonged to which chromosome. They then cut those large segments into even smaller pieces and sequenced the nucleotides of

The map of sequences grew, and as sequences were assigned to specific chromosomal locations, it became much easier for scientists to find a home for gene sequences that had already been identified. Since the information was uploaded to an online database, any scientist could view the map to see where exactly his or her gene was located and apply that information to further research. But after a year, the project wasn't progressing as quickly as planned, and some estimated that if things kept plodding along at the same pace, the total cost could reach more than $100 billion.

Criticism began to mount over the costs and delays of the HGP. In 1991, Craig Venter, who had helped develop automated gene sequenc-

ing techniques at the NIH, publicly proposed an alternative plan. He suggested breaking the entire genome into small fragments and sequencing them simultaneously. Scientists including Venter had successfully used this approach on smaller, bacterial genomes. But the human genome was much larger and contained many repetitive sequences. Critics countered that computer software would not be able to piece all the sequences together accurately. Venter's approach–called whole genome shotgun sequencing–would be akin to trying to put a jigsaw puzzle together without a photo of the finished puzzle as a guide.

Nevertheless, this approach would be faster. Venter claimed that this method could find up

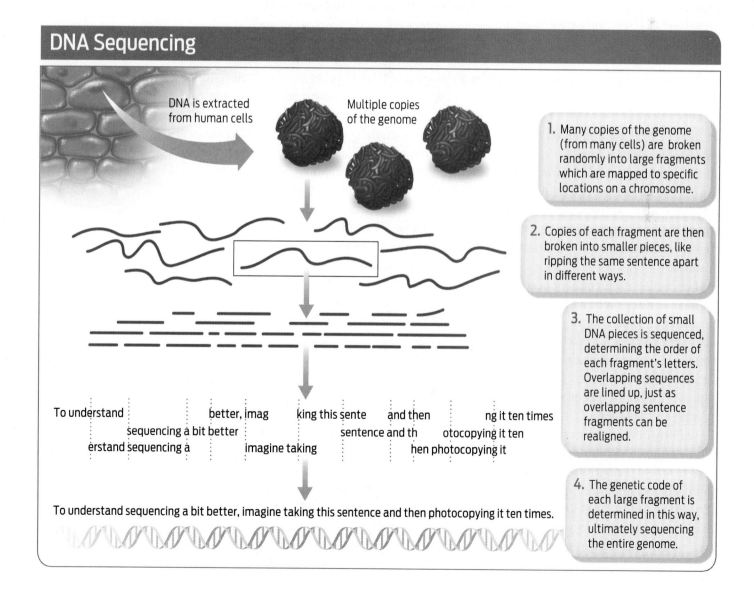

DNA Sequencing

DNA is extracted from human cells

Multiple copies of the genome

1. Many copies of the genome (from many cells) are broken randomly into large fragments which are mapped to specific locations on a chromosome.

2. Copies of each fragment are then broken into smaller pieces, like ripping the same sentence apart in different ways.

3. The collection of small DNA pieces is sequenced, determining the order of each fragment's letters. Overlapping sequences are lined up, just as overlapping sentence fragments can be realigned.

To understand   better, imag   king this sente   and then   ng it ten times
     sequencing a bit better   sentence and th   otocopying it ten
   erstand sequencing a   imagine taking   hen photocopying it

4. The genetic code of each large fragment is determined in this way, ultimately sequencing the entire genome.

To understand sequencing a bit better, imagine taking this sentence and then photocopying it ten times.

A visitor to the American Museum of Natural History in New York City in 2001 looks at a digital representation of the human genome.

to 90% of human genes within a few years. He further asserted that the approach would be a bargain compared to the cost of the HGP.

In 1992 Watson resigned as head of the HGP, and in early 1993 NIH appointed a new head, the geneticist Francis Collins. The idea of speeding up the sequencing didn't sit well with either Watson or Collins. In fact, Collins claimed publicly that Venter's idea wouldn't work. Both men argued that although it seemed feasible, it would create sequence data that might be riddled with mistakes.

Then, in 1998, Venter announced that he had made an arrangement with the Perkin-Elmer Corporation, which was about to unveil a new automated sequencing machine. Together they would create a new company, to be called Celera Genomics, that intended single-handedly to sequence the human genome in just three years for a mere $300 million—a fraction of the cost of the publicly funded consortium

Collins and other leaders of the public project were troubled. Congress might favor Celera's approach and stop funding the public project altogether. Collins was also concerned that Celera was going to try to patent their sequence data, which would have restricted public access to it.

The race was on. Collins and his colleagues stepped up the pace. Venter wasn't the only one who had access to new automated sequencing machines and powerful computers that could process large amounts of data. Publicly financed scientists, too, had access to these and other new tools. Such technological advances dramatically cut the amount of time it took to sequence each nucleotide, and cut costs, too.

About six months after Venter's announcement, Collins announced that the public consortium would complete sequencing the genome by 2003–two years ahead of schedule. The consortium also planned to produce a rough draft of the genome by 2001, which was about the same time that Venter planned to finish his draft. Collins justified his decision by stating that scientists were clamoring for the data even in rough form.

For a few years the contest between the privately funded Celera and the publicly funded HGP was bitter, each side criticizing the other's methods. The two sides eventually agreed to share the glory and appeared at a White House press briefing on June 26, 2000, together with U.S. President Bill Clinton and British Prime Minister Tony Blair to publicly announce that they had completed a rough draft of the human genome sequence. In February 2001 both groups published their drafts of the human genome simultaneously in the journals *Nature* and *Science*.

Just whose genome sequence was in fact published? Geneticists working on the publicly funded project had collected blood samples from anonymous donors. The ultimate sequence is thus a composite pieced together from the gene sequences of several individuals. Celera's sequence data comes from the DNA of Venter himself.

When the HGP was completed in 2003, the achievement was hailed as one of the greatest scientific accomplishments of the 20th century. Some even consider it the greatest achievement ever in biology. Not only did it reveal new characteristics of the human genome, it also shed light on how we differ from other organisms.

The human genome may encode more than 1 million proteins from fewer than 25,000 genes.

While the human genome was being sequenced, scientists had also finished sequencing the genomes of some other organisms. Scientists compared our genome to the genomes of other organisms, and what they found astounded them.

Before the sequencing was complete, scientists thought that what made humans such complex creatures was gene number–the more complex the creature, the more genes it should have. But the Human Genome Project showed that humans carry a mere 20,000 to 25,000 genes–about the same number as a lowly roundworm. This evidence suggested that gene number wasn't as important as how those genes produced proteins. Before the project, scientists didn't think noncoding regions of the genome were important. Now they know that noncoding regions actually regulate genes and consequently contribute to the complexity of higher organisms. Moreover, the number of genes says nothing about the number of proteins that are

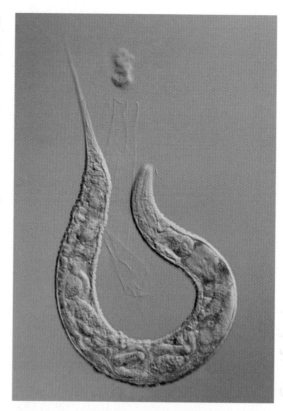

The roundworm *Caenorhabditis elegans* carries about the same number of genes as a human: a mere 20,000 to 25,000.

produced in an organism. A relatively small number of genes produce an enormous number of diverse proteins. For example, the human genome may encode more than 1 million proteins from fewer than 25,000 genes.

By mining the HGP data, scientists have identified genes that confer susceptibility to different types of cancers, heart disease, and neurological disorders such as Alzheimer disease. Without its road map of sequences, we might not today have diagnostic tests that can predict susceptibility to various cancers, or be able to develop cancer medicines that are tailored to a tumor's genotype. Scientists around the globe access the HGP data daily to help them link diseases, traits, and behaviors to specific genes, to study gene function, and to help them understand more about human evolution. The human genome sequence will remain a crucial scientific tool for years to come. ■

Paramedic Plants

Paramedic Plants

Will herbs be the next cancer therapy?

In the late nineties, online chat rooms began buzzing that an herbal supplement called PC-SPES could shrink prostate tumors. This was a "natural" approach that caused fewer side effects than conventional prostate cancer therapy, so it was claimed. PC-SPES was introduced to the U.S. market in 1996, and within a few years, as many as 10,000 men in the United States with prostate cancer were taking the supplement and seeing their tumors shrivel—or so the rumors went.

Although some studies had shown that PC-SPES could kill cancer cells, no one had ever studied the supplement in men with prostate cancer. Concerned that so many cancer patients were taking an unproven therapy, a group of scientists at the National Center for Alternative and Complementary Medicine decided to test the herbal mixture in clinical trials.

To their surprise, initial studies seemed to support the rumors. PC-SPES, which was marketed as a mixture of eight herbs known in China since ancient times, appeared to fight prostate tumors.

But now community chat rooms were buzzing again, this time with suspicions that the supplement was contaminated with toxic metals and perhaps even harmful drugs. These stories made their way to the California Department of Health Services, which decided to investigate. The department's analyses were shocking: in many instances, the quantity of each herb varied considerably from bottle to bottle. Moreover, some bottles were laced with three different prescription drugs, including an estrogen-like drug and a blood-thinner. The FDA—the U.S. Food and Drug Administration—issued a warning to all consumers to stop taking PC-SPES. By late 2002, the company that made PC-SPES had voluntarily taken the product off the market and went bankrupt soon after.

PC-SPES isn't the only herbal supplement that's gotten into trouble with the law. Authorities have found that many supplements are contaminated with dangerous heavy metals or bacteria. While contamination with prescription drugs or other substances appears to

be a rare phenomenon, investigators have found that the amount of active ingredient in many supplements commonly varies from pill to pill and from bottle to bottle. Some melatonin pills, for example, which some people take to ward off jet lag, have been found to contain very little melatonin, and the quantity can vary from pill to pill within the same bottle, according to a 2003 study by researchers at the University of Colorado, Denver.

This lack of consistency makes it difficult for consumers to know if they are getting what they are paying for. In the United States, supplements like echinacea, ginseng, and St. John's wort aren't stringently regulated. Although manufacturers are prohibited from making false statements on their labels, there is no government agency that *certifies* that a supplement actually contains what is listed on its label or does what it claims to do. More significantly, very few herbal supplements have been rigorously studied, and few have been shown to contain consistent levels of active compounds from batch to batch, or proved to prevent or treat illnesses. Authorities typically discover that a product has been falsely advertised or contaminated only after investigating complaints from consumers.

But some of this is changing. In 2004, the FDA put procedures in place that allow companies to apply to the agency to market an herbal supplement almost like a pharmaceutical drug. Not only must a company prove with clinical trial data that a supplement works, it must also prove that it can produce the supplement with consistent quality and quantity from batch to batch. While the new guidelines don't require sellers of supplements to apply to the FDA, they do offer manufacturers a financial incentive to cooperate: a company can sell an FDA-approved supplement by prescription for 5 years without competition from a similar product. In 2006, the FDA approved the first such herbal, a topical green tea extract used to treat genital warts, available by prescription only.

"The payoff will be huge," says Robert Tilton, vice president of science and technology at PhytoCeutica, Inc., a company based in New Haven, Connecticut, that is investigating the use of Chinese herbs in the treatment of cancer. There are so many diseases or afflictions for which available treatments are inadequate, Tilton points out. While researchers are investigating herbs to treat everything from alcohol addiction to heart disease, the next to be offered to patients by prescription may be herbs that help treat cancer. At a time when doctors are actively searching for new cancer therapies, herbal supplements may offer a way to improve existing cancer therapies and make them less toxic.

Cell Division and Cancer

The founders of PhytoCeutica sought to study herbs that might help improve the treatment of **cancer**–a disease of unregulated cell division, for which most existing therapies cause severe side effects. They decided on a mixture of four herbs developed in China more than 1,800 years ago to treat gastrointestinal distress. They dubbed the mixture PHY906.

The notion that an herb can have a medicinal benefit certainly is not new. As the origins of PHY906 attest, traditional cultures have relied on herbs to treat disease for thousands of years. Many modern medicines are also derived from plant sources. Salicylic acid, for example, the primary ingredient in aspirin, was initially extracted from the bark of the willow tree. And the cancer drug paclitaxel was originally extracted from the bark of the Pacific yew tree.

TABLE 9.1

Drugs from Plants

The well-established drugs listed below are among dozens that were developed after scientists began to analyze the chemical constituents of plants used by traditional peoples for medicinal or other purposes.

DRUG	MEDICAL USE	PLANT SOURCE	COMMON PLANT NAME
Aspirin	Reduces pain and inflammation	*Filipendula ulmana*	Meadowsweet
Codeine	Eases pain, suppresses coughing	*Papaver somnifenum*	Opium poppy
Ipecac	Induces vomiting	*Psychotria ipecacuanha*	Ipecacuanha
Pilocarpine	Reduces pressure in the eye	*Pilocarpus jaborandi*	Jaborandi
Pseudoephedrine	Reduces nasal congestion	*Ephedra sinica*	Ephedra, ma huang
Quinine	Combats malaria	*Cinchona pubescens*	Quinine tree
Reserpine	Lowers blood pressure	*Rauvolia serpentina*	Serpentine wood, snakeroot
Scopolamine	Eases motion sickness	*Datura stramonium*	Jimson weed
Theophyline	Opens bronchial passages	*Camellia sinensis*	Tea plant
Paclitaxel	Chemotherapeutic drug	*Taxus brevitolin*	Pacific yew tree

Source: Cox, PA et al. 1994. *Scientific American* 82–87.

Traditional cultures have relied on herbs to treat disease for thousands of years.

Modern drugs tend to contain a single active ingredient that has been highly purified and tested in clinical trials and shown to be safe and effective (**Table 9.1**).

But herbal supplements aren't nearly as well studied or developed as conventional prescription drugs. Herbal supplements are dried pieces of a plant or fungus that are typically ground up

CANCER
A disease of unregulated cell division: cells divide inappropriately and accumulate, in some instances forming a tumor.

How Conventional Drugs Differ from Herbal Supplements

Conventional Drugs

One specific active ingredient is purified from a plant or fungus or synthesized in the laboratory and concentrated into pill, capsule, or injectable form.

The bark of the Pacific yew, *Taxus brevifolia*

Paclitaxel

Paclitaxel, commercially known as Taxol, is a single ingredient originally purified from the bark and needles of the Pacific yew tree, *Taxus brevifolia*.

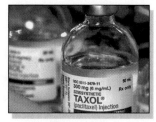

Mandatorily tested in clinical trials and approved by the FDA

Herbal Supplements

Plants, algae, fungi, and combinations of these are used as a tea, an extract, or ground into powder and sold in capsule form. They contain complex mixtures of many different unpurified plant molecules.

Baikal skullcap
Scutellaria baicalensis

PC-SPES Herbal Supplement contains extracts from specific parts of the following eight plants and fungus:

Baikal skullcap (Root)	*Scutellaria baicalensis*
Reishi (Stem)	*Ganoderma lucidum*
Rabdosia (Leaf)	*Rabdosia rubescens*
Dyer's woad (Leaf)	*Isatis indigotica*
Chrysanthemum (Flower)	*Dendrathema morifolium*
Saw palmetto (Berry)	*Serenoa repens*
San-Qi ginseng (Root)	*Panax notoginseng*
Licorice (Root)	*Glycyrrhiza uralensis*

PC-SPES contains baicalein extracted from the plant *Scutellaria baicalensis*. The supplement also contains dozens of unknown ingredients from seven other plants and several pharmaceutical compounds.

Not mandatorily tested in clinical trials or approved by the FDA

"Whether botanicals will provide a source of products to treat cancer is hard to say for sure."
–K. Simon Yeung

and packaged into pills or capsules, or brewed into tea. Botanical herbs used for supplements often contain a number of different compounds rather than one single active ingredient **(Infographic 9.1)**.

Unlike drugs, herbal supplements typically contain several compounds, and they can target diverse biochemical pathways. This is precisely why PhytoCeutica founders believed that PHY906 might prove helpful in treating cancer. Many cancer therapies, especially chemotherapy, cause vomiting and diarrhea as side effects. Because the herbal mixture in PHY906 works on different biochemical pathways, it seemed plausible that the herbs in PHY906 might not only lessen the side effects of chemotherapy but also have other, positive, effects. And indeed, early studies with cancer patients suggest that PHY906 can reduce vomiting and diarrhea and may even make chemotherapy more effective.

"Whether botanicals will provide a source of products to treat cancer is hard to say for sure," says K. Simon Yeung, a research pharmacist

Why Do Cells Divide?

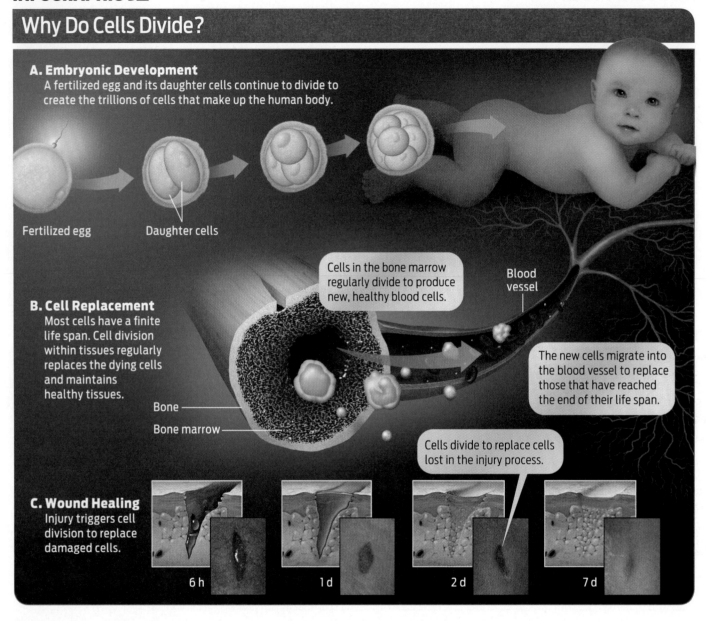

A. Embryonic Development
A fertilized egg and its daughter cells continue to divide to create the trillions of cells that make up the human body.

Fertilized egg

Daughter cells

Cells in the bone marrow regularly divide to produce new, healthy blood cells.

Blood vessel

B. Cell Replacement
Most cells have a finite life span. Cell division within tissues regularly replaces the dying cells and maintains healthy tissues.

Bone

Bone marrow

The new cells migrate into the blood vessel to replace those that have reached the end of their life span.

Cells divide to replace cells lost in the injury process.

C. Wound Healing
Injury triggers cell division to replace damaged cells.

6 h

1 d

2 d

7 d

and certified herbalist at Memorial Sloan-Kettering Cancer Center in New York City. "But it definitely holds promise in the future." In fact, researchers at Sloan-Kettering are also studying different Chinese herbs to treat cancer. In particular, they are investigating whether herbals combined with other drugs, such as those used in chemotherapy or other types of immune boosting treatments, may make those treatments more effective.

> **Although we may think of our bodies as relatively fixed structures, most of our tissues are in a state of constant flux.**

Generally speaking, **chemotherapy** refers to treatment of disease by the use of chemicals. Most commonly, the term refers to drugs used to treat cancer. There are different classes of chemotherapeutic drugs, most of which attack cancer by interfering with a fundamental part of a cell's life: **cell division.**

Although we may think of our bodies as relatively fixed structures, most of our tissues are in a state of constant flux as cells divide periodically to replace cells that have

CHEMOTHERAPY
The treatment of disease, specifically cancer, by the use of chemicals.

CELL DIVISION
The process by which a cell reproduces itself; cell division is important for normal growth, development, and repair of an organism.

reached the end of their life span. In fact, cell division in our bodies begins long before we are even born. During embryonic development, a single fertilized egg cell divides, and its daughter cells divide again and again, eventually forming trillions of cells by the time a baby is born. As we age, our tissues continually discard old cells and generate new ones in their place. And when we cut or injure ourselves, cells in the area divide to heal the wound **(Infographic 9.2)**.

To produce new cells, each cell passes through a series of stages collectively known as the **cell cycle.** During the cell cycle, one cell becomes two. A cell doesn't simply split in half to form two new cells, however. If it did, each resulting cell would be smaller than the original, and with each division, each cell would lose half its contents. So before a cell divides, it first makes a copy of its contents so that each new cell has the same amount of organelles, DNA, and cytoplasm as the original cell. This preparatory stage of the cell cycle, known as **interphase**, is divided into separate subphases: G_1 phase, when the cell grows and prepares to divide both its DNA and its organelles; synthesis phase (S), when DNA is replicated; and G_2 phase, when the cell is ready to divide.

INFOGRAPHIC 9.3

The Cell Cycle: How Cells Reproduce

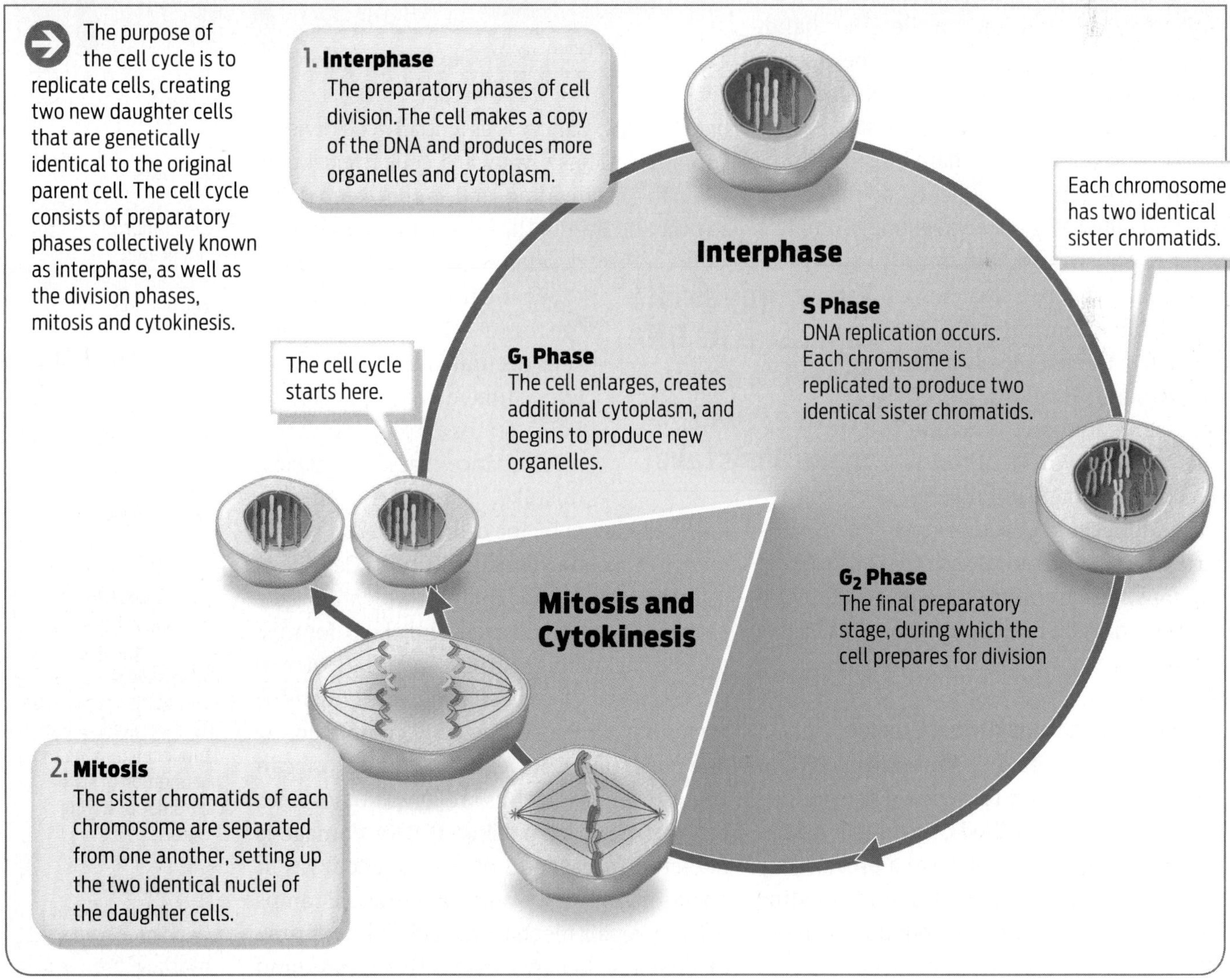

The purpose of the cell cycle is to replicate cells, creating two new daughter cells that are genetically identical to the original parent cell. The cell cycle consists of preparatory phases collectively known as interphase, as well as the division phases, mitosis and cytokinesis.

1. Interphase
The preparatory phases of cell division. The cell makes a copy of the DNA and produces more organelles and cytoplasm.

Each chromosome has two identical sister chromatids.

Interphase

S Phase
DNA replication occurs. Each chromsome is replicated to produce two identical sister chromatids.

G_1 Phase
The cell enlarges, creates additional cytoplasm, and begins to produce new organelles.

The cell cycle starts here.

G_2 Phase
The final preparatory stage, during which the cell prepares for division

Mitosis and Cytokinesis

2. Mitosis
The sister chromatids of each chromosome are separated from one another, setting up the two identical nuclei of the daughter cells.

Once the cell duplicates its contents, it enters the division phases of the cell cycle: mitosis and cytokinesis. During **mitosis,** the chromosomes line up along the midline of the cell. The two **sister chromatids** of each replicated chromosome are connected at a region of the chromosome known as the **centromere.** The sister chromatids then pull apart from each other. Each chromatid will form one of two genetically identical chromosomes. During **cytokinesis,** the cytoplasm divides into two separate cells, each containing a full complement of organelles and DNA. In this way, one parent cell divides into two new daughter cells, each of which is identical to the original parent cell (Infographic 9.3).

Cell division is not like making a photocopy. When you photocopy a photo, you start with the photo and end with the original photo and a copy of it. Cell division, on the other hand, begins with a parent cell that undergoes a series of steps and then splits into two *new* daughter cells. The parent cell no longer exists.

The cell cycle has two main purposes: producing sufficient "ingredients" to make two new daughter cells, and segregating a complete copy of those ingredients to each daughter cell to create two genetically identical cells. In particular, the cell's 46 chromosomes must be carefully copied and segregated into daughter cells (Infographic 9.4). Mitosis—the cell's mechanism for separating duplicated chromosomes—occurs in a series of coordinated phases that ensure that one copy of each chromosome will make it into each new daughter cell (see **Up Close: The Phases of Mitosis**).

When Division Runs Amok: Cancer

Normal cells divide only periodically. When cells no longer need to divide—for example, when a wound has healed or worn-out tissues have been replaced—cells take a break. They pause in their life cycle and stop dividing (although they still carry out other normal cellular functions). By contrast, cancer cells divide

INFOGRAPHIC 9.4

Cell Division: The Chromosome Perspective

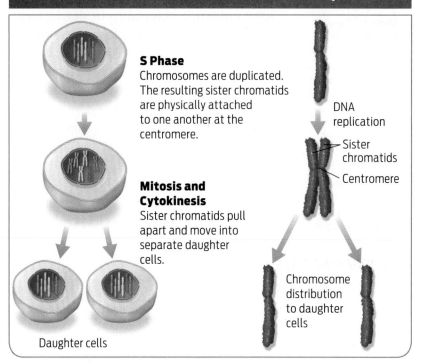

S Phase
Chromosomes are duplicated. The resulting sister chromatids are physically attached to one another at the centromere.

Mitosis and Cytokinesis
Sister chromatids pull apart and move into separate daughter cells.

Daughter cells

DNA replication

Sister chromatids

Centromere

Chromosome distribution to daughter cells

haphazardly. Cancer is essentially cell division run amok. These wayward cells don't know when to stop dividing, and so they keep doing so over and over until, eventually, a tumor may develop.

What causes certain cells to "go rogue"? Cancer results when cells accumulate DNA damage. Every time a cell replicates its DNA, for example, there is a small chance that it will make a mistake. Normally, such mistakes are caught by the cell and fixed at what's known as a **cell cycle checkpoint.** Cells have a series of such checkpoints, which monitor each stage of the cell cycle and check for mistakes. Checkpoints also prevent progression of the cell through the cell cycle until previous stages have been successfully completed. At one checkpoint, for example, proteins scan DNA for damage such as broken chromosomes or incorrect base pairing. If DNA damage is detected, a cascade of events occurs that results in one of two outcomes: the cell ramps up DNA repair mechanisms, giving itself time to repair the damage. Or, in cases of severe and

> **Every time a cell replicates its DNA, there is a small chance it will make a mistake.**

MITOSIS
The segregation and separation of duplicated chromosomes during cell division.

SISTER CHROMATID
One of the two identical DNA molecules that make up a duplicated chromosome following DNA replication.

CENTROMERE
The specialized region of a chromosome where the sister chromatids are joined. This site is critical for proper alignment and separation of sister chromatids during mitosis.

CYTOKINESIS
The physical division of a cell into two daughter cells.

The Phases of Mitosis

 Mitosis, the process of separating duplicated chromosomes, occurs in a series of phases that are part of the cell cycle. A dividing cell passes through each of these phases in sequence. The cellular machinery that actually pulls chromosomes apart is known as the mitotic spindle.

Dividing animal cell

Interphase
· Each chromosome replicates in interphase, resulting in two sister chromatids connected at the centromere.
· Chromosomes are loosely gathered in the nucleus.

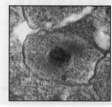

Prophase
· Replicated chromosomes begin to coil up.
· The nuclear membrane begins to disassemble.
· Protein fibers of the mitotic spindle begin to form.

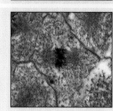

Prometaphase
· Chromosomes condense (shorten) so they are easier to separate.
· Spindle fibers attach to chromosomes on both sides at the centromere region.

Spindle fibers

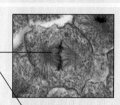

Metaphase
· Spindle fibers from opposite ends of the cell pull on chromosomes.
· Chromosomes are aligned along the middle of the cell.

Anaphase
· Spindle fibers shorten and pull sister chromatids to opposite ends of the cell.

Taxol interferes with the action of spindle fibers during anaphase.

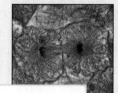

Telophase
· An identical set of chromosomes reaches each pole.
· Spindle fibers dissemble.
· Nuclear membrane forms around each set of chromosomes, forming the daughter cell nuclei.

Cytokinesis
· Cell membrane pinches in to completely surround each new daughter cell.

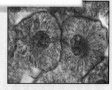

Interphase
· Two identical daughter cells are formed, each with the same number of chromosomes as the parent cell.

Cell Division Is Tightly Regulated

 Normal cells have mechanisms to ensure that cell division is carried out precisely and only when necessary. Regulated cell division ensures that adequate cell number and healthy tissue structure are maintained in the body.

Cell Cycle Checkpoints
During the cell cycle, a system of checkpoints regulates a cell's progress. Checkpoints prevent a cell from progressing to the next stage until it accurately finishes the current stage.

Apoptosis
When a normal cell sustains irreparable damage, it undergoes programmed cell death. This cellular suicide prevents cells from producing more damaged daughter cells.

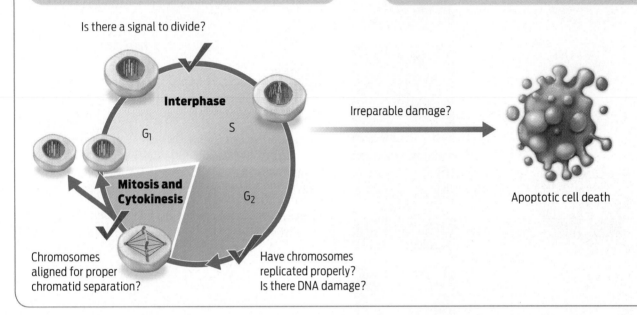

Is there a signal to divide?

Interphase

G₁   S

Mitosis and Cytokinesis

G₂

Irreparable damage?

Apoptotic cell death

Chromosomes aligned for proper chromatid separation?

Have chromosomes replicated properly? Is there DNA damage?

---

irreparable damage, the checkpoints direct a cell to commit suicide in a process called **apoptosis.** Apoptosis is programmed cell death, a biochemical pathway in which the cell's DNA is degraded into small pieces and the cell breaks apart and dies. Other cells in the area then engulf these remnants. Checkpoint mechanisms ensure that cells divide accurately and only when necessary **(Infographic 9.5).**

Even with these repair mechanisms, however, cells with DNA damage do occasionally manage to complete the cell cycle and divide because some of the damage includes injury to the very proteins that function as checkpoints. When cells accumulate enough DNA damage, the result is cancer. Cancer cells plow through the cell cycle uninhibited. Because their checkpoint functions are damaged, they have no stop signals. With nothing to tell them to stop, these damaged cells survive and can divide again and again, uncontrollably. And with every

round of cell division, cellular and chromosomal defects go unrepaired while additional defects occur, which causes DNA damage to accumulate **(Infographic 9.6).**

Fighting Cancer

For many types of cancer, the first line of treatment is often surgery to remove the lump of rogue cells completely. Surgery is effective for certain solid tumors, but not for blood cancers, or cancers that have undergone **metastasis,** that is, that have spread to other parts of the body. In these cases, the best option is usually chemotherapy, treatment to target cancer cells circulating throughout the body.

Most drugs used in chemotherapy work by interfering with the cell cycle. Several plant-derived drugs do just this. Paclitaxel, a drug originally extracted from the needles of Pacific yew trees, for example, interferes with a cell's ability to separate sister chromatids during mito-

CELL CYCLE CHECKPOINT
A cellular mechanism that ensures that each stage of the cell cycle is completed accurately.

APOPTOSIS
Programmed cell death; often referred to as cellular suicide.

METASTASIS
The spread of cancer cells from one location in the body to another.

Cancer: When Checkpoints Fail

➔ Cancer cells have damaged checkpoint mechanisms, which enable them to divide when they should not. This means that DNA damage or errors in chromosome separation are passed on to daughter cells. These damaged cells also bypass apoptosis. With each cell division, the damage is perpetuated and additional errors in DNA accumulate.

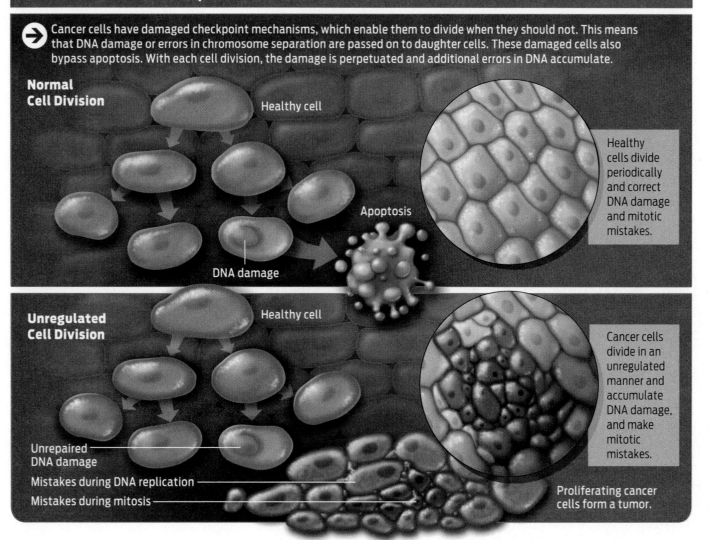

Normal Cell Division

Healthy cell

Apoptosis

DNA damage

Healthy cells divide periodically and correct DNA damage and mitotic mistakes.

Unregulated Cell Division

Healthy cell

Unrepaired DNA damage

Mistakes during DNA replication

Mistakes during mitosis

Cancer cells divide in an unregulated manner and accumulate DNA damage, and make mitotic mistakes.

Proliferating cancer cells form a tumor.

sis. Unable to properly segregate chromosomes, the cells fail to divide. Vinblastine, another plant-derived anticancer drug–it comes from the Madagascar periwinkle plant–also interferes with chromosome separation during mitosis. Other cancer drugs interrupt other parts of the cell cycle. The common chemotherapeutic drug irinotecan, for example, inhibits an enzyme that helps DNA replicate during S phase. Cells treated with irinotecan are unable to copy their DNA and are thus prevented from reproducing themselves. Other types of anticancer drugs work by inhibiting progression through the cell cycle in other ways.

If the cancer has not yet spread throughout the body, doctors may treat a tumor with **radia-tion therapy** (often in conjunction with surgery and chemotherapy). In radiation therapy, beams of high-energy electrons kill dividing cells. Such radiation severely damages molecules and causes rampant DNA damage. This DNA damage can trigger apoptosis, causing the cells to die. Both chemotherapy and radiation therapy interfere with or kill all dividing cells. While most normal cells are able to repair the damage inflicted by these therapies, cancer cells typically have hobbled repair mechanisms and can't effectively do so, and consequently die.

Although cancer remains a leading cause of death in Western countries, researchers have developed an arsenal of chemotherapeutic

RADIATION THERAPY
The use of ionizing (high-energy) radiation to treat cancer.

Conventional Cancer Therapy

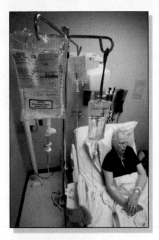

Chemotherapy

Mode of Action:
Chemicals interfere with different parts of cell division. For example, Taxol interferes with the separation of sister chromatids during mitosis; irinotecan interferes with DNA replication.

Side Effects:
Because chemotherapy targets all dividing cells, including healthy ones, the treatment causes side effects. Patients may experience nausea, vomiting, diarrhea, hair loss and a high risk of infection when the treatment interferes with cell division in the digestive tract, hair follicles, and bone marrow.

Radiation

Mode of Action:
High-energy radiation severely damages DNA. Excessive DNA damage will cause cells to die, either by apoptosis or by interrupting DNA replication.

Side Effects:
Because radiation targets all dividing cells in the irradiated area, it causes side effects. If cells in the digestive tract, hair follicles, or bone marrow are irradiated, patients may experience nausea, vomiting, diarrhea, hair loss, and susceptibility to infection, as the radiation interferes with cell division in these locations.

drugs to aid in the fight against this deadly disease, and cancer patients have been surviving longer and longer. Cancer death rates have been slowly declining over the years; the latest research shows that deaths from all cancers dropped 15.8% between 1991 and 2006. Part of the decline in deaths is due to more effective chemotherapeutic drugs.

The downside of both radiation and chemotherapy is that they can cause severe side effects. That's because neither therapy is very specific—both radiation and chemotherapy damage all rapidly dividing cells, including healthy ones. These cancer treatments kill healthy cells lining the intestinal tract, the cells in hair follicles, and cells in bone marrow (which divide rapidly throughout our lives to replace worn-out blood cells), often leading to side effects such as hair loss, vomiting, bruising, and susceptibility to infections. Since healthy cells can repair DNA damage, they aren't as severely affected as cancer cells, which have dysfunctional checkpoints and hobbled repair mechanisms. While scientists are trying to develop cancer therapies that target only cancerous cells, chemotherapy and radiation remain the mainstays of cancer therapy today. Anything that might lessen side effects without hindering a drug's effectiveness would be a boon to patients **(Infographic 9.7)**.

The herbal supplement PHY906 is meant to be taken in addition to chemotherapy. So far, it has been tested in people with colon, liver, and pancreatic cancer who are also being treated

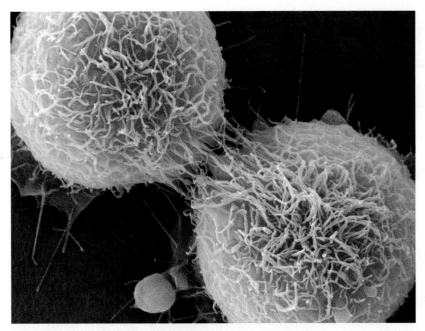
Cervical cancer cells dividing unchecked.

with the chemotherapy drug irinotecan. These initial studies suggest that PHY906 can reduce the side effects of chemotherapy.

Although researchers aren't sure exactly how PHY906 works, they do have some clues. Research in animals suggests that the herb mixture reduces inflammation in the gut. Because chemotherapy kills cells lining the stomach and intestines, gut tissue becomes inflamed. The ability of PHY906 to calm inflammation might account for its ability to reduce nausea, vomiting, and diarrhea during chemotherapy. And while PHY906 does not by itself kill cancer cells, animal studies by the company suggest that the supplement enhances the effect of chemotherapy by making cancer cells more permeable to drugs. **(Infographic 9.8).** That PHY906 can reduce side

INFOGRAPHIC 9.8

Herbal Supplements May Complement Cancer Therapy

Experiments show that the herbal supplement PHY906 can improve traditional cancer therapy. While PHY906 doesn't appear to kill cancer cells when used alone, it enhanced the ability of a chemotherapeutic agent called irinotecan to shrink colorectal tumors in mice. In other words, the drug and herbal supplement worked synergistically. Since PHY906 does not directly kill cancer, it will likely only complement existing cancer therapy regimens.

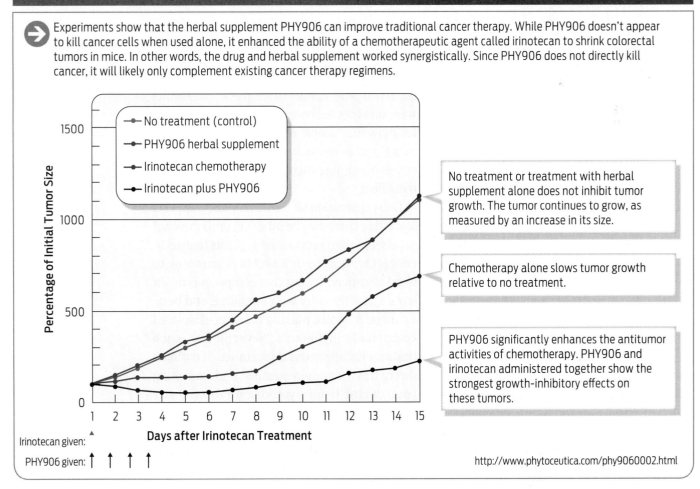

Legend:
- No treatment (control)
- PHY906 herbal supplement
- Irinotecan chemotherapy
- Irinotecan plus PHY906

y-axis: Percentage of Initial Tumor Size (0, 500, 1000, 1500)
x-axis: Days after Irinotecan Treatment (1–15)

Irinotecan given: ▲
PHY906 given: ↑ ↑ ↑ ↑

No treatment or treatment with herbal supplement alone does not inhibit tumor growth. The tumor continues to grow, as measured by an increase in its size.

Chemotherapy alone slows tumor growth relative to no treatment.

PHY906 significantly enhances the antitumor activities of chemotherapy. PHY906 and irinotecan administered together show the strongest growth-inhibitory effects on these tumors.

http://www.phytoceutica.com/phy9060002.html

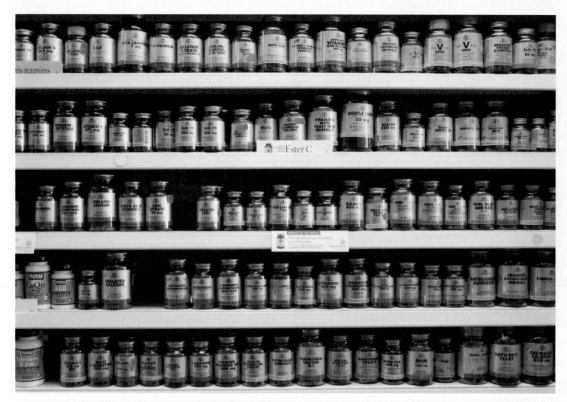

Thanks to recent FDA incentives, herbal supplements may soon join the ranks of some of our most valued prescription drugs.

effects from chemotherapy also raises the possibility that higher doses of chemotherapy could be administered without a corresponding increase in side effects.

PHY906 isn't the only herbal supplement that is showing promise in cancer treatment: some common plant ingredients may also aid the fight against cancer. For instance, scientists at the University of California, Los Angeles, have found in laboratory experiments that green tea extracts, which have been popularly thought to be cancer-fighting agents, can in fact slow down aberrant cell division and increase the likelihood that a damaged cell will go through apoptosis. Studies on the spice turmeric, commonly used in Asian cooking, suggest that it can help fight cancer: in animals, turmeric has been shown to protect the liver, inhibit tumors, and reduce inflammation. A few studies have also shown that the Chinese herb astragalus, combined with another herb extracted from the berries of the glossy privet tree native to Japan and Korea, can boost the immune system and help it fight cancer.

But the news on supplements isn't all good. One of the largest studies to date, which followed almost 30,000 smokers for 8 years, found that beta-carotene supplements actually increased the risk of lung cancer in smokers. The same study found that vitamin E supplements appeared to have no effect in reducing the risk of lung cancer in smokers. The results were unexpected because epidemiologic studies show that people who eat diets rich in vitamin E and beta-carotene have a lower risk of developing lung cancer, suggesting a preventative effect.

One explanation for the discrepancy, experts say, is that there may be other yet undiscovered cancer-fighting compounds in plants that act in concert with vitamin E and beta-carotene to ward off cancer. These other compounds might work synergistically with vitamin E and beta-carotene in foods, making extracts of isolated compounds ineffective. Nevertheless, such findings highlight the importance of eating a diet rich in fruits and vegetables and consulting your doctor before you decide to take any supplement.

While it's too soon to be certain that any herbal supplement will join the ranks of our most valuable prescription drugs, most

In the next few years, herbs might join conventional drugs as therapeutics.

experts are optimistic. After all, many of our most important drugs started their pharmaceutical life as plant extracts, so it's reasonable to think that some of today's herbals will eventually prove useful, too. In the next few years, herbs might join conventional drugs as therapeutics, according to Mary Hardy, associate director of UCLA's Center for Dietary Supplements Research. And with the FDA's new initiative that offers the agency's approval to herbal supplements that have been shown to be effective through clinical trials, we can expect to see greater integration of these supplements into medical treatments in the United States. A similar procedure already is in place in Europe: Germany and the Netherlands require clinical evidence of an herb's efficacy stated on its label, and herbal drugs are prescribed by doctors in those countries.

Some herbs have valid medicinal uses, says Hardy, and studying them will help doctors better advise their patients. "Eventually," she says, "herbs will be integrated into the broader medical paradigm." ■

▶ Summary

■ Cell division is a fundamental feature of life, necessary for normal growth, development, and repair of the body.

■ The cell cycle is the sequence of steps that a cell undergoes in division. Stages of the cell cycle include interphase, mitosis, and cytokinesis.

■ During mitosis, replicated chromosomes segregate to opposite poles of the dividing cell; during cytokinesis, the cell physically divides into two daughter cells.

■ Cell cycle checkpoints ensure accurate progression through the cell cycle; repair mechanisms at each checkpoint can fix mistakes that occur, such as DNA damage.

■ In the absence of proper checkpoint function, cells can acquire DNA damage during cell division and pass these DNA defects on to daughter cells.

■ Mistakes in the course of cell division can lead to cancer, which is unregulated cell division.

■ Cancer cells have lost the ability to regulate cell division and reproduce uncontrollably, often eventually forming a tumor.

■ Conventional cancer treatments—chemotherapy and radiation—kill all rapidly dividing cells, both cancer cells and healthy cells.

■ Many drugs, including some of those used to treat cancer, are extracted from plants.

■ Herbal supplements may enhance conventional cancer treatments.

BASICS OF CELL DIVISION AND THE CELL CYCLE

Dividing cells progress through a series of stages known as the cell cycle. Checkpoints monitor passage through the cell cycle.

HINT See Infographics 9.2–9.5.

❯ KNOW IT

1. Following mitosis and cytokinesis, daughter cells are
   a. genetically unique.
   b. genetically identical to each other.
   c. genetically identical to the parent cell.
   d. contain half of the parent cell's chromosomes.
   e. b and c

2. During the cell cycle, DNA is replicated during
   a. mitosis.
   b. G_1.
   c. S.
   d. G_2.
   e. in cytokinesis

3. Explain how embryonic development, wound healing, and replacement of blood cells are related.

❯ USE IT

4. If a cell fails to replicate its DNA completely, what will happen?
   a. It will progress through G_2 and mitosis.
   b. It will die by apoptosis.
   c. It will pause to allow DNA replication to complete.
   d. It will stop in S phase and never progress further through the cell cycle.
   e. It will stay in interphase indefinitely.

5. Many drugs interfere with cell division. Why shouldn't pregnant women take these drugs?

CANCER AND CANCER THERAPIES

When cells fail to accurately progress through the cell cycle, cancer may arise. Cancer cells may have lost checkpoint function, or may divide even without a signal to do so.

HINT See Infographics 9.1 and 9.6–9.8.

❯ KNOW IT

6. A normal cell that sustains irreparable amounts of DNA damage will most likely
   a. divide out of control.
   b. die by apoptosis.
   c. arrest in G_2.
   d. immediately go back to S phase.
   e. stop in S phase and never progress through the cell cycle.

7. Explain why traditional chemotherapy can cause nausea, diarrhea, and hair loss.

8. Which type of cancer treatment relies on purified chemicals to kill rapidly dividing cells?
   a. chemotherapy
   b. radiation therapy
   c. herbal supplement therapy
   d. a combination of chemotherapy and herbal supplement therapy
   e. none of the above

❯ USE IT

9. After a bad sunburn, skin usually peels. What process best describes what happens to the burned skin cells?
   a. skin cancer
   b. metastasis
   c. apoptosis
   d. checkpoint failure
   e. cytokinesis

10. Liver cells and neurons rarely, if ever, divide in normal circumstances. The cells lining the digestive tract are replaced by cell division on a regular basis. Explain why chemotherapy frequently causes digestive symptoms but less frequently causes cognitive symptoms.

11. Your pet mouse has developed colon cancer. Which of the following treatments will likely be most effective?
   a. PHY906
   b. irinotecan chemotherapy
   c. PHY906 plus irinotecan
   d. radiation therapy
   e. There is no treatment for colon cancer.

12. Look at Infographic 9.8.
   a. Does irinotecan actually shrink tumors in the colon? Explain your answer.
   b. Does PHY906 plus irinotecan shrink colon tumors? Explain your answer.

13. Why might a beta-carotene supplement not have the same effect on cancer as a diet with lots of food rich in beta-carotene?

A CLOSER LOOK AT MITOSIS AND THE CELL CYCLE

Mitosis is a critical stage of cell division. It ensures that chromosomes accurately separate into daughter cells.

HINT See Infographic 9.4 and Up Close: The Phases of Mitosis.

⊙ KNOW IT

14. During which stage of the cell cycle do sister chromatids separate from each other?

15. During which stage of the cell cycle are sister chromatids initially produced?

⊙ USE IT

16. What would be the result if a cell completed interphase and mitosis but failed to complete cytokinesis? (That is, how many cells would there be, and how many chromosomes relative to the parent cell would those cells have?)

17. Looking at **Up Close: The Phases of Mitosis**, would you say that a drug that stabilizes spindle fibers, preventing them from shortening, would be a valuable chemotherapy drug? Why or why not?

SCIENCE AND ETHICS

18. What are some of the risks of taking an over-the-counter herbal supplement as an alternative to conventional cancer therapy?

19. PHY906 has been tested in mice. What steps would you take to establish its efficacy in humans ethically and safely?

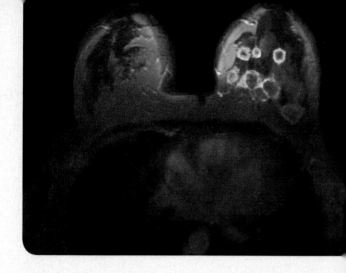

Fighting Fate

Fighting Fate

Some are genetically predisposed to cancer—but surgery may cut their risk

Lorene Ahern wasn't totally surprised when she tested positive for breast cancer. "Half of me was expecting it all my life and part of me was saying, 'No, this won't happen to me,'" says the 47-year-old mother of two in Twinsberg, OH. She knew that her risk of cancer might be higher than average— her mother had died of cancer at 49. But until the day she learned the test result, Ahern, who took good care of herself and lived a healthy lifestyle, had never fully believed she would develop cancer.

There was more bad news in store for Ahern. About a year after she received the diagnosis of breast cancer, Ahern had DNA extracted from her blood and tested for **mutations** in two genes— *BRCA1,* located on chromosome 17, and *BRCA2,* located on chromosome 13 ("BRCA" stands for "breast cancer susceptibility"). Women who are born with mutations in either of these two genes have an exceptionally high risk of developing breast and ovarian cancers. Men with these alleles are at higher risk for breast and prostate cancers. This test, too, was positive: Ahern had a mutation in one of her copies of the *BRCA1* gene, which meant that she was at high risk for other cancers as well. Moreover, she could have passed on this mutation to her children.

> ### "Half of me was expecting it all my life . . . ."
>
> —Lorene Ahern

Aside from nonmelanoma skin cancer, breast cancer is the most common cancer to affect women. Breast cancer affects nearly 200,000 women in the United States a year, according to a 2006 report by the U.S. Centers for Disease Control and Prevention. For most women, the risk of developing breast cancer ranges between 12% and 15%, or 1 in every 7 women. For women with mutations in *BRCA1* or *BRCA2,* however, the risk is much higher: a 40% to 80% lifetime

MUTATION
A change in the nucleotide sequence of DNA.

risk of developing breast cancer and a 20% to 50% risk of developing ovarian cancer, depending on the particular *BRCA* alleles they carry.

The good news is that studies have shown diet and lifestyle changes can dramatically cut a woman's risk of getting cancer–just quitting smoking cuts the risk by 30%. The bad news is that prevention is not that simple for women with inherited predispositions to breast cancer– for this group, diet and lifestyle changes don't necessarily make a difference. "Their cancers just behave differently," says Thomas Sellers, executive vice president of the H. Lee Moffitt Cancer Center and Research Institute in Tampa, Florida. Even with traditional treatment like chemotherapy and radiation, hereditary breast cancers are more likely to recur in the same tissue or other tissue in the body. But these women do have treatment options that can drastically reduce their risk of getting cancer or of having it recur.

Inherited Mutations

What is hereditary cancer? And how does it differ from other forms of cancer?

A woman who has hereditary breast cancer has a genetic predisposition to the disease. This predisposition is caused by a mutation in a gene she inherited from one or both of her parents. In Ahern's case, an inherited mutation in one of the copies of her *BRCA1* gene predisposed her to cancer. The mutation causes the *BRCA1* gene to make a dysfunctional BRCA1 protein, which in its normal form helps to regulate the cell cycle so that a cell can repair DNA damage (Chapter 9).

Recall that alleles are alternate versions of the same gene. In fact, an allele and a mutated gene are essentially the same thing; scientists often use the terms interchangeably. Ahern's mutation is just one of more than 600 mutations found in the *BRCA1* gene. Another way of saying this is that there are more than 600 alleles of the *BRCA1* gene in the population. Each allele has a different nucleotide sequence. These different alleles arose because mutations caused nucleotides to be substituted, deleted, or inserted within the gene (**Infographic 10.1**).

To understand how there can be so many versions of a single gene, remember that every time a cell replicates its DNA, mistakes can occur. About once every 10,000 to 100,000 times, DNA polymerase, the enzyme responsible for adding nucleotides to a new strand of DNA, will add the wrong nucleotide, choosing for instance a guanine (G) instead of a thymine (T) to pair with an adenine (A), or add too many or too few bases in a specific location.

Most of the time, mistakes in the nucleotide order are repaired by a cell's error correction machinery. Groups of enzymes "proofread" the nucleotides added to the new DNA strand. If a newly added nucleotide is not complementary to its partner on the template strand—a G incorrectly paired with an A, for example—these enzymes replace that nucleotide with the correct one. DNA damage is usually repaired at cell cycle checkpoints, points in the cycle at which the cell is monitored for such mistakes, and where if necessary the cell's progress through the cycle is delayed until DNA repair has taken place. If the damage is irreparable, the cell self-destructs. On average, fewer than one mistake in a billion nucleotides makes it through this system of checkpoints.

But think how often a cell replicates its DNA, and you can see how over time a rate of one in a billion can produce quite a number of mistakes. Such "mistakes" are called mutations. But not all mutations are passed on to offspring. Mutations are only inherited by offspring if they occur in sperm or egg cells during meiosis. This is the only way they can be

passed to the next generation. A new allele of any gene can also arise if a gene in an embryo in its very early stages develops a mutation. This mutation will be present in every cell of the adult individual and will be passed on to future generations. So each uncorrected mutation that occurs in a gene of an early embryo or of a sperm or egg cell can produce a new allele of that gene (**Infographic 10.2**).

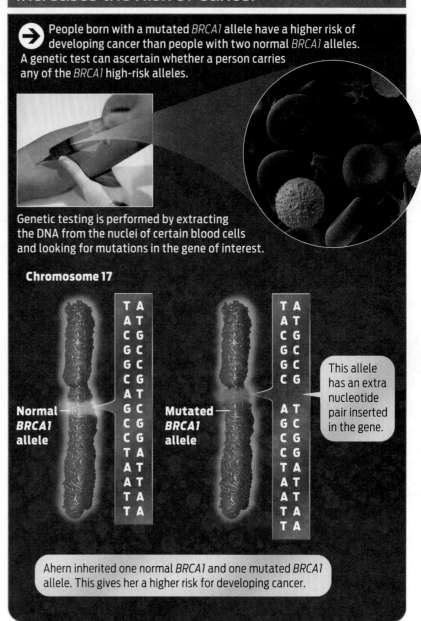

Inheriting One Mutated *BRCA1* Allele Increases the Risk of Cancer

People born with a mutated *BRCA1* allele have a higher risk of developing cancer than people with two normal *BRCA1* alleles. A genetic test can ascertain whether a person carries any of the *BRCA1* high-risk alleles.

Genetic testing is performed by extracting the DNA from the nuclei of certain blood cells and looking for mutations in the gene of interest.

Chromosome 17

Normal *BRCA1* allele

Mutated *BRCA1* allele

This allele has an extra nucleotide pair inserted in the gene.

Ahern inherited one normal *BRCA1* and one mutated *BRCA1* allele. This gives her a higher risk for developing cancer.

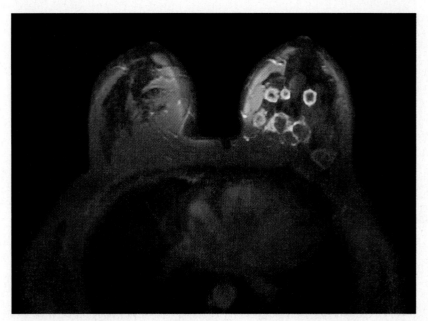

An MRI showing several malignant tumors in the breast of a 32-year-old woman.

their effects; they may change a nucleotide or an amino acid here and there, but don't seriously affect a person's health. And some mutations are actually beneficial: a mutation that enables the blood to carry more oxygen, for example, might be an advantage to someone who lives in high altitudes.

Regardless of whether a mutation is harmful or not, a nucleotide change is significant if it alters the amino acid sequence of the corresponding protein. Because an altered amino acid sequence can also alter the shape of the protein, it could disable the protein and make it unable to perform its usual job. BRCA proteins produced from mutated alleles, for example, do not perform their job as cell cycle regulators and make cells more likely to divide uncontrollably and become cancerous (Infographic 10.3).

Some of these mutations are so detrimental that they aren't compatible with life, and the embryo may spontaneously abort. Others aren't severe enough to harm a fetus or prevent birth, but they impair health after birth—as with diseases like cystic fibrosis and Huntington Disease. Some mutations are neutral in

Inherited mutations are present in all the cells of the body. Moreover, such mutations are faithfully copied every time body cells divide. Hereditary mutations are also called germ-line mutations because the gene changes are in

INFOGRAPHIC 10.2

Mistakes in DNA Replication Can Produce Mutations

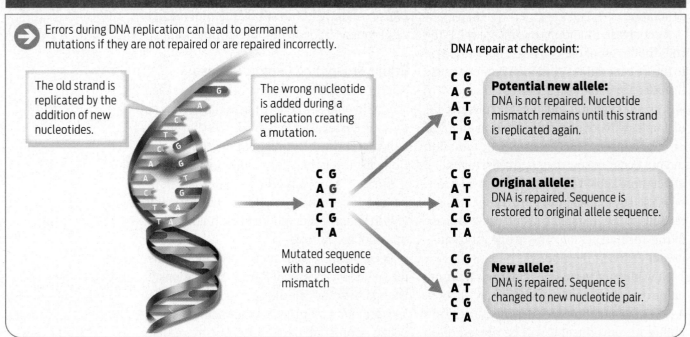

Errors during DNA replication can lead to permanent mutations if they are not repaired or are repaired incorrectly.

The old strand is replicated by the addition of new nucleotides.

The wrong nucleotide is added during a replication creating a mutation.

Mutated sequence with a nucleotide mismatch

C G
A G
A T
C G
T A

DNA repair at checkpoint:

C G
A G
A T
C G
T A

Potential new allele:
DNA is not repaired. Nucleotide mismatch remains until this strand is replicated again.

C G
A T
A T
C G
T A

Original allele:
DNA is repaired. Sequence is restored to original allele sequence.

C G
C G
A T
C G
T A

New allele:
DNA is repaired. Sequence is changed to new nucleotide pair.

Mutations in DNA Can Alter Protein Function and Cause Cancer

Mutations alter the nucleotide sequence of DNA. If a mutation changes the coding region of any gene, the corresponding protein may be dysfunctional. When the protein in question helps regulate the cell cycle, cancer may result.

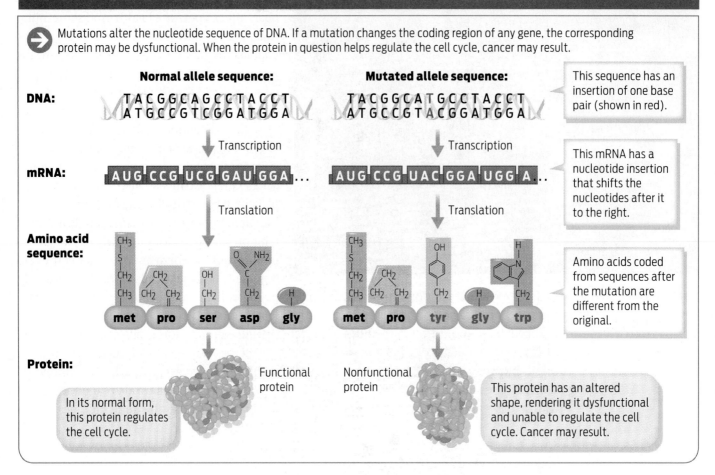

Normal allele sequence:

DNA: TACGGCAGCCTACCT
ATGCCGTCGGATGGA

Transcription

mRNA: AUG CCG UCG GAU GGA ...

Translation

Amino acid sequence:
met pro ser asp gly

Protein:
In its normal form, this protein regulates the cell cycle.
Functional protein

Mutated allele sequence:

TACGGCATGCCTACCT
ATGCCGTACGGATGGA

This sequence has an insertion of one base pair (shown in red).

Transcription

mRNA: AUG CCG UAC GGA UGG A ...

This mRNA has a nucleotide insertion that shifts the nucleotides after it to the right.

Translation

met pro tyr gly trp

Amino acids coded from sequences after the mutation are different from the original.

Nonfunctional protein

This protein has an altered shape, rendering it dysfunctional and unable to regulate the cell cycle. Cancer may result.

---

the sperm and egg cells–the germ cells–and can be passed from parent to child each generation.

By contrast, mutations in somatic cells–the cells in the rest of the body–are not passed on to future generations, although they can cause disease. A person who acquires a mutation in a skin cell from too much sun exposure, for example, will not pass this mutation on to his or her children. That's because the mutation did not occur in sperm or egg cells, nor will it affect those cells. This mutation can, however, be passed by mitosis and cell division to daughter cells of the mutated cell and cause disease in the affected person. This is one way *nonhereditary* cancers develop.

Now imagine germ-line mutations accumulating over thousands of years in a population. As long as a mutation does not affect a person's ability to reproduce, it will be passed on to

future generations through sexual reproduction. The result is that a single gene such as *BRCA1* can have hundreds of different nucleotide sequences, or alleles, in a population.

Ethnic Groups and Genetic Disease

Ahern descends from a subgroup of Jews called the "Ashkenazi"–the term generally refers to Jews of Eastern European descent. Ahern's father was born in Germany, immigrating to the United States in 1939; her mother was born in the United States; but Ahern's maternal grandfather was born in Russia. But the history of this Jewish subgroup extends much further back than modern Europe.

Recent gene studies support the biblical history of Jews as descended from populations in what is now the Middle East. The Ashkenazi Jews are a subgroup that left the Middle East and began populating parts of Europe more than

TABLE 10.1

Incidence of Hereditary Diseases in Different Populations

HEREDITARY DISEASE	CARRIER RATE IN ASHKENAZI JEWISH POPULATION	CARRIER RATE IN GENERAL POPULATION
Tay-Sachs disease	1 in 25	1 in 250
Canavan disease	1 in 40	Rare/unknown
Niemann-Pick disease, type A	1 in 90	1 in 40,000
Gaucher disease, type 1	1 in 14	1 in 100
Bloom syndrome	1 in 100	Rare/unknown
BCRA mutation	1 in 40	1 in 350–1,000
Familial dysautonomia	1 in 30	Rare/unknown

2,000 years ago. The majority of Ashkenazis, however, migrated into Europe in the 10th century from the region of present-day Israel, settling in the Rhineland, the valley of the Rhine River, in Germany.

A number of historical factors have made the Ashekenazi Jewish population more susceptible to genetic diseases. First, they descend from a small group of people. Second, that population has expanded and contracted over time. Third, and most important, members of the population usually marry within the community. In other words, Ashkenazi Jews have many of the characteristics of an isolated population—new alleles are not frequently introduced by people immigrating into the population.

Consequently, Ashkenazi Jews are an example of an ethnic group that has a more homogeneous genetic background than the general population, and is more likely to pass on certain genetic diseases to future generations. Scientists have discovered more than 1,000 recessive diseases in the general population, but most of them are rare. In Ashkenazi Jews, however, the prevalence of some recessive diseases is increased 100-fold or more. Tay-Sachs disease, Gaucher disease, and Bloom syndrome are genetic diseases that all occur more frequently

in this ethnic group than in the general population; approximately 1 in 25 Ashkenazi Jews carry disease alleles for at least one of these disorders (Table 10.1).

Ashkenazi Jews are not the only ethnic group to have a higher incidence of certain genetic diseases than occurs in the general population. For example, people from Mediterranean, African, and Asian countries have higher rates of thalassemias, blood disorders that cause anemia. Sickle-cell anemia, another type of hereditary anemia, is more common among people of African descent.

Ashkenazi Jews are also more likely than the general population to carry mutations in *BRCA1* and *BRCA2*. Some studies have found that more than 8% of Ashkenazi women carry a mutated *BRCA1* gene, compared to only 2.2% of other women. These alleles can take the form of changes in one DNA base pair, or in several. In some cases, large DNA segments are rearranged. In mutated *BRCA2* genes, a small number of additional DNA base pairs is inserted into or deleted from the gene. These mutations, or alleles, of these genes arose and became prevalent over thousands of years.

Cancer Genetics

Inheriting a gene that carries a predisposition to a disease such as cancer doesn't mean you will automatically get the disease. Inherited predispositions increase the risk, but they don't definitively determine that the disease will occur. In most cases, there are several other contributing factors. Cancer often occurs only when additional, nonhereditary, mutations in a cell accumulate.

Environmental insults such as chemicals, ultraviolet light, radiation, and other factors can damage our DNA and cause it to mutate. Exposure to ultraviolet light for example, impairs the DNA in our skin cells and can lead to skin cancer. Physical or chemical agents that cause mutations with either positive, negative, or neutral outcomes are known as **mutagens.** Chemicals and other factors such as pesticides and pollutants that can cause cancer are a class of mutagens known as **carcinogens** because

MUTAGEN
Any chemical or physical agent that can damage DNA by changing its nucleotide sequence.

CARCINOGEN
Any chemical agent that causes cancer by damaging DNA. Carcinogens are a type of mutagen.

What Causes Mutations?

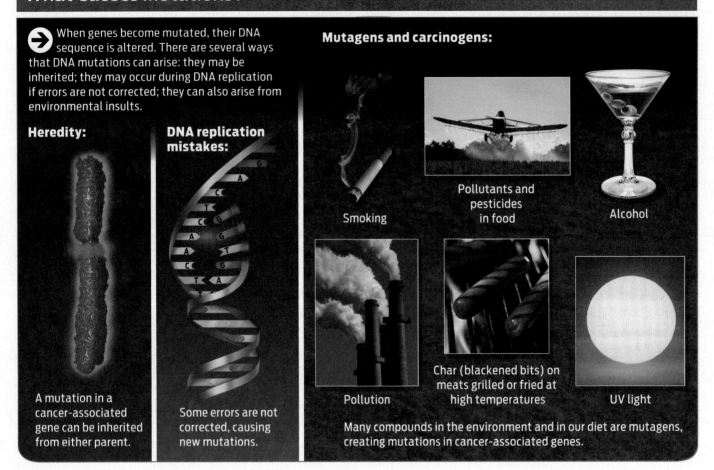

➡ When genes become mutated, their DNA sequence is altered. There are several ways that DNA mutations can arise: they may be inherited; they may occur during DNA replication if errors are not corrected; they can also arise from environmental insults.

Heredity:

A mutation in a cancer-associated gene can be inherited from either parent.

DNA replication mistakes:

Some errors are not corrected, causing new mutations.

Mutagens and carcinogens:

Smoking

Pollutants and pesticides in food

Alcohol

Pollution

Char (blackened bits) on meats grilled or fried at high temperatures

UV light

Many compounds in the environment and in our diet are mutagens, creating mutations in cancer-associated genes.

---

they damage DNA in a harmful way that can lead to cancer (**Infographic 10.4**).

Environmental insults such as chemicals, ultraviolet light, radiation, and other factors can damage our DNA and cause it to mutate.

Normally our cells are able to repair such DNA damage. But very rarely a mistake may remain uncorrected; over time and with age, if enough mutations accumulate in the same cell, that cell may begin to divide abnormally and become cancerous. Such acquired somatic mutations can develop throughout a person's life as he or she is exposed to carcinogenic environmental insults and as cells divide.

Mutations that cause cancer typically occur in two categories of genes that influence the cell cycle: **proto-oncogenes** and **tumor suppressor genes.** Normal proto-oncogenes promote cell division and cell differentiation, but only in response to appropriate signals. But mutated proto-oncogenes can become permanently "turned on" or activated, stimulating cells to divide all the time. In this state they are called **oncogenes**—genes that cause cancer. In other words, oncogenes are proto-oncogenes that have been mutated to become overexpressed or permanently activated. *Her2,* a gene overexpressed in certain types of breast cancer, is an example of a proto-oncogene.

Tumor suppressor genes, or tumor suppressors, normally pause cell division, repair damaged DNA, and tell cells when to die. Tumor suppressor genes cause cancer when they are inactivated by mutation. "You can think of

PROTO-ONCOGENE
A gene that codes for a protein that helps cells divide normally.

TUMOR SUPPRESSOR GENES
Genes that code for proteins that monitor and check cell cycle progression. When these genes mutate, tumor suppressor proteins lose normal function.

ONCOGENE
A mutated and overactive form of a proto-oncogene. Oncogenes drive cells to divide continually.

Mutations in Two Types of Cell Cycle Genes Cause Most Types of Cancer

During the cell cycle, proteins regulate whether the cell is ready to continue to the next stage or if the cell requires additional time to repair DNA damage before progressing. The proteins that regulate these checkpoints are made by proto-oncogenes and tumor suppressor genes. Accumulated mutations in these types of genes cause cancer.

Proto-oncogenes signal cells to progress through the cell cycle at the appropriate time. Mutations in these genes cause them to be overstimulated, causing too much cell division.

Tumor suppressor genes signal cells to pause the cell cycle to fix mistakes. Mutations in these genes cause them to be underexpressed, allowing damaged cells to divide inappropriately.

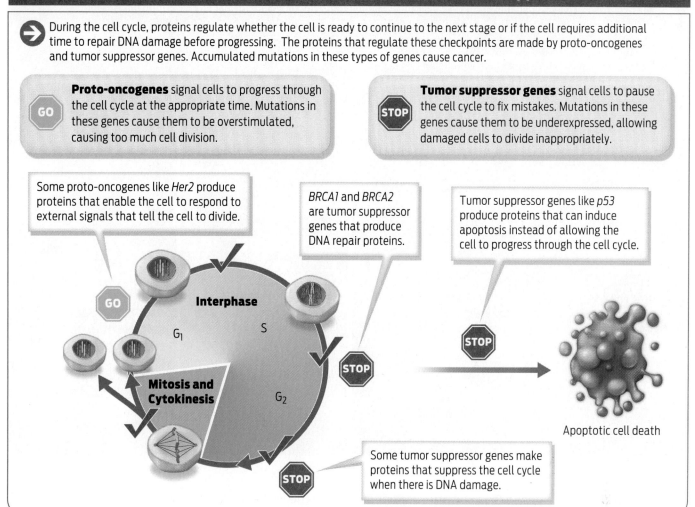

Some proto-oncogenes like *Her2* produce proteins that enable the cell to respond to external signals that tell the cell to divide.

BRCA1 and *BRCA2* are tumor suppressor genes that produce DNA repair proteins.

Tumor suppressor genes like *p53* produce proteins that can induce apoptosis instead of allowing the cell to progress through the cell cycle.

Interphase

GO

G₁

S

Mitosis and Cytokinesis

G₂

STOP

STOP

STOP

Apoptotic cell death

Some tumor suppressor genes make proteins that suppress the cell cycle when there is DNA damage.

the suppressors as brakes and the oncogenes as the accelerators," Thomas Sellers of the H. Lee Moffitt Cancer Center and Research Institute explains. "They are sort of the yin and yang of each other." Both *BRCA1* and *BRCA2* are tumor suppressors that, when normally expressed, code for proteins that help the cell progress normally through the cell cycle and respond to DNA damage **(Infographic 10.5)**.

It usually takes more than a single mutation in a cell to cause cancer. In most cases, a cell will become cancerous only after it has acquired mutations in several genes that regulate the cell cycle or repair DNA damage. The collective mutated genes can include a combination of

> **"You can think of the suppressors as brakes and the oncogenes as the accelerators."**
>
> **–Thomas Sellers**

tumor suppressor genes that have lost their function *and* proto-oncogenes that have been activated to oncogenes. This is one reason why cancer affects people as they age: as cells accumulate mutations over time through exposure to carcinogens and repeated rounds of cell division, the chances increase that a cell will

Tumors Develop in Stages as Mutations Accumulate in a Cell

➔ It takes more than a single mutation to cause cancer. Individuals who have inherited high-risk mutations require fewer additional mutations to get to cancer, and therefore get cancer at a much earlier age.

One possible scenario for cancer progression (from A to D):

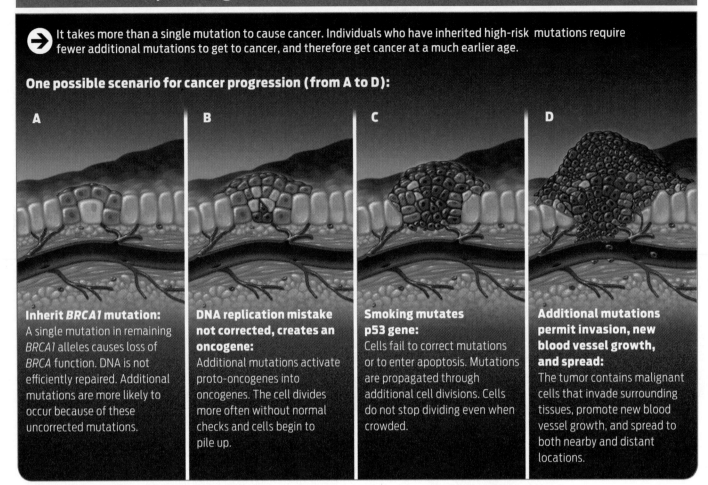

A

Inherit *BRCA1* mutation:
A single mutation in remaining *BRCA1* alleles causes loss of *BRCA* function. DNA is not efficiently repaired. Additional mutations are more likely to occur because of these uncorrected mutations.

B

DNA replication mistake not corrected, creates an oncogene:
Additional mutations activate proto-oncogenes into oncogenes. The cell divides more often without normal checks and cells begin to pile up.

C

Smoking mutates p53 gene:
Cells fail to correct mutations or to enter apoptosis. Mutations are propagated through additional cell divisions. Cells do not stop dividing even when crowded.

D

Additional mutations permit invasion, new blood vessel growth, and spread:
The tumor contains malignant cells that invade surrounding tissues, promote new blood vessel growth, and spread to both nearby and distant locations.

accumulate enough mutations to become cancerous **(Infographic 10.6)**.

People who have inherited high-risk mutations start life with at least one cancer-predisposing mutation, so they require fewer additional mutations to get cancer. For example, Ahern was born with a predisposing mutation in one of her *BRCA1* alleles. If a second mutation in one of her somatic cells disables her second *BRCA1* allele, that cell and all its descendants will no longer be able to respond effectively to DNA damage. Consequently, the cells of women with *BRCA* mutations accumulate DNA damage at a faster rate, which is why hereditary breast cancer often strikes women who are in their 30s and 40s—much younger than women who have no inherited predisposition to cancer.

Since *BRCA* genes are expressed in many cell types in addition to breast tissue, mutations in either gene raise the risk of cancers in other organs, too. Scientists have linked mutations in both genes to a higher than average risk of prostate, colon, and pancreatic cancers, among others. But the breasts and ovaries are at especially high risk of developing cancers because they respond to the hormone estrogen, which causes cells in these organs to divide more often. In breast tissue, for example, the rise in estrogen during a woman's monthly cycle signals cells lining the milk glands to divide to prepare to produce milk should a woman become pregnant.

The *BRCA* genes aren't the only ones that predispose women to breast cancer. Scientists now think that genes other than *BRCA* cause up to half of all hereditary breast cancers. Other inherited mutations in tumor suppressor genes and proto-oncogenes have been linked to various other cancers as well **(Infographic 10.7)**.

BRCA Mutation Increases the Risk of Breast Cancer

→ People with one copy of particular *BRCA1* alleles ("carriers") are at higher risk of developing breast cancer at earlier ages.

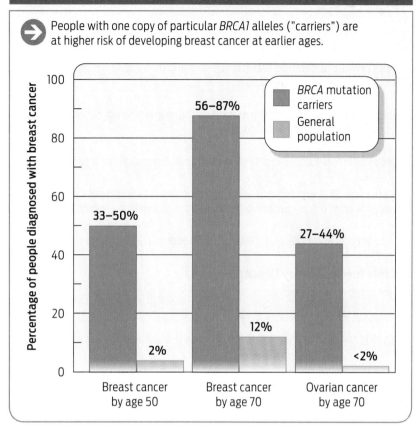

Legend:
- *BRCA* mutation carriers
- General population

Percentage of people diagnosed with breast cancer

Breast cancer by age 50	**Breast cancer by age 70**	**Ovarian cancer by age 70**
33–50% vs 2%	56–87% vs 12%	27–44% vs <2%

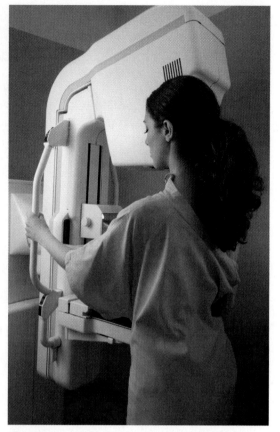

Mammograms (X-rays of the breast) are the best way to detect cancer.

Preventative Measures

A decade ago, women diagnosed with hereditary breast cancer had to face the continuing fear of developing new cancers. Their breast cancer could be treated with chemotherapy and radiation, but because they were born with genetic predispositions that make their cells less able to repair DNA damage from the start, the likelihood that the cancer would recur or that a new cancer would develop was high. To reduce the risk, they could choose to have their breasts or ovaries surgically removed. But that option seemed drastic, and the evidence supporting surgery as means to reduce the risk of repeat cancers was slim.

But over the years, studies have followed women after prophylactic preventative surgery to remove both breasts or the ovaries and have shown a reduced risk of cancers in those organs by as much as 90%. "The difference is that now we have empirical data," Sellers comments. There is a small risk that the cancer will recur because breast tissue is distributed across the chest wall and can be found near the armpit, above the collarbone, and as far down as the abdomen. Therefore it is impossible for a surgeon to remove all breast tissue, and there is a small chance that breast cancer can still recur. And removal of the breasts or ovaries will not reduce the increased risk of developing cancer in other areas of the body.

Ahern consulted a genetic counselor and decided to have both ovaries removed in 2007, a year after doctors diagnosed her breast can-

> A decade ago, women diagnosed with hereditary breast cancer had to face the continuing fear of developing new cancers.

TABLE 10.2

Reducing the Risk of Cancer

Cancer Risk Reduction Measures for Everyone

WHAT?	HOW?
Wear sunscreen.	Sunscreen helps prevent UV-induced DNA damage in skin cells. Women who use tanning beds increase their risk of skin cancer by 55%.
Avoid tobacco (both smoking and chewing).	Agents in tobacco can break DNA, causing mutations and many different cancers. Smoking causes 80–90% of lung cancer deaths.
Avoid or reduce alcohol consumption.	Excessive consumption of alcohol increases the risk of oral, liver, and breast cancers.
Maintain a healthy weight.	Obesity and lack of physical exercise increases the risk of breast cancer in postmenopausal women. Between 14 and 20% of cancer deaths result from overweight and obesity.
Get screened.	Screening helps detect cancers early and improves the odds of successful treatment.

Special Measures for Hereditary Cancers

WHAT?	HOW?
Consider genetic counseling and testing.	Genetic counseling and testing enables better-informed decisions about prevention and treatment.
Screen early.	Cancer screening at an earlier age than recommended for the general population can aid in prevention.
Consider prophylactic surgery.	Removal of tissue, mastectomy, or removal of the ovaries, for example, reduces the risk of developing cancer.
Involve other members of the family.	Genetic testing may help others in the family also make better-informed decisions.

cer. Since ovarian cells also have estrogen receptors and, because of a woman's monthly cycle, divide more frequently than other tissues, ovarian cells are also at high risk of turning cancerous. Ahern is also considering a mastectomy to remove her breasts.

She feels "pretty good" right now, Ahern says, though there was a time when she was visiting online breast cancer discussion groups every evening after work and all weekend long. They not only helped her cope emotionally but also helped to inform her about her disease and her treatment options. However, this may not be the best route to support for everyone.

But unfortunately, according to Sue Friedman, executive director of Facing Our Risk of Cancer Empowered (FORCE), there is still a lot of misinformation out there about hereditary breast cancer. People still assume that diet and lifestyle changes will cut the risk of cancer in people with hereditary cancer. "Those factors may help, but not enough in our community," says Friedman, who has had cancer herself. Women also have options regarding prophylactic surgery—when to have it and how much is necessary—that aren't always effectively communicated by health professionals (Table 10.2).

Scientists admit that surgery isn't the most palatable treatment. "Surgery cuts your risk substantially, but it's still pretty traumatic," says Sellars. "It would be nice to say we've got a medication you can take and you'll have the same effect. But we just don't have that kind of treatment right now." ∎

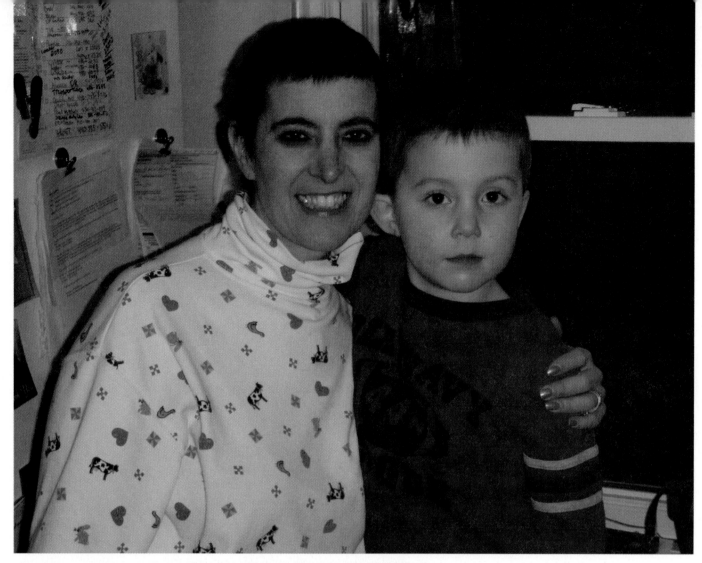

Lorene Ahern with her son. Her hair was just growing back after chemotherapy.

▶ Summary

■ Cancer is uncontrolled cell division caused by mutations in DNA.

■ Mutations occur spontaneously during DNA replication. They can also be caused by environmental triggers such as tobacco or UV radiation.

■ Mutations in certain genes can lead to cancer if they damage the normal function of the proteins those genes code for.

■ Mutations in two types of genes, proto-oncogenes and tumor suppressors, cause most cancers.

■ Multiple mutations must occur in the same cell for it to become cancerous.

■ Mutations that occur in somatic (that is, body) cells, for example skin or breast cells, *are not* inherited by offspring.

Mutations that occur in sperm or egg cells *are* inherited by offspring.

■ People with "hereditary" cancer inherit predispositions to the disease in the form of specific genetic mutations. These mutations are present in all body cells and can serve as the first mutation that may lead to cancer.

■ Women with *BRCA* mutations have a much higher risk of developing cancer, and at an earlier age, than women without these mutations.

■ Mutations introduce new alleles into the population. These alleles may code for proteins that have advantageous, neutral, or harmful affects on an individual.

■ Certain alleles are more common in specific ethnic groups that have been reproductively isolated.

MUTATIONS AND CANCER

Cancer occurs when cells accumulate several DNA mutations that enable the cells to divide uncontrollably. People with inherited predispositions to cancer develop the disease at an earlier age than others because their cells already have one mutation that hinders their cells' ability to divide normally.

HINT See Infographics 10.1–10.7.

➔ KNOW IT

1. What are some differences and some similarities between tumor suppressor genes and oncogenes?

2. What is the role of *BRCA1* in normal cells?

3. In an otherwise normal cell, what happens if one mistake is made during DNA replication?
 a. Nothing; mistakes just happen.
 b. A cell cycle checkpoint detects the damage and pauses the cell cycle so the error can be corrected.
 c. The cell will begin to divide out of control, forming a malignant tumor.
 d. A checkpoint will force the cell to carry out apoptosis, a form of cellular suicide.
 e. The mutation will be inherited by the individual's offspring.

4. Which of the following can cause cancer to develop and progress?
 a. a proto-oncogene
 b. an oncogene
 c. a tumor suppressor gene
 d. a mutated tumor suppressor gene
 e. b and d
 f. b and c

5. Someone with a *BRCA1* mutation
 a. will definitely develop breast cancer.
 b. is at increased risk of developing breast cancer.
 c. must have inherited it from her mother, because of the link to breast cancer.
 d. will also have a mutation in *BRCA2*.
 e. b and c

6. Why does wearing sunscreen reduce cancer risk?
 a. Sunscreen can repair damaged DNA.
 b. Sunscreen can activate checkpoints in skin cells.
 c. Sunscreen can reduce the chance of mutations caused by exposure to UV radiation present in sunlight
 d. It doesn't; sunscreen causes mutation and actually increases cancer risk.
 e. Sunscreen can prevent cells with mutations from being destroyed.

➔ USE IT

7. Lorene Ahern was born with an inherited predisposition to cancer. At the cellular and genetic level, what was she born with? At birth, were cells in her breast genetically identical to cells in her liver? Now that she has breast cancer, are her cancer cells genetically identical to her normal breast cells? Explain your answers.

8. What would you say to a niece if she asked you how she could reduce her risk of getting breast cancer? Assume there is no family history of breast cancer. How might each of your suggestions reduce her risk?

9. If you wanted to change your lifestyle to reduce your risk of developing cancer, which of the following behaviors would be important?
 a. limiting alcohol consumption
 b. wearing sunscreen
 c. avoiding exposure to tobacco
 d. by avoiding exposure to pesticides
 e. all of the above

10. Who of the following women would be most likely to benefit from genetic testing for breast cancer?
 a. a 25-year-old woman whose mother, aunt, and grandmother had breast cancer
 b. a healthy 75-year-old woman with no family history of breast cancer
 c. a 40-year-old woman who has a cousin with breast cancer
 d. a 55-year-old woman whose older sister was just diagnosed with breast cancer
 e. All women can benefit from genetic testing for breast cancer.

11. People like Lorene Ahern have inherited a mutated version of *BRCA1*. Why does this mutation pose a problem? Why are these people at high risk of developing breast cancer when they still have a functional *BRCA1* allele? Describe how the protein encoded by normal *BRCA1* compares to that encoded by mutant alleles of *BRCA1*.

SCIENCE AND ETHICS

12. Nellie has a family history similar to Lorene Ahern's. Nellie's mother died at an early age from breast cancer, as did her maternal aunt (her mother's sister). Nellie is not yet 35 but has started having annual mammograms. She has also been tested for *BRCA1* and *BRCA2* mutations. She has a *BRCA2* mutation and is considering prophylactic surgery. Her younger sister, Anne, doesn't want to know the results of Nellie's genetic testing because if Nellie has a *BRCA2* mutation, then there is a chance that Anne could have inherited the same mutation from their mother. Does Nellie or Nellie's doctor have an obligation to tell Anne about the test results? What about Nellie's older brother? Should he be told? There are personal and medical benefits and risks to consider here.

Rock for a Cause

Rock for a Cause

Research lightens the load of cystic fibrosis

Emily Schaller had no idea that her Detroit-based band would be strumming its way to fame one day. A few years ago she and her girlfriends were goofing around singing songs when someone floated the idea of forming a band. None of them could play guitar, bass, or drums, but that didn't stop them. "We just went out and bought a bunch of instruments," says Emily, who chose to play drums. The friends practiced in Emily's parents' basement. Her older brother overheard the original mix of punk and classic rock songs that the five girls put together and was so impressed he asked them to open for his own band on New Year's Eve. The five-girl rock 'n' roll band called Hellen was born.

"It's getting really huge," says Emily excitedly. Hellen performs almost every weekend in Detroit and is now taking its show on the road,

> **Approximately 2,500 babies are born with the disease every year, making CF the most common fatal genetic disease in the United States.**

with gigs scheduled in Cleveland, Chicago, and soon, Emily hopes, New York City.

Emily may look like a typical rock 'n' roller, with her bleached blond hair and tattooed forearms; but her carefree appearance masks a serious underlying condition. Emily has cystic fibrosis (CF)—a genetic disease she inherited from her parents—and each day she takes a cocktail of drugs and vitamins. CF has many symptoms, the most dangerous of which is mucus that clogs airways in the lungs and makes it difficult to breathe. People with CF also can't digest food well—mucus blocks the passageways through which the necessary enzymes travel to the intestines. So Emily must swallow enzymes before each meal to ensure that her body gets enough nutrients. She's grown accustomed to the schedule, but having to take such meticulous care of her health is hardly routine for most people her age.

In 2000, Emily and her friends launched an all-girl band called Hellen. From left to right: Katie, Charmain, Becca, Emily, and Amanda.

Approximately 2,500 babies are born with the disease every year, making CF the most common fatal genetic disease in the United States. In 1989, a team of scientists led by Lap Chee Tsui, at Toronto's Hospital for Sick Children, and Francis Collins, then at the University of Michigan, discovered that the disease is caused by genetic mutations in a specific gene that sits on chromosome 7. A **mutation** is a change in the nucleotide sequence of DNA, which creates alternative alleles of a gene. As we saw in Chapter 8, alleles are alternative nucleotide sequences of the same gene. Most genes have not just a single allele but several,

MUTATION
A change in the nucleotide sequence of DNA.

each created by genetic mutation. Tsui and Collins discovered that CF is caused by mutations in a gene called *CFTR*, which codes for the protein known as the cystic fibrosis transmembrane regulator.

The discovery was a milestone. Now that they knew the gene responsible, scientists could study how mutations in it make people sick. Because genes provide instructions for making proteins, a change in gene sequence can change the function or shape of a protein. In the most common CF allele, three nucleotides within the *CFTR* gene are deleted. People who carry this allele produce a defective CFTR protein. This slight change wreaks havoc on victim's bodies—their lungs, sweat glands, and pancreas no longer function normally **(Infographic 11.1)**.

Today, almost 20 years later, scientists understand the disease better, and this has led to better drugs and therapies to treat symptoms; victims of CF are living longer than ever. But despite scientific advances, there is still much to learn. One aspect of the disease that scientists are studying intensively is that people with identical CF alleles vary in the course of their disease—some have worse symptoms and live shorter lives than others. In recent years, scientists have discovered that there are other genes that contribute to a patient's overall health—so-called modifier genes. That discovery is leading to exciting new therapies that may extend Emily's life and the lives of thousands of other people with CF.

How Is CF Inherited?

When Emily's mother, Debbie, learned that her daughter had CF, she was shocked. She and her husband, Lowell, were both healthy, and they already had two healthy sons. How did their daughter Emily develop a disease that neither Debbie nor her husband had?

The answer is inheritance. Genes, which provide instructions for making proteins, are the units of inheritance, physically transmitted from parents to children. The particular alleles of genes you received from your parents are the reason you resemble your mother and father, and possibly also an uncle or a grandparent. But not every child of a couple receives the exact same parental genes, and so children can and do differ from their parents and from each other.

Consider Emily's parents. Because they are **diploid** organisms, each of their body cells carries two copies of each chromosome—one inherited from mom, the other from dad. Such paired chromosomes are called **homologous chromosomes**. Because chromosomes come in pairs, we have two copies of nearly every gene in our body cells. Genes located on the X and Y chromosome in males do not have a second copy. While the two gene copies have the

same general function, the nucleotide sequence of each copy can differ. In other words, a person can carry two different alleles of the same gene, one of which functions differently from the other. In the case of the gene *CFTR*, a person can have one CF-associated allele and remain healthy if his or her other

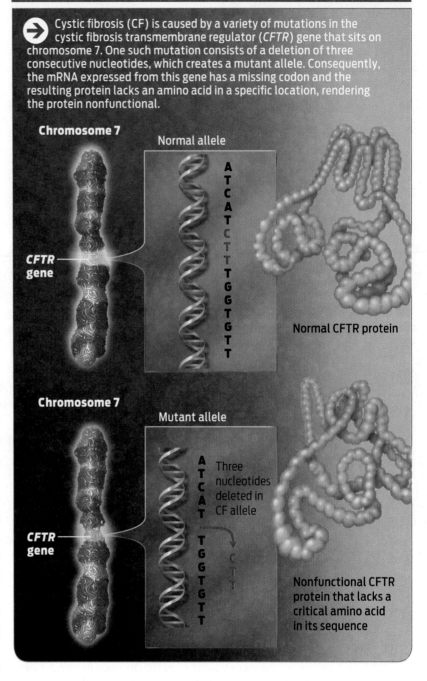

INFOGRAPHIC 11.1

CF Is Caused by Mutations in the *CFTR* Gene

→ Cystic fibrosis (CF) is caused by a variety of mutations in the cystic fibrosis transmembrane regulator (*CFTR*) gene that sits on chromosome 7. One such mutation consists of a deletion of three consecutive nucleotides, which creates a mutant allele. Consequently, the mRNA expressed from this gene has a missing codon and the resulting protein lacks an amino acid in a specific location, rendering the protein nonfunctional.

Chromosome 7

CFTR gene

Normal allele

ATCATCTTTGGTGTT

Normal CFTR protein

Chromosome 7

CFTR gene

Mutant allele

ATCAT TGGTGTT

Three nucleotides deleted in CF allele

CTT

Nonfunctional CFTR protein that lacks a critical amino acid in its sequence

DIPLOID
Having two copies of every chromosome.

HOMOLOGOUS CHROMOSOMES
The two copies of each chromosome in a diploid cell. One chromosome in the pair is inherited from the mother, the other is inherited from the father.

Humans Have Two Copies of Nearly Every Gene

Human cells have 23 pairs of homologous chromosomes. One chromosome of each pair is inherited from mom, one from dad. This makes us diploid, as virtually every cell in the body carries two copies of every gene. Each copy of each gene has two alleles that can either be identical to each other or different. In the case of CF, carrying at least one normal allele is enough to remain healthy.

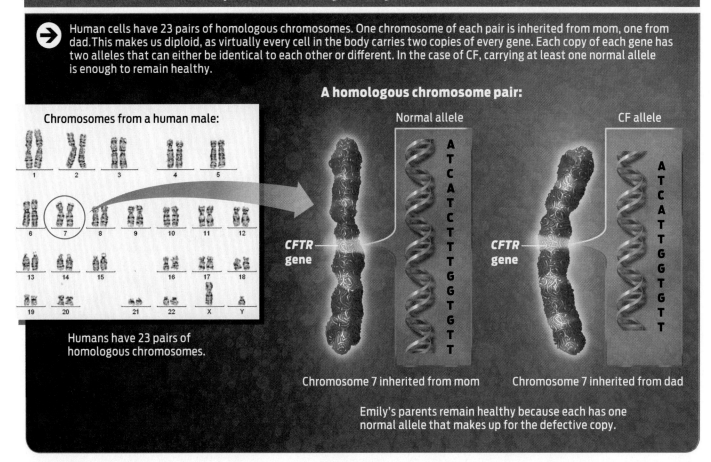

Chromosomes from a human male:

Humans have 23 pairs of homologous chromosomes.

A homologous chromosome pair:

Normal allele

CF allele

CFTR gene

ATCATCTTGGTGTT

CFTR gene

ATCATTGGTGTT

Chromosome 7 inherited from mom

Chromosome 7 inherited from dad

Emily's parents remain healthy because each has one normal allele that makes up for the defective copy.

PHENOTYPE
The visible or measurable features of an individual.

GENOTYPE
The particular genetic makeup of an individual.

GAMETES
Specialized reproductive cells that carry one copy of each chromosome (that is, they are haploid). Sperm are male gametes; eggs are female gametes.

HAPLOID
Having only one copy of every chromosome.

chromosome has a normal allele to make up for the defective copy (Infographic 11.2). That's why Emily's parents, even if each of them had a CF-associated gene, could be healthy.

Geneticists make a distinction between a person's observable or measurable traits, or **phenotype,** and his or her genes, or **genotype.** As in the case of Debbie and Lowell, one cannot always determine genotype from phenotype. Both Debbie and Lowell have normal phenotypes, but they both also carry a disease allele as part of their genotype. They each inherited one CF allele from one of their parents and therefore can pass that defective allele along to their children—as they did to Emily.

But not all the Schaller children have the disease—Debbie and Lowell also have two healthy boys. Why didn't these children inherit CF? Sexual reproduction is a bit like shuffling cards. Before parents pass their genes to their offspring, those genes are first mixed up and then the two copies of each gene are separated from each other, so that not every child receives the same combination of alleles. It is the unique combination of maternal and paternal alleles that come together during fertilization that determines a person's genotype and contributes to his or her phenotype.

To reproduce sexually, organisms must first create sex cells called **gametes.** In humans, these are the egg and sperm cells. Unlike the rest of the body's cells, which are diploid, gametes carry only one copy of each chromosome, which makes them **haploid.** To become haploid, the cells that form gametes go through a unique kind of cell division, called

Gametes Pass Genetic Information to the Next Generation

→ To reproduce sexually, diploid organisms produce specialized sex cells called gametes, which are haploid — they carry only one copy of each chromosome. When a sperm fertilizes an egg the resulting diploid zygote divides by mitotic cell division, eventually generating enough cells to form a baby. The baby is diploid.

The process of cell division that creates gametes is known as meiosis. Men produce sperm, the male gametes, and women produce eggs, the female gametes.

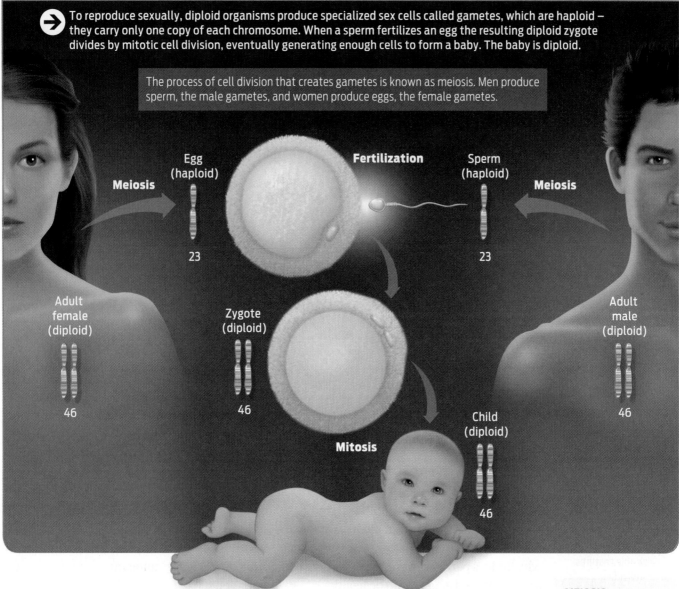

meiosis, which halves the number of chromosomes from 46 to 23. When a haploid sperm fertilizes a haploid egg, the result is a diploid **zygote** that now carries two copies of every gene on 46 chromosomes. In turn, this zygote will divide by mitosis to become an **embryo,** which will eventually grow into a human child (**Infographic 11.3**).

Meiosis, the cell division that creates sperm and egg, is similar to mitotic cell division (Chapter 9), except that in meiosis there are two separate divisions. The first division separates homologous chromosomes; the second division separates sister chromatids (**Infographic 11.4**).

Because it unites haploid egg and sperm from two people, sexual reproduction is the primary reason why children don't look and behave exactly like one parent in particular; they inherit alleles from both parents and consequently are genetically a combination of the two.

Besides forming haploid sex cells, meiosis contributes to the genetic diversity of offspring in other ways as well. No two gametes produced by the same parent are identical, and that is because of two major events during meiosis that contribute to the huge variation we see among parents, children, and their siblings. The first is **recombination,** in which homologous

MEIOSIS
A specialized type of cell division that generates genetically unique haploid gametes.

ZYGOTE
A cell that is capable of developing into an adult organism. The zygote is formed when an egg is fertilized by a sperm.

EMBRYO
An early stage of development reached when a zygote undergoes cell division to form a multicellular structure.

Meiosis Produces Haploid Egg and Sperm

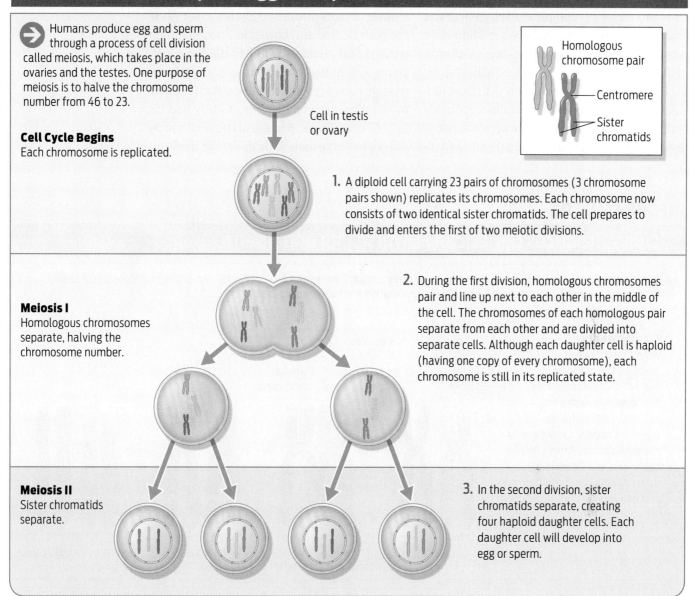

Humans produce egg and sperm through a process of cell division called meiosis, which takes place in the ovaries and the testes. One purpose of meiosis is to halve the chromosome number from 46 to 23.

Cell Cycle Begins
Each chromosome is replicated.

Cell in testis or ovary

Homologous chromosome pair

Centromere

Sister chromatids

1. A diploid cell carrying 23 pairs of chromosomes (3 chromosome pairs shown) replicates its chromosomes. Each chromosome now consists of two identical sister chromatids. The cell prepares to divide and enters the first of two meiotic divisions.

Meiosis I
Homologous chromosomes separate, halving the chromosome number.

2. During the first division, homologous chromosomes pair and line up next to each other in the middle of the cell. The chromosomes of each homologous pair separate from each other and are divided into separate cells. Although each daughter cell is haploid (having one copy of every chromosome), each chromosome is still in its replicated state.

Meiosis II
Sister chromatids separate.

3. In the second division, sister chromatids separate, creating four haploid daughter cells. Each daughter cell will develop into egg or sperm.

RECOMBINATION
The stage of meiosis in which maternal and paternal chromosomes pair and physically exchange DNA segments.

INDEPENDENT ASSORTMENT
The principle that alleles of different genes are distributed independently of one another during meiosis.

maternal and paternal chromosomes pair up and swap genetic information. As a result of recombination, maternal chromosomes actually contain segments (and therefore alleles) from paternal chromosomes and vice versa.

The second vitally important aspect of meiosis is **independent assortment,** which means that alleles of different genes are distributed independently of one another, not as a package. Because the number of possible combinations of alleles is therefore huge, a unique

combination of maternal and paternal chromosomes is distributed into each sperm and each egg cell. This distribution occurs at the first division of meiosis (known as meiosis I), when maternal and paternal chromosomes line up along the midline of the cell and segregate into newly forming cells. Because maternal and paternal chromosomes line up randomly (sometimes on the "left," sometimes on the "right,"), the exact combination of maternal and paternal chromosomes that each sperm or egg inherits differs every time meiosis occurs **(Infographic 11.5).**

When meiosis is complete, each gamete has only 23 chromosomes that are a mishmash of maternal and paternal alleles–and this is the reason that not everyone in the Schaller family has CF. Because alleles randomly distribute into each gamete, some of the Schallers' gametes will carry the CF allele while others will not. If by chance a sperm that carries a CF allele fertil-izes an egg that also carries a CF allele, the resulting child will have CF.

After doctors diagnosed Emily's cystic fibrosis, both Debbie and Lowell learned that their parents had relatives who had died at a very young age. At the time, the cause of death was thought to be a respiratory illness such as pneumonia. But these relatives most likely had CF, Debbie now thinks; doctors at the time simply did not have the tools to diagnose the disease.

INFOGRAPHIC 11.5

Meiosis Produces Genetically Diverse Egg and Sperm

→ Meiosis produces haploid gametes that are genetically unique. Each egg and sperm has its own distinct combination of alleles. The two events that create this diversity are recombination and independent assortment.

1. Recombination

Before separating at meiosis I, the maternal and paternal chromosomes line up next to each other and physically exchange segments of DNA. Consequently, maternal chromosomes contain segments (and thus alleles) from paternal chromosomes, and vice versa.

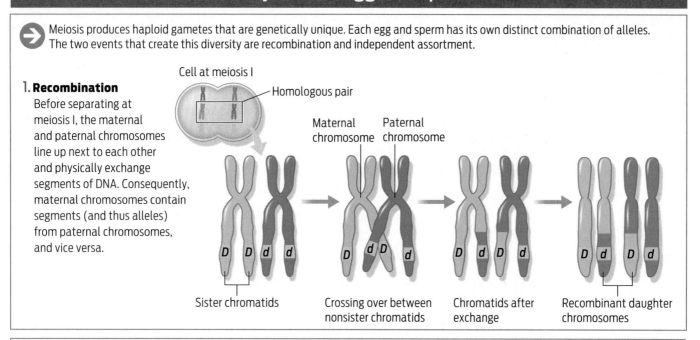

Cell at meiosis I

Homologous pair

Maternal chromosome   Paternal chromosome

Sister chromatids   Crossing over between nonsister chromatids   Chromatids after exchange   Recombinant daughter chromosomes

2. Independent Assortment

Maternal and paternal chromosome pairs separate according to how they have randomly lined up in the cell. Each time meiosis occurs, the chromosome pairs line up differently, and thus a different chromosome combination is produced in the resulting gametes. When all 23 chromosome pairs are considered, there are more than 8 million unique chromosome combinations possible.

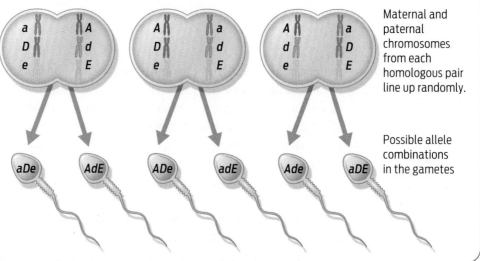

Maternal and paternal chromosomes from each homologous pair line up randomly.

Possible allele combinations in the gametes

Cystic fibrosis patients like Emily wear vibrating vests to loosen the mucus in their lungs while inhaling a saltwater solution to thin out the mucus.

membrane that allows certain ions in and out of the cell, keeping the cell's chemistry in balance. But in people with CF, the channel is distorted or dysfunctional and the mechanism goes awry. The result is that mucus—a slippery substance that lubricates and protects the linings of the airways, digestive system, reproductive system, and other tissues—becomes abnormally thick (**Infographic 11.6**).

This abnormal mucus blocks ducts throughout the body. The most problematic symptom, however, is that thick mucus builds up in the lungs. Patients have trouble breathing, and the mucus provides fertile ground for bacteria and other organisms. Over time, repeated infections permanently damage the lungs. Suffocation often kills CF victims as they slowly lose their ability to breathe.

To avoid lung damage, every morning Emily dons an inflatable vest that vibrates to loosen mucus in her lungs. During this 30-minute therapy she inhales a saltwater solution and another medication to thin her mucus, which she then coughs out periodically. To that regime she adds two other medications three times a week to keep her lungs from becoming inflamed and to kill off infections. But despite her best efforts, Emily has been hospitalized more frequently in recent years because of serious lung infections that hinder her ability to breathe.

The Schallers now understood that the disease ran in both their families. But they could still not help Emily. "They told us she would only live to be about 12 years old," Debbie recalls, adding, "We just put ourselves in the hands of medical professionals."

Living with the Disease

Growing up, Emily was scarcely aware of her own disability. The visits to doctors and periodic stays in the hospital were just a part of life. All her teachers and friends knew that she had CF. "My family and friends were all so supportive," she says. In high school she played volleyball, basketball, and soccer, and participated in many walkathons to raise money for CF

Emily remains undaunted. "I just live each day at a time," she says. She works about 30 hours a week at a retail shop in downtown Detroit and spends her evenings practicing with her band, performing at concerts, playing guitar, or hanging out with friends. While she doesn't plan too far ahead into the future, she hopes her band's fame and success will grow. If the band's following expands beyond Detroit, she hopes to tour Europe. Emily hasn't ruled out having a family of her own one day. Even though Emily has CF, her children will not necessarily have the disease.

Why not? Remember that since Emily has CF, her parents, Lowell and Debbie, both must carry disease alleles. But as neither of them has the disease, the CF alleles must be "hidden." When one allele masks the effect of

---

"They told us she would only live to be about 12 years old. We just put ourselves in the hands of medical professionals." —Debbie Schaller

---

research. Thanks to medical progress, Emily has outlived doctors' original expectations by more than a decade.

But she deals daily with the legacy of her genetic inheritance. In healthy people, the CFTR protein acts as a channel within a cell's

INFOGRAPHIC 11.6

The CFTR Protein and Cystic Fibrosis

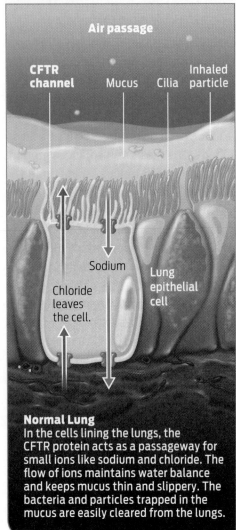

Air passage

CFTR channel · Mucus · Cilia · Inhaled particle

Sodium

Chloride leaves the cell.

Lung epithelial cell

Normal Lung
In the cells lining the lungs, the CFTR protein acts as a passageway for small ions like sodium and chloride. The flow of ions maintains water balance and keeps mucus thin and slippery. The bacteria and particles trapped in the mucus are easily cleared from the lungs.

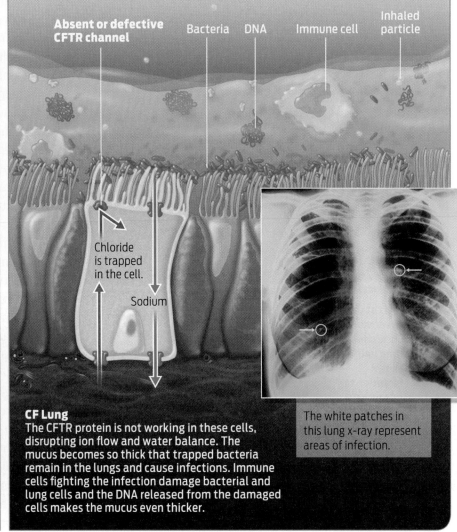

Absent or defective CFTR channel · Bacteria · DNA · Immune cell · Inhaled particle

Chloride is trapped in the cell.

Sodium

CF Lung
The CFTR protein is not working in these cells, disrupting ion flow and water balance. The mucus becomes so thick that trapped bacteria remain in the lungs and cause infections. Immune cells fighting the infection damage bacterial and lung cells and the DNA released from the damaged cells makes the mucus even thicker.

The white patches in this lung x-ray represent areas of infection.

---

another, the hidden allele is described as **recessive** (designated by a lower-case letter, e.g., *a*). The normal allele, which conceals the effect of the recessive allele, is known as the **dominant allele** (designated by a capital letter, e.g., *A*). Debbie and Lowell are healthy because they each have a dominant normal allele that compensates for their defective recessive CF allele. Geneticists call their genotype **heterozygous.** Their two healthy sons either are heterozygous like their parents, or have two normal alleles–that is, their genotype is **homozygous**. A genotype made up of two dominant alleles is known as homozygous dominant. Emily's genotype, however, is

homozygous recessive: she inherited one recessive CF allele from each parent, which is why she has the disease.

What were the chances that Debbie and Lowell would have a child with CF? To figure out the likelihood that parents will have a child with a particular trait, we can plot the possibilities on a **Punnett square,** a tool named for the British geneticist Reginald C. Punnett, who devised it. A Punnett square matches up the possible parental gametes and shows the likelihood that particular parental alleles will combine. As heterozygous individuals, Debbie and Lowell each have a 50% chance of passing on their CF allele to a child, which means they have a

RECESSIVE ALLELE
An allele that reveals itself in the phenotype only if the organism has two copies of that allele.

DOMINANT ALLELE
An allele that can mask the presence of a recessive allele.

HETEROZYGOUS
Having two different alleles.

HOMOZYGOUS
Having two identical alleles.

25% chance of having a child with CF and a 75% chance of having a healthy child. The chance that a child will be a heterozygous **carrier**– that is, that the child will carry the recessive allele for CF but will not have the disease because the allele's effect is masked by the dominant allele–is 50% **(Infographic 11.7)**.

Just as Emily's genotype is different from her parents' genotype, Emily's children will have different genotypes from her own. Whether or not her children develop CF depends on the father's genotype. Since Emily is homozygous, she can contribute only recessive CF alleles to her children. If Emily were to have children with a man who had two normal alleles, for example,

none of her children would have the disease– they would all have a heterozygous genotype but a normal phenotype. But as carriers they could pass on the disease to their children. If she had children with a man who was heterozygous for the CF gene, then her children would have a 1 in 2, or 50%, chance of having CF.

Not all recessive alleles cause disease. Physical traits such a blue eyes, for example, result from the inheritance of two recessive alleles of the same gene. And not all genetic diseases are caused by recessive alleles; some, such as the neurodegenerative disorder called Huntington disease, are determined by dominant alleles. Diseases caused by dominant alleles, however,

INFOGRAPHIC 11.7

How Recessive Traits Are Inherited

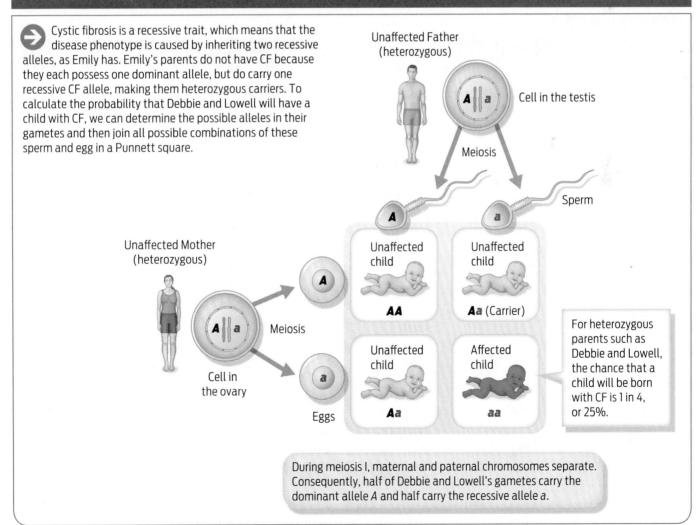

Cystic fibrosis is a recessive trait, which means that the disease phenotype is caused by inheriting two recessive alleles, as Emily has. Emily's parents do not have CF because they each possess one dominant allele, but do carry one recessive CF allele, making them heterozygous carriers. To calculate the probability that Debbie and Lowell will have a child with CF, we can determine the possible alleles in their gametes and then join all possible combinations of these sperm and egg in a Punnett square.

Unaffected Father (heterozygous)

Cell in the testis

Meiosis

Sperm

Unaffected Mother (heterozygous)

Meiosis

Cell in the ovary

Eggs

Unaffected child
AA

Unaffected child
Aa (Carrier)

Unaffected child
Aa

Affected child
aa

For heterozygous parents such as Debbie and Lowell, the chance that a child will be born with CF is 1 in 4, or 25%.

During meiosis I, maternal and paternal chromosomes separate. Consequently, half of Debbie and Lowell's gametes carry the dominant allele *A* and half carry the recessive allele *a*.

have a high probability of being passed to the next generation (Infographic 11. 8).

In all cases, anyone with a genetic disease is at risk for passing it on to his or her children. The risk merely varies, depending on whether the alleles are dominant or recessive and on the genotype of the partner (Table 11.1).

Couples who carry disease genes needn't feel that having children is a roll of the dice. There are ways to ensure that their children won't develop the diseases they could otherwise inherit. Many couples in this situation use a technology called pre-implantation genetic diagnosis to detect and select embryos that do not carry defective alleles. Through in vitro fertilization, a man's sperm

> **Couples who carry disease genes needn't feel that having children is a roll of the dice.**

can fertilize a woman's eggs outside the body. The genes of each resulting embryo are then examined for specific alleles, and then only embryos that don't contain defective alleles are implanted into the mother. Hundreds of thousands of babies have already been born by this technique.

New Research in the Pipeline
Some couples, however, may choose not to undergo assisted reproduction because of religious or other reasons. In the case of CF, there are new treatments in the pipeline that could help Emily, and her children and grandchildren, too.

Furthest along are a class of medications that, when inhaled, can restore the balance of ions inside affected cells. Scientists are presently

INFOGRAPHIC 11.8

How Dominant Traits Are Inherited

→ Some genetic conditions, such as Huntington disease, a degenerative neurological disease, and polydactyly, having more than five fingers or toes per limb, are caused by dominant alleles. Plenty of common traits such as dark eyes and dimples are also determined by dominant alleles. In these cases, inheriting one copy of the dominant allele is sufficient to display the trait.

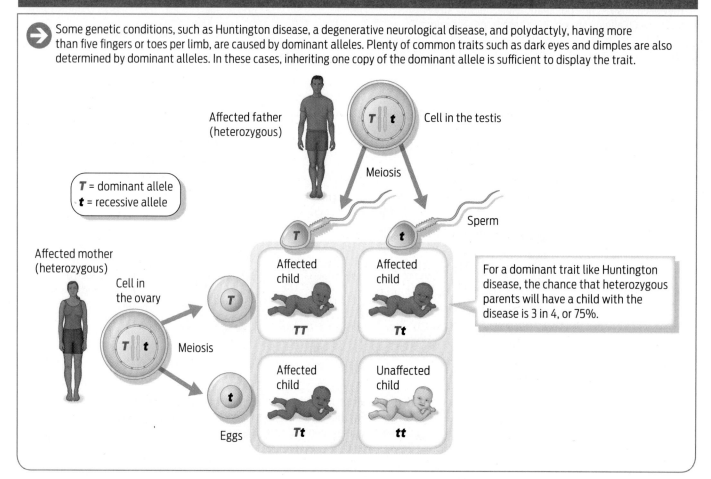

Affected father (heterozygous)

Cell in the testis

Meiosis

T = dominant allele
t = recessive allele

Sperm

Affected mother (heterozygous)

Cell in the ovary

Meiosis

Eggs

Affected child
TT

Affected child
Tt

Affected child
Tt

Unaffected child
tt

For a dominant trait like Huntington disease, the chance that heterozygous parents will have a child with the disease is 3 in 4, or 75%.

TABLE 11.1

Inherited Genetic Conditions in Humans

RECESSIVE TRAITS	PHENOTYPE
Albinism	Lack of pigment in skin, hair, and eyes
Cystic fibrosis	Excess mucus in lungs, digestive tract, and liver; increased susceptibility to infections
Sickle-cell disease	Sickled red blood cells; damage to tissues
Tongue rolling	Ability to curl tongue into a U-shape
Tay-Sachs disease	Lipid accumulation in brain cells; mental deficiency, blindness, and death in childhood

DOMINANT TRAITS	PHENOTYPE
Huntington disease	Mental deterioration and uncontrollable movements; onset at middle age
Freckles	Pigmented spots on skin, particularly on face and arms
Polydactyly	More than five digits on hands or feet
Dimples	Indentation in the skin of the cheeks
Chin cleft	Indentation in chin

But in recent years, scientists have learned that there is more to the story. Researchers have discovered other genes on different chromosomes that contribute to the severity of CF symptoms. The genes so far discovered predominantly influence a person's immune system, which helps the body fight off infections.

For example, scientists have found that one allele of a gene called *TGFB1,* located on chromosome 19, is associated with more severe lung disease in CF patients. This gene influences a person's immune response to infection. Scientists suspect that CF patients with certain *TGFB1* alleles mount a more vigorous response to infections than those with other alleles. Such a heightened immune response can cause lung tissue to scar. So if a CF patient also inherited this specific allele of *TGFB1,* his or her lungs are more likely to scar in response to infections. The impact of such modifier genes on the CF phenotype makes it more complicated to assess how disabling any particular person's CF disease will be–but it is not impossible.

Parents who are heterozygous carriers of CF, for example, have a 1 in 4, or 25%, chance of having a child who has CF. If these two parents are also heterozygous for *TGFB1,* then the probability that their child will be homozygous recessive for *TGFB1* is also 1 in 4 (25%). The chance of two independent events occurring together is calculated by multiplying the two independent chances together. So the probability of being homozygous recessive for both *CFTR* and *TGFB1* is $\frac{1}{4} \times \frac{1}{4}$, or 1 in 16. This probability can also be calculated using a Punnett square **(Infographic 11.9).**

Understanding how these modifier genes contribute to the disease may point the way to even more therapies. In some cases, existing drugs may prove useful. Drugs that reduce inflammation by targeting the TGFβ1 protein, for example, may help reduce scarring in the lungs.

testing at least six different experimental drugs in humans.

Through basic research, scientists continue to learn more about the disease. Over the past 20 years, scientists have discovered more than 1,000 different alleles of the *CFTR* gene. The most common is *ΔF508,* which accounts for about 70% of all CF alleles. This particular CF allele is associated with more severe disease. But researchers have long puzzled over why the disease varies in two people with identical CF alleles–even two people homozygous for the *ΔF508* allele will vary in how their disease progresses. Researchers long thought that perhaps environmental factors such as diet, social relationships, and exercise might be responsible.

> **Over the past 20 years, scientists have discovered more than 1,000 different alleles of the *CFTR* gene.**

Emily recently had her genotype tested. While she doesn't carry *ΔF508*, she does carry another allele associated with severe disease. Her second CF allele, however, is much rarer, and scientists can't assess how these two particular alleles along with alleles of Emily's other genes will interact as she gets older. So they can't predict how quickly her disease will progress.

Keenly aware of how medical progress has extended her life, Emily conducts her own share of fund raising and education. After that fateful New Year's Eve when she and her friends opened for her brother's band, Self Normal, Hellen's following grew. The band wrote more songs and refined their sound. "I favor the old stuff, AC-DC, Led Zeppelin, the Ramones," Emily remarks. And fans just

INFOGRAPHIC 11.9

Tracking the Inheritance of Two Genes

People with CF differ in the severity of their disease. Some of this variability is influenced by alleles of other genes that sit on other chromosomes. One such gene, called *TGFB1*, is located on chromosome 19, shown here with symbol *D*. We can also use a Punnett square to follow the inheritance of two genes, as in the example below.

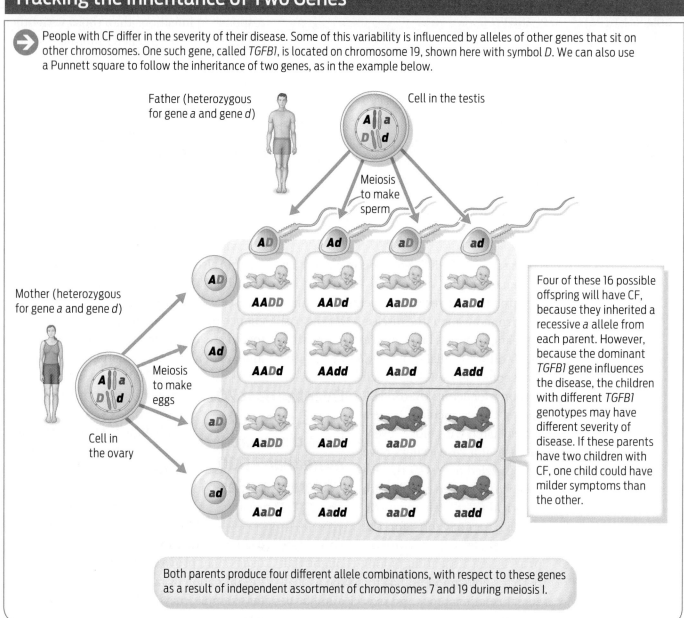

Father (heterozygous for gene *a* and gene *d*)

Cell in the testis

Meiosis to make sperm

AD   Ad   aD   ad

Mother (heterozygous for gene *a* and gene *d*)

Meiosis to make eggs

Cell in the ovary

	AD	Ad	aD	ad
AD	AADD	AADd	AaDD	AaDd
Ad	AADd	AAdd	AaDd	Aadd
aD	AaDD	AaDd	aaDD	aaDd
ad	AaDd	Aadd	aaDd	aadd

Four of these 16 possible offspring will have CF, because they inherited a recessive *a* allele from each parent. However, because the dominant *TGFB1* gene influences the disease, the children with different *TGFB1* genotypes may have different severity of disease. If these parents have two children with CF, one child could have milder symptoms than the other.

Both parents produce four different allele combinations, with respect to these genes as a result of independent assortment of chromosomes 7 and 19 during meiosis I.

Every year Emily plays a concert to benefit CF.

couldn't seem to get enough of the girls' music. Each year, the number of fans has grown.

Sparked by a Self Normal hit single called "Just Breathe," Emily had the idea to organize a concert to benefit CF research. The song has nothing to do with CF, but Emily thought that it might be a good theme song for a concert. Besides, she says, "We were tired of walk-athons and black tie events with tickets that cost $300 each."

Emily and her brother organized the first benefit concert in 2004. Called "Just Let Me Breathe," it featured four Detroit bands. The concert sold out and raised about $9,000. With all of her fund-raising activities, Emily has raised more than $150,000 so far. But she doesn't plan on stopping there. The concert was so successful that Emily and her brother have planned others at bigger venues in coming years. She hopes to draw chart topping Detroit bands, raise even more money, and "rock CF." ■

▶ Summary

■ An organism's physical traits constitute its phenotype, while its genes constitute its genotype. A person's genotype can't always be determined from his or her phenotype.

■ Genes, which code for proteins, are the units of inheritance, physically passed down from parents to offspring.

■ Different versions of the same gene are known as alleles. Alleles arise from mutations that change the nucleotide sequence of a gene.

■ Alleles may be dominant or recessive. Dominant alleles can mask the effects of recessive alleles, which can be hidden.

■ Many traits result from carrying two recessive alleles, while others result from carrying one dominant allele.

■ Meiosis is a type of cell division that produces genetically distinct sperm and egg.

■ Homologous chromosomes recombine and assort independently during meiosis to generate genetically diverse sperm and eggs.

■ Haploid gametes fuse randomly during fertilization, generating genetically unique zygotes.

■ Cystic fibrosis (CF) is a recessively inherited genetic disease. Alterations in the gene *CFTR* cause disease by interfering with ion and water balance.

■ A Punnett square can help predict a child's genotype and phenotype when the pattern of inheritance, dominant or recessive, is known.

GENES, CHROMOSOMES, ALLELES

Humans have two copies of nearly every gene, located on pairs of chromosomes.

HINT See Infographics 11.1 to 11.3.

➲ KNOW IT

1. How many chromosomes are present in one of your liver cells?

2. How many chromosomes are present in one of your gametes?

➲ USE IT

3. How many copies of the CF-associated allele does a person with CF have in one of his or her lung cells? How does this compare to someone who is a carrier for CF? How does it compare to someone who is homozygous dominant for the gene *CFTR?*

4. Strictly on the basis of the following *CFTR* genotypes, what do you predict the phenotype of each to be?
   a. heterozygous
   b. homozygous dominant
   c. homozygous recessive

5. From the discussion in this chapter, why might a person with a homozygous recessive *CFTR* genotype have a somewhat different phenotype from someone with a homozygous recessive *CFTR* genotype?

MEIOSIS

Meiotic cell division is critical for making gametes. Two separate events, recombination and independent assortment, occur during meiosis to produce genetic diversity.

HINT See Infographics 11.4 and 11.5.

6. A human female has _____ chromosomes in each skin cell and _____ chromosomes in each egg.
   a. 46; 46
   b. 23; 46
   c. 46; 23
   d. 23; 23
   e. 92; 46

7. A woman is heterozygous for the CF-associated gene (the alleles are represented here by the letters *A* and *a*). Assuming that meiosis occurs normally, which of the following represent eggs that she can produce?
   a. *A*
   b. *a*
   c. *Aa*
   d. *AA*
   e. *aa*
   f. either *A* or *a*
   g. any of *A*, *a*, or *Aa*

8. Draw a maternal version of chromosome 7 in one color and a paternal version of chromosome 7 in another color. Maintaining this color distinction, now draw a possible version of chromosome 7 that could end up in a gamete following meiotic division.

➲ USE IT

9. An alien has 82 total chromosomes in each of its body cells. The chromosomes are paired, making 41 pairs. If the alien's gametes undergo meiosis, what are the number and arrangement (paired or not) of chromosomes in one of its gametes? Give the reason for your answer.

10. Describe at least two major differences between mitosis (discussed in Chapter 9) and meiosis.

11. If meiosis were to fail and a cell skipped meiosis I, so that meiosis II was the only meiotic division, how would you describe the resulting gametes?

PREDICTING PATTERNS OF INHERITANCE

Cystic fibrosis is a genetic disease with one pattern of inheritance; other genetic diseases have different inheritance patterns.

HINT See Infographics 11.6 to 11.9.

➲ KNOW IT

12. What is the genotype of a person with CF?
   a. homozygous dominant
   b. homozygous recessive
   c. heterozygous
   d. any of the above
   e. none of the above

13. A person has a heterozygous genotype for a disease gene and no disease phenotype. Does this disease have a dominant or a recessive inheritance pattern?

14. Women can inherit alleles of a gene called *BRCA1* that makes them susceptible to breast cancer. The alleles associated with elevated cancer risk are dominant. Of the genotypes listed below, which has the lowest genetic risk of developing breast cancer?
   a. *BB*
   b. *Bb*
   c. *bb*
   d. *BB* and *Bb* have less risk than *bb*.
   e. All have equal risk.

➔ USE IT

15. Assume that Emily (who has CF) decides to have children with a man who does not have CF and who has no family history of CF.
   a. What combination of gametes can each of them produce?
   b. Place these gametes on a Punnett square and fill in the results of the cross.
   c. On the basis of the Punnett square results, what is the probability that they will have a child with CF?
   d. On the basis of the Punnett square results, what is the probability that they will have a child who is a carrier for CF?

16. Your friend's mother has Huntington disease and her mother's mother does not have Huntington disease. If your friend's father does not have Huntington disease, what is the probability that your friend will develop Huntington disease? (Hint: draw a Punnett square.)

SCIENCE AND ETHICS

17. Emily took a genetic test to determine which CF alleles she inherited. The results revealed she has one allele about which very little is known. Although genetic testing can predict whether a person will develop CF and drugs can prolong life, for some other genetic diseases, such as Huntington disease, treatment is limited and there is no cure. If you were faced with the decision to take a genetic test, especially for a disease for which there is no cure, would you take the test? Why or why not?

Mendel's Garden

→ What You Will Be Learning

By studying the inheritance of traits in pea plants, Mendel unknowingly discovered genes and the chromosomal basis of inheritance.

Mendel's Garden

An Austrian priest lays the foundation for modern genetics

The garden outside the Augustinian Abbey in Brno, where Mendel performed his experiments.

Johann Gregor Mendel was an unlikely father of genetics. He was a melancholy Austrian priest who by all accounts suffered from debilitating test-taking anxiety, failing his teaching exam twice. Mendel nevertheless collected the first research suggesting that each parent passes discrete "elements" to each child that determine specific traits. These "elements" remain intact and can be passed on indefinitely to future generations without being diluted. Although he couldn't define these "elements," in fact Mendel had discovered what came to be called genes.

Mendel's idea was new and radical, with implications for how human traits were understood. Some contemporary scientists in the mid-19th century believed that parental traits were blended—for example, a tall mother and short father would have a son or daughter of medium height. Other scientists thought that a sperm or egg contained a miniature adult waiting to be born. But through a series of simple yet elegant experiments conducted in a monastery garden, Mendel revolutionized our understanding of heredity decades before the word "genetics" was coined.

In 1843, Mendel became a monk at the Augustinian Abbey of St. Thomas in Brünn (now Brno) in what was then the Austro-Hungarian Empire. He studied theology and was ordained a priest in 1847. When he failed his teaching exam (the Augustinians were a teaching order), the abbot at St. Thomas, who encouraged intellectual pursuits among the members of the abbey, sent Mendel to the University of Vienna, where for 2 years he studied math, physics, zoology, botany, and plant physiology.

Mendel returned to the monastery in 1853, and a year later began experiments to study hybrids—the offspring of different breeds or varieties. He began by breeding mice but, as Robin Marantz Henig writes in her book *The Monk in the Garden* (2001), the local bishop found "toying with the reproduction of

> **Although he couldn't define these "elements," in fact Mendel had discovered what came to be called genes.**

Theories of Inheritance Before Mendel

→ "Preformation theories" of inheritance, popular in the 1800s, posited that the next generation of life already existed fully formed in miniature inside the egg or sperm. It was thought that these tiny individuals only needed to grow prior to being born. Other ideas of inheritance speculated that substances from the mother and father blend together during conception to create the traits of the offspring.

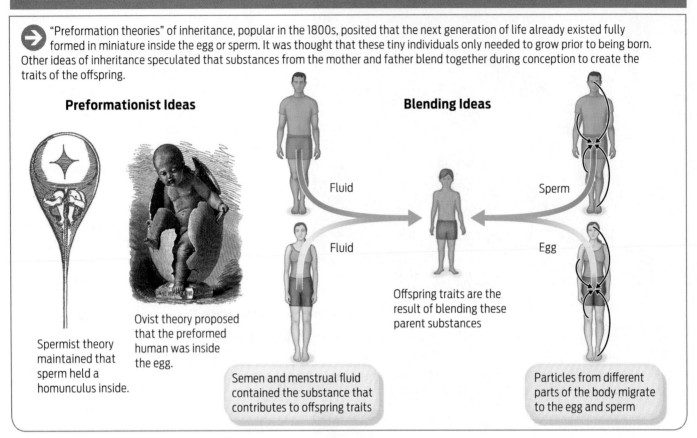

Preformationist Ideas

Spermist theory maintained that sperm held a homunculus inside.

Ovist theory proposed that the preformed human was inside the egg.

Blending Ideas

Fluid

Fluid

Sperm

Egg

Offspring traits are the result of blending these parent substances

Semen and menstrual fluid contained the substance that contributes to offspring traits

Particles from different parts of the body migrate to the egg and sperm

View through a window of the Abbey of St. Thomas to the garden used by Gregor Mendel for his experiments with pea plants.

animals simply too vulgar an undertaking for a priest." So Mendel decided to work instead with pea plants, which proved a better model organism anyway. The plants grew quickly, and he could better control their environment and breeding.

Mendel didn't start out with the goal of understanding heredity, however. He was interested in how hybrids form, and he hoped to explain what he and many others had observed: that physical traits (size, color, etc.) can skip a generation.

Mendel began by choosing specific traits that he could see and study, among them seed shape, seed color, pod shape, pod color, flower color, and stem length. Each of the traits he chose to study appeared in two forms. For example, seed shape was either round or wrinkled; seed color was either green or yellow. Because, as he and others had observed, in several types of organisms some traits seemed to disappear in one generation only to show up again in the next, he started his breeding experiments with plants that "bred true"– plants with offspring that carried the same traits as the parents, generation after generation. Only then could he study what happened to particular traits when purebred plants of one variety were mated with purebred plants of another variety.

Pea plants can self-pollinate, which means that the pea flower contains both male and female sexual organs, and a single plant can fertilize itself to produce offspring. To create true-breeding plants, Mendel covered pea flowers with a small bag so that he could control fertilization, manually fertilizing plants with their own pollen and preventing pollen from another plant from entering. Once he had established true-breeding plants, he could then set up a cross between two different plants. What would happen if he crossed a green-seeded plant with one that produced yellow seeds? Or a purple-flowered plant with a white-flowered plant? For each cross, Mendel painstakingly pollinated individual flowers from the two plants by hand. He also prevented self-pollination by removing the male reproductive parts from the plants to be fertilized.

Mendel noticed that when he bred a true-breeding white-flowering plant with a true-breeding purple-flowering plant, the first generation of offspring (the F_1 generation) all had purple flowers. That the flowers were true purple rather than pale purple suggested that

Mendel's Experiments

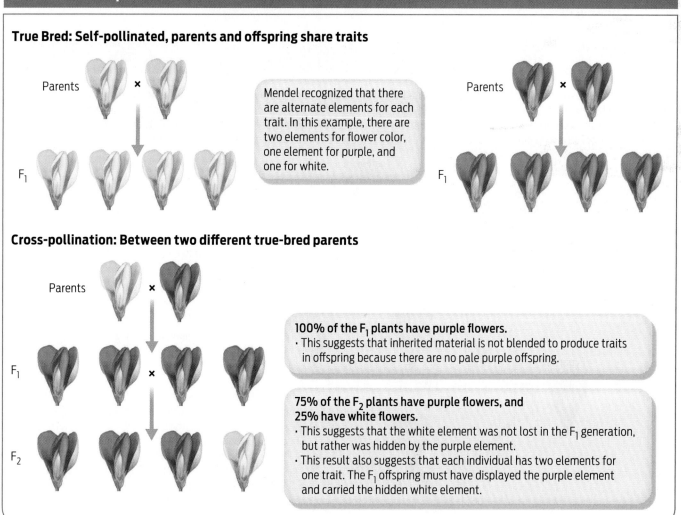

True Bred: Self-pollinated, parents and offspring share traits

Parents

F_1

Mendel recognized that there are alternate elements for each trait. In this example, there are two elements for flower color, one element for purple, and one for white.

Parents

F_1

Cross-pollination: Between two different true-bred parents

Parents

F_1

F_2

100% of the F_1 plants have purple flowers.
· This suggests that inherited material is not blended to produce traits in offspring because there are no pale purple offspring.

75% of the F_2 plants have purple flowers, and 25% have white flowers.
· This suggests that the white element was not lost in the F_1 generation, but rather was hidden by the purple element.
· This result also suggests that each individual has two elements for one trait. The F_1 offspring must have displayed the purple element and carried the hidden white element.

parental traits were not blended, as earlier theories of inheritance would have predicted. But the trait for white flowers did not disappear completely, either. When Mendel randomly selected two F_1 purple-flowering plants to breed, he found that on average 1 out of every 4 plants of the second generation of offspring (the F_2 generation) had white flowers. Mendel reasoned that a hidden white element must be present in the purple F_1 plants. So each F_1 plant must have two such elements, one representing purple (the trait that appeared) and the other representing white (the hidden trait).

These results sound familiar, don't they? They reflect dominant and recessive patterns of inheritance. Purple flower color is dominant

Mendel revolutionized our understanding of heredity decades before the word "genetics" was coined.

over white, which is recessive. Mendel was the first to gather evidence showing that traits could be inherited in a dominant or recessive fashion (although it was only later that the terms "dominant" and "recessive" were used). While earlier scientists had noticed that traits could disappear in one generation and reappear in later generations, Mendel used simple ratios to explain why traits appeared in the frequency they did.

Over 7 years, Mendel studied thousands of pea plant crosses and came up with the basic principles of inheritance. He published his results in 1866.

Today we know that Mendel's "elements" are alleles of genes, and that genes are located on chromosomes. The principles he discovered

Mendel's Law of Segregation

Mendel's experiments enabled him to formulate the Law of Segregation. This law has held up over time, although today we call Mendel's "elements" alleles. Mendel's Law of Segregation states that when an organism produces gametes, the two alleles for any given trait separate so that each gamete receives only one allele. Consequently, each parent donates only one of any two alleles to any offspring. The alleles don't blend, but remain as discrete pieces of information as they pass from one generation to the next.

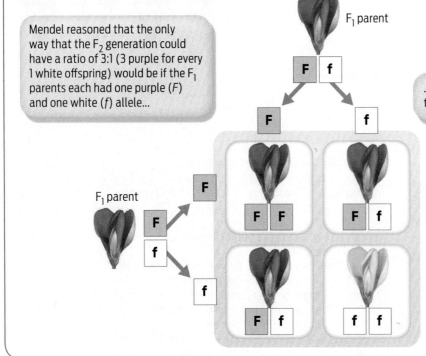

Mendel reasoned that the only way that the F_2 generation could have a ratio of 3:1 (3 purple for every 1 white offspring) would be if the F_1 parents each had one purple (F) and one white (f) allele...

F_1 parent

...and contributed only one of these two alleles randomly to each offspring.

As egg and sperm join during fertilization, the resulting offspring has two alleles, just like each parent.

When an offspring has one of each allele (Ff) it displays the trait of the "dominant" allele (F), and the "recessive" allele (f) is masked.

have been formalized into two laws. The first of these is Mendel's law of segregation, which states that for any diploid organism, the two alleles of each gene segregate separately into gametes. That is, every gamete receives only one of the two alleles and the specific allele that any one gamete receives is random.

Over 7 years, Mendel studied thousands of pea plant crosses and came up with the basic principles of inheritance.

The second law, the law of independent assortment, states that the two alleles of any given gene segregate independently from any two alleles of a second gene. Because of independent assortment, offspring can display any and all combinations of the different traits,

Gregor Mendel at work in his laboratory.

The pea plant, *Pisum sativum*.

rather than inheriting the traits together. We now know this holds true only for genes that are located on different chromosomes, or far enough away from each other to recombine. It was mere happenstance that Mendel chose traits for which genes are located on different chromosomes.

Despite Mendel's groundbreaking research, no one realized the significance of his results at the time—not even Charles Darwin, whose *Origin of Species* was published in 1859. In 1868, Mendel was elected abbot of St. Thomas, and largely shifted his focus from science to monastic life and the administration of the abbey. Although Mendel's research was cited by other scientists, he didn't receive much notice until botanists who were also studying how traits are inherited in plants rediscovered his work 30 years later. They used Mendel's work with pea plants to inform their own research. Mendel was finally recognized as the researcher who had solved a crucial mystery of inheritance many years before. ■

Mendel's Law of Independent Assortment

Mendel went on to study how multiple traits are inherited. For example, he studied plants that had different seed color and seed texture and how those traits passed to the next generation. Tracing two traits at a time helped him form the Law of Independent Assortment. This law posits that two alleles for any given trait will segregate independently from any other alleles when passed on to gametes. Consequently, each gamete may acquire all possible allele combinations and traits.

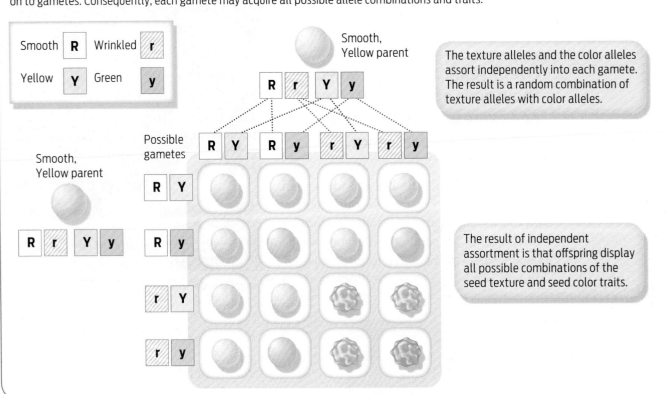

Smooth **R** Wrinkled **r**

Yellow **Y** Green **y**

Smooth, Yellow parent

Smooth, Yellow parent

Possible gametes

The texture alleles and the color alleles assort independently into each gamete. The result is a random combination of texture alleles with color alleles.

The result of independent assortment is that offspring display all possible combinations of the seed texture and seed color traits.

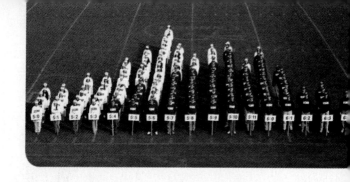

Q & A: Genetics

> ➔ **What You Will Be Learning**

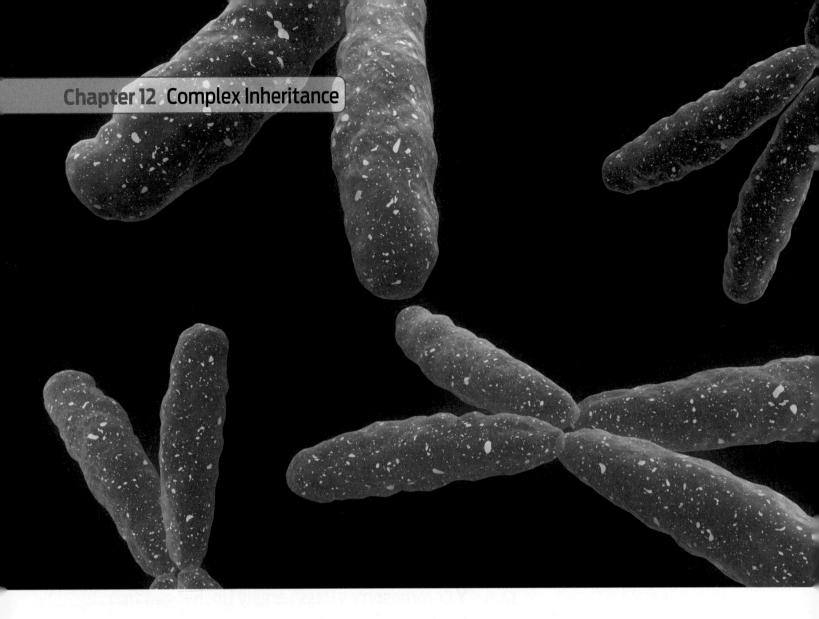

Q & A: Genetics

Complexities of human genetics, from sex to depression

In Chapter 11, we saw how dominant and recessive traits are inherited. Not all traits are inherited so simply, however. Other factors—including sex and the number of genes that influence a particular trait—can alter patterns of inheritance. In this chapter, we consider more complex patterns of inheritance. Although these types of inheritance differ from simple dominant and recessive patterns, they still rely on the underlying genetic variation resulting from meiotic division.

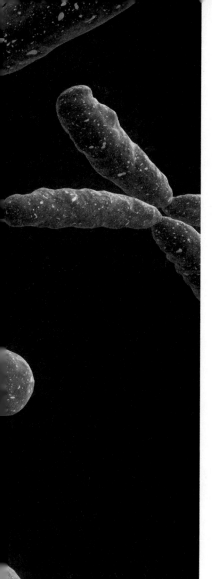

AUTOSOMES
Paired chromosomes present in both males and females; all chromosomes except the X and Y chromosomes.

SEX CHROMOSOMES
Paired chromosomes that differ between males and females, XX in females, XY in males

Y CHROMOSOME
One of two sex chromosomes in humans. The presence of a Y chromosome signals the male developmental pathway during fetal development.

X CHROMOSOME
One of the two sex chromosomes in humans.

Q What determines a person's sex?

A A botched circumcision in 1965 on a little boy named Bruce Reimer in time became a landmark example of how nature determines a person's sex. Doctors at the hospital where Reimer was circumcised used an experimental procedure that involved burning off the foreskin. The procedure went awry, burning most of Bruce's penis. With the limited surgical techniques available at the time, his penis would never be completely functional again. On the advice of John Money, a well-known and respected doctor at Johns Hopkins University in Baltimore, who had written much about the importance of environment in determining a person's sexual identity, Bruce's parents decided to have their little boy surgically turned into a little girl and rear him as "Brenda."

But Brenda never behaved like a girl. She didn't like playing with other girls and often got into fistfights at school. By the time Brenda reached puberty, her behavior became so troublesome that her father broke down and told her what had happened to her.

Brenda eventually had reconstructive surgery to recreate a penis, and she changed her name to David. David told his story in a book published in 2000, *As Nature Made Him: The Boy Who Was Raised as a Girl,* by John Colapinto. By that time, it had become increasingly clear that sexual identity is largely influenced by biology. Studies had shown that prenatal exposure to fetal sex hormones such as testosterone not only determine whether a fetus will develop female or male genitalia, but also that these sex hormones act on a developing baby's brain. Male hormones like testosterone promote masculine behaviors, whereas female hormones such as estrogen promote feminine behaviors.

Sex hormones are produced by sex organs known as the gonads—ovaries in females, testes in males. In a developing fetus, these hormones shape the development of both internal and external sexual anatomy. It is the external sexual anatomy, or genitalia, that doctors typically use to categorize someone as male or female: people with male genitalia are male and people with female genitalia are female. Whether a fetus develops male or female gonads, and produces male or female hormones, depends on the set of chromosomes it receives from its parents.

Humans have 23 pairs of chromosomes; 22 pairs are **autosomes** and 1 pair consists of the **sex chromosomes,** X and Y. Sons inherit one **Y chromosome** from their father and one **X chromosome** from their mother. Daughters inherit two X chromosomes, one each from mother and father. So males are XY and females are XX. In the absence of a Y chromosome, a fetus will develop into a female. Thus, fathers determine the sex of a baby based on whether the sperm fertilizing a mother's egg carries an X or a Y sex chromosome. More specifically, the Y chromosome contains genes that masculinize a developing fetus **(Infographic 12.1).**

Although the vast majority of men carry an XY chromosome pair, and the vast majority of women carry an XX chromosome pair, there are exceptions to this rule. Each year about 1 in every 1,600 babies born in America falls into an intermediate sex category termed "intersex," which groups people who have a "disorder of sex development." An intersexual person is someone whose external genitalia do not match his or her internal sex organs—for example, a person with an XX chromosome pair who has internal ovaries but external genitalia that

appear male. Often intersex babies are born with ambiguous genitalia.

Debate over the case of David Reimer, and similar cases, as well as research showing how strongly biology influences sexual identity, has changed the care of intersex babies. Today, such babies are often assigned a sex by parents and doctors only after a period of observation to assess behavior patterns. Surgeons then perform surgery to create either male or female genitalia. Or, parents may forgo surgery, preferring that their child remain as is.

Genetically speaking, disorders of sex development can arise from a number of genetic mutations. For example, if the Y chromosome has a mutation in a gene called *SRY*, the embryo is likely to have undeveloped gonads with external female genitalia, even though it carries an XY chromosome pair.

There are also cases of people with XY sex chromosomes who are missing genes that code for androgen (a male hormone) receptors. So even though they carry a functional *SRY* gene and have internal testes, a mutation in a gene carried on the X chromosome causes a failure of their cells to respond to male hormones like testosterone. As a result, complete male external genitalia do not develop, and these people appear to be female.

Similarly, there are people with XX sex chromosomes who have male genitalia. In some cases, this is caused by a condition called congenital adrenal hyperplasia. These individuals have one or more mutations in genes on autosomal chromosomes. One result is excessive production of male hormones. These people have ovaries but may have genitals that appear more male than female.

INFOGRAPHIC 12.1

X and Y Chromosomes Determine Human Sex

Males and females differ by virtue of a pair of sex chromosomes. Females have two X chromosomes and males have a single X and a single Y chromosome. Every person must have at least one X chromosome, but it's the presence of a gene on the Y chromosome that initiates male development.

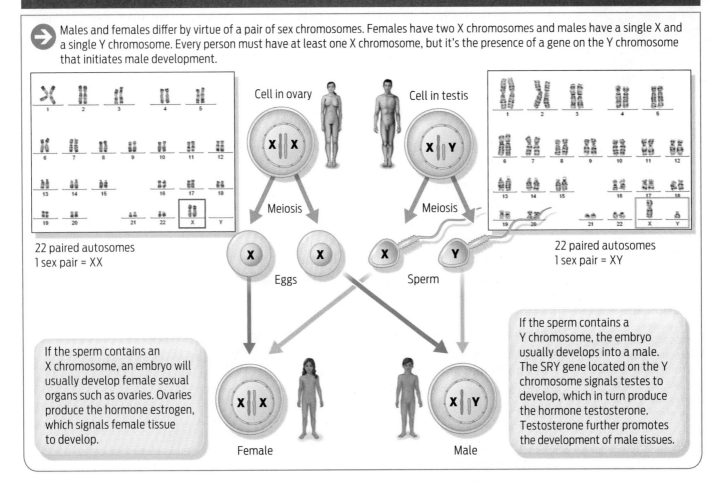

22 paired autosomes
1 sex pair = XX

Cell in ovary

Cell in testis

Meiosis

Meiosis

22 paired autosomes
1 sex pair = XY

Eggs

Sperm

If the sperm contains an X chromosome, an embryo will usually develop female sexual organs such as ovaries. Ovaries produce the hormone estrogen, which signals female tissue to develop.

Female

Male

If the sperm contains a Y chromosome, the embryo usually develops into a male. The SRY gene located on the Y chromosome signals testes to develop, which in turn produce the hormone testosterone. Testosterone further promotes the development of male tissues.

TABLE 12.1

How Many Sexes Are There?

Each of the following individuals has 22 pairs of autosomal chromosomes. Discrepancies in the sexual phenotype may result from environmental factors, hormone imbalance, or having too many or too few sex chromosomes.

SEX CATEGORY	CHROMOSOMES	GONADS	GENITALIA	OTHER CHARACTERISTICS
Female	XX	Ovaries	Female	
Male	XY	Testes	Male	
Female pseudo-hermaphroditism	XX	Ovaries	Male	Infertile
Male pseudo-hermaphroditism	XY	Testes	Female or Ambiguous	Infertile
True gonadal intersex	XX and/or XY	Ovaries and testes	Male or Female or ambiguous	Infertile; historically called true hermaphrodites
Triple X syndrome	XXX	Ovaries	Female	Fertile, taller than average, learning disabilities
Klinefelter syndrome	XXY	Testes	Male	Infertile, enlarged breast tissue
47, XYY syndrome	XYY	Testes	Male	Fertile, taller, learning and emotional disabilities
Turner syndrome	X	Ovaries	Female	Infertile, broad chest, webbed neck

Some people have only one sex chromosome, and others may have three sex chromosomes. Every person must have at least one X chromosome (the only sex chromosome the mother can contribute). Because of errors in chromosome segregation during meiosis, a variety of other X and Y combinations are possible: XXY men, women with only a single X chromosome, XXX females, and XYY males. In many of these cases, a person's physical traits and genitalia reveal that they do not have the usual makeup of sex chromosomes, but not always. These are just a few examples of the many chromosomal possibilities that determine sex. Environmental factors, like exposure to chemicals or abnormal levels of hormones during sexual development, can also play a role in defining a person's sex in terms of external genitalia (Table 12.1).

The question of sex goes beyond a person's genitalia or genotype, however. Further, defining what counts as "masculine" and "feminine" can be even more complicated. For example, some men have characteristics that we typically identify as female, such as a high voice and sparse body hair, yet they are genetically and anatomically male. And many women have what are considered to be more masculine features, such as angular faces and more muscle as compared to body fat. Yet, they, correspondingly, are genetically and anatomically female. In other words, there is no set of physical or mental characteristics that is entirely male or entirely female. In addition, some people may mentally identify with one sex even though their genitalia and chromosomal makeup classify them as the other. Consequently, "sexual identity" is more complicated than simply having an X or a Y chromosome, or male or female genitalia.

SEX-LINKED INHERITANCE

⍰ Why do some genetic conditions affect sons more often than daughters?

A Some 10 million American men—about 7% of the male population—either cannot distinguish red from green, or see these two colors as different hues from the ones most other people perceive. But such red-green color blindness affects only 0.4% of women. Similarly, 1 in 5,000 boys worldwide is born with hemophilia, a blood-clotting disorder; yet hemophilia rarely afflicts girls.

Why this disparity? Some genetic conditions are more common in boys than in girls. These conditions are caused by genes found on the X chromosome. When a gene is located on either of the sex chromosomes, daughters and sons don't share the same probability of inheriting it.

Take the neuromuscular condition Duchenne muscular dystrophy (DMD), for example. DMD is a disease in which muscles slowly degenerate, leading to paralysis. About 1 in 2,400 boys worldwide is born with the condition each year. Most affected boys are in wheelchairs by the time they become teenagers, and they rarely live longer than 30 years. Why does DMD primarily affect men? Recall that a female has two X chromosomes. For a recessive trait like DMD, a normal copy on one X chromosome masks the recessive disease allele on the other X chromosome. A male, on the other hand, has a single X chromosome, and so will show the effects of any recessive alleles located on his X chromosome. Because females can carry the disease allele without showing it, they may not even know they are at risk of passing it on to their sons.

So, for example, if you are female, and DMD runs in your family—say, a male cousin has the disease—you could in theory pass the disease allele on to your children. Whether or not your children will have the disease depends on which X chromosome they inherit from you. Note that a woman always passes one of her two X chromosomes to each of her children. A man, on the other hand, passes his single X chromosome to his daughters and his Y chromosome to his sons. So if a male carries a disease allele on his X chromosome, he can't pass it to his sons. He can, however, pass the disease allele to his daughters. Diseases and other traits such as DMD that are inherited on X chromosomes are called **X-linked traits.** By contrast, boys and girls share the same probability of inheriting diseases that, like cystic fibrosis, are carried on autosomes **(Infographic 12.2).**

If you are female, and your male cousin has DMD, your aunt was likely a DMD carrier–a person who has a recessive gene for a particular disease, in this case on one of her X chromosomes. Her son, your cousin, inherited her DMD-carrying X chromosome. Because males have only one X chromosome, your cousin doesn't have another allele to mask his defective one.

Your children's risk depends on whether or not you carry a defective DMD gene, which you might have inherited from your mother if she, too, carries a defective DMD gene. If you are a carrier and your husband is healthy, a son who inherits a diseased DMD gene from you would have the disease. This pattern is typical of X-linked traits, which pass down through generations to boys via their mothers **(Infographic 12.3).**

CHROMOSOME ANALYSIS

⍰ Did Thomas Jefferson father children with a slave?

A Thomas Jefferson was the third president of the United States, the principal architect of the Declaration of Independence, and founder of the University of Virginia. He was

X-linked traits are passed from mothers to children on their X chromosome.

X-LINKED TRAIT
A phenotype determined by an allele on an X chromosome.

X-linked Traits Are Inherited on X Chromosomes

➔ Duchenne Muscular Dystrophy (DMD) is an example of an X-linked trait. Recessive mutations of the dystrophin gene on the X chromosome cause the disease. DMD primarily affects males because they inherit only one copy of the X chromosome (from their mothers). Therefore, the single DMD allele they inherit determines their phenotype. Since females have two X chromosomes, they may carry the DMD allele, but have a healthy phenotype.

Mother has a DMD allele:

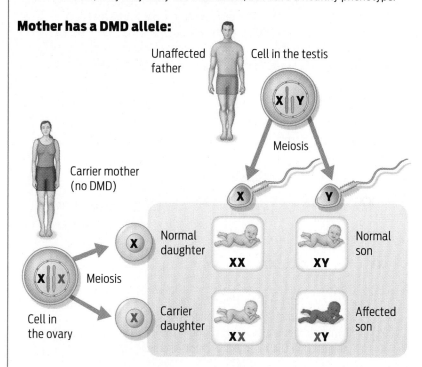

X has a normal dystrophin allele.

X has a DMD allele.

Mother does not have DMD, but passes the DMD allele to half her sons and daughters. Only the sons have DMD.

Father has a DMD allele:

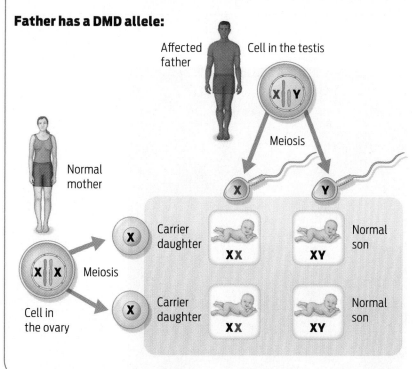

While most males with DMD do not survive long enough to have children, IF a male with DMD did have children, none of his sons would be affected and all of his daughters would carry the DMD allele.

Female Carriers Can Pass Disease Alleles to Their Children

→ The following diagram, known as a pedigree, shows how an X-linked trait passes through generations.

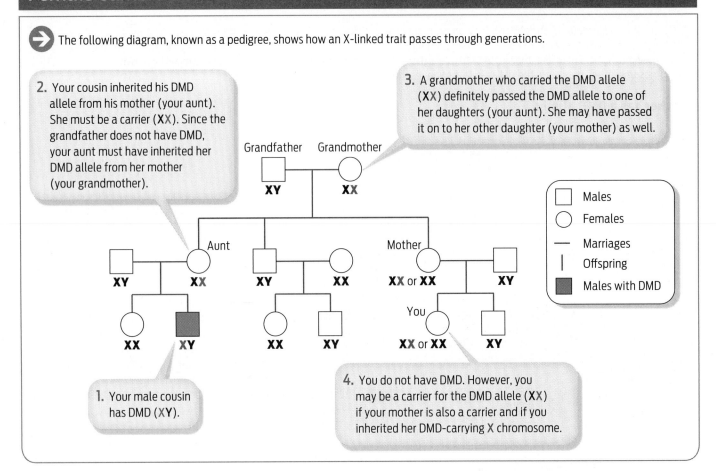

2. Your cousin inherited his DMD allele from his mother (your aunt). She must be a carrier (**XX**). Since the grandfather does not have DMD, your aunt must have inherited her DMD allele from her mother (your grandmother).

3. A grandmother who carried the DMD allele (**XX**) definitely passed the DMD allele to one of her daughters (your aunt). She may have passed it on to her other daughter (your mother) as well.

Grandfather **XY** — Grandmother **XX**

Aunt **XX** — **XY**

Mother **XX or XX** — **XY**

XX **XY**   **XX** **XY**   You **XX or XX** **XY**

| Males |
| Females |
| Marriages |
| Offspring |
| Males with DMD |

1. Your male cousin has DMD (**XY**).

4. You do not have DMD. However, you may be a carrier for the DMD allele (**XX**) if your mother is also a carrier and if you inherited her DMD-carrying X chromosome.

also a slave holder. Historians have long debated the meaning of these and other seeming contradictions in the founding father's life and politics. For example, although Jefferson's writings clearly show that he did not believe in the institution of slavery, he owned at least 200 slaves. He made disparaging comments about slaves, yet maintained close relationships with those living in his home. In fact, Jefferson was rumored to have fathered at least six children with Sally Hemings, a slave who tended to his family. For decades, historians discredited the rumor as unreliable oral history. But in 1997, DNA supported what many historians had discounted. Scientists tested the DNA of both Hemings's and Jefferson's descendants using a technique called Y-chromosome analysis. The results: Jefferson could have fathered at least one of Hemings's children.

Third U.S. president Thomas Jefferson.

Are these people descendants of Thomas Jefferson and Sally Hemings?

Y-chromosome analysis is commonly used to study ancestry and to identify paternity. It is just one of several ways that science can complement history. Scientists can use it to verify, discredit, or fill in missing pieces of historical information.

For example, scientists have used Y-chromosome analysis to show that 90% of the Cohanim–members of the Jewish priesthood–are related, supporting oral and written histories claiming the Cohanim are all descended from Aron, brother of Moses. They've also used Y-chromosome analysis to support the oral history of the Lemba, an African tribe, who claim they are descended from Jews. And Y-chromosome analysis suggests that about 8% of Eastern European and Asian men are descended from Genghis Khan or his family.

How does Y-chromosome analysis work? Sons inherit their Y chromosome from their bio-logical fathers. These Y chromosomes are passed through generations largely intact. That's because Y chromosomes have no homologous partner chromosome with which to pair and exchange DNA during meiosis. In other words, the Y chromosome rarely undergoes genetic recombination. Consequently, the Y chromosome that a son inherits from his father is almost identical to the Y chromosome that his father inherited from his father. In this way, Y chromosomes are transmitted essentially unchanged from fathers to sons.

Comparing DNA sequences on Y chromosomes can reliably establish paternity. In 1998, a team led by Eugene A. Foster, a pathologist, compared the Y chromosomes of four groups of men: descendants of Thomas Jefferson's grandfather Field Jefferson; descendants of Thomas Woodson, a man who claimed he was Sally Hemings and Thomas Jefferson's

Y Chromosomes Pass Largely Unchanged from Fathers to Sons

Paternity testing (Y-chromosome analysis) relies upon the fact that the Y chromosome does not undergo recombination during meiosis and so passes unchanged from the father to his sons.

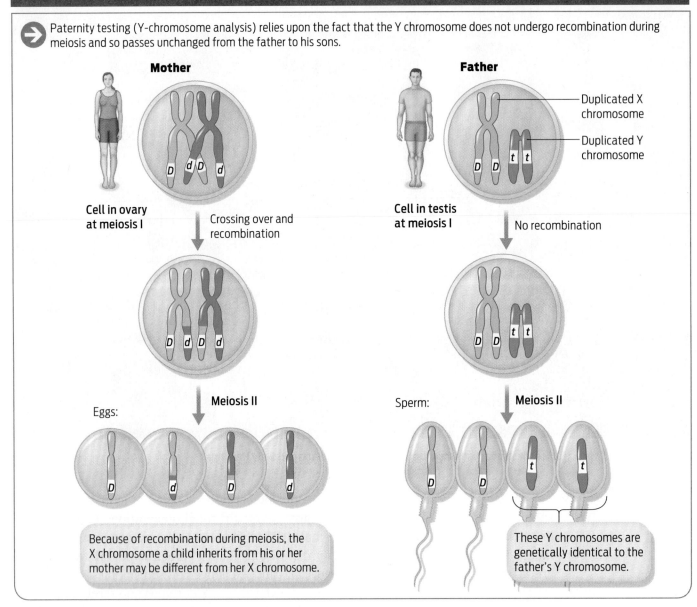

Mother

Cell in ovary at meiosis I

Crossing over and recombination

Meiosis II

Eggs:

Because of recombination during meiosis, the X chromosome a child inherits from his or her mother may be different from her X chromosome.

Father

Duplicated X chromosome

Duplicated Y chromosome

Cell in testis at meiosis I

No recombination

Meiosis II

Sperm:

These Y chromosomes are genetically identical to the father's Y chromosome.

first child; descendants of Eston Hemings, Sally Hemings's son; and descendants of John Carr, Jefferson's sister's son. Since Jefferson's only surviving child from his wife was a daughter, he did not have any direct male descendants, which is why scientists tested descendants of Jefferson's grandfather **(Infographic 12.4)**.

The study analyzed 11 short tandem repeats (STRs) on the Y chromosome. (Recall from Chapter 7 that STRs are regions of noncoding DNA that show differences in the number of times a short DNA sequence is repeated among different people.) The results showed that the man descended from Eston Hemings has the same Y chromosome as the descendants of Field Jefferson. Consequently, Thomas Jefferson could have fathered Eston Hemings. However, any male Jefferson could have fathered Hemings's son Eston **(Infographic 12.5)**.

In fact, some historians have argued that Thomas's brother Randolph Jefferson fathered Eston. But other experts have argued that historical evidence places Thomas himself rather than Randolph under the same roof as Sally at the time of her conceptions.

DNA Links Sally Hemings's Son to Jefferson

➡ Scientists compared DNA sequences on the Y chromosome of Sally Hemings's and Thomas Jefferson's grandfather's descendants. The DNA sequences match at the eleven different STR locations analyzed.

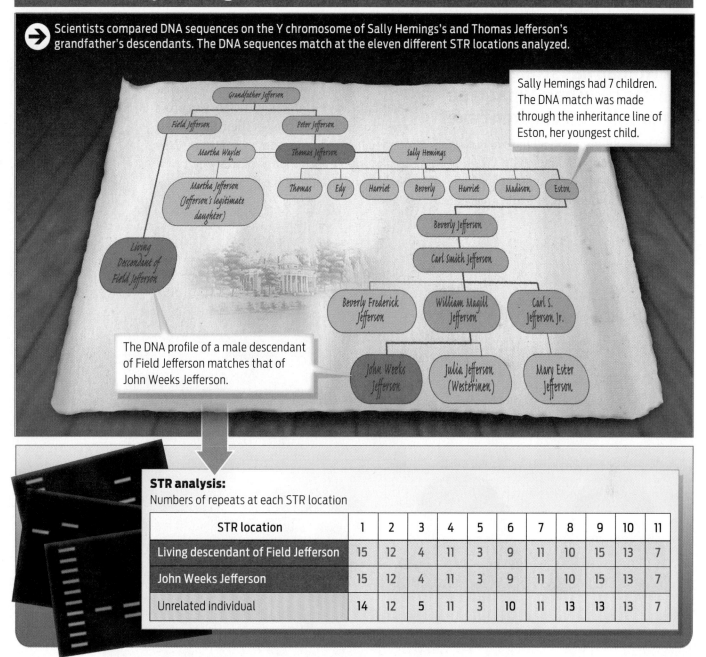

Sally Hemings had 7 children. The DNA match was made through the inheritance line of Eston, her youngest child.

The DNA profile of a male descendant of Field Jefferson matches that of John Weeks Jefferson.

STR analysis:
Numbers of repeats at each STR location

STR location	1	2	3	4	5	6	7	8	9	10	11
Living descendant of Field Jefferson	15	12	4	11	3	9	11	10	15	13	7
John Weeks Jefferson	15	12	4	11	3	9	11	10	15	13	7
Unrelated individual	14	12	5	11	3	10	11	13	13	13	7

For the descendants of Eston Hemings, the DNA study was powerful vindication. They had long argued that they were descended from Thomas Jefferson, but without hard evidence, most historians disregarded their claims. "I feel wonderful about it," Julia Jefferson Westerinen, a Staten Island artist and Eston's great-great-granddaughter told the *New York Times* when the study results were published. "I feel honored."

As for the relationship between Jefferson and Sally Hemings, historians continue to debate whether it was consensual or forced. "I was a history major," said Jefferson Westerinen, "And we learned not to say, 'I feel this, I think that,' without knowing the facts. They had a relationship of 38 years. I would like to think they were in love, but how would I know?"

We'll likely never know the truth. Neither Thomas Jefferson nor Sally Hemings left any written evidence of their relationship.

Q A woman has straight hair, a man has curly hair. What type of hair will their children have?

A These children will all have an intermediate phenotype: wavy hair. Like flower color in some plants, hair type is an example of **incomplete dominance,** a form of inheritance in which heterozygotes have a phenotype intermediate between homozygous dominant and homozygous recessive.

There are two versions of the gene that specify hair texture, straight (*h*) and curly (*H*). People with straight hair are homozygous recessive (*hh*); they don't produce any protein that makes hair wavy. Heterozygotes (*Hh*) express some protein, which makes their hair wavy. People who are homozygous dominant (*HH*) express double the amount of wavy hair protein and consequently have curly hair (**Infographic 12.6**).

There are two versions of the gene that specify hair texture, but three phenotypes: curly, straight, and wavy.

Hair Texture Exhibits Incomplete Dominance

Incomplete dominance means that heterozygotes display a phenotype intermediate between homozygous dominant homozygous recessive. Hair texture is an example. There are two alleles of the gene that specify hair texture, and curly (*H*) and straight (*h*).

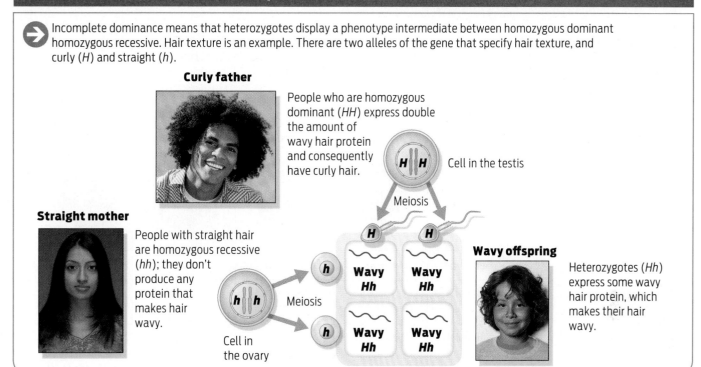

Curly father

People who are homozygous dominant (*HH*) express double the amount of wavy hair protein and consequently have curly hair.

Cell in the testis

Meiosis

Straight mother

People with straight hair are homozygous recessive (*hh*); they don't produce any protein that makes hair wavy.

Meiosis

Cell in the ovary

Wavy offspring

Heterozygotes (*Hh*) express some wavy hair protein, which makes their hair wavy.

Wavy *Hh* · Wavy *Hh* · Wavy *Hh* · Wavy *Hh*

CODOMINANCE

Q Who can be a universal blood donor?

A When a person needs a blood transfusion, the donated blood cannot come from just anyone. The transfused blood must match the recipient in ways that are determined by genetics. The two most important genetic attributes are ABO blood type and Rhesus (Rh) factor. Your blood type indicates the presence of specific molecules on the surface of your red blood cells. Your Rh status, (+) or (-), indicates the presence or absence of Rh proteins on the surface of your red blood cells. Both ABO blood type and Rh factor must match between donor and recipient; mixing incompatible blood causes blood cells to clump, which is life threatening for people receiving transfusions.

There are three basic blood type alleles: *A*, *B*, and *O*. Since we inherit one allele from each parent, the possible combinations of the three alleles are *OO, AO, BO, AB, AA,* and *BB*. Blood type is an example of **codominance**–both

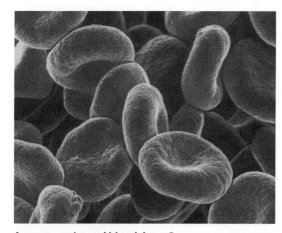

Are you a universal blood donor?

maternal and paternal alleles contribute equally and separately to the phenotype. Unlike incomplete dominance in which heterozygotes carry an intermediate phenotype, codominant traits share the limelight: heterozygotes express both phenotypes.

Blood type alleles *A* and *B* are codominant, while *O* is recessive to both *A* and *B*. Consequently, if you have blood type A, your genotype will either be *AA* homozygous or *AO*

CODOMINANCE
A form of inheritance in which both alleles contribute equally to the phenotype.

Human Blood Type Is a Codominant Trait

→ In codominant inheritance, heterozygotes display the effects of both alleles in their phenotype. Human blood type is an example. Alleles for blood type code for different surface markers on red blood cells. A person with type AB blood, for example, displays both A and B markers, while type O blood displays no surface markers. A person's blood type must be considered when he or she gives or receives blood.

Blood Transfusions:
The ability to donate or receive blood is based on immune rejection. If two people have the same surface markers, then their blood will be compatible. People with type O blood have no surface markers to provoke an immune response in a recipient (so O is the universal donor). People with type AB blood will not recognize either marker as foreign, so can receive blood from any donor.

	Type A markers	Type B markers	Type A and B markers	No blood group markers
Red blood cell type:	A	B	AB	O
Genotype:	*AA* or *AO*	*BB* or *BO*	*AB*	*OO*
Can donate to:	Type A or AB recipient	Type B or AB recipient	Type AB recipient	Type A, B, AB, or O recipient
Can receive from:	Type A or O donor	Type B or O donor	Type A, B, AB, or O donor	Type O donor

heterozygous. The same goes for blood type B **(Infographic 12.7)**.

By contrast, Rh factor genes are inherited in a dominant and recessive fashion. The positive Rh factor allele (Rh+) is dominant over the negative Rh factor allele (Rh-). So if a person carries one positive and one negative allele, the positive allele will dominate and the person will have an Rh-positive phenotype.

Type O Rh negative donors are known as universal donors because their blood can be transfused to patients of any other blood type without causing an immune reaction. Because any patient can receive O Rh negative blood, O negative donors are always in demand. Blood banks can fall short of type O Rh negative blood during such disasters as earthquakes or hurricanes in which many people are hurt and require blood **(Infographic 12.8)**.

MULTIFACTORIAL INHERITANCE

Q How much of human height is inherited?

A The short answer is, a lot. Experts estimate that height is 60% to 80% inherited. In other words, genes determine 60% to 80% of the difference in height you see from person to person; the rest is determined by environmental factors such as nutrition. But there isn't one single gene that determines height—there are several. You may have noticed that your own height differs from that of your parents, or that your siblings are all of different heights. That's because height is an example of a **polygenic trait**—a single trait that is determined by more than one gene. In fact, more than 20 different parts of the genome have been implicated in influencing a person's height, which is why we

POLYGENIC TRAIT
A trait whose phenotype is determined by the interaction among alleles of more than one gene.

INFOGRAPHIC 12.8

A Mismatched Blood Transfusion Causes Immune Rejection

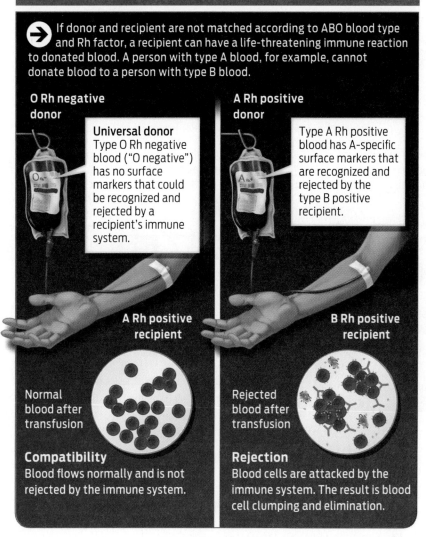

→ If donor and recipient are not matched according to ABO blood type and Rh factor, a recipient can have a life-threatening immune reaction to donated blood. A person with type A blood, for example, cannot donate blood to a person with type B blood.

O Rh negative donor

Universal donor
Type O Rh negative blood ("O negative") has no surface markers that could be recognized and rejected by a recipient's immune system.

A Rh positive recipient

Normal blood after transfusion

Compatibility
Blood flows normally and is not rejected by the immune system.

A Rh positive donor

Type A Rh positive blood has A-specific surface markers that are recognized and rejected by the type B positive recipient.

B Rh positive recipient

Rejected blood after transfusion

Rejection
Blood cells are attacked by the immune system. The result is blood cell clumping and elimination.

see such a range of heights among us. When many genes act together, their effects are cumulative—they add up to determine, for example, a person's height.

Height is an example of a trait that shows continuous variation in any given population. In the United States, most people fall between 5 feet and 6.2 feet tall, and women tend to be shorter than men. This means that if one were to plot height on a graph, the result would resemble a bell curve, with most people falling between these two heights. This is in contrast to the discrete traits we've encountered, in which individuals have one of only two or three possible phenotypes for a given trait—round or wrinkled pea plants, or AB blood type, for example. With traits that vary continuously, such as height, there are many possible phenotypes in the population, and individuals can vary by as little as half an inch. Other examples of polygenic traits include skin color and eye color.

Even though height is largely determined by genes (60% to 80%), another 20% to 40% is determined by environmental factors such as nutrition. Why such large variation? In developed countries, where most people have access to adequate nutrition, height is more than 80% heritable. This means that when scientists compare the height of a person to his or her relatives, they find that height varies only about 20% among direct relatives, that is, relatives in a direct line of descent—grandparents, parents,

Height is an example of a trait that shows continuous variation in any given population.

Human Height Is Both Polygenic and Multifactorial

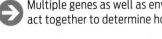

Multiple genes as well as environmental factors such as diet, nutrition, and overall health act together to determine how tall we become.

Polygenic (blue bars on graph):
Many genes contribute to determine one's height. The combination of alleles a person inherits (*aabbcc*, *AabbCc*, etc.) predicts a distinct height phenotype.

Multifactorial (red line on graph):
Human populations show a continuous range of heights, however, rather than the genetically predicted number of distinct phenotypes. This is due to environmental influences interacting with one's genetically determined potential for height.

A mating between two people with medium height (where three genes control height): *AaBbCc* × *AaBbCc* produces seven distinct phenotypes determined by the number of dominant genes inherited.

Male gametes

	ABC	*ABc*	*AbC*	*aBC*	*Abc*	*aBc*	*abC*	*abc*
ABC	*AABBCC*	*AABBCc*	*AABbCC*	*AaBBCC*	*AABbCc*	*AaBBCc*	*AaBbCC*	*AaBbCc*
ABc	*AABBCc*	*AABBcc*	*AABbCc*	*AaBBCc*	*AABbcc*	*AaBBcc*	*AaBbCc*	*AaBbcc*
AbC	*AAbBCC*	*AAbBCc*	*AAbbCC*	*AabBCC*	*AAbbCc*	*AabBCc*	*AabbCC*	*AabbCc*
aBC	*aABBCC*	*aABBCc*	*aABbCC*	*aaBBCC*	*aABbCc*	*aaBBCc*	*aaBbCC*	*aaBbCc*
Abc	*AABBCc*	*AABbCc*	*AAbbCc*	*AabBCc*	*AAbbcc*	*AabBcc*	*AabbcC*	*Aabbcc*
aBc	*aABBCc*	*aABBcc*	*aABbCc*	*aaBBCc*	*aABbcc*	*aaBBcc*	*aaBbcC*	*aaBbcc*
abC	*aAbBCC*	*aAbBCc*	*aAbbCC*	*aabBCC*	*aAbbCc*	*aabBcC*	*aabbCC*	*aabbCc*
abc	*aAbBcC*	*aAbBcc*	*aAbbcC*	*aabBcC*	*aAbbcc*	*aabBcc*	*aabbcC*	*aabbcc*

Female gametes

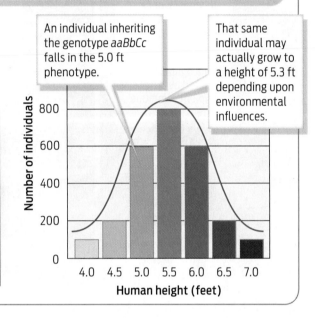

An individual inheriting the genotype *aaBbCc* falls in the 5.0 ft phenotype.

That same individual may actually grow to a height of 5.3 ft depending upon environmental influences.

Number of individuals / Human height (feet)

children, and so on. In developing countries, where many people are still malnourished, environment plays a larger role. Another way of looking at this is that more people in developed countries have reached their genetic potential than people in developing countries because most of us in the developed world have access to adequate nutrition. In developing countries, access to nutrition varies much more and this variation is reflected in larger variations in height between a person and his or her direct relatives. In fact, the average height of the U.S. population has almost leveled off in the past decade, suggesting that the environment has almost maximized the genetic potential of height in this country.

When both genes and environment work together to influence a given trait, the trait is described as multifactorial. So height is both polygenic and multifactorial (**Infographic 12.9**).

Multifactorial inheritance is a common pattern of inheritance. Diseases such as asthma, diabetes, and heart disease are all caused by a combination of several genes and their interaction with the environment. For example, some studies have found that cigarette smoke, air pollution, and ozone can exacerbate asthma. Other studies have shown that people who carry the *E4* allele of a gene called *APO* have an increased chance of developing heart disease if they smoke and don't exercise, compared to people with other *APO E* alleles who smoke and don't exercise. Even for traits that are largely genetically determined, the environment plays a very important role in influencing our phenotypes.

MULTIFACTORIAL INHERITANCE
An interaction between genes and the environment that contributes to a phenotype or trait.

Depression cannot be explained by genetic or environmental factors alone, but by an interaction between the two.

Q Can people be genetically predisposed to depression?

A In the early 1990s, Stephen Suomi and Dee Higley, researchers at the National Institute of Child Health and Human Development, were studying how stress affected the mental development of infant monkeys. More specifically, they were looking at whether certain alleles of a specific gene, called serotonin transporter, made infant monkeys more vulnerable to stress early in life.

The serotonin transporter gene is located on chromosome 17; it exists as two alleles, a short version and a long version. The long version contains about 44 extra base pairs. Previous research had suggested that people who had at least one copy of a short version of this gene were much more likely to have an anxiety disorder.

Higley and Suomi showed that infant monkeys exposed to stress, such as being deprived of their mothers, and who carried short versions of this allele, behaved differently from their counterparts: they were more anxious, aggressive, and some even became alcoholics as adults.

Despite this finding in monkeys, it quickly became clear that having short versus long alleles could not explain why some people become severely depressed while others are more resilient. Researchers could not find a clear association between any particular allele and depression in people.

Taking their cue from Higley and Suomi, in 2003 Terrie Moffitt and Avshalom Caspi, a husband-and-wife team of psychologists at King's College London, decided to test whether environmental influences might also contribute to depression in people. Moffitt and Caspi turned to a long-term study of almost 900 New Zealanders, identified these subjects' serotonin transporter alleles, and interviewed them about traumatic experiences in early adulthood–experiences such as a major breakup, a death in the family, or serious injury–to see if these difficulties brought out an underlying genetic tendency toward depression.

The results were striking: clinical depression was diagnosed in 43% of subjects who had two copies of the short allele and who had experienced four or more tumultuous events. By contrast, only 17% of subjects who had two copies of the long allele and who had endured four or more stressful events had become depressed–this was no more than the rate of depression in the general population. Subjects with the short allele who experienced no stressful events fared pretty well, too–they also became depressed at the average rate. Clearly, it was the combination of hard knocks

Serotonin Transporter Function Is Linked to Depression

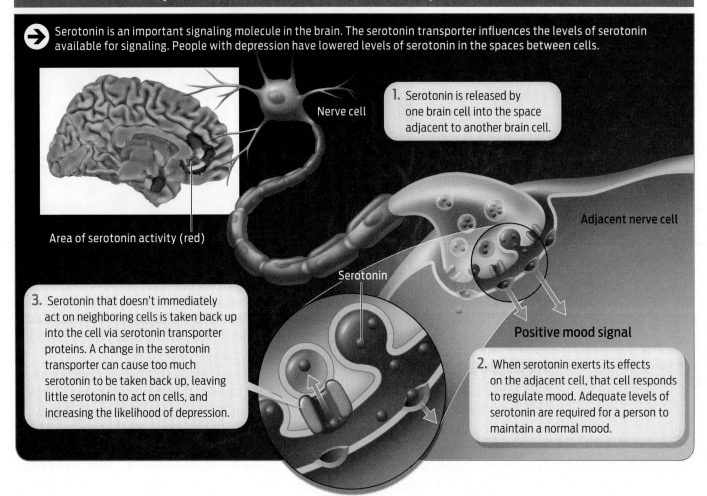

→ Serotonin is an important signaling molecule in the brain. The serotonin transporter influences the levels of serotonin available for signaling. People with depression have lowered levels of serotonin in the spaces between cells.

Nerve cell

Area of serotonin activity (red)

1. Serotonin is released by one brain cell into the space adjacent to another brain cell.

Adjacent nerve cell

Serotonin

3. Serotonin that doesn't immediately act on neighboring cells is taken back up into the cell via serotonin transporter proteins. A change in the serotonin transporter can cause too much serotonin to be taken back up, leaving little serotonin to act on cells, and increasing the likelihood of depression.

Positive mood signal

2. When serotonin exerts its effects on the adjacent cell, that cell responds to regulate mood. Adequate levels of serotonin are required for a person to maintain a normal mood.

and short alleles that more than doubled the risk of depression.

Since the early 1990s, researchers have shown that the serotonin transporter gene influences the levels of serotonin present in the spaces between brain cells in humans and other animals, and that low levels of serotonin in these spaces is one biological hallmark of depression in people (**Infographic 12.10**). But there are likely other factors that contribute to a person's risk of depression—there are many people who carry long alleles and who suffer from depression, as well as people who carry short alleles who do not, despite having gone through distressing experiences.

Nevertheless, Caspi and Moffitt's study was one of the first to examine the combined effects of genetic predisposition and experience on a specific trait—psychiatrists were delighted. While scientists had been trying to

tease apart environmental from genetic influences on physical diseases like cancer, this was the first study to investigate this relationship in a mental disorder. Moreover, the findings reinforced the emerging view that the majority of mental illnesses and other complex diseases cannot be explained by genetic or environmental factors alone, but often arise from an interaction between the two. That is, mental illnesses exhibit multifactorial inheritance (**Infographic 12.11**).

NONDISJUNCTION

Q Why does the risk of having a baby with Down syndrome go up as a woman ages?

A At age 25, a woman's risk of having a baby with Down syndrome is 1 in 1,250 births. At age 40 her risk skyrockets to 1 in 100 births.

INFOGRAPHIC 12.11

Depression Is a Multifactorial Trait

→ In 2003, Terrie Moffitt and Avshalom Caspi showed that a specific allele of the serotonin transporter gene—a gene that influences levels of the signaling molecule serotonin in the brain—in combination with stressful life events can cause depression. The gene comes in long and short versions.

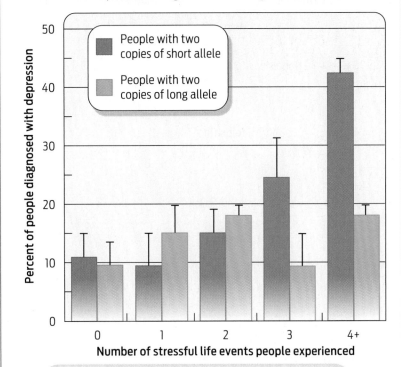

Legend:
- People with two copies of short allele
- People with two copies of long allele

Y-axis: Percent of people diagnosed with depression (0 to 50)

X-axis: Number of stressful life events people experienced (0, 1, 2, 3, 4+)

People with two copies of short allele who also experienced four or more stressful events were more than twice as likely to become depressed as those with two copies of the long allele who also experienced stressful events.

Source: Caspi, A. et al. 2003. *Science.* 301(5631):386–389.

ANEUPLOIDY
An abnormal number of one or more chromosomes (either extra or missing copies).

NONDISJUNCTION
Failure of chromosomes to separate accurately during cell division; nondisjunction in meiosis leads to aneuploid gametes.

TRISOMY 21
Carrying an extra copy of chromosome 21; also known as Down syndrome.

an error in chromosome segregation, leading to a chromosomal abnormality.

A chromosomal abnormality means that a developing fetus carries a chromosome number that differs from the usual 46. The most common abnormalities in humans are called **aneuploidies,** deviations from the normal number of chromosomes because single chromosomes are either duplicated or deleted. Most aneuploidies arise during meiosis in the parents' sex cells. If a gamete makes a mistake when chromosomes segregate, an occurrence called **nondisjunction,** it will either lack a chromosome or carry an extra copy. When that egg is fertilized by a normal gamete, the resulting zygote can have an abnormal number of chromosomes. In most cases, the abnormality is so severe the zygote spontaneously aborts.

There are, however, cases in which the abnormality is not life threatening but does cause severe disability–the most common is **trisomy 21,** also called Down syndrome. Trisomy 21 results when an embryo inherits an extra copy of chromosome 21. Anyone can conceive a child with the abnormality, but older women are at exceptionally high risk **(Infographic 12.12).**

Most Down syndrome children have learning disabilities that range from mild to moderate, but some have profound mental disability. They

In fact, as women age, the risk of giving birth to a baby with any chromosomal abnormality increases. That's because as a woman ages, so do her eggs. All the eggs that a woman will ever have were formed before she was born, and they have been aging like the rest of the cells in her body. Until puberty, a woman's eggs are "paused" in the middle of meiosis (at meiosis I); they haven't yet completed their cell division. During a menstrual cycle, one egg resumes meiosis and is ovulated. In older women, when these eggs complete meiosis and are ovulated, they are more likely to have

Chromosomal Abnormalities: Aneuploidy

→ Birth defects can arise when chromosomes fail to separate normally during meiosis, a phenomenon called nondisjunction. The resulting gametes carry an abnormal number of chromosomes, a condition called aneuploidy.

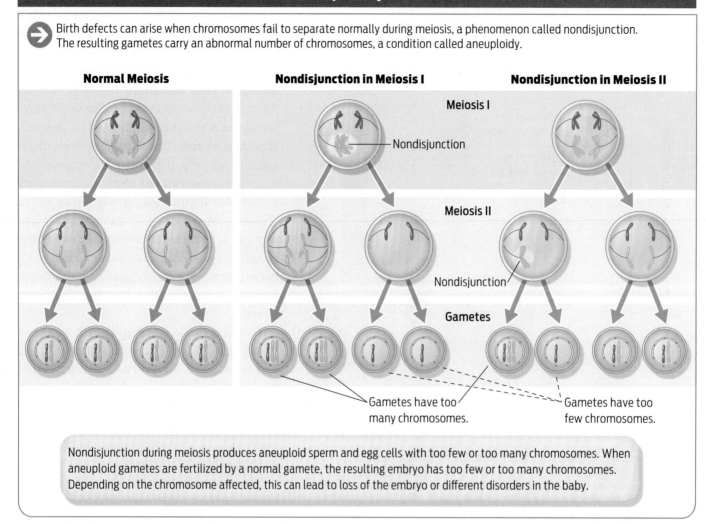

Normal Meiosis **Nondisjunction in Meiosis I** **Nondisjunction in Meiosis II**

Meiosis I

Nondisjunction

Meiosis II

Nondisjunction

Gametes

Gametes have too many chromosomes.

Gametes have too few chromosomes.

Nondisjunction during meiosis produces aneuploid sperm and egg cells with too few or too many chromosomes. When aneuploid gametes are fertilized by a normal gamete, the resulting embryo has too few or too many chromosomes. Depending on the chromosome affected, this can lead to loss of the embryo or different disorders in the baby.

are also at higher risk for other diseases and typically don't live beyond 50 years of age.

Down syndrome, as well as other chromosomal abnormalities, can be diagnosed by **amniocentesis.** This procedure is usually performed between 14 and 20 weeks of pregnancy, although some medical centers may perform it as early as 11 weeks. The procedure is quick. A long, thin, hollow needle is inserted through a woman's abdominal wall and into her uterus. Through the needle, the equivalent of 2 to 4 teaspoons of amniotic fluid, which surrounds the growing fetus, is removed. This fluid contains fetal cells that contain the fetus's DNA. From that fluid, technicians analyze the fetal **karyotype**—that is, the chromosomal makeup in its cells **(Infographic 12.13).**

The reasons to undergo amniocentesis vary from couple to couple. But if a test comes back positive, couples have options: they can begin to plan for a disabled child, or make the decision not to carry the child to term.

Although scientists have linked some of the most obvious birth defects to the age of a woman's eggs, recent research also shows that a man's age affects his sperm quality. Men who father children after age 45 are more likely to have children with cognitive disorders such as autism, for example. Male fertility declines over time, too, although much more gradually than does female fertility. Research shows that the older the man, the more likely he is to produce sperm with genetic defects. ■

AMNIOCENTESIS
A procedure that removes fluid surrounding the fetus to obtain and analyze fetal cells to diagnose genetic disorders.

KARYOTYPE
The chromosomal makeup of cells. Karyotype analysis can be used to detect trisomy 21 prenatally.

Amniocentesis Provides a Fetal Karyotype

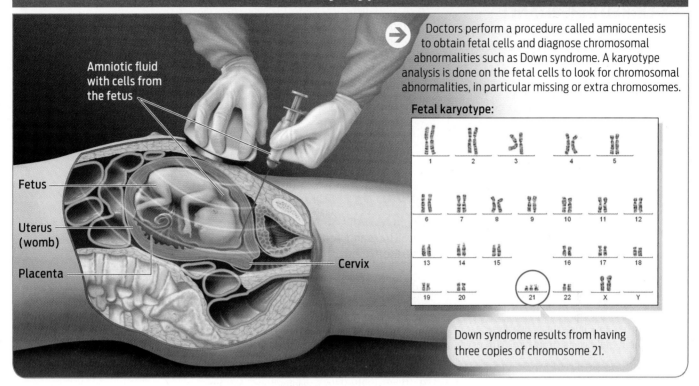

Amniotic fluid with cells from the fetus

Fetus

Uterus (womb)

Placenta

Cervix

Doctors perform a procedure called amniocentesis to obtain fetal cells and diagnose chromosomal abnormalities such as Down syndrome. A karyotype analysis is done on the fetal cells to look for chromosomal abnormalities, in particular missing or extra chromosomes.

Fetal karyotype:

Down syndrome results from having three copies of chromosome 21.

▶ Summary

■ Humans have 23 pairs of chromosomes. One of these pairs is the sex chromosomes: XX in females and XY in males. It is the presence of the Y chromosome that determines maleness.

■ Because the Y chromosome in a male does not have a homologous partner, it does not experience recombination during meiosis. The Y chromosome a son inherits from his father is essentially identical to the Y chromosome his father inherited from his father (the grandfather), a fact that can be used to establish paternity.

■ Some genes are located on the X chromosome; these are known as X-linked genes. Disorders inherited on X chromosomes are called X-linked disorders, and are more common in males than in females.

■ Hair type is an example of incomplete dominance, a form of inheritance in which heterozygotes have a phenotype intermediate between homozygous dominant and homozygous recessive.

■ ABO blood type is an example of a codominant trait—both maternal and paternal alleles contribute equally and separately to the phenotype.

■ Many traits are polygenic—that is, they are influenced by the additive effects of multiple genes. Polygenic traits show a normal distribution in the population.

■ In many cases, a person's phenotype is determined by both his or her genotype at a number of different genes as well as by environmental influences; this type of inheritance is described as multifactorial. Human height, cardiovascular disease, and depression are examples of multifactorial inheritance.

■ Some genetic disorders result from having a chromosome number that differs from the usual 46. Down syndrome, or trisomy 21, is caused by having an extra copy of chromosome 21.

SEX-LINKED INHERITANCE

The two human sex chromosomes are X and Y. Genes located on the sex chromosomes are said to be sex linked.

HINT **See Infographics 12.1–12.5.**

⊙ KNOW IT

1. Which of the following most influences the development of a female fetus?
   a. the presence of any two sex chromosomes
   b. the presence of two X chromosomes
   c. the absence of a Y chromosome
   d. the presence of a Y chromosome
   e. either b or c

2. Why are more males than females affected by X-linked recessive genetic diseases?

3. If a man has an X-linked recessive disease, can his sons inherit that disease from him? Why or why not?

⊙ USE IT

4. Which of the following couples could have a boy with Duchenne muscular dystrophy?
   a. a male with Duchenne muscular dystrophy and a homozygous dominant female
   b. a male without Duchenne muscular dystrophy and a homozygous dominant female
   c. a male without Duchenne muscular dystrophy and a carrier female
   d. a and c
   e. none of the above

5. Predict the sex of a baby with each of the following pairs of sex chromosomes. (You may want to use this question to go back and check your answer to Question 1.)
   a. XX
   b. XXY
   c. XY
   d. X

6. Consider your brother and your son.
   a. If you are female, will your brother and your son have essentially identical Y chromosomes? Explain your answer.
   b. If you are male, will your brother and your son have essentially identical Y chromosomes? Explain your answer.

7. A wife is heterozygous for Duchenne muscular dystrophy alleles and her husband has a dominant allele on his X chromosome. What percentage of their sons, and what percentage of their daughters, will have:
   a. Duchenne muscular dystrophy (which is determined by a recessive allele on the X chromosome)
   b. an X-linked dominant form of rickets (a bone disease)

OTHER PATTERNS OF INHERITANCE

Not all traits are inherited in simple dominant and recessive patterns.

HINT **See Infographics 12.6–12.11.**

⊙ KNOW IT

8. What aspects of height make it a polygenic trait?

9. Which of the following inheritance patterns includes an environmental contribution?
   a. polygenic
   b. X-linked recessive
   c. X-linked dominant
   d. multifactorial
   e. none of the above

10. What is the difference between polygenic inheritance and multifactorial inheritance?

11. How does incomplete dominance differ from codominance?

12. If you are blood type A-positive, to whom can you safely donate blood? Who can safely donate blood to you? List all possible recipients and donors and explain your answer.

⊙ USE IT

13. If two women have identical alleles of the suspected 20 height-associated genes, why might one of those women be 5 feet 5 inches tall and the other 5 feet 8 inches tall?

14. Look at Infographic 12.11. How do the data given support the hypothesis that both genes and the environment influence at least some cases of clinical depression?

15. Look at Infographic 12.11. At approximately how many stressful experiences does the homozygous short genotype begin to influence the depression phenotype?

16. From information in this chapter, how can you account for two people with the same genotype for a predisposing disease allele having different phenotypes?

CHROMOSOMAL ABNORMALITIES

Improper chromosome segregation during cell division can lead to birth defects.

HINT See Infographics 12.12 and 12.13.

➔ KNOW IT

17. What is the normal chromosome number for each of the following:
- **a.** a human egg
- **b.** a human sperm
- **c.** a human zygote

18. When looking at a karyotype, for example to diagnose trisomy 21 in a fetus, is it possible to use that analysis also to tell if the fetus has inherited a cystic fibrosis allele from a carrier mother?

➔ USE IT

19. Which of the following can result in trisomy 21?
- **a.** an egg with 23 chromosomes fertilized by a sperm with 23 chromosomes
- **b.** an egg with 22 chromosomes fertilized by a sperm with 23 chromosomes
- **c.** an egg with 24 chromosomes, two of which are chromosome 21, fertilized by a sperm with 23 chromosomes
- **d.** an egg with 23 chromosomes fertilized by a sperm with 24 chromosomes, two of which are chromosome 21

20. From information in this chapter, which of the possibilities in Question 19 is most likely? Explain your answer.

SCIENCE AND ETHICS

21. What factors would lead you to consider prenatal genetic testing? In your opinion, what is the value of having this information?

Grow Your Own

Grow Your Own

Stem cells could be the key to engineering organs

In 1995, Charles Vacanti, an anesthesiologist, and Linda Griffith-Cima, then an assistant professor of chemical engineering at the Massachusetts Institute of Technology (MIT), amazed the world with an unusual and important experiment. Under the skin of a laboratory mouse, they injected cow cartilage cells into an implanted and biodegradable mold shaped like a human ear. The result? A structure of cartilage, shaped like a human ear, grew on the mouse's back. The sensational image splashed across tabloids, and many hailed the feat as a great scientific accomplishment–but some likened it to creating Frankenstein's monster. The experiment had a serious purpose, however: the mouse's body nurtured the ear as it grew, and once the ear was large enough, a surgeon could remove it and attach it to someone whose ear was missing. Doctors never actually transplanted the ear; the scientists merely intended to demonstrate the possibilities of tissue engineering. Lose an ear because of an accident? Doctors can grow you a new one.

Today, more than a decade later, we know that the mouse experiment helped pave the way for an entirely new kind of transplanted organ, one grown from a patient's own cells. In 2006, Anthony Atala, director of the Wake Forest University Institute for Regenerative Medicine,

Cartilage cells grow on a mouse back into the shape of a human ear.

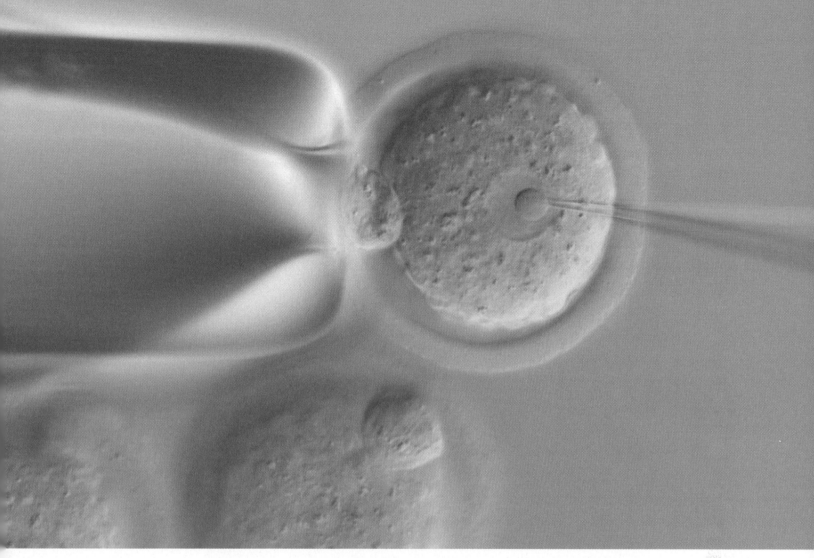

Scientists can manipulate the nucleus of a single cell to unlock its therapeutic potential.

The mouse experiment helped pave the way for an entirely new kind of transplanted organ.

announced that he and his colleagues had successfully transplanted engineered human bladders into several children and teenagers. Although scientists had been growing human tissue outside the body for years, the bladder transplants were the first time that a damaged organ was repaired with a person's own cells.

"It was very significant work," says William Wagner, deputy director of the McGowan Institute for Regenerative Medicine at the University of Pittsburgh. "He's overcome a huge number of challenges."

The potential applications of the technique are enormous. For years transplant surgeons have worked to help patients suffering from organ failure who are in need of an organ donor. But each year, the demand for organs such as hearts, livers, and kidneys vastly exceeds supply. Last year, for example, surgeons transplanted about 30,000 organs, according to the Organ Procurement and Transplantation Network. Meanwhile, there are about 100,000 people waiting for an organ transplant. And even when an organ does become available, the recipient's body may reject the organ because the donor and recipient immune systems are not compatible–leaving the patient sicker than before the transplant.

Growing organs from a person's own cells would not only sidestep organ rejection, it would also eliminate the need for donors. A decade ago, most scientists considered such a

Cells Are Organized into Tissues, Organs, and Systems

> Tissues are integrated groups of specialized cells working together. Multiple tissues combine to form organs, which in turn cooperate as part of a single functioning organ system.

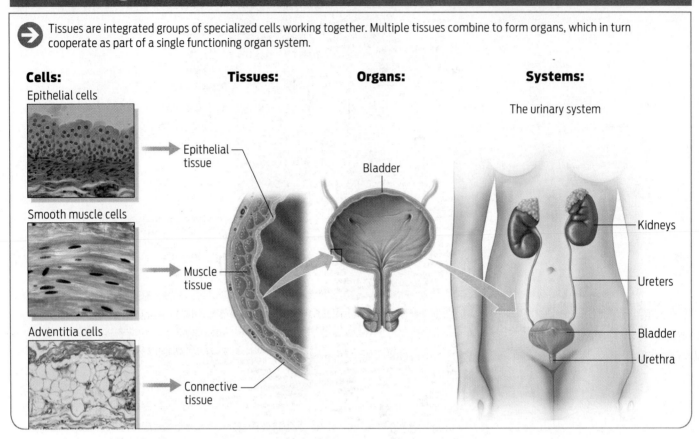

Cells:
Epithelial cells

Smooth muscle cells

Adventitia cells

Tissues:
Epithelial tissue

Muscle tissue

Connective tissue

Organs:
Bladder

Systems:
The urinary system

Kidneys

Ureters

Bladder

Urethra

---

feat a pipe dream. They knew that human **tissues** consisted of integrated groups of different and specialized cells that together perform a specific function **(Infographic 13.1)**. But controlling tissues to repair organs seemed impossible.

More recently, researchers discovered that most body tissues contain pools of **stem cells**—immature cells that can spontaneously divide repeatedly and give rise to more specialized cell types in the body. For example, stem cells found in bone, heart, and brain tissues help regenerate those tissues and organs **(Infographic 13.2)**.

The discovery has fueled a search for ways to harness the regenerative potential of stem cells to treat ailing patients. In addition to using stem cells to create organs for transplant, scientists also hope one day to tap

> ## Lose an ear because of an accident? Doctors can grow you a new one.

into the body's natural healing processes and coax stem cells into healing damaged tissue.

From Ears to Bladders

The effort to engineer human tissue for transplants dates back almost 40 years and is based on the knowledge that the majority of our cells continuously die and are replaced by new cells. Without such cell division, an organism would neither grow nor heal. During early development, for example, a single fertilized egg cell divides to begin to form an embryo, and these cells divide again and again to form millions of cells by the time the embryo becomes a fetus. As we age, the body discards old cells and generates new ones in their place. And when we cut or injure ourselves, cells in the area undergo cell division to heal the injury. Transplant science tries to

TISSUE
An organized group of different cell types that work together to carry out a particular function.

STEM CELLS
Immature cells that can divide and differentiate into specialized cell types.

Stem Cells in Tissues Have Regenerative Properties

→ Stem cells in various tissues divide to produce more stem cells and the specialized cells that make up that tissue. In this way, stem cells help keep the tissues in which they reside healthy.

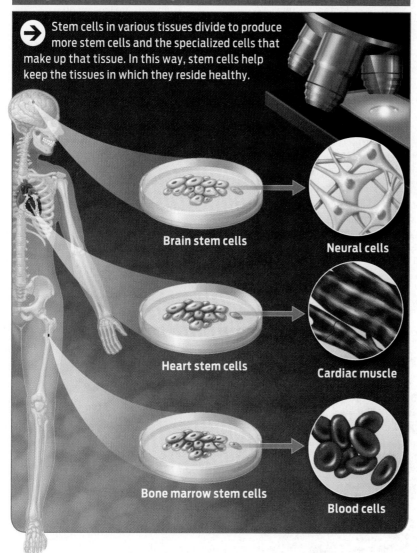

Brain stem cells

Neural cells

Heart stem cells

Cardiac muscle

Bone marrow stem cells

Blood cells

skin to grow to cover the wound. Their method is still used today to treat burn victims, patients who undergo plastic surgery, and patients with recurrent skin wounds.

But the field made its largest strides in the late 1980s, when Joseph Vacanti of Boston's Children's Hospital teamed up with Robert Langer at MIT to engineer tissues. The pair wanted to design synthetic biodegradable scaffolds that could be molded into particular shapes—a human ear, for example—and then coat the mold with cells that would grow into a tissue. The scaffold would never need to be removed—it would in time dissolve. Vacanti's brother, Charles, and Linda Griffith-Cima used this technique to grow a "human" ear on a mouse's back.

About the time of Joseph Vacanti and Langer's achievement, Anthony Atala, who had collaborated with Joseph Vacanti and Langer, applied some of this research on biodegradable scaffolds to his own work on engineered bladders. Atala, a urologist, sought to help his patients whose bladders were not functioning normally because of cancer, injury, or an inborn defect. For about a century, doctors have treated such patients by using pieces of their stomach or bowel to reconstruct their bladders. But because the procedure requires surgically removing pieces of healthy tissue, it is not an ideal treatment. A better option would be to grow a piece of new bladder tissue to repair the organ.

harness this natural ability of the body to grow and heal.

So far, progress has been incremental but steady. In the early 1970s, W. T. Green, an orthopedist, tried to grow cartilage tissue outside the body. He placed cartilage cells onto scaffolds made out of bone to try to get the cells to grow in specific formations. Although he was unsuccessful, his experiments set the stage for growing cells on scaffolds. Several years later, John Burke of Harvard Medical School and Ioannis Yannas of MIT developed a method using scaffolds transplanted into wound areas that stimulated both the dermis and the epidermis of the

Langer was one of the first to devise a way to grow tissue on biodegradable scaffold.

Engineering an Organ Using Stem Cells

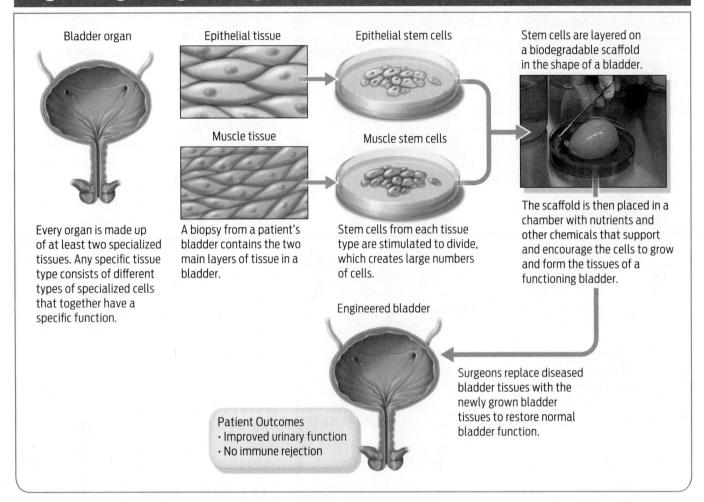

Bladder organ

Every organ is made up of at least two specialized tissues. Any specific tissue type consists of different types of specialized cells that together have a specific function.

Epithelial tissue

Muscle tissue

A biopsy from a patient's bladder contains the two main layers of tissue in a bladder.

Epithelial stem cells

Muscle stem cells

Stem cells from each tissue type are stimulated to divide, which creates large numbers of cells.

Stem cells are layered on a biodegradable scaffold in the shape of a bladder.

The scaffold is then placed in a chamber with nutrients and other chemicals that support and encourage the cells to grow and form the tissues of a functioning bladder.

Engineered bladder

Surgeons replace diseased bladder tissues with the newly grown bladder tissues to restore normal bladder function.

Patient Outcomes
· Improved urinary function
· No immune rejection

Although it has been possible for decades to grow human skin outside the body to treat burn victims, growing more complex organs like bladders has been challenging. Scientists typically grow only one or two skin layers for grafting. To grow a bladder, scientists must grow several layers of tissue, including muscle and connective tissue. Moreover, the thicker the tissue, the more blood vessels required to nourish it.

It took Atala 17 years of research to achieve success. He spent the bulk of those years devising a biodegradable scaffold and a mechanical incubator that could mimic the conditions found in the human body and grow bladder tissue in three dimensions. Once he had devised these tools, the next step was to test the bladder tissue in patients.

For each patient, Atala excised a small piece of muscle tissue less than half an inch wide from inside the bladder and extracted two types of cells—muscle stem cells and urothelial stem cells. He then mixed these stem cells with chemicals that encouraged them to divide. Next he placed the stem cells onto the biodegradable scaffold, which he had sculpted to resemble a human bladder. He bathed the scaffold in nutrients to encourage the cells to grow, and then placed the scaffold with nutrients and other growth factors in an incubator to simulate conditions inside the human body. The cells went through several cell divisions and turned into the tissue layers that make up a bladder. Two months later, surgeons reconstructed the patient's bladders using the new bladder tissue **(Infographic 13.3)**.

The treatment, so far, seems remarkably successful. Not only does the technique restore partial to complete bladder function, it avoids the complications of using tissues from other organs–like the bowel–for bladder repair. "Doing bowel-for-bladder replacements in children really got to me. It's one thing to put them into an adult, but putting them in a child with a 70-plus life expectancy didn't make sense when you knew there would be trouble down the line," Atala told the *New York Times* in 2006, soon after he had published the results of transplants he performed on seven children whose repaired bladders were still functioning well six years after transplant.

Atala's technique is also safer than bladder tissue transplanted from a donor. Just as a food allergy can cause shock when the offending food is ingested, a person's immune system can react similarly to a donated organ and reject it. To prevent tissue rejection, transplant patients must take powerful immune-suppressing drugs. But because the drugs suppress the body's natural immune defenses, they also make that person vulnerable to infection. By contrast, tissue grown from a person's own cells poses no such risk of rejection because the tissue is genetically related to the donor.

Since 2006, scientists have implanted engineered bladders in 10 more children with a congenital condition called spina bifida in which the spinal cord does not completely close before birth. Often the nerves of the lower spine are compromised, including those that control urination, which can lead to bladder damage. Of the 10 children who received the implants, 6 had significantly improved bladder function a year after the surgery, scientists reported in 2009.

Anthony Atala, whose lab at Wake Forest University Institute for Regenerative Medicine grows human organs—including bladders—from just a few human cells.

Specialized Cells Express Different Genes

→ Every cell in your body has the same genes, or genome. What distinguishes one cell type from another is the pattern of gene expression and, consequently, the proteins each cell makes. A muscle cell makes a different set of proteins than a B cell, a type of immune system cell. Muscle cells, for example, express large amounts of actin and myosin proteins, which help muscles contract, whereas B cells express high levels of antibody proteins, which help the body fight infections.

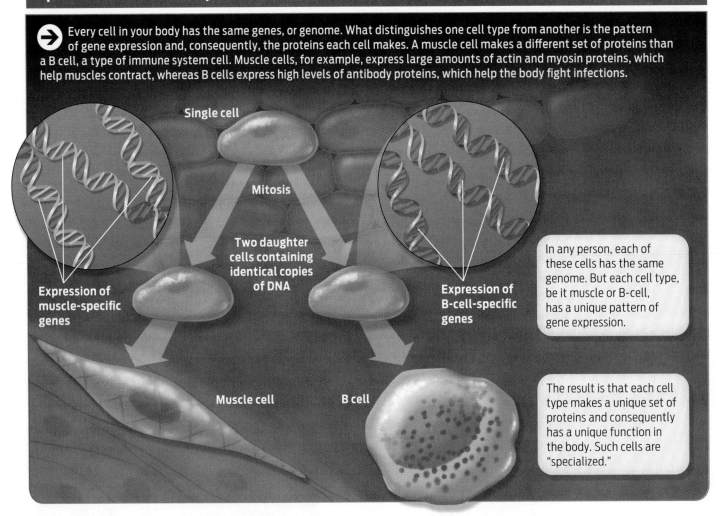

Single cell

Mitosis

Two daughter cells containing identical copies of DNA

Expression of muscle-specific genes

Expression of B-cell-specific genes

Muscle cell

B cell

In any person, each of these cells has the same genome. But each cell type, be it muscle or B-cell, has a unique pattern of gene expression.

The result is that each cell type makes a unique set of proteins and consequently has a unique function in the body. Such cells are "specialized."

Regenerate Instead of Engineer

A major drawback of the engineering technique is that surgeons must operate on patients twice—once to obtain stem cells, and then again to repair the damaged organ. It's a disadvantage that has scientists studying other methods to grow spare tissues from a person's own cells that could sidestep double surgery. What if, instead, doctors could coax the body's existing natural repair mechanisms into healing damaged organs?

Since cells in the body continually divide in response to damage or to replace aging cells, your body is actually many years younger than your chronological age (Table 13.1). In fact, parts of you may be just 10 years old or less. According to Jonas Frisen, a biologist at the Karolinska Institutet in Stockholm, who dated many of the cell populations in the human body, the average age of all the cells in an adult's body is 7 to 10 years—some cells, like cells on the surface of the gut, are replenished about every 5 days, while muscle cells in tissues around the ribs are replenished about every 15 years.

It has been known for decades that some body cells, like skin and blood cells, divide continually. But only within the past 20 years have scientists discovered that there are specific stem cells responsible for regenerating specific tissue types: for example, one type of stem cell may regenerate skin cells, while another type of stem cell may regenerate heart cells. Stem cells found in adult tissues are known as **adult stem cells** or **somatic stem cells** ("somatic" means "referring to the body"). Scientists are still searching for exactly where adult stem cells are located in each type of tissue. They suspect these cells reside in a specific tissue

ADULT STEM CELLS (SOMATIC STEM CELLS)
Stem cells located in tissues that help maintain and regenerate those tissues.

TABLE 13.1

How Old Are You?

 Your body is younger than you think. Each kind of tissue has its own turnover time, depending in part on the workload endured by its cells. But the lens cells of the eye, the neurons of the cerebral cortex, and perhaps the muscle cells of the heart last a lifetime.

TISSUE TYPE	TURNOVER RATE
Epidermis (skin surface)	2 to 3 weeks
Red blood cells	120 days
Liver	300 to 500 days
Bones	More than 10 years
Gut	15.9 years
Rib muscle	15.1 years
Lens of the eye	Never replaced
Neurons of the cerebral cortex	Never replaced

CELLULAR DIFFERENTIATION
The process by which a cell specializes to carry out a specific role.

DIFFERENTIAL GENE EXPRESSION
The process by which genes are "turned on" (that is, expressed) in different cell types.

area where they may remain quiescent (that is, nondividing) for many years until disease or injury triggers them to divide.

To heal tissue damaged by injury or disease, stem cells must do more than simply divide repeatedly. The new cells must also go through a process of specialization to develop into the specific cell types appropriate to the tissue in need of healing. Remember that during embryonic development a single cell becomes millions as the embryo grows. These dividing cells eventually become specialized as muscle cells, kidney cells, heart cells, and more than 200 other cell types in the body by the time we are born. This process, in which a cell develops from an immature cell type into a more specialized one, is called **cellular differentiation.** Cells become specialized by turning some genes "on" and others "off," in what's known as **differential gene expression.** So while every cell in our body carries the exact same DNA, it is a cell's pattern of gene expression—and therefore the proteins produced from those

genes—that defines it as one cell type or another.

Take, for example, two cell types with very distinct characteristics: muscle cells and B-cells of the immune system. Muscle cells are long and slender, and allow the body to move by contracting and releasing. B-cells are round with antibody receptors protruding from their surfaces that detect foreign objects like viruses and bacteria, thus helping the body fight off infection. A cell's physical shape and function are dictated by the kinds of proteins found within it. Muscle cells and B-cells each contain a different collection of proteins; the set of proteins found in muscle cells allow them to contract and cause body movement, while the set of proteins found in B-cells allow them to display surface receptors and fight infection. Muscle cells and B-cells both contain the exact same DNA, which provides instructions (genes) for making every protein for every cell type in the body. But only a subset of those proteins is required by each cell type. Put another way, because each cell type only expresses a subset of a person's genes, each cell type has a unique pattern of gene expression. As a result, each cell type produces a unique set of proteins that distinguish one cell type from another **(Infographic 13.4).**

When scientists discovered that each tissue had its own stem cell type, the search began for ways to coax quiescent stem cells to divide and differentiate when they otherwise would not, so that damaged organs could be repaired from within. This field of research is called regenerative medicine.

One approach to regenerative medicine is to use therapeutic drugs to stimulate specific stem cells in the body to grow and differentiate. Another involves cell therapy to remove stem cells from the body, chemically induce them to reproduce and differentiate, and then re-implant a small sample of differentiated cells into a patient with a damaged tissue or organ. The differentiated cells would repair the existing damaged organ inside the body **(Infographic 13.5).**

The idea of transplanting cells isn't entirely new. In fact, doctors have been treating leuke-

Regenerative Medicine

Therapeutic Drugs:
Somatic stem cells in the brain can be stimulated by therapeutic medicines to differentiate into nerve cells. This induces the tissue to repair itself.

Patients can take the chemicals in the form of medicines that stimulate somatic stem cells in the brain to differentiate into nerve cells and repair damaged tissue.

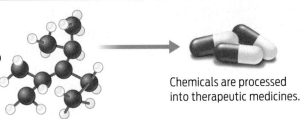

Specific chemicals signal stem cells to differentiate into nerve cells.

Chemicals are processed into therapeutic medicines.

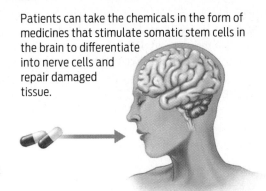

Cell Therapy:
In the lab, scientists can isolate stem cells and then expose them to chemicals that stimulate specific genes to be expressed. This triggers the stem cells to differentiate into a specific tissue type of interest, a neuron for example.

Nerve cells are injected to repair injury or damage.

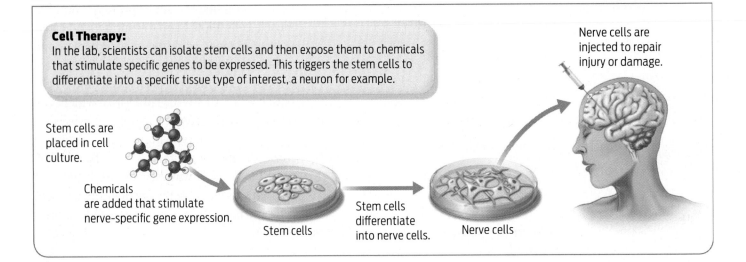

Stem cells are placed in cell culture.

Chemicals are added that stimulate nerve-specific gene expression.

Stem cells

Stem cells differentiate into nerve cells.

Nerve cells

---

mia—a type of white blood cell cancer—with stem cell transplants for decades. Like other blood cells, white blood cells, known as leukocytes, are derived from stem cells in the bone marrow. When signaled by disease or foreign bodies (viruses or bacteria, for example), these stem cells undergo cellular differentiation and become mature leukocytes that pass into the bloodstream, where they fight off infection and invaders. Leukemia is caused by defective leukocytes that divide uncontrollably. To treat leukemia, doctors administer chemotherapy to kill the patient's marrow cells and then replace those cells with marrow cells from an immune-matched donor. Stem cells in the new marrow undergo cel-

> **It turns out that not all stem cells are created equal.**

lular differentiation and repopulate the patient's bloodstream with healthy blood cells. The goal of regenerative medicine is similar: scientists want to use stem cells to heal damaged or diseased tissues. The difference is that regenerative medicine seeks to prod stem cells to differentiate into cell types that they wouldn't differentiate into on their own. Most bone marrow stem cells, for example, in normal circumstances differentiate only into blood cells—but not neurons or other unrelated cell types. One goal of regenerative medicine is to broaden adult stem cells' potential fates and get them to differentiate into a type of cell that they wouldn't otherwise differentiate into.

MULTIPOTENT
Describes a cell with the ability to differentiate into a limited number of cell types in the body.

EMBRYONIC STEM CELLS
Stem cells that make up an early embryo, which can differentiate into nearly every cell type in the body.

Embryonic vs. Adult Stem Cells

→ Embryonic stem cells (ESC) come from embryos and have different characteristics than adult cells. ESCs can develop into almost any cell type, and therefore may have a greater therapeutic potential compared to adult stem cells.

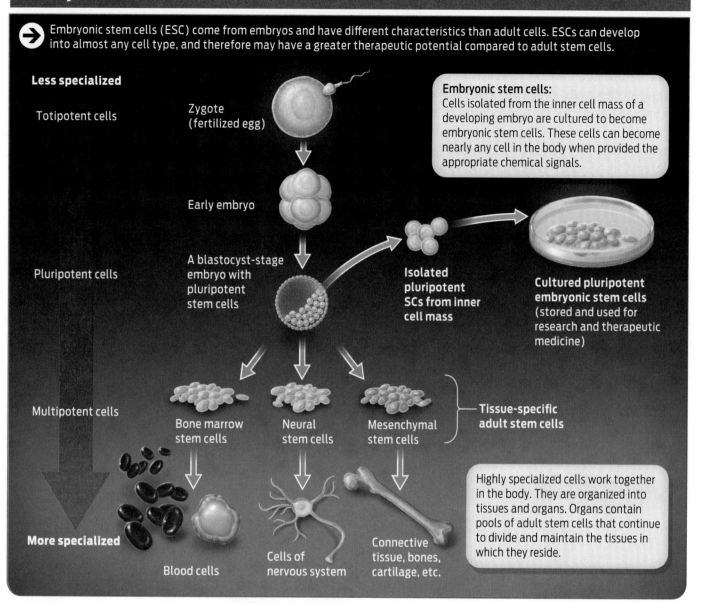

Less specialized

Totipotent cells

Zygote (fertilized egg)

Embryonic stem cells:
Cells isolated from the inner cell mass of a developing embryo are cultured to become embryonic stem cells. These cells can become nearly any cell in the body when provided the appropriate chemical signals.

Early embryo

Pluripotent cells

A blastocyst-stage embryo with pluripotent stem cells

Isolated pluripotent SCs from inner cell mass

Cultured pluripotent embryonic stem cells (stored and used for research and therapeutic medicine)

Multipotent cells

Bone marrow stem cells

Neural stem cells

Mesenchymal stem cells

Tissue-specific adult stem cells

Highly specialized cells work together in the body. They are organized into tissues and organs. Organs contain pools of adult stem cells that continue to divide and maintain the tissues in which they reside.

More specialized

Blood cells

Cells of nervous system

Connective tissue, bones, cartilage, etc.

BLASTOCYST
The stage of embryonic development in which the embryo is a hollow ball of cells. Researchers can derive embryonic stem cell lines during the blastocyst stage.

PLURIPOTENT
Describes a cell with the ability to differentiate into nearly any cell type in the body.

Embryonic Stem Cells

While stem cells show great promise in regenerative medicine, there are limitations. It turns out that not all stem cells are created equal. Adult stem cells typically can differentiate only into one or a few cell types. Such cells with restricted ability to differentiate are described as **multipotent** and can give rise to a limited number of cell types.

But there are some stem cells that can differentiate into *any* of the body's cell types. Scientists are especially interested in stem cells that

are active during early embryonic development, the so-called **embryonic stem cells.** Embryonic stem cells are found in an early embryo at what's known as the **blastocyst** stage, when the embryo is mostly a hollow ball of cells. Unlike adult stem cells, which differentiate only into certain cell types, embryonic stem cells can give rise to nearly any cell type in the body. For this reason, they are referred to as **pluripotent**—they can differentiate into most of the body's cell types. At even earlier stages of embryonic development, embryonic cells can differentiate into any cell type, and these are described as **totipotent (Infographic 13.6).**

Some scientists argue that pluripotent and totipotent cells hold greater potential in treating disease because they are not as specialized as multipotent stem cells, which can differentiate only into tissue-specific cell types. Bone marrow stem cells, for example, typically differentiate only into blood cells. Similarly, the immature cells that Atala used to generate bladders are already committed to differentiate into bladder

cells; their fates are determined by the particular genes they express–that is, by their differential gene expression. Embryonic stem cells have much more developmental flexibility. With embryonic stem cells, scientists might be able to develop ready-to-use therapeutic cell populations that doctors could access as needed.

While embryonic stem cells are potentially valuable in medicine, the challenge has always been how to obtain them. One source of embryonic stem cells is discarded human embryos from fertility clinics. Scientists can extract cells

INFOGRAPHIC 13.7

Somatic Cell Nuclear Transfer Creates Cloned Embryonic Stem Cells

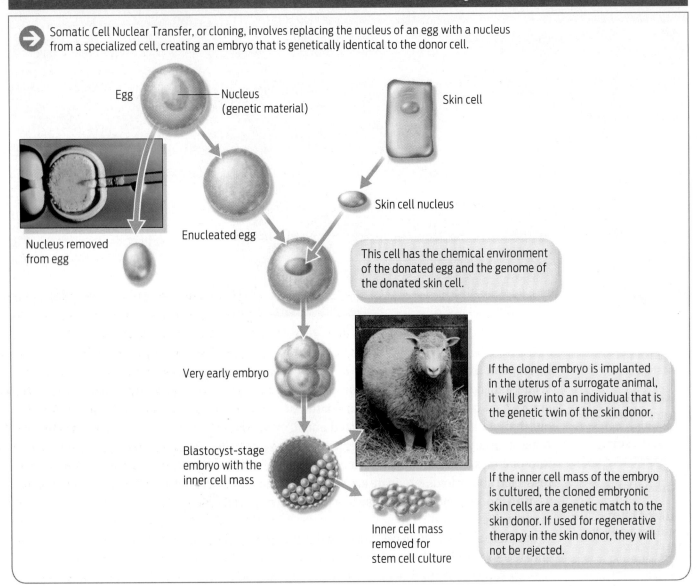

Somatic Cell Nuclear Transfer, or cloning, involves replacing the nucleus of an egg with a nucleus from a specialized cell, creating an embryo that is genetically identical to the donor cell.

Egg — Nucleus (genetic material)

Skin cell

Nucleus removed from egg

Enucleated egg

Skin cell nucleus

This cell has the chemical environment of the donated egg and the genome of the donated skin cell.

Very early embryo

If the cloned embryo is implanted in the uterus of a surrogate animal, it will grow into an individual that is the genetic twin of the skin donor.

Blastocyst-stage embryo with the inner cell mass

Inner cell mass removed for stem cell culture

If the inner cell mass of the embryo is cultured, the cloned embryonic skin cells are a genetic match to the skin donor. If used for regenerative therapy in the skin donor, they will not be rejected.

from an embryo and place them in special media that encourages them to divide. These cells are then stored, used in research, and could potentially be used in treatment.

Another way to obtain embryonic stem cells is by a technique called cloning. In this method, scientists replace the nucleus of a haploid unfertilized human egg with the diploid nucleus taken from another cell, a skin cell, for example. The chemical soup inside the egg turns on specific genes in the donated nucleus to reset them to an embryonic cell state. This technique, known technically as somatic cell nuclear transfer (SCNT), creates a new embryo with the same genes as the donor cell. This is how Dolly, the first cloned sheep, was created in 1997 (**Infographic 13.7**).

If scientists were to implant a cloned embryo into a woman's womb, it would, in theory, develop into a fetus that has the same nuclear DNA as the person who donated the cells from which the diploid nucleus was taken. While such "reproductive cloning," as it is called, is prohibited in the United States, scientists funded by nongovernmental sources are allowed to create cloned embryos for research–this is called "therapeutic cloning," to distinguish the process from reproductive cloning. Embryonic stem cells created by SCNT can be extracted from the new embryo and grown in a petri dish to create a population of stem cells that are essentially genetically identical to the donor. Consequently, any differentiated cells derived from these stem cells could be transplanted back into the donor without fear of an immune response against the transplanted cells.

The main difficulty with both techniques is that they destroy embryos, either frozen embryos from fertility clinics, or cloned embryos generated in the lab. To date, most embryonic stem cell lines in the United States have been derived from embryos stored at fertility clinics and subsequently donated to research. If not donated to science, these extra embryos would be destroyed after a period of time. And although SCNT embryos are not intended to become a human, they may potentially become one if implanted in a wom-

INDUCED PLURIPOTENT STEM CELL
A pluripotent stem cell that was generated by manipulation of a differentiated somatic cell.

an's uterus. Consequently, many people find the idea of deliberately interfering with an embryo, regardless how it was made, ethically troubling.

Scientific Progress

Keenly aware of the ethical challenges, the scientific community has been searching for other methods to generate embryonic stem cells. In late 2007, scientists accomplished a feat that could in fact make the ethical controversy obsolete. They discovered a way to create embryonic-like stem cells without touching an embryo. James Thompson at the University of Wisconsin and Shinya Yamanaka of Kyoto University in Japan independently showed that they could turn a mature human cell–one that is already differentiated–into an "embryonic" stem cell by adding a few genes to its genome. They called the new cells **induced pluripotent stem cells** to reflect the fact that mature cells had been genetically manipulated to become embryonic-like stem cells. Thompson, for example, took an adult skin cell and inserted four genes into it–genes that are normally switched off in skin cells. The additional genes expressed proteins that were able to "de-differentiate" the skin cell–to turn back the clock, in a sense– and return the cell to its pluripotent embryonic state. The cell regained the ability to turn into nearly any cell type of the body. It was a major technological breakthrough. This promising technique may offer a way to create transplantable cells that are genetically matched to a patient without depending on embryos (**Infographic 13.8**).

Since Thompson and Yamanaka's discovery, scientists have improved on the method and developed ways to create embryonic stemlike cells by adding fewer genes or even just proteins. Whether or not these cells will prove to have the same potential to treat disease as embryonic stem cells created by other methods is an active area of research.

Given the unknowns, scientists continue to pursue all avenues of research. Research on both embryonic and adult stem cells, including investigation of ways to control their differentiation,

Induced Pluripotent Stem Cells

→ One method of creating embryonic stem cells is to induce adult stem cells to de-differentiate. Scientists have done just that by inserting either specific genes or proteins into adult cells. The expression of these genes causes the differentiated cells to act like pluripotent stem cells. This technology offers the potential to create immune matched stem cells for therapy, just like cloning, except it does not involve destroying an embryo.

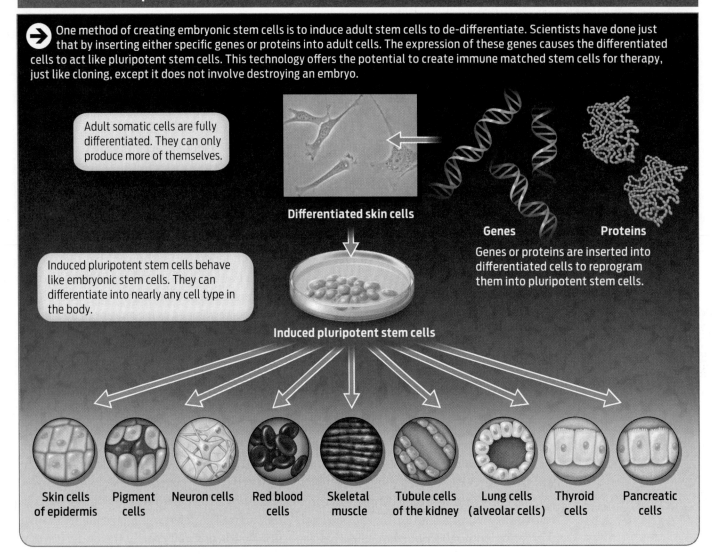

Adult somatic cells are fully differentiated. They can only produce more of themselves.

Differentiated skin cells

Genes   **Proteins**

Genes or proteins are inserted into differentiated cells to reprogram them into pluripotent stem cells.

Induced pluripotent stem cells behave like embryonic stem cells. They can differentiate into nearly any cell type in the body.

Induced pluripotent stem cells

| Skin cells of epidermis | Pigment cells | Neuron cells | Red blood cells | Skeletal muscle | Tubule cells of the kidney | Lung cells (alveolar cells) | Thyroid cells | Pancreatic cells |

is forging ahead. The first clinical trial to test whether immature nerve cells can help heal spinal cord injuries is already under way. Scientists at Geron, a Menlo Park, California, biotech company, are transplanting immature nerve cells obtained from human embryonic stem cells into a small group of patients with recent spinal cord injuries. Other research aims to manipulate adult stem cells into becoming totipotent by supplying them with specific proteins, or converting one multipotent cell type into another by similar methods.

"Even the dumbest stem cell is smarter than the smartest neuroscientist"
—Evan Snyder

While the research is promising, hurdles remain. Scientists still have trouble identifying and extracting adult stem cells. And even when they can extract or produce stem cells, they understand little about how to nudge them down a particular path to differentiation. There are countless growth factors and biochemical signals that cue any stem cell to become a kidney cell, a muscle cell, or a heart cell, for example.

"Even the dumbest stem cell is smarter than the smartest neuroscientist," says Evan Snyder, program director at the Burnham Institute for Medical Research in La Jolla, California. "The cells are making stuff that we might not be able to identify for centuries."

All the more reason, says Snyder, for scientists to pursue several methods to achieve the same goal. Anthony Atala, for one, is continuing to study how to engineer an entire bladder. While induced pluripotent stem cells may provide another source of cells to grow bladder tissue or an entire bladder from a person's own cells without the need for a tissue biopsy, getting a whole engineered organ to function normally once transplanted will require much more research, he says. To replace the organ completely would require a full set of nervous connections, so that when the bladder is full it would send a message of urgency to the brain. Also, during urination the sphincter muscles would have to relax while the wall of the bladder contracted. To get all this working properly is an immense challenge.

"Ultimately we're depending on advances in molecular and cell biology," says William Wagner of the University of Pittsburgh's McGowan Institute for Regenerative Medicine. "There is still a lot that we don't know."

Growing a solid organ such as a kidney or heart would be even more complicated. Such organs have a very complex structure and carry out very complicated bodily functions. It took Atala 17 years of small steps before he succeeded in engineering bladder tissue for repairs. It may take another 17 years or even longer to engineer an entire organ. But there is reason to be optimistic, he says: "There has been lots of steady progress." ■

▶ Summary

■ Tissues are integrated groups of specialized cells that perform specific functions.

■ Stem cells are relatively unspecialized cells that can divide and specialize (that is, differentiate) into different cell types.

■ Adult stem cells, also known as somatic stem cells, are found in tissues; embryonic stem cells make up early embryos.

■ Stem cells can be used therapeutically to engineer or regenerate tissues and organs.

■ Making new tissues requires both cell division and cell differentiation. Cell differentiation is the process in which an unspecialized cell becomes a specialized cell with a unique function.

■ All cells in any individual have the same genome but express different genes. Such differential gene expression causes each cell type to produce different proteins and to have different functions.

■ Adult stem cells are multipotent, able to differentiate into a limited number of different cell types.

■ Embryonic stem cells are pluripotent, able to differentiate into nearly any cell type in the body.

■ Embryonic stem cells can be obtained from human embryos or from cloned embryos. They may also be created by inducing adult stem cells to "de-differentiate."

■ Both adult and embryonic stem cells are being investigated as possible therapies to restore damaged tissue in humans.

TISSUES AND CELL DIFFERENTIATION

Tissues are made up of a variety of specialized types of cells. Each cell type differentiates from an unspecialized stem cell.

HINT See Infographics 13.1–13.4 and Table 13.1.

➔ KNOW IT

1. Does a 5-year-old child have adult stem cells in his or her tissues? Explain your answer

2. Relative to one of your liver cells, one of your skin cells
   a. has the same genome (that is, the same genetic material).
   b. has the same function.
   c. has a different pattern of gene expression.
   d. a and c
   e. b and c

3. You shed dead skin cells every day. How are those cells replaced?
   a. by mitotic division and specialization of embryonic stem cells
   b. by differentiation of neighboring neurons into skin cells
   c. by differentiation of red blood cells that leave the circulation and migrate into deeper layers of the skin
   d. by mitotic division and differentiation of skin stem cells

4. The brain and spinal cord are made up of nervous tissue. This tissue includes neurons, cells that fire electrical impulses and communicate information in the brain. Nervous tissue also includes glial cells, cells that support neurons. Some glial cells enable the electrical impulse to travel faster. What characteristics of glial cells and neurons tell you that they both make up nervous tissue?

➔ USE IT

5. From information in Question 4, would it be sufficient to replace the damaged neurons in someone who had suffered nervous-tissue damage? Why or why not?

6. Different cells have different functions: muscles contract because of the sliding action of actin and myosin proteins in muscle cell fibers; a protein known as retinal makes up the light detecting photoreceptor cells in the retina of the eye; helper T cells of the immune system have a protein on their surface known as CD4 which helps in mounting an immune response. From this information, complete the following table.

	Photo-receptor cells of the retina	Heart muscle fibers	Helper T cells
Myosin gene present?			
Myosin mRNA present?			
Myosin protein present?			
Retinal gene present?			
Retinal mRNA present?			
Retinal protein present?			
CD4 gene present?			
CD4 mRNA present?			
CD4 protein present			

7. Do stem cells have a larger genome than specialized cells?
   a. yes, because they need the genes found in every cell type, whereas specialized cells need only a subset of all the genes
   b. yes, because they express more genes than do specialized cells
   c. no, because all cells in a person have the identical set of genes in their genome
   d. no, they have a smaller genome, because stem cells are equivalent to gametes (which are haploid) in that they can potentially create an entire individual
   e. no, they have a smaller genome because stem cells only express a subset of genes

STEM CELLS AND REGENERATIVE MEDICINE

Stem cells can potentially repair damaged tissue. The challenges are how to stimulate existing stem cells to divide, or how to transfer stem cells to the area of damage to promote repair.

HINT See Infographics 13.2 and 13.5–13.8.

KNOW IT

8. List and then describe several advantages of using one's own cells to regenerate an organ over receiving a transplant from an organ donor.

9. Describe at least two differences between embryonic stem cells and somatic (that is, adult) stem cells.

10. Why does the recipient of a liver transplant have a high risk of bacterial infections?

    a. because the liver plays a critical role in the immune response

    b. because donor livers are often contaminated with disease-causing bacteria

    c. because transplant recipients have to take drugs that suppress their immune systems

    d. because the surgery poses a high risk for introducing bacteria into the recipient

    e. because the immune system may reject the liver

USE IT

11. Why is engineering a bladder more challenging than engineering skin?

12. Which of the following populations of adult stem cells (if any) could be stimulated to divide to treat each of the following conditions?

    a. heart attack (damage to the heart muscle)

    b. cancer (unregulated cell growth)

    c. type I diabetes (destruction of insulin-producing cells in the pancreas)

    d. Parkinson disease (loss of specific neurons in the brain)

13. List and then describe some of the successes and challenges associated with using adult stem cells in comparison with embryonic stem cells for stem cell therapy.

14. If all specialized cells have the same genes in their genomes (including stem cell genes), why did James Thompson have to add genes into a mature cell in order to get it turn into a stem cell?

SCIENCE AND ETHICS

15. If you were head of the National Institutes of Health and responsible for allocating research funds to different avenues of research, which line(s) of stem cell research would you fund? Why?

16. Most people make a distinction between therapeutic cloning and reproductive cloning, at least for humans. There have been several reports of cloned pets and other animals. How does the science differ between reproductive cloning and therapeutic cloning? Do you think reproductive cloning should be legal or illegal?

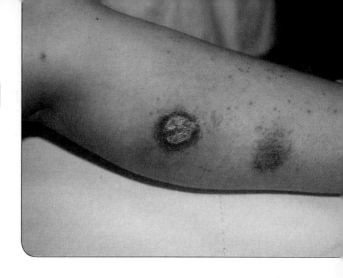

Bugs That Resist Drugs

Bugs That Resist Drugs

Drug-resistant bacteria are on the rise. Can we stop them?

In January 2008, sixth-grader Carlos Don, an active footballer and skateboarder, boarded a bus headed for a class trip, happy and healthy. A month later Carlos was dead.

In April 2006, 17-year-old Rebecca Lohsen was a model student at her high school; she was on the honor roll and was a member of the swim team. Four months later Rebecca was dead.

In December 2003, Ricky Lannetti was a college senior, a star football player and all-around athlete. A few weeks later Ricky was dead.

The list of surprising deaths like these goes on and on. But contrary to what you might think, these young people weren't killed in accidents, nor by violence; they were all killed by methicillin-resistant *Staphylococcus aureus* (MRSA)—an infectious bacterium that has become widespread in recent years and that is difficult to treat with most existing antimicrobial drugs.

MRSA sickens some 94,000 people in the United States each year and kills almost 19,000, according to a 2007 study by Monina Klevens and her colleagues at the Centers for Disease Control and Prevention (CDC). Formerly, outbreaks of MRSA were confined mainly to hospitals. But since the late 1990s, growing numbers of healthy people are becoming infected outside hospitals. In addition, there are new high-risk groups that never had high rates of infection before: day care attendees, the prison population, men who have sex with men, and certain ethnic groups now are showing MRSA infections at a higher rate than

> **MRSA sickens some 94,000 people in the United States each year and kills almost 19,000.**

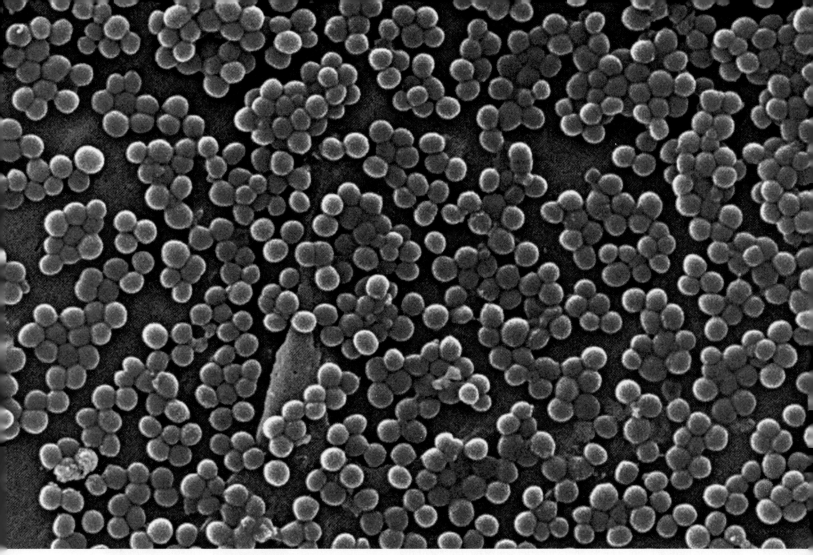

A resistant strain of *Staphylococcus aureus* at 4780x magnification.

the general population. Schools nationwide have been reporting outbreaks and young, healthy people are getting sick.

"This is a major public health imperative," says Robert Daum, professor of microbiology at the University of Chicago and a member of the Infectious Diseases Society of America. "We need a plan of attack now."

Staph the Microbe

MRSA infection is caused by the *Staphylococcus aureus* bacterium–often simply called "staph." Although several species of staph bacteria can cause human disease, the medical community is especially concerned about those, such as *S. aureus,* that have developed resistance to antibiotic drugs that once effectively killed them. "MRSA" is actually a misnomer because

methicillin is no longer used to treat staph infections. In fact, drug-resistant strains of staph are usually resistant to several different types of antibiotics. Some people use the terms "MRSA" and "drug-resistant staph" interchangeably to refer to staph strains that are resistant to the common classes of antibiotics–penicillins and cephalosporins–that are used to treat staph infections.

Staph bacteria are harmless to most people who carry them. Between 30% and 40% of the population carries staph on their skin or in their noses, and about 1% of the population carries drug-resistant strains, according to the Centers for Disease Control and Prevention. If you carry staph of any strain but aren't sick, you are "colonized" but not infected. Healthy people can be colonized with any staph strain, including

MRSA, and not become ill. However, they can pass the staph to others via skin-to-skin contact or through shared, contaminated items such as towels and bars of soap. Infections typically occur when the bacteria come into contact with a wound. Athletes who have cuts and scrapes may acquire a staph infection in locker rooms or during contact sports **(Infographic 14.1)**.

"Every one of us has probably had a staph infection at some point," explains Daum. "Staph ranges from the commonest cause of infected fingernails all the way to a severe syndrome with rapid death, and everything in between. Most staph infections don't even result in a medical encounter."

In otherwise healthy people, staph infection usually causes only minor skin eruptions such as boils or pustules that can resemble spider bites. A healthy person can, however, become more severely ill if he or she undergoes a medical procedure that either weakens the immune system or creates a break in the skin that becomes infected with staph. The elderly, who may have weakened immune systems, and children, whose immune systems are still immature, are at especially high risk of developing severe diseases such as pneumonia, infections of the bloodstream, or infections of surgical wounds caused by staph. When bacteria such as staph do cause illness, they do so by multiplying on or in human tissues. They can also secrete toxic substances that harm human cells or interfere with essential cellular processes.

Staph bacteria can cause such a range of disease because the bacteria exist as many different strains. Each strain differs from all others in

INFOGRAPHIC 14.1

The Bacterium *Staphylococcus aureus*

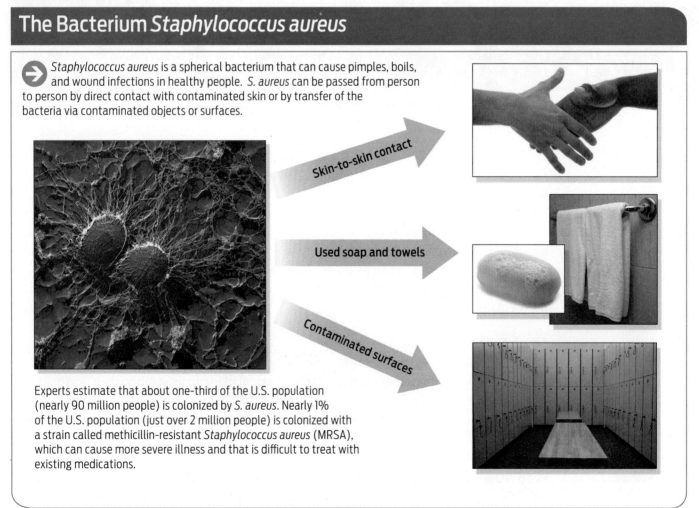

→ *Staphylococcus aureus* is a spherical bacterium that can cause pimples, boils, and wound infections in healthy people. *S. aureus* can be passed from person to person by direct contact with contaminated skin or by transfer of the bacteria via contaminated objects or surfaces.

Skin-to-skin contact

Used soap and towels

Contaminated surfaces

Experts estimate that about one-third of the U.S. population (nearly 90 million people) is colonized by *S. aureus*. Nearly 1% of the U.S. population (just over 2 million people) is colonized with a strain called methicillin-resistant *Staphylococcus aureus* (MRSA), which can cause more severe illness and that is difficult to treat with existing medications.

its genetic makeup. MRSA, for example, is composed of a number of unique strains of staph bacteria, and some cause more serious disease than others. In recent years there have been several cases of healthy people becoming severely ill from MRSA infection, most likely because they were infected by an especially deadly strain of drug-resistant staph.

Ricky Lannetti, for example, was a perfectly healthy 21-year-old football player at Lycoming College in Williamsport, Pennsylvania. "He was strong as an ox and he ran like a deer," says his mother, Theresa Drew. A few days before Ricky died, he had come down with a bout of flu. Ricky wasn't recovering, however, and on the morning of December 6, 2003, Drew drove her son to Williamsport Hospital. By the time he was admitted, his blood pressure had dropped dangerously low and his body temperature was erratic. As each hour passed, his condition worsened. His lungs began to fail. Doctors tried five different antibiotics, in vain. When his heart began to weaken, his doctors prepared him to be flown to the cardiac center at a bigger hospital in Philadelphia. But it was too late. Ricky died that night.

It was only after an autopsy was performed that it was known what had killed him: MRSA that had infected Ricky's bloodstream. Although doctors couldn't be sure how Ricky contracted MRSA, they reasoned that he had inhaled it—the fact that his lungs were so damaged suggested the lungs as the first place of infection. Since MRSA can colonize people's noses, it can travel to their respiratory systems, where it can cause severe damage.

"Doctors tried every antibiotic imaginable, including vancomycin," says his father, Rick Lannetti. But the treatment was too late. Ricky's immune system was already weak because of the flu. When he contracted MRSA at the same time, his body was unable to fight back as well as it otherwise would have. "In the end," his father says, "MRSA had broken every one of his organs beyond repair."

The Antibiotic Revolution

Bacterial infections were a common cause of death before the 1940s, when scientists developed the first antibiotics. **Antibiotics** are chemicals that either kill bacteria or slow their growth by interfering with the function of essential bacterial cell structures. Research in the early twentieth century had revealed that certain microorganisms, such as the fungus *Penicillium*, produce compounds that can kill bacteria. In 1928, the Scottish biologist Alexander Fleming isolated the antibiotic penicillin, although it took more than a decade of research by Fleming and others to develop it into a usable drug. The 1940s saw a major search for other drugs to treat infections, and within a few years scientists successfully purified more antibiotics. Though the original antibiotics were derived from microorganisms, many are now synthesized in the laboratory.

Over the decades, antibiotics have been effective in treating most common bacterial infections, including staph, and have saved thousands of lives. But soon after antibiotics were in general use, microorganisms that could survive antibiotics—drug-resistant "bugs"—began to emerge. Within the last decade drug-resistant bacterial strains have become much more common. Although people infected with drug-resistant bacterial strains are treatable, they have fewer treatment options. And sometimes—as in Ricky Lannetti's case—existing drugs are completely ineffective.

Drug-resistant strains of staph, for example, are typically resistant to an entire class of antibiotic drugs called the beta-lactams. Beta-lactams include penicillin and the cephalosporin antibiotics, such as methicillin and cephalexin. Beta-lactams are the most commonly prescribed class of antibiotics. They work by interfering with a bacterium's ability to synthesize cell walls. A variety of non-beta-lactam classes of antibiotics can treat MRSA infections, and vancomycin, an intravenous non-beta-lactam drug, is the antibiotic of choice when a serious or severe MRSA infection is confirmed. But even vancomycin isn't always effective; there are now staph strains resistant to vancomycin, too (**Infographic 14.2**).

Ricky Lannetti did not respond to vancomycin. Nor did Rebecca Lohsen, the 17-year-old

ANTIBIOTICS
Chemicals that either kill bacteria or slow their growth by interfering with the function of essential bacterial cell structures.

How Beta-lactam Antibiotics Work

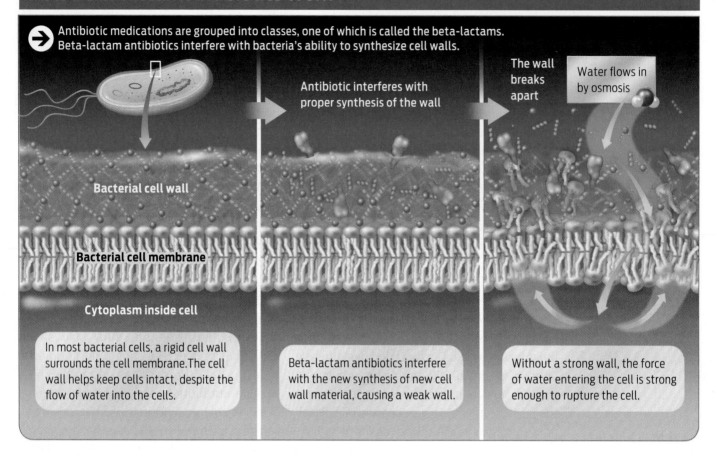

➡ Antibiotic medications are grouped into classes, one of which is called the beta-lactams. Beta-lactam antibiotics interfere with bacteria's ability to synthesize cell walls.

Antibiotic interferes with proper synthesis of the wall

The wall breaks apart

Water flows in by osmosis

Bacterial cell wall

Bacterial cell membrane

Cytoplasm inside cell

In most bacterial cells, a rigid cell wall surrounds the cell membrane. The cell wall helps keep cells intact, despite the flow of water into the cells.

Beta-lactam antibiotics interfere with the new synthesis of new cell wall material, causing a weak wall.

Without a strong wall, the force of water entering the cell is strong enough to rupture the cell.

high school swimmer. She was diagnosed with MRSA two days after she was admitted to the hospital for pneumonia. But antibiotics were ineffective in controlling the MRSA that attacked her lungs and then her heart. She died in August 2006.

Twelve-year-old Carlos Don, who skateboarded and played football, suffered a similar fate. Carlos was first diagnosed with pneumonia that was likely caused by inhaling MRSA, although doctors did not initially identify the organism causing his lung infection. He was sent home with several different antibiotics, including vancomycin, only to return to the hospital the next morning, his condition worsening. He died two weeks later, on February 4, 2007, after his lungs, heart, and kidneys were too damaged to function on their own.

Scientists don't entirely understand all the reasons antibiotics might not work even if the appropriate one is used. "It's complicated and depends upon the interactions between the person and the bacteria," says Ruth Lynfield, state epidemiologist and medical director of the Minnesota Health Department. Personal factors include the individual's general health and state of the immune system, and which body sites are infected. Bacterial factors include whether or not the strain is making toxins or has ways to avoid the immune system. There are treatment factors as well: whether the antibiotics used are active against the strain, for instance, and whether the antibiotic achieves a high enough concentration in the body site that is infected.

The very severe and fatal cases are the "tip of the iceberg," Lynfield says. Of the 94,000 invasive MRSA infections each year in the United States, the Centers for Disease Control and Prevention estimates that about 20% are fatal. And 85% of these invasive infections occur in

How Bacteria Reproduce

Bacteria reproduce through a process called binary fission. Binary fission is a form of asexual reproduction in which a single parent cell replicates its contents and then divides into two daughter cells. Note that each daughter cell inherits all its DNA from the single parent cell.

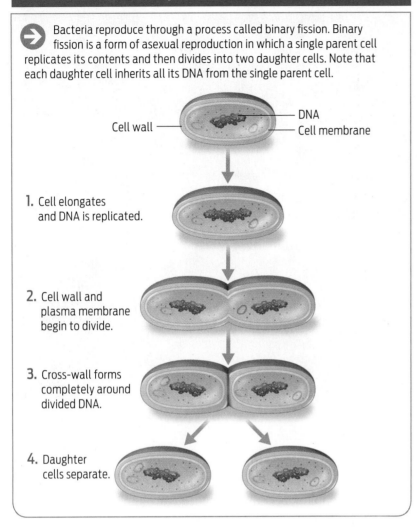

Cell wall — DNA
— Cell membrane

1. Cell elongates and DNA is replicated.

2. Cell wall and plasma membrane begin to divide.

3. Cross-wall forms completely around divided DNA.

4. Daughter cells separate.

binary fission, a single parental cell simply replicates its single chromosome, grows in size, and then splits into two daughter cells, each with a copy of the parental DNA. Each time DNA is replicated, however, there is a chance that genetic mutations will occur, and the new alleles will then be carried into each daughter cell. And because bacteria reproduce much more rapidly than other organisms—one generation of bacteria can reproduce itself in as little as 20 minutes—they accumulate mutations at a relatively high rate. An entire population of bacteria that is genetically different from the original cell can arise very quickly (Infographic 14.3).

Bacteria might never grow resistant to drugs if not for random mutations that create new alleles and generate genetic diversity.

Mutation isn't the only way a bacterium's genetic material can change. A bacterium can also acquire new alleles through a mechanism called gene swapping, in which bacteria can swap pieces of DNA with other bacteria. (By contrast, sexual organisms, which are also subject to mutation, become genetically diverse primarily through meiosis—cell division producing sperm and egg cells that randomly mixes up maternal and paternal genes.)

Staph bacteria, for example, became resistant to drugs either by mutations in their own genes or by picking up "resistance" genes from other drug-resistant bacteria. The genetic changes ultimately alter bacterial proteins in ways that helped staph dodge antibiotic drugs. Specifically, the altered or acquired genes code for proteins that can disable antibiotics, or they code for proteins with altered shapes to which antibiotics can no longer bind. Some bacteria produce enzymes called beta-lactamases that chew up beta-lactam antibiotics and render them ineffective. Because different strains of

patients who are hospitalized or in people with underlying illnesses or who are exposed to the health care system (health care workers, for example). The other 15% occur in healthy people in the community.

Acquiring Resistance

Bacteria might never grow resistant to drugs if not for random mutations that create new alleles and generate genetic diversity. Like all organisms, bacteria can acquire mutations when their DNA replicates during reproduction. Bacteria reproduce asexually by a process called **binary fission.** Unlike sexual reproduction, in which gametes from two parents fuse, asexual reproduction does not require a partner. During

BINARY FISSION
A type of asexual reproduction in which one parental cell divides into two.

INFOGRAPHIC 14.4

How Bacterial Populations Acquire Genetic Variation

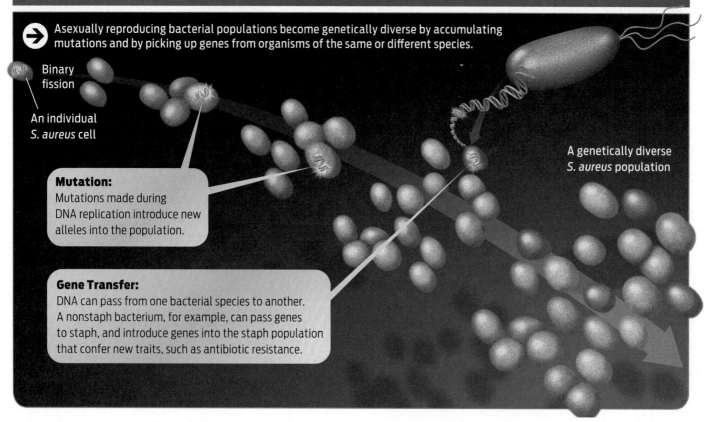

→ Asexually reproducing bacterial populations become genetically diverse by accumulating mutations and by picking up genes from organisms of the same or different species.

Binary fission

An individual *S. aureus* cell

A genetically diverse *S. aureus* population

Mutation:
Mutations made during DNA replication introduce new alleles into the population.

Gene Transfer:
DNA can pass from one bacterial species to another. A nonstaph bacterium, for example, can pass genes to staph, and introduce genes into the staph population that confer new traits, such as antibiotic resistance.

bacteria have developed antibiotic resistance independently, several strains of genetically unique drug-resistant staph circulate through human communities at the same time (**Infographic 14.4**).

How Populations Evolve

While an individual bacterium—or any individual organism, from mushroom to manatee—can undergo genetic changes that may give it new traits, this doesn't entirely explain how bacterial populations such as staph develop resistance to drugs. An entire population of organisms with a new trait can arise only when the environment favors that trait—that is, when carrying the specific trait is advantageous to the organisms carrying it. A **population** is a group of individuals of the same species living together in the same geographic area. Geographic area is relative; it could be an open prairie, or a drop of pond water. A population of bacteria can exist anywhere. In the case of staph, populations exist in people's noses, and on other parts of their skin.

When a population's environment favors some traits over others, the frequencies of the alleles that code for those traits in the population change over time. Take the trait for drug resistance, for example. Bacterial populations, like populations of any organism, consist of genetically varied individuals. In an environment free of antibiotics, individual bacteria would have about an equal chance of reproducing, whether or not they carried a resistance gene. In other words, the ability to resist antibiotics would confer neither an advantage nor a disadvantage. In the presence of an antibiotic, however, bacteria with an allele for resistance survive, whereas most other bacteria die. The surviving bacteria, which are drug resistant, reproduce more prolifically, and more alleles for drug resistance are passed on to future generations. Consequently, the frequency of the resistance trait increases. This is how popula-

POPULATION
A group of organisms of the same species living together in the same geographic area.

EVOLUTION
Change in allele frequencies in a population over time.

INFOGRAPHIC 14.5

An Organism's Fitness Depends on Its Environment

→ The term "fitness" describes how successfully an organism is able to reproduce in a particular environment. Fitness is determined by the interaction between phenotype and environment. Antibiotic-resistant bacteria, for example, have high fitness in the presence of antibiotics.

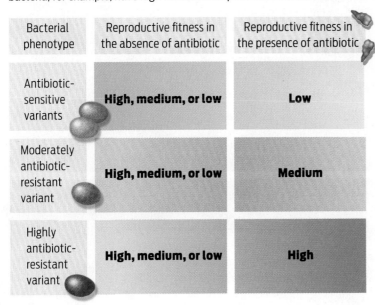

Bacterial phenotype	Reproductive fitness in the absence of antibiotic	Reproductive fitness in the presence of antibiotic
Antibiotic-sensitive variants	High, medium, or low	Low
Moderately antibiotic-resistant variant	High, medium, or low	Medium
Highly antibiotic-resistant variant	High, medium, or low	High

FITNESS
The relative ability of an organism to survive and reproduce in a particular environment.

NATURAL SELECTION
Differential survival and reproduction of individuals in response to environmental pressure that leads to change in allele frequencies in a population over time.

ADAPTATION
The response of a population to environmental pressure, so that advantageous traits become more common in the population over time.

tions evolve. **Evolution** is defined as a change in the frequency of alleles in the population over time.

An organism's ability to survive and reproduce in a particular environment is called that organism's **fitness.** The greater an organism's fitness, the more likely that alleles carried by that organism will be passed on to future generations and increase in frequency. In an environment in which antibiotics are abundant, drug-resistant bacteria are more fit than nonresistant bacteria (**Infographic 14.5**).

In the case of staph, the antibiotic-resistance trait has become so common because it gave bacteria with the trait a reproductive edge in an environment in which antibiotic use has been rampant. These antibiotic-resistant variants might not do well in another environment, however. For example, some resistant bacteria survive well in hospitals, but outside the hospital environment they perish. That's because muta-

tions can come at a cost. An altered protein may enable a bacterium to withstand an antibiotic assault, but at the same time it might render that bacterium weaker in other ways; for instance, it might not be able to reproduce as fast as a bacterium without the mutation. Consequently, in an antibiotic-free environment the bacterium without the resistance allele might be able to out-reproduce the drug-resistant one. On the other hand, some antibiotic-resistant bacteria have been shown to develop secondary mutations that render them just as fit as antibiotic-susceptible bacteria in an antibiotic-free environment.

Patterns of Natural Selection
Ultimately, this interplay between an organism's traits, or its phenotype (which is largely determined by its genes or genotype), and its environment is what determines what traits will predominate in any population. Organisms can be fit in one environment and not in another. This process of differential survival and reproduction of individuals within a population in response to environmental pressure is known as **natural selection.**

The interplay between an organism's traits and its environment determines what traits will predominate in any population.

When natural selection favors some traits over others, the population shows **adaptation** to its environment. In other words, specific advantageous traits become more common in a population over time. This is what we see with antibiotic-resistant bacteria—the population has become better suited, or adapted, to an environment in which antibiotics are abundant because individual bacteria carrying resistance genes are more fit in this environment. The various finch

Evolution by Natural Selection

→ In any genetically diverse population, individual fitness varies. When an organism's environment favors specific genetic variants to survive and reproduce over others, natural selection occurs. Those with high fitness tend to reproduce more successfully. Over generations, the frequency of alleles that confer higher fitness increase while those that confer lower fitness decrease. This non-random change in allele frequencies over generations is called evolution by natural selection.

For example, in the *absence* of antibiotic:

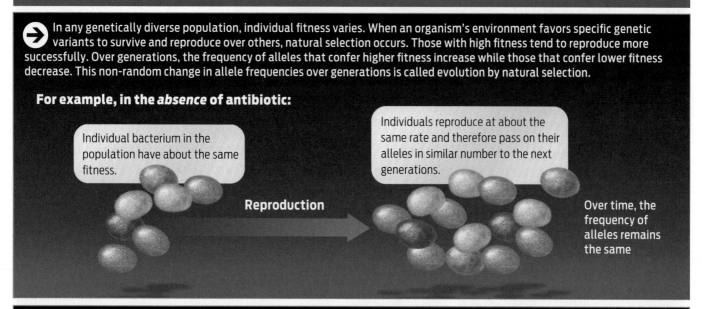

Individual bacterium in the population have about the same fitness.

Individuals reproduce at about the same rate and therefore pass on their alleles in similar number to the next generations.

Reproduction

Over time, the frequency of alleles remains the same

In the *presence* of antibiotic:

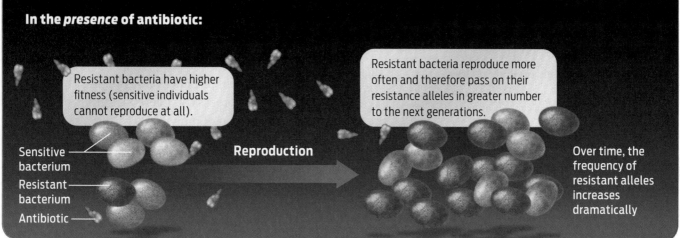

Resistant bacteria have higher fitness (sensitive individuals cannot reproduce at all).

Resistant bacteria reproduce more often and therefore pass on their resistance alleles in greater number to the next generations.

Sensitive bacterium

Resistant bacterium

Antibiotic

Reproduction

Over time, the frequency of resistant alleles increases dramatically

---

species that Charles Darwin observed on the Galápagos Islands had evolved different types of beaks as adaptations to different food sources.

Note that evolution by natural selection occurs in populations, not individuals. Individual organisms do not experience a change in allele frequencies over time. Therefore, individual organisms do not evolve **(Infographic 14.6)**.

By studying how populations have evolved in the past, scientists have defined three major patterns of natural selection. When the predominant phenotype in the population has shifted in one particular direction, we say

that **directional selection** has occurred. For example, when bacterial populations evolve from populations sensitive to drugs into ones that resist drugs–that is, toward antibiotic resistance–they are exhibiting directional selection.

When the phenotype of the population settles around the middle of the phenotypic spectrum, we call this **stabilizing selection.** Or, a population can also "spread out," so that the population shows extremes of the phenotypic spectrum; this pattern is known as **diversifying selection.** The particular pattern of natural selection a population follows depends on the interaction of phenotypes with

DIRECTIONAL SELECTION
A type of natural selection in which organisms with phenotypes at one end of a spectrum are favored by the environment.

STABILIZING SELECTION
A type of natural selection in which organisms near the middle of the phenotypic range of variation are favored.

Natural Selection Occurs in Patterns

Directional selection occurs when a single phenotype predominates in a particular environment.

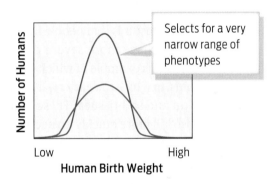

Selects for phenotypes toward one end of the spectrum

— Before natural selection
— After natural selection

Number of Bacteria

Low   High
Antibiotic Resistance

Example:
Antibiotic-containing environments favor resistant strains of bacteria.

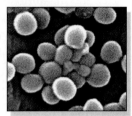

Stabilizing selection occurs when phenotypes at each end of the spectrum are less suited to the environment than organisms in the middle of the phenotypic range.

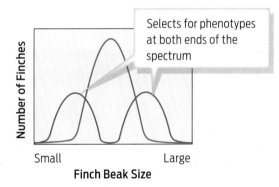

Selects for a very narrow range of phenotypes

Number of Humans

Low   High
Human Birth Weight

Example:
Human babies with very low birth weights do not survive as well as larger babies, and very large human babies are not easily delivered through the birth canal. Mid-range babies are favored.

Diversifying selection typically occurs in a "patchy" environment, in which extremes of the phenotypic range do better than middle range individuals.

Selects for phenotypes at both ends of the spectrum

Number of Finches

Small   Large
Finch Beak Size

Example:
The African finch *Pyrenestes* lives in an environment where only large, hard seeds and small, softer seeds are available. Birds with either large or small beak sizes are selected for, while medium beaks, which are not as successful at cracking either type of seed, are selected against.

DIVERSIFYING SELECTION
A type of natural selection in which organisms with phenotypes at both extremes of the phenotypic range are favored by the environment.

the environment. So, for example, in the absence of antibiotics, populations of staph bacteria might have followed stabilizing or diversifying selection. Instead, directional selection led to the MRSA that killed Carlos Don, Rebecca Lohsen, and Ricky Lannetti (**Infographic 14.7**).

MRSA in the Community
Drug-resistant staph strains first emerged in hospitals during the early 1960s. Since then, hospitals have remained hot spots for staph infections. Surgical procedures create wounds that are vulnerable to infection, and certain medications can weaken the immune system,

making a patient less able to fight off staph than a healthier person. Patients also can pick up staph directly from other patients, from infected health care workers, or in some cases from contaminated objects. In addition, a drug-resistant strain of any bacterium can emerge in a patient taking antibiotics by mutating or by grabbing genes from another bacterial strain that is merely colonizing the patient. While this new resistant bacterial strain may not sicken the patient, the patient may transmit it to other people, who may become ill.

In response, several hospitals have implemented systems to reduce infections. For example, studies have shown that measures as simple as requiring all health care workers to wash their hands before handling each patient can dramatically reduce the number of infections. However, the rate of hand washing among health care workers remains dismally low. A 2010 study published by researchers at the Atlanta Veterans' Affairs Medical Center found that 50% of health care workers or less followed guidelines for hand hygiene in the hospitals observed in the study. Other studies have shown similar numbers. "It is really important that people do low tech things that make a high difference," Lynfield, from the Minnesota Health Department, says. "Washing hands well and often is absolutely critical."

More alarming than MRSA infections in hospitals are MRSA infections in the community at large. Though MRSA has been around for over 40 years, in the mid 1990s the rate of infections in the United States began to soar, explains Daum from the University of Chicago. In some groups, such as day care attendees and prisoners, 60% to 90% of those who show up at a hospital or clinic because of a skin infection are infected by some strain of drug-resistant staph, he says.

What happened during that time? A new strain of *Staphylococcus aureus* emerged. Daum thinks that drug-resistant staph strains circulating in the community evolved separately and more recently than other resistant strains. In addition, he and his colleagues recently showed that a strain called USA300 is more virulent than other MRSA strains. "It appears to be juiced up," he says. Many USA300 genes

Requiring all health care workers to wash their hands before handling each patient can dramatically reduce the number of infections.

are expressed at high levels. One gene in particular that controls expression of a number of staph toxins is turned on all the time.

The result is that people infected with such strains have more severe disease, the most infamous example of which is necrotizing fasciitis, in which the bacteria literally eat through skin and soft tissues. These "superbugs" can also kill more quickly. In necrotizing pneumonia, the bacteria eat through lung tissue and kill the victim. Symptoms can appear so suddenly that, according to Daum, "you could be healthy at 1:00 in the afternoon and be dead by 1:00 in the morning."

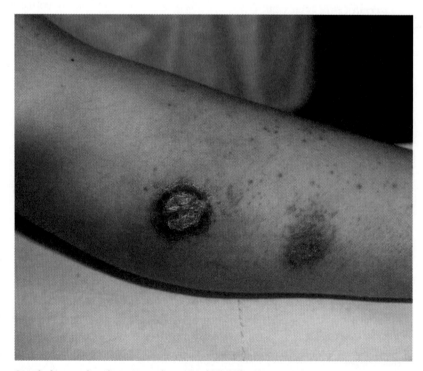

Staphylococcal ecthyma, an ulcerative skin infection, on a leg.

"I've been an infectious disease guy for over 20 years now and we didn't talk about staph necrotizing pneumonia like we do now." It was likely USA300 or a related strain that killed Carlos Don, Rebecca Lohsen, and Ricky Lannetti.

Even more troubling, staph is continuing to evolve. There is evidence that when strains that are prevalent in the community mix with strains that are prevalent in hospitals, the risk that an even more virulent staph strain will emerge increases.

Stopping Superbugs

Staph aren't the only bacteria that can cause disease or that have grown resistant to antibiotics. About 200 species of bacteria are known to cause human diseases, including *Mycobacterium tuberculosis, Salmonella, Neisseria gonorrhoeae,* and *Klebsiella pneumoniae.* And although it gets the most attention, MRSA is by no means the only superbug out in the community. It is getting harder to treat patients with severe *Salmonella* food poisoning caused by drug-resistant strains. Gonorrhea has become resistant to another important group of antibiotics, the fluoroquinolones. In hospitalized pneumonia patients, infections by *Klebsiella* strains that are resistant to every available antibiotic are now emerging.

Because the very use of antibiotics can drive bacterial populations to evolve resistance, antibiotic resistance is inevitable. But humans have hastened the emergence of drug-resistant strains of bacteria by the haphazard use and overuse of antibiotics. For more than 40 years, physicians have typically prescribed antibiotics for colds, coughs, and earaches, most of which are caused by viruses that aren't killed by antibiotics anyway. Antibiotics are frequently overused or misused for many other ailments as well.

Doctors aren't the only culprits. Agricultural practices are also to blame. Antibiotics used in low doses to promote growth are often given to poultry, swine, and beef. This practice can cause food-borne pathogens such as *Salmonella* or *Campylobacter* to develop

antibiotic resistance. Undigested antibiotics in animal manure can contaminate the environment through groundwater or when manure is used as fertilizer. In this environment drug-resistant bacteria are more fit, and will therefore be selected for and become more prevalent over time.

"We really have to be careful about how we use antibiotics because antibiotic use is the biggest driver of antibiotic resistance," says Lynfield.

Clearly, creating stronger antibiotics isn't the only or the best solution because bacteria will ultimately adapt to those, too. Perhaps the best way to control resistance, say experts, is to change practices that enable resistant strains to thrive. Careful hygiene and prevention of infection through vaccination are important tools. It is also important that when an antibiotic is prescribed it is taken precisely as prescribed, for the full course of treatment, no matter how much better the patient may be feeling. If bacteria are exposed to antibiotics at low levels or for short durations, the entire population may not be eradicated. Remaining bacteria may be resistant to the antibiotic and emerge as the dominant population. And anyone taking antibiotics exposes all the bacteria in his or her body to the antibiotics, which may enable other drug resistant bacteria to emerge. These drug-resistant bacteria might then be transmitted to others.

At the community level, the more antibiotics that are used, the more resistance will emerge. So doctors are heavily discouraged from prescribing antibiotics unnecessarily. And efforts are being made to crack down on the practice of feeding livestock low levels of antibiotics (**Infographic 14.8**).

Of course, these measures won't fight resistant strains that are already circulating. But there are ways to reduce and perhaps prevent infections in this case, too. Because MRSA is more prevalent in certain populations, they present opportunities for health care workers to intervene. Prisons, for example, are a hot spot for infection because so many people pass through. "People go in, they pick up MRSA, they take it home, and then I see the kids come in sick," says Daum. By preventing transmission in prisons, health care workers may be able to prevent infections in the larger community.

A vaccine would be another way to prevent staph infections. For example, a vaccine for children against *Streptococcus pneumoniae* introduced in 2000 caused the rate of infection—and especially the rate of drug-resistant infections—to drop dramatically. And not only did the rate of infection drop in vaccinated children, but other age groups benefited as well because the bacteria were not being transmitted as frequently. As another example of the impact that vaccines have on infections, Daum points to the bacterium *Haemophilus influenzae*, which frequently caused pneumonia, meningitis, and other serious diseases in children. Today, children are vaccinated against it. "When I was an intern we used to see 60 to 80 *Haemophilus* infections a month," he says. "Today we see none, it's gone. And MRSA needs to be gone too." ∎

INFOGRAPHIC 14.8

Treating and Preventing Infection by Antibiotic-Resistant Bacteria

Reduce antibiotics in livestock feed.
Excessive antibiotics in the environment create continuous selective pressure for all bacteria.

Keep locker rooms and sports equipment clean.
Protect young people from contact with contaminated surfaces.

Disinfect common surfaces.
Avoid transmitting infection by contact with contaminated surfaces, especially in facilities that serve a lot of people.

Wash hands frequently.
This is especially important for people in close contact with other people.

Research new vaccines.
Prevent resistant bacterial strains from making people ill.

Do not take antibiotics for viral infections.
Viruses are not killed by antibiotics. Overuse of antibiotics causes resistant bacterial strains to become widespread.

◗ Summary

■ Populations are groups of individuals of the same species living together in the same geographic area.

■ Within any population, genetic variation exists among individuals.

■ Bacterial populations, which reproduce asexually, develop genetic variation by mutation and gene exchange; populations of sexually reproducing organisms generate genetic variation by meiosis and fusion of gametes as well as by mutation.

■ Genetic variation in a population gives rise to corresponding phenotypic variation in the population.

■ Different phenotypes can affect the survival and reproduction, or fitness, of individuals in the population.

■ The differential survival and reproduction of individuals in a population over time in response to environmental pressure is known as natural selection.

■ Natural selection is one cause of evolution, defined as a change in the allele frequency of a population over time.

■ Individuals with higher fitness in a given environment reproduce and pass on their alleles more frequently than do individuals with lower fitness, resulting in evolution.

■ Natural selection can shift the allele frequencies in a population in one of several patterns: directional selection, diversifying selection, or stabilizing selection.

■ Antibiotic-resistant populations of bacteria emerge by directional selection in the presence of antibiotics.

■ Over time, natural selection leads to adaptation: advantageous traits become more common in the population, which becomes more suited to its environment as a result.

BACTERIA AND DISEASE

Bacteria live in and on humans and may or may not cause disease.

HINT See Infographics 14.1 and 14.2.

➤ KNOW IT

1. The term "MRSA" as it is used today refers to
a. *Staphylococcus aureus* bacteria that are resistant to many antibiotics.
b. a collection of skin and other infections, caused by a type of bacteria.
c. *Staphylococcus aureus* bacteria that are found only in humans with certain types of skin infections.
d. *Staphylococcus aureus* bacteria that are normal residents of human skin in the vast majority of the human population.
e. all bacteria that are resistant to antibiotics.

2. What is the difference between a *Staphylococcus aureus* colonization and a *Staphylococcus aureus* infection?

3. Where is MRSA most likely to be a problem?
a. on the surface of the skin
b. in nasal passages
c. in the bloodstream
d. on your fingernails
e. The presence of MRSA in any of those locations indicates a serious infection.

➤ USE IT

4. A young athlete has a nasty skin infection caused by MRSA. How might he have contracted this infection?

5. For the patient in Question 4, which antibiotic would you choose to treat the infection? What other measures would you recommend to prevent spread of MRSA to the athlete's teammates and family? Explain your answer.

6. Why do the beta-lactam antibiotics affect sensitive bacterial cells but not eukaryotic cells? (You may need to review cell structure to answer this question.)

EVOLUTION BY NATURAL SELECTION

Evolution is a change in allele frequencies in a population over time. When genetically diverse individuals differ in their ability to survive and reproduce, evolution by natural selection occurs.

HINT See Infographics 14.3–14.8.

➤ KNOW IT

7. What are the two major mechanisms by which bacterial populations generate genetic diversity?
a. mutation and meiosis
b. binary fission and evolution by natural selection
c. gene exchange and mutation
d. mutation and binary fission
e. gene exchange and replication

8. What is the environmental pressure in the case of antibiotic resistance?
a. the growth rate of the bacteria
b. how strong or weak the bacterial cell walls are
c. the relative fitness of different bacteria
d. the presence or absence of antibiotics in the environment
e. the temperature of the environment

9. What is the evolutionary meaning of the term "fitness"?

10. The evolution of antibiotic resistance is an example of
a. directional selection.
b. diversifying selection.
c. stabilizing selection.
d. random selection.
e. steady selection

11. In humans, very-large-birth-weight babies and very tiny babies do not survive as well as midrange babies. What kind of selection is acting on human birth weight?
a. directional selection
b. diversifying selection
c. stabilizing selection
d. random selection

➤ USE IT

12. Binary fission is asexual. What does this mean? How could two daughter cells end up with different genomes at the end of one round of binary fission?

13. In what sense do bacteria "evolve faster" than other species?

14. If we take the most fit bacterium from one environment—one in which the antibiotic amoxicillin is abundant, for example—and place it in an environment in which a different antibiotic is abundant, will it retain its high degree of fitness?

a. yes, fitness is fitness, regardless of the environment

b. yes, once a bacterium is resistant to one antibiotic it is resistant to all antibiotics

c. not necessarily; fitness depends on the ability of an organism to survive and reproduce, and it may not do this as well in a different environment

d. no, what is fit in one environment will never be fit in another environment

15. If a single bacterial cell that is sensitive to an antibiotic—for example, vancomycin—is placed in a growth medium that contains vancomycin, it will die. Now consider another single bacterial cell, also sensitive to vancomycin, that is allowed to divide for many generations to become a larger population. If this population is placed into vancomycin-containing growth medium, some bacteria will grow. Why do you see growth in this case, but not with the transferred single cell?

16. If evolution by natural selection is a change in allele frequencies in a population, then why do we detect the process of evolution by natural selection as a change in *phenotype* frequencies in the population?

17. Imagine that a genetically diverse population of garden snails occupies your backyard, in which the vegetation is a variety of shades of green with some brown patches of dry grass.

a. If birds like to eat snails, but they can see only the snails that stand out from their background and don't blend in, what do you think the population of snails in your backyard will look like? Explain your answer.

b. Suppose you move the population of snails to a new environment, one with patches of dark brown pebbles and patches of yellow ground cover. Will individual snails mutate to change their color immediately? As the population evolves and adapts to the new environment, what do you predict will happen to the phenotypes in your population of snails after several generations in this new environment? How did this occur? Include the terms *gametes, mutation, fitness, phenotype,* and *environmental selective pressure* in your answer.

SCIENCE AND ETHICS

18. Your friend has had a virus-caused cold for 3 days and is still so stuffy and hoarse that he is hard to understand. He seems to be telling you that his doctor called in a prescription for an antibiotic for him to pickup at his pharmacy. You hope that you misunderstood him, but you realize that you heard him perfectly well.

a. Why are you dismayed to hear his story?

b. Will the antibiotic help your friend's cold?

c. What are the risks to your friend if he takes the antibiotic? (Think about what might happen if he should develops a wound infection.) What are the risks to you, as his friend?

19. Your roommate has been prescribed an antibiotic for bacterial pneumonia. She is feeling better and stops taking her antibiotic before finishing the prescribed dose, telling you that she will save them to take the next time she becomes sick. What can you tell your roommate to convince her that this is not a good plan?

Adventures in Evolution

➔ What You Will Be Learning

Informed by extensive reading, travel, and observations of the natural world, Darwin and Wallace independently proposed the mechanism for evolution.

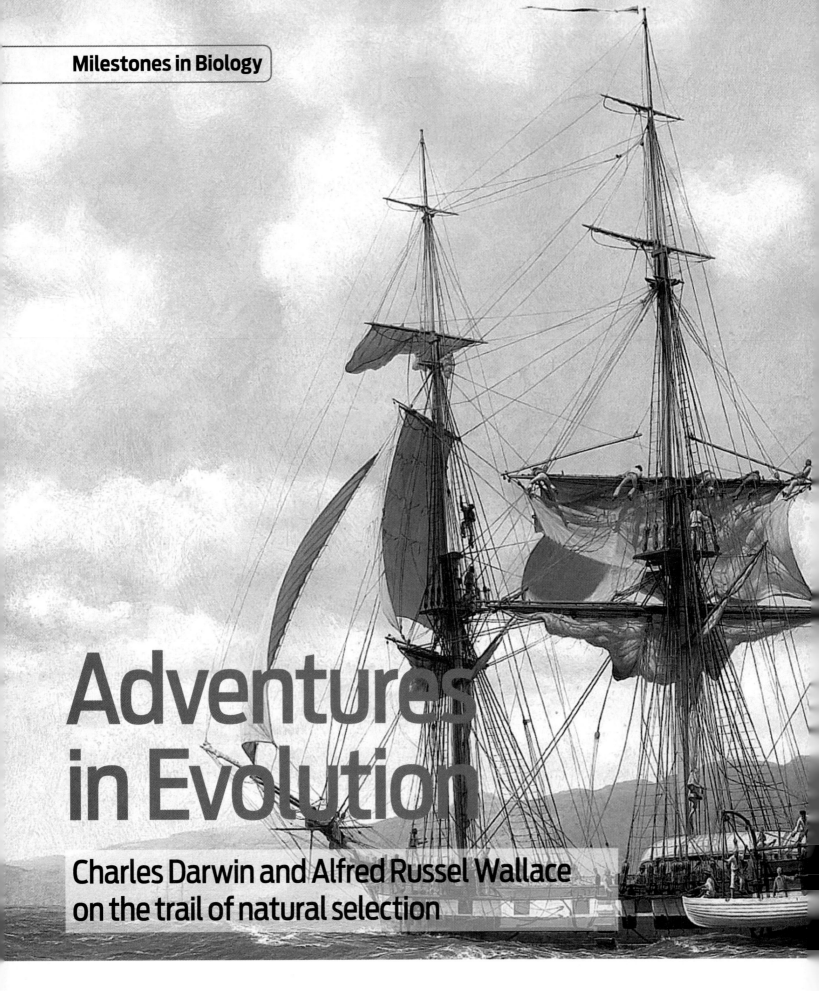

Adventures in Evolution

Charles Darwin and Alfred Russel Wallace on the trail of natural selection

It was the unpaid internship of a lifetime: a 5-year, around-the-world trip as a naturalist aboard a British surveying ship. The ship's captain, Robert Fitzroy, wanted a travel companion who would also collect specimens along the way. He approached a professor at Cambridge University, who nominated one of his best students—a good-natured 22-year-old bug collector named Charles Darwin. Unsure what he wanted to do with his life but eager to see the world, Darwin jumped at the chance to travel on the HMS *Beagle*. He later said of the trip, "The voyage of the *Beagle* has been by far the most important event in my life and has determined my whole career."

Yet he almost didn't go. His father, Dr. Robert Darwin, wanted his son to become a physician, like himself. But young Charles was more interested in spending time outdoors than studying medicine. "You care for nothing but shooting, dogs, and rat-catching, and you will be a disgrace to yourself and all your family," his father told him. When Charles's aversion to rote memorization and queasiness at the sight of blood effectively ruled out a career in medicine, they agreed that Charles would instead do the next best thing: become an Anglican minister. Far from furthering that goal, an impetuous sea voyage seemed to Robert Darwin a useless distraction—a "wild scheme," he called it—and at first he refused to let his son go. But eventually, at the cajoling of his family, he relented. Charles packed his bags, said goodbye to his girlfriend, Emma, and set sail for South America. It was December 1831.

The passage aboard the 90-foot vessel was frequently harrowing, and Darwin suffered debilitating bouts of seasickness, but his journey aboard the *Beagle* set in motion one of the greatest revolutions in science. What he saw on that trip planted the seeds of ideas that have completely changed the way we view the world and our place in it. As the evolutionary biologist Stephen Jay Gould put it, "The world has been different ever since Darwin."

> **"The world has been different ever since Darwin."**
> —Stephen Jay Gould

Journey to an Idea

Though Darwin is the most famous figure associated with the theory of evolution, he did not invent the idea. Nor was he alone among his contemporaries in studying it. In fact, the notion that species change gradually over time had been around for generations. To be sure, most people in the 1830s (Darwin included) still assumed that species were fixed and unchanging, created perfectly by God. But evidence to the contrary had been accumulating for some time. Explorers and naturalists were traveling to faraway lands, discovering a host of never-before-seen plants and animals. Fossils were being unearthed, providing evidence that some species no longer seen on earth had lived in the past. And anatomists were noting uncanny physical resemblances between different species, including chimpanzees and humans. Evolution was in the air when Darwin began thinking about it. His own grandfather, Erasmus Darwin, had even written a book about evolution in the 1790s.

However, the ideas that people in Darwin's time had proposed to explain *how* species changed were flawed. One common misconception was Lamarckianism, named after the French naturalist Jean-Baptiste Lamarck, who suggested that species could change through the inheritance of acquired characteristics. In the Lamarckian view, giraffes, for example, developed their long necks by continually stretching them to feed on tall trees. Once it acquired its long neck, a giraffe could then pass that advantageous trait on to its offspring. This idea of the inheritance of acquired character-

Lamarckianism: An Early (and Incorrect) Idea about Evolution

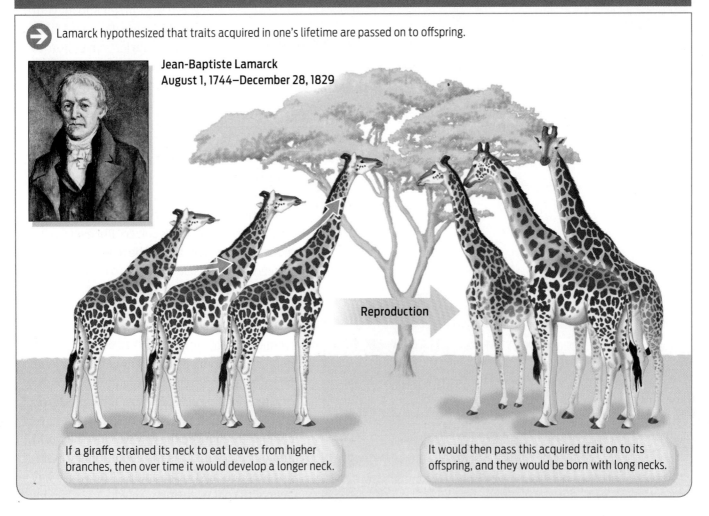

→ Lamarck hypothesized that traits acquired in one's lifetime are passed on to offspring.

Jean-Baptiste Lamarck
August 1, 1744–December 28, 1829

Reproduction

If a giraffe strained its neck to eat leaves from higher branches, then over time it would develop a longer neck.

It would then pass this acquired trait on to its offspring, and they would be born with long necks.

The Evolution of Darwin's Thought

 Darwin was influenced by the work of others, which informed the way he interpreted his own research and collections.

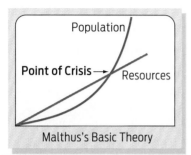

Malthus's Basic Theory

Population

Point of Crisis → Resources

**Charles Darwin
(February 12, 1809–April 19, 1882)**

Malthus's work (1798):
· Populations are limited by a number of factors, including food, water, and disease.
· Some individuals die and some survive.

Barnacle research (1846–1854):
· Darwin characterized differences between different groups of barnacles, and developed ideas about how these differences arose.

SKELETON OF THE MEGATHERIUM. Page 106.

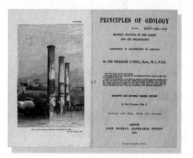

PRINCIPLES OF GEOLOGY

Fossil of giant land sloth (1833):
· Buried in a sediment layer below a deposit of shells.
· No modern-day South American animals resemble this fossilized creature. What happened to it?

***Beagle* voyage (1831–1836):**
· Darwin collected plants, animals, and fossils from across the globe.
· Darwin observed similarities and differences and attempted to explain these characteristics.

Lyell's work (1833):
· The earth's geology is formed by slow-moving forces.
· The earth is much older than thought at the time.

istics, while incorrect, was a popular one in Darwin's time–one that even Darwin himself found it hard to fully shake off in his writings.

While at sea, Darwin had plenty of time to read and think about the ideas then being discussed in scientific circles. He read, for instance, the work of the geologist Charles Lyell. Lyell's *Principles of Geology* (1833) argued that the earth was much older than the 6,000 years popularly accepted at the time (a figure based on a literal reading of the Bible), and that its geology had been shaped entirely by incre-

mental forces operating over a vast expanse of time. "No vestige of a beginning, no prospect of an end," was how James Hutton, Lyell's mentor, had put it. With such thoughts percolating through his mind, Darwin studied the plants, animals, and geology at each stop on his trip, collecting fossils and specimens of local flora and fauna wherever he went.

While exploring the shore of Argentina in August 1833, Darwin unearthed a particularly prized find: the complete fossil of a giant sloth embedded in a cliff, below a layer of

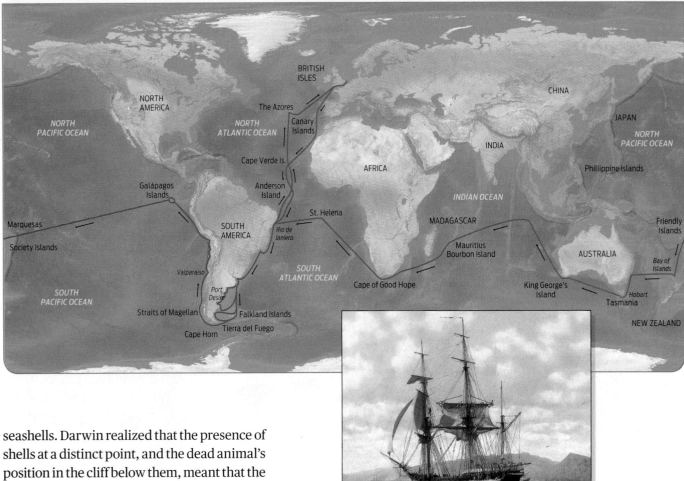

seashells. Darwin realized that the presence of shells at a distinct point, and the dead animal's position in the cliff below them, meant that the animal had lived in the area before it was an ocean environment. The animal had also clearly lived a very long time ago, since many layers of earth sat on top of it. More mysterious was the fact that there were no living animals in the present-day region that looked remotely like the creature he found. Where had they gone? Perhaps, Darwin reasoned, when the landscape changed the animals had been unable to adapt and had become extinct. Darwin also noticed that some South American species resembled European ones, despite living in an entirely different environment. Taken together, these pieces of evidence seemed to conflict with the notion that God had created each species perfectly adapted to its environment. Darwin was beginning to question the common wisdom of his day.

In 1835, the young naturalist stepped ashore on the Galápagos Islands, off the coast of Ecua-dor. On this archipelago, Darwin observed and collected many creatures, among them a variety of small birds. Months later, while studying the specimens when he was back in England, he realized they were all closely related species of finch. Each species was distinguishable by a different size and shape of beak. He later wrote in the second edition of *The Voyage of the Beagle* (1845), "One might really fancy that, from an original paucity of birds in this archipelago, one species had been taken and modified for different ends." His ideas were taking shape.

After returning home from his eye-opening voyage, Darwin made detailed notes of the evidence supporting his speculations. A key insight came to him in September 1838 while reading the work of the political economist Thomas Malthus, whose pessimistic book, *An Essay on the Principle of Population* (1798), described how hunger, starvation, and disease would ultimately limit human population growth. Darwin realized that, for animals, such limitations would lead to competition for resources that would put weaker individuals at a disadvantage. In these circumstances, Darwin wrote in his notebook, "favourable variations would tend to be preserved, and unfavourable ones to be destroyed." Competition for survival and reproduction among members of a species, he realized, would lead gradually to the species becoming more adapted to its surroundings. In effect, the environment was "selecting" for favorable traits, much as plant and animal breeders selected and perpetuated desirable varietals–a plant with especially large fruit, for

---

"One might really fancy that . . . one species had been taken and modified for different ends."
–Charles Darwin

---

instance. This idea of "natural selection" was Darwin's original contribution to the theory of evolution–what he called "descent with modification." (Darwin avoided using the term "evolution," which he thought gave a mistaken idea of progressive development, preferring instead this more descriptive phrase.) Others had speculated at length about species change–most notably Robert Chambers in *Vestiges of the Natural History of Creation* (1844)–but Darwin was the first to provide a clear *mechanism* of evolution. The philosopher of science Daniel Dennett has called the theory of natural selection "the single best idea anyone has ever had."

By 1844, Darwin had developed his ideas into a 200-page manuscript that he hoped would be the definitive word on the subject. He did not rush his ideas about natural selection into print, however. He knew that his ideas would be controversial, contradicting as they did strongly held beliefs about God and the special creation of all animals, including

humans. Other scientists with evolutionary ideas were causing quite a stir in England and being openly ridiculed (for this reason Robert Chambers had published his book anonymously). Even sharing his theory of evolution by natural selection with trusted colleagues, Darwin said, was "like confessing a murder." To withstand challenges, he knew he would need more detailed evidence.

And so, at age 37, Darwin began to investigate closely one group of animals: barnacles, the small invertebrates that cling to ships or marine life. Darwin spent 8 years, from 1846 to 1854, carefully analyzing the barnacles' tiny adaptations. It was tedious work, leading Darwin to write, "I hate a Barnacle as no man ever did before." Yet the work proved valuable, putting detailed meat on the bones of his skeletal idea. "What he found in barnacles," wrote Janet Browne, a professor of the history of science at Harvard and the author of *Darwin's Origin of Species: A Biography* (2007), "brought important shifts in his biological understanding, strengthened his belief in evolution and provided an essential backdrop to *Origin of Species*."

Darwin was hard at work fleshing out his idea in painstaking detail when he received a letter from a young naturalist with whom he had a casual acquaintance, a collector named Alfred Wallace who made a living selling rare butterflies and birds to other collectors and museums. The envelope was postmarked from an island in Indonesia. Inside was a 20-page manuscript describing the author's bold new idea about how species change over time, which he wanted Darwin to read and have published. Darwin, it seemed, had been scooped.

In Darwin's Shadow
Although we often credit Charles Darwin with the discovery of evolution by natural selection, he was not alone in charting this intellectual territory. Another British naturalist was also hot on the trail. Like Darwin, Wallace was fascinated by natural history and had a thirst for adventure. In other ways, though, the two men couldn't have been more different. Darwin came from a

wealthy, upper-class family and had received a prestigious Cambridge education. He was greeted as a minor celebrity when he returned from his trip around the world. Wallace, on the other hand, was a man of much more humble origins, for whom nothing in life had come easily.

The eighth of nine children, Wallace could not afford a university education, and he was plagued by financial difficulties his whole life. He attended night school and supported himself as a builder and railroad surveyor. His budding fascination with natural history, though, led him to read widely. Like Darwin, he read Lyell's work on geology, Malthus's work on human population, and Chambers's *Vestiges*. And, of course, he devoured Darwin's travel account, *The Voyage of the Beagle* (1839), which kindled his sense of adventure.

In 1848, having scrimped and saved, the 25-year-old Wallace set sail for Brazil, to the

The Evolution of Wallace's Thought

 Like Darwin, Wallace was influenced by the writings and work of others, which shaped his interpretations of his own observations and research.

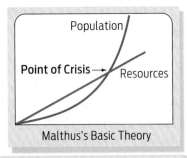

Malthus's Basic Theory

Malthus's work (1798):
· Populations are limited by a number of factors, including food, water, and disease.
· Some individuals die and some survive.

Alfred Russel Wallace (January 8, 1823–November 7, 1913)

Wallace–Darwin correspondence (185?–1858):
· Wallace began corresponding with Darwin several years before sending Darwin his completed manuscript.
· The two men were clearly developing very similar ideas about the nature of evolutionary change.

Amazon trip (1848–1852):
· Wallace observed that related (or "closely allied") species occupied neighboring geographic areas.
· He noted the role of physical barriers (such as the Amazon River) in separating related species from one another.

First publication (1855):
· Wallace's publication dealt with the physical distribution of species.
· He introduced the idea that new species are temporally and spatially connected to a related species.

Disease and famine (1858):
· While suffering from malaria, Wallace pondered the role of disease and famine in keeping human populations in check.
· He wondered how these factors could apply to the evolution of animal species.

mouth of the Amazon River. There he hoped to earn his reputation as a respectable scientist. Wallace was an unusually keen observer of the natural world. Exploring the rain forest of the Amazon, he was struck by the distribution of distinct yet similar-looking ("closely allied") species, which were often separated by a geographic barrier, such as a canyon or river. For example, he noted that different species of sloth monkey were found on different banks of the Amazon River. Over the course of his 4-year trip, Wallace scoured the Amazon and collected thousands of specimens.

Wallace was on his way home to London with his specimens in 1852 when disaster struck: his ship caught fire and sank. Wallace survived, but he lost everything—his notes, sketches, journals, and all his specimens. In spite of this catastrophe, Wallace was undeterred. He was, as his biographer Michael Shermer noted in his book *In Darwin's Shadow* (2002), "a veritable scientific and literary engine," a man who was singularly devoted to his research. Less than 2 years later, he was off on another collecting expedition, this time to the islands of Southeast Asia.

Wallace's first paper, "On the Law Which Has Regulated the Introduction of New Species," was published in September 1855. Based on his island work, it focused on the similar geographical distribution of closely allied species. For example, he wrote, "the Galápagos Islands . . . contain little groups of plants and animals peculiar to themselves, but most nearly allied to those of South America." From these observations, Wallace deduced this law, as he called it: "Every species has come into existence coincident both in space and time with a pre-existing closely allied species."

Wallace's article was groundbreaking, foreshadowing Darwin in a number of ways, but it lacked an explanation—a mechanism—of exactly how one species might have evolved from another.

> "Every species has come into existence coincident both in space and time with a pre-existing closely allied species." —Alfred Russel Wallace

Wallace continued his research, but in early 1858, disaster struck again: while traveling in what is now Malaysia, he contracted malaria. Not one to waste a perfectly good research opportunity, Wallace turned his convalescence into a sabbatical. As he later recalled, "I had nothing to do but to think over any subjects then particularly interesting me." Acutely aware of his own illness, he thought about what Malthus had written about disease and how it kept human populations in check. He also ruminated on Lyell's recent discoveries, which suggested that the earth was much older than previously thought. How might these forces of disease and death, multiplied over time, influence the composition of different populations, he wondered? As his fever waned, inspiration struck—like "friction upon the specially-prepared match," he later said: in every generation, weaker individuals will die and those with the fittest variations will remain and reproduce; as a result the species will become better adapted to its environment. Wallace had worked out the mechanism for evolution that was missing from his earlier work. He quickly wrote out his idea and sent it to the one naturalist he thought might be able to appreciate it. Wallace's paper arrived on Darwin's doorstep on June 18, 1858.

Darwin was stunned. For 20 years he had been working diligently on the same idea and now it seemed someone else might get credit for it. "All my originality will be smashed," he wailed to Lyell, asking him what he thought he should do. Recognizing the delicacy of Darwin's situation, Lyell and other colleagues devised a plan that would clearly establish Darwin's intellectual precedence: they would arrange to have papers by both men presented at a meeting of the Linnaean Society in London. The meeting took place on July 1, 1858. The papers were dutifully read, but there was no discussion or fanfare. In fact, neither of the authors was even present: Wallace was still traveling in Malaysia and Darwin was mourning the recent

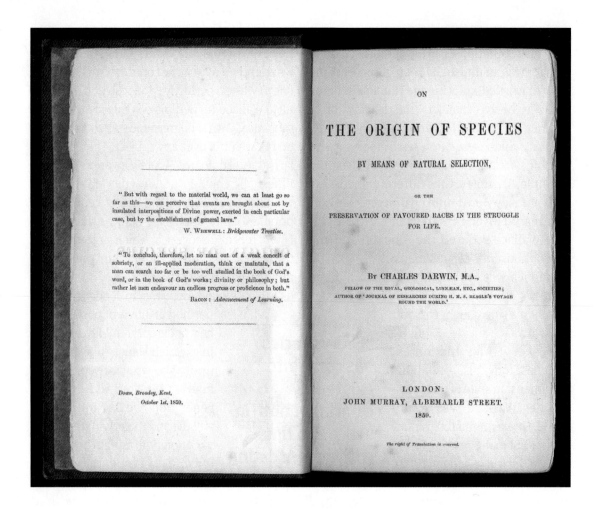

On

THE ORIGIN OF SPECIES

BY MEANS OF NATURAL SELECTION,

OR THE

PRESERVATION OF FAVOURED RACES IN THE STRUGGLE
FOR LIFE.

By CHARLES DARWIN, M.A.,
FELLOW OF THE ROYAL, GEOLOGICAL, LINNÆAN, ETC., SOCIETIES;
AUTHOR OF 'JOURNAL OF RESEARCHES DURING H. M. S. BEAGLE'S VOYAGE
ROUND THE WORLD.'

LONDON:
JOHN MURRAY, ALBEMARLE STREET.
1859.

The right of Translation is reserved.

" But with regard to the material world, we can at least go so far as this—we can perceive that events are brought about not by insulated interpositions of Divine power, exerted in each particular case, but by the establishment of general laws."

W. WHEWELL: *Bridgewater Treatise.*

"To conclude, therefore, let no man out of a weak conceit of sobriety, or an ill-applied moderation, think or maintain, that a man can search too far or be too well studied in the book of God's word, or in the book of God's works; divinity or philosophy; but rather let men endeavour an endless progress or proficience in both."

BACON: *Advancement of Learning.*

Down, Bromley, Kent,
October 1st, 1859.

death of his young son and too distraught to attend.

The scientific meeting secured Darwin's reputation, but still he was unsettled. Wallace's communication had lit a fire under his feet. He needed to finish his book. That work, *On the Origin of Species by Means of Natural Selection,* was published in November 1859. It would become one of the most famous books of all time, going through six editions by 1872. Just as every species is a product of its predecessors, so are ideas: in his book, Darwin credited Malthus and Lamarck, as well as Wallace.

Although it may seem that Wallace was cheated of his rightful recognition as a discoverer of evolution by natural selection, he was never bitter. On the contrary, he was delighted when he heard about his copublication with Darwin. He fully accepted that Darwin had for-mulated a more complete theory of natural selection before he did, and there is no trace of resentment in his later writings. In fact, Wallace titled his major work *Darwinism,* in recognition of the other man's intellectual influence.

After the presentation of 1858, Wallace stayed in the Malay archipelago for 4 more years, systematically recording its fauna and flora, and securing his reputation as both the greatest living authority on the region and an expert on speciation. In fact, Wallace is responsible for our modern-day definition of "species." In work on butterflies, he defined "species" as groups of individuals capable of interbreeding with other members of the group but not with individuals from outside the group. This idea–known today as the biological species concept–remains one of the most important in evolutionary theory. ■

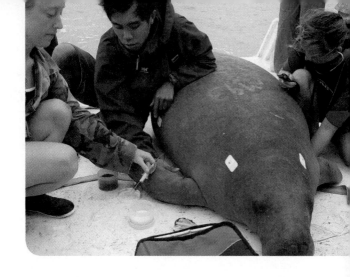

Evolution in the Fast Lane

Evolution in the Fast Lane

Can Florida's manatees cope with a rapidly changing environment?

Close your eyes. Picture a world in which you must eat, sleep, and breathe on a highway filled with cars. Now imagine the number of cars increasing every day, driving faster and faster.

"Strange as this scenario seems," says biologist Robert Bonde, that's essentially the predicament that Florida's manatees face today. These large marine mammals live in the congested waterways of Florida's panhandle, sharing their habitat with an ever-increasing number of speedboats. Frequent run-ins with boat propellers mean slashed backs, severed flippers, mangled tails—and even death—for the manatees, whose maximum speed of 5 miles per hour is eclipsed by that of the boats whizzing by. In fact, boat-manatee collisions are so common that researchers often use a manatee's distinctive battle scars as identification.

"If manatees weren't air breathers it wouldn't be much of a problem," explains Bonde, who works with the U.S. Geological Survey in Gainesville. "But because they have to go to the surface to get that breath of air, they're very susceptible to being struck."

Boat-related deaths reached an all-time high in 2009, when 97 animals were killed; in 2006, 92 manatees perished in collisions. And boats aren't the only threat to the manatees. Other causes of death include being caught in fishing lines and crushed in locks and flood dams. Every year, more than 100 manatees die from human-related causes.

Recognizing the plight of the manatee, the state of Florida has created speed zones in some manatee habitats and restricted boat access to others. Such measures have certainly helped, "but to completely protect the animals from

speedboats," says Bonde, "you'd almost have to take speedboats completely out of the picture," and people just aren't willing to do that.

Manatees belong to a group of mammals known as sirenians, a category that also includes the manatees' closest living relative, the dugong. There are four living species of sirenian. A fifth sirenian species, Steller's sea cow, was driven to extinction by hunting in the 1770s. Manatees, too, were once routinely hunted for their meat and hide and were in danger of becoming extinct by the late 1800s. To protect the creatures, Florida passed legislation in 1893 prohibiting the hunting of manatees, but

Manatees were once routinely hunted for their meat and hide and were in danger of becoming extinct by the late 1800s.

enforcement was lax, and their numbers continued to decline throughout the next century.

Not until 1967, when the species was officially listed as "endangered," did manatees receive protection from the federal government. Today, manatees are protected by both federal and state laws, and they are now Florida's official state marine mammal. Because of these efforts, the Florida manatee population has increased to more than 5,000 individuals, up from only 1,000 30 years ago. While the manatee is no longer on the brink of extinction and its numbers are increasing, biologists are still worried about the endangered

Left: A sign in a Florida marina alerts boaters to manatees.
Right: Damage to a manatee fin from a boat propeller.

species' long-term survival. Part of the problem is the continued threat from human influences such as coastal development, which forces manatees out of their preferred habitats. Equally worrisome is the fact that Florida's population of manatees is geographically isolated from manatees in other regions and thus in danger of being subdivided further. Small, isolated populations tend to have less genetic diversity than larger populations made up of many interbreeding individuals. Limited genetic diversity can seriously impair a population's evolutionary success in the face of a changing environment.

A Swim in the Manatee Gene Pool

Robert Bonde and his colleagues work with the Sirenia Project of the U.S. Geological Survey, which studies the life history, genetics, and ecology of the manatee. Bonde himself has been studying the creatures for more than 30 years; they are clearly his passion. The most fun, he says, is swimming with them. When you "get in the water with an animal the size of an elephant and look eye-to-eye with it, you come out of the water a different person." It's a kind of "enchantment," he says, that inspires his work and led him to become a world expert in manatee population genetics.

How did Florida get its famous population of manatees? DNA and fossil evidence suggests that Florida's manatees are descended from ancestors

> When you "get in the water with an animal the size of an elephant and look eye-to-eye with it, you come out of the water a different person."
> –Robert Bonde

that originally lived in South America. Two million years ago, a group of manatees living in the waters of the Amazon Basin journeyed north, slowly making their way up the coast of South America to the Caribbean. A few eventually swam all the way to Florida. What we know as the Florida manatee (*Trichechus manatus latirostris*) is actually one of two subspecies of the West Indian manatee, which ranges from South America to the southeastern United States. A third species of manatee lives along the coast of western Africa, and a fourth sirenian relative lives along the coasts of eastern Africa, India, Asia, and northern Australia. No one knows exactly how many manatees journeyed north to Florida; manatees don't make the trek today, so it's impossible to say. But whatever the number, it was enough to establish a successful population of manatees in the region, one that has existed for the last 15,000 years (Infographic 15.1).

From a genetic perspective, each population of manatees (or of any organism, for that matter) has its own particular collection of alleles, known as its **gene pool.** Within the gene pool, each allele is present in a certain proportion, or **allele frequency,** relative to the total num-

GENE POOL
The total collection of alleles in a population.

ALLELE FREQUENCY
The relative proportion of an allele in a population.

ber of alleles (for example, 50 out of 1,000, or .05). Over time, several forces can change the frequency of each allele—that is, how common it is in the population. When alleles change in frequency over time, a population evolves. (As discussed in Chapter 14, that's the definition of evolution.)

Evolutionary changes in a gene pool can have lasting consequences for a population—sometimes good, sometimes bad, sometimes neutral. For example, changes in the gene pool can result in a species becoming more adapted to its environment—think of antibiotic-resistant bacteria. The gene-pool-altering force that results in adaptation is natural selection, which is discussed in Chapter 14.

Natural selection isn't the only force that alters allele frequencies, although it is the only one that results in adaptation. Mutation, which introduces new alleles into a population, also alters allele frequencies, but since the process is random, it does not by itself lead to adaptation. In other words, mutation is a type of **nonadaptive evo-**lution. Nonadaptive evolution isn't necessarily "bad" or *maladaptive*. Without mutation introducing variation into a population, for example, there would be no evolution at all. But by itself, mutation does not lead to a population becoming more adapted. Another such nonadaptive cause of evolution—one of particular relevance to manatees—is **genetic drift,** change in allele frequencies between generations that occurs purely by chance.

Changing by Chance

Genetic drift is a bit like rolling the evolutionary dice. By simple chance, some individuals just happen to survive and reproduce, while others do not. Those that pass on their genes weren't necessarily more fit or better adapted; they just got lucky. Because only a subset of the population (with a subset of the total alleles) reproduces, only a subset of alleles is represented in the next generation.

Over time, genetic drift decreases the genetic diversity of a population. Genetic drift tends to

INFOGRAPHIC 15.1

Geographic Ranges of Manatees and Dugongs

This map shows the geographic ranges of the four living species of manatees and dugongs.

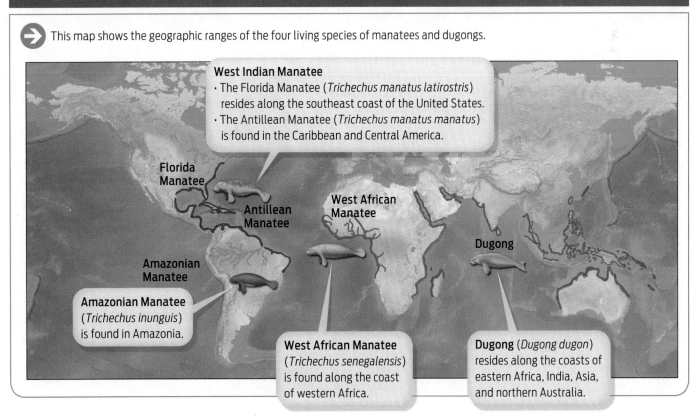

West Indian Manatee
· The Florida Manatee (*Trichechus manatus latirostris*) resides along the southeast coast of the United States.
· The Antillean Manatee (*Trichechus manatus manatus*) is found in the Caribbean and Central America.

Florida Manatee

Antillean Manatee

West African Manatee

Dugong

Amazonian Manatee

Amazonian Manatee (*Trichechus inunguis*) is found in Amazonia.

West African Manatee (*Trichechus senegalensis*) is found along the coast of western Africa.

Dugong (*Dugong dugon*) resides along the coasts of eastern Africa, India, Asia, and northern Australia.

have more dramatic effects in smaller populations than in larger ones, since in a population with few individuals any single individual that does not reproduce could spell the loss of alleles from the population.

Manatees are believed to have experienced just such a chance loss of genetic diversity through a type of genetic drift known as the **founder effect.** The population of manatees that settled in Florida was a group of colonists that emigrated from the larger population living in the Caribbean. As a founder population, Florida's original manatees likely contained only a subset of the total alleles present in the original Caribbean gene pool. "Manatees came in, in a smaller population, and started to grow," explains Bonde. The reduced level of genetic diversity that Florida manatees have now reflects the reduced diversity of the original founders **(Infographic 15.2).**

Genetic drift has affected manatees in other ways as well. In some parts of their range, for example, manatees were hunted almost to extinction. When a population is cut down sharply, and often suddenly, there's a good chance that the remaining population will possess a greatly impoverished gene pool—a type of drift known as the **bottleneck effect.** Bottlenecks can also occur from natural causes—say, an extremely cold winter that causes half the population to die. Populations forced through a genetic bottleneck contain only a small fraction of the original starting diversity in the population **(Infographic 15.3).**

As another example of a genetic bottleneck, consider the cheetah (*Acinonyx jubatus*), the

FOUNDER EFFECT
A type of genetic drift in which a small number of individuals leaves one population and establishes a new population; by chance, the newly established population may have lower genetic diversity than the original population.

BOTTLENECK EFFECT
A type of genetic drift that occurs when a population is suddenly reduced to a small number of individuals, and alleles are lost from the population as a result.

INFOGRAPHIC 15.2

The Founder Effect Reduces Genetic Diversity

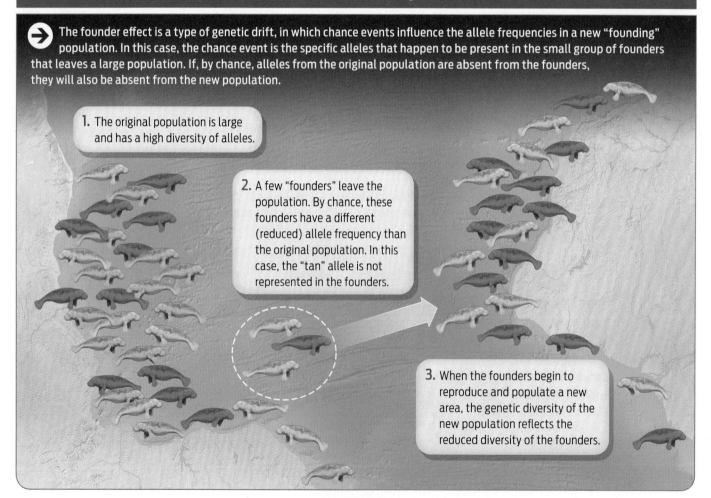

The founder effect is a type of genetic drift, in which chance events influence the allele frequencies in a new "founding" population. In this case, the chance event is the specific alleles that happen to be present in the small group of founders that leaves a large population. If, by chance, alleles from the original population are absent from the founders, they will also be absent from the new population.

1. The original population is large and has a high diversity of alleles.

2. A few "founders" leave the population. By chance, these founders have a different (reduced) allele frequency than the original population. In this case, the "tan" allele is not represented in the founders.

3. When the founders begin to reproduce and populate a new area, the genetic diversity of the new population reflects the reduced diversity of the founders.

Bottlenecks Can Reduce Genetic Diversity

→ Genetic bottlenecks occur when a population loses a large proportion of its members. Bottlenecks are more consequential when the starting population is small. In a large population, the reduced population is still likely to retain the same alleles present in the original population (top panel). In a small population (lower panel), the loss of population members is more likely to result in loss of alleles, and therefore a dramatic change in the allele frequency and genetic diversity of the "restored" population.

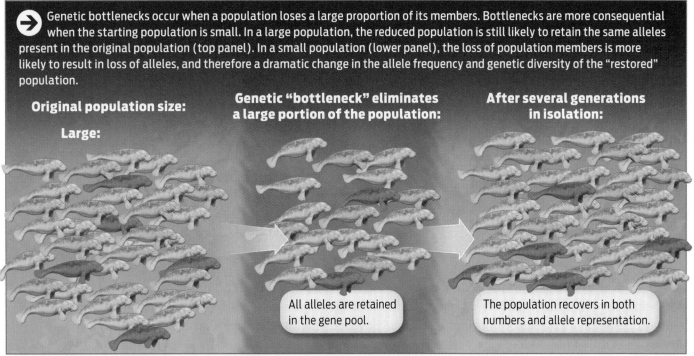

Original population size:

Large:

Genetic "bottleneck" eliminates a large portion of the population:

After several generations in isolation:

All alleles are retained in the gene pool.

The population recovers in both numbers and allele representation.

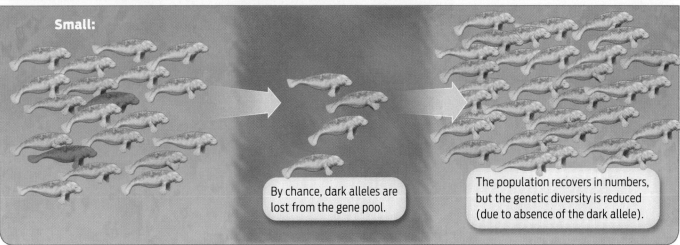

Small:

By chance, dark alleles are lost from the gene pool.

The population recovers in numbers, but the genetic diversity is reduced (due to absence of the dark allele).

fastest land animal. Cheetahs nearly became extinct 10,000 years ago, when harsh conditions of the last ice age claimed the lives of many large vertebrates on several continents. Ultimately, a few cheetahs survived and reproduced, but the more than 12,000 individuals alive today are now so genetically similar that skin grafts between unrelated individuals do not cause immune rejection; the animals are essentially genetically identical.

Why is genetic diversity important? In essence, a diverse gene pool gives a population more flexibility to survive in a changing environment: the more genetically diverse a population is, the more ways it has to adapt. Different alleles produce variation in inherited characteristics such as skin color or body shape. In the right environment, some genetic traits can greatly enhance survival and reproduction.

You can think of a gene pool as a population's portfolio of assets. Having a diverse array of investments is a better strategy for long-term success than having all your money tied up in one kind of stock–especially if that stock loses value in changed economic times.

For example, say a population of manatees suddenly finds itself in a colder environment (as indeed many did during the last ice age). If the population carries with it a rich variety of alleles, then some of these alleles (ones for more insulating blubber, for example) may help that population survive and reproduce in the altered conditions. Individuals with these alleles will be more fit in this environment, and the population will adapt by natural selection. With less diversity in the population, adaptation will be more limited, and the population may shrink. "Generally, if your allelic diversity is low, you're not as fit to survive and persist over a long period of time in the environment," explains Bonde. That's why it's so important to keep an eye on the manatee gene pool.

Analyzing Evolution

As part of their work with the Sirenia Project, Bonde and his colleagues regularly set out to capture, tag, and analyze the slow-moving manatees. Capturing a 1-ton, 10-foot-long, slippery animal is no small feat. It requires a special boat with a flat cargo bed that can be submerged in water like a boat ramp. The researchers lower the ramp into the water, beneath the manatee, and then hoist the animal on deck. Though the manatee at first tries to wriggle off the boat, once out of the water it becomes fairly complacent.

Assessing genetic diversity in populations of threatened or endangered species is a key part of conservation biology.

With the manatee safely on board, the biologists take the animal's vital signs and photograph it. They also collect blood and tissue samples for genetic studies. Manatees, like all mammals, breathe air, but they can dry out while in the boat, so the team pours water on them to keep their skin moist during the procedures. Finally, before a manatee is released back into the water, a radio tag is attached so the team can monitor its movements.

One of the main things that Bonde and his team want to measure is the level of genetic

Members of the Sirenia Project draw blood from a manatee for genetic studies.

diversity in the manatee population. Assessing genetic diversity in populations of threatened or endangered species is a key part of conservation biology; it's one way that biologists measure the genetic health of populations. If a population is shown to have low levels of genetic diversity, which could threaten its ability to adapt to a changing environment, then preventive measures might be taken, such as attempting to reintroduce genetic diversity and protecting habitat. The hope is that with early intervention it may be possible to avoid the more serious threat of irreversible damage–even extinction.

How do biologists measure the genetic health of populations? Put another way, how can you tell if a population is evolving in ways that could be detrimental to its long-term success? One way would be to measure allele frequencies in a population over generations to see if total genetic diversity is going down. Since the life span of a manatee can be 70 years, however, it is not practical to wait around to witness such evolution happening. A shortcut used by population biologists is, in essence, to take a "snapshot" of the gene pool at a given time, and compare it to the picture of a population that is known *not* to be evolving. If these two pictures differ, then you know your population is evolving, and you can begin to investigate why.

What does a nonevolving population look like? Allele frequencies in a nonevolving population behave in a predictable way: by definition, they do not change over time. Furthermore, in a nonevolving population, genotype frequencies remain unchanged from one generation to the next, a condition known as **Hardy-Weinberg equilibrium.** The Hardy-Weinberg equilibrium provides a baseline from which to judge if a population is evolving or not; it describes the default pattern of genotypes in nonevolving populations (see **Up Close: Calculating Hardy-Weinberg Equilibrium**).

How is Hardy-Weinberg equilibrium useful to conservation biologists? Suppose we have a population of manatees with two possible alleles for hide thickness, thick skin (T) and thin skin (t). Let's say we sample this population of manatees and find that the frequency of heterozygote individuals (Tt) is lower than their expected frequency based on Hardy-Weinberg equilibrium. In this case, we know that something is causing heterozygotes to become less common. This could happen if, for example, natural selection were favoring homozygotes to preferentially survive and reproduce, which would be a type of adaptive evolution.

Heterozygotes can also become less common in a population if closely related members of the population are mating with each other, a phenomenon known as **inbreeding.** Inbreeding is a type of nonrandom mating that will cause genotype frequencies to differ from predicted Hardy-Weinberg values. Specifically, inbreeding will cause homozygous individuals to become more common. Because closely related individuals are more likely to share the same alleles, the chance of two recessive alleles coming together during a mating is high. When that happens, homozygous recessive genotypes are created, and previously hidden recessive alleles start to affect phenotypes.

The accumulation of harmful recessive phenotypes can lower fitness (and thus fertility), a phenomenon known as **inbreeding depression,** which can threaten a species' long-term survival. One species that has suffered from inbreeding depression is the Florida panther, which is actually a subspecies of puma. In the past, Florida panthers mated with puma populations from neighboring states, where their ranges overlapped. This interbreeding–that is, breeding between populations–fostered an exchange of alleles that continually enriched the local populations' genetic diversity.

By the mid-20th century, however, hunting and development had squeezed the Florida panther population into an isolated region at the state's southernmost tip. By 1967, only 30 panthers remained, and the U.S. Fish and Wildlife Service listed them as endangered. By 1980, the panthers showed unmistakable signs of inbreeding depression–birth defects, low sperm count, missing testes, and bent tails.

HARDY-WEINBERG EQUILIBRIUM
The principle that, in a non-evolving population, both allele and genotype frequencies remain constant from one generation to the next.

INBREEDING
Mating between closely related individuals. Inbreeding does not change the allele frequency within a population, but it does increase the proportion of homozygous individuals to heterozygotes.

INBREEDING DEPRESSION
The negative reproductive consequences for a population associated with having a high frequency of homozygous individuals possessing harmful recessive alleles.

How do we know if a population is evolving? To find out, we can use a mathematical formula called the **Hardy-Weinberg equation**, which calculates the frequency of genotypes you would expect to find in a nonevolving population. For a gene with one dominant and one recessive allele, p and q, this formula can be written as:

$$p^2 \quad + \quad 2pq \quad + \quad q^2 \quad = 1$$

Frequency of homozygous dominants	Frequency of heterozygotes	Frequency of homozygous recessives

By definition, a population is not evolving (and is therefore in Hardy-Weinberg equilibrium) when it has stable allele frequencies and, therefore, stable genotype frequencies from generation to generation. This can only be achieved when *all five* of the following conditions are met:

1. No mutation introducing new alleles into the population
2. No natural selection favoring some alleles over others
3. An infinitely large population size (and therefore no genetic drift)
4. No influx of alleles from neighboring populations (i.e., no gene flow)
5. Random mating of individuals

In nature, no population can ever be in strict Hardy-Weinberg equilibrium, since it will never meet all five conditions. In particular, because no real population is infinitely large, genetic drift will always occur. In other words, all natural populations are evolving. Nevertheless, by describing the pattern of genotypes in a nonevolving population, Hardy-Weinberg equilibrium provides a baseline from which to measure evolution.

To see how the Hardy-Weinberg equation can be used to detect evolutionary change, consider the following example. Say you have a population of manatees with two possible phenotypes for hide color, gray and white. The allele for gray hide color, G, is dominant; the allele for white hide color, g, is recessive. As every individual in the population has two alleles for the hide-color gene (one maternal and one paternal), there are twice as many alleles as there are members of the population. So a population of 500 manatees has 1,000 alleles of the gene for hide color.

In this population, assume there are 800 G alleles, and 200 g alleles. We would then say that the frequency of the dominant allele is 0.8 (800/1,000) and the frequency of the recessive allele is 0.2 (200/1,000). Since there are only two alleles in the population, their combined frequencies must add up to 1. If we use p to denote the frequency of the dominant allele and q to denote the frequency of the recessive allele, then we can say that $p + q = 1$.

Suppose we want to use those allele frequencies to calculate the expected frequency of white-hided (gg) individuals in the population. If the frequency of g in the population is q, then we know from the Hardy-Weinberg equation that the frequency of gg is $q^2 = (.2)(.2) = .04$. Thus, in our population of manatees, 4%, or 20 manatees, will have white hides. If we find out that the *actual* number of white manatees in the population is appreciably more or less than this number, then we know that our population is evolving, and we can begin to investigate why.

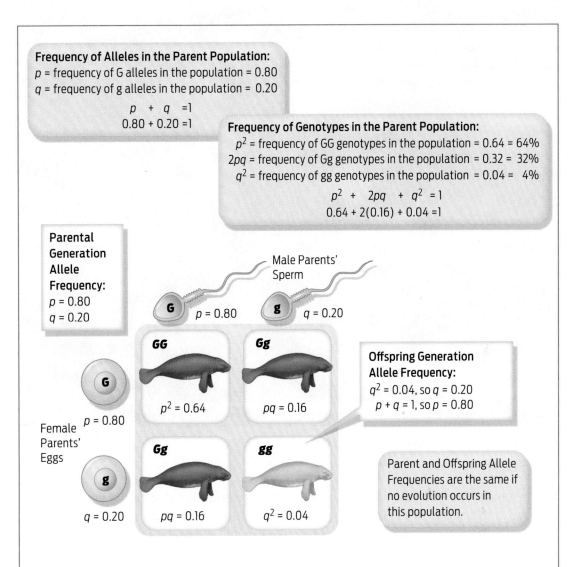

Frequency of Alleles in the Parent Population:
p = frequency of G alleles in the population = 0.80
q = frequency of g alleles in the population = 0.20

$$p + q = 1$$
$$0.80 + 0.20 = 1$$

Frequency of Genotypes in the Parent Population:
p^2 = frequency of GG genotypes in the population = 0.64 = 64%
$2pq$ = frequency of Gg genotypes in the population = 0.32 = 32%
q^2 = frequency of gg genotypes in the population = 0.04 = 4%

$$p^2 + 2pq + q^2 = 1$$
$$0.64 + 2(0.16) + 0.04 = 1$$

Parental Generation Allele Frequency:
$p = 0.80$
$q = 0.20$

Male Parents' Sperm

G $p = 0.80$ g $q = 0.20$

GG   Gg

Offspring Generation Allele Frequency:
$q^2 = 0.04$, so $q = 0.20$
$p + q = 1$, so $p = 0.80$

G

$p^2 = 0.64$ $pq = 0.16$

Female Parents' Eggs $p = 0.80$

Gg   gg

g

$q = 0.20$ $pq = 0.16$ $q^2 = 0.04$

Parent and Offspring Allele Frequencies are the same if no evolution occurs in this population.

Hardy-Weinberg also has important applications in public health. For example, the Hardy-Weinberg equation can be used to estimate the frequency of carriers (heterozygotes) of rare recessive diseases, such as cystic fibrosis. Since we know that the frequency of CF is approximately 1 in 3,000 Caucasian babies in the United States, we know that the frequency of homozygous recessive individuals, q^2, is 1 in 3,000 (or 0.0003). This means that the frequency of the recessive allele, q, must equal 0.018 (the square root of 0.0003). And if $q = 0.018$, then $p = 1 − 0.018 = 0.982$ (since $p + q = 1$). Therefore, the frequency of heterozygotes, $2pq$, is 0.035, or 3.5%. Knowing the frequency of the disease and carriers in the population helps health workers offer genetic counseling and plan for interventions.

A manatee rescue in Homosassa Springs, Florida.

In response, the Fish and Wildlife Service took active measures. In 1995, it brought in female pumas from Texas to mate with Florida's male panthers. The program was successful: the hybrid kittens—30 in all—showed none of the symptoms of inbreeding depression. Today, more than 100 healthy panthers roam the swamps and grasslands of Florida.

Some researchers were concerned that inbreeding might be occurring in the geographically isolated Florida manatee. Despite the manatee's ability to swim large distances, populations tend to stay close to protected coastal waters and rivers. "It's very much against their well-being to travel across open water," explains conservation geneticist Brian Bowen, a colleague of Bonde's who conducted genetic studies of manatees when he worked at the University of Florida.

"Out there, they are big, slow, and tasty. They are just a big shark egg roll, and that's one of the reasons they don't move much between their different habitats in the West Atlantic," Bowen told reporters at the University of Florida. When populations are isolated—when they don't mix—they're stuck with the limited amount of genetic variation that each already contains, and the likelihood of inbreeding goes up. Moreover, the small size of their populations makes genetic drift more likely, compounding the problem.

Acutely aware of the importance of genetic diversity to populations, Bonde and others have analyzed the genetic makeup of a number of Florida manatees. This work has revealed some good news. While the overall allelic diversity of Florida manatees is indeed relatively low—as you would expect of a founder population—it is not nearly as low as, for example, that of the cheetah population that went through a sudden bottleneck. Researchers are still able to find numerous subtle DNA differ-

Gene Flow Between Populations Increases Genetic Diversity

 Mutation introduces new alleles into a population at low frequency. Migration and interbreeding of individuals moves alleles between populations. DNA analysis of Florida manatee populations shows genetic diversity in nuclear DNA, consistent with healthy gene flow between Florida populations, as shown in the example below.

Inbred, or reproductively isolated, populations have limited diversity:

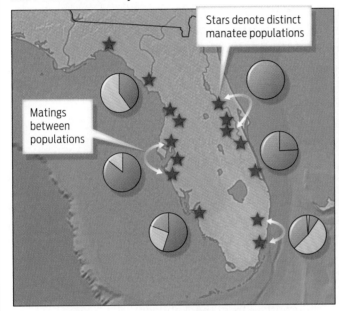

Stars denote distinct manatee populations

Matings between populations

Populations that mix their alleles with other populations are more diverse:

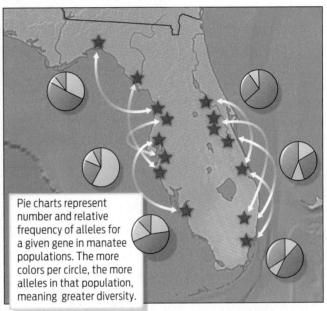

Pie charts represent number and relative frequency of alleles for a given gene in manatee populations. The more colors per circle, the more alleles in that population, meaning greater diversity.

These populations have a limited genetic diversity because they only share alleles with themselves. Many of these populations have a limited number of the possible alleles available.

These populations have higher levels of genetic diversity, as there is a flow of alleles between gene pools. Each population has all possible alleles but carries them in different frequency.

GENE FLOW
The movement of alleles from one population to another, which may increase the genetic diversity of a population.

ences among manatee individuals. Moreover, when researchers compare the frequency of genotypes in the population to those predicted by Hardy-Weinberg equilibrium, they find no significant differences, suggesting that inbreeding is not occurring.

These results have given conservationists a reason to be optimistic. Manatees seem to be preserving their genetic diversity. However, as researchers are quick to point out, the continued health of the gene pool clearly hinges upon the population's being large enough to avoid loss of genetic variation through genetic drift, which in turn depends upon reducing human-related deaths as well as protecting habitat. Especially crucial is preserving migratory cor-

ridors that manatees use to encounter each other and mate.

Manatees Mixing It Up, or Gene Flow

Though they don't travel in the open ocean, manatees still manage to migrate great distances, especially during the breeding season. "In the course of a year, they could go over several hundred if not thousands of miles," says Bonde. Through this annual mixing and mating of wandering manatees, alleles are continually shared between neighboring gene pools in a process known as **gene flow**.

Like genetic drift, gene flow is a type of evolution that does not lead to adaptation. But unlike genetic drift, which *decreases* the genetic diversity of a population, gene flow works in the opposite direction: it *increases* the genetic diversity of a local population by introducing alleles from its neighbors **(Infographic 15.4)**.

Thanks to high levels of gene flow, manatees in Florida have access to all the available alleles in the statewide population. To date, no region-specific alleles have been found. This is good news for the population as whole: "There's evidence now that the populations are mixing, and that there's enough mixture in the population that it's keeping gene flow channels open and the populations have been healthy," explains Bonde. In fact, he and his colleagues have even detected a flow of alleles between populations on the east and west coasts of Florida, even though east-west coast migrations of individuals are rarely observed; somehow, the individuals are mixing.

Although manatees are now doing well–for the moment–there is concern that continued habitat destruction or increased run-ins with boats could result in a fragmented population with isolated groups that are not able to interbreed. Conservation biologists are therefore trying to ensure that the Florida manatees are able to keep moving from place to place.

While genetic isolation can be dangerous for small populations, it isn't always a detrimental force in evolution. In fact, genetic isolation has a lot to do with how manatees became manatees in the first place.

How One Manatee Species Became Three

Though they are water-living mammals, manatees are actually more closely related to elephants than to whales and dolphins. DNA and fossil evidence shows that manatees first evolved in freshwater regions of South America, such as the Amazon River basin, and subsequently branched out into the Caribbean. The first group of Caribbean migrants included those that would later evolve into the West

> **Though they are water-living mammals, manatees are actually more closely related to elephants than to whales and dolphins.**

African manatee when they made the journey across the Atlantic and adapted to a new environment. The ancestors that remained in the Caribbean would evolve into the West Indian manatee. All of this took place within the past 2 million years.

The current population of manatees in Florida likely arrived there about 15,000 years ago, between ice ages, when the climate warmed and ice caps receded north, making migration between landmasses easier. When the climate again cooled, migration routes were closed, and the Florida population was thereby isolated from the Caribbean population by ecological barriers. "Today, the cool winters of the northern Gulf Coast, on the one hand, and the deep water and strong currents of the Straits of Florida, on the other, are believed to be the barriers . . . to gene flow that have allowed the Florida subspecies to differentiate from the Antillean manatees inhabiting Mexico and Cuba," explains Daryl Domning, a professor of anatomy at Howard University in Washington, D.C., and an expert on sirenian evolution.

Biologists recognize three living species of manatees: the Amazonian manatee (*Trichechus inunguis*), the West Indian manatee (*Trichechus manatus*), of which Florida manatees are a subspecies (*Trichechus manatus latirostris*), and the West African manatee (*Trichechus senegalensis*). But what, exactly, is a species? The term *species* comes from the Latin word for "kind" or "appearance." To define "species" more precisely, evolutionary biologists rely on the **biological species concept,** which defines a species as a population of individuals whose members can interbreed and produce fertile offspring.

Members of different species cannot mate with each other because their populations are reproductively isolated. Such **reproductive isolation** can be caused by a number of factors. For example, the two species may have a different mating time, location, or mating rit-

BIOLOGICAL SPECIES CONCEPT
The definition of a species as a population whose members can interbreed to produce fertile offspring.

REPRODUCTIVE ISOLATION
Mechanisms that prevent mating (and therefore gene flow) between members of different species.

SPECIATION
The genetic divergence of populations owing to a barrier to gene flow between them, leading over time to reproductive isolation and the formation of new species.

INFOGRAPHIC 15.5

Species Are Reproductively Isolated

 Species maintain their reproductive isolation in a variety of ways.

Ecological Isolation:
Different environments.
The Arctic Fox and the Desert Fox live in such different places, they never encounter one another.

Temporal Isolation:
Mating behavior or fertility at different times.
The Leopard Frog mates in early spring and the Bullfrog mates in early summer.

Behavioral Isolation:
Different mating activities.
The Prairie Chicken is not attracted to the mating display of the Ring-necked Pheasant.

Mechanical Isolation:
Mating organs are incompatible.
Plants pollinated by the hummingbird do not receive pollen from plants pollinated by the Black Bee.

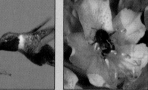

Gametic Isolation:
Gametes cannot unite.
The gametes from a dog and a cat cannot unite to form a zygote.

Hybrid Inviability:
Gametes unite but viable offspring cannot form.
The goat and sheep can mate, but the zygote formed does not survive.

Hybrid Infertility:
Viable hybrid offspring cannot reproduce.
Zebras and horses are different species because their hybrid offspring, zebroids, cannot make offspring of their own.

ual—so, like ships that pass in the night, they may never have the opportunity to meet. Or, anatomical differences between the two species may make the physical act of mating impossible (**Infographic 15.5**). Over evolutionary time, two closely related but reproductively isolated species—say, manatees and dugongs—will gradually become more and more dissimilar as genetic differences between them accumulate.

Genetic differences can also occur and accumulate within a single species, *provided some barrier to gene flow occurs between populations*. Once this barrier forms, the separated gene pools will evolve independently of each other. The specific alleles present in each separated gene pool will then depend on the precise balance among the four main forces of evolution acting on it: mutation, natural selection, genetic drift, and gene flow. Eventually, if enough genetic changes accumulate between populations of the same species, the two populations may diverge into separate species, a process known as **speciation.** This is what happened among manatees.

Some manatee species are geographically as well as reproductively isolated (for example, the West Indian manatee and the Amazonian manatee, which have different numbers of chromosomes, will not produce fertile offspring); and each has evolved by natural selection to become adapted to a different environment. You can see the results of natural selection in the different populations of manatees living today.

For instance, snout shape: West Indian manatees, including those living in Florida, have a pronounced bend in their snout, an adaptation that allows them to take better advantage of the kinds of aquatic vegetation found in marine environments, particularly the rooted plants that grow on riverbeds. Amazonian manatees, by contrast, have a flatter snout, better for scooping up overhanging freshwater vegetation near the water's surface. West Indian manatees are also about twice as large as Amazonian manatees. With all that extra padding and insulation, the West Indian manatee is able to tolerate the cooler water temperatures found in their northern range. The three manatee species–West Indian, Amazonian, and West African–acquired their distinct adaptations over hundreds of thousands of years through the process of natural selection **(Infographic 15.6)**.

Numerous genetic studies have confirmed that Antillean and Florida manatees are genetically distinct from their manatee cousins living in the Amazon, that both are distinct from their West African counterparts, and that there is very little, if any, gene flow between the populations. These studies have also helped to untangle the evolutionary history of the three species. To study ancestry, researchers often rely on sequences of mitochondrial DNA, which exists in the mitochondria of cells (Chapter 20 discusses how mitochondrial DNA can be used to trace ancestry). In Florida, only one sequence of mitochondrial DNA has been found in the entire

INFOGRAPHIC 15.6

Physical Traits in the Order Sirenia

 All sirenia species have limited amounts of body hair and long whiskers on either side of the snout. The whiskers are very sensitive and help the animal identify food. The upper snout of all Sirenia is split so that each side can move independently to push food into the mouth.

Family Dugongidae: Dugongs and Sea Cows (extinct)
Dugongs have a more streamlined body than manatees. The body tapers into a tail, which flares out into two distinct flukes. The dugong snout points downward and is flat on the bottom to help it dig up underground plants and roots from the sea bottom. They also have a set of tusks that emerge from the upper jaw. Dugongs can spend their entire lives in saltwater.

Dugong
Dugong dugon

Steller's Sea Cow
Hydrodamalis gigas (extinct in the 1770s due to hunting)

Family Trichechidae: Manatees
The manatee's body is distinctly more rotund than that of the dugong. It narrows slightly before it flares into a broad, paddle-shaped tail. Manatees have three or four fingernails at the ends of their flippers, which aid in feeding. Manatees must periodically find freshwater to drink and cannot live exclusively in saltwater.

West Indian Manatee
· Florida Manatee, *Trichechus manatus latirostris*
· Antillean Manatee, *Trichechus manatus manatus*

Amazonian Manatee
Trichechus inunguis

West African Manatee
Trichechus senegalensis

The Mitochondrial DNA of Florida Manatees Lacks Genetic Diversity

→ Looking at the distribution of mitochondrial DNA (mtDNA) variants in the manatee range is one way to assess genetic diversity.

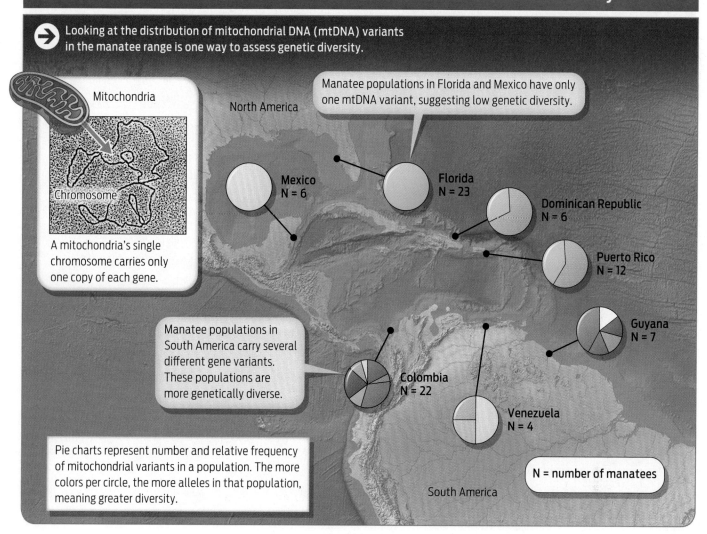

Mitochondria

Chromosome

A mitochondria's single chromosome carries only one copy of each gene.

North America

Manatee populations in Florida and Mexico have only one mtDNA variant, suggesting low genetic diversity.

Mexico
N = 6

Florida
N = 23

Dominican Republic
N = 6

Puerto Rico
N = 12

Guyana
N = 7

Manatee populations in South America carry several different gene variants. These populations are more genetically diverse.

Colombia
N = 22

Venezuela
N = 4

Pie charts represent number and relative frequency of mitochondrial variants in a population. The more colors per circle, the more alleles in that population, meaning greater diversity.

South America

N = number of manatees

manatee population, which indicates that this population has remained isolated from those populations in other parts of the Americas that have different mitochondrial DNA sequences **(Infographic 15.7).**

Speciation that occurs because of geographic or climatic separation is known as **allopatry.** According to evolutionary ecologist Juliana Vianna of Andrés Bello University in Chile, that is how the three species of manatees evolved: "Manatee species diversification occurred mainly by isolation [that is, allopatry], followed by local selection pressures, such as the freshwater adaptation of Amazonian manatees." In

ALLOPATRY
Speciation that occurs because of geographic or climatic barriers to gene flow.

that sense, manatee speciation resembles the path to speciation take by another group of animals with famously distinctive faces: the more than 13 species of finches living on the Galápagos Islands, near Ecuador, that Charles Darwin encountered on his voyage aboard the *Beagle* (see **Milestones in Biology: Adventures in Evolution**); **(Infographic 15.8).**

Left Out in the Cold?

Like all species, manatees are adapted to survive in only certain environments. Their ability to move into new habitats is limited by their range of physical adaptations, which they

Allopatric Speciation: How One Species Can Become Many

→ The Galápagos archipelago is a series of islands off the coast of South America. Finches first came to the Galápagos from a population on the mainland of South America. As they spread from island to island, they encountered different environments, including available food sources, which influenced bill size and shape in each new island population. As separated finch populations evolved in different food environments, they diverged from their ancestral population to have smaller, pointed bills for insects, longer bills for cactus fruit and flowers, or thick, strong bills for hard seeds. In addition, they evolved such that each separated population could no longer interbreed. At least 13 finch species have diverged from the original South American species.

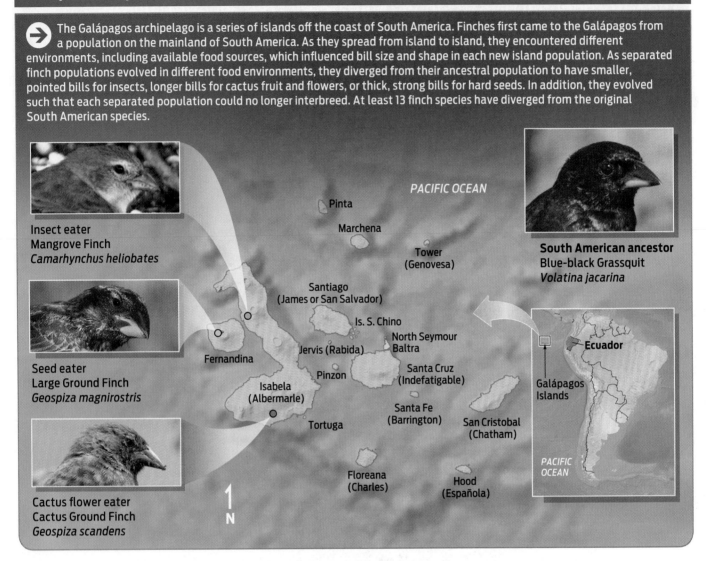

Insect eater
Mangrove Finch
Camarhynchus heliobates

Seed eater
Large Ground Finch
Geospiza magnirostris

Cactus flower eater
Cactus Ground Finch
Geospiza scandens

PACIFIC OCEAN

Pinta

Marchena

Tower
(Genovesa)

Santiago
(James or San Salvador)

Is. S. Chino

North Seymour
Baltra

Jervis (Rabida)

Fernandina

Pinzon

Santa Cruz
(Indefatigable)

Isabela
(Albermarle)

Santa Fe
(Barrington)

San Cristobal
(Chatham)

Tortuga

Floreana
(Charles)

Hood
(Española)

South American ancestor
Blue-black Grassquit
Volatina jacarina

Ecuador

Galápagos
Islands

PACIFIC
OCEAN

N

evolved over thousands of years. For example, as creatures of the tropics manatees cannot survive long exposures to temperatures below 68°F (20°C). Florida generally marks the northern tip of their range (though during summer months manatees can sometimes be found swimming as far north as Rhode Island).

Historically, during the winter months manatees would congregate in the warm waters of Florida's natural springs and shallow, sun-warmed Everglades. Development has since blocked access to those warm areas or destroyed them altogether, so manatees have been forced to find alternative hot spots—for example, in the warm water emitted by coal-fired power plants.

Basking in these human-made hot tubs helps thaw the chilled creatures. But this is at best a short-term solution: as coal-fired power plants are closed or decommissioned and replaced with less-polluting forms of energy, Florida manatees will have few places to turn for warmth. Many, undoubtedly, will be left out in the cold and die. In 2010 alone, 245 manatees died from cold stress.

Whether or not manatees are able to adapt and survive depends not only on our efforts to safeguard their habitat but also on their intrinsic genetic diversity. Will they make it? No one knows for sure. "But give 'em credit," says Bonde, "they're in it for the fight." ∎

▶ Summary

■ From a genetic perspective, a population is identified by the particular collection of alleles in its gene pool.

■ Genetic diversity, as reflected in the number of different alleles in a population's gene pool, is important for the continued survival of populations, especially in the face of changing environments.

■ Evolution is a change in allele frequencies in a population over time. Evolution can be adaptive or nonadaptive. Mutation, genetic drift, and gene flow are nonadaptive forms of evolution.

■ The founder effect is a type of genetic drift in which a small number of individuals establishes a new population in a new location, with reduced genetic diversity as a possible result.

■ The bottleneck effect is a type of genetic drift that occurs when the size of a population is reduced, often by a natural disaster, and the genetic diversity of the remaining population is reduced.

■ Inbreeding of closely related individuals may occur in small, isolated populations, posing a threat to the health of a species.

■ Gene flow is the movement of alleles between different populations of the same species, often resulting in increased genetic diversity of a population.

■ Genetic diversity can be measured by using DNA sequences to assess allele frequency.

■ Hardy-Weinberg equilibrium describes the frequency of genotypes in a nonevolving population. The Hardy-Weinberg equation can be used to detect evolutionary change in a population.

■ According to the biological species concept, species are reproductively isolated populations of individuals that can interbreed to produce fertile offspring.

■ Speciation can occur when gene pools are separated, gene flow is restricted, and populations diverge genetically over time.

POPULATIONS AND GENE POOLS

The genetic diversity of a population is reflected in its collective bank of alleles, or gene pool. The amount of genetic diversity in a population has implications for its evolution.

HINT See Infographic 15.2 and Up Close: Calculating Hardy-Weinberg Equilibrium.

➲ KNOW IT

1. Genetic diversity is measured in terms of allele frequencies (the relative proportion of different alleles in a gene pool). A population of 3,200 manatees has 4,200 dominant *G* alleles and 2,200 recessive *g* alleles. What is the frequency of *g* alleles in the gene pool?

2. Of the three populations described below, each of which has 1,000 members, which population has the highest genetic diversity? Note that only one gene is being presented, and that this gene has three possible alleles: *A1*, *A2*, and *a*.

   Population A: 70% have an *A1/A1* genotype, 25% have an *A1/A2* genotype, and 5% have an *A1/a* genotype.
   Population B: 50% have an *A1/A1* genotype, 20% have an *A2/A2* genotype, 10% have an *A1/A2* genotype, 10% have an *A2/a* genotype, and 10% have an *a/a* genotype.
   Population C: 80% have an *A1/A1* genotype and 20% have an *A1/a* genotype.

3. A starting population of bacteria has two alleles of the *TUB* gene: *T* and *t*. The frequency of *T* is 0.8 and the frequency of *t* is 0.2. The local environment undergoes an elevated temperature for many generations of bacterial reproduction. After 50 generations of reproduction at the elevated temperature, the frequency of *T* is 0.4 and the frequency of *t* is 0.6. Has evolution occurred? Explain your answer.

➲ USE IT

4. Question 2 looked at the allele frequencies of populations A, B, and C. From your answer to that question, which population would you predict to have the greatest chance of surviving an environmental change? Explain your answer.

5. Which of the four populations in the table below would you be concerned about from a conservation perspective? Why would you be concerned?

Popu-lation	Number of individ-uals	Number of alleles, gene 1	Number of alleles, gene 2	Number of alleles, gene 3
1	50	1	7	5
2	1,000	1	5	7
3	50	3	2	2
4	1,000	1	1	2

6. Phenylketonuria (PKU) is a rare, recessive genetic condition that affects approximately 1 in 15,000 babies born in the United States. (You may have noticed on products that contain aspartame the statement "Phenylketonurics: contains phenylalanine," a warning for people with PKU that they should avoid consuming that product.) Calculate the expected frequency of carriers (that is, of heterozygotes) in the U.S. population, based on the information provided about rates of PKU among U.S. births. (Remember the Hardy-Weinberg equation.)

7. Assume a population of 100 individuals. Five are homozygous dominant (*AA*), 80 are heterozygous (*Aa*), and 15 are homozygous recessive (*aa*) for the *A* gene. Determine *p* and *q* for this population. Now use those values for *p* and *q* and plug them into the Hardy-Weinberg equation. Is this population in Hardy-Weinberg equilibrium? Why or why not?

GENETIC DRIFT

Genetic drift can alter the allele frequency of a population. Genetic drift tends to lower the genetic diversity of a population.

HINT See Infographics 15.2 and 15.3.

➲ KNOW IT
8. Which of the following are examples of genetic drift?
   a. founder effect
   b. bottleneck effect
   c. inbreeding
   d. a and b
   e. a, b, and c

9. A bottleneck is best described as
   a. an expansion of a population from a small group of founders.
   b. a small number of individuals leaving a population.
   c. a reduction in the size of an original population followed by an expansion in its size as the surviving members reproduce.
   d. the mixing and mingling of alleles by mating between members of different populations.
   e. an example of natural selection.

10. A population of manatees has 12 different alleles, *A* through *L,* of a particular gene. A drunk motorboat driver recklessly tears through the water where the manatees live, killing 90% of them. The surviving manatees are all homozygous for allele *B.*
    a. What is the impact of this event on the frequency of alleles *A* through *L*?
    b. What type of event is this?

◉ USE IT
11. Why is genetic drift considered to be a form of evolution? How does it differ from evolution by natural selection?

12. In humans, founder effects may occur when a small group of founders immigrates to a new country, for example to establish a religious community. In this situation, why might the allele frequencies in succeeding generations remain similar to those of the founding population rather than gradually becoming more similar to the allele frequencies of the population of the country to which they immigrated?

GENE FLOW AND SPECIATION
Gene flow can alter the genetic diversity of a population as individuals from neighboring populations mix and mate with the original population. Barriers to gene flow contribute to speciation.

HINT See Infographics 15.5, 15.7, and 15.8.

13. The biological species concept defines a species
    a. on the basis of similar physical appearance.
    b. on the basis of close genetic relationships.
    c. on the basis of similar levels of genetic diversity.
    d. on the basis of the ability to mate and produce fertile offspring.
    e. on the basis of recognizing one another's mating behaviors.

14. How does geographic isolation contribute to speciation events?

◉ USE IT
15. Two populations of rodents have been physically separated by a large lake for many generations. The shore on one side of the lake is drier and has very different vegetation from that on the other side. The lake is drained by humans to irrigate crops, and now the rodent populations are reunited. How could you assess if they are still members of the same species?

16. Why is inbreeding detrimental to a population?

17. If geographically dispersed groups all converge at a common location during breeding season, then return to their home sites to bear and rear their young, what might happen to the gene pools of the different groups over time?

18. A small population of 25 individuals has five alleles, *A* through *E,* for a particular gene. The *E* allele is only represented in one homozygous individual:

   5 individuals are *D/A* heterozygotes
   5 individuals are *A/A* homozygotes
   5 individuals are *A/B* heterozygotes
   5 individuals are *C/D* heterozygotes
   4 individuals are *C/C* homozygotes
   1 individual is an *E/E* homozygote.

If five *A/E* heterozygotes migrate into the population, what will be the impact on the allele frequencies of each of the five alleles?

SCIENCE AND ETHICS
19. Consider the situation of Florida manatees.
    a. What is the difference between an endangered and a threatened species, according to the classification established by the U.S. Endangered Species Act? At the present time, what is the status of the Florida manatee?
    b. For a species like the Florida panther, why is a habitat conservation approach not sufficient to ensure a healthy recovery?
    c. What approach could be taken to try to restore genetic diversity to a species such as the cheetah, given that all cheetahs are survivors of a bottleneck and are essentially genetically identical?

A Fish with Fingers?

A Fish with Fingers?

A transitional fossil sheds light on how evolution works

For 5 years, biologists Neil Shubin and Ted Daeschler spent their summers trekking through one of the most desolate regions on earth. They were fossil hunting on the remote island of Ellesmere, in the Canadian Arctic, about 600 miles from the north pole. Even in summer, Ellesmere is a forbidding place: a windswept, frozen desert where sparse vegetation grows no more than a few inches tall, where sleet and snow fall in the middle of July, and where the sun never sets. Only a handful of wild animals survive here, but those that do make for dangerous working conditions. Hungry polar bears and charging herds of muskoxen are hazards of working in the Arctic, says Daeschler. The team carried shotguns for protection.

Braving these conditions, the researchers drilled, chiseled, and hammered their way through countless tons of rock looking for fossils. Not just any rocks and fossils, but ones dating from 375 million years ago, when animals were taking their first tentative steps on land. For three summers, they scoured the site of what was once an active streambed but found little of interest, mostly pieces of ancient fish.

> **_Tiktaalik_ "splits the difference between something we think of as a fish and something we think of as a limbed animal."** —Ted Daeschler

Then, in 2004, the team made a tantalizing discovery: the snout of a curious-looking creature protruding from a slab of pink rock. Further excavation revealed the well-preserved remains of several flat-headed animals between 4 and 9 feet long. In some ways, the creatures resembled giant fish—they had fins and scales. But they also had traits that resembled those of

VERTEBRATE
An animal with a bony or cartilaginous backbone.

A model of *Tiktaalik roseae*, the fossil discovery that represents a critical phase in the evolution of four-legged, land-dwelling animals.

The *Tiktaalik roseae* fossil.

land-dwelling amphibians–notably, a neck, wrists, and fingerlike bones. They named the new species *Tiktaalik roseae*; *tiktaalik* (pronounced tic-TAH-lick) is a native word meaning "large freshwater fish." This hybrid animal no longer exists today, but it represents a critical phase in the evolution of four-legged, land-dwelling **vertebrates**–including humans.

Tiktaalik "splits the difference between something we think of as a fish and something we think of as a limbed animal," says Daeschler, a curator of vertebrate paleontology at the Academy of Natural Sciences in Philadelphia. "In that sense, it is a wonderful transitional fossil between two major groups of vertebrates."

Today, of course, four-legged animals are found far and wide across the land. But 400 million years ago it was a different story. Life was mostly aquatic then, restricted to oceans and freshwater streams. How life made the jump from water to land is a question that has long intrigued evolutionary biologists. In fact, scientists have been searching for evidence of this milestone ever since Charles Darwin first speculated that all life on the planet is related by a tree of common descent. According to Shubin, a professor of biology at the University of Chicago and

INFOGRAPHIC 16.1

Fossils Form Only in Certain Circumstances

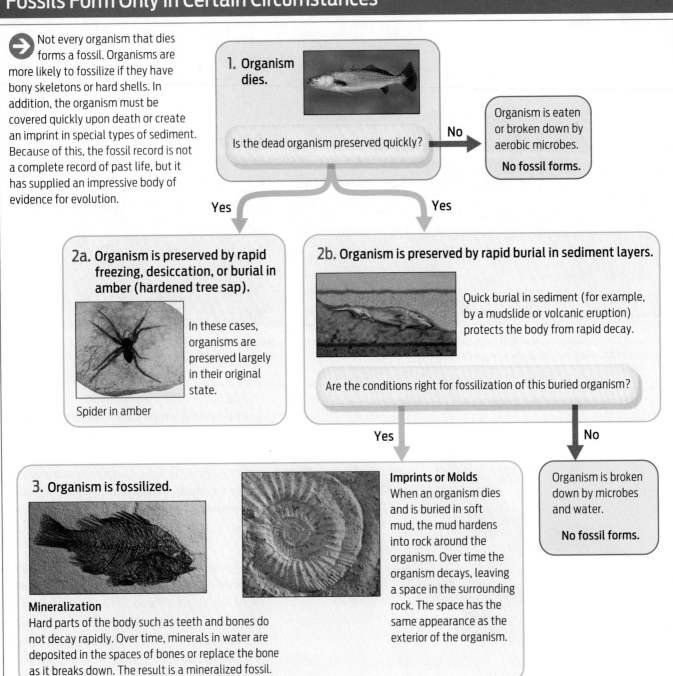

Not every organism that dies forms a fossil. Organisms are more likely to fossilize if they have bony skeletons or hard shells. In addition, the organism must be covered quickly upon death or create an imprint in special types of sediment. Because of this, the fossil record is not a complete record of past life, but it has supplied an impressive body of evidence for evolution.

1. Organism dies.

Is the dead organism preserved quickly?

No → Organism is eaten or broken down by aerobic microbes. **No fossil forms.**

Yes ↓ **Yes** ↓

2a. Organism is preserved by rapid freezing, desiccation, or burial in amber (hardened tree sap).

In these cases, organisms are preserved largely in their original state.

Spider in amber

2b. Organism is preserved by rapid burial in sediment layers.

Quick burial in sediment (for example, by a mudslide or volcanic eruption) protects the body from rapid decay.

Are the conditions right for fossilization of this buried organism?

Yes ↓ **No** ↓

3. Organism is fossilized.

Mineralization
Hard parts of the body such as teeth and bones do not decay rapidly. Over time, minerals in water are deposited in the spaces of bones or replace the bone as it breaks down. The result is a mineralized fossil.

Imprints or Molds
When an organism dies and is buried in soft mud, the mud hardens into rock around the organism. Over time the organism decays, leaving a space in the surrounding rock. The space has the same appearance as the exterior of the organism.

Organism is broken down by microbes and water. **No fossil forms.**

The Shubin and Daeschler expedition looking for fossils on Ellesmere Island.

the Field Museum of Natural History, *Tiktaalik* is the most compelling example yet of an animal that lived at the cusp of this important transition. Not only does it fill a gap in our knowledge, the discovery also provides persuasive evidence in support of Darwin's great idea.

Reading the Fossil Record

The theory of evolution–what Darwin called **descent with modification**–draws two main conclusions about life on earth: that all living things are related, and that the different species we see today have emerged over time as a result of natural selection operating over millions of years. Many lines of evidence support this theory (remember that "theory" in science means it is considered to be an established fact). The most direct evidence of evolution comes from **fossils**, the preserved remains or impressions of once-living organisms. Fossils are like "snapshots" of past life, capturing particular moments in time.

They are formed in a number of ways: an animal or plant may be frozen in ice, trapped in amber, or buried in a thick layer of mud. Rapid decay is thereby prevented and the organism's shape is preserved. Not all organisms are equally likely to form fossils, though: animals with bones or shells are more likely to be pre-

served than animals without such hard parts (think earthworms or jellyfish) that decay quickly. And conditions permitting fossilization are rare: the organism has to be in just the right place at just the right time. Still, the **fossil record** is remarkably rich and offers an exciting window onto the past. **Paleontologists**, scientists who study ancient life, have uncovered hundreds of thousands of fossils throughout the world, from many evolutionary time periods **(Infographic 16.1)**.

When fossils are arranged in order of age, they provide a tangible history of life on earth. Because not all organisms are preserved, the fossil record is not a complete record of past life. Nevertheless, it is extensive enough to show the overall arc of life, and provides compelling evidence in support of Darwin's theory. For example, if all organisms have descended from a single common ancestor billions of years ago as the theory of evolution concludes they did, then we would expect the fossil record to show an ordered succession of evolutionary stages as organisms evolved and diversified. And, indeed,

---

DESCENT WITH MODIFICATION
Darwin's term for evolution, combining the ideas that all living things are related and that organisms have changed over time.

FOSSILS
The preserved remains or impressions of once-living organisms.

FOSSIL RECORD
An assemblage of fossils arranged in order of age, providing evidence of changes in species over time.

PALEONTOLOGIST
A scientist who studies ancient life by means of the fossil record.

that is exactly what you see: prokaryotes appear before eukaryotes, single-cell organisms before multicellular ones, water-dwelling organisms before land-dwelling ones, fish before amphibians, reptiles before birds, and so on.

Moreover, we would expect to see changes over time in a family of organisms, and we do. To use one exceptionally well studied example, consider horses. Comparisons of modern-day horse bones with fossils of horse ancestors reveal how in the course of evolution horses have lost most of their toes. Overall, the fossils show a continual series of change, with the most recent fossils being the most similar to modern organisms, and the more ancient fossils being the most different. But they all clearly share a family resemblance (**Infographic 16.2**).

> **When fossils are arranged in order of age, they provide a tangible history of life on earth.**

Descent with modification also predicts that the fossil record should contain evidence of intermediate organisms–those with a mixture of "old" and "new" traits. Darwin acknowledged in *On the Origin of Species* that the fossil record of his day did not provide many examples of such intermediate organisms–a state of affairs he described as "probably the gravest and most obvious of all the many objections which may be urged against my views." Yet Darwin knew that if his hypothesis was correct, then such intermediate fossils would eventually be found. And indeed they have been. Scientists have discovered animals with mixtures of reptile and bird characteristics, and animals with mixtures of reptile and mammal characteristics, for instance. But the transition between fish and amphibians has remained fuzzy.

The Fossil Hunt

Shubin and Daeschler began their hunt for fossils in the Canadian Arctic in 1999, after stumbling upon a map in an old geology textbook. The map showed that the region contained large swaths of exposed rock from just the period the researchers were interested in, roughly 380–375 million years ago. Why was this period so important to the scientists? They knew that no land-dwelling vertebrates are present in the fossil record before 385 million years ago. By 365 million years ago, organisms easily recognizable as amphibians are well documented in the fossil record. The paleontologists hypothesized that if they looked at rocks sandwiched in between these two time periods–around 375 million years old–they might find one of Darwin's elusive "intermediates." Moreover, Ellesmere is one of only three places on earth where rocks of this time period are exposed (for example, not covered with a shopping mall), yet to Shubin and Daeschler's knowledge, no other paleontologists had explored the area, which meant it was a potential fossil gold mine.

Knowing exactly where to look for fossils was a tricky proposition, since Ellesmere Island covers 75,000 square miles. To locate the most promising dig site, the scientists first studied aerial photographs. Once on the ground, the team, with six to nine members at various times, split up and spent the first two seasons just walking the rocky exposures, prospecting for bits and pieces of fossils that had eroded out from the rock. When they found something interesting on the surface, they would start to dig.

It was while walking these rocky exposures in 2002 that Daeschler and his team found the first piece of what would turn out to be a *Tiktaalik* fossil–"basically part of the snout," he says. At first, they didn't think much of the find, but collected it anyway along with other fossil pieces. Back in Philadelphia, researchers cleaned the fossil, removing the remaining rock. Even then, says Daeschler, it wasn't clear what the snout belonged to. It wasn't until a visiting graduate student remarked on the resemblance of the skull to one from the earliest known amphibians that the researchers realized what they had found. If ever there was a "lightbulb" moment, he notes, this was it. But, alas, they had only one small piece of the creature.

The team returned to Ellesmere in 2004 for another round of hunting and digging. It didn't

INFOGRAPHIC 16.2

Fossils Reveal Changes in Species over Time

The fossil record of horses supports the theory of descent with modification. Forelimb fossils are similar to one another, but show changes over time from the earliest horse ancestors to modern-day horses, as species diverged from a common ancestor. In the fossil record we can observe over time a reduction in toe number, as the central toe became dominant, allowing horses to move more rapidly in new prairie-like environments.

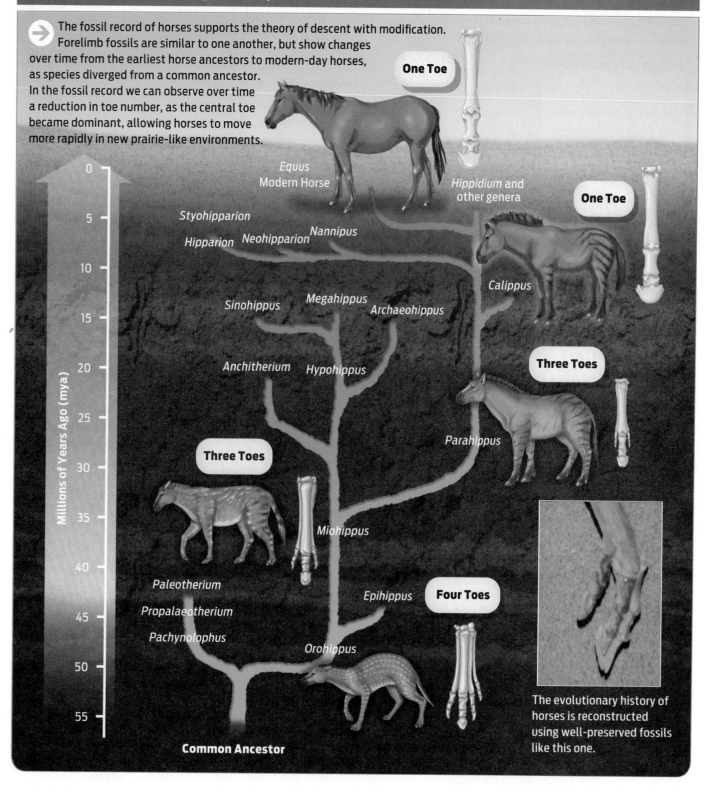

One Toe

Equus
Modern Horse

Hippidium and other genera

One Toe

Styohipparion

Hipparion   Neohipparion   Nannipus

Calippus

Sinohippus   Megahippus   Archaeohippus

Three Toes

Anchitherium   Hypohippus

Parahippus

Three Toes

Miohippus

Paleotherium

Propalaeotherium

Pachynolophus

Epihippus   **Four Toes**

Orohippus

Millions of Years Ago (mya)

0
5
10
15
20
25
30
35
40
45
50
55

Common Ancestor

The evolutionary history of horses is reconstructed using well-preserved fossils like this one.

How Fossils Are Dated

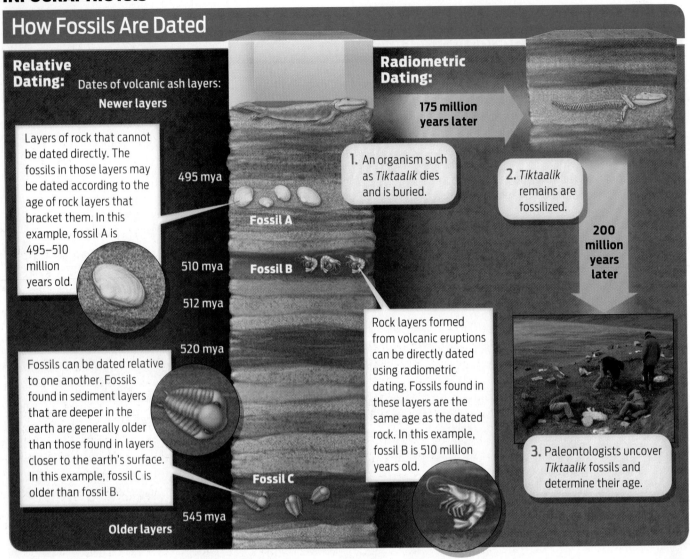

Relative Dating: Dates of volcanic ash layers:

Newer layers

Layers of rock that cannot be dated directly. The fossils in those layers may be dated according to the age of rock layers that bracket them. In this example, fossil A is 495–510 million years old.

495 mya

Fossil A

510 mya **Fossil B**

512 mya

520 mya

Fossils can be dated relative to one another. Fossils found in sediment layers that are deeper in the earth are generally older than those found in layers closer to the earth's surface. In this example, fossil C is older than fossil B.

Fossil C

545 mya

Older layers

Radiometric Dating:

175 million years later

1. An organism such as *Tiktaalik* dies and is buried.

2. *Tiktaalik* remains are fossilized.

200 million years later

Rock layers formed from volcanic eruptions can be directly dated using radiometric dating. Fossils found in these layers are the same age as the dated rock. In this example, fossil B is 510 million years old.

3. Paleontologists uncover *Tiktaalik* fossils and determine their age.

take long for their patience to be rewarded: "Literally inches," Daeschler says, from where they'd been excavating before, they hit pay dirt.

The researchers determined that the fossils they found were 375 million years old—just the right age to show transitional features—but how did they know? Fossils are at least as old as the rocks that encase them. Since some types of rocks can be dated directly by a method known as radiometric dating (described in Chapter 17), it is possible to determine the age of fossils embedded within them. If fossils are found sandwiched in rock layers that cannot be directly dated, they can be dated indirectly by their position with respect to rocks or fossils of known age, a technique called **relative dating.** Generally speaking, the deeper the fossils, the older they are. Using a combination of both methods, scientists have determined that the rocks where *Tiktaalik* was found are 375 million years old, which means *Tiktaalik* is that old as well **(Infographic 16.3).**

Setting the Stage for Life on Land

The geologic time period that Shubin and Daeschler are interested in is known as the Devonian—roughly 400–350 million years ago. Great transformations were occurring during the Devonian: jawed fishes, sharks, land plants, and insects all diversified in this period. Because sea levels were high worldwide, and much of the land lay submerged under water, the Devonian period has been called the age of fishes.

RELATIVE DATING
Determining the age of a fossil from its position relative to layers of rock or fossils of known age.

Back then, what is now the Canadian Arctic had a warm, wet climate and a landscape veined by shallow, meandering streams. Early in the Devonian period there was little plant growth, and the world would have looked fairly brown and empty. By the middle of the Devonian, if you were standing on the bank of a stream you would have seen some of the first land plants, the first forests, as well as the first **invertebrates**–spiderlike creatures and millipedes, for example–crawling on land. Still, there would have been no land-dwelling vertebrates at this time: nothing with bony limbs, nothing with a backbone or skull.

By the late Devonian, things were changing quickly. By then, says Daeschler, "you had a green flood plain, a green world." It was this green world–a rich and productive ecosystem, with energy-rich leaf litter flowing into shallow streams–that set the stage for the move of vertebrates onto land.

The physical challenges of living on land are very different from those in water. Water is dense and difficult to move through, but fish glide smoothly through water thanks to a sleek shape, a muscular body, and flexible fins. By contrast, animals that walk on land have to cope with gravity. Air doesn't support animals as they move, so the bodies of land animals need a sturdier structure. Animals on land can also dry out, which is dangerous for them because cells need water to function. And, of course, taking in oxygen is different in land and in water.

Of the many features that distinguish land animals from fish, biologists have singled out one as a key evolutionary milestone: no living fish has true limbs, that is, bony appendages with wrists, ankles, and digits. Instead, they have webbed fins. In most fishes, the fin bones are thin and splayed like the rays of a fan. These so-called ray-finned fishes include the familiar modern-day perch, trout, and bass. By contrast, amphibians, birds, most reptiles, and mammals all have two pairs of limbs, defining them as **tetrapods** (from the Greek for "four-footed").

While having limbs is a key feature distinguishing tetrapods from fish, one small group of fish–the lobe-finned fish–seems to blur this distinction. First appearing in the fossil record about 400 million years ago, lobe-finned fish have sturdy fins with bony supports that resemble primitive limb bones.

Lobe-finned fish are thought to have evolved in shallow streams, where rich plant material lured small fish and other creatures close to the water's edge. The lobe-finned fish likely used their strong fins to touch the bottom of the streambed while maneuvering to catch prey. As Daeschler explains, it was the unique ecological opportunity afforded by shallow streams that enabled the lobe-finned fish to start developing features that were adaptive in shallow water. But lobe-finned fish were still very far from being true tetrapods. *Tiktaalik,* on the other hand, is inching closer: "It looks like a fish in that it has scales and fins," says Shubin, "but when you look inside the skeleton you see how special it really is."

The Fish That Did Pushups

Shubin and Daeschler were lucky: the fossils they found were so well preserved that they were able to study *Tiktaalik*'s skeletal anatomy in detail, even seeing how the bones interacted and where muscles attached. From these fossil bones, they determined that *Tiktaalik* was a predatory fish with sharp teeth, scales, and fins. In addition to these fishy attributes, it had a flat skull reminiscent of a crocodile head and a flexible neck (in other words, the skull was not rigidly fixed to the shoulders as it is in most fishes). To Shubin and Daeschler, the neck was one of the most surprising finds. Having a flexible neck meant that *Tiktaalik* could swivel its head independently of its body, perhaps enabling it to catch a glimpse of predators sneaking up on it from behind or to hunt its own prey. It also had the full-fledged ribs of a modern land animal, sturdy enough to support the animal's trunk out of water even against the force of gravity.

But it is *Tiktaalik*'s fins that have made it famous. While possessing many features of a lobe-finned fish, *Tiktaalik* appears also to have had a jointed wrist and fingerlike bones. From the fossil pieces, Shubin and Daeschler were

INVERTEBRATE
An animal without a backbone.

TETRAPOD
An organism with four true limbs, that is, bony appendages with jointed wrists, ankles, and digits—i.e., mammals, amphibians, birds, and reptiles.

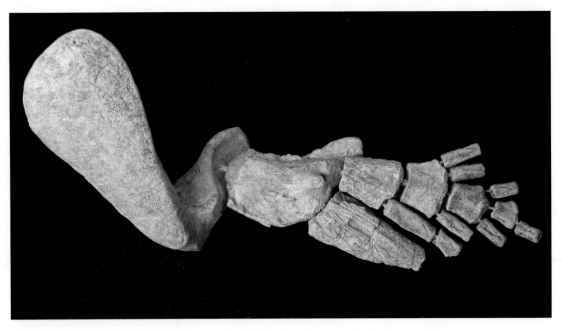

Shubin and Daeschler created a digital model of Tiktaalik's bones and how they would have moved relative to one another.

able to create a model of how the bones would have moved relative to one another, and they are now modeling these movements digitally. The models show that the bones and joints were strong enough to support the body and worked like those of the earliest known tetrapods–the early amphibians. "This animal was able to hold its fin below its body, bend the fin out toward what we think of as a wrist, and bend the elbow," explains Daeschler. In other words, it was a fish that could do a push-up.

With this hybrid anatomy, *Tiktaalik* was not galloping on land, of course. It probably lived most of the time in water, but Shubin and Daeschler suspect that *Tiktaalik* may have used its supportive fins to pull itself out of the water for brief periods. "This is a fish that can live in the shallows or even make short excursions onto land," says Shubin. The ability to crawl onto land would certainly have been a useful trait in the Devonian, when open water was a brutal fish-eat-fish world, whereas land was a predator-free paradise, full of nourishing bugs.

There was, of course, no forethought involved in this process. Fish did not develop limbs for the *purpose* of walking on land. Rather, limbs first evolved in shallow water, where they proved adaptive and were thus retained in the descendants of the organisms who first developed them. Then, when there was an opportu-

"This is a fish that can live in the shallows or even make short excursions onto land." –Neil Shubin

nity to take advantage of a tantalizing new habitat–the land–the amphibious creatures already had the skeletal "toolkit."

For all its amphibian-like adaptations, *Tiktaalik* is still a fish because its limbs lack the true jointed fingers and toes that define tetrapods. But it's by far the most tetrapod-like of all the fishes so far discovered. Scientists have jokingly referred to it as a "fishapod" (Infographic 16.4).

And that's what makes *Tiktaalik* such an important find: it embodies a previously unknown midpoint between fish and tetrapods. Such intermediate, or transitional, fossils document important steps in the evolution of life on earth. They help biologists understand how groups of organisms evolved, through natural selection, from one form into

HOMOLOGY
Anatomical, genetic, or developmental similarity among organisms due to common ancestry.

another. And they confirm that Darwin's theory of descent with modification–which predicts such intermediate forms–is correct.

A Fin Is a Paw Is an Arm Is a Wing

In *On the Origin of Species*, Darwin asked, "What can be more curious than that the hand of a man, formed for grasping, that of a mole for digging, the leg of the horse, the paddle of the porpoise, and the wing of the bat, should all be constructed on the same pattern, and should include similar bones, in the same relative positions?" To Darwin, this uncanny similarity was evidence that all these organisms were related–that they share a common ancestor in the ancient past.

The fact that all tetrapods share the same forelimb bones, arranged in the same order, is an example of **homology**–a similarity due to common ancestry. Before Darwin, comparative anatomists had identified many such similarities in anatomy; what they lacked was a satisfactory explanation for why such similarity should exist. Darwin provided that explanation:

INFOGRAPHIC 16.4

Tiktaalik, an Intermediate Fossilized Organism

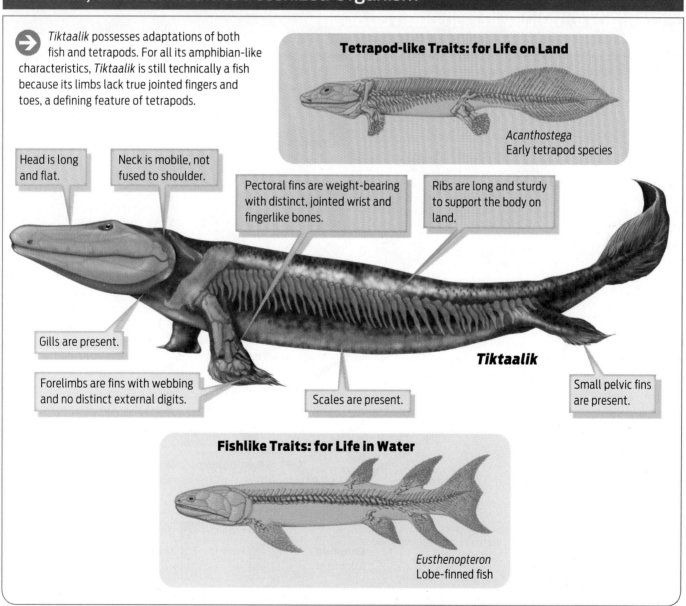

Tiktaalik possesses adaptations of both fish and tetrapods. For all its amphibian-like characteristics, *Tiktaalik* is still technically a fish because its limbs lack true jointed fingers and toes, a defining feature of tetrapods.

Tetrapod-like Traits: for Life on Land

Acanthostega
Early tetrapod species

Head is long and flat.

Neck is mobile, not fused to shoulder.

Pectoral fins are weight-bearing with distinct, jointed wrist and fingerlike bones.

Ribs are long and sturdy to support the body on land.

Gills are present.

Forelimbs are fins with webbing and no distinct external digits.

Scales are present.

Tiktaalik

Small pelvic fins are present.

Fishlike Traits: for Life in Water

Eusthenopteron
Lobe-finned fish

homologous structures are ones that are similar because they are inherited from the same ancestor–in this case, an amphibious creature like *Tiktaalik*. Why is this significant? Think of it this way: every time you bend your wrist back and forth–to swipe a paint brush or hold a cell phone to your ear, for example–you are using structures that first evolved 375 million years ago in fish. In many ways, you have a fish to thank for some of your most useful adaptations. As Shubin points out, "This is not just some archaic, weird branch of evolution; this is *our*

branch of evolution." (For more on this, check out Shubin's book *Your Inner Fish*.) **(Infographic 16.5)**.

If they have the same bones, why then do a human arm and a bird wing look so different? Remember that during the process of inheritance mutations are continually introduced into the DNA of genes. Such mutations can produce subtle changes in the proteins encoded by those genes–proteins involved in constructing the bones that make up an arm or a wing, for example. Changes in bone proteins can result in

INFOGRAPHIC 16.5

Forelimb Homology in Fish and Tetrapods

The number, order, and underlying structure of the forelimb bones are similar in all the groups illustrated below. The differences in the relative width, length, and strength of each bone contribute to the specialized function of each forelimb. This anatomical homology is strong evidence that these organisms all shared a common ancestor at some time in the distant past. The variations in bone shape and function reflect evolutionary adaptations to different environments.

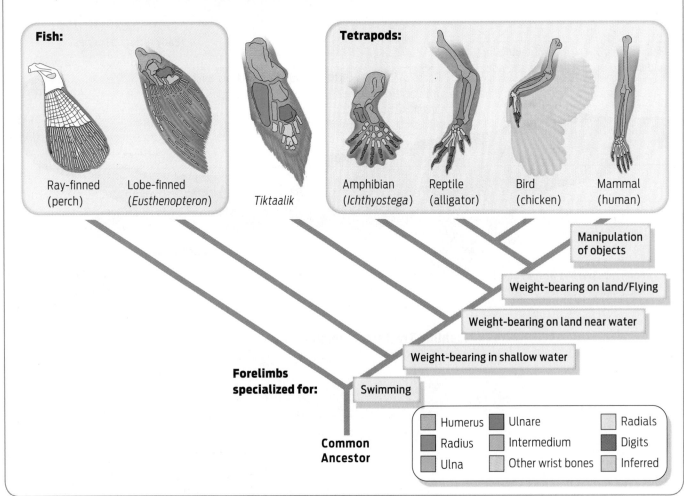

Vertebrate Animals Share a Similar Pattern of Early Development

→ We can identify homologous structures by tracing their embryological development. Some of our middle ear bones, for example, are homologous with the jaw bones of reptiles and bones supporting gills in fish. We know this because all of these structures develop from the pharyngeal pouches that appear in all vertebrate embryos early in development. This developmental homology is strong evidence that all vertebrate animals are related by common ancestry. Genetic changes over time have introduced modifications in later stages that give rise to distinct species with vast physical differences.

Early Embryos:

| Human | Cat | Fish | Snake | Chick |

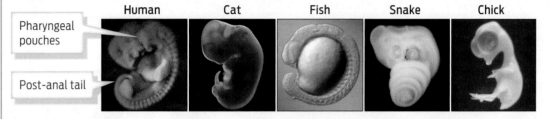

Pharyngeal pouches

Post-anal tail

Early-stage embryos of related organisms share common structures.

Late Development:

Later in development, these structures take on species-specific shape and function.

slightly altered bones, making them longer or thinner, for instance. When these modified bones are helpful to an organism's survival and reproduction, the advantageous traits are passed on to the next generation, and populations emerge that have these adaptations. This "descent with modification" (Darwin's phrase

> **"This is not just some archaic, weird branch of evolution; this is *our* branch of evolution."**
> —Neil Shubin

VESTIGIAL STRUCTURES
A structure inherited from an ancestor that no longer serves a clear function in the organism that possesses it.

again) results in diverse organisms sharing common—homologous—structures and putting them to different uses.

We can see homology not only in adult anatomy, but in early development as well. Take a look at early embryos of vertebrate animals as diverse as humans, fish, and chickens and you'll see that they all look remarkably similar **(Infographic 16.6)**. Why should the embryonic stage of a human resemble the embryonic stage of a fish when the adults of each species look so dif-

ferent? Similar embryological structures are further evidence that all vertebrates shared a common ancestor.

Development helps us solve other evolutionary conundrums as well, such as why reptiles like snakes don't have legs like other tetrapods. In fact, snake embryos *do* possess the beginnings of limbs, but these limb "buds" remain rudimentary and do not develop into full-fledged limbs (although you can still see stubby hindlimbs in some species of snake today). Such **vestigial structures**, which serve no apparent function in an organism, are strong evidence for evolution: these "useless" features are inherited from an ancestor in whom they *did* serve a function.

Zooming in even further, to the molecular level, we find still more examples of homology—yet more evidence of common ancestry. Scientists have known since the 1960s that DNA is the molecule of heredity, and that it is shared by all living organisms on earth. Every molecule of DNA—whether from fish, maple tree, bacterium, or human—is made of the same four nucleotides

DNA Sequences Are Shared among Related Organisms

→ Related organisms share DNA sequences inherited from a common ancestor. Over time, the sequence in each species acquires independent mutations. The more time that has passed, the greater the number of sequence differences that will be present. Thus, the percentage of nucleotides that differ between two species gives an indication of the evolutionary distance between them.

Sequence homology between species

Species A	GGTATCGAGGTTCTACATTGCAACTTCTAC
Close relative	GGAAACGAGGTTCTACATTGCCACTTCTAC
Distant relative	GGAAACGAGGTTCGACATAGCCACTTCTAC

3 differences in 30 nucleotides
3/30 = 10%; or 90% similarity

5 differences in 30 nucleotides
5/30 = 17%; or 83% similarity

Similarity to human DNA sequences*

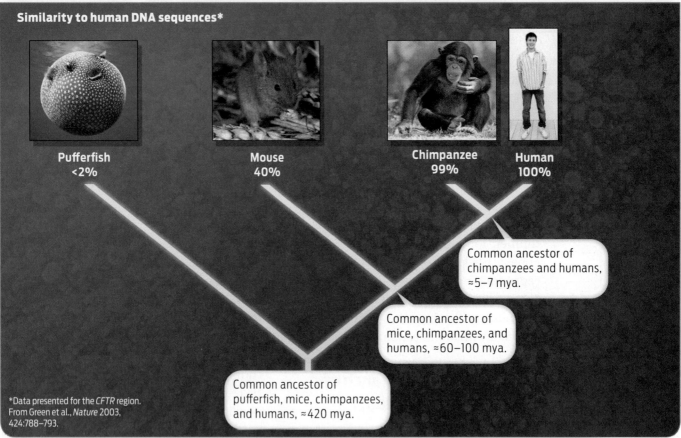

Pufferfish
<2%

Mouse
40%

Chimpanzee
99%

Human
100%

Common ancestor of chimpanzees and humans, ≈5–7 mya.

Common ancestor of mice, chimpanzees, and humans, ≈60–100 mya.

Common ancestor of pufferfish, mice, chimpanzees, and humans, ≈420 mya.

*Data presented for the *CFTR* region. From Green et al., *Nature* 2003, 424:788–793.

(A, C, T, and G), and all organisms use the information encoded by those nucleotides to make proteins in the same basic way, using the universal genetic code (discussed in Chapter 8). Why should all living things use the same system of decoding genetic information? The best explanation is that this system was the one used by the ancient ancestor of all living organisms, passed on to all of its descendants, and preserved throughout billions of years of evolution.

DNA and Descent

While all living organisms share DNA and the genetic code, no two species will share the exact same sequence of DNA nucleotides. That's because (as described in Chapter 10) errors in DNA replication and other mutations are continually introducing variation into DNA sequences (and the proteins they encode). Over time, neutral and advantageous mutations will tend to be preserved, while harmful mutations will tend to be selected against and eliminated. In addition, much of our DNA is noncoding. Because mutations in noncoding DNA have no effect on an organism, they accumulate over time. As mutations are passed on to descendants, the number of sequence differences between the ancestor and its descendants grows–slowly in the case of sequences coding for critical proteins whose structures are well adapted to their functions, and more rapidly in the case of noncoding DNA. Closely related species will therefore have fewer DNA sequence differences than species that are more distantly related.

For example, when scientists looked at one specific region of DNA–the cystic fibrosis transmembrane region, which contains both coding and noncoding regions–they discovered that human DNA in this region is 99% identical to chimpanzee DNA. The fact that the DNA of the two species is nearly identical reflects the fact that humans and chimps share a common ancestor that lived relatively recently–just 5–7 million years ago. By contrast, human DNA is only 40% identical to the DNA of a mouse at this same region, which makes sense given that humans and mice share a common ancestor that lived between 60 and 100 million years ago. Even less sequence identity would be seen between a human and a toad, whose common ancestor–a lobed-finned fish–lived roughly 375 million years ago. The more distantly related two species are, the more sequence differences in DNA sequences you will see. In essence, DNA serves as a kind of molecular clock: each additional sequence difference is like a tick of the clock, showing the amount of time that has elapsed since the two species' common ancestor **(Infographic 16.7)**.

When combined with evidence from the fossil record, anatomy, and development, molecular data become a powerful tool for understanding evolution. As we'll see in Chapter 17, DNA evidence is often a more reliable clue to common ancestry than physical appearance, and can serve as a check on conclusions derived from the fossil record or anatomy. As well, DNA is deepening our knowledge of how limbs evolved. Scientists have discovered that even species that are only very distantly related share some of the same genes. Animals as seemingly different as humans and fruit flies, for example, use some of the very same genes to get their heads on straight and their limbs in the right place. Learning how these genes work and how changes in their DNA sequences can produce large-scale changes in body plan or limb structure is a hot area of biology right now, familiarly known as "evo-devo."

Filling in the Gaps

Asked what he thinks is most interesting about the discovery of *Tiktaalik,* Ted Daeschler homes in on what he sees as a popular misconception about the fossil record–that it's "spotty" and "chaotic." But that's simply not true, he says. Despite the fact that it does not record *all* past life, the fossil record is still "very good"–so good, in fact, that you can use it to make and test predictions. You can, for example, look at the fossil record of fish and tetrapods and–suspecting on the basis of anatomy that the two groups are related–hypothesize that an intermediate-looking animal must have existed at some point. Then you can go look for it. Daeschler refers to this process as "filling in the gaps," and it's exactly what he and Neil Shubin did with *Tiktaalik.* They knew, based on the existing fossil record, *when* such a creature was likely to have existed, so then it was just a question of *where* to look for it.

For Shubin and Daeschler, *Tiktaalik* is exciting mostly because it shows that our understanding of evolution is correct: "It confirms that we have a very good understanding of the framework of the history of life," says Daeschler. "We predicted something like *Tiktaalik,* and sure enough, with a little time and effort, we found it." ■

▶ Summary

■ The theory of evolution—what Darwin called "descent with modification"—draws two main conclusions about life: that all living things are related, sharing a common ancestor in the distant past; and that the species we see today are the result of natural selection operating over millions of years.

■ The theory of evolution is supported by a wealth of evidence, including fossil, anatomical, and DNA evidence.

■ Fossils are preserved remains or impressions of once-living organisms that provide a record of past life on earth. Not all organisms are equally likely to form fossils.

■ Fossils can be dated directly or indirectly based on the age of the rocks they are found in, or on their position relative to rocks or fossils of known ages.

■ When fossils are dated and placed in sequence, they show how life on earth has changed over time.

■ As predicted by descent with modification, the fossil record shows the same overall pattern for all lines of descent: younger fossils are more similar to modern organisms than are older fossils.

■ Descent with modification also predicts the existence of "intermediate" organisms, such as *Tiktaalik*, that possess mixtures of "old" and "new" traits.

■ An organism's anatomy reflects adaptation to its ecological environment. Changed ecological circumstances provide opportunities for new adaptations to evolve by natural selection.

■ Homology—the anatomical, developmental, or genetic similarities shared among groups of organisms—is strong evidence that those groups descend from a single common ancestor that lived many millions of years ago.

■ Homology can be seen in the common bone structure of the forelimbs of tetrapods, the similar embryonic development of all vertebrate animals, and the universal genetic code.

■ Many genes, including those controlling body plans, are shared among distantly related species, an example of molecular homology owing to common ancestry.

■ More-closely related species show greater DNA sequence homology than do more-distantly related species.

THE FOSSIL RECORD

While incomplete, the record of past life preserved in the fossil record gives us valuable insight into evolutionary changes in organisms over time and the history of life.

HINT See Infographics 16.1–16.3.

➲ KNOW IT

1. Generally speaking, if you are looking at layers of rock, at what level would you expect to find the newest—that is, the youngest—fossils?

2. Which of the following is most likely to leave a fossil?
- **a.** a jellyfish
- **b.** a worm
- **c.** a wolf
- **d.** a sea sponge (an organism that lacks a skeleton)
- **e.** All of the above are equally likely to leave a fossil.

3. What can the fossil shown below tell us about the structure and lifestyle of the organism that left it? Describe your observations.

➲ USE IT

4. You have molecular evidence that leads you to hypothesize that a particular group of soft-bodied sea cucumbers evolved at a certain time. You have found a fossil bed with many hard-shelled mollusks dating to the critical time, but no fossil evidence to support your hypothesis about the sea cucumbers. Does this find cause you to reject your hypothesis? Why or why not?

5. A specific type of oyster is found in North American fossil beds dated to 100 million years ago. If similar oyster fossils are found in European rock, in layers along with a novel type of barnacle fossil, what can be concluded about the age of the barnacles? Explain your answer.

TIKTAALIK AND ITS SIGNIFICANCE

Tiktaalik provides a glimpse into the adaptations of vertebrate animals as they moved from the water onto land.

HINT See Infographics 16.4 and 16.5.

➲ KNOW IT

6. Which of the following features of *Tiktaalik* is not shared with other bony fishes?
- **a.** scales
- **b.** teeth
- **c.** a mobile neck
- **d.** fins
- **e.** none of the above

7. *Tiktaalik* fossils have both fishlike and tetrapod-like characteristics. Which characteristics are related to supporting the body out of the water?

➲ USE IT

8. *Tiktaalik* fossils are described as "intermediate" or "transitional" fossils. What does this mean? Why are transitional organisms so significant in the history of life?

9. *Tiktaalik* has been called a "fishapod"—part fish, part tetrapod. Speculate on the fossil appearance of its first true tetrapod descendant—what features would distinguish it from *Tiktaalik*? How old would you expect those fossils to be, relative to *Titkaalik*?

10. If some fish acquired modifications that allowed them to be successful on land, why didn't fish just disappear? In other words, why are there still plenty of fish in the sea if the land presented so many favorable opportunities?

COMPARATIVE ANATOMY

Strong evidence for evolution can be seen in the anatomical and developmental structures shared among diverse species of vertebrate animals.

HINT See Infographics 16.5 and 16.6.

11. Compare and contrast the structure and function of a chicken wing with the structure and function of a human arm.

12. Vertebrate embryos have structures called pharyngeal pouches. What do these structures develop into in an adult human? In an adult bony fish?

13. What is the evolutionary explanation for the fact that both human hands and otter paws have five digits?

14. Could you use the presence of a tail to distinguish a human embryo from a chicken embryo? Why or why not?

MOLECULAR EVIDENCE

All living organisms use DNA as their hereditary molecule and make proteins using the same genetic code, a reflection of the fact that all life on earth shares a common ancestor that lived in the distant past.

HINT See Infographic 16.7.

15. You have three sequences of a given gene from three different organisms. How could you determine how closely the three organisms are related to one another?

16. If, in humans, the DNA sequence TTTCTAGGAATA encodes the amino acid sequence phenylalanine–leucine–glycine–isoleucine, what amino acid sequence will that same DNA sequence specify in bacteria?

17. Gene *X* is present in yeast and in sea urchins. Both produce protein X, but the yeast protein is slightly different from the sea urchin protein. What explains this difference? How might you use this information to judge whether humans are closer evolutionarily to yeast or to sea urchins?

SCIENCE AND ETHICS

18. Fossils allow us to understand the evolution of many lineages of plants and animals. They therefore represent a valuable scientific resource. What if *Tiktaalik* (or an equally important transitional fossil) had been found by amateur fossil hunters and sold to a private collector? Do you think there should be any regulation of fossil hunting to prevent the loss of valuable scientific information from the public domain?

Q & A: Evolution

339

Q & A: Evolution

The history, classification, and phylogeny of life on earth

The modern theory of evolution draws two main conclusions about life on earth: that all living things are related, and that the different species we see today have emerged over millions of years as a result of genetic change. We can use evidence from geology, chemistry, paleontology, biogeography, comparative anatomy, and genetics to reconstruct the details of that evolutionary history.

Q How old is the earth, and how do we know?

A When the *Apollo 11* astronauts Neil Armstrong and Buzz Aldrin returned to earth from their historic 1969 moon walk, they carried with them a cargo of lunar rock chipped from the moon's surface. Embedded within these hunks of shimmering anorthosite lay clues to the earliest history of our solar system, including the planet we call home.

According to the nebular hypothesis, the planetary objects in our solar system are the result of a single event: the collapse of a swirling solar nebula, which formed both the sun and the planets out of cosmic dust. Since all the planets were formed at roughly the same time, we can date the age of the solar system by dating any planetary object within it. Of the many moon rocks obtained over the course of the six *Apollo* missions, the oldest have been calculated to be some 4.4 to 4.5 billion years old, which means that the earth is at least that old as well.

Why go to the moon to date the earth? With the exception of a few meteorite battle scars, the moon's surface has remained largely intact over the course of its existence. In contrast, the earth is a swirling ball of molten lava that continuously churns and digests its rocky outer crust. Because of this perpetual churning, it is difficult to find original, undisturbed rocks from earth's earliest period. The oldest known intact piece of earth's land surface, the Acasta Gneiss in a remote region of northern Canada, dates from 3.9 billion years ago. Some of these ancient rocks contain minerals as old as 4.1 to 4.2 billion years.

While these values do not establish an absolute age of the earth, they do provide a lower limit: the earth is at least as old as the materials that make it up. From these earth minerals and moon rocks, as well as material from meteorites that have fallen to earth, scientists estimate

The Genesis Rock, a sample of lunar crust from about the time the moon was formed, was retrieved by *Apollo 15* astronauts James Irwin and David Scott.

that the age of the earth—and of the solar system more generally—is 4.54 billion years, give or take a few million years.

How are such rocks, extraterrestrial or earthly, dated? The most important method is **radiometric dating,** in which the amount of radioactivity present in a rock is used as a kind of geologic clock. When rocks form, the minerals in them contain a certain amount of **radioactive isotopes**—atoms of elements such as uranium-238, potassium-40, and rubidium-87—that are unstable and decay into other atoms.

Radioactive isotopes decay by releasing high-energy particles from the nucleus, a change that causes one element literally to transform into another. For example, an atom of the radioactive isotope uranium-238 decays in a stepwise fashion into a stable atom of lead-206. The time it takes for half the isotope in a sample to break down is called its **half-life**.

Different radioactive elements decay at different rates. Uranium-238 has a half-life of 4.5 billion years, whereas potassium-40 has a half-life of 1.3 billion years. The half-life of carbon-14 (which is used to date once-living, organic remains rather than rocks) is relatively short: it decays to nitrogen-14 in just 5,730 years.

> Scientists estimate that the age of the earth—and of the solar system more generally—is 4.54 billion years.

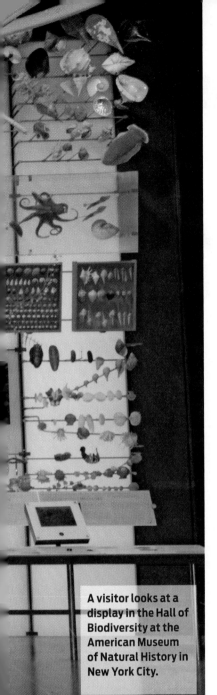

A visitor looks at a display in the Hall of Biodiversity at the American Museum of Natural History in New York City.

RADIOMETRIC DATING
The use of radioactive isotopes as a measure for determining the age of a rock or fossil.

RADIOACTIVE ISOTOPE
An unstable form of an element that decays into another element by radiation, that is, by emitting energetic particles.

HALF-LIFE
The time it takes for one-half of a substance to decay.

Unstable Elements Undergo Radioactive Decay

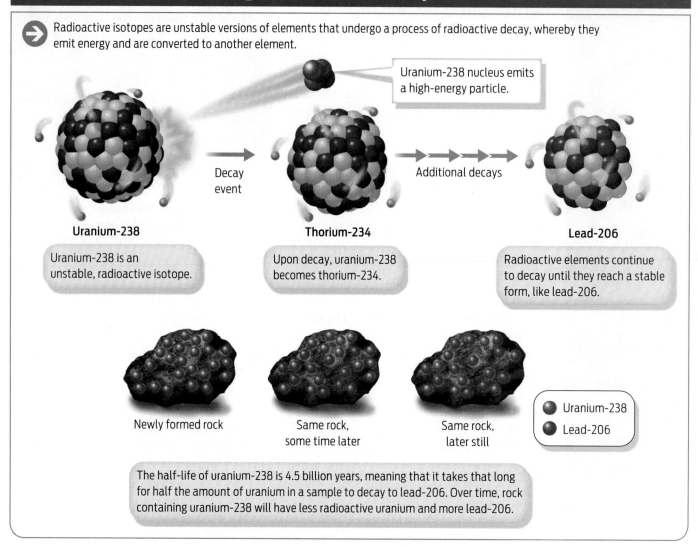

Radioactive isotopes are unstable versions of elements that undergo a process of radioactive decay, whereby they emit energy and are converted to another element.

Uranium-238 nucleus emits a high-energy particle.

Decay event

Additional decays

Uranium-238

Thorium-234

Lead-206

Uranium-238 is an unstable, radioactive isotope.

Upon decay, uranium-238 becomes thorium-234.

Radioactive elements continue to decay until they reach a stable form, like lead-206.

Newly formed rock

Same rock, some time later

Same rock, later still

● Uranium-238
● Lead-206

The half-life of uranium-238 is 4.5 billion years, meaning that it takes that long for half the amount of uranium in a sample to decay to lead-206. Over time, rock containing uranium-238 will have less radioactive uranium and more lead-206.

Because the isotopes decay at a known rate, they can be used to determine the age of the materials in which they're found **(Infographic 17.1)**.

As wind and water washed over rocks throughout earth's history, they stripped off, or eroded, particles and carried them to other places. Sometimes the deposited particles were compressed over many years into new rock layers by water or by additional particles. Such rock, called sedimentary rock, can be seen in the distinctive striations, or stripes, marking successive layers of sandstone and limestone found in former riverbanks like those surround-

ing the Grand Canyon in Arizona. Most fossils are found in sedimentary rocks.

Rocks can also form suddenly as erupting volcanoes spew lava and ash over an area. When this molten debris cools and hardens, it forms what is called igneous rock ("igneous" is from the Latin word for "fire"). Radiometric dating is performed on igneous rocks. When the rocks form, the radioactive clock is set to zero; no products of radioactive decay are present. Over time, more and more radioactive decay will occur, and more and more stable product will be present. By measuring the ratio of a radioactive isotope to stable product present in a layer

INFOGRAPHIC 17.2

Radioactive Decay Is Used to Date Some Rock Types

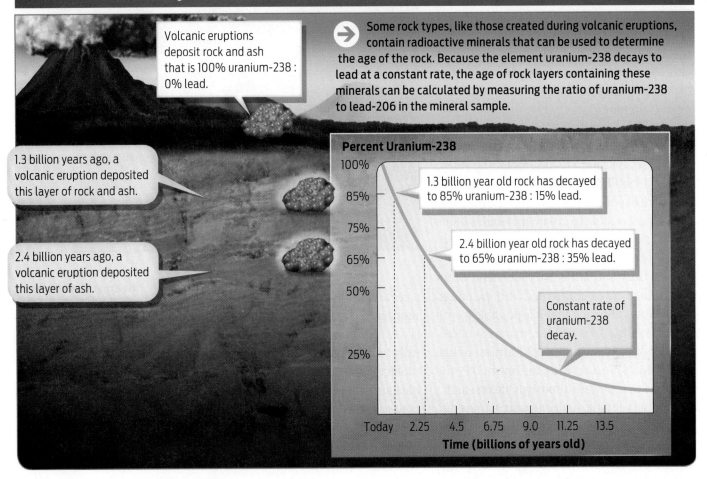

Volcanic eruptions deposit rock and ash that is 100% uranium-238 : 0% lead.

Some rock types, like those created during volcanic eruptions, contain radioactive minerals that can be used to determine the age of the rock. Because the element uranium-238 decays to lead at a constant rate, the age of rock layers containing these minerals can be calculated by measuring the ratio of uranium-238 to lead-206 in the mineral sample.

1.3 billion years ago, a volcanic eruption deposited this layer of rock and ash.

2.4 billion years ago, a volcanic eruption deposited this layer of ash.

Percent Uranium-238

1.3 billion year old rock has decayed to 85% uranium-238 : 15% lead.

2.4 billion year old rock has decayed to 65% uranium-238 : 35% lead.

Constant rate of uranium-238 decay.

Time (billions of years old)

---

of igneous rock, scientists can calculate its age **(Infographic 17.2)**. Sedimentary rocks, on the other hand, cannot be dated by radiometric methods because they are made up of particles from rocks of various ages.

Dating rocks by radioactive isotopes is quite precise and can be confirmed by a number of methods. For example, minerals taken from layers of rock in Saskatchewan, Canada, were dated by three methods: the potassium-argon method yielded an age of 72.5 million years; the uranium-lead method, an age of 72.4 million years; and the rubidium-strontium method, an age of 72.54 million years.

BIOCHEMISTRY

Q When and how did life begin?

A At some point in the earth's distant past, life did not exist. Then, at a later point, it

did. Where did this life come from? How did it start? The precise details of the transition from nonliving to living are lost in the mists of time. We can now only hypothesize how that transition might have occurred.

Scientists have offered a number of hypotheses to explain how life began on earth, including the idea that it arrived here fully formed on an asteroid or meteorite from outer space. Others hypothesize that life emerged in stages over time, as inorganic chemicals combined into successively more-complex molecules, including ones that were capable of self-replicating—that is, of copying themselves. A landmark experiment lending support to this hypothesis was performed by University of Chicago chemist Harold Urey and his 23-year-old graduate student Stanley Miller in 1953.

Urey and Miller hypothesized that they could synthesize organic molecules—the building blocks of life–by replicating the chemical environment of the early earth. To simulate the early atmosphere, they combined the gases hydrogen (H_2), methane (CH_4), ammonia (NH_3), and water vapor (H_2O) in a flask filled with warm water–the "primordial sea." They then mimicked lightning by discharging sparks into the chamber.

As the gases condensed and rained into the sea, a host of new molecules was produced from these basic ingredients, including amino acids, the building materials of proteins. This landmark experiment showed for the first time that it was possible to create molecules of life from the inorganic materials found in the primordial soup.

Since Urey and Miller's experiment, other researchers have confirmed and extended their results, showing that it is possible to create essentially all the building materials necessary for life, including all 20 amino acids, sugars, lipids, nucleic acids, and even ATP–the molecule that powers almost all life on earth.

Although organic molecules are a prerequisite for life, they are not themselves alive. To be alive, something must be able to grow, reproduce, and metabolize, among other things. Today, of course, cells carry out these life-sustaining functions. How then did living cells come about? Recall from Chapter 2 that one of the major components of cells are their lipid membranes. Researchers hypothesize that at

Stanley Miller recreates the experiment he first performed in 1953 with Harold Urey.

a certain point the lipid molecules in the primordial soup formed bubbles–which makes sense since lipids are hydrophobic and naturally form bubbles in water. The other organic molecules were incorporated into these bubbles and, as researchers speculate, over the course of millions of years, such membrane-bound bubbles filled with self-replicating molecules eventually became cells, capable of reproducing. While these ideas are highly speculative, research on microorganisms living today in such unlikely places as hydrothermal vents at the bottom of the ocean are giving us concrete insights into how life might have begun (see Chapter 18).

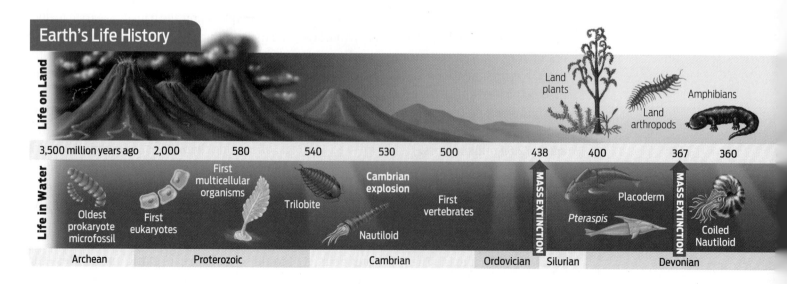

Earth's Life History

Life on Land

Land plants

Land arthropods

Amphibians

| 3,500 million years ago | 2,000 | 580 | 540 | 530 | 500 | 438 | 400 | 367 | 360 |

Life in Water

Oldest prokaryote microfossil

First eukaryotes

First multicellular organisms

Trilobite

Nautiloid

Cambrian explosion

First vertebrates

MASS EXTINCTION

Pteraspis

Placoderm

MASS EXTINCTION

Coiled Nautiloid

Archean   Proterozoic   Cambrian   Ordovician   Silurian   Devonian

What was life like millions of years ago?

A Humans weren't around millions of years ago, so we have no cave paintings or other records to help us picture what life on earth was like. Most of what we know about past life on earth comes from fossils—the preserved remains of once-living organisms, such as *Tiktaalik,* which is discussed in Chapter 16.

While each fossil find is a treasure, any single specimen reveals only a tiny slice of geologic history. What paleontologists really want to understand is how each fossil fits into the larger story told by the fossil record. By dating the rock layers, or strata, near where fossils are buried, scientists can determine when different organisms lived on the earth. Combined with knowledge from geology, chemistry, and biology, the fossil record has enabled scientists to construct a geologic timeline of life on earth **(Infographic 17.3)**.

The geologic timeline shows that during the 4.6 billion years or so that the earth has been around, its geography and climate have gone through dramatic changes. For the first few million years or so it was a molten ball of lava continually bombarded by meteorites. Not until the

> **The oldest known fossils date from some 3.5 billion years ago.**

surface cooled down about 3.8 billion years ago–to a balmy 45°C to 85°C (113°F to 185°F)– could it support life.

The oldest known fossils date from some 3.5 billion years ago, when earth's climate was very different from what it is today. The atmosphere lacked substantial oxygen (O_2), churning instead with ammonia, methane, and hydrogen. In this oxygenless world, the only organisms that could thrive were unicellular prokaryotes that used these other gases as a fuel source. Only with the emergence and proliferation of unicellular photosynthetic organisms, between 3.0 to 2.5 billion years ago, did oxygen begin to accumulate in the atmosphere, opening the door for more-complex eukaryotic organisms to evolve.

The first multicellular, eukaryotic organisms to make use of this oxygen were green algae, which appeared 1.2 billion years ago. Soft-bodied aquatic animals followed, about 600 million years ago, but it is only from 545 million years ago, during the Cambrian period, that we see fossil evidence of a truly diverse animal world. During the Cambrian explosion, as this event is known, ocean life swelled with a mind-boggling array of strange-looking creatures, including *Opabinia,* an organism with five eyes and a snout resembling a vacuum-cleaner hose, discovered in fossils from this period.

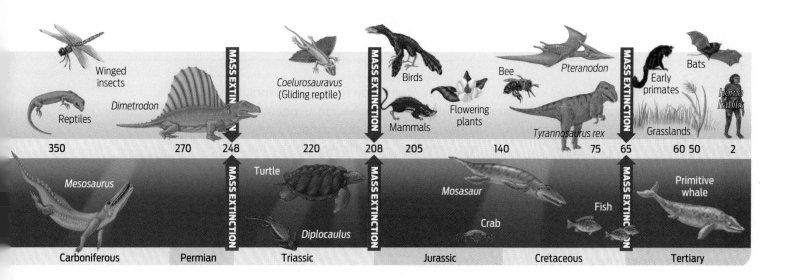

Geologic Timeline of the Earth

 Based on fossil evidence and radiometric dating of rock layers from around the world, scientists have produced a geologic timescale of the earth. The geologic timescale provides a chronological history of the main periods of the earth and its inhabitants.

Era	Period	Millions of Years Ago	Notable Events in the History of Life
Cenozoic	Quaternary	1.8	Many large mammals go extinct. Modern organisms are present today.
	Tertiary	65	Mammals, birds, and flowering plants diversify. Grasses appear. The first primates and early humans.
Mesozoic	Cretaceous	144	Dinosaurs diversify. Cone-bearing plants dominate, while flowering plants take over in many habitats. Era ends in mass extinction of dinosaurs.
	Jurassic	206	The first flowering plants and bird species appear. Large dinosaurs are plant-eaters.
	Triassic	251	Dinosaurs and mammals appear on land. Ocean life diversifies in recovery from Permian extinction.
Paleozoic	Permian	290	Reptiles appear on land. Oceans abundant in coral species. Era ends with mass extinction of 95% of living organisms.
	Carboniferous	354	Forests of seedless plants dominate land. Amphibians appear and begin to diversify.
	Devonian	408	Fish species diversify. The first insects and seed-bearing plants appear on land.
	Silurian	439	Seedless plants, primitive insects, and soft-bodied animals appear on land.
	Ordovician	495	Diverse plant and animal life in the oceans. The first fungal species appear.
	Cambrian	543	Expansion of ocean animal diversity. Ancestors of vertebrates appear.
Pre-Cambrian			Single-celled organisms in the ocean dominate life. Some soft-bodied invertebrates develop.

The first organisms to colonize land were primitive plants, appearing roughly 450 million years ago. By 350 million years ago, forests of seedless plants covered the globe.

Then, 290 million years ago, life was drastically cut down: roughly 95% of living species were suddenly extinguished in a mass die-off known as the Permian **extinction.** The extinction wasn't bad for all organisms, though; some flourished as space and resources opened up for the survivors, who spread and diversified in a phenomenon known as **adaptive radiation.** Among these were reptiles, who thrived in the hot, dry climate of the Triassic period. The most famous group of reptiles, the dinosaurs, dominated the land for nearly 200 million years, until they died out in another mass extinction at the end of the Cretaceous period, 65 million years ago.

EXTINCTION
The elimination of all individuals in a species; extinction may occur over time or in a sudden mass die-off.

The Burgess Shale, in the Canadian Rockies, is famous for the fine state of preservation of the soft parts of its fossils.

Q Why are there no penguins at the north pole, and no polar bears at the south pole?

A In terms of habitat, the north pole and the south pole are quite similar: cold, snowy, and surrounded by ocean. Yet each place is home to different creatures. Why? We can get some clues from **biogeography,** the study of the natural geographic distribution of species. Biogeography seeks to explain why islands and isolated land areas–such as the Arctic and the Antarctic–have evolved their own distinct flora and fauna.

Penguins make their home in the southern hemisphere, especially in the coastal regions of Antarctica. According to fossil evidence, penguins first appeared about 65 million years ago near southern New Zealand and Antarctica. Polar bears, on the other hand, live only in the Arctic. From fossil and DNA evidence, polar bears likely evolved from brown bears roughly 150,000 years ago in an area in Siberia, when the region became isolated by glaciers. Both penguins and polar bears have thus lived in their respective northern or southern habitat for a long time, and it's easy to understand why they haven't migrated from one pole to the other–obstacles included great distance and the warmer oceans ringing their icy habitat **(Infographic 17.4)**. But how did they get to their homes in the first place?

Though today they are at opposite ends of the earth, the Arctic and Antarctic landmasses weren't always so far apart. In fact, 250 million years ago, the continents we currently see on earth were bound together in one large landmass that geologists call Pangaea. At that time it was theoretically possible for populations of land-dwelling animals to roam far and wide over the entire land surface. But because of a geologic process known as **plate tectonics**, over time this giant landmass split and split again, forming the continents of the northern and southern hemispheres. In the process, the ancestors of penguins and polar bears were

---

ADAPTIVE RADIATION
The spreading and diversification of organisms that occur when they colonize a new habitat.

PUNCTUATED EQUILIBRIUM
The theory that most species change occurs in periodic bursts as a result of sudden environmental change.

BIOGEOGRAPHY
The study of how organisms are distributed in geographical space.

PLATE TECTONICS
The movement of the earth's upper mantle and crust, which influences the geographical distribution of landmasses and organisms.

The reason for the extinction of the dinosaurs was a mystery for many years; evidence now suggests that what killed off the dinosaurs (and 60% of the other species living at the time) was a massive 6-mile-wide asteroid that plowed into earth with almost unimaginable force, sending a thick layer of soot and ash into the atmosphere and blocking out the sun for months. A crater 110 miles wide in Mexico's Yucatán peninsula, near the town of Chicxulub, is the likely impact site.

Evidence now suggests that what killed off the dinosaurs (and 60% of the other species living at the time) was a massive 6-mile-wide asteroid.

With the extinction of the dinosaurs, it was mammals' chance to spread and diversify on land and thus give rise to many of the species of organisms we see on the planet today. This pattern of sudden change–extinctions followed by adaptive radiations–is seen in the fossil record, and is an example of **punctuated equilibrium,** the theory that most evolutionary change occurs in sudden bursts related to environmental change rather than taking place gradually.

The Geographic Distribution of Species Reflects Their Evolutionary History

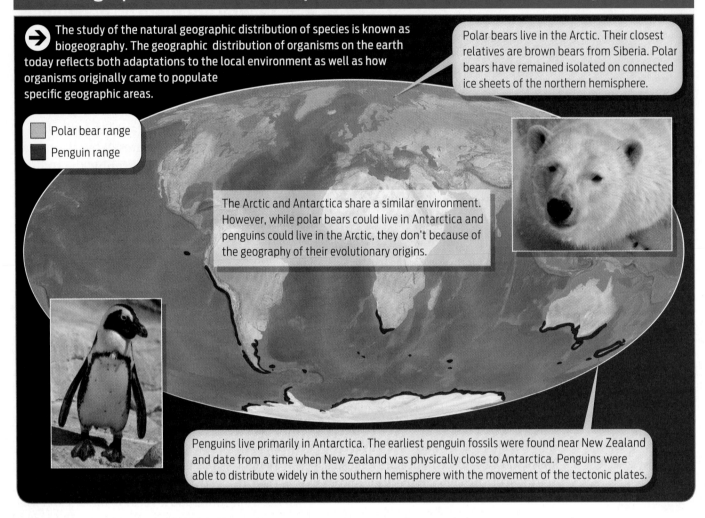

→ The study of the natural geographic distribution of species is known as biogeography. The geographic distribution of organisms on the earth today reflects both adaptations to the local environment as well as how organisms originally came to populate specific geographic areas.

Polar bears live in the Arctic. Their closest relatives are brown bears from Siberia. Polar bears have remained isolated on connected ice sheets of the northern hemisphere.

Polar bear range
Penguin range

The Arctic and Antarctica share a similar environment. However, while polar bears could live in Antarctica and penguins could live in the Arctic, they don't because of the geography of their evolutionary origins.

Penguins live primarily in Antarctica. The earliest penguin fossils were found near New Zealand and date from a time when New Zealand was physically close to Antarctica. Penguins were able to distribute widely in the southern hemisphere with the movement of the tectonic plates.

isolated from each other, as if on different life-boats cast out to sea. Because the animals we know today as penguins and polar bears evolved from their ancestors after the split of the northern and southern landmasses, they are found today at different ends of the earth (Infographic 17.5).

Q Are creatures that look alike always closely related?

A Polar bears share many traits with their brown-bear cousins—both species are recognizable as bears despite obvious differences in color. The fact that polar bears resemble brown bears in important respects is

persuasive evidence that the two species share a recent common ancestor. But common ancestry is not the only reason that two species might appear similar. Even species that are not closely related may share similar adaptations as a result of independent episodes of natural selection, a phenomenon called **convergent evolution.**

Cold-dwelling fish provide a good example. In the frigid waters of the Antarctic Ocean, fish have a unique adaptation that keeps them from becoming ice cubes: their blood is pumped full of "antifreeze." Fish antifreeze is actually molecules called glycoproteins that lower the temperature at which body fluids would otherwise freeze by surrounding tiny ice crystals and keeping them from growing. Arctic fish, at the

CONVERGENT EVOLUTION
The process by which organisms that are not closely related evolve similar adaptations as a result of independent episodes of natural selection.

earth's other pole, also have antifreeze proteins, but the genes that code for them are different.

In the Antarctic Ocean, fish have a unique adaptation that keeps them from becoming ice cubes: their blood is pumped full of "antifreeze."

Arctic and Antarctic fish diverged from their common ancestor long before each species developed antifreeze proteins, which means these adaptations must have evolved more than once. In other words, at least two separate, independent episodes of evolution occurred with the same functional results. Sometimes similar environmental challenges will favor the same adaptations time and time again.

SYSTEMATICS

Q How many species are there on earth, and how do scientists keep track of them?

A Current estimates of the total number of species on earth range anywhere from 5 to 30 million, of which 1.8 million or so have been formally described. Many of these species are found in diversity hot spots such as rain

INFOGRAPHIC 17.5

Movement of the Earth's Plates Influences Climate and Biogeography

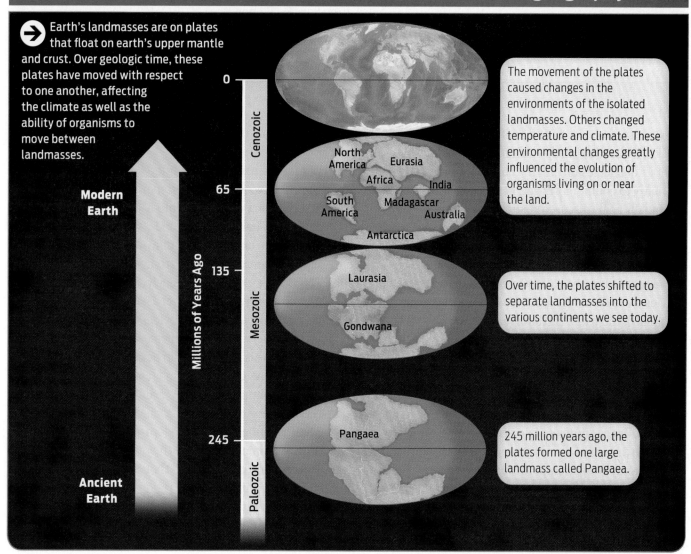

→ Earth's landmasses are on plates that float on earth's upper mantle and crust. Over geologic time, these plates have moved with respect to one another, affecting the climate as well as the ability of organisms to move between landmasses.

Modern Earth

Millions of Years Ago

Ancient Earth

Cenozoic

Mesozoic

Paleozoic

0

65

135

245

North America
Eurasia
Africa
India
South America
Madagascar
Australia
Antarctica

Laurasia

Gondwana

Pangaea

The movement of the plates caused changes in the environments of the isolated landmasses. Others changed temperature and climate. These environmental changes greatly influenced the evolution of organisms living on or near the land.

Over time, the plates shifted to separate landmasses into the various continents we see today.

245 million years ago, the plates formed one large landmass called Pangaea.

How Many Species Are There?

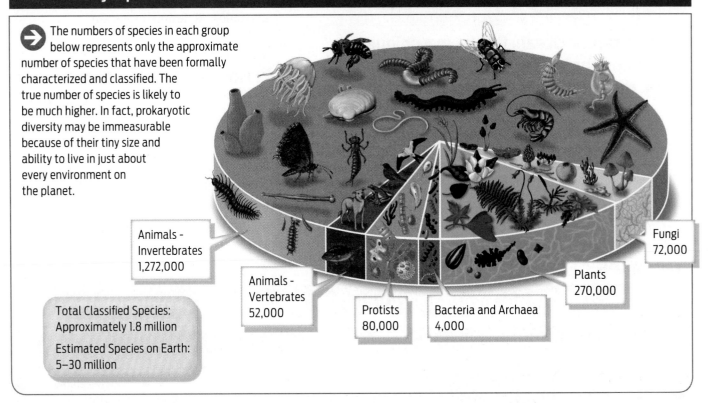

→ The numbers of species in each group below represents only the approximate number of species that have been formally characterized and classified. The true number of species is likely to be much higher. In fact, prokaryotic diversity may be immeasurable because of their tiny size and ability to live in just about every environment on the planet.

Animals - Invertebrates
1,272,000

Animals - Vertebrates
52,000

Protists
80,000

Bacteria and Archaea
4,000

Plants
270,000

Fungi
72,000

Total Classified Species:
Approximately 1.8 million

Estimated Species on Earth:
5–30 million

---

forests. But as the wide range of the estimate implies, it's hard to put an exact number on the number of species on earth–there are simply too many to count. Moreover, new species are continually being discovered. In 2007, for example, scientists identified 11 new species of plants and animals in a remote part of Vietnam. And a recent study by researchers at Arizona State University found that 17,000 new eukaryotic species were discovered in 2006 alone, more than half of them insects **(Infographic 17.6)**.

With so many species out there, how do scientists keep track of them all? The process by which scientists systematically identify, name, and classify organisms is called **taxonomy.** (Taxonomy is part of the broader study of systematics, or the study of biological diversity of life on earth.)

Taxonomy is an attempt to impose a human sense of order on this vast array of species, categorizing them on the basis of features they have in common, such as whether their cells are eukaryotic or prokaryotic, whether they photosynthesize, whether or not they have four legs and fur.

By studying the many similarities and differences among organisms, taxonomists have come up with a system for sorting organisms into a series of eight progressively narrower categories: domain, kingdom, phylum, class, order, family, genus, species. As you move down the list, from domain to species, the categories get increasingly exclusive, until finally only one member is included. The genus and species names provide the scientific name for every living organism. Because the scientific name is in Latin, it can be easily recognized in many languages.

Take humans, for example. Humans are eukaryotes, members of the domain Eukarya. They are also animals, members of the kingdom Animalia. Within the animal kingdom, they belong to the phylum Chordata, a group that includes the **vertebrates**, animals with a rigid backbone. Further, humans are **mammals**, members of the class Mammalia; they share with all members of this class mammary

TAXONOMY
The process of identifying, naming, and classifying organisms on the basis of shared traits.

VERTEBRATES
Animals with a rigid backbone.

MAMMALS
Members of the class Mammalia; all members of this class have mammary glands and a fur-covered body.

glands and a body that is covered with fur. Humans belong to the Primate order, which also includes monkeys, apes, and lemurs. And humans are members of the Hominidae family, and so are closely related to their fellow hominids: chimpanzees, gorillas, and orangutans.

Our scientific name–made up of our genus and species names–is *Homo sapiens* ("wise man") **(Infographic 17.7)**.

Classification would seem to be a simple matter–just observe, measure, and sort. But deciding which category an organism belongs

INFOGRAPHIC 17.7

Classification of Species

→ Organisms are classified into groups that are increasingly exclusive. In the broadest category (Animal Kingdom), all animals are included. Closely related organisms are grouped based on morphological, nutritional, and genetic characteristics. There are far fewer organisms in an order than in a phylum.

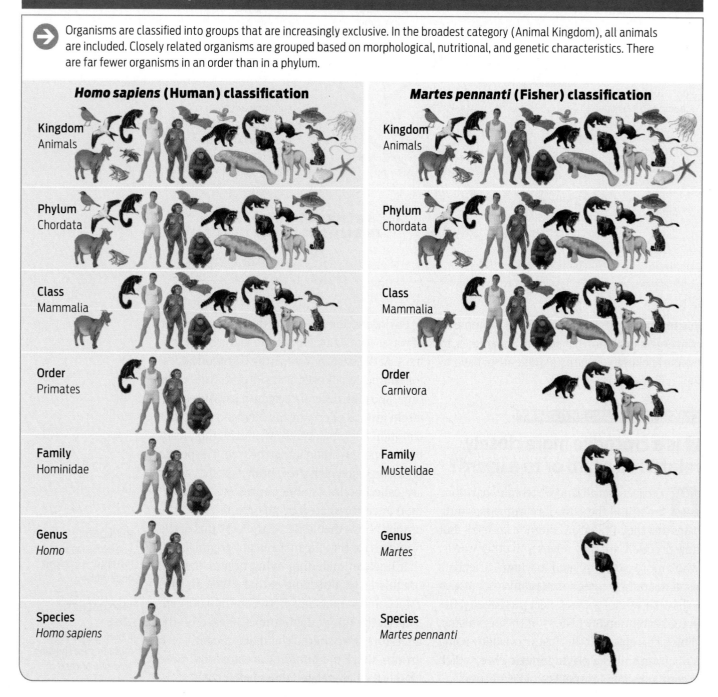

Homo sapiens (Human) classification

Kingdom
Animals

Phylum
Chordata

Class
Mammalia

Order
Primates

Family
Hominidae

Genus
Homo

Species
Homo sapiens

Martes pennanti (Fisher) classification

Kingdom
Animals

Phylum
Chordata

Class
Mammalia

Order
Carnivora

Family
Mustelidae

Genus
Martes

Species
Martes pennanti

How to Read an Evolutionary Tree

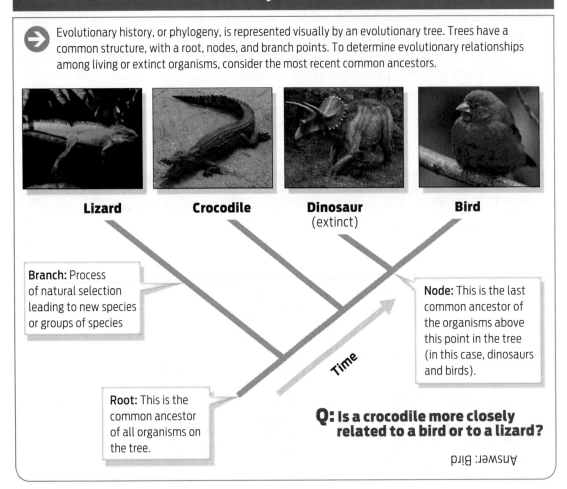

→ Evolutionary history, or phylogeny, is represented visually by an evolutionary tree. Trees have a common structure, with a root, nodes, and branch points. To determine evolutionary relationships among living or extinct organisms, consider the most recent common ancestors.

Lizard   **Crocodile**   **Dinosaur** (extinct)   **Bird**

Branch: Process of natural selection leading to new species or groups of species

Node: This is the last common ancestor of the organisms above this point in the tree (in this case, dinosaurs and birds).

Time

Root: This is the common ancestor of all organisms on the tree.

Q: Is a crocodile more closely related to a bird or to a lizard?

Answer: Bird

---

in can sometimes be tricky, as the example of convergent evolution has shown. Sometimes, to properly classify organisms, scientists have to look a little deeper.

CLASSIFICATION AND PHYLOGENY

Ⓠ Is a crocodile more closely related to a bird or to a lizard?

Ⓐ The fact that all land vertebrates have four limbs and the same forelimb bones indicates that they all share a common ancestor. But how precisely are they related? In other words, who's more closely related to whom? Scientists want not only to categorize organisms, but also to have those categories reflect **phylogeny,** the actual evolutionary history of the organisms. Biologists represent this history visually using a diagram called a **phylogenetic tree**, which is similar in some respects to a family tree.

Phylogenetic trees can be drawn in a number of ways, but most have certain features in common. At the base, or root, is the common ancestor shared by all organisms on the tree. Over time, and with different selective pressures, different groups of organisms diverged from that common ancestor and from one another, leading to separate branches on the tree. The points on the tree at which these branch points occur are called nodes. A node represents the common ancestor shared by all organisms on the branch above that node. At the very tips of the branches we find the most recent organisms in that lineage, including living organisms and organisms that became extinct. We can thus establish relationships between living organisms (at the tips of the branches) based on the ancestors they share. The more recently two groups share a common ancestor, the more closely they are related **(Infographic 17.8)**.

PHYLOGENY
The evolutionary history of a group of organisms.

PHYLOGENETIC TREE
A branching tree of relationships showing common ancestry.

A phylogenetic tree is a visual representation of the best hypothesis we currently have for how species are related. The evidence for a phylogenetic tree comes from many sources, including the fossil record, physical traits, and shared DNA sequences. For many years, biologists relied solely on observable physical or behavioral features to construct evolutionary trees. But with the genetic revolution, it's become common to include DNA evidence. Typically, researchers compare sequence differences in a gene that is found in all living organisms, such as the ribosomal RNA (rRNA) genes.

Sometimes the new genetic information yields surprises. Modern genetic evidence shows, for example, that crocodiles are more closely related to birds than they are to lizards, appearances notwithstanding. Genetics, you might say, is shaking the evolutionary tree.

CLASSIFICATION AND PHYLOGENY

Q How many branches does the tree of life have?

A Since each living species sits on its own branch in a phylogenetic tree, the complete tree of life has as many branches as there are species in the world. Today's species are like thin twigs in the upper branches of an enormous tree. Closer to the bottom of the tree, nearer to the ancient trunk, however, we find significant forks. Just how many forks there are at the bottom of the tree is a question that has been debated for decades.

Before the 18th century, biologists divided living things into just two main categories: animals and plants. This classification was based on whether an organism moved around and ate or did not move around and eat. By the mid-19th century, use of the microscope had revealed a whole new world of microscopic organisms, and so a third branch was added to life's tree: protists.

By the 1960s, taxonomists realized that even three such branches did not fully capture the diversity of life; many organisms—such as fungi—didn't fit neatly into any of these groups, and so another classification scheme was proposed. This one grouped all living organisms into five large kingdoms on the basis of how they looked (both anatomically and microscopically) and how they obtained their food. The five kingdoms were Animalia, Plantae, Fungi, Protista, and Monera. Protista comprised mostly single-cell eukaryotic organisms (such as the amoeba), and Monera included all prokaryotic organisms (such as bacteria).

Yet even this revised classification scheme eventually had to be overhauled as more infor-

An early version of the tree of life drawn in 1866 by E. H. P. A. Haeckel

Monophyletischer Stammbaum der Organismen, entworfen und gezeichnet von Ernst Haeckel. Jena, 1866.

I, Feld: p m n q (19 Stämme)
II, Feld: p x y q (3 Stämme)
III, Feld: p s t q (1 Stamm)
stellen 3 mögliche Fälle der universalen Genealogie dar

Plantae — Cormophyta; Anthophyta (Angiospermae); Pteridophyta (Lepidophyta, Rhizocarpeae, Filices, Calamophyta); Gymnospermae; Bryophyta (Phyllobrya, Thallobrya); Fucoideae (Sargassaceae, Laminariaceae, Chordariaceae); Florideae (Sphaerococceae, Ceramiaceae); Characeae; Jnophyta (Lichenes, Fungi); Archephyta (Ulva, Conferva, Desmidium, Nostoc, Codiolum)

Protista — Myxomycetes (Physarum, Stemonitis, Lycogala, Trichia); Spongiae; Petrospongiae (Siphonida, Ocellarida, Lymnorida, Bothroconida, Turonida); Autospongiae (Calcispongiae, Silicispongiae, Ceratospongiae, Myxospongiae); Rhizopoda (Radiolaria, Actinophryida, Acyttaria, Polythalamia); Flagellata (Peridinium, Euglena, Volvox); Myxocytoda (Noctilucae); Protoplasta (Arcellae, Gregarinae, Autamoeba, Amoebae); Diatomeae (Areolatae, Vittatae, Striatae); Moneres (Protogenes, Protamoeba, Vampyrella, Protomonas, Vibrio)

Animalia — Vertebrata; Amniota; Articulata; Arthropoda (Tracheata, Crustacea); Aves; Mammalia; Reptilia; Vermes (Annelida, Rotatoria); Anamnia; Amphibia; Pisces; Scolecida; Infusoria; Amphirhina; Monorrhina; Echinodermata (Holothuriae, Echinida, Crinoida, Asterida); Leptocardia; Mollusca (Otocardia); Himatega; Coelenterata (Nectacalephae, Petracalephae)

Archephylum vegetabile — Archephylum protisticum — Archephylum animale

Protista — Plantae — Animalia

Radix communis Organismorum — Moneres autogonum

DNA Defines Three Domains of Life: Bacteria, Archaea, Eukarya

All living organisms have evolved from a common ancestor. Based on DNA evidence, we can group living things into one of three domains of life, each with a distinct evolutionary history. While the Bacteria and Archaea both have prokaryotic cells, they have distinct evolutionary histories, with Archaea being more genetically related to Eukarya than Bacteria. The domain Eukarya encompasses protists, plants, fungi, and animals, including humans (see Chapter 19).

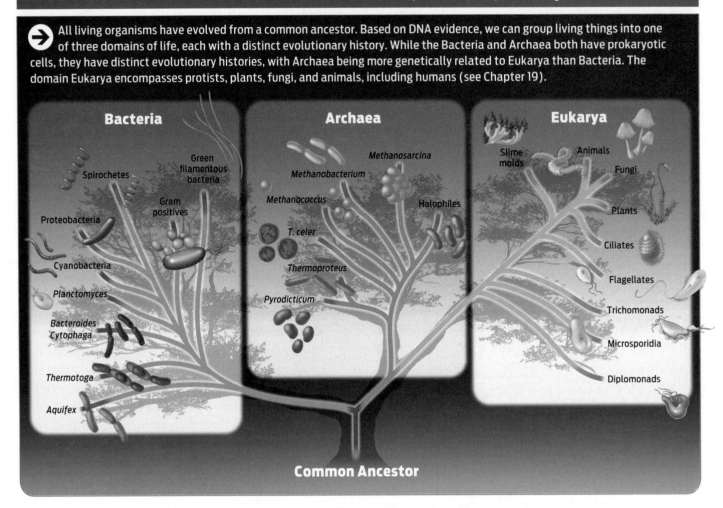

mation became available. In the 1970s, genetic studies by Carl Woese revealed that, on the basis of on genetic relatedness, not all prokaryotes could be lumped together; likewise, protists were too genetically diverse to be put in one category. Consequently, scientists now group organisms into one of three large **domains**—Bacteria, Archaea, and Eukarya—which represent three fundamental branch points in the trunk of the evolutionary tree. The original kingdom Monera is now divided into two domains, Archaea and Bacteria. Within the domain Eukarya, Animalia, Plantae, and Fungi remain recognized kingdoms, but the protists (former members of the kingdom Protista) are dispersed across the domains of life, on the basis of DNA evidence (**Infographic 17.9**). ∎

DOMAIN
The highest category in the modern system of classification; there are three domains—Bacteria, Archaea, and Eukarya.

▶ Summary

■ The age of the earth and its rock layers can be determined by measuring the amount of radioactive isotopes present in certain types of rocks, a method known as radiometric dating.

■ Life on earth may have emerged in stages, as inorganic molecules combined to form organic ones in the primordial soup, and as these were incorporated into lipid bubbles to form cells.

■ Using geological evidence and the fossil record, paleontologists have been able to construct a geologic timeline of life on earth.

■ Earth's history can be divided into important eras and periods. Dinosaurs, for example, lived primarily from 250 to 65 million years ago, during the Mesozoic era, from the Triassic through the Cretaceous periods.

■ The history of life on earth is marked by repeated extinctions and adaptive radiations, a phenomenon of intermittent rather than steady change known as punctuated equilibrium.

■ Ancient movement of earth's major landmasses affected the eventual distribution of species around the globe, the study of which is known as biogeography.

■ Convergent evolution is the evolution of similar adaptations in response to similar environmental challenges in groups of organisms that are not closely related.

■ Life is astoundingly diverse. Current estimates of the total number of species on earth range anywhere from 5 to 30 million, of which 1.8 million have been formally described.

■ Biologists sort organisms into a series of nested categories based on shared anatomical and genetic features: domain, kingdom, phylum, class, order, family, genus, species.

■ The scientific name of an organism is given by its genus and species names (for humans it is *Homo sapiens*).

■ Both physical evidence and genetic evidence are used to understand evolutionary history, or phylogeny. Branching trees of common ancestry are used to represent that history visually.

■ On the basis of DNA evidence, all living organisms can be classified into one of three domains: Bacteria, Archaea, or Eukarya.

HISTORY OF LIFE

Life on earth has changed dramatically since it first emerged. We can learn a great deal about the history of life on earth by studying rock layers and the fossils found within them.

HINT See Infographics 17.1–17.3.

KNOW IT

1. What do uranium-238, carbon-14, and potassium-40 have in common?

2. To date what you suspect to be the very earliest life on earth, which isotope would you use: uranium-238, carbon-14, or potassium-40? Explain your answer.

3. Place the following evolutionary milestones in order from earliest to most recent (numbering them from 1 to 7), providing approximate dates to support your answer.

the first multicellular eukaryotes _____

the first prokaryotes _____

the Permian extinction _____

the Cambrian explosion _____

the first animals _____

the extinction of dinosaurs _____

an increase in oxygen in the atmosphere

_____

USE IT

4. Consider a rock formed at about the same time as the earth was formed.
   a. How old is this rock?
   b. How much of the original uranium-238 is likely to be left today in that rock?

5. If an igneous rock contains 75% lead, how old is that rock? (Look at Infographic 17.2.)

6. Diverse animal fossils are found dating from the Cambrian period, not earlier. Why might these organisms have made their first appearance in the fossil record only then, even though their ancestors may have been living, and evolving, for a long time before the Cambrian? (Think about what kinds of new structures might have evolved during the Cambrian period that would have allowed these organisms to leave fossils.)

7. Along the banks of a river, some sedimentary rock strata have been revealed by erosion. If the sedimentary layers had been deposited in the Carboniferous period, would you expect paleontologists to find fossils of amphibians in these strata? Reptiles? Sharks? Justify your answers.

BIOGEOGRAPHY AND PLATE TECTONICS

The current distribution of organisms on earth reflects both evolutionary history and accidents of geology.

HINT See Infographics 17.4 and 17.5.

KNOW IT

8. If two organisms strongly resemble each other in terms of their physical traits, can you necessarily conclude that they are closely related? Explain your answer.

9. What did the arrangement of landmasses on earth look like between 135 and 65 million years ago? What happened to these landmasses, and how does this change help explain the distribution of organisms found on the planet?

USE IT

10. A cactus called ocotillo (*Fouquieria splendens*), which grows in New Mexico, looks very much like *Alluaudia procera*, a species of plant that grows in the deserts of Madagascar. These two plant species are not closely related—why then do they look so alike?

11. If penguins and polar bears had evolved before Pangaea split into northern and southern continents, what might you predict about their geographic distribution today?

12. Both bats and insects fly, but bat wings have bones and insect wings do not. Would you consider bat and insect wings to be a result of convergent evolution, or of homology—evolution based on inheritance of similar structures from a common ancestor? Explain your answer.

CLASSIFICATION AND PHYLOGENY

Categorizing the many species on earth and understanding how they are related is a challenging task, made easier by genetic information.

HINT See Infographics 17.6–17.9.

➔ KNOW IT

13. Which of the following is *not* a domain of life?

  a. Animalia
  b. Eukarya
  c. Bacteria
  d. Archaea
  e. Plantae
  f. Neither a nor e is a domain of life.

14. Put the following terms in order, from most inclusive (1) to least inclusive (5).

  domain _____
  species _____
  kingdom _____
  genus _____
  phylum _____

15. A phylogenetic tree represents

  a. a grouping of organisms on the basis of their shared structural features.
  b. a grouping of organisms on the basis of their cell type.
  c. a grouping of organisms on the basis of their complexity.
  d. a grouping of organisms on the basis of their evolutionary history.
  e. a grouping of organisms on the basis of where they are found.

➔ USE IT

16. Why was the classification of the kingdom Monera split into two domains? What are these two domains?

17. Which number on the tree below shows the most recent common ancestor of humans and corn?

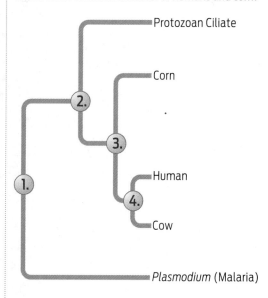

  a. 1
  b. 2
  c. 3
  d. 4
  e. Humans and corn do not share any ancestors.

SCIENCE AND ETHICS

18. How might knowledge of the evolutionary history of organisms affect human health? How might such knowledge affect decisions regarding the environment?

Lost City

Lost City

Scientists probe life's origins in an undersea world of extreme-loving microbes

Gretchen Früh-Green's heart had never beat so fast. Late on a December evening in 2000, she was hunkered in the control room of the research ship *Atlantis*, monitoring live video streaming up to the ship from a camera swimming 2,100 feet below. Strange white shapes began to appear in the blackness of her screen. As the camera panned, an underwater landscape of ghostly white towers suddenly came into focus. The huge limestone structures, which resembled the stalagmites found in caves, loomed above an otherwise empty seafloor. Früh-Green, a geologist with the Institute for Mineralogy and Petrology in Zurich, Switzerland, knew immediately that she was looking at something special, and raced to tell her colleagues. "It was really quite exciting," she said. "It was late at night and kind of woke us all up."

> **A dense "cityscape" of rocky spires, Lost City stretches across a distance of roughly six football fields of dark seafloor.**

The next day, the mission's chief scientist, Deborah Kelley and two members of the team dove in a submersible to the site, which lies at a depth of 2,600 feet on an undersea mountain in the middle of the Atlantic Ocean. There, they found a dense "cityscape" of rocky spires, stretching across roughly six football fields of dark seafloor. It was an uncharted world no one had known existed. Kelley named the undersea world Lost City. The tallest tower, measuring 200 feet, she dubbed Poseidon, after the Greek god of the sea.

Subsequent research has shown that each tower is a type of underwater chimney, or spring, known as a hydrothermal vent. As rocks from the earth's crust come in contact with seawater, they react chemically, giving off a steady stream of heat and combustible gas that seeps out of

each vent. The resulting fluid is highly caustic, with a pH of 9-11–similar to that of drain cleaner–and temperatures as hot as 90°C (194°F).

Though deep-sea hydrothermal vents had been discovered before, the ones at Lost City were unique in their chemistry and in the type of life they support. Nothing like them had ever been seen. "Rarely does something like this come along that drives home how much we still have to learn about our own planet," said Kelley, an oceanographer at the University of Washington in Seattle, who now leads the international team of scientists who are exploring Lost City's mysteries.

An extreme and seemingly inhospitable environment, the towers of Lost City are nonetheless home to a surprising number of life forms, some of them unlike any seen anywhere else on earth. Most prevalent are dense layers of unicellular microbes that coat the towers, inside and out. Microbial life exists pretty much everywhere on earth–in frozen glaciers, in radioactive dirt, in the intestines of animals–yet the microbes at Lost City have earned the fascination of scientists.

Left: Gretchen Früh-Green works on samples collected by *Atlantis*. Right: Researchers monitor camera feed from underwater dives.

"They're living in boiling toilet bowl cleaner full of flammable gas," says Bill Brazelton, a postgraduate researcher with NASA and the University of Washington, who studies the microbes. The extreme conditions in which these microbes live resemble what researchers think the earth was like 3.8 billion years ago, when life was getting started. Scientists have long wondered how life managed to emerge and flourish in such inhospitable conditions. At Lost City, they are finding some tantalizing clues **(Infographic 18.1)**.

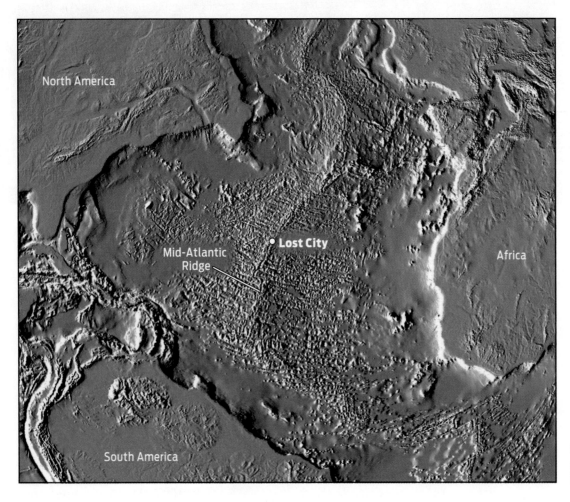

Lost City: 2,600 feet down on the Mid-Atlantic Ridge.

Lost City Hosts Unique Microscopic Life

The hydrothermal vents of Lost City are huge rock chimneys. Scalding fluid with an extremely basic pH flows out of the tops of these chimneys. Temperatures cool farther down each tower and are cooler yet at their bases. The towering size and range of temperatures within a single Lost City spire mean it can host a variety of unique microbial communities.

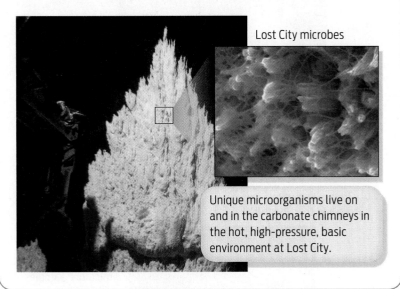

Lost City microbes

Unique microorganisms live on and in the carbonate chimneys in the hot, high-pressure, basic environment at Lost City.

Exploring the Deep

Since Lost City was first discovered in 2000, researchers have organized three exploratory trips to the site. During each of these month-long expeditions, a team of more than 50 people–scientists and researchers, pilots, engineers, and ship crew–works around the clock to orchestrate dives and obtain specimens for research. A stable of sophisticated robotic assistants aids the effort.

Researchers dive to the site in *Alvin,* a tiny three-person submersible craft. Each trip to the murky depths takes about 30 minutes and is a risky descent into dark, uncharted waters. This isn't flat seafloor, after all; the Lost City towers are like tall buildings, some as high as 18 stories. Members of the team compare the journey to flying through New York City in a helicopter at night with no lights. It's well worth the effort, though: "All the time you're looking at something that nobody's ever looked at before, and that's really cool," says Brazelton.

Upon reaching their destination, the researchers begin the work of collecting specimens, using a pair of remotely operated mechanical vehicles, *Jason* and *Hercules.* Both *Hercules* and *Jason* have robotic arms that are used to collect rock samples. But that's easier said than done: think of the arcade game in which you try to grab a toy with a shaky mechanical claw. Now imagine doing that under water, remotely, while strong ocean currents blow your claw around. The chalky limestone prizes can also be quite

The *Alvin* is deployed from the *Atlantis*.

Investigating Life in Lost City

Collecting Samples:

Various types of life are collected from the surface of the Lost City spires and surrounding fluid. Collection is painstakingly completed with the use of robotic claws and suction devices that move samples into collection boxes for transport to the surface.

Robotic arms grasp rock samples for further analysis on *Atlantis*.

A "slurp gun" sucks up samples containing smaller organisms.

Processing Microbes:

Once Lost City microbe samples arrive in the laboratory on board *Atlantis* they are processed so that each organism can be grown in a laboratory culture and then identified.

Lost City microbes are placed in tubes with specialized energy sources, like hydrogen or methane, and placed in the incubator on board *Atlantis*.

A scientist processes microbial samples in an anaerobic glove bag that eliminates oxygen that may be harmful to Lost City microbes.

brittle, crumbling if squeezed too tightly. Often the researchers would grab something only to have it fall down between the spires.

To collect living specimens, the researchers use what they call a "slurp gun." A robotic arm aims the gun, which then gulps a sample from the surface of the spires. Everything is caught on camera by *Hercules*. Eight hours later, the vehicles return to *Atlantis* with their cargo of treasures.

Because many of the Lost City microbes cannot tolerate oxygen, the biologists have to be careful not to expose them to air. Using a special airtight bag with built-in gloves, the researchers transfer samples of microbes into test tubes without introducing oxygen into their environment. The microbes are then put in warm incubators and coaxed to grow (**Infographic 18.2**).

What Are They?

The microbial life in Lost City poses many riddles for the scientists trying to understand this mysterious deep-sea world: What are these creatures? How are they related to known organisms? What adaptations allow them to survive?

Lost City houses a community of life forms ranging from mats of microbes to translucent 1-centimeter-long animals and the larger fish that eat them. But most of the living things at Lost City are **prokaryotes**—unicellular organisms whose single cell lacks internal membrane-bound organelles, and whose ribbon of DNA floats freely in the cytoplasm (rather than being housed in a nucleus, as in eurkaryotes; see Chapter 3). Most prokaryotic organisms are unicellular and microscopic, on the order of 1-10 microns, which is about 1/10 the thickness of a human hair (**Infographic 18.3**).

PROKARYOTE
A usually unicellular organism whose cell lacks internal membrane-bound organelles and whose DNA is not contained within a nucleus.

Jason, a remotely operated mechanical assistant used on the exploratory trips to Lost City.

What prokaryotes lack in size they make up for in numbers. There are more prokaryotes in a handful of dirt than there are plants and animals in a rain forest. More prokaryotic organisms live on and in you right now than there are human cells in your body. At Lost City, up to 1 billion such prokaryotic organisms inhabit each gram of chimney rock, forming mucus-like biofilms several centimeters thick. It looks like the chimneys got sneezed on, says Brazelton.

As their numbers testify, prokaryotes are an extraordinarily successful product of evolution. From fossil evidence, we know that prokaryotes were the first colonizers of our planet, and for nearly 2 billion years were its only life form. Having first evolved nearly 4 billion years ago, prokaryotic organisms have had plenty of time to adapt to a wide range of environments, including many that would kill most eukaryotes. In fact, prokaryotes are almost endlessly adaptive and can thrive just about anywhere. At another

INFOGRAPHIC 18.3

Prokaryotic Cells Are Small and Lack Organelles

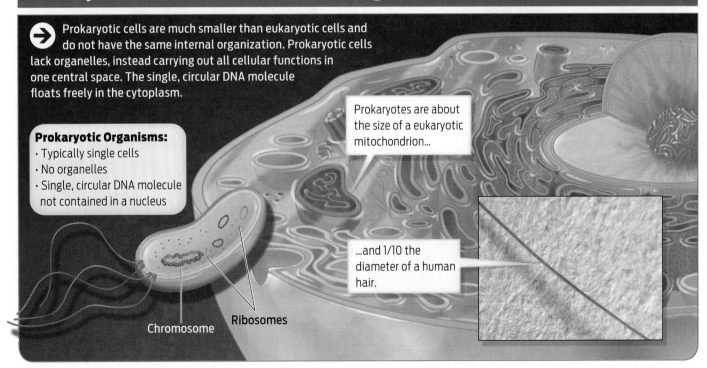

→ Prokaryotic cells are much smaller than eukaryotic cells and do not have the same internal organization. Prokaryotic cells lack organelles, instead carrying out all cellular functions in one central space. The single, circular DNA molecule floats freely in the cytoplasm.

Prokaryotic Organisms:
· Typically single cells
· No organelles
· Single, circular DNA molecule not contained in a nucleus

Prokaryotes are about the size of a eukaryotic mitochondrion...

...and 1/10 the diameter of a human hair.

Chromosome

Ribosomes

type of deep-sea hydrothermal vent, off the Oregon coast—one with very acidic, extremely hot fluids—scientists have discovered more than 40,000 different kinds of prokaryotes (Infographic 18.4).

How do the prokaryotic organisms found at Lost City compare to those living elsewhere? When biologists want to identify a prokaryote, they can't always rely on physical appearance, since many prokaryotic organisms look similar under a microscope. Nor can prokaryotes necessarily be grown in the laboratory. Many of the unusual prokaryotes at Lost City were almost impossible to culture in the lab, making it hard to study their features

By looking at DNA sequences, researchers have discovered several new species of prokaryotic organisms living at Lost City

and behavior. Instead, biologists generally rely on DNA to identify prokaryotic organisms. Finding a unique DNA sequence in a sample means the researchers have discovered a new organism. The number of DNA sequence similarities between the new species and known ones establishes their degree of relatedness.

By looking at DNA sequences, researchers have discovered several new species of prokaryotic organisms living at Lost City. Two of these species fall into a group of prokaryotes known as the domain Archaea. Archaea aren't the only prokaryotes present at Lost City—the site is rich in bacterial populations, too—but

INFOGRAPHIC 18.4

Prokaryotes Are Abundant and Diverse

→ Even in seemingly inhospitable environments, there can be large numbers and many different types of prokaryotic microorganisms.

Lost City Microbes

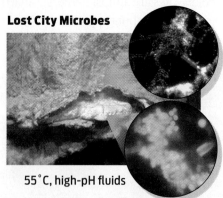

1 billion filamentous microbes per gram in the outer carbonate crust

Methane-metabolizing microbes in fluids rich in volatile gases.

55˚C, high-pH fluids

Oregon Vent Microbes

> 40,000 different types of microbes

Hydrothermal vent on the Pacific deep-sea volcano, Axial Seamount

A Variety of Microbial Environments

Glaciers

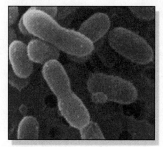

Environment: freezing, high-pressure, low-nutrient, low-oxygen

Salt Lakes

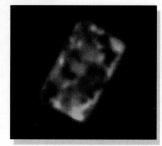

Environment: high-salt, low-nutrient
Organism trait: photosynthetic

Intestines

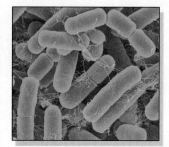

Environment: no oxygen, moist, strict 37˚C
Organism trait: metabolize sugars

Miles Underground

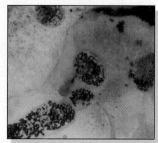

Environment: dry, high-pressure, low-nutrient
Organism trait: metabolize coal

Bacteria and Archaea, Life's Prokaryotic Domains

 Two of the domains of life, Bacteria and Archaea, have prokaryotic cells, but they each have distinct evolutionary histories, with Archaea being genetically closer to Eukarya than to Bacteria. The genetic differences between Bacteria and Archaea translate into a variety of structural and functional adaptations.

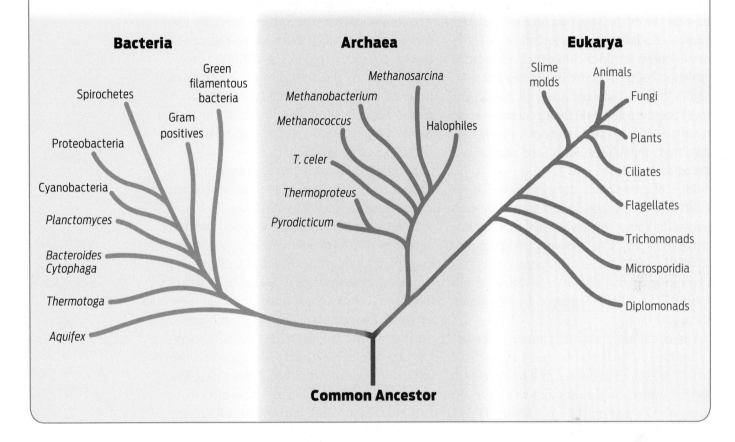

it's the archaea that are most interesting to researchers. That's because the archaea are doing things that even bacteria can't do.

At Lost City, bacterial populations congregate on the outsides of the vents, where temperatures are relatively mild and where oxygen is present in the seawater. Archaea, by contrast, are found only inside the vents, where temperatures are hottest and where there is no oxygen. So far, just two species of archaea have been detected in this environment. "The conditions are so extreme that they're the only thing that has been able to survive," says Brazelton. Because of their preference for such extreme environments, archaea have been nicknamed "extremophiles."

As intriguing as these extreme-loving organisms are, however, they weren't even recognized as a distinct evolutionary group until quite recently.

For many years all prokaryotes were lumped together into one large group—the kingdom Monera—a classification based largely on their cell structure. Then, in the late 1970s, Carl Woese and his colleagues at the University of Illinois made the surprising discovery that not all prokaryotic organisms are genetically similar enough to be classified as a single group. His work established an entirely new branch of prokaryotic organisms, the Archaea. While most archaea don't look that different from bacteria under the microscope—both are unicellular prokaryotes—genetically they are as different from bacteria as humans are. In other words, they represent a distinct evolutionary domain of life. Together, the domains Bacteria and Archaea represent a very large slice of the total diversity of life on earth—they are two of life's three domains (**Infographic 18.5**).

Bounteous and (Mostly) Beneficial: Bacteria

Bacteria are everywhere. They include commonly encountered microbes such as the beneficial strain of *Escherichia coli* present in the human gut or the *Staphylococcus aureus* that causes skin infections, as well as not so commonly encountered ones like the fluorescence-emitting bacteria living in the head of a sea squid. While all bacteria are prokaryotic, and most possess a cell wall, their genetic diversity translates into a wide variety of differences in nutrition, metabolism, structure, and lifestyle (**Infographic 18.6**).

Like all organisms, bacteria can be categorized by what they eat. Some bacteria are autotrophs (literally, "self-feeders"): they are able to make their own food directly, using material from the nonliving environment. Others are heterotrophs (literally, "other feeders"): they must rely on other living organisms to provide them with food.

One of the largest and most important groups of autotrophic bacteria are the cyanobacteria, which are found in oceans and freshwater as well as on exposed rocks and soil–virtually everywhere that sunlight can reach them. Cyanobacteria use the energy of sunlight to carry out photosynthesis in a manner similar to plants, taking in carbon dioxide and generating much of the oxygen that other organisms, including humans, rely on. Many cyanobacteria also perform the ecologically useful task of converting nitrogen from the atmosphere into a form that plants can use to grow. This process, called **nitrogen fixation,** is indispensable for the survival of life on earth (Chapter 23). Cyanobacteria are thought to be the oldest photosynthetic organisms on earth, dating back roughly 2.5 billion years and playing a pivotal role in making the atmosphere breathable for the rest of us.

Bacteria can feed on more than just sunlight and carbon dioxide. Certain autotrophic bacteria, including those living at Lost City, can obtain energy directly from geological sources such as inorganic gases pouring out of hydrothermal vents–making them among the few organisms on earth that do not rely on the sun's energy to survive. Heterotrophic bacteria include those that obtain food by consuming material from living or dead organisms. Such heterotrophic bacteria play an important role in decomposition, allowing carbon and other elements–which would otherwise be trapped in dead organisms, sewage, or landfills–to be recycled. They are also useful in bioremediation projects. For example, some types of bacteria metabolize droplets of oil, much the way humans digest butter, so they can be used to help clean up an oil spill.

Bacteria break down their food molecules via a variety of metabolic pathways, some of which require oxygen, some of which do not. For example, many bacteria employ the anaerobic process known as fermentation (Chapter 6) to get energy from food. The products of fermentation can be valuable (and tasty) to humans. You may have seen "L. bulgaricus" listed as an ingredient of yogurt; live *Lactobacillus bulgaricus* bacteria are present and at work in the yogurt, fermenting sugars into lactic acid. Other bacteria use oxygen to break down organic molecules, like the aerobic bacteria that feast on an oil spill.

Highly resourceful, bacteria have a diverse range of living arrangements with other creatures. Many live in close association, or **symbiosis,** with other organisms–often to the benefit of one or both partners. Naturally occuring lactobacilli in the female vaginal tract, for example, obtain nourishment by fermenting naturally occurring sugars to lactic acid. The resulting acidity of the vaginal tract suppresses the growth of yeast, keeping women free of yeast infections. Antibiotics taken for a bacterial infection are likely to kill the resident lactobacilli as well as the invaders, and a yeast infection is often an unhappy side effect.

Another example of beneficial bacterial symbiosis is *Vibrio fischeri,* a bioluminescent bacterium that lives and feeds inside the light organs of certain species of squid. The glow-in-the-dark *Vibrio* creates light beneath the squid and helps obscure the shadow that the squid might cast on a moonlit night, making it less noticeable to its prey as it hunts.

BACTERIA
One of the two domains of prokaryotic life; the other is Archaea.

NITROGEN FIXATION
The process of converting atmospheric nitrogen into a form that plants can use to grow.

SYMBIOSIS
A relationship in which two different organisms live together, often interdependently.

INFOGRAPHIC 18.6

Exploring Bacterial Diversity

→ Bacteria live in every imaginable place on earth and have a diverse array of lifestyles. Within this domain, subgroups can be established based on shared evolutionary history, as reflected by genetic relationships.

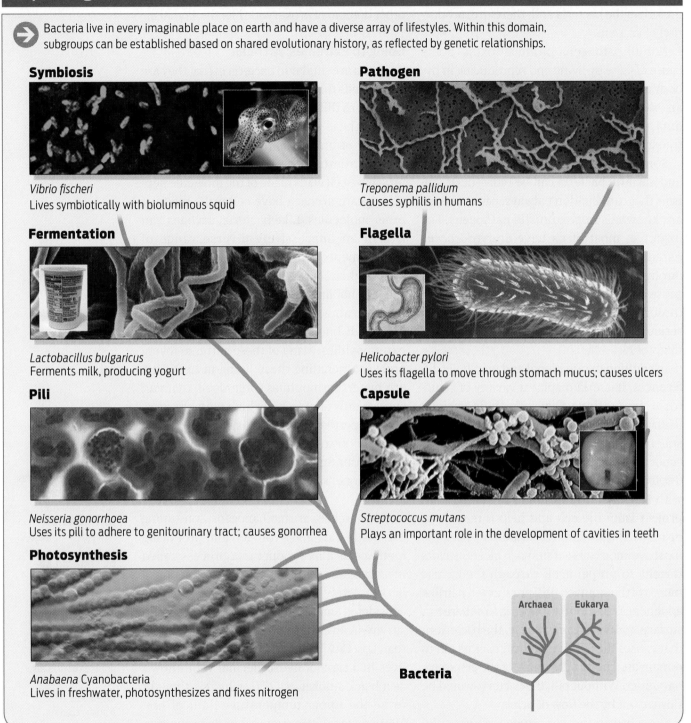

Symbiosis

Vibrio fischeri
Lives symbiotically with bioluminous squid

Fermentation

Lactobacillus bulgaricus
Ferments milk, producing yogurt

Pili

Neisseria gonorrhoea
Uses its pili to adhere to genitourinary tract; causes gonorrhea

Photosynthesis

Anabaena Cyanobacteria
Lives in freshwater, photosynthesizes and fixes nitrogen

Pathogen

Treponema pallidum
Causes syphilis in humans

Flagella

Helicobacter pylori
Uses its flagella to move through stomach mucus; causes ulcers

Capsule

Streptococcus mutans
Plays an important role in the development of cavities in teeth

Archaea   Eukarya

Bacteria

PATHOGEN
A disease-causing agent, usually an organism.

Unfortunately, not all bacteria are beneficial to the host. While the vast majority of bacteria do not cause human disease, some do. Bacteria and other organisms that cause disease are known as **pathogens.** Many pathogenic bacteria cause disease by producing toxins that harm their hosts. Such toxins can either be part of the bacterial cell itself or secreted by the bacterium. For example, the bacterium *Staphylococcus aureus* secretes a potent toxin that causes severe gastrointestinal discomfort in its host (as anyone who has had food poisoning knows). Keeping food refrigerated helps prevent food poisoning by slowing the growth of the bacteria and, therefore, production of its toxin.

Not all pathogens produce toxins. Some cause disease by living and reproducing in the body and intefering with its normal processes–an example is the bacterium *Treponema pallidum*, which causes syphilis, a sexually transmitted disease (STD).

Sometimes the line between harmless and harmful bacteria can be blurred. Organisms that can, but don't always, cause disease are known as opportunistic pathogens. For example, most of us have *Staphylococcus aureus* on our skin at many times during our lives. Most of the time, *S. aureus* does not cause any harm, but if it penetrates the skin–through a wound, for example–it can cause a serious infection and even death, as related in Chapter 14.

In addition to nutritional and metabolic differences, bacteria display a variety of structural adaptations that suit their various lifestyles. They come in different shapes: spherical (in which case they are known as "cocci"), rod-shaped ("bacilli"), and spiral ("spirochetes"). Many bacteria are equipped with **flagella,** tiny whiplike structures that project from the cell and help it move. For example, the bactrium *Helicobacter pylori,* the most common cause of stomach ulcers, uses its flagella to propel itself through the gastric mucus of the stomach. **Pili** are shorter, hairlike appendages that enable bacteria to adhere to a surface. *Neisseria gonorrhoeae,* the bacterium that causes the STD gonorrhea, uses its pili to remain attached to the lining of the genitourinary tract. Without pili, the bacteria would be flushed out by the flow of urine.

Other bacteria are surrounded by a **capsule,** a sticky outer layer that helps the cell adhere to surfaces and to avoid the defenses of the host. *Streptococcus mutans,* for example, produces a capsule that allows it to adhere to teeth, where it forms the plaque that can lead to cavities.

For all their impressive diversity and abundance, bacteria are far from the totality of pro-karyotic life. Moreover, their lifestyles and adaptations may seem tame next to those of the other domain of prokaryotic life: the Archaea.

Going to Extremes: Archaea

Archaea are similar to bacteria in that they are simple cells that lack a nucleus, but genetically they are as different from bacteria as humans are. All those genetic differences add up to a number of unique features that distinguish archaea from bacteria. For example, while bacteria have cell walls made of the molecule peptidoglycan, archaea have cell walls made of other molecules. Like bacteria, archaea can live in an impressively diverse range of environments.

Though archaea are found in many run-of-the-mill habitats such as rice paddies, forest soils, ocean waters, and lake sediments, the most well known species are the so-called extremophiles. Many of these extreme-loving archaea, including those living in Lost City, are hyperthermophiles–organisms that can survive only at extremely high temperatures. Many hyperthermophilic archaea are anaerobic and rely on sulfur instead of oxygen in their metabolism. Sulfur-rich hot springs like those in Yellowstone National Park are home to these archaea.

Other archaea are methanogens, consuming carbon dioxide and hydrogen and producing methane as a by-product in a process called methanogenesis. Because this gaseous meal is completely inorganic, these archaea are considered autotrophs. A methanogen that can survive in an even more extreme environment than Lost City is *Methanopyrus kandleri*, which lives in a type of hydrothermal vent known as a black smoker at 121°C, which is thought to be the upper temperature limit of life. Methanogens occupy more mundane environments as well, including the digestive systems of methane-belching cows.

Some archaea are halophiles, or "salt lovers," and prefer a home saturated in salt, which would shrivel most other living things. Their presence is detectable by the colorful pigments they produce–bright reds, yellows, and pur-

FLAGELLA (SINGULAR: FLAGELLUM)
Whiplike appendages extending from the surface of some bacteria, used in movement of the cell.

PILI (SINGULAR: PILUS)
Short, hairlike appendages extending from the surface of some bacteria, used to adhere to surfaces.

CAPSULE
A sticky coating surrounding some bacterial cells that adheres to surfaces.

ARCHAEA
One of the two domains of prokaryotic life; the other is Bacteria.

ples—as seen in salt ponds in San Francisco Bay (Infographic 18.7).

Archaea's affinity for such extreme environments is suggestive of their evolutionarily ancient roots. The early earth was a lot warmer than it is now, and it's long been a question how living things could withstand the heat that prevailed. If the archaea at Lost City are any indication, they would have done just fine.

In addition, some researchers suspect that methanogenesis may be among the most ancient forms of metabolism on earth. Scientists who study the origin of life believe that the first organisms were likely autotrophs that obtained their carbon from carbon dioxide, using hydrogen as an energy source and emitting methane as a by-product—in other words, doing exactly what certain archaea do today.

More recently, scientists studying Lost City have suggested that the archaea found there may be able to survive by consuming rather

INFOGRAPHIC 18.7

Exploring Archaeal Diversity

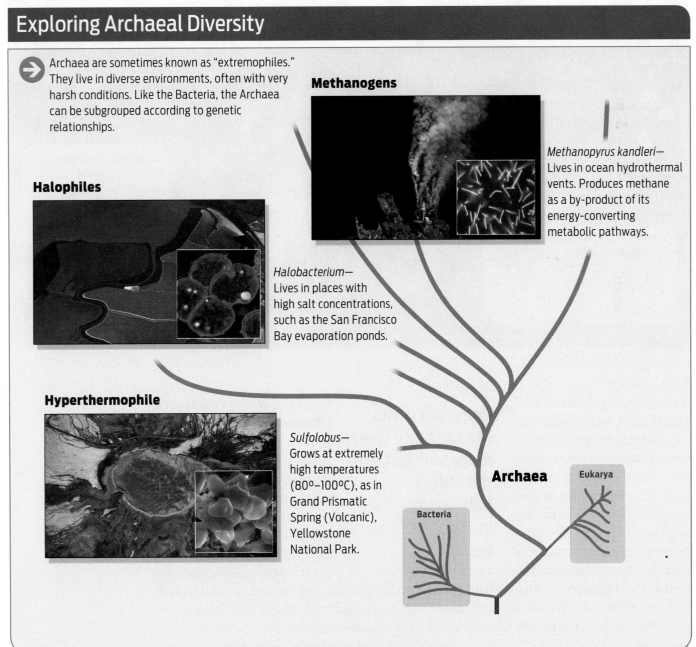

Archaea are sometimes known as "extremophiles." They live in diverse environments, often with very harsh conditions. Like the Bacteria, the Archaea can be subgrouped according to genetic relationships.

Methanogens

Methanopyrus kandleri— Lives in ocean hydrothermal vents. Produces methane as a by-product of its energy-converting metabolic pathways.

Halophiles

Halobacterium— Lives in places with high salt concentrations, such as the San Francisco Bay evaporation ponds.

Hyperthermophile

Sulfolobus— Grows at extremely high temperatures (80°–100°C), as in Grand Prismatic Spring (Volcanic), Yellowstone National Park.

Archaea

Eukarya

Bacteria

Energy from the Earth Fuels Life at Lost City

→ The environment deep on the ocean floor is anaerobic (that is, lacking in oxygen) and hostile to most life. But many prokaryotes thrive here, living off of energy and carbon molecules produced abiotically from chemical reactions occurring between rocks and seawater. These reactions do not involve sunlight, and may be similar to ones that sparked life on the early earth, when photosynthetic organisms did not yet exist.

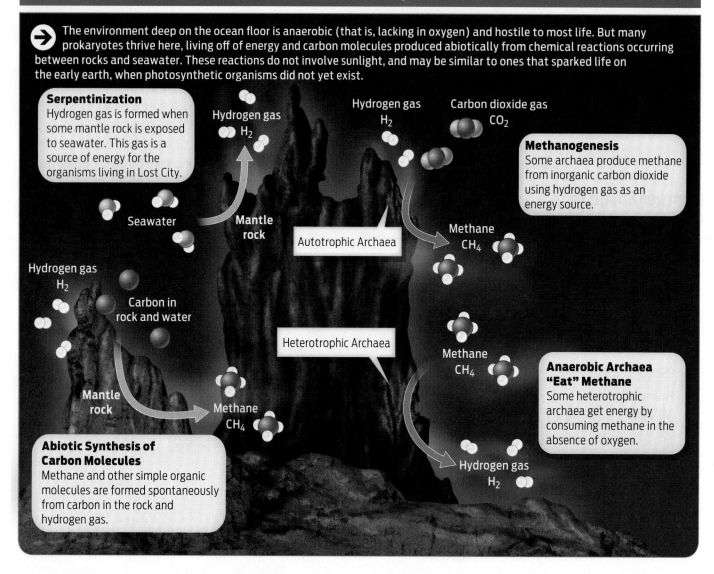

Serpentinization
Hydrogen gas is formed when some mantle rock is exposed to seawater. This gas is a source of energy for the organisms living in Lost City.

Methanogenesis
Some archaea produce methane from inorganic carbon dioxide using hydrogen gas as an energy source.

Abiotic Synthesis of Carbon Molecules
Methane and other simple organic molecules are formed spontaneously from carbon in the rock and hydrogen gas.

Anaerobic Archaea "Eat" Methane
Some heterotrophic archaea get energy by consuming methane in the absence of oxygen.

than producing methane. For a long time, it was thought that the only organisms capable of "eating" methane were bacteria that required oxygen in order to do so, much as we require oxygen to perform aerobic respiration (Chapter 6). Only recently have scientists learned that archaea can consume methane without oxygen. That's a handy trait to have when you live in an oxygen-free environment–which is exactly what the early earth was.

Deep-sea vents like Lost City likely represent some of the oldest habitats for microbial life on earth.

Lost City: Where Prokaryotic Life Began?

In its early days, more than 4 billion years ago, earth was mostly warm ocean, and its atmosphere lacked significant oxygen. Photosynthetic organisms did not exist, and thus there were no living things to harness the energy of sunlight to make organic molecules. Such conditions pose a conundrum for those who study the origin of life. With no photosynthetic organisms to make organic molecules, where did the energy and building blocks necessary to assemble living things come from? Lost City provides an important clue.

The hydrothermal vents at Lost City are driven by a geochemical process called serpentinization, which occurs any time a particular type of rock from the earth's mantle comes in contact with seawater. The reaction generates a lot of heat, releases hydrogen gas, and–when this hydrogen reacts with carbon from rocks or seawater–produces hydrocarbons such as methane and simple organic molecules. All this happens completely abiotically–that is, without the help of living things. Such an environment would have been an ideal place for life to begin, says Brazelton. "You have energy and organic compounds and liquid water all in a warm spot . . . that's a great place where you might imagine life could have got started."

"You have energy and organic compounds and liquid water all in a warm spot . . . that's a great place where you might imagine life could have got started." –Bill Brazelton

Lost City may also help to explain the chicken-and-egg problem posed by the origin of life: organic molecules are needed to build living things, but living things are generally the source of organic molecules. So which came first, the chicken or the egg? "I think a big clue to the chicken-and-egg problem is that you don't need life to make organic compounds," says Brazelton. "We are studying environments right now where the organic compounds are literally pouring out of these chimneys, and they're being made without the help of life" (Infographic 18.8).

Many microbiologists would agree that deep-sea vents like Lost City likely represent some of the oldest habitats for microbial life on earth. Radiometric dating of the rock layers indicates that Lost City's vents have been pumping strong for at least 100,000 years, and likely much longer: Lost City sits on a layer of earth's crust that is at least 1.5 million years old. And geological reactions like those at Lost City were likely more common and more widespread during earth's early history than they are today. Back then, rocks from the interior of the earth–the mantle–were much closer to the surface than they are now, which means their reaction with seawater would have been more common. A journey to Lost City is thus like a journey back in time, to earth's primordial past.

Lost City may even provide a clue to life beyond earth. The rocks that cause serpentinization are quite common in the solar system. They are all over the surface of Mars, for example, and researchers suspect that the chemical reaction might be occurring right now beneath the surface of Mars, where recent evidence suggests that methane is being produced. NASA is therefore extremely interested in Lost City as a way to understand potential life on Mars (Chapter 2).

Ironically, scientists know more about the surface of Mars than they do about the ocean floor of our own planet. Ocean covers 70% of earth's surface, yet much of it remains unexplored. If Lost City is any indication, many scientific treasures await the patient explorer. "[Lost City] is a good example of what we really don't know and what there is to still discover on the sea floor," says geologist Früh-Green.

If life on earth did begin at hydrothermal vents like those at Lost City, then it could mean that these extreme-loving prokaryotes are the descendants of the most ancient form of life on earth. For nearly 2 billion years, these earliest prokaryotic organisms reined supreme, with no challengers. Not until photosynthetic prokaryotes evolved, some 2.5 billion years ago, did they meet their match. Then, in an instant, geologically speaking, life on our planet underwent a radical and unprecedented change: 2 billion years ago, one of these early prokaryotes engulfed another and the two cells began a symbiotic relationship. That was the birth of the first eukaryote. ∎

▶ Summary

■ Prokaryotes are unicellular organisms that lack internal organelles and whose DNA is not contained in a nucleus.

■ Prokaryotes are found in virtually every environment on earth, even those with seemingly inhospitable conditions, such as hydrothermal vents on the ocean floor.

■ Genetic analysis has led to the categorization of life into three domains: Bacteria, Archaea, and Eukarya. Each domain of life has a distinct evolutionary history.

■ Both bacteria and archaea have prokaryotic cells, but they otherwise differ in their structure, biochemistry, and lifestyles.

■ Bacteria are a diverse group of prokaryotic organisms with many unique adaptations such as flagella and capsules that allow them to live and thrive in many environments.

■ Some bacteria are disease-causing pathogens, but most are harmless and even beneficial. Cyanobacteria, for example, are responsible for much of the photosynthesis that supports life on earth.

■ Often known as "extremophiles," many archaea live in some of the most inhospitable conditions on earth, such as hydrothermal vents. Many archaea flourish in less extreme environments.

■ The harsh conditions of Lost City may resemble the conditions of early earth. The prokaryotic inhabitants of Lost City may be metabolically similar to the earliest known life.

■ The energy that fuels life in Lost City comes from a geological source, rather than from sunlight, making it one of the few communities on earth that is not powered by photosynthesis.

DOMAINS OF LIFE

All living organisms can be grouped into three domains on the basis of genetic relatedness. Each domain has a distinct evolutionary history.

HINT See Infographics 18.3 and 18.5.

➲ KNOW IT

1. Organisms are placed into one or another of the three domains of life on the basis of
   a. cell type.
   b. physical appearance.
   c. evolutionary history assessed by genetic relatedness.
   d. ability to cause disease.
   e. degree of sophistication, that is, how evolutionarily advanced they are.

2. Describe the major difference(s) between prokaryotic and eukaryotic organisms.

3. The absence of membrane-bound organelles in a cell tells you that the cell must be
   a. from a member of the domain Bacteria.
   b. from a member of the domain Archaea.
   c. from a member of the domain Eukarya.
   d. either a or b.
   e. either b or c.

➲ USE IT

4. Why were the bacteria and archaea originally grouped together?

5. When first discovered, the archaea were called "archaeabacteria." Why do you suppose this was? What are the strengths and weaknesses of this earlier term?

BACTERIAL AND ARCHAEAL DIVERSITY

Both bacteria and archaea have prokaryotic cells but are distinguished by a number of genetic differences and unique adaptations.

HINT See Infographics 18.4, 18.6, and 18.7.

➲ KNOW IT

6. The term *prokaryotic* refers to
   a. a type of cell structure.
   b. a domain of life.
   c. a group with a shared evolutionary history.
   d. a type of bacteria.
   e. a type of archaea.

7. If you were looking for a bacterium, where would expect to find one?
   a. on your skin
   b. in the soil
   c. in the ocean
   d. associated with plants
   e. any of the above

8. What is the function of flagella?
   a. production of methane
   b. sticking to a surface
   c. motility
   d. luminescence
   e. metabolism

9. If you are unable to culture archaea from an environmental sample, is it safe to conclude that there are no archaea present? Why or why not?

➲ USE IT

10. Can you use cell structure to classify a cell as either bacterial or archaeal? Explain your answer.

11. Many prokaryotic organisms can carry out both photosynthesis and nitrogen fixation. Why are these processes important to humans?

12. If *Neisseria gonorrhoeae* had no pili, would it still be a successful pathogen? Explain your answer.

LIFE AT LOST CITY

The hydrothermal vents at Lost City are an extreme environment that may resemble the conditions of the early earth. By studying life at Lost City, we can test hypotheses about the earliest living organisms on earth.

HINT See Infographics 18.1, 18.2, and 18.8.

➲ KNOW IT

13. List the features that make Lost City a particularly harsh environment. For each feature, give a brief explanation of why that environment is inhospitable for many organisms.

14. If you were a prokaryotic organism and wanted to be successful at Lost City, what energy source must you be able to use?
   a. sunlight
   b. oxygen
   c. hydrogen gas
   d. electricity
   e. None of the above is available at Lost City.

USE IT

15. What is the significance of methane and other hydrocarbons at Lost City? (Think of both the origin of life and the sustenance of early life.)

16. If methane were not produced abiotically at Lost City, what would be the implications for early life?

17. Would you expect to find photosynthetic organisms at Lost City? Explain your answer.

SCIENCE AND ETHICS

18. Do you think that the scientists studying Lost City should be concerned about introducing microbial contaminants from their submersibles onto the towers of Lost City? How probable is this, given the conditions at Lost City and on the surface? If such an event could happen, what would be the implications?

19. Do you think that, instead of spending time studying microbes from extreme and remote environments, scientists should be studying microbes that are more apparently relevant to humans, such as ones that cause disease? In what ways might understanding the organisms at Lost City might be useful to humans?

Rain Forest Riches

→ What You Will Be Learning

Rain Forest Riches

Exploring eukaryotic diversity in Olympic National Park

On a chilly January afternoon in 2008, a crowd of eager onlookers gathered at a snowy campground in Washington State to watch natural history being made. The stars of the show were a trio of rarely seen animals held inside small wooden crates. As photographers craned for a good look and schoolchildren held their breath, the door to the first crate was opened. With a flash of whiskers and brown fur, a weasel-like animal bolted from the box and made a break for the forest. The fisher was finally home.

"There were lots of oooohs and aaaahhs and clapping and cheering," recalls Jeffrey Lewis of the Washington Department of Fish and Wildlife, who helped coordinate the

The release of these fishers, imported from British Columbia, Canada, are part of a coordinated effort to bring the species back to Washington State.

EUKARYOTE
Any organism of the domain Eukarya; eukaryotic cells are characterized by the presence of a membrane-enclosed nucleus and organelles.

release of the fishers into the Washington forest that day.

It was a long-awaited homecoming. Once plentiful in the state, fishers have been hunted and trapped nearly to extinction; they have not been seen in Washington since the early twentieth century. The animal was named a Washington State endangered species in 1998, after a careful investigation failed to find any evidence of a local population. The three animals released that afternoon—two females and one male—were imported from British Columbia, Canada, and are part of a coordinated effort to bring fishers back to Washington.

Olympic National Park is a microcosm of the planet's eukaryotic diversity.

They are the first of 90 fishers scheduled to be released over 3 years into their new home: Olympic National Park.

A nearly 1-million-acre plot of wilderness occupying the northwest corner of the state, Olympic National Park is attractive to more than just fishers. It is home to a mind-boggling array of different species, in numbers uncommon in most other parks. "It's got amazing biological diversity in a very compressed area," says Patti Happe, who is Wildlife Branch Chief at the park and helped orchestrate the fisher release. Among this biodiversity are a seemingly endless variety of **eukaryotes**—the

Tree of Life: Domain Eukarya

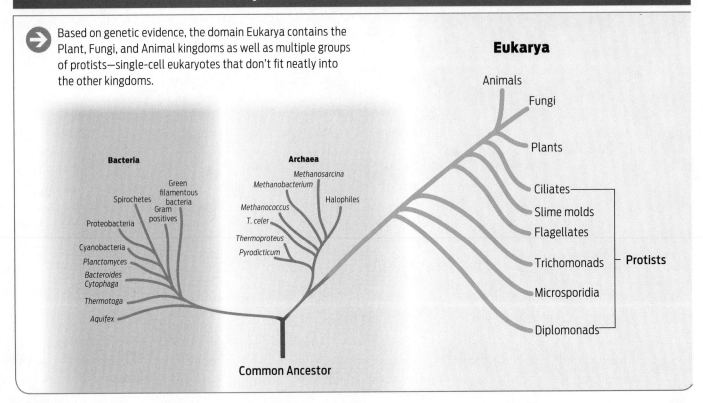

→ Based on genetic evidence, the domain Eukarya contains the Plant, Fungi, and Animal kingdoms as well as multiple groups of protists—single-cell eukaryotes that don't fit neatly into the other kingdoms.

Eukarya

- Animals
- Fungi
- Plants
- Ciliates ⎤
- Slime molds │
- Flagellates │
- Trichomonads ⎬ **Protists**
- Microsporidia │
- Diplomonads ⎦

Bacteria

- Green filamentous bacteria
- Spirochetes
- Gram positives
- Proteobacteria
- Cyanobacteria
- *Planctomyces*
- *Bacteroides Cytophaga*
- *Thermotoga*
- *Aquifex*

Archaea

- *Methanosarcina*
- *Methanobacterium*
- Halophiles
- *Methanococcus*
- *T. celer*
- *Thermoproteus*
- *Pyrodicticum*

Common Ancestor

plants, animals, fungi, and unicellular protists making up one of the three main branches of life **(Infographic 19.1).**

Visitors to the park can find some of the world's oldest and tallest Douglas fir and Sitka spruce trees, the country's largest herd of Roosevelt elk, and a number of eukaryotic species found here and nowhere else, including the Olympic marmot, the Olympic pocket gopher, and the Olympic torrent salamander. A designated UNESCO Biosphere Reserve and a World Heritage Site, Olympic National Park is a microcosm of the planet's eukaryotic diversity, and biologists are eager to protect it.

"National parks are here to serve as our national treasures, where we preserve and protect both our cultural and our natural heritage for future generations," says Happe. "Really what we're doing is wise stewardship of our nation's resources."

A Green World

Olympic's unique collection of wildlife reflects the geological history and ecology of the region. During the last ice age, approximately 20,000 years ago, the Olympic Peninsula was isolated by glaciers and largely separated from the rest of what is now the United States. Today, it is surrounded by saltwater on three sides, and is essentially an ecological island, distinctive in its geography and topography.

Far from being a single landscape, however, Olympic National Park is more like three parks in one: a glacier-topped mountain region with flowering subalpine meadows, valleys of temperate rain forest coursing with freshwater rivers and lakes, and nearly 60 miles of jagged Pacific coastline. Central to the park's ecology is the towering presence of Mount Olympus, a giant landmass that traps warm air blowing in from the Pacific, making the western slope one of the wettest spots in the United States: it's doused in 12 feet of rain each year. Taken all together, the park is a mosaic of physical, geographic, and climatic conditions providing numerous habitats for the many species of eukaryotes that live here **(Infographic 19.2).**

PLANT
A multicellular eukaryote that has cell walls, carries out photosynthesis, and is adapted to living on land.

BRYOPHYTE
A nonvascular plant that does not produce seeds.

Each segment of the park is marked by its distinct form of vegetation, or plant life. Within the low-elevation rain forests, for example, stands of giant Sitka spruce and Douglas fir trees form a thick canopy of green, steadily dripping moisture. In dense forest understory, 300 feet below the canopy, plants form a junglelike tangle, growing on and in other plants: nearly every log and tree trunk is coated with a shaggy carpet of mosses, ferns, and lichens (a partnership between a fungus and a photosynthetic organism), while hanging plants drape branches like luxurious scarves with long ground-reaching stems.

In all this variety, what exactly defines a plant, scientifically speaking? At the most basic level, a **plant** is a multicellular eukaryote that possesses cells with cell walls and carries out photosynthesis. Land plants such as those found in Olympic first evolved from water-dwelling algae about 450 million years ago, when life on earth was confined primarily to the seas. As plants radiated and diversified on land, they evolved a number of adaptations that made them increasingly independent of water.

The earliest plants to make the transition from water to land were small, seedless plants called **bryophytes.** Bryophytes lack roots and tissue for transporting water and nutrients throughout their bodies, and therefore can grow only in damp environments, where they

INFOGRAPHIC 19.2

The Landscape of Olympic National Park

 Olympic National Park has an enormous diversity of physical, geological, and climatic conditions, all of which contribute to the huge biological diversity in the park.

Old Growth Forest
The Olympic National Park temperate rain forest has tremendous plant diversity, providing habitat and resources for other living organisms.

Pacific Coastline
The park includes saltwater beaches.

Olympic National Park

WASHINGTON

Olympic Mountains
Mount Olympus has a major influence on local climate.

Hart Lake
A variety of lakes provides a type of aquatic habitat for living organisms.

Elwha River
Rivers running through the park provide a specific type of aquatic habitat for many species.

Evolution of Plant Diversity

All plants have evolved from an ancient protist ancestor. Different groups of plants have developed different specializations to allow them to be successful on land.

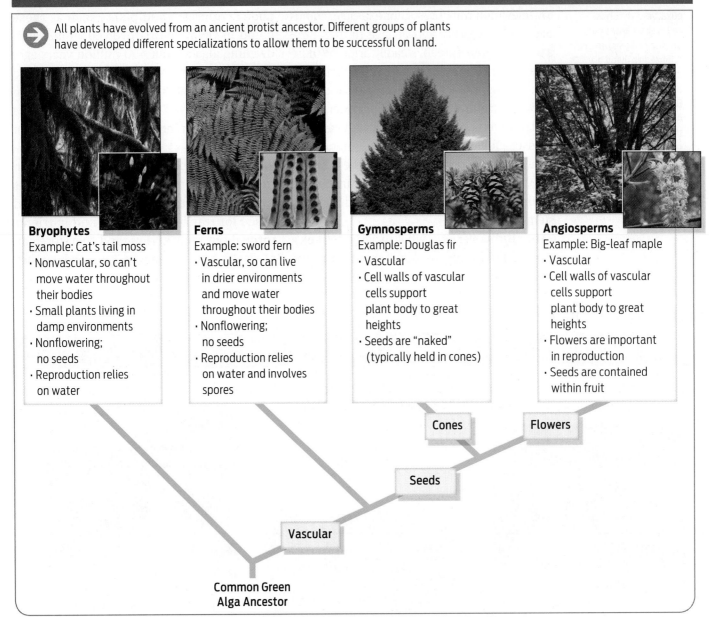

Bryophytes
Example: Cat's tail moss
- Nonvascular, so can't move water throughout their bodies
- Small plants living in damp environments
- Nonflowering; no seeds
- Reproduction relies on water

Ferns
Example: sword fern
- Vascular, so can live in drier environments and move water throughout their bodies
- Nonflowering; no seeds
- Reproduction relies on water and involves spores

Gymnosperms
Example: Douglas fir
- Vascular
- Cell walls of vascular cells support plant body to great heights
- Seeds are "naked" (typically held in cones)

Angiosperms
Example: Big-leaf maple
- Vascular
- Cell walls of vascular cells support plant body to great heights
- Flowers are important in reproduction
- Seeds are contained within fruit

Cones

Flowers

Seeds

Vascular

Common Green
Alga Ancestor

can easily absorb water. One of the wettest places on earth, the Olympic rain forest is a soggy paradise for bryophytes, such as mosses and liverworts, which appear as squat, spongy mats.

The rain forest is also home to many **vascular plants**—those with specialized tissues for transporting nutrients and water through the plant body. The first true vascular plants were **ferns,** such as the hip-high sword ferns that cover a good portion of the forest. Like bryo-

phytes, ferns do not produce seeds. Yet unlike those vertically challenged relatives, ferns can stand upright and grow tall, thanks to the vascular tissue that keeps stems rigid and transports water and nutrients from one end of the plant to the other. At one time, ferns ruled the plant world, spreading their massive fronds across the entire landscape in the Carboniferous period. But their reign was short lived. Soon, another kind of plant evolved to challenge the ferns' dominance—those with seeds.

VASCULAR PLANT
A plant with tissues that transport water and nutrients through the plant body.

FERN
The first true vascular plants; ferns do not produce seeds.

GYMNOSPERM
A seed-bearing plant with "naked" seeds typically held in cones.

ANGIOSPERM
A seed-bearing flowering plant with seeds typically contained within a fruit.

Seed plants first emerged about 360 million years ago, during the late Devonian period. A seed, which envelopes a plant's embryo, is an ideal package for withstanding harsh conditions and traveling to a location where it can grow into a new plant. Seed plants were so successful that they quickly came to dominate forests by the time of the dinosaurs in the Mesozoic era.

Today, more than 90% of all living plants are seed plants. Plants with largely exposed seeds, as in a pinecone, are known as **gymnosperms**–spruce, pine, redwood, fir, and other conifers, for example. ("Gymnos" is Greek for "naked," so the name literally means "naked seeds.") **Angiosperms** are flowering plants with seeds contained in a fruit–an apple, say, or an acorn ("angio" is from the Greek for "vessel" or "container"). Olympic National Park is home to many species of angiosperms, including oaks, maples, huckleberry bushes, and willows, as well as hundreds of species of flowers (**Infographic 19.3**).

One species of angiosperm in the park, *Toxicodendron diversilobum,* is quite versatile, growing as a vine stretching up the sides of trees or as bushes hugging trails near water. It looks attractive, with shiny green leaves arranged in threes, white flowers, and white berries, but it secretes a powerful toxin that causes a painful rash in unsuspecting admirers who get too close. Its popular name is poison oak.

A Forest for Fishers (and More)

With its dense population of trees more than 200 years old, Olympic National Park is an especially good home for the tree-loving fisher (*Martes pennanti*). "Because the park is 95% wilderness area, little of it has been logged, and it contains great expanses of older forest, which provide the large trees, snags, and logs that fishers need," says Lewis. Fishers rest in nooks within the trees, and females use tree cavities as dens in which to birth and nurse their kits. The somewhat shy fishers are also attracted to places with dense canopy cover, woody debris, and understory vegetation, all of which provide plentiful hiding places. Many of the tree species found here make for prime fisher habitat, including western Hemlock, Sitka spruce, and Pacific silver fir, which also provide a reliable food source for seed- and insect-eating mammals that fishers stalk as prey, such as squirrels, mice, and shrews.

A Pacific fisher (*Martes pennanti*).

The fisher is an animal, of course, but what defines an animal? Scientifically, an **animal** is a multicellular eukaryotic heterotroph that obtains nutrients by ingestion—that is, by eating. When we humans think of "animal," we tend to picture mammals, such as the fur-covered fisher. But the definition applies to a great many kinds of creatures, ranging from sponges to worms to insects to humans. To help bring some order to this diversity, biologists sort animals into smaller groups on the basis of shared characteristics and ancestry.

Many features can be used to group and sort animals. Historically, anatomical and embryo-logical evidence were relied upon most, but in recent years it has become more common to use DNA. From DNA evidence, it is clear that all animals descended from the same common ancestor that lived nearly 1 billion years ago and subsequently diversified into the different forms we see today **(Infographic 19.4)**.

Early in their history, animals branched into three main lineages, the legacy of which can be seen in three distinct animal body plans in existence today. The simplest living animals, such as sponges, lack defined tissues or organs and have an amorphous shape. These asymmetrical organisms are likely similar to the earliest ani-

ANIMAL
A eukaryotic, multicellular, organism that obtains nutrients by ingesting other organisms.

INFOGRAPHIC 19.4

Evolution of Animal Diversity

All animals have descended from an ancestral protist. Many features can be used to classify animals, including body symmetry, type of body support, and the presence or absence of a spinal cord and backbone. In addition, genetic sequencing and the study of embryonic development have informed this phylogenetic tree.

The arthropods are the most successful animal group.

The chordates are the only group that includes vertebrates.

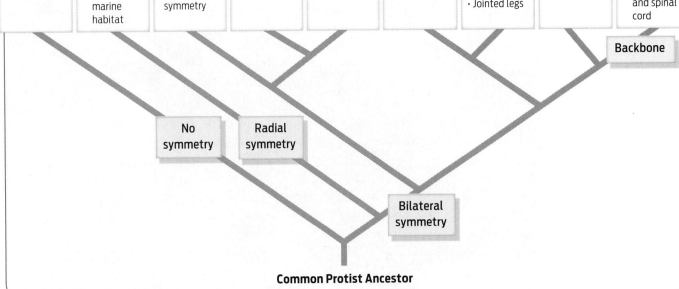

Sponges:
E.g., sea sponge
· No organized tissues
· No symmetry

Cnidarians:
E.g., jellyfish
· Radial symmetry
· Aquatic and marine habitat

Flatworms:
E.g., flatworm
· Simplest animal with bilateral symmetry

Molluscs:
E.g., clam
· Soft body
· Single, hard outer shell

Annelids:
E.g., earthworm
· Long, segmented body

Nematodes:
E.g., roundworm
· Long, unsegmented body

Arthropods:
E.g., insect
· Exoskeleton
· Segmented body
· Jointed legs

Echinoderms:
E.g., starfish
· Endoskeleton
· Spiny outer skin

Chordates:
E.g., dog
· Vertebrates have a backbone and spinal cord

Backbone

No symmetry

Radial symmetry

Bilateral symmetry

Common Protist Ancestor

384 **UNIT 3:** HOW DOES LIFE CHANGE OVER TIME? EVOLUTION AND DIVERSITY

mals to have populated the oceans. All other animals have defined tissues and fall into one of two broad categories based on the type of body symmetry they possess.

Animals such as jellyfish and coral exhibit **radial symmetry**, meaning that they're shaped like a pizza–circular, with no clear left and right sides. All other animals–everything from worms and insects to fishers and humans–exhibit **bilateral symmetry:** if you cut them down the middle you'd produce clear left and right halves that are mirror images of each other.

Bilateral symmetry has become as prevalent as it is in the animal kingdom because it is a useful adaptation for seeking out food, stalking prey, and avoiding predators. For instance, bilaterally symmetrical animals have eyes on both sides of their face, enabling them to look straight ahead. In the fisher's case, such bilateral symmetry aids in its ability to climb down trees head first in search of prey.

Fishers are **vertebrates,** meaning they are animals with a backbone. A fisher's backbone is made of bony vertebrae, a feature they share with most other vertebrates. (A few vertebrates, primarily sharks and several other fish, have backbones made of cartilage.) While vertebrates are some of the most easily recognized animals–including humans–they represent but a sliver of the total animal world (note they are found on only one branch of the animal tree, the chordates). In fact, most animals lack a backbone and are therefore called **invertebrates.** While invertebrates are often lumped together on the basis of what they lack, the division of the animal world into those with and those without backbones makes about as much sense as dividing the world into sponges and nonsponges–it camouflages a lot of differences, and obscures the fact that most animals–an astounding 95%–are invertebrates.

Olympic National Park hosts a squirming, wriggling, buzzing swarm of invertebrates. If you were hiking or camping in the park, you would easily encounter invertebrates from several major phyla–some more welcome than others, perhaps.

The banana slug, one of the forest's most voracious inhabitants, is distasteful to predators.

Sliding quietly amid leaf litter on forest trails are many specimens of the Pacific coast's best known **mollusk,** the brightly colored banana slug (perhaps the only mollusk to serve as a university mascot, as it does for the University of California at Santa Cruz). A squishy yellow creature that can grow nearly to the size of its namesake fruit, the banana slug is one of the forest's most voracious inhabitants, eating its way through just about everything in its path, from animal carcasses and droppings to mushrooms, lichens, and leaves–including those of poison oak. This mollusk is so successful in the forest in part because it is distasteful to would-be predators, who know by its unmistakable color to avoid eating it.

Slugs–and their shelled cousins, the snails–are often considered garden pests. Yet by digesting dead plant material, these mollusks help recycle nutrients. And with their calcium-rich shells, snails provide this valuable mineral to the creatures that feast on them, such as rodents and birds. Some humans find mollusks a tasty treat as well: if you have enjoyed clams, oysters, or squid, then you have eaten some aquatic varieties of mollusk.

Move a rock while you're pitching your tent or digging a hole in the ground and you'll likely uncover numerous squirmy **annelids,** or segmented worms. Annelids such as earthworms perform a critical ecological service by creating passageways in the soil as they move around. The passageways allow air and water to enter the soil, which is important for plants

RADIAL SYMMETRY
The pattern exhibited by a body plan that is circular, with no clear left and right sides.

BILATERAL SYMMETRY
The pattern exhibited by a body plan with clear right and left halves that are mirror images of each other.

VERTEBRATE
An animal with a bony or cartilaginous backbone.

INVERTEBRATE
An animal lacking a backbone.

MOLLUSK
A soft-bodied invertebrate, generally with a hard shell (which may be tiny, internal, or absent in some mollusks).

ANNELID
A segmented worm, such as an earthworm.

and other aerobic organisms that require water and oxygen. By eating and digesting leaf and other plant litter, earthworms also make nutrients available for other plants.

When you look at a dragonfly or a water strider, you are observing a member of the park's enormous collection of **arthropods.** Arthropods are the most abundant invertebrates in the park, and on earth in general. There are an estimated 2-4 million species of arthropods, of which 855,000 have been officially described so far. The number of individual arthropods on the planet is estimated to be over 10^{18} (that's 1 with 18 zeros after it). They include animals as diverse as water-dwelling crustaceans like crabs and lobsters to terrestrial spiders, millipedes, and flying insects.

Despite their abundance and diversity, all arthropods share some common physical characteristics. They have segmented bodies with jointed appendages such as legs, antennae or pincers, and a hard **exoskeleton,** or external skeleton, made up of proteins and chitin (a type of polysaccharide). An arthropod's exoskeleton serves multiple functions: it protects the organism from predators, keeps it from drying out, and affords structure and support for movement, just as our internal **endoskeleton** does.

Most arthropods are harmless or even helpful to humans, but some are not. A few produce powerful venoms that can be deadly when they are conveyed to victims through bites or stings. When camping, it's always a good idea to check your shoes before putting them on; a few shakes will dislodge any scorpions or black widow spiders that may have crawled in.

The vast majority of all arthropods are **insects**–arthropods with three pairs of jointed legs and a three-part body consisting of head, thorax, and abdomen. Insects include animals such as the honey bees and butterflies that pollinate flowers, the termites and cockroaches that infest our walls, and those blood-sucking insects hated by campers everywhere: mosqui-toes. Insects are evolution's great success story. Having first evolved some 400 million years ago, there are now more insect species on the planet than all other animals species combined.

Insect bodies boast an array of useful adaptations, including three-pronged mouthparts that are used variously for biting, chewing, or sucking; but what really sent insect diversity soaring was the evolution of wings. Wings enable insects to fly away from predators, access distant food sources, and travel to find mates. Among the most successful of all flying insects are beetles, which have two sets of wings and mouthparts specialized for biting, mincing, and chewing. Taxonomists have catalogued approximately 350,000 beetle species so far, and some estimates put the total number of species in the millions.

But even insects with six feet firmly planted on the ground can be remarkably successful– just look at the ants. They can't fly, but they can communicate, split up tasks, solve problems, and shape their local environment–your picnic, for example–for their own needs. Given their complex social behavior and adaptations, it's not surprising to find ants nearly everywhere on the planet; Antarctica, Greenland, Iceland, and Hawaii are some of the few places on earth believed to harbor no native ant species.

A Variety of Vertebrates

Amid the enormous crowd of invertebrates in the park live plenty of animals *with* a backbone, the vertebrates. Fishers fall into a class of vertebrates called **mammals,** animals with mammary glands and a body covered with fur. Like many mammals, fishers are predators–hunting mostly other small and midsize mammals such as snowshoe hares, squirrels, mice, and beaver. With their keen sense of sight and smell, sharp teeth, and non-retractable claws, these nocturnal creatures are among the most effective predators on the ground and in trees.

Fishers' biggest foe are humans, especially fur traders who trap and kill the animals for their soft pelts.

ARTHROPOD
An invertebrate having a segmented body, a hard exoskeleton, and jointed appendages.

EXOSKELETON
A hard external skeleton covering the body of many animals, such as arthropods.

ENDOSKELETON
A solid internal skeleton found in many animals, including humans.

INSECT
A six-legged arthropod with three body segments: head, thorax, and abdomen.

MAMMAL
An animal having mammary glands and a body covered with fur.

Fishers themselves have few natural predators. They are occasionally killed by cougars, coyotes, and eagles, but by far their biggest foe are humans, especially fur traders who trap and kill the animals for their soft pelts. Fishers were also an unintended target of a massive predator control program in Washington State that sought to reduce the number of wolves and cougars in state forests. The unsuspecting fishers ate the poison bait and perished.

Because humans are largely responsible for the decline of the fisher in Washington, many conservationists believe it is our duty to help undo that damage. There is an ecological, as well as a moral, rationale for such action. "When you start getting up into the mammalian species, going up the phylogenetic tree," says wildlife manager Happe, "there's fewer species, but one animal has a big effect on the ecosystem." That is especially true of predators, such as the fisher, that act as a natural population control on other species.

For example, in some parts of their range, fishers are one of the few natural predators of porcupines. A healthy fisher presence in a forest helps keep the porcupine population under control and prevents what would quickly become a prickly problem for the trees that the porcupines eat and the loggers who cut them for lumber. Restoring fishers to the ecosystem is therefore necessary to keep the natural web of the environment intact. Exactly what effect the reintroduced fishers are having in Olympic, researchers can't say for certain. "I'm sure that it's having repercussions in the ecosystem, it's just that we can't interview the squirrels and the rabbits to find out what they think of all this," says Happe.

Besides the fisher, Olympic is home to many other vertebrates, including other mammals–cougar, black-tailed deer, mountain goat, black bear, river otter, and Douglas squirrel, for example, as well as hundreds of species of fish, amphibians, birds, and reptiles. These various vertebrates are easily recognized by the unique adaptations

Olympic National Park is home to many vertebrates, including mammals such as river otter and cougar, as well as hundreds of species of fish, reptiles, birds, and amphibians.

that allow them to survive and flourish in their particular habitats within the park. Fish can live in the ponds and rivers because of their scales, fins, and gills; amphibians metamorphose from a water-dwelling juvenile form to an air-breathing adult; birds have hollow bones and feathers that enable many of them to fly; and reptiles have a body covered in water-tight scales that equip them for life on dry land.

Cycles of Life in the Rain Forest

The rain forest is a place of irrepressible life; it is also one of death. Coyotes kill fishers. Fishers hunt squirrels. Insects eat trees. Death casts a long shadow over life in the park, yet without it there would be no life at all.

When organisms are alive, they store nutrients and chemical building blocks in the fabric of their bodies. When the organisms die and decompose, these nutrients and building blocks are returned to the earth and eventually taken up into new life. Crucial to this cycle of life and growth, death and decomposition, are **fungi,** a third major branch on the eukaryotic tree.

By breaking down organic matter into smaller particles, fungi help release trapped nutrients. Without fungi, dead trees and animal carcasses would pile up in the forest and smother everything in it. Thanks to the action of fungi, however, the organisms decompose and the elements they contained will nourish many organisms throughout the web of life. Many decomposing organisms even provide shelter, such as the tree holes that fishers and other animals use as dens.

Fungi come in many forms. There are unicellular species, such as molds and yeasts, and multicellular species, such as mushrooms and the shelf fungus you sometimes see growing on a tree trunk. Underlying this physical diversity is a method of obtaining nutrients common to all fungi: they secrete digestive enzymes onto their food source and then absorb the digested products. As one of nature's **decomposers,** fungi can break down just about anything that has organic components, including plant parts that are indigestible to many bacterial decomposers. All fungi also have cell walls made of chitin (the same molecule that makes up the exoskeleton of the arthropods), and all modern fungi evolved from a common unicellular ancestor—the same one that gave rise to animals—approximately 1 billion years ago.

Multicellular fungi, such as the mushrooms poking up through leaves in the forest, have a body composed of threadlike structures known as **hyphae.** Each individual hypha is a chain of many cells, capable of absorbing nutrients. Fungal hyphae interweave to form a spreading mass known as a **mycelium.** The mushrooms you see on the forest floor are merely one aboveground part of what can be a huge, underground fungal mycelium; the mold on a slice of bread, by contrast, has a mycelium that is all right there for you to see.

There are fungi everywhere in Olympic National Park, growing on, in, and under the abundant vegetation. Some species play an especially critical role in the soil, where they form a symbiotic relationship with the roots of many trees. Their slender hyphae grow into microscopic spaces in the soil where the tree's roots can't fit, greatly enhancing a root's ability to absorb water and nutrients. In return, trees supply nutrients to the fungi, which do not photosynthesize. Many fungi also live in and on animals, as you probably know if you have ever had athlete's foot or a yeast infection **(Infographic 19.5).**

A Microcosm in a Drop of Water

The remaining eukaryotes in Olympic National Park are the ones you rarely see—swimming in drops of water or hiding in puddles under the leaf cover. Informally known as **protists,** these varied members of what used to be called kingdom Protista do not fit neatly into one group and are tricky to classify. They comprise several side branches on the eukaryotic tree shown in Infographic 19.3.

Most protists are unicellular, but there are also multicellular varieties, such as some types of **algae.** Multicellular algae share with plants the ability to photosynthesize, but they differ from plants in lacking specialized adaptations for living on land, such as roots, stems, and leaves. Other protists are similar to animals in that they are heterotrophic, eating other

FUNGUS (PLURAL: FUNGI)
A unicellular or multicellular eukaryotic organism that obtains nutrients by secreting digestive enzymes onto organic matter and absorbing the digested product.

DECOMPOSER
An organism such as a fungus or bacterium that digests and uses the organic molecules in dead organisms as sources of nutrients and energy.

HYPHA (PLURAL: HYPHAE)
A long, threadlike structure through which fungi absorb nutrients.

MYCELIUM (PLURAL: MYCELIA)
A spreading mass of interwoven hyphae that forms the often subterranean body of multicellular fungi.

PROTIST
A eukaryote that cannot be classified as a plant, animal, or fungus; usually unicellular.

Fungi, the Decomposers

Fungi are a diverse group of organisms with a variety of reproductive strategies. However, all fungi share the way that they obtain their nutrition: they all secrete digestive enzymes onto their food, then absorb the digested products.

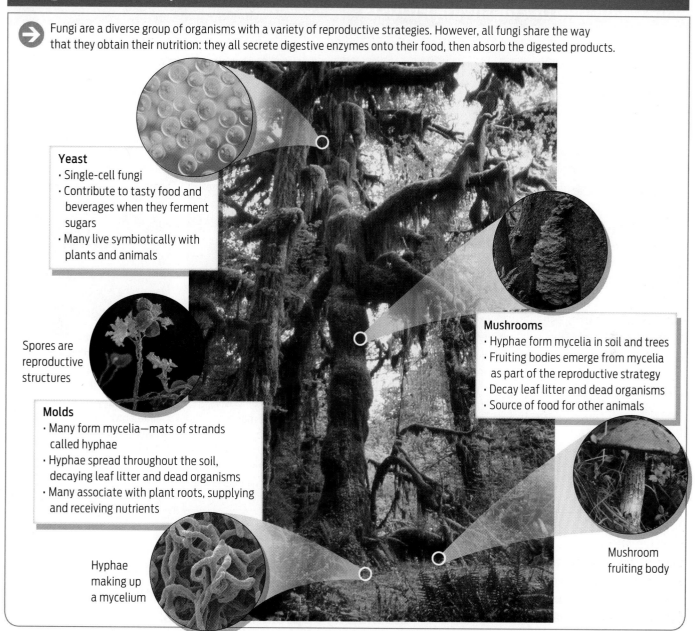

Yeast
· Single-cell fungi
· Contribute to tasty food and beverages when they ferment sugars
· Many live symbiotically with plants and animals

Spores are reproductive structures

Molds
· Many form mycelia—mats of strands called hyphae
· Hyphae spread throughout the soil, decaying leaf litter and dead organisms
· Many associate with plant roots, supplying and receiving nutrients

Hyphae making up a mycelium

Mushrooms
· Hyphae form mycelia in soil and trees
· Fruiting bodies emerge from mycelia as part of the reproductive strategy
· Decay leaf litter and dead organisms
· Source of food for other animals

Mushroom fruiting body

organisms, but since they are unicellular they are not technically animals. Some protist species have long filamentous bodies resembling fungi, but they are no more related to fungi than animals are. In fact, genetic evidence shows that protists do not form a cohesive evolutionary group; some may be as distinct from each other as plants are from animals. Still a work in progress, our understanding of these diverse organisms is likely to evolve in coming years as we continue to learn more about them (Infographic 19.6).

ALGA (PLURAL: ALGAE)
A uni- or multicellular photosynthetic protist.

Despite their diversity, protists do share some common traits. They are all susceptible to drying out, so they are typically found in wet environments: lakes, oceans, ponds, moist soils, and living hosts. Many disease-causing protists, for example, must spread directly from host to host because otherwise they would dry out. This explains why trichomoniasis, a sexually transmitted infection caused by the protist *Trichomonas vaginalis*, can be spread only by direct sexual contact.

Other protists live in the gastrointestinal systems of animals such as beavers and can be found

The Challenge of Classifying Protists

Protists are a diverse group of organisms that are difficult to classify. They share features with animals, plants, and fungi, but are not classified as any one of these. Nor do they have a single unifying characteristic that places them within a single evolutionary group. Most protists are unicellular.

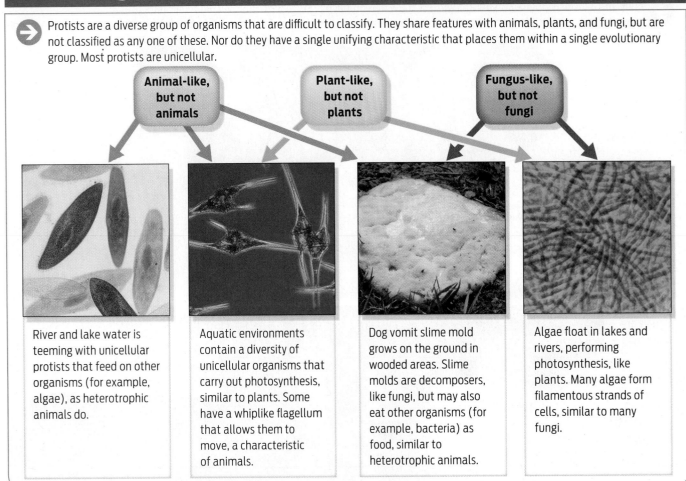

Animal-like, but not animals

Plant-like, but not plants

Fungus-like, but not fungi

River and lake water is teeming with unicellular protists that feed on other organisms (for example, algae), as heterotrophic animals do.

Aquatic environments contain a diversity of unicellular organisms that carry out photosynthesis, similar to plants. Some have a whiplike flagellum that allows them to move, a characteristic of animals.

Dog vomit slime mold grows on the ground in wooded areas. Slime molds are decomposers, like fungi, but may also eat other organisms (for example, bacteria) as food, similar to heterotrophic animals.

Algae float in lakes and rivers, performing photosynthesis, like plants. Many algae form filamentous strands of cells, similar to many fungi.

---

in pond water where those animals defecate. Unwary campers who drink from a pond may find themselves stricken with an unpleasant diarrheal disease called giardiasis (aka "beaver fever") caused by the protist *Giardia lamblia*.

However small and difficult they are to classify, protists can rightfully claim an important position in the eukaryotic family tree. According to the theory of **endosymbiosis** put forward by biologist Lynn Margulis (Chapter 3), it was a single-cell protist that gave rise, some 2 billion years ago, to the ancestor of all living eukaryotes, from fungi to flowers to fishers (Infographic 19.7).

Protecting Diversity

As richly diverse as the Olympic forest is, it is not as rich as it once was. The fisher is but one

Between the 1930s and the early 1990s, the total area of old-growth forest in Washington State was slashed by approximately 70%.

species whose existence in Washington has been threatened by human actions. Between the 1930s and the early 1990s, the total area of old-growth forest in Washington State was slashed by approximately 70%–down to 3 million acres from more than 9 million–most of it used for lumber. Much of the remaining habitat is fragmented by highways, power lines, railroads, and residential development, leaving no place for many species to call home.

The challenges facing Washington are not unique. The United Nations Food and Agricultural Organization estimates that the total global

ENDOSYMBIOSIS
The theory that free-living prokaryotic cells engulfed other free-living prokaryotic cells billions of years ago, forming eukaryotic organelles such as mitochondria and chloroplasts.

The First Eukaryotes Were Products of Endosymbiosis

➡️ All eukaryotes are characterized by the presence of a membrane-enclosed nucleus and organelles. While the origin of the nucleus is not yet completely understood, there is good evidence that mitochondria and chloroplasts arose by the engulfment of ancient prokaryotes by the earliest eukaryotes.

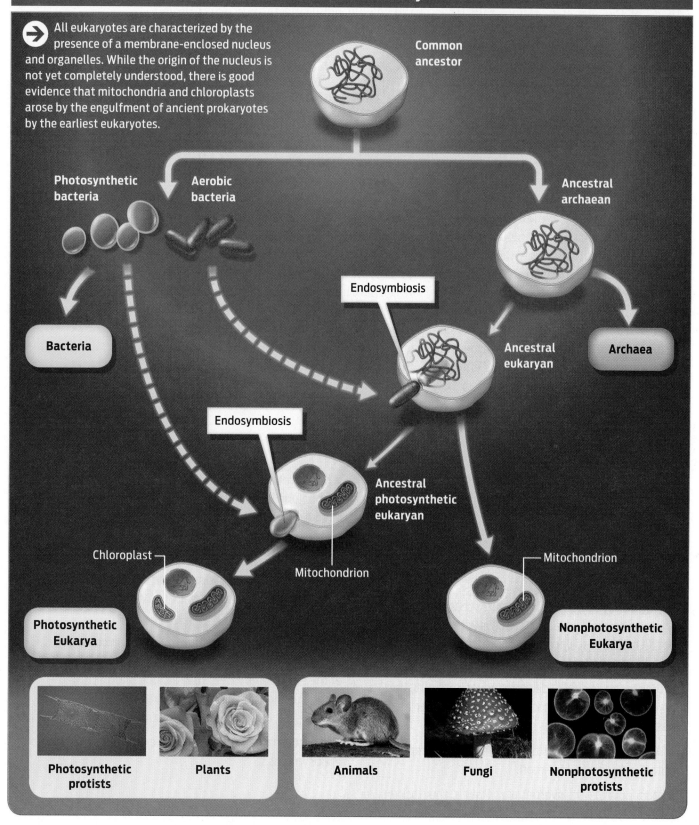

Common ancestor

Photosynthetic bacteria

Aerobic bacteria

Ancestral archaean

Endosymbiosis

Bacteria

Ancestral eukaryan

Archaea

Endosymbiosis

Ancestral photosynthetic eukaryan

Chloroplast

Mitochondrion

Mitochondrion

Photosynthetic Eukarya

Nonphotosynthetic Eukarya

Photosynthetic protists

Plants

Animals

Fungi

Nonphotosynthetic protists

area covered by forests shrank by 23 million acres a year during the 1990s–most of it cleared for agriculture. Some experts estimate that only half the acreage of the planet's original rain forest remains.

Forests are only one place biodiversity is in danger. Habitat destruction in ecosystems around the globe–wetlands, ice caps, coral reefs–poses a grave threat to countless species. If current rates of habitat destruction continue, we may witness levels of extinction rivaling the greatest extinction events of geological history. Can anything be done to reverse the trend of dwindling biodiversity around the globe?

Though the rapidly expanding human population (Chapter 24) is gobbling up resources faster than the earth can restore them, there are things we can do to mitigate the destruction. One conservation strategy is to protect those areas that are known to be especially diverse, ensuring that they remain so. That means safeguarding habitat and forbidding overhunting. Where possible, it also means taking efforts to restore missing diversity, in an effort to keep ecosystems whole. "We have an obligation to try to keep an ecosystem intact if we can," says Lewis, from Washington Department of Fish and Wildlife, noting that the best way to protect the environment may be to keep all of its parts in place.

"We have an obligation to try to keep an ecosystem intact if we can." –Jeffrey Lewis

---

For the fishers, at least, things seem to be looking up. They are dispersing and reproducing in the forest–at least seven females so far have had kits–and park manager Happe says she is "guardedly optimistic" about their chance of survival. Over the next few years, Happe and Lewis will continue to monitor the fishers, which have been equipped with radio collars, to make sure they are adapting to and surviving in their new home. Only then will they be able to label the restoration project a success. Fishers may have returned to the forest, says Happe, but "they're not out of the woods." ∎

▶ Summary

■ Rain forests are sites of great biological diversity, as measured by the number and variety of different species present.

■ The domain Eukarya encompasses all eukaryotic organisms—plants, animals, fungi, and the many types of protists.

■ Plants are multicellular eukaryotes that carry out photosynthesis. All plants have cells with cell walls, but not all have a vascular system, not all produce seeds, and not all produce flowers.

■ Plants can be subdivided into groups, including the bryophytes, ferns, gymnosperms, and angiosperms, on the basis of their terrestrial adaptations.

■ Animals are multicellular eukaryotic heterotrophs that obtain nutrients by ingestion.

■ Most animals are invertebrates (that is, they lack a backbone). The most abundant invertebrates by far are arthropods, and especially insects.

■ Vertebrates (animals with a backbone) are members of the phylum Chordata. Common vertebrates include mammals such as the fisher, as well as amphibians, reptiles, birds, and fish.

■ Fungi are decomposers, acquiring their nutrition and energy by breaking down dead organic matter and absorbing the results. There are unicellular and multicellular fungi.

■ Protists are a diverse group of mostly unicellular eukaryotic organisms that do not cluster on a single branch of the evolutionary tree. They include photosynthetic plantlike algae and animal-like parasites.

■ All eukaryotes are descendants of a unicellular protist that first emerged some 2 billion years ago as the result of endosymbiosis.

DOMAIN EUKARYA

The domain Eukarya encompasses all eukaryotic organisms, including plants, animals, fungi, and protists. Eukaryotic organisms display great diversity in their many evolutionary adaptations.

HINT See Infographics 19.1 and 19.3.

● KNOW IT

1. How does the physical landscape diversity of Olympic National Park affect biodiversity in the park?

2. What are the defining features of the domain Eukarya?

3. What do a fisher and a fir tree have in common?

● USE IT

4. How do you think the diversity of eukaryotic organisms in each of the following areas would compare to the diversity in Olympic National Park—would there be more or less? Explain the reasons for your answers.
    a. Lake Michigan
    b. the Sonoran Desert in Arizona
    c. the prairies of Kansas

5. If a fungicide were applied throughout Olympic National Park, how might it affect eukaryotes in the park? Explain your answer.

PLANT DIVERSITY

Plants are photosynthetic, multicellular eukaryotes. Plants have evolved many different structures and adaptations for living and reproducing on land.

HINT See Infographic 19.6.

● KNOW IT

6. Which group of plants was the first to live on land? Why do we find these plants only in particular environments (after all, if they were first, shouldn't they have spread everywhere by now)?

7. A major difference between a fern and a moss is
    a. the presence of seeds.
    b. the presence of flowers.
    c. the presence of cones.
    d. the presence of a vascular system.
    e. the ability to carry out photosynthesis.

● USE IT

8. What is an advantage of having seeds? (Think about spreading to new locations and whether or not reproduction relies on water.)

9. What type of seed plant is likely to rely on hungry animals to spread its seeds? Explain your answer.

10. How did the evolution of vascular systems in plants change the landscape?

ANIMAL DIVERSITY

Animals are multicellular eukaryotes that obtain nutrients through ingestion. Animals exhibit a wide variety of body shapes and structures.

HINT See Infographic 19.5.

● KNOW IT

11. A sand dollar gets its name from its body shape—it resembles a large coin. What type of body symmetry does a sand dollar have?
    a. bilateral
    b. radial
    c. none (sand dollars are amorphous)
    d. hyphae
    e. mycelium

12. What do a backbone and an exoskeleton have in common?
    a. They are found in closely related groups of animals.
    b. They are made of the same substance.
    c. They both help provide support to an animal's body.
    d. They both require an animal to molt in order to be able to grow.
    e. all of the above

13. You and a fisher are both mammals; as such, what are some characteristics you and the fisher have in common?

14. Which of the following statement(s) is/are true about both cockroaches and lobsters?
    a. They are invertebrate insects with bilateral symmetry.
    b. They are mollusks with an exoskeleton.
    c. They are arthropods with segmented bodies and no symmetry.
    d. They are arthropods with an exoskeleton.
    e. They are mollusks with a segmented body.

● USE IT

15. Many characteristics are used to classify animals. Why do we need to use so many different characteristics? Consider the following five animals: woodpecker, human, wasp, ant, and fisher; and the following three characteristics: ability to fly, two-legged, bearing feathers
    a. Which of the five animals could be grouped by each characteristic?
    b. Would this grouping reflect their real taxonomic relationship?

c. What feature(s) would you use to put wasps and ants together in their own group? What about human and fisher?

16. Judging from their numbers, arthropods are a tremendously successful group. What traits do you think have enabled them to be so successful? Justify your answer with examples.

FUNGAL DIVERSITY

Fungi are unicellular and multicellular eukaryotes that obtain nutrients by secreting digestive enzymes onto organic matter and absorbing the digested product.

HINT See Infographic 19.7

➔ KNOW IT

17. Consider the "eating habits" of fungi.
  a. Can fungi carry out photosynthesis?
  b. Can fungi ingest their food?
  c. How do fungi obtain their nutrients and energy?

18. Which of the following meals include fungi as food?
  a. a bread and blue cheese platter with fruit
  b. mushroom risotto
  c. a and b
  d. a fruit salad
  e. yogurt

➔ USE IT

19. A very early classification scheme placed the fungi together with the plants. Why do you think fungi were grouped with plants? What features distinguish them from plants?

A PLETHORA OF PROTISTS

Protists are a diverse group of primarily unicellular eukaryotic organisms that are considered together only because they do not sort neatly into any other single evolutionary category.

HINT See Infographics 19.8 and 19.9.

➔ KNOW IT

20. What do members of the informal group known as protists have in common?
  a. nothing
  b. They are all eukaryotic.
  c. They all carry out photosynthesis.
  d. They are all human parasites.
  e. They are all decomposers.

21. Endosymbiosis describes the process by which:
  a. protists diverged from plants.
  b. eukaryotic cells acquired certain organelles.
  c. protists became multicellular.
  d. eukaryotes diverged from prokaryotes.
  e. a and c
  f. b and d

➔ USE IT

22. Why do scientists no longer consider protists a separate kingdom? How might scientists find new taxonomic "homes" for the protists? Do you think structural features (for example, chloroplasts) or genetic information will be more useful in their classification?

23. Many protists have an organelle called the contractile vacuole that pumps out water that enters the cell by osmosis. Why is this a useful adaptation for a protist? What might happen to a protist if its contractile vacuole stopped working? (Think about where many protists live, and what happens to bacteria whose cell walls are disrupted by antibiotics.)

SCIENCE AND ETHICS

24. Reintroducing species to their native habitats is sometimes controversial. One reintroduction effort in particular that has caused quite a stir is the reintroduction of the Mexican gray wolf (*Canis lupus baileyi*) into New Mexico and Arizona. You can read about this project at http://www.fws.gov/southwest/es/mexicanwolf/.
  a. Why might it be important to reintroduce species into their native habitats? Answer first in general terms, then specifically for the Mexican gray wolf.
  b. What factors could impede the success of such reintroductions? Again, answer in general terms first, then specifically for the Mexican gray wolf.

25. Many species reintroductions are being carried out across the United States. Do some research to learn about at least one such effort. For the species you research, address the following questions:
  a. What caused it to be lost from its native habitat?
  b. Is its reintroduction important?
  c. Are there are controversies about its reintroduction?
  d. What made you interested in this particular species and its reintroduction? Is it an "attractive" species? Is it being reintroduced near where you live?

What Is Race?

What You Will Be Learning

20.1 How Do We Define Race?

20.2 Melanin Influences Skin Color

20.3 Folate and Vitamin D Are Necessary for Reproductive Health

20.4 Human Skin Color Correlates with UV Light Intensity

20.5 Modern Human Populations Are Descendants of “Eve”

20.6 Mitochondrial DNA Is Inherited from Mothers

20.7 Out of Africa: Human Migration

20.8 Traits of Modern Humans Reflect Evolutionary History

20.9 Natural Selection Influences Human Evolution

20.10 The Evolution of Skin Color

What Is Race?

Science redefines the meaning of racial categories

When Barack Obama was elected in 2008, he was hailed as America's first black president. When Tiger Woods won the Masters Golf Tournament in 1997, he was lauded as the first black man to win. When Halle Berry won an Oscar in 2001 for best actress, she was commended as the first black woman to win in that category.

Why was skin color so significant? A 250-year history of slavery and racial discrimination in the United States has left a bitter legacy. Almost 150 years after slavery was legally abolished in the United States, people of color are still underrepresented in positions of power and prestige. Although the reasons for this underrepresentation are complex, the recognition of the achievements of Obama, Woods, and Berry was important because it signaled a major change: barriers to social advancement were beginning to come down.

To shoehorn any of these three people into a simple racial category, however, is misleading: Barack Obama was born to a white mother and an African father; Tiger Woods's background includes African, Chinese, Dutch, and Thai forebears; Halle Berry was born to a white mother and a black father. So what does the term "black" mean?

Historically, racial categories were employed primarily by one group to maintain power over another and to justify forms of oppression, including slavery. In the United States, racial categories were reinforced by laws such as the "one drop" rule adopted by several states in the 1920s, which held that any American with one drop of African blood was to be considered black. People then continued to use these categories and their connotations to justify racial discrimination and, in some places, racial segregation.

Though social and political attitudes have changed, people continue to invoke racial categories like "black" or "white" for various reasons, including simple physical description. Regardless of the reason, it is increasingly clear that from a biological perspective racial categories are meaningless. Research on the evolution of humans increasingly shows that race is a

social, not a biological, category. Groups of people can and do share similar physical characteristics, such as skin color and other features, but all humans are members of a single biological species, *Homo sapiens*. The only thing that skin color might accurately identify is the geographical location where a person's ancestors lived (**Infographic 20.1**).

Humans are a recent species, first walking the earth a mere 200,000 years ago.

Why is there so much variation in human skin tone? And how did the geographical variation come about? The answers lie in the evolution and migration of our earliest ancestors. Humans are a recent species, first walking the earth a mere 200,000 years ago. The physical differences we see among people today have all emerged in the very recent past. And while the physical differences between, say, an African from Senegal and a European from Sweden may appear large, biologically speaking such differences are actually quite small. In fact, genetic studies comparing regions of the human genome from person to person show that each person's DNA is 99.9% identical to any other unrelated person. Nevertheless, this 0.1% difference holds clues that help explain how our varying skin tones and other physical traits evolved.

How Do We Define Race?

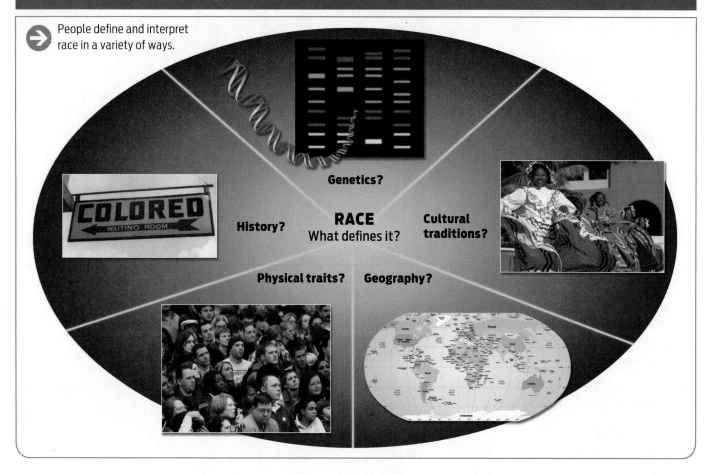

People define and interpret race in a variety of ways.

Genetics?

History?

RACE
What defines it?

Cultural traditions?

Physical traits?

Geography?

COLORED
← WAITING ROOM →

Over our 200,000-year history humans have evolved many traits that helped them survive the environments they encountered as they migrated around the globe. People with African ancestry, for example, carry with high frequency an allele that helps them resist malaria. An allele common among Northern Europeans enables them as adults to digest milk better than other populations do, an indication that at some point in history, dairy products provided an important source of nutrition and those who could digest them were more likely than others to survive. Tibetans carry in high frequency an allele that helps their red blood cells compensate for the low oxygen level in their high-altitude environment.

Skin color is another example of human evolution. The reason for the adaptation, however, wasn't clear until Nina Jablonski, an anthropologist at Pennsylvania State University, studied human skin color variation in depth.

In certain environments, is there an evolutionary advantage to light or dark skin?

MELANIN
Pigment produced by a specific type of skin cell that gives skin its color.

FOLATE
A B vitamin also known as folic acid, folate is an essential nutrient, necessary for basic bodily processes such as DNA replication and cell division.

The Evolution of Skin Color

More than a decade ago, Jablonski and her husband, George Chaplin, a geographic information systems specialist, set out to understand why human populations had evolved varying skin tones. Skin tone largely reflects the amount of **melanin** present in the skin; people naturally produce different levels of melanin, resulting in different skin tones. Skin also responds to sunlight by producing more melanin and becoming darker temporarily (**Infographic 20.2**).

In general, skin tone correlates with geography: people from northern climates tend to be fair and those from areas close to the equator tend to have dark skin. Jablonski wanted to understand this, so she searched the scientific literature. Might there be an evolutionary advantage to having light or dark skin in different environments?

She found her first clue in a 1978 study showing that an hour of intense sunlight can halve the level of an important vitamin known as **folate** in light-skinned people. Folate, also called folic acid, is an essential nutrient, necessary for basic bodily processes like DNA replication and cell division.

Then, at a seminar, Jablonski learned that low folate levels can cause severe birth defects such as spina bifida, in which the spinal column does not close, and anencephaly, the absence at birth of all or most of the brain. She subsequently came across three case studies that linked such birth defects to the mothers' visits to tanning studios, where the women would have been exposed to ultraviolet light. She also learned

INFOGRAPHIC 20.2

Melanin Influences Skin Color

Melanocytes are a type of cell located in the epidermis, the outermost layer of skin. Melanocytes make the pigment melanin, and deposit it into other cells in the skin. A person's skin color depends largely on the amount and type of melanin that his or her skin melanocytes produce. Sunlight can also temporarily increase the amount of melanin in a person's skin.

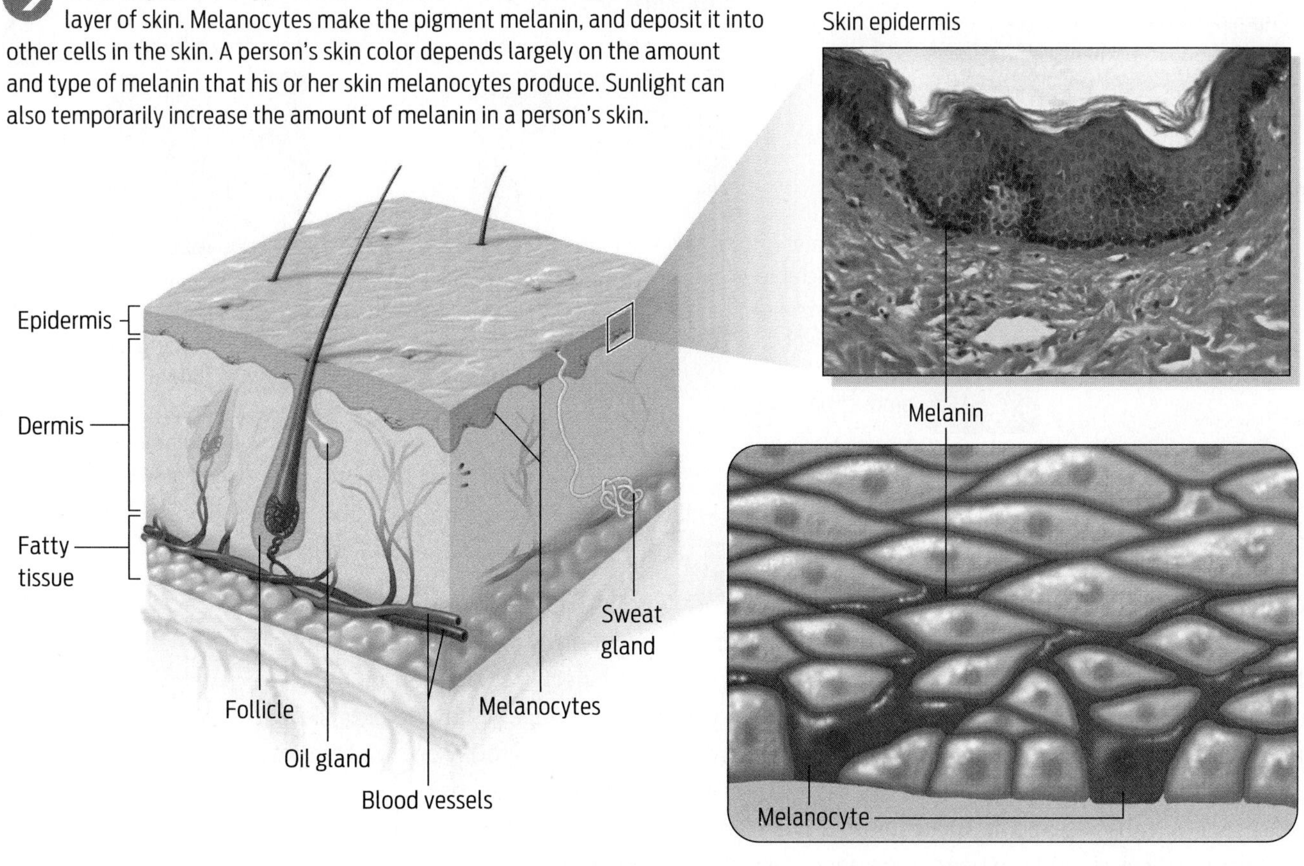
Skin epidermis
Epidermis
Dermis
Fatty tissue
Follicle
Oil gland
Blood vessels
Melanocytes
Sweat gland
Melanin
Melanocyte

Folate and Vitamin D Are Necessary for Reproductive Health

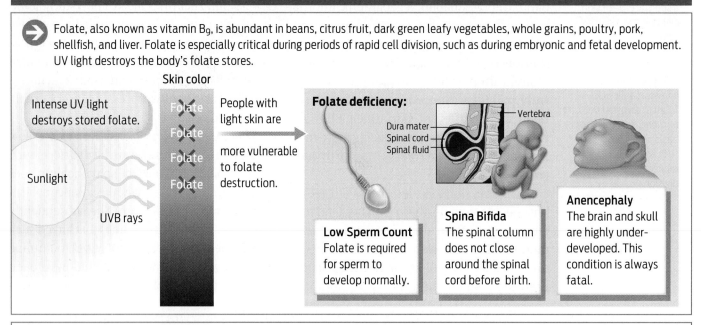

Folate, also known as vitamin B₉, is abundant in beans, citrus fruit, dark green leafy vegetables, whole grains, poultry, pork, shellfish, and liver. Folate is especially critical during periods of rapid cell division, such as during embryonic and fetal development. UV light destroys the body's folate stores.

Skin color

Intense UV light destroys stored folate.

Sunlight

UVB rays

Folate
Folate
Folate
Folate

People with light skin are more vulnerable to folate destruction.

Folate deficiency:

Vertebra
Dura mater
Spinal cord
Spinal fluid

Low Sperm Count
Folate is required for sperm to develop normally.

Spina Bifida
The spinal column does not close around the spinal cord before birth.

Anencephaly
The brain and skull are highly under-developed. This condition is always fatal.

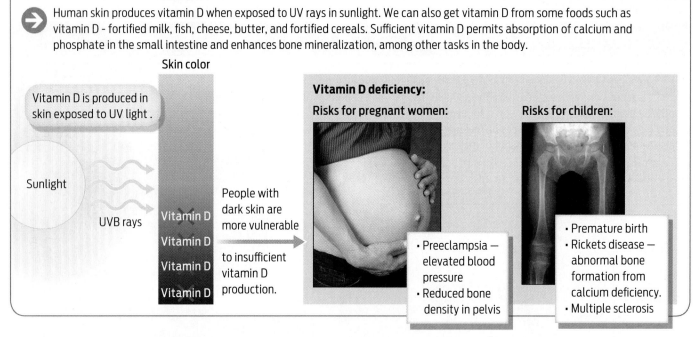

Human skin produces vitamin D when exposed to UV rays in sunlight. We can also get vitamin D from some foods such as vitamin D - fortified milk, fish, cheese, butter, and fortified cereals. Sufficient vitamin D permits absorption of calcium and phosphate in the small intestine and enhances bone mineralization, among other tasks in the body.

Skin color

Vitamin D is produced in skin exposed to UV light .

Sunlight

UVB rays

Vitamin D
Vitamin D
Vitamin D
Vitamin D

People with dark skin are more vulnerable to insufficient vitamin D production.

Vitamin D deficiency:

Risks for pregnant women:

Risks for children:

- Preeclampsia — elevated blood pressure
- Reduced bone density in pelvis

- Premature birth
- Rickets disease — abnormal bone formation from calcium deficiency.
- Multiple sclerosis

that folate is necessary for sperm to develop normally.

Taken together, the results of Jablonski's literature search suggested that people with light skin are more vulnerable to folate destruction than are darker-skinned people–presumably because melanin absorbs and dissipates damaging UV light as heat. Could the need to protect the body's folate stores from UV light have driven the evolution of darker skin shades? The supporting evidence was compelling. But then what was the advantage of having light skin at all, as many populations today do? And why are there geographical differences?

Jablonski's work built on a hypothesis first proposed in the 1960s by biochemist W. Farnsworth Loomis, who suggested that vitamin D might play a role in the evolution of skin color. Unlike folate, which is destroyed by excess sunlight, the production of vitamin D *requires* ultraviolet light. **Vitamin D** is crucial for good health: it helps the body absorb calcium and deposit it in bones. During pregnancy women need extra vitamin D to nourish the growing embryo. In addition, since vitamin D is so important for healthy bone growth, too little might also cause bone distortion, and a distorted pelvis would make it difficult for a woman to bear children (**Infographic 20.3**).

In 2000, Jablonski and Chaplin published a study in the *Journal of Human Evolution* that compared data on skin color in indigenous populations from more than 50 countries to levels of global ultraviolet light. They found a clear correlation: the weaker the ultraviolet light, the fairer the skin, a compelling suggestion that both dark and light skin are linked to levels of global sunlight. They now had a complete hypothesis: light skin evolved because in sun-poor parts of the world it helped the body produce vitamin D, while dark skin evolved because it helped protect the body's folate stores in people who lived in sunny climates. The body's need to balance levels of these two important nutrients explains why there is so much variation in skin tone around the globe (**Infographic 20.4**).

Since the publication of Jablonski and Chaplin's work, many other scientists have tested their hypothesis, and it is now the most widely

> **The body's need to balance levels of vitamin D and folate explains why there is so much variation in skin tone around the globe.**

INFOGRAPHIC 20.4

Human Skin Color Correlates with UV Light Intensity

Nina Jablonski and George Chaplin used NASA satellite measurements of UVB intensity to predict the amount of skin pigment that would best block harmful UV rays yet still enable the body to produce sufficient vitamin D in populations around the globe. Their predictions closely match actual skin color variations around the world.

Predicted pigmentation: skin-color prediction based on UVB intensity

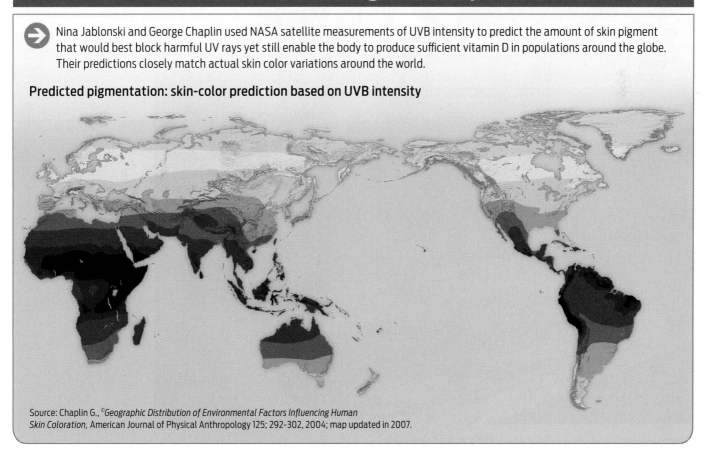

Source: Chaplin G., *"Geographic Distribution of Environmental Factors Influencing Human Skin Coloration,* American Journal of Physical Anthropology 125; 292-302, 2004; map updated in 2007.

accepted explanation for the evolution of human skin color. As Jablonski points out, "It synthesizes the available information on the biology of skin from anatomy, physiology, genetics, and epidemiology, and has not been contradicted by any subsequent data."

Out of Africa

If light and dark skin tones developed over time, then at some point in history, all humans likely had the same skin tone. This scenario, accepted by most scientists, is supported by genetic studies suggesting that anatomically modern humans first evolved in Africa.

In 1987, a team led by Allan Wilson of the University of California at Berkeley used **mitochondrial DNA (mtDNA)**—genetic material we inherit solely from our mothers—to construct an evolutionary tree of humanity. Wilson and his two colleagues Rebecca Cann and Mark Stoneking determined that all humans can trace their ancestry back to a single woman who lived in eastern Africa some 200,000 to 150,000 years ago, a woman he called the mitochondrial Eve.

In other words, if every person on the planet were to construct a family tree that listed every relative for thousands of generations back in time, they would all eventually converge at a single common female ancestor and a single common male ancestor; this female is the one Wilson dubbed Eve **(Infographic 20.5)**. Note that Eve wasn't the only female living at the time; she was merely one female in a population of many ancient humans. But her mitochondrial DNA is the only DNA that modern humans still carry today. In other words, other females living at the time also had descendants, but the lines of these descendants died off over time. Eve's descendants—and only her descendants—populate the earth today.

MITOCHONDRIAL DNA (mtDNA)
The DNA in mitochondria that is inherited solely from mothers.

Jablonski (right) and Chaplin examine a world map of skin color.

Modern Human Populations Are Descendants of "Eve"

→ Many women of Eve's generation left descendants, but the mtDNA data suggest that Eve's descendants are the ones who went on to become the modern human populations across the globe we know today.

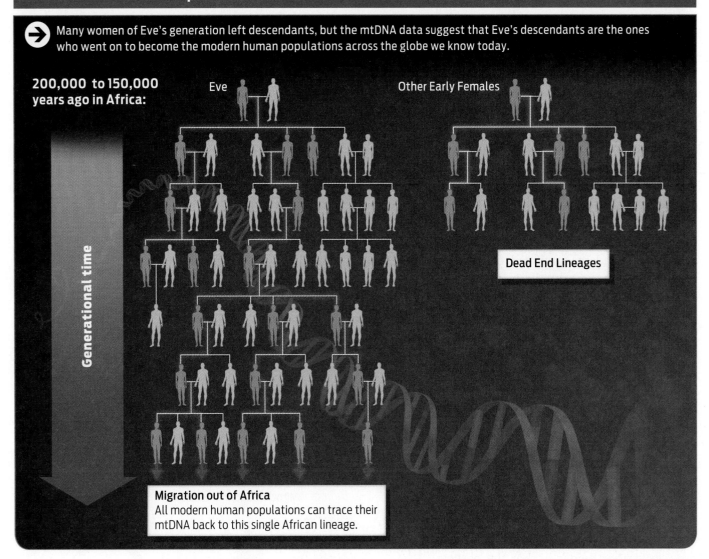

200,000 to 150,000 years ago in Africa:

Eve

Other Early Females

Generational time

Dead End Lineages

Migration out of Africa
All modern human populations can trace their mtDNA back to this single African lineage.

Mitochondrial DNA is DNA located in the mitochondria in all our cells. Unlike nuclear DNA, which is inherited from both parents in most multicellular organisms (including humans and other animals), and which undergoes recombination during meiosis, mtDNA passes from mothers to offspring essentially unchanged. That's because sperm do not con-

All humans can trace their ancestry back to a single woman who lived in eastern Africa some 200,000 to 150,000 years ago.

tribute their mitochondria to the newly formed zygote (**Infographic 20.6**).

Like nuclear DNA, mtDNA mutates at a fairly regular rate, although it appears to mutate faster

than nuclear DNA. A mother with a mutation in her mtDNA will pass it to all her children, and her daughters will pass it to their children in turn. Because these mutations pass down without being combined and rearranged with paternal mitochondrial DNA, mtDNA is a powerful tool by which to track human ancestry back through hundreds of generations.

To conduct the Eve study, Wilson and his colleagues collected mtDNA from 147 contemporary individuals from Africa, Asia, Australia, Europe, and New Guinea. On the basis of the mtDNA sequence patterns, the researchers created an evolutionary tree. Branches of the tree from all five areas could be traced back to Eve. However, the tree had two major evolutionary branches: one that included the ancestors of populations now living in Asia, Australia, Europe, and New Guinea, and one that included the ancestors of modern-day Africans.

The mtDNA of people on the African branch had acquired twice as many mutations as the mtDNA of people on the rest of the tree. The most likely interpretation of these data, the scientists reasoned, was that the African mtDNA had had more time to accumulate mutations, and was consequently older, evolutionarily speaking. This would mean that humans likely originated in Africa, where they formed several ancestral populations. After some period of time, one group of Africans left the continent, and their descendants continued to migrate to other continents, eventually becoming the ancestors of modern-day Asians, Australians, and Europeans.

Since Wilson's study, additional evidence continues to back the "out of Africa" hypothesis. Fossil discoveries in Ethiopia in 2003 and 2005 represent the oldest known fossils of modern humans—160,000 and 195,000 years old, respectively—and plug a major gap in the human fossil record. Both sets of remains date precisely from the time when Wilson and his colleagues think that a genetic Eve lived in eastern Africa. The fossil discoveries provide evidence that anatomically modern humans were living in that region around the same time that Eve lived, and provide further evidence that the earliest humans originated in Africa.

This hypothesis is also supported by research that sampled genetic diversity from nuclear DNA. In a 2008 study, for example, Richard Myers, of the Stanford University School of Medicine, and colleagues found less and less genetic variation in people the farther away from Africa they lived—the same pattern of variation that scientists have found in human mtDNA sequences. This finding suggests that as each small group of people broke away to explore a new region, it took only a sample of the parent population's genes. Consequently, genetic diversity decreased in tandem with the distance people traveled away from Africa—a classic example of the founder effect described in Chapter 15 **(Infographic 20.7)**.

INFOGRAPHIC 20.6

Mitochondrial DNA Is Inherited from Mothers

→ When egg and sperm fuse during fertilization, sperm contribute only nuclear DNA to the nucleus of the newly formed zygote. The egg provides all other organelles, including mitochondria. Consequently, only mothers contribute mitochondrial DNA to their children.

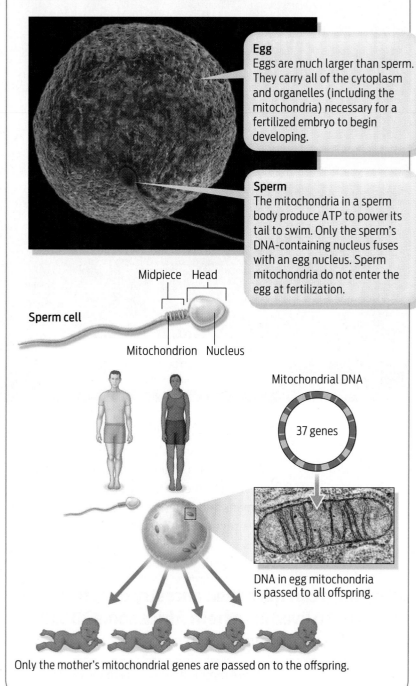

Egg
Eggs are much larger than sperm. They carry all of the cytoplasm and organelles (including the mitochondria) necessary for a fertilized embryo to begin developing.

Sperm
The mitochondria in a sperm body produce ATP to power its tail to swim. Only the sperm's DNA-containing nucleus fuses with an egg nucleus. Sperm mitochondria do not enter the egg at fertilization.

Midpiece   Head

Sperm cell

Mitochondrion   Nucleus

Mitochondrial DNA

37 genes

DNA in egg mitochondria is passed to all offspring.

Only the mother's mitochondrial genes are passed on to the offspring.

Becoming Human

A number of lines of evidence peg Eve as the likely common ancestor of all humans living today. However, she represents merely one branch on the evolutionary tree that includes our species; this tree has several other branches representing other hominid species that came before her (Infographic 20.8). A **hominid** is any member of the biological family Hominidae, which includes living and extinct humans and apes.

Humans and apes are grouped together because the fossil evidence shows that modern humans and present-day apes evolved from a common ancestor that lived 13 million years ago. Of the living primates, humans and chimpanzees are the most closely related, although it has been more than 6 million years since their shared ancestor lived. During those 6 million years, both humans and chimps have undergone a tremendous amount of evolutionary change, which is why living humans look and behave so differently from chimps—or any other primate species living today.

Scientists haven't yet discovered fossil remains of the last common ancestor between chimps and humans. However, in October 2009

INFOGRAPHIC 20.7

Out of Africa: Human Migration

→ Genetic evidence suggests that the earliest modern humans originated and evolved for thousands of years in Africa before a group of them and their descendants migrated to the other continents.

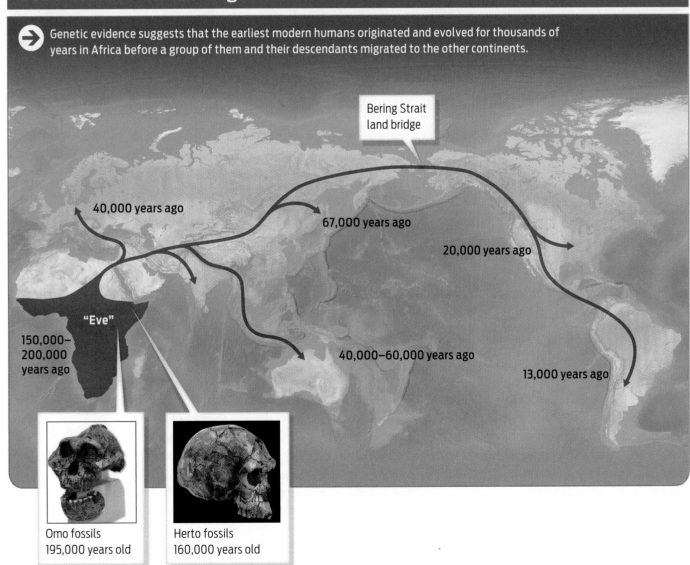

Bering Strait
land bridge

40,000 years ago

67,000 years ago

20,000 years ago

"Eve"

150,000–
200,000
years ago

40,000–60,000 years ago

13,000 years ago

Omo fossils
195,000 years old

Herto fossils
160,000 years old

Traits of Modern Humans Reflect Evolutionary History

→ *Homo sapiens* is the only surviving lineage in the evolutionary history of humans. In other words, several hominids have existed or coexisted as related but distinct species in the past. The physical traits that modern-day humans have, such as skin color and body hair, evolved in response to selective pressures. A species with less hair could better regulate body temperature in hot and sunny environments, for example, but would require darker skin to protect it from high UV light exposure.

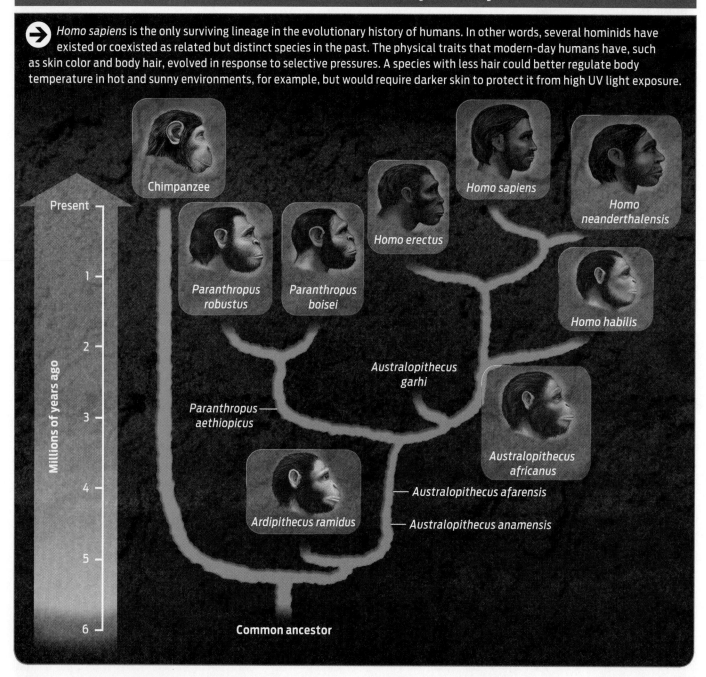

the first analyses of fossil remains of a 4.4-million-year-old hominid, *Ardipithecus ramidus*, nicknamed Ardi, were published. Ardi's remains are among the oldest hominid fossils so far discovered and, as such, give tantalizing clues to early human origins.

Among the defining characteristics of *Homo sapiens* are the ability to walk upright and a big brain. An upright gait meant the hands were free to make and use tools. A big brain enabled

H. sapiens to develop complex language. Ardi helped scientists discover that the ability to walk upright evolved first. Ardi had a small brain, suggesting that it could not use complex language. By studying Ardi's bones, scientists also know that it could maneuver on all fours in trees, but it could also walk upright without dragging its knuckles.

The fossil record after Ardi has also helped show us some of the major milestones in human evolution. For example, artifacts found at various archeological sites indicate that simple tool use began approximately 2.6 million years ago, most likely when our hominid ancestors began eating meat from large animals. The first tool-users were members of the genus *Australopithecus*. This genus walked upright and appears to have lost the ability to live in trees, as evidenced by the lack of an opposable big toe, which had helped the early hominids grip branches.

Another milestone was the ability to use and control fire, which appeared about 800,000 years ago. Artifacts such as clay shards found at various fossil sites show that *Homo erectus* was likely the first species able to control fire. Fire use enabled *Homo erectus* to cook meat and bone marrow, to stay warm, and probably to fight off predators.

Finally, at some point between 800,000 and 200,000 years ago, hominid brain size began to expand rapidly. Geological studies show that this was also a time of rapid and dramatic climate change. Scientists hypothesize that a larger brain would have enabled better communication and problem-solving, which would have been very useful as our hominid ancestors had to cope with climate change. This was also around the time that anatomically modern humans like Eve and our own species, *Homo sapiens*, appeared.

Selection for Skin Color

That anatomically modern humans evolved in Africa suggests that the first humans likely had dark skin. But then how might varying skin tones have later developed? Nina Jablonski's research has revealed that environmental factors likely played a role in the evolution of different skin tones. Environment alone doesn't cause evolution, however. Rather, the environment acts on traits, or phenotypes, increasing or decreasing the frequency of alleles in a population by natural selection. Where did these alleles come from?

Recall that each time a cell replicates, mutations—errors in replication—can occur. If these mutations occur in germ cells during meiosis (Chapter 10), they will permanently change the genome of the next generation. This process continually introduces new alleles into the population. Some of these alleles can be negative or harmful, as in the case of hereditary cancer or cystic fibrosis. But new alleles can also be benign or even beneficial. Indeed, sometimes alleles can be so positive and confer such a survival advantage that they become more common in succeeding generations and can eventually become fixed in a population **(Infographic 20.9)**.

Sometimes alleles that are harmful in one environmental context may be beneficial in another. For example, the recessive allele responsible for cystic fibrosis (CF) can cause this serious disease when it occurs in homozygotes, who have two copies of the allele. However, research has suggested that being heterozygous for CF—that is, having only one CF allele—may have reduced the severity of diarrhea caused by cholera or some other infection. Consequently, carrying a CF allele provided an advantage during epidemics. This would help explain why the CF allele became relatively common.

Skin color is another example of a trait that likely conferred an advantage to humans and underwent natural selection at some point in human history. Otherwise, dark or light skin color wouldn't be so common among specific populations. In fact, the dark skin of those *Homo sapiens* who evolved in Africa was probably an early adaptation; it is likely that before evolving dark skin, our earliest ancestors had light skin, just as chimpanzees do today.

Fossil and genetic evidence suggests that about 2 million years ago hominids became "bipedal striders, long distance walkers and possibly even runners," according to Nina Jablonski. But to sustain such activities, they needed an effective cooling system, a feat they could have accomplished only by losing excessive body hair and gaining more sweat glands. In contrast, hairy chimpanzees, our closest living animal relatives, can sustain only short bouts of activity without getting overheated. "It's like sweating in a wool blanket," Jablonski explains. "After that blanket gets saturated, you can't lose very much heat."

At some point, some factor—food scarcity, perhaps—forced ancient hominids out of the forests

Natural Selection Influences Human Evolution

The environment selects for specific genetically determined traits. Different environments will select for different traits, and therefore different alleles.

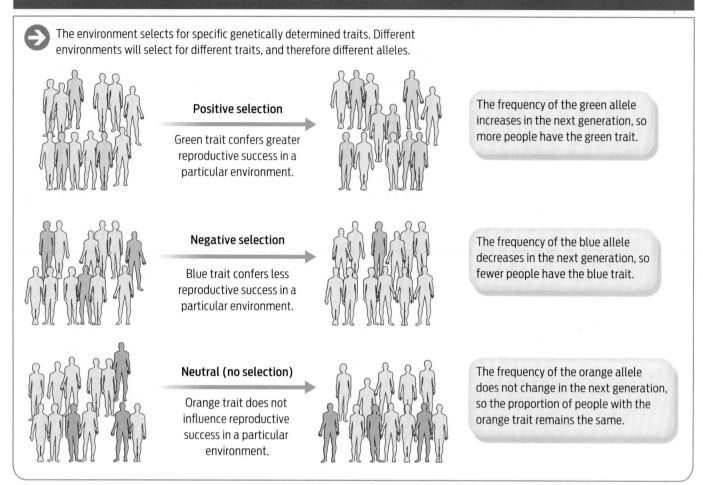

Positive selection

Green trait confers greater reproductive success in a particular environment.

The frequency of the green allele increases in the next generation, so more people have the green trait.

Negative selection

Blue trait confers less reproductive success in a particular environment.

The frequency of the blue allele decreases in the next generation, so fewer people have the blue trait.

Neutral (no selection)

Orange trait does not influence reproductive success in a particular environment.

The frequency of the orange allele does not change in the next generation, so the proportion of people with the orange trait remains the same.

and into the open savannahs to hunt for food. Hominids with less hair and more sweat glands were likely better hunters because they could sustain long bouts of activity without getting overheated. Like modern-day chimpanzees, these hominids likely had fair skin under their hair. Without hair to protect their light skin, they were exposed to the intense African sun. And, scientists hypothesize, exposure to the sun would have reduced their folate levels and thus their fitness in the sun-drenched environment.

Any of these ancient hominids that carried an allele or developed or inherited a mutation that increased their ability to produce more melanin would have been able to spend more time in the sun without the detrimental effects. Darker coloration would have protected their skin, and consequently their folate levels, from the sun, enabling these prehistoric humans to hunt and travel in the open fields.

Evidence to support this hypothesis comes from genetics. In 2004, Alan Rogers and his colleagues studied a gene that influences skin shade. They discovered that more than a million years ago, an allele that contributes to dark skin became fixed—that is, its frequency approached 100%—in the African population. "This is critical," Jablonski says. "It shows that darkly pigmented skin became extremely important to us" around the time that hominids became more humanlike.

The allele for darker skin was such an advantage in terms of survival and reproduction that hominids with darker skin left more offspring than their lighter-skinned relatives. Though some hominids were certainly born with rare mutations that gave them light skin, they weren't able to survive and reproduce in great enough numbers for the trait to persist in the population. The allele for darker skin eventually increased in the population until it reached 100%.

Populations that migrated north, away from the African sun, however, faced a different environment. Folate was not as easily destroyed in this lower-UV-light environment. But the high levels of melanin present in dark skin were a disadvantage; they prevented bodies from producing enough vitamin D. In this low-UV-light environment, fair skin allowed the body to soak up more ultraviolet light and produce essential vitamin D. In these environments fair-skinned people thus were more fit and left more descendants than dark-skinned people. Consequently, the frequency of light skin in northern climates increased with each generation.

Genetic studies show that the frequency of alleles for light skin increased and swept through populations as they migrated north—most likely more than once. The fact that light skin in people from Northwestern Europe and light skin in people from Eastern Asia is determined by at least three different genes suggests that mutations for light skin arose independently and spread through those two populations separately.

There are a number of other hypotheses for the evolution of skin tone, but of all of them, the folate-vitamin D hypothesis has the most evidence supporting it and is consequently "the most reasonable," says Mark Stoneking of the Max Planck Institute for Evolutionary Anthropology in Leipzig, Germany. In fact, Stoneking says, "skin tone is one of the best examples of human evolution." It's an example in which we can see that genes have definite phenotypic effects on skin pigmentation and for which scientists can also see that the trait was selected, he explains.

There are at least a dozen different genes that interact to determine skin color, maybe more than 100 genes in total, says Stoneking. The dozen or so that have been identified are the ones that are known to have very strong effects.

> **"Skin tone is one of the best examples of human evolution."**
> —Mark Stoneking

Scientists also know that these skin color genes have been favored by specific environments because they carry genetic signatures of natural selection. To study whether natural selection favored any particular trait, scientists typically look at the amount of sequence variation that exists in a gene of interest. Less variation than average means that there was some environmental pressure that selected the alleles for that trait to be conserved over time.

Indeed, skin-color genes show this very pattern—they show less sequence variation than genes for other traits. Consequently, we know that the amount of melanin in the skin represents a compromise, or evolutionary trade-off, most likely between the need to protect folate levels from excess sunlight and the need to absorb sunlight to make vitamin D—and the way the trade-off was resolved depended on the environment **(Infographic 20.10)**. Skin color is thus a proxy for the geographic origin of our ancestors, but not much else.

Throughout human history, the lines between what we have come to call races have been fluid. Genetic studies show that hardly any population is pure in the way that many have thought. As people moved around the globe, they settled and often bore children with people they met along the way, introducing their alleles into the local gene pool. The particular environment people encountered favored some traits over others, and that is why populations that live in similar environments share similar features.

Though people tend to create racial groupings based on obvious physical characteristics, such features can be shared with other groups, says Jablonski. Not all Africans have equally dark skin and not all Europeans are fair-skinned, for example. And as humans travel more, settle in different areas, and intermarry, Jablonski says, "racial categories will get messier and messier." Perhaps the concept of race itself in time will disappear. ■

The Evolution of Skin Color

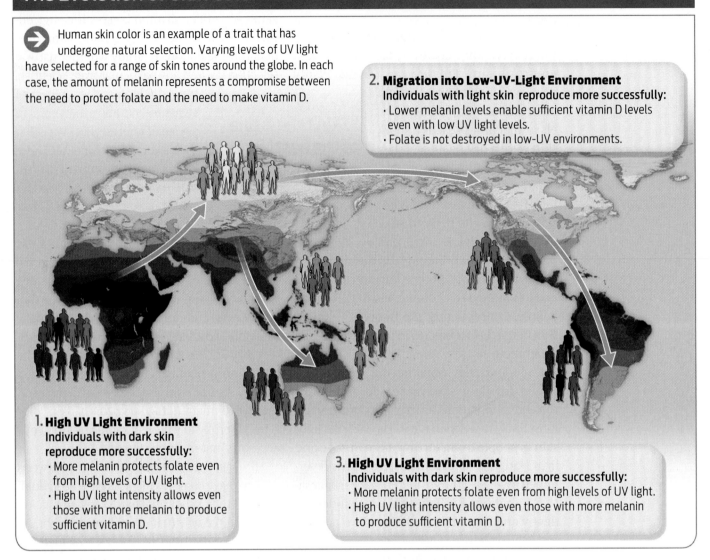

→ Human skin color is an example of a trait that has undergone natural selection. Varying levels of UV light have selected for a range of skin tones around the globe. In each case, the amount of melanin represents a compromise between the need to protect folate and the need to make vitamin D.

2. Migration into Low-UV-Light Environment
Individuals with light skin reproduce more successfully:
· Lower melanin levels enable sufficient vitamin D levels even with low UV light levels.
· Folate is not destroyed in low-UV environments.

1. High UV Light Environment
Individuals with dark skin reproduce more successfully:
· More melanin protects folate even from high levels of UV light.
· High UV light intensity allows even those with more melanin to produce sufficient vitamin D.

3. High UV Light Environment
Individuals with dark skin reproduce more successfully:
· More melanin protects folate even from high levels of UV light.
· High UV light intensity allows even those with more melanin to produce sufficient vitamin D.

◗ Summary

■ Physical features shared by people within populations reflect adaptations to specific environments.

■ Alleles can be harmful, beneficial, or neutral in their effect on survival and reproduction.

■ Skin color most likely evolved in response to environmental UV levels, an example of human adaptation by natural selection. Alleles for darker skin conferred an advantage in sunnier environments, while alleles for lighter skin conferred an advantage in regions that receive weak sunlight.

■ Skin color represents an evolutionary trade-off between the need for vitamin D, which requires adequate sunlight for its production, and the need for folate, which is destroyed by too much sunlight.

■ Fossil evidence shows that humans and apes descended from a common ancestor and that walking upright preceded development of a big brain. There were many species that could walk upright before *Homo sapiens* appeared.

■ Mitochondrial DNA evidence shows that modern-day humans first emerged in Africa, approximately 200,000 years ago, and subsequently spread to other continents.

■ Humans evolved from apelike primate ancestors who likely had fair skin. Darker skin emerged in tandem with loss of body hair as our hominid ancestors ventured into the hot savannah.

■ Biologically distinct human races do not exist. All humans are members of the same biological species.

THE EVOLUTION OF SKIN COLOR

Skin color most likely evolved in response to environmental UV levels, and represents an evolutionary trade-off between a need for vitamin D and folate.

HINT See Infographics 20.1 to 20.4, 20.9, and 20.11.

➔ KNOW IT

1. In the course of human evolution, which of the following environmental factors likely influenced whether populations had mostly light-skinned individuals or mostly dark-skinned individuals?
   - **a.** average annual temperature
   - **b.** average annual rainfall
   - **c.** levels of UV light
   - **d.** the vitamin D content of the typical diet
   - **e.** mitochondrial DNA inheritance

2. As hypothesized by Jablonski and Chaplin, darker skin is advantageous in _____ UV environments because darker skin _____.
   - **a.** high-; reduces Vitamin D production
   - **b.** high-; protects folate from degradation
   - **c.** high-; increases the rate of folate synthesis
   - **d.** low-; allows more vitamin D to be produced
   - **e.** low-; allows more folate to be produced

➔ USE IT

3. If folate is *not* destroyed by UV radiation, predict the skin color you might find in each of the following populations. Explain your answers.
   - **a.** populations living at the equator
   - **b.** populations living in Greenland

4. Which of the following would help darker-skinned people who live in low-UV environments remain healthy?
   - **a.** folate supplementation
   - **b.** sunscreen
   - **c.** reduced production of melanin
   - **d.** vitamin D supplementation
   - **e.** calcium supplements

5. Our closest primate relatives, chimpanzees, have light-colored skin yet live in tropical (high-UV) environments. How would the Jablonski-Chaplin hypothesis explain this observation?
   - **a.** Chimpanzees don't need folate for successful reproduction.
   - **b.** Chimpanzees are not susceptible to skin cancer.
   - **c.** The hair of chimpanzees protects their light skin from UV light.
   - **d.** Chimpanzees require much higher levels of vitamin D than humans do.
   - **e.** Chimpanzees use a light-colored pigment as their UV protection.

EARLY HUMAN ORIGINS

Fossil and mtDNA evidence suggests that humans and apes evolved from a single ancestor, likely in Africa.

HINT See Infographics 20.5 to 20.8, and 20.10.

➔ KNOW IT

6. Why is mtDNA a useful tool in the study of human evolution?

7. According to the "out of Africa" hypothesis of human origins and migration, which group of people should show the highest level of genetic diversity?
   - **a.** Africans
   - **b.** Europeans
   - **c.** Asians
   - **d.** South Americans
   - **e.** Australians

➔ USE IT

8. Of the following traits that are associated with being human, which evolved most recently?
   - **a.** upright walking
   - **b.** ability to control fire
   - **c.** social communication
   - **d.** tool use
   - **e.** big brain

9. Rank the levels of genetic diversity you would expect to find within the four populations listed in Question 7 from highest to lowest. Justify your ranking.

10. Why would individual Australopithicines who could make and use tools have had a selective advantage (that is, higher fitness) over individuals who could not make or use tools?

11. Ardi was partially arboreal (that is, the species could live in trees). The ability to move around in trees was facilitated by an opposable big toe that would help grip branches. Once ancient hominids moved permanently to a grounded lifestyle, would there have been any selective pressure to maintain an opposable big toe? Explain your answer.

12. Members of the genus *Australopithecus* walked upright, and their fossilized footprints show no evidence of an opposable big toe.

  a. What foot structure and lifestyle might have been selected if early hominid evolution occurred in a forested environment? In a grasslands environment? Would you predict any differences because of the selective pressures in each environment? Why or why not?

  b. What other traits would you expect to be favored in a forested environment? In open grasslands?

SCIENCE AND SOCIETY

13. Vitiligo is a disease in which melanocytes are destroyed, causing loss of pigmentation. If a dark-skinned person develops vitiligo and therefore lighter-colored skin, would his or her race change? What factors have led people to classify (or misclassify) themselves or others as members of one race or another?

14. Visit the 2010 Census Constituent FAQs page, http://2010.census.gov/partners/pdf/ConstituentFAQ.pdf. What is the U.S. Census definition of race? Why do you think the Census asks people to specify their race? What factors do you think go into a person's choice of a particular race on the U.S. Census form?

On the Tracks of Wolves and Moose

On the Tracks of Wolves and Moose

Ecologists are learning big lessons from a small island

> Teeth, hooves, blood, bruises, adrenalin, exhaustion. Romeo killed a moose. Very likely, this is the first moose he'd ever killed. He'd seen his parents, the alpha pair of Chippewa Harbor Pack, do it many times. He would have even helped his parents kill moose. He'd wounded moose a couple of times this winter, but never killed one. His pride heightened because he killed this moose with the help of a girlfriend. By early morning they slept with full bellies while a dozen ravens celebrated the accomplishment with a feast of their own."

That's an entry, made on February 20, 2010, in biologist John Vucetich's field journal, describing the exploits of a young gray wolf researchers have named Romeo. For almost 20 years, Vucetich has been shadowing wolves like Romeo and his kin on Isle Royale, a remote island about 15 miles off the Canadian shore in the northwest corner of Lake Superior. A 200-square-mile slice of roadless wilderness that is accessible only by boat and seaplane, Isle Royale may seem an unlikely place for a scientific laboratory, but that's exactly what it is for Vucetich and his colleagues. Every summer, and for a few weeks every winter, they investigate the island's packs of gray wolves (*Canis lupis*) and the herd of moose (*Alces alces*) that are their lifeblood.

Begun in 1958, the Isle Royale wolf and moose study is the longest-running predator-prey study in the world. For more than 50 years, researchers have studied how these two island inhabitants have interacted and co-existed. They are motivated by a simple goal: "to observe and understand the dynamic fluctuations of Isle Royale's wolves and moose, in the hope that such knowledge will inspire a new, flourishing relationship with nature." And all the effort may finally be paying off.

ECOLOGY
The study of the interactions between organisms, and between organisms and their nonliving environment.

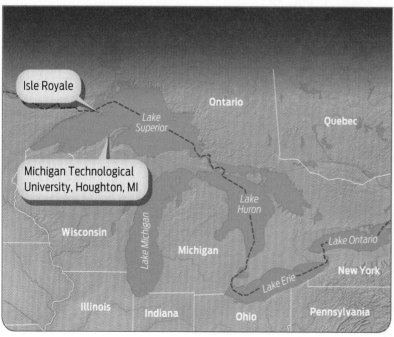

Vucetich began studying wolves as a college student at Michigan Technological University in the early 1990s. In 2001, he became coleader of the Isle Royale study, working alongside his former teacher and mentor, Rolf Peterson. It's challenging work at times, but Vucetich says he may have been destined for this career path: "*Vuk*"–the root of his last name–"is the Croatian word for wolf."

In Nature's Laboratory

Isle Royale essentially functions as a natural laboratory in which biologists can learn about **ecology,** the interactions between organisms and between organisms and their nonliving environment. A number of features make Isle Royale a good place for ecological research.

Because the island is uninhabited by humans and is protected as a national park, scientists can study moose-wolf interactions in a nearly natural environment, undisturbed by settlement, hunting, or logging.

Isle Royale is also an ideal distance from shore–close enough to the mainland for moose and wolves to have got there, but far enough away that other animals do not migrate easily to it. Because there are no other predators or prey on the island, the only things eating moose are wolves, and moose are just about the only thing wolves eat. These simplified conditions allow scientists a good look at the two residents' behavior and ecological impact.

Another thing that makes Isle Royale good for research is its size. The island is not so big as to have an unmanageably large population of moose, and not so small as to be unsupportive of a wolf population. "It's a little bit of the Goldilocks thing," says Vucetich, now a professor of ecology at Michigan Tech. "Isle Royale is not too big and it's not too small and it's not too close and not too far. It's just the right size to have a population of wolves and moose that we can study."

Ecologists study organisms at a number of levels: They can look at an individual organism, such as a single moose or wolf, studying how it fares in its surroundings. They may also look at a group of individuals of the same species living in the same place–a herd of moose, or a pack of wolves, for example–watching what happens to this **population** over time. Two or more interacting populations constitute a **community**. Isle Royale, for example, is home to a community of wolves, moose, and the plants the moose feed on.

Finally, ecologists may want to understand the functioning of an entire **ecosystem,** all the living organisms in an area and the nonliving components of the environment with which they interact. When moose eat trees, for example, they reduce the available habitat for other animals, such as birds. However, the heat of summer can reduce the ability of moose to feed, which in turn improves tree growth (Infographic 21.1).

Ecology draws not only on many areas of biology but also on many other branches of science, including geography and meteorology as well as mathematics.

A multidisciplinary science, ecology draws not only on many areas of biology but also on many other branches of science, including geography and meteorology as well as mathematics. Vucetich was initially drawn to ecology as a way to experience the outdoors, and being good at math was not something he could have

POPULATION
A group of organisms of the same species living and interacting in a particular area.

INFOGRAPHIC 21.1

Ecology of Isle Royale

Individual – a single organism of a particular species
· one wolf

Population – a group of individuals of the same species living and interacting in the same region
· a pack of wolves

Community – interacting populations of different species
· wolves prey on moose
· ticks infest moose
· moose feed on trees

Ecosystem – species interacting with other species and the environment
· moose eat the trees, changing the vegetation, which in turn changes the landscape for other animals
· hot summers reduce the ability of moose to feed, affecting their winter survival

predicted. "As a high school student, I didn't like math at all," he says. Only when he saw that math allowed him to spend more time outdoors doing what he loved did he become "interested and inspired to learn a great deal about math."

Vucetich is a population ecologist, and population ecology is all about numbers. On Isle Royale, the main numbers the researchers are interested in year after year are the numbers of wolves and moose. "In any given season there are more or less of those species and we want to understand why," says Vucetich. Answering the "why" involves a lot of time, patience, and, of course, counting.

Much of the counting is done from the air. Sitting one in front of the other inside a tiny two-person plane, pilot and observer circle the island scanning for evidence of wolves and moose. Wolves are relatively easy to find and count because their tracks are easy to follow in the snow. "You follow the wolf tracks until you find the wolves," says Vucetich. The other thing that makes counting wolves easy is that they live in packs: if you find one wolf, you've generally found the others. And since there are usually only a couple dozen wolves on the island at any time, it's possible to count every one.

It's a different story with moose. There can be more than a thousand moose on the island—too many to count all at once. Besides, moose are relatively solitary creatures, and their brown coloring makes them harder to spot against the backdrop of dark evergreen trees. When moose are feeding in the forest—which is much of the time—counting them is, according to Vucetich, "like trying to count fleas on a dog from across the room." It's simply not possible to count them all.

Instead, the team uses a shortcut: they count all the moose in a series of square-kilometer plots representing about 20% of the island, average the number of moose per plot, and then extrapolate to the rest of the island. But even this shortcut requires many careful hours of study in the plane, straining to see the moose through the trees. To help himself concentrate,

Vucetich recites a sort of mantra: "Think moose, think moose, look for the moose." That's the only way to make sure he doesn't miss one (Infographic 21.2).

The somewhat random dispersion of individually roaming moose represents one type of **distribution pattern** found in nature. Distribution patterns generally reflect behavioral or ecological adaptation. For moose, being solitary and randomly distributed may help protect them from predation, since single moose are harder to spot in the forest than a large group would be. A random distribution may also allow individuals to maximize their access to resources. Pine trees, for example, have air-blown seeds that are spread far and wide by gusty winds, resulting in a random distribution of trees in the forests on the island.

> When moose are feeding in the forest, counting them is "like trying to count fleas on a dog from across the room."
> —John Vucetich

A truly random distribution is rare in nature; even wind-blown seeds must fall on fertile soil to grow, and this does not always happen. More common is a clustered, or clumped, distribution, which results when resources are unevenly distributed across the landscape, or when social behavior dictates grouping, as it does with the highly social wolf. Clumping has its advantages: for wolves, clumping helps them to gang up on moose; they circle their prey and close in for the kill. Clumping can also be a defense against predation, as it is for a school of fish.

A third distribution pattern found in nature is uniform distribution. In this case, individuals keep apart from one another at regular distances, usually because of some kind of territorial behavior. Birds such as penguins that nest

COMMUNITY
Interacting populations of different species in a defined habitat.

ECOSYSTEM
All the living organisms in an area and the nonliving components of the environment with which they interact.

DISTRIBUTION PATTERN
The way that organisms are distributed in geographic space, which depends on resources and interactions with other members of the population.

Distribution Patterns Influence Population Sampling Methods

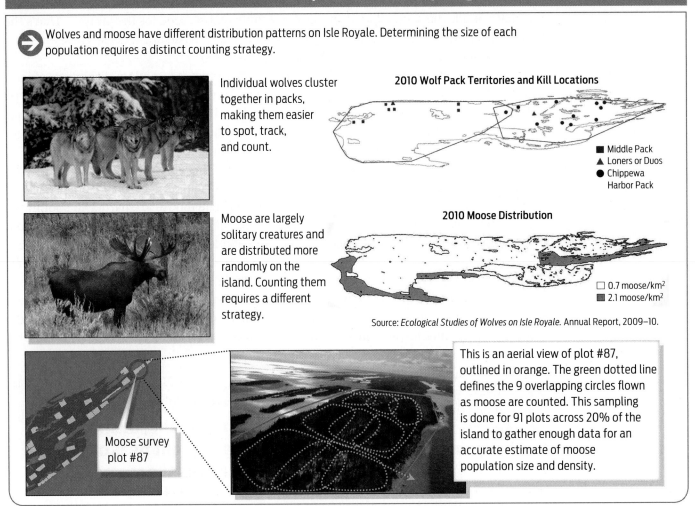

Wolves and moose have different distribution patterns on Isle Royale. Determining the size of each population requires a distinct counting strategy.

Individual wolves cluster together in packs, making them easier to spot, track, and count.

2010 Wolf Pack Territories and Kill Locations

■ Middle Pack
▲ Loners or Duos
● Chippewa Harbor Pack

Moose are largely solitary creatures and are distributed more randomly on the island. Counting them requires a different strategy.

2010 Moose Distribution

☐ 0.7 moose/km²
■ 2.1 moose/km²

Source: *Ecological Studies of Wolves on Isle Royale.* Annual Report, 2009–10.

Moose survey plot #87

This is an aerial view of plot #87, outlined in orange. The green dotted line defines the 9 overlapping circles flown as moose are counted. This sampling is done for 91 plots across 20% of the island to gather enough data for an accurate estimate of moose population size and density.

in defined spaces a few feet away from each other are a good example **(Infographic 21.3).**

Population Boom and Bust

Moose have not always roamed Isle Royale. The first antlered settlers likely arrived around 1900, when a few especially hardy individuals swam across the 15-mile-wide channel from Canada. With an abundant food supply and no natural predators on the island, the moose population exploded, growing from a handful of individuals around the turn of the century to more than a thousand by 1920.

This rapid increase reflected the population's high **growth rate,** a rate defined as the birth rate minus the death rate. Because it denotes the simple balance between birth and death, the growth rate is also known as the rate of natural increase. When the birth rate of a pop-

ulation is greater than the death rate, the population grows; when the death rate is greater than the birth rate, the population declines; and when the two rates are equal, the result is zero population growth.

In many populations, immigration and emigration make substantial contributions to population growth. But because the moose and wolves of Isle Royale are isolated, and individuals neither come to nor go from the island on a regular basis, their population growth rates are due only to births and deaths.

Ecologists describe two general types of population growth. The rapid and unrestricted increase of a population growing at a constant rate is called **exponential growth.** When a population is growing exponentially, it increases by a certain fixed percentage every generation. Thus, instead of a constant number

GROWTH RATE
The difference between the birth rate and the death rate of a given population; also known as the rate of natural increase.

EXPONENTIAL GROWTH
The unrestricted growth of a population increasing at a constant growth rate.

A bull moose walking across a waterway.

of individuals being added at each generation—say, the population going from 100 to 120 to 140 to 160—the increase is more like credit card interest, with each increase added to the principal (the population) before the percentage is applied. And so, with an exponential growth rate of 20%, that population would increase at each generation from 100 to 120 to 144 to 173 to 207. If the population continued to grow exponentially, it would quickly get out of control, not unlike a credit card bill you don't pay on time.

Such unrestricted growth is rarely if ever found unchecked in nature. As populations increase in numbers, various environmental factors such as food availability and access to habitat limit an organism's ability to reproduce.

INFOGRAPHIC 21.3

Population Distribution Patterns

 Different organisms have different distribution patterns. There are three main types, but few organisms in nature fall into strictly one category.

Random

Individuals are equally likely to be anywhere within the area.

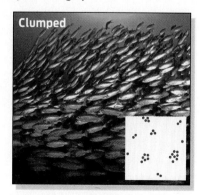
Clumped

High-density clumps are separated by areas of low abundance.

Uniform

Individuals maximize space between them by being uniformly spaced.

When population-limiting factors slow the growth rate, the result is **logistic growth**, a pattern of growth that starts rapidly and then slows.

Eventually, after a period of rapid growth, the size of the population may level off and stop growing. At this point, the population has reached the environment's **carrying capacity**–the maximum number of individuals that an environment can support given its space and resources. Carrying capacity places an upper limit on the size of any population; no natural population can grow exponentially forever without eventually reaching a point at which resource scarcity and other factors limit population growth. This is true even of the human population, as discussed in Chapter 24 (Infographic 21.4).

Note that the size of a population may fluctuate around the environment's carrying capacity, briefly exceeding it and then dropping back.

After an initial overshoot of carrying capacity, factors such as disease or food shortage will cause the population to shrink. This drop in turn may allow the environment time to recover its food supply, at which point the population may begin to grow again, briefly exceeding carrying capacity, and so on, in a cycle of boom and bust.

When moose first arrived on Isle Royale, their population grew exponentially. This unchecked proliferation of hungry mouths took a severe toll on the island; by 1929, the moose had munched their way through most of its vegetation. In turn, the reduction of the island's food supply caused the moose population to crash. The moose population had exceeded the island's carrying capacity, and by 1935, they had dwindled to a few hundred starving individuals.

The herd got lucky, though. The next summer, fire consumed 20% of the island, and the scorched areas provided space for new trees to

LOGISTIC GROWTH
A pattern of growth that starts off fast and then levels off as the population reaches the carrying capacity of the environment.

CARRYING CAPACITY
The maximum population size that a given environment or habitat can support given its food supply and other natural resources.

INFOGRAPHIC 21.4

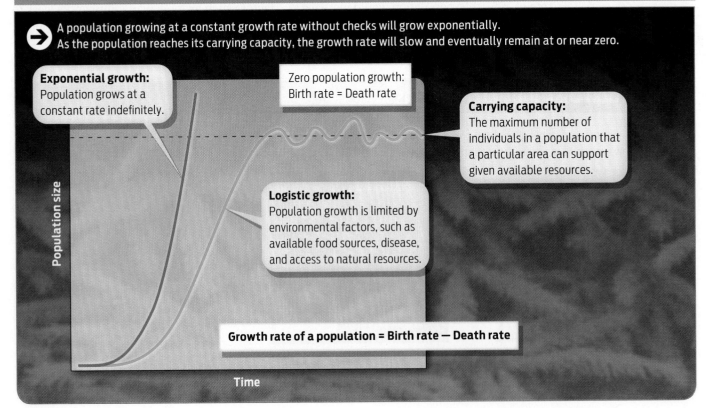

Population Growth and Carrying Capacity

→ A population growing at a constant growth rate without checks will grow exponentially. As the population reaches its carrying capacity, the growth rate will slow and eventually remain at or near zero.

Exponential growth:
Population grows at a constant rate indefinitely.

Zero population growth:
Birth rate = Death rate

Carrying capacity:
The maximum number of individuals in a population that a particular area can support given available resources.

Logistic growth:
Population growth is limited by environmental factors, such as available food sources, disease, and access to natural resources.

Growth rate of a population = Birth rate — Death rate

Population size (y-axis)

Time (x-axis)

John Vucetich with the carcass of a dead moose.

INFOGRAPHIC 21.5

Population Cycles of Predator and Prey

→ The wolf and moose populations are intimately linked. The wolf population peaks about 10 years after a peak in the moose population, and then declines, following the decline in the moose.

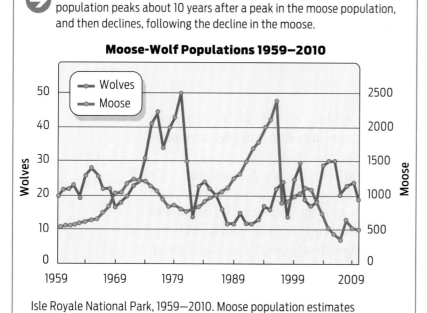

Moose-Wolf Populations 1959–2010

Isle Royale National Park, 1959–2010. Moose population estimates during 1959–2001 are based on population reconstruction from recoveries of dead moose. Estimates from 2002–2010 are based on aerial surveys.

grow. But as soon as the forest recovered, moose numbers again began to explode, ravaging the forests once more.

Then, around 1950, everything changed. One especially cold winter, a pair of gray wolves crossed an ice bridge connecting Canada to Isle Royale, forever altering the ecology of the island. Since then, the fates of the wolves and moose have been inextricably linked, with the size of one population influencing the size of the other.

Near the beginning of the Isle Royale study, in 1959, there were about 550 moose and 20 wolves. Moose numbers climbed for about 15 years, reaching a peak of approximately 1,200 animals in 1972, and then declined rapidly, to a low of approximately 700 moose in 1980. As moose numbers fell, wolf numbers rose—from a low of 17 wolves in 1969 to a high of 50 animals in 1980. These two trends were linked: the wolves were feeding themselves well enough to increase their own population, but by hunting and killing so many moose they caused the moose death rate to exceed the birth rate. With a negative growth rate, the moose population shrank.

What would happen next? Would the wolf predators simply drive their moose prey to extinction? No one knew. The only thing to do was watch and wait. Eventually, it became clear that the two populations were rising and falling together in a specific pattern, with the size of the wolf population peaking several years after the size of the moose herd and then dropping.

Why does the wolf population fall? Because even for wolves, there's no such thing as a free lunch: they pay a price for predation in the form of a declining food supply. The result is a repeating cycle in the number of predator and prey. Rather than growing exponentially and leveling off, the populations cycle through repeated rounds of boom and bust (Infographic 21.5).

Ecological Detectives

One pattern to emerge in the decades of data collected on Isle Royale is a correlation between

Patterns of Population Growth

Wolf, moose, and tree populations are all interconnected. Trees provide food for moose, and moose provide food for wolves. Anything that impacts the size of one population will impact the size of the others.

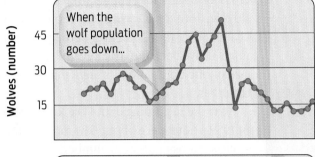

When the wolf population goes down...

The main diet of Isle Royale wolves is moose. Wolf populations grow and diminish in response to the availability of this food resource.

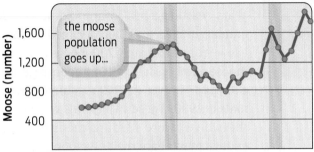

the moose population goes up...

The main diet of moose is trees. A larger moose population means that more tree material is eaten, so tree growth slows.

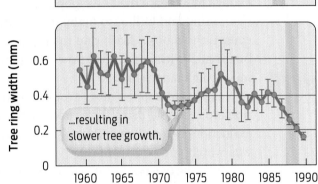

...resulting in slower tree growth.

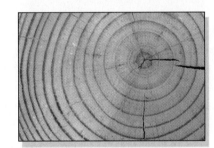

Tree growth can be measured by the width of each tree ring. One ring represents the amount of growth in 1 year: the wider the ring, the more growth in that year.

a large wolf population and vigorous tree growth. When wolves are plentiful, they keep the moose population in check. Because trees are the primary food source for moose, they grow more when fewer moose are eating them. It's therefore possible to follow the rise and fall of the wolf population by monitoring the state of the forest.

One way ecologists can determine forest growth and health is to count and measure the width of tree rings, which reflect how much trees have grown season by season. They also measure how tall the trees are. Taller and bigger trees mean that fewer moose have been forag-ing on them, which in turn indicates that more wolves have been keeping the moose popula-tion in check.

Wolves affect tree growth in another, indi-rect, way as well. Because the wolves don't always consume the entire carcass of a moose they kill, the remains decay and fertilize the ground where they lie, enriching the soil with nutrients for plant growth. In fact, researchers have found that nitrogen levels are between 25% and 50% higher in these hot spots compared to controls. This work shows that predators–in this case, wolves–are an important component of a balanced and healthy ecosystem, and "illustrates what can be protected or lost when predators are preserved or exterminated," according to Joseph Bump, an assistant professor at Michigan Tech's School of Forest Resources and Environmental Science (Infographic 21.6).

INFOGRAPHIC 21.7

Moose and Wolf Health is Monitored Using a Variety of Data

 In addition to information about population size, researchers collect other data that are essential for monitoring the physical health of moose and wolf populations.

Moose droppings can reveal the vegetation preferences of the moose populations.

Moose bones provide information on the presence of arthritis and osteoporosis, as well as bone marrow fat content for nutritional health.

Wolf scat is a source of DNA which provides a genetic profile for each wolf on the island.

Urine-soaked snow is tested to determine the ratio of urea and creatinine in the urine, an indication of a moose's nutritional status.

Another clue the ecological detectives look at in determining the population patterns on Isle Royale is urine-soaked snow and droppings, also known as scat. Urine and scat may seem crude objects of scientific study, but to the trained eye aided by a microscope they reveal a host of information about the animal that produced them. For example, by analyzing scat samples under the microscope, researchers can tell exactly what moose have been eating. During the winter months, for example, moose eat mostly twigs from deciduous (leaf-shedding) plants and needles from balsam fir and cedar trees.

Scat also provides important information about an animal's genetics. It is used to obtain DNA profiles, for example, which can be used to confirm population counts and to track which wolves were involved in killing which moose. DNA can also be used to look for diseases or signs of inbreeding. "Through the DNA we can get a good sense of individual wolves—how they live and how they die," says Vucetich. (For more information on DNA profiling, see Chapter 7.)

Yet more clues can come from studying a moose kill site, which is a bit like analyzing a crime scene. Researchers can tell if a moose was killed by wolves because in that case there will often be blood spattered on nearby trees and signs of struggle in the form of broken branches. Wolves also typically scatter bones as they feast, whereas the carcasses of moose that die of starvation may be relatively intact.

At the kill site, researchers gather moose bones. From these bones, the researchers can tell how old a moose was when it died, as well as learn about other aspects of the animal's health, such as whether it had arthritis or osteoporosis. The value of this information goes beyond understanding individual animals. It allows researchers to know whether wolves are targeting healthy moose or sickly ones. Killing a healthy moose has a bigger effect on moose population dynamics than killing one that is already near death, because a young, healthy moose might have gone on to reproduce had it lived (Infographic 21.7).

Too Close for Comfort?

Moose are formidable foes of their wolf predators. At 900 pounds and 10 times the weight of a wolf, an adult moose can successfully defend itself against an aggressive pack of wolves with its powerful front legs. For that reason, wolves often attack older and weaker or young moose. They typically target the nose and rear, where they bite and latch onto the flesh like a steel trap. When enough wolves are attached, their collective weight brings down the moose, and the feeding begins.

A number of factors can influence the likelihood that wolves will kill moose. One of the simplest is **population density,** the number of organisms per given area. Because the area of Isle Royale stays the same, as the size of the moose population increases so does its density. At high population density, moose are easier for wolves to locate and kill. Further, when the moose population is at high density, food scarcity can also be a problem, leaving moose hungry and weak and therefore more vulnerable to attack.

Because wolf predation and plant abundance have a greater effect on moose when the moose population is large, these are examples of **density-dependent factors** influencing population size. As living organisms, they are also examples of **biotic** factors influencing growth. Not all density-dependent factors are biotic. Nonliving, or **abiotic,** factors like weather, habitat, and breeding places can also influence population size in a density-dependent manner.

Some environmental pressures take a toll on a population no matter how large or how small it is. In an exceptionally cold winter with deep snow, for example, moose can die of cold or starvation. The weather can also weaken them so they are easier to hunt and kill. Since cold weather affects moose regardless of population size, it is considered a **density-independent factor**: whether 10 moose or 1,000, a harsh winter affects them all.

Conversely, harsh winters tend to benefit wolves, since moose are easier to catch in deep snow. "A mild winter is always tough on the

At 900 pounds and 10 times the weight of a wolf, an adult moose can successfully defend itself against an aggressive pack of wolves with its powerful front legs.

wolves," says Peterson, who notes that moose can more easily escape wolves when snow cover is light. Other common density-independent factors influencing population growth include rainfall, drought, and fire. Density-independent factors can be nature's form of bad luck, often striking without warning. Most, but not all, density-independent factors are abiotic (Info-graphic 21.8).

Watching and Waiting

For the scientists on Isle Royale, population ecology is full of unexpected twists and turns. There is often no sure way to know how various environmental factors will influence the growth of a population. Even on an isolated island with only one large predator and one large prey, population dynamics are never simple. Scientists gather data, look for patterns, and form hypotheses, but predicting what will happen next is much more difficult. "[What] Isle Royale has shown us . . . convincingly for the past 50 years, is that we're lousy at predicting the future," says Vucetich. "What we're a fair bit better at is explaining the past."

For example, beginning around 1980, a disease known as canine parvovirus (CPV) infected Isle Royale's population of wolves. The disease typically affects domestic dogs and was likely brought to the island on the boots of unsuspecting hikers. The disease killed all but 14 of the island's wolves, and over the next 10 years the moose population skyrocketed, demonstrating that wolves exert a strong influence on the abundance of their prey. The event was useful from a scientific standpoint—but entirely unexpected. "There's no way that anyone could have predicted that. Not in a million years," says Vucetich.

That wasn't the end of the surprises. In the last 15 years, it's become apparent that a

POPULATION DENSITY
The number of organisms per given area.

DENSITY-DEPENDENT FACTOR
A factor whose influence on population size and growth depends on the number and crowding of individuals in the population (for example, predation).

BIOTIC
Refers to the living components of an environment.

ABIOTIC
Refers to the nonliving components of an environment, such as temperature and precipitation.

DENSITY-INDEPENDENT FACTOR
A factor that can influence population size and growth regardless of the numbers and crowding within a population (for example, weather).

INFOGRAPHIC 21.8

Abiotic and Biotic Influences on Population Growth

→ Both living (biotic) and nonliving (abiotic) environmental factors influence the size and growth of populations.

Abiotic Factors:

Climate:
Harsh winter temperatures and snowfall stress moose. Food-seeking and hiding from predators become difficult.

Temperature:
High summer temperatures can cause heat stress in moose. Changes in climate may increase insect parasitism.

Biotic Factors:

Food:
The availability of trees as a food source influences moose population growth.

Predators:
The presence of a wolf predator limits moose population growth.

Disease:
Tick infestations weaken moose.

warming climate, not just predation by wolves, is influencing moose population size. The first decade of the 21st century was one of the hottest on record. Sweltering summer temperatures hit moose especially hard. The large herbivores get hot easily, and they don't perspire; they escape the heat by resting in the shade. A lot of time spent resting means less time for eating, and a moose who's been dieting all summer has less insulation for winter.

Warmer temperatures have affected moose in a more insidious way as well. About 10 years ago, Vucetich and his colleagues began to notice that a tick parasite was bothering the moose, and that warm weather seems to favor ticks. Ticks suck the moose's blood and cause them to itch. The moose scratch themselves against trees and chew their hair out trying to rid themselves of the itchy freeloaders. Since a single moose may host many thousands of ticks, the combination of tick-related blood loss and heat-induced weight loss can be deadly. In 2004, the average moose had lost more than 70% of its body hair, the result of carrying more than 70,000 ticks.

By 2007, the deadly combination of blood-sucking ticks, hot summers, and relentless predation from wolves had driven the moose population to its lowest point in at least 50 years–385, down from 1,100 in 2002. Predictably, the wolf population followed suit, declining from 30 individuals in 2005 to 21 in 2007. As of 2010, the moose population has remained low, at about 510 individuals–half their typical abundance–while the wolves declined to just 19 individuals (Infographic 21.9).

Hunted by wolves, preyed on by ticks, dogged by oppressive heat, moose certainly do not have it easy. They can live to be 17 years old, but most moose die before reaching their tenth birthday. To paraphrase philosopher Thomas Hobbes, a moose's life is often nasty, brutish, and short.

It's no picnic for wolves, either. While the wolf lifespan–the longest they can live–is about 12 years, most die by age 4. The most common cause of death is starvation. With few available food sources, a wolf may go 10 days without eating. Obtaining a meal on the eleventh day may mean having to wrestle a 900-pound moose on an empty stomach.

As of 2010, the moose population has remained low, at about 510 individuals—half their typical abundance—while the wolves declined to just 19 individuals.

The difficulty of finding food is just one obstacle for wolves. They also have a very high incidence of bone deformities, which cause back pain and partial paralysis of the hind legs. In the early years of the Isle Royale study, such deformities were rare, but they've become more common in recent years, almost certainly as a result of inbreeding. For the last 12 years, every dead wolf on Isle Royale has had such deformities.

It's not the first time wolf populations have been in trouble. When colonial settlers first arrived in North America, the gray wolf roamed throughout all of the future 48 contiguous U.S. states. By 1914, hunting and trapping had greatly reduced the population, and survivors were limited to remote wooded regions of Michigan, Wisconsin, and Minnesota. The federal government officially listed the species as endangered

INFOGRAPHIC 21.9

Warming Climate Influences Moose and Wolf Population Size

➔ In recent years, climate change has become a significant influence on moose and wolf populations on Isle Royale. Warmer temperatures lead to increased tick infestations of moose, resulting in a weakened and depleted population.

One moose may be home to tens of thousands of ticks at a time.

Ticks cause moose to lose their hair, their appetite, and a good deal of blood.

Ticks make moose weak and vulnerable to predation and starvation. So, while ticks have been increasing...

...the moose population has been decreasing.

Moose weakened by ticks are easier for wolves to catch. After an initial population increase in response to an abundance of moose, the wolf population begins to suffer (2007) as the moose population continues its decline.

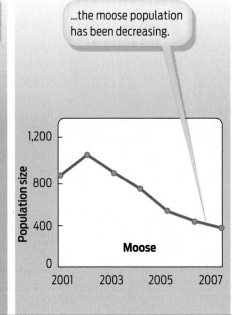

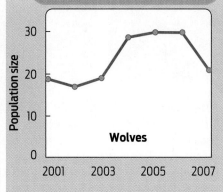

Chippewa Harbor, Isle Royale. © JOHN AND ANN MAHAN

in the early 1970s, when it seemed on the verge of extinction.

The wolves' latest plight poses an ethical dilemma: should scientists intervene on their behalf–say, by importing wolves from another population to reintroduce genetic diversity–or let nature take its course? It's a question that Vucetich thinks about a lot. The answer, he says, will require balancing a number of competing values–not just the value of individual animals, but the values of population and ecosystem health in addition to the values of scientific knowledge and the value of wilderness. Without wolves, for example, would the moose population once again explode and decimate the island's forest? Would healthier wolves be able to completely overwhelm moose, and drive them to extinction on the island? These are the sorts of difficult questions that wildlife manag-

ers will need to consider when debating whether and how to intervene.

If only one value mattered, Vucetich notes, it would be easier to make a decision, but here the values are often competing. The dilemma is a familiar one to conservation biologists. According to Vucetich, these competing values show up in varying degrees in almost any management question that we have in any part of the world. They represent, he says, "this grand question of How should humans relate to nature?" To this question, there are no easy or obvious answers. Nevertheless, he believes it is important for people to debate and discuss these issues–not just scientists and experts, but lay people, too, because "every citizen has a stake in this question of how we relate to nature." ■

▶ Summary

■ Ecology is the study of the interactions between organisms and between organisms and their nonliving environment.

■ Ecologists study these interactions at a number of levels, including population, community, and ecosystem.

■ Living organisms can have a clumped, random, or uniform distribution pattern, depending on ecological and behavioral adaptations. Few organisms fall into strictly one category.

■ Population growth is an increase in the number of individuals in a population. The growth rate of a population is defined as the birth rate minus the death rate. When immigration and emigration are excluded, it is also known as the rate of natural increase.

■ Exponential growth is the unrestricted growth experienced by a population growing at a constant rate. Logistic growth is the slowing of the growth of a population due to environmental factors such as crowding and lack of food.

■ Carrying capacity is the maximum population size that an area can support, given its food supply and other life-sustaining resources. Populations cannot grow exponentially forever; eventually, they hit the carrying capacity for the region and stop growing.

■ Population growth can be limited by a variety of factors, including biotic (living) and abiotic (nonliving) parts of the environment.

■ Density-independent factors, such as a severely cold winter, can affect a population of any size.

■ Density-dependent factors, such as the presence of predators, have different impacts on the population, depending on the size and crowding of individuals in the population.

■ Populations in a community are interconnected, with the fate of one often influencing the fate of the others.

STUDYING ECOLOGY

Ecology is the study of the ways organisms interact with one another and the environment. Ecologists study these interactions at a number of levels.

HINT See Infographics 21.1–21.3.

➔ KNOW IT

1. What is the difference between a community and a population?

2. An ecosystem ecologist might study
   a. plant populations.
   b. herbivores that eat the plants.
   c. predators in the population.
   d. the impact of precipitation patterns on the plant populations.
   e. all of the above

3. Why do the researchers collect scat as part of their study on Isle Royale?

➔ USE IT

4. Which of the following is an example of population growth?
   a. The average weight of Americans has increased substantially in the past decade.
   b. Tropical fish have been found in more northern waters than their usual habitat.
   c. The number of people in a town has increased by 25% in the past 5 years.
   d. The number of butterflies in a region have stayed the same from 1950 to 2010.
   e. all of the above

5. How would you explain to a 10-year-old what ecologists do?

6. A local environmental group wants to determine the population size of squirrels in a nearby nature preserve. What are some methods you could use to estimate the size of the squirrel population? Would the same approaches be as useful in determining the population size of maple trees in the same area? Why or why not?

7. Why is it important for researchers to determine the cause of death of moose on Isle Royale? Can this information be used to help make predictions about moose and wolf populations? Explain your answer.

8. How would you use scat analysis to determine whether an herbivore had a preference for a particular type of vegetation? Be specific about both the type of analysis, and what the analysis would reveal for herbivores with or without a preference for a particular type of vegetation.

POPULATION GROWTH

Ecologists analyze the growth of populations with the help of concepts from population ecology. A number of different environmental factors can affect the growth of populations.

HINT See Infographics 21.4–21.9.

➔ KNOW IT

9. Which of the following would cause a population to grow?
   a. identical increases in both the birth rate and the death rate of a population
   b. a decrease in the birth rate and an increase in the death rate of a population
   c. an increase in the birth rate and a decrease in the death rate of a population
   d. an increase in the birth rate and no change in the death rate of a population
   e. an identical decrease in both the birth rate and the death rate of the population

10. When a population reaches its carrying capacity, what happens to its growth rate?

➔ USE IT

11. You are studying a group of predatory fish that live in a school in a large lake. If a parasite were introduced to the lake by a vacationing fisherman, would you expect it to have a greater impact on the population if the fish were at high density or low density (assume the parasite is passed from one fish to another through the water, but can only remain alive in the water for a very short period of time)? What would happen to this same population if there was a severe drought and very hot summer?

12. Classify each of the following as a biotic or an abiotic factor in an ecosystem. Then predict the impact of each on the moose population of Isle Royale. Explain your answers, keeping in mind possible interactions between the various factors and between the moose and wolf populations.
   a. hot summer temperatures
   b. ticks that parasitize moose
   c. declining numbers of balsam fir trees
   d. a parvovirus in wolves
   e. deep winter snowfall

13. Assume that a new herbivore is added to Isle Royale that is not a prey for wolves. Predict the effect of this introduction on

    a. the populations of trees.

    b. the moose population.

    c. the wolf population.

14. If the moose population remains stable, what other factors could influence the wolf population on Isle Royale?

15. Population Q has 100 members. Population R has 10,000 members. Both are growing exponentially at a 5% annual growth rate.

    a. Which population will add more individuals in 1 year? Explain your answer.

    b. After 5 years, what will be the size of each population?

    c. If the larger population reaches its carrying capacity at the end of the third year, what will its size be after 5 years?

SCIENCE AND ETHICS

16. The wolves of Isle Royale are suffering from bone deformities, probably as a result of inbreeding in their small population.

    a. Do you think that humans should intervene to save the wolves? Would your answer be different if the wolves were near human populations or agricultural centers?

    b. If humans were to intervene, what kinds of strategies might help stabilize or increase the wolf population? Explain your answer.

17. In the 1960s the Asian carp was introduced into U.S. waterways, where it now consumes massive amounts of plankton (including photosynthetic algae). This species is becoming a concern to ecologists and sport fishermen; why do you think this is the case? Think about possible consequences for the communities in which this species now resides. What if the Asian carp invades waterways with recreational or commercial fisheries? What management strategies can you suggest?

What's Happening to Honey Bees?

What's Happening to Honey Bees?

A mysterious ailment threatens to unravel the human food chain

Dave Hackenberg has been keeping bees for more than 40 years. Every spring, as flowering plants start to bloom, he trucks bees from his home in central Pennsylvania to farms around the country, where they help farmers pollinate local crops—everything from California almonds to Florida melons. In November 2006, as he had done for years, Hackenberg brought his buzzing cargo to his winter base in central Florida. When he dropped them off, his 400 healthy hives were "boiling over" with bees. Three weeks later, when he returned to check on them, the bees had essentially vanished; only 40 healthy hives remained.

Mysteriously, there were no dead bees lying in or near the hives. Nor were there any signs of intruders who might have destroyed the hives in search of honey. The bees were simply gone. It was, as Hackenberg said, a bee ghost town.

"I literally got down on my hands and knees and looked between the stones for dead bees," says Hackenberg, but the beekeeper found none. "I was kind of speechless. And people know I'm not speechless."

Hackenberg's was the first reported case of what has since become known as colony collapse disorder, or CCD. But Hackenberg was not alone. Surveys conducted in 2007 and 2008 by the U.S. Department of Agriculture and the Apiary Inspectors of America found that one quarter of all beekeepers across the United States had suffered similar unexplained devastation, losing anywhere from 30% to 90% of their colonies.

Since that first case, some 3 million honey bee colonies across the United States have reportedly been wiped out, with American beekeepers losing an average of 30% to 40% of their colonies every year from 2007 through 2010. To

In three short years, American bee keepers have lost 3 million colonies.

POLLEN
Small, thick-walled plant structures that contain cells that will develop into sperm.

POLLINATION
The transfer of pollen from male to female plant structures so that fertilization can occur.

date, CCD has been documented in 27 states and in Canada, as well as in Europe and Asia.

What's happening to honey bees? No one knows for sure, but it's a predicament that has beekeepers, farmers, and scientists racing to understand and combat the plight of the precious pollinator. At stake are not only billions of dollars worth of agricultural crops, but also the health and diversity of natural ecosystems that rely on bees for their valuable services.

The Dance of Pollination

A hundred million years ago, when dinosaurs roamed the earth, the plant world was dominated by cone-bearing conifers such as pine and redwood trees, which spread their **pollen** via the wind. But a new type of relationship was evolving at this time, one that would forever change the terrestrial landscape. The relationship was the dance of **pollination;** the dance partners, flowering plants and insects.

With the arrival of pollinating insects, including bees, on the evolutionary scene, flowering plants blossomed, diversified, and radiated around the globe. Their great success owes everything to the reproductive advantage of relying on insects, rather than wind, to deliver pollen.

Wind pollination is like junk mail: you need to send a million letters to hook just one receptive customer—or in this case, to fertilize just one plant. All those wasted pollen grains represent a huge energy loss to the plant. Pollinating insects, on the other hand, are like FedEx: they deliver a pollen package directly to the appropriate recipient.

Of the 250,000 species of flowering plants, or angiosperms, that exist worldwide, more than 75% are dependent on insect pollinators. And of the many types of insect pollinators, bees are by far the most important ecologically. In fact, in many natural environments, bees species are **keystone species,** meaning they play a central role in holding the **community** together.

You can think of a keystone species as analogous to the keystone in an archway–it doesn't support as much weight as the other stones, but if it is removed, the doorway collapses (**Infographic 22.1**).

The bee species cultivated by the majority of beekeepers in the United States and Europe is *Apis mellifera,* the Western honey bee. It is hard to overestimate the importance of this tiny pollinator to modern agriculture. In the United States alone, more than 100 different crops–worth an estimated $15 billion annually–are dependent on honey bee pollination, including apples, oranges, blueberries, melons, pears, pumpkins, cucumbers, cherries, raspberries, broccoli, avocados, asparagus, clover, alfalfa, and almonds. A 2007 study published in the *Proceedings of the Royal Society B* (London), found that 87 of the leading global food crops, accounting for 35% of global crop production, are dependent upon pollinators, the most important of which are honey bees.

INFOGRAPHIC 22.1

Bees Are Keystone Species

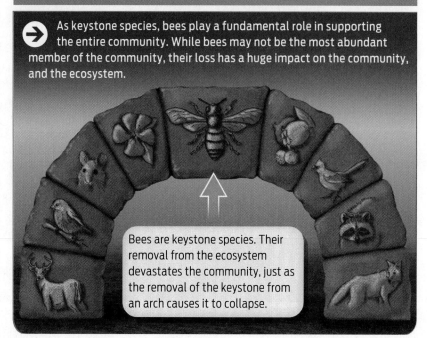

→ As keystone species, bees play a fundamental role in supporting the entire community. While bees may not be the most abundant member of the community, their loss has a huge impact on the community, and the ecosystem.

Bees are keystone species. Their removal from the ecosystem devastates the community, just as the removal of the keystone from an arch causes it to collapse.

"One in every three bites of food we eat is pollinated directly or indirectly by honey bees," says Dennis vanEngelsdorp, State Apiarist for Pennsylvania's Department of Agriculture. Without honey bees, he says, we wouldn't starve–we would still have wheat, rice, corn, and other crops that are either wind- or self-pollinated–but many of our favorite foods might no longer grace our tables (**Infographic 22.2**).

For bees, flowers are food: they contain the protein-rich pollen and sugary nectar that bees need to nourish themselves and their hives. With their long retractable proboscis, a tongue-like organ, bees are able to reach deep into a flower to draw out the nectar. Being fuzzy and having a slight electrical charge, bees attract pollen as they snuggle up to a flower the way warm socks attract other clothes as they come out of the dryer. The bees can then transfer this pollen to other plants as they continue their hunt for food.

Honey bees are more efficient at pollination than many other types of pollinators, which is why farmers have come to depend on them. The average honey bee will make 12 or more foraging trips a day, visiting several thousand flowers. On each trip, she (the foragers are

KEYSTONE SPECIES
Species on which other species depend, and whose removal has a dramatic impact on the community.

COMMUNITY
A group of interacting populations of different species living together in the same area.

A beekeeper with his hives.

> **"One in every three bites of food we eat is pollinated directly or indirectly by honey bees."**
> —Dennis vanEngelsdorp

all female) will confine herself to flowers from a single plant species, thus ensuring delivery of the proper pollen. Because they come and go from a central home base–their hive–honey bees can be counted on to stay in a fixed area around a crop. And with roughly 40,000 individual bees per hive, this is a versatile workforce that's also easy to transport from crop to crop.

Of course, pollination isn't just about feeding bees and humans; it's also how flowering plants have sex. Masters of seduction, flowers have evolved countless colorful and fragrant adaptations that lure their pollinators to the blossom. In the words of poet Kahlil Gibran, "For to the bee a flower is a fountain of life, and to the flower a bee is a messenger of love."

In order for a flowering plant to reproduce, male pollen must find its way to the female eggs of a plant of the same species. A flower is the reproductive hub of an angiosperm–it is where its reproductive organs are located, male and female in the same flower in some species, separate flowers in others. The male reproductive

INFOGRAPHIC 22.2

Commercial Crops Require Bees

→ Many of the crops that we rely on for food, fuel and fiber rely on bees for their pollination and reproduction.

Bee hives are placed in the field. Bees leave the hive in search of nectar and pollen, which they use to make food for themselves and their hive.

	Value attributed to honeybees (in millions, 2000 estimates)	Percentage of crop pollinated by… Honeybees / Other insects / Other
Alfalfa, hay & seed	$4,654.2	
Apples	1,352.3	
Almonds	959.2	
Citrus	834.1	
Cotton (lint & seed)	857.7	
Soybeans	824.5	
Onions	661.7	
Broccoli	435.4	
Carrots	420.7	
Sunflower	409.9	
Cantaloupe/honeydew	350.9	
Other fruits & nuts	1,633.4	
Other vegetables/melons	1,099.2	
Other field crops	70.4	
Total	**14,564**	

Besides insects, other means of pollination include birds, wind, and rainwater.

Source: Compiled by CRS using values reported in R.A. Morse and N.W. Calderone. *The Value of Honeybees as Pollinators of U.S. Crops in 2000*, March 2000. Cornell University.

Bees gather nectar from blossoms. During this process, they transfer pollen between flowers.

Flowering Plant Reproduction Relies on Pollinators

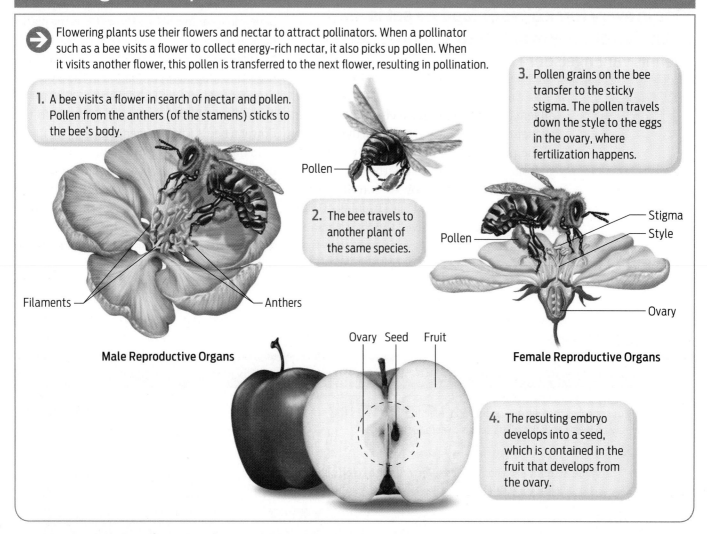

Flowering plants use their flowers and nectar to attract pollinators. When a pollinator such as a bee visits a flower to collect energy-rich nectar, it also picks up pollen. When it visits another flower, this pollen is transferred to the next flower, resulting in pollination.

1. A bee visits a flower in search of nectar and pollen. Pollen from the anthers (of the stamens) sticks to the bee's body.

Pollen

2. The bee travels to another plant of the same species.

Pollen

3. Pollen grains on the bee transfer to the sticky stigma. The pollen travels down the style to the eggs in the ovary, where fertilization happens.

Stigma
Style
Ovary

Filaments

Anthers

Male Reproductive Organs

Female Reproductive Organs

Ovary Seed Fruit

4. The resulting embryo develops into a seed, which is contained in the fruit that develops from the ovary.

---

organ, called a **stamen,** consists of a stemlike filament topped with a pollen-saturated anther. When a bee lands on or brushes against an anther during her pursuit of nectar and pollen, her furry body picks up pollen grains.

As the bee continues to forage, she carries the pollen to the female reproductive organ of the flower—the **pistil.** The pistil is topped with a sticky "landing pad" called a stigma. When a bee lands on the stigma, pollen grains are deposited, and from there travel down a tubelike style into the ovary, where they fertilize the eggs. A fertilized egg will eventually develop into an embryo-containing **seed,** while the surrounding ovary eventually becomes the fruit (Infographic 22.3).

> **As keystone species, bees play a crucial role in the food chain.**

A Critical Link in the Food Chain

By helping plants reproduce, bees not only help sustain human food production, they also maintain the integrity and productivity of many natural communities. Without these miniature matchmakers, many flowering plants would become extinct, and many birds and mammals would go hungry. That's because, as keystone species, bees play a crucial role in **food chains**—the linked sequences of feeding relationships in a community.

Take the little-known southeastern blueberry bee (*Habropoda laboriosa*), which feeds mainly on the flowers of blueberry plants. This speedy pollinator will visit 50,000 blueberry flowers over the course of a few

STAMEN
The male reproductive structure of a flower, made up of a filament and an anther.

PISTIL
The female reproductive structure of a flower, made up of a stigma, style, and ovary.

SEED
The embryo of a plant, together with a starting supply of food, all encased in a protective covering.

weeks in spring, contributing to the production of more than 6,000 blueberries. All those blueberries are an important food source for many wild animals, including bluebirds, robins, wild turkeys, fox, mice, rabbits, skunks, chipmunks, and deer. In turn, the small animals that eat the blueberries are fed upon by larger carnivorous animals, such as hawks and coyotes. Without the blueberry bee—the keystone in this community—the chain is broken.

Organisms in a food chain can be categorized by who eats whom. At the base of the food chain are **producers**—autotrophs such as plants and algae, which obtain energy directly from the sun and supply it to the rest of the food chain. Organisms higher up the food chain are **consumers**—heterotrophic organisms that eat the producers or eat other organisms lower on the chain to obtain energy.

When one organism feasts on another, that's called **predation.** Usually when we think of predators, we think of large, fierce animals such as wolves hunting moose (Chapter 21). Pre-

dation can also take more subtle forms, however. Feeding on plants, for example, is a type of predation known as **herbivory,** an activity that may or may not kill the plant. Predators that feed on plants are known as herbivores—bees feasting on nectar and pollen, for example, or a deer nibbling on a blueberry bush. Herbivores are an important food source for carnivorous consumers higher up the chain.

At the top of the food chain are those animals known as top consumers—animals such as coyotes, hawks, and wolves (as well as meat-eating humans), who have no natural predators and are not generally eaten by anything else in the community.

While it might seem preferable to be at the top of a food chain rather than at the bottom, there are downsides to being last in line to eat. For one, it's harder to obtain the energy necessary to live. As consumers prey on organisms below them in the chain, energy is transferred up the chain through what are known as **trophic levels** (from the Greek *trophe*, meaning "food"). But not all the energy stored in a lower level

Wolves are predators at the top of the food chain.

Energy Flows Up a Food Chain

➔ In a food chain, energy flows in one direction: from producers to consumers. Producers obtain energy from the sun. Consumers obtain energy by eating producers. As consumers eat other consumers, the flow of energy continues up the food chain. The passage of energy is not efficient, however, as only 10% of energy makes it from one trophic level to the next. The result is an energy "pyramid."

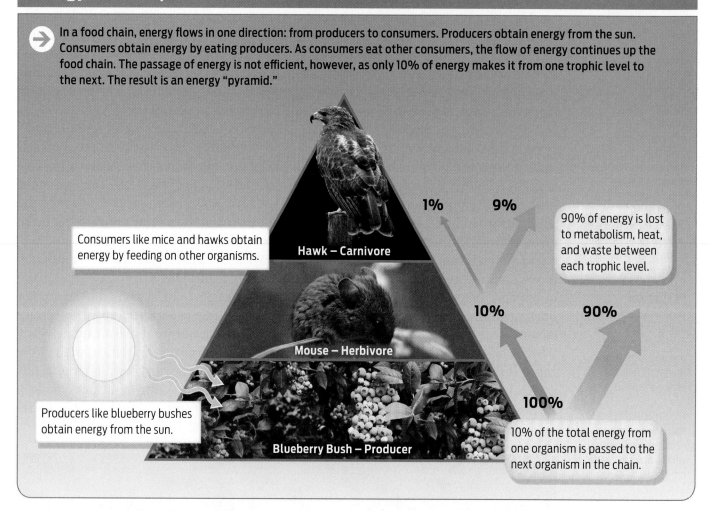

Consumers like mice and hawks obtain energy by feeding on other organisms.

Hawk – Carnivore

Mouse – Herbivore

Producers like blueberry bushes obtain energy from the sun.

Blueberry Bush – Producer

1%   9%

90% of energy is lost to metabolism, heat, and waste between each trophic level.

10%   90%

100%

10% of the total energy from one organism is passed to the next organism in the chain.

---

makes it to the level above it; at each step, much of this energy is lost from the chain (**Infographic 22.4**).

When a deer feeds on a blueberry bush, for example, most of the energy in the blueberries is either burned as fuel (in aerobic respiration) and given off as heat, or passed through the deer as indigestible plant fiber. Only a very small portion (about 10%) of the energy stored in the blueberries goes to putting weight on the deer.

This is the main reason why top carnivores like coyotes, hawks, and wolves are scarce on earth, and why there are no predators of these creatures: there's simply not enough energy left in the chain to sustain more of them. (It's also why vegetarianism is more energetically effi-

cient than meat-eating: the same amount of a crop can feed many more vegetarians than meat-eaters who eat the animals that eat the crop.)

While it's helpful to think of the food chain as a stepwise series, the food-chain concept is an oversimplification. Many organisms are omnivores (that is, they eat both plants and animals) and so occupy more than one position in the chain. The result is a complex, intertwined **food web**—like the one that links bees to the food on your breakfast table. This food web includes not just the fruits and vegetables that bees pollinate directly, but the animals that eat bee-pollinated crops such as alfalfa and provide us with meat and dairy products (**Infographic 22.5**).

FOOD WEB
A complex interconnection of feeding relationships in a community.

PARASITISM
A type of symbiotic relationship in which one member benefits at the expense of the other.

MUTUALISM
A type of symbiotic relationship in which both members benefit; a "win–win" relationship.

A Swarm of Problems

The United States is home to approximately 1,000 commercial beekeepers, who together cultivate about 2.5 million bee colonies. To a beekeeper–essentially a small business owner– losing even 30% of his or her colonies represents an unsustainable financial loss. The sudden disappearances were a serious worry for Dave Hackenberg and other beekeepers.

While the losses from CCD have indeed been devastating, this was actually not the first time that beekeepers' livelihood has been hit hard. Since 1987, beekeepers have had to battle significant annual losses from an aggressive pest: the blood-sucking varroa mite.

An invasive species, the varroa mite was likely introduced into the United States on the backs of imported bees. The sesame seed-size freeloader is a parasite that feeds on bees' blood, weakening their immune systems and making them more susceptible to disease. **Parasitism** is a type of symbiosis in which one species (in this case, the mite) clearly benefits, and one species (the honey bee) clearly loses. Because it involves one species feeding on another, parasitism is also a form of predation.

Not all symbiotic relationships are harmful to one of the partners; they can sometimes be mutually beneficial. Bees and flowering plants are a perfect example of one such **mutualism.**

INFOGRAPHIC 22.5

A Honey Bee Food Web

The intersection of multiple food chains in a community results in a complex food web. Individual organisms in the food web have multiple important roles that keep the community healthy.

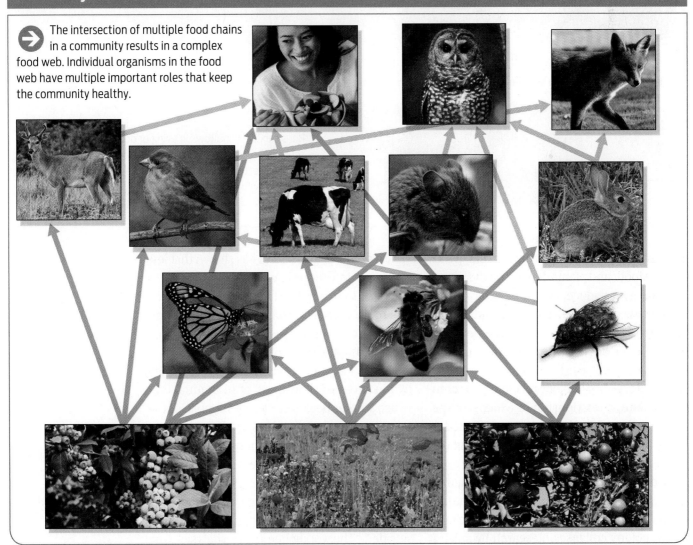

Organisms Live Together in Symbioses

 Symbioses are cases of different species living together in close association. These associations can provide benefits or do harm to the partners involved.

a. Mutualism – Both species benefit from the interaction

 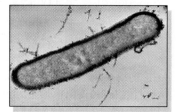

Mutualistic bacteria

Bees and flowering plants represent a mutualism: Bees get nectar and pollen which they use for food. Pollination allows successful reproduction for the plant. Bacteria living in the bee gut provide protection against bee pathogens. In turn, the bacteria get a safe place to live and a constant source of food.

b. Parasitism – One species benefits and the other is harmed

 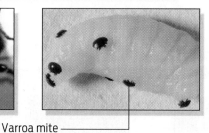

—— Varroa mite ——

The varroa mite has historically been the primary pathogen of bees. It parasitizes both larvae and adult bees, obtaining nutrition, and leaving the bee immune system suppressed. Hives infected with the mite are susceptible to fatal infections caused by bacteria and viruses.

c. Commensalism – One species benefits and the other is unharmed

Bees can live in hollows in living or downed trees. In this case, the bees benefit from the shelter provided by the tree, and the tree is not harmed by the symbiotic association with the bees.

Bees can't survive without the flowers, which provide food, and plants depend on the bees to help them reproduce. Honey bees have other mutualistic symbioses, as well, including with bacteria that live safely inside the bees and benefit their hosts by helping them combat disease.

A third type of symbiotic relationship is **commensalism,** a relationship in which one species benefits while the other is unaffected or unharmed–bees living in a hollowed-out oak tree, for example (Infographic 22.6).

As devastating as the parasitic varroa mite infestation has been, it is unlikely to be the sole or even primary factor responsible for the most recent colony collapses. Research by apiarist vanEngelsdorp, and others has shown that levels of mite infections in collapsing colonies are no higher than they had been in previous years. Moreover, a mite infestation does not explain the most curious aspect of the condition: the sudden disappearance of entire colonies.

Honey bees are a colonial species: they live in hives of thousands of individuals, in which worker bees collectively support all the juvenile larvae. In collapsing colonies, the worker bees (all female) abandon the hive. With no workers to help larvae reach maturity, the colony dies.

According to vanEngelsdorp, the worker bees are likely practicing something called "altruistic suicide." "The worker bee knows she's sick," he

COMMENSALISM
A type of symbiotic relationship in which one member benefits and the other is unharmed.

explains. "She knows, 'Well, I better fly out of here and die away from the hive and maybe preserve my nest mates.'" But that altruistic practice can spiral out of control, he says, and the result is a collapsing colony.

Not surprisingly, the sudden disappearances

The United States is home to approximately 1,000 commercial beekeepers, who together cultivate about 2.5 million bee colonies.

have fueled intense speculation among beekeepers and laypeople alike about what's going on. Hypotheses have included everything from pesticides, viruses, and genetically modified crops to cell phone radiation, global warming, and even alien abduction. It's a baffling who-done-it with many suspects but no smoking gun.

Honey Bee Forensics

Among the first to investigate the die-offs was a team of Pennsylvania State University biologists headed by vanEngelsdorp and Diana Cox-Foster. It was Cox-Foster whom beekeeper Hackenberg called the day his bees went missing.

The team started their investigation by performing autopsies on the few remaining bees in Hackenberg's colonies. When vanEngelsdorp looked through his microscope, he was shocked by what he saw: "a lot of different scar tissue, and [what] looked like foreign organs," he says. There were also signs of multiple infections, including a parasitic fungus called *Nosema ceranae*. The bees' insides were overrun with pathogens.

Though the bees were clearly sick, each colony seemed to suffer from a different spectrum of ailments. "The bees are getting the flu," says vanEngelsdorp. "What we don't understand is the fact that it's not always the same strain of flu." The researchers hypothesized that something had compromised the bees' immune system, making them vulnerable to infections that a healthy colony could normally fend off. Some observers have even likened the condition to "bee AIDS."

The analogy certainly seems fitting. But early attempts to identify a bee-equivalent of HIV (the virus that causes AIDS) were unsuccessful. The initial prime suspect, the varroa mite, was not present at high enough levels to cause a crippled immune system. And all of the other parasites and infections had previously been documented in healthy bee populations, and so were unlikely to have caused the catastrophic losses.

Hoping to isolate a previously unidentified culprit, Cox-Foster and her colleagues enlisted genomics experts from Columbia University to scour genetic material from the hives for evidence of a new invader. After months of intensive work, their efforts seemed to pay off: genetic tests revealed that a virus called Israeli acute paralysis virus (IAPV) was present in 96% of the hives affected with CCD. The researchers thought they had found the smoking gun. Subsequent research, however, showed that not all honey bee colonies that are infected with IAPV have symptoms of CCD, suggesting that the virus alone is not the source of the problem.

More recently, in 2010, researchers from the University of Montana and the U.S. Army's Edgewood Chemical Biological Center presented evidence that another viral culprit–invertebrate iridescent virus (IIV)–was present in essentially all collapsing hives. Whether this virus proves to be the decisive factor in CCD remains to be seen. But since IIV is present in noncollapsing hives as well, it is unlikely to be acting alone.

In fact, it may be that there is no single cause of CCD, but rather a complex combination of causes. "All the evidence so far has really supported the idea that it's likely a combination of factors that are stressing the bees beyond their ability to cope," says Maryann Frazier, a bee researcher at Penn State who is part of Cox-Foster's team.

One factor that almost certainly plays a role in exacerbating the condition is poor nutrition. Just like humans, bees need a well-balanced diet that contains all the essential nutrients to remain healthy. For a number of reasons,

Pollinators Have Different Ecological Niches

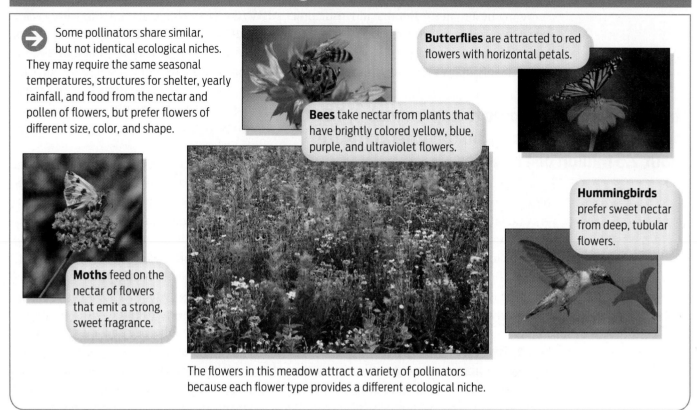

Some pollinators share similar, but not identical ecological niches. They may require the same seasonal temperatures, structures for shelter, yearly rainfall, and food from the nectar and pollen of flowers, but prefer flowers of different size, color, and shape.

Butterflies are attracted to red flowers with horizontal petals.

Bees take nectar from plants that have brightly colored yellow, blue, purple, and ultraviolet flowers.

Hummingbirds prefer sweet nectar from deep, tubular flowers.

Moths feed on the nectar of flowers that emit a strong, sweet fragrance.

The flowers in this meadow attract a variety of pollinators because each flower type provides a different ecological niche.

honey bees are finding it harder and harder to obtain a nutritious diet.

Competing for Resources

When European settlers first brought the Western honey bee to the United States in the 1600s, the bees quickly spread from managed colonies into the wild, in some cases displacing native bee species. Honey bees have been so successful at colonizing new habitats because they are largely generalists when it comes to flower choice.

Though they tend to visit a single species of flower on each foraging trip, honey bees may visit more than 100 species of flowers within a single geographic region over the course of a season. In warm climates, they are active year round, and tend to feed throughout the day and start foraging earlier in the morning than many native bee species. In other words, honey bees have a broad ecological **niche**—the space, environmental conditions, and resources (including other living species) that a species needs in order to survive and reproduce.

Different pollinators generally have different niches, owing to their varying sizes and preferences for different flower types. Bees, for instance, are attracted to brightly colored blossoms—those with yellow, blue, and purple petals, for example, but not red. Butterflies, by contrast, are commonly attracted to red flowers that are large and easy to land on. Moth-pollinated flowers tend to have pale or white petals with no distinctive color pattern but with strong fragrance (**Infographic 22.7**).

When two or more species rely on the same limited resources—that is, when their niches overlap—the result is competition. Competition tends to limit the size of competing populations and may even drive one out. In theory, no two species can successfully coexist in identical niches in a community because one would eventually out-compete the other—a concept described by the **competitive exclusion principle.** In reality, however, very few species share *exactly* the same niche, so different species may find a competitive balance by

NICHE
The space, environmental conditions, and resources that a species needs in order to survive and reproduce.

COMPETITIVE EXCLUSION PRINCIPLE
The concept that when two species compete for resources in an identical niche, one is inevitably driven to extinction.

subdividing resources (so-called resource partitioning).

Some species compete through behavior. African honey bees (*Apis mellifera scutellata*), for example, are a subspecies that was brought from Africa to Brazil in 1956 and which quickly expanded its range to include Central America and the southern United States. So-called africanized honey bees are much more aggressive than Western honey bees. They will chase away other pollinators from food sources, and even swarm and sting animals that get near their hives—behaviors earning them the colloquial name "killer bee" (Infographic 22.8).

Some scientists were worried that killer bees might displace or even interbreed with populations of Western honey bees, with potentially disastrous results for beekeeping and agriculture. (Imagine trying to convince a farmer to let you pollinate his almond grove with killer bees!) This hasn't happened. The much more serious problem for the Western honey bee, it seems, is competition from an even more dangerous species: humans. Humans and their activities have limited the resources used by bees and other pollinators. Agriculture, suburban sprawl, and development, for example, have all decreased bees' natural forage areas, and fragmented their habitat into nonoverlapping zones. Unable to access as many resources in a single foraging trip, bees must compete with each other in the patches that remain.

Take the well-manicured lawns that many of us are so proud of. An immense stretch of green grass and no flowers, a lawn is "basically a desert to pollinators," says Frazier. There is literally nothing for them to eat. Likewise, many agricultural areas are planted with monocultures (that is, single crops), which all bloom at the same time, leaving no flowers for the rest of the year. Worse yet, certain genetically engineered pollen-free crops trick bees into thinking they'll find food, only to leave them hungry. And certain non-native plants have floral structures that are inaccessible to indigenous pollinating insects.

As a consequence of these and other human actions, in some geographic regions bees must compete both with one another and with other pollinators for a dwindling supply of food-providing flowers. Many are going hungry and thus are left vulnerable to conditions that can lead to CCD.

INFOGRAPHIC 22.8

Bees Compete for Resources

→ Species with similar niches compete for resources that may be limited due to natural or human influences. Species may outcompete one another or find a balance, depending on their foraging abilities and behaviors.

Blueberry Bee
(*Osmia ribifloris*)

Food Partitioning
Native bees like the blueberry bee have had to share their food resources with non-native species brought to North America. In some cases, their populations have declined, as they have had to partition, or split up, the resources to share with other bee species.

Honey Bee
(*Apis mellifera*)

Generalist Foraging Patterns
Imported honeybees can forage over great distances and feed on a wide variety of flowers. As they feed year round in warmer climates, they are very successful in their competition with other bee species, which may have more limited niches.

Africanized Honey Bee
(*Apis mellifera scutellata*)

Defensive Behavior
Killer bees compete successfully due to their aggressive defense of food resources. They chase other pollinators away from available food.

Well-manicured lawns are "a desert to pollinators," as are areas planted with a single crop (which blooms once, leaving no flowers for the rest of the year).

Honey Bee in the Coal Mine?

Honey bees aren't the only pollinator in peril. According to a report published in 2007 by the National Research Council, the number and abundance of pollinator species have declined greatly over the last several years. In fact, several bumblebee species are becoming or have become extinct in North America. The concern among researchers and beekeepers is that honey bees may be the "canary in the coal mine," forecasting what's in store for other pollinators. "It's not only the honey bees that are in trouble," says Hackenberg. "All the beneficial insects are in a bad situation."

What's ailing these insects? In addition to a shrinking and fragmented habitat, a disquieting possibility is that they are being poisoned by pesticides. Penn State researcher Frazier and her colleagues have looked at pollen and wax from beehives and found large amounts of many different kinds of pesticides, some of which are approaching toxic levels for the bees. "Pesticides are definitely in the mix and we think they are definitely a player in the stresses that bees are experiencing," says Frazier.

Of particular concern to beekeepers is a class of pesticides known as neonicotinoids, or neonics for short. Neonics are an artificial form of nicotine used in commercial agriculture. (Nicotine, made by tobacco plants, is a natural deterrent to plant-eating insects.) Research has shown that neonics can impair honey bees' ability to find and return to their hives, and the U.S. Environmental Protection Agency acknowl-

edges that neonics are "highly toxic to honey bees." As of 2008, both France and Germany had banned the use of neonicotinoid products after severe bee losses occurred in those countries, but the ban does not seem to have prevented further die-offs.

Although researchers have not yet been able to prove that neonics are playing a role in CCD– "the jury is still out," says Frazier–beekeepers like Hackenberg are understandably cautious about what pesticides they will allow their colonies to be exposed to (Infographic 22.9).

Because it may involve a complex combination of triggers, there is no easy remedy to CCD. It may require making fundamental changes to our beekeeping and agricultural practices. In particular, we could break up fields of monocul-

A more bee-friendly awareness is emerging.

What is Causing Colony Collapse Disorder?

→ Bees live in a social colony with a single queen and her offspring. The collapse of colonies all over the world is of great concern. The cause of this disorder is likely to be complex and to involve an interplay of several factors.

A healthy bee colony is full of busy adult bees.

A "collapsing" colony has very few adults, so the developing larvae that depend on them will not survive.

Likely causes →

Stress and nutrition
Bees need both nectar and pollen for a complete, nutritional diet. When blossoms are scarce, bee keepers feed their colonies sugar mixtures, but not pollen supplements. The lack of essential nutrients leaves the colony stressed, which weakens their defenses.

Pesticides
The pesticides sprayed on crops can make their way into pollen particles. The amounts measured in pollen have reached toxic levels, which may be affecting the health of the bees that eat the pollen.

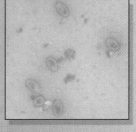

***Nosema ceranae* parasite**
This intestinal parasite prevents bees from processing food properly which can weaken and kill bees.

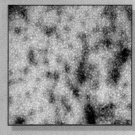

Israeli acute paralysis virus
The virus causes paralysis and death in bees.

tures with varied bee-friendly plants: red clover, foxglove, and bee balm, for example. We could also use pesticides sparingly and avoid spraying at times of day when bees are actively foraging.

While these individual steps would certainly help matters, apiarist Dennis vanEngelsdorp diagnoses a more systemic problem. In his estimation, we suffer from NDD–"nature deficit disorder." To help bees, he says, we need also to cure ourselves. As treatment, he prescribes reconnecting to nature in a more immediate and local way–"having a meadow or living by a meadow," for example, or becoming a beekeeper oneself.

In addition, says bee expert Frazier, "people need to take more time to understand where their food comes from, what it takes to produce food and have this incredible supply of food available to us."

While the fate of the honey bees remains uncertain, there are signs that a more bee-friendly awareness is beginning to emerge, thanks in part to the concerns raised by CCD. Häagen-Dazs, the ice-cream maker, has recently

> "People need to take more time to understand where their food comes from, what it takes to produce food and have this incredible supply of food available to us." —Maryann Frazier

launched a "Help the Honey Bee" campaign, noting that honey bee-dependent products are used in 25 of its 60 flavors. The company also introduced a new flavor called Vanilla Honey Bee, the proceeds of which are being used to fund CCD research. Even ordinary citizens are catching the bee buzz. From city-dwellers becoming amateur rooftop beekeepers to suburbanites letting more flowers grow in their yards, the ranks of people wanting to make the environment pollinator-friendly has swelled. And that's a cause that just about everyone can get behind–because, as more and more people are coming to realize, a world without honey bees just wouldn't be as sweet. ■

▶ Summary

■ An ecological community is made up of interacting populations of different species.

■ Bees are keystone species in that they play a fundamental role in supporting the entire community, much like the keystone in an arch.

■ Bees are the primary pollinators for many species of flowering plants, which depend on the pollinators to transfer pollen between plants of the same species.

■ Flowers are the reproductive hub of a plant, containing male and female reproductive structures. Pollination, the transfer of pollen from male to female structures, results in fertilization.

■ The organisms in a community are connected by a food chain. Each player in the chain is an important ecological link in the chain.

■ Organisms at the base of the food chain are producers—they obtain energy directly from the sun and supply it to the rest of the food chain; organisms higher up the food chain are consumers— they obtain energy by eating organisms lower on the chain.

■ In predation, one organism eats another. Herbivory—eating plants—is one type of predation.

■ As energy flows through different trophic levels in the food chain, some of it is lost to the environment.

■ Organisms can have different types of symbiotic relationships. In mutualistic symbioses, both members benefit; in parasitism, one member benefits while the other suffers; and in commensalism, one member benefits while the other is unharmed.

■ The space and resources, including other members of the community, that a species uses to survive and reproduce define its ecological niche. Some species have overlapping niches, leading to competition for resources.

■ Bees are not the only pollinators in peril. Human development and agriculture have decreased habitat and foraging areas for many natural pollinators, resulting in increased competition among them.

Chapter 22 Test Your Knowledge

KEYSTONE SPECIES AND POLLINATION

Keystone species are critical in community structure. Insect pollinators play key roles by ensuring reproduction of many species of flowering plants.

HINT See Infographics 22.1–22.3.

KNOW IT

1. How does a community differ from a population?

2. What are keystone species?

3. A rocky shoreline that is covered at high tide but exposed at low tide supports a community of mussels, algae, barnacles, and starfish. An ecologist systematically removes species from different areas of the beach. Removing the mussels doesn't substantially change the community, but removing the starfish dramatically changes the mix of species in the area. Which is the keystone species?
   a. mussels
   b. barnacles
   c. algae
   d. starfish
   e. all of the above

4. Bees transfer pollen from the _____ to the _____.
   a. anther; stigma
   b. stigma; style
   c. filament; ovary
   d. anther; ovary
   e. stigma; anther

USE IT

5. Think about a community of organisms that you are familiar with. From what you know about this community, choose what you think might be a keystone species and defend your choice.

6. If you have pollen allergies, are you more likely to be suffering from the effects of bee-carried pollen or wind-carried pollen? Explain your answer.

FOOD CHAINS AND ENERGY FLOW

Energy is initially captured by autotrophs and flows through organisms in food chains. As energy flows from producers through consumers, some of it is lost to the environment as heat.

HINT See Infographic 22.4.

KNOW IT

7. In relation to a food chain, what do plants and photosynthetic algae have in common?
   a. nothing
   b. they are both producers
   c. they are both first level consumers
   d. they are both top level consumers
   e. their numbers are limited by the energy they take in from heterotrophic food sources

8. A bear who eats both blueberries and fish from a river can be referred to as
   a. an omnivore
   b. a heterotroph
   c. a consumer
   d. a producer
   e. all of the above
   f. a, b and c
   g. a and c

USE IT

9. Describe a natural food web that includes a terrestrial food chain (including honeybees) and at least one aquatic organism from an aquatic food chain.

10. Explain how a cow can eat so many kilograms of grain but not produce the equivalent amount of energy in the form of meat. What happens to the energy stored in the grain once it is ingested by the cow?

11. Compare the diet of a human who is an herbivore with that of a human who is a top consumer.

COMMUNITY INTERACTIONS

Organisms in a community interact in many different ways, which are sometimes helpful to one another, sometimes not. The precise role that each species plays in a community defines its niche.

HINT See Infographics 22.5–22.7.

KNOW IT

12. What are some important features of a honey bee niche? How is it that other nectar-feeding organisms can coexist with bees as part of a community?

13. Competition is most likely to occur
   a. when one species eats another.
   b. when two species occupy different niches.
   c. when one species helps another.
   d. when two species occupy overlapping niches.
   e. when two species help each other.

> USE IT

14. On a rocky intertidal shoreline (the area between the highest and lowest tidelines, so the intertidal zone is alternately exposed and covered by seawater), mussels and barnacles live together attached to rocks where they obtain food by filtering it from ocean water. Since these two species coexist in the same habitat, we predict that they do not have identical niches. What might be separating their niches enough to allow them to occupy the same rocky intertidal zone?

15. If a meadow of wildflowers were converted to a field of corn, would you predict the number and diversity of bees in the community to increase or decrease? Explain your answer.

> KNOW IT

16. Which of the following characterizations best defines a symbiotic relationship?
   a. Both organisms benefit.
   b. The organisms live in close association.
   c. Only one organism benefits.
   d. The relationship is mutually harmful.
   e. Neither organism benefits.

17. Would you characterize the relationship between the bacteria that live symbiotically within bees and their bee hosts as a type of competition, parasitism, mutualism, or commensalism? Explain your answer.

> USE IT

18. What is the evidence for and against varroa mites and IAPV being responsible for colony collapse disorder (CCD)?

19. We all have *E. coli* bacteria living in our intestinal tracts. Occasionally these *E. coli* can cause urinary tract infections. From this information, which of the following terms would you say describe(s) the relationship between us and our intestinal *E. coli*? Why did you choose the term(s) you did?
   a. competition
   b. mutualism
   c. parasitism
   d. symbiosis
   e. predator–prey

SCIENCE AND ETHICS

20. Many people consider bees a stinging nuisance. What could you say to such people to dissuade them from killing all the bees in their backyards?

21. Farmers often plant large acreage of a single crop in order to maximize yield and simplify harvesting. From what you have read in this chapter, what arguments can you make for *not* growing acre after acre of almonds, even for almond lovers?

The Heat Is On

The Heat Is On

From migrating maples to shrinking sea ice, signs of a warming planet

For more than two centuries, Burr Morse's family has collected sap from Vermont's maple trees and boiled it to sweetened perfection. If you pour maple syrup over your breakfast pancakes or eat maple-cured ham, you've likely enjoyed the results of their careful craft, or that of other Vermont sugar farmers. About one in four trees in the state of Vermont is a sugar maple (*Acer saccharum*), and each year the state produces between half a million and a million gallons of syrup, making Vermont the number one maple syrup producer in the United States. Yet what has been a proud family tradition and the economic lifeblood for generations of sugar farmers could very well be in jeopardy.

"In the last 20 years we have had a number of bad seasons and most of those I would attribute to temperature that is a little too warm," says Morse. "For maple sugaring to work right, the nights have to freeze down into the mid 20s, and the days have to thaw up into the 40s. And the nights for those 20 years, it seemed, were not quite getting cold enough."

Morse isn't the only one to notice the shift. Sugar farmers across New England have noted the changes in temperature and are leery about their long-term effects.

Warmer winters in New England could have a large economic impact on the region. As

One out of 4 trees in Vermont is a sugar maple.

ecologist Tim Perkins, director of the Proctor Maple Research Center at the University of Vermont, testified to Congress in 2007, "If the northeast regional climate continues to warm as projected, we expect that the maple industry in the U.S. will become economically untenable during the next 50-100 years." This is not just icing on the cake; according to Perkins, "the total economic impact of maple in Vermont alone is nearly $200 million each year."

Before 1900, 80% of the world's maple syrup came from trees in the United States, the rest from Canada. Today, the pattern is reversed, with Canada greatly out-producing the United States. Canada now accounts for about 80% of world maple syrup production. While part of this reversal has to do with marketing, Canadian government subsidies, and improved technologies,

> **"The total economic impact of maple in Vermont alone is nearly $200 million each year."**
> **—Tim Perkins**

scientists believe that climate change is a significant contributing factor, putting New England sugar farmers at a competitive disadvantage (Infographic 23.1).

New England's maples are not the only ones feeling the heat. Plant and animal species throughout the world–from herbs in Switzerland to starfish in California–are being affected by rising temperatures. Some are shifting their ranges as a result: many historically subtropical aquatic animals, such as seahorses and turtles, are drifting toward the coasts of northern England and Scotland, where ocean temperatures are warmer than they used to be. And fish that were once wholly tropical are turning up in North Atlantic waters. Other organisms that cannot easily relocate, such as plants and mountain-dwelling animals, are being driven to extinction.

Climate change is a natural part of the environment, of course, and nothing new in the long history of earth. But scientists are finding increasing and compelling evidence that humans are accelerating the pace of change, with potentially dire consequences for life on our planet.

To Everything a Season

In nature, timing is everything. And for many species temperature is nature's clock, cueing their seasonally appropriate tasks such as mating or producing flowers in the springtime. Rising temperatures around the globe are interfering with these natural rhythms. Many plants are flowering earlier now than they once did; animals—an example is the yellow-bellied marmot—are emerging from hibernation earlier; and many bird and butterfly species are migrating north and breeding earlier in the spring than they did a few decades ago. It's a pattern of change that scientists are seeing around the globe (Infographic 23.2).

So what, you might ask, if flowers bloom earlier or marmots shake off their wintry slumber earlier in the season? Because living things are exquisitely adapted to their environments, a change in one part of an ecosystem may upset others.

As the name implies, an **ecosystem** is a complex, interwoven system of interacting components. It includes both the community of living organisms present in an area and features of the nonliving environment–physical conditions such as temperature and moisture and chemical resources found in soil, water, and

ECOSYSTEM
The living and nonliving components of an environment, including the communities of organisms present and the physical and chemical environment with which they interact.

INFOGRAPHIC 23.1

Vermont Maple Syrup: A Thing of the Past?

The amount of maple syrup produced in Vermont has been declining, in part because of a shortening of the maple syrup season in Vermont. Meanwhile, Canadian production has been increasing because of increased marketing, government subsidies, improved technologies, and, likely, climate change.

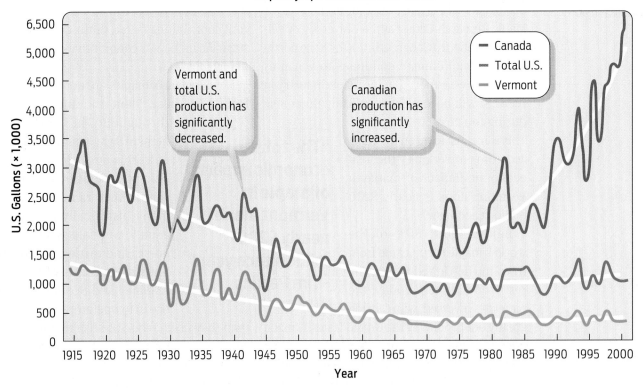

Maple Syrup Production 1916–2000

Vermont and total U.S. production has significantly decreased.

Canadian production has significantly increased.

Legend: Canada, Total U.S., Vermont

y-axis: U.S. Gallons (× 1,000): 0, 500, 1,000, 1,500, 2,000, 2,500, 3,000, 3,500, 4,000, 4,500, 5,000, 6,500

x-axis: Year: 1915, 1920, 1925, 1930, 1935, 1940, 1945, 1950, 1955, 1960, 1965, 1970, 1975, 1980, 1985, 1990, 1995, 2000

Source: *New England Regional Assessment*, Barrett Rock and Shannon Spencer.

Rising Temperatures Affect Plant Behavior

 Global warming is changing the seasonal behavior of plants and animals. Near Oxford, England, many plants are flowering earlier now than they did between 1954 and 1990. In fact, the average first date of flowering is now 4.5 days earlier than the long-term historic average (1954–1990). One species studied is now flowering 55 days sooner than it did in previous decades.

Plants are flowering earlier:

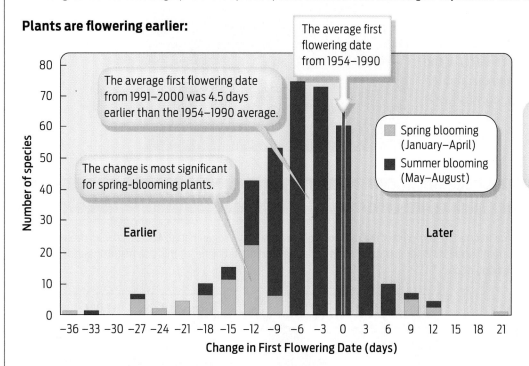

Early blooming correlates with temperature increase:

Month of average first flowering day	Number of species	Number of days flowering was advanced by a 1° C increase in average monthly temperature
February	12	6.0 days earlier
March	22	4.5 days earlier
April	63	4.3 days earlier
May	105	2.0 days earlier
June	108	1.7 days earlier
July	67	2.8 days earlier

Rapid Changes in Flowering Time in British Plants. A. H. Fitter and R. S. R. Fitter (2002) *Science*, Vol 296, p. 1689–1691.

air. Because the biotic and abiotic parts of an ecosystem can and do change, ecosystems are not static entities but dynamic systems. And because the parts of an ecosystem are so interconnected, a small change in one part of an ecosystem can have a domino effect.

No one knows this better than sugar farmers. "The flow of sap from maple trees during the spring season is controlled almost entirely by the daily fluctuation in temperature," explains ecologist Perkins. "Small changes in the day-to-day temperature pattern will have large consequences on sap flow."

Historically, trees have been tapped in early March when the sap began to flow; the sap was then collected for the next 6 weeks. But about 10 years ago, Perkins started getting calls from sugar producers saying that they were tapping earlier and making syrup earlier. Curious, he and his colleagues decided to investigate. They scoured historical records and surveyed thousands of maple sugar producers in New England. Their results were startling: over a mere 40 years, between 1963 and 2003, the start of the tapping season had moved forward by about 8 days. Even more significant, the end of the season, when maples begin to leaf out and the sap is no longer good for syrup, now comes 11 days earlier.

Collecting sap from a sugar maple tree (*Acer saccharum*) to make maple syrup.

INFOGRAPHIC 23.3

Maple Tree Range Is Affected by Increasing Temperature

→ Models including 25 environmental parameters predict the rapid disappearance of the sugar maple, *Acer saccharum*, from the United States with even small increases in temperature. As the ideal environmental niche for this tree migrates north into Canada, so does the tree population.

1 23 46 69 92 115 138

Colors indicate woody production (m³/ha/yr) of the sugar maple

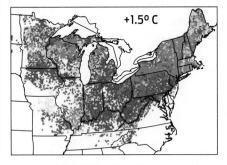

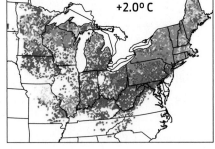

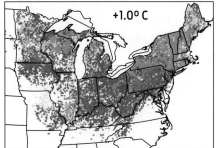

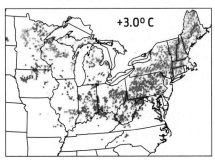

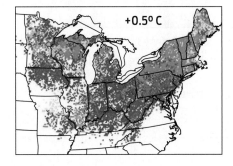

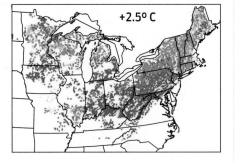

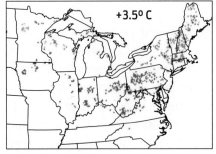

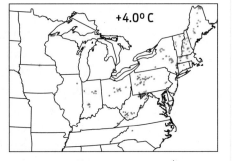

4th International Conference on Integrating GIS and Environmental Modeling (GI/EM4):
W. H. Hargrove and F. M. Hoffman *Problems, Prospects and Research Needs*.
Banff, Alberta, Canada, September 2–8, 2000.

HABITAT
The physical environment where an organism lives and to which it is adapted.

BIOME
A large geographic area defined by its characteristic plant life, which in turn is determined by temperature and levels of moisture.

"Over that 40-year time period we've lost about 3 days of the season," Perkins told Vermont Public Radio. "That doesn't seem like a lot until you realize that the maple production season averages about 30 days in length. So we've lost about 10% of the season."

To some extent, losses from a shortened tapping season have been offset by improved sap-removal technologies that make it possible to extract sap even under poor conditions. The bigger problem is what will happen if climate changes so much that New England no longer provides a suitable **habitat** for maple trees.

As Perkins testified to Congress, current climate models predict that by the end of the century New England's forests will more closely resemble those of present-day Virginia, North Carolina, and Tennessee, dominated by hickory, oak, and pine rather than maple, beech, and birch **(Infographic 23.3)**. If that happens, not only maple syrup but the brilliant fall foliage New England is famous for will be a thing of the past.

New England's colorful foliage is part of a **biome** known as temperate deciduous forest. Biomes are large, geographically cohesive regions whose defining vegetation–its plant life–is determined principally by climatic factors such as temperature and rainfall (see **Up Close: Biomes**). Climate change is beginning to alter the boundaries and plant composition of

UP-CLOSE Biomes

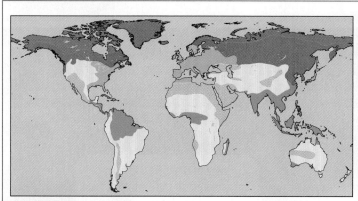

Desert:
A biome characterized by extreme dryness. Cold deserts experience cold winters and hot summers, while hot deserts are uniformly warm throughout the year.

Coniferous Forest:
A biome characterized by evergreen trees, with long and cold winters and only short summers.

Tundra:
A biome that occurs in the Arctic and mountain regions. Tundra is characterized by low-growing vegetation and a layer of permafrost (frozen all year long) very close to the surface of the soil.

Temperate Deciduous Forest:
A biome characterized by trees that drop their leaves in winter. Winters are much colder than summers.

Tropical Forest:
Tropical forests are biomes characterized by warm temperatures and sufficient rainfall to support the growth of trees. Tropical forests may be deciduous or evergreen, depending on the presence or absence of a dry season.

Grassland:
A biome characterized by perennial grasses and other nonwoody plants. In North America, the prairies are examples of grasslands.

Aquatic: Marine
This biome covers about three-fourths of the earth and includes the oceans, coral reefs, and estuaries.

Aquatic: Freshwater
A biome characterized by having a low salt concentration. Freshwater biomes include ponds and lakes, rivers and streams, and wetlands.

temperate deciduous forests, and may one day push sugar maples north into Canada.

Such changes are not uncommon—climate change is beginning to redraw the map of biomes around the world. In northern Alaska, where once there was only sparsely vegetated tundra, woody shrubs now grow. When Montana's Glacier National Park was opened in 1910, it held approximately 150 large glaciers; in 2010, there were only 25. As the vegetation in these landscapes changes, so will the community of organisms that rely on it for food and habitat.

> **When Montana's Glacier National Park was opened in 1910, it held approximately 150 large glaciers; in 2010, there were only 25.**

Warming Planet, Diminishing Biodiversity
Although temperature swings and shifts in the ranges of organisms are natural phenomena, the amount of warming in recent years is unprecedented, and evidence suggests that the change is not merely part of a natural cycle. From 1880 until 2010, the earth's surface has warmed, on average, by about 0.8°C (1.4°F), according to a 2010 study by NASA's Goddard Institute for Space Studies. That may not sound like a lot. But consider this: the difference in global average temperatures between today and the last ice age—when much of North America was buried under ice—is only about 5°C (9°F). Where global temperatures are concerned, even a 1-degree change is significant.

The rate of warming has increased as well. Eighteen of the warmest years on record occurred in just the past 20 years. The last decade, from 2000 through 2010, was the hottest decade so far, with 2010 tying 2005 for the title of hottest year on record. Much of this warming is attributable to the **greenhouse effect,** the trapping of heat in earth's atmosphere. As sunlight shines on our planet, it warms the earth's surface. This heat radiates back to the atmosphere, where it is absorbed by **greenhouse gases** such as carbon dioxide. The heat trapped by greenhouse gases raises the temperature of the atmosphere, and in turn, the surface of the earth (Infographic 23.4).

The greenhouse effect is a natural process that helps maintain life-supporting temperatures on earth. Without this greenhouse effect, the average surface temperature of the planet would be a frigid –18°C (0°F). In recent years, however, rising levels of greenhouse gases have increased the strength of the greenhouse effect, a phenomenon known as the enhanced greenhouse effect. As the amount of greenhouse gases in the atmosphere has increased, so have temperatures. The result is **global warming**, an overall

GREENHOUSE EFFECT
The normal process by which heat is radiated from the earth's surface and trapped by gases in the atmosphere, helping to maintain the earth at a temperature that can support life.

GREENHOUSE GAS
Any of the gases in earth's atmosphere that absorb heat radiated from the earth's surface and contribute to the greenhouse effect, for example carbon dioxide and methane.

GLOBAL WARMING
An increase in the earth's average temperature.

INFOGRAPHIC 23.4

The Greenhouse Effect

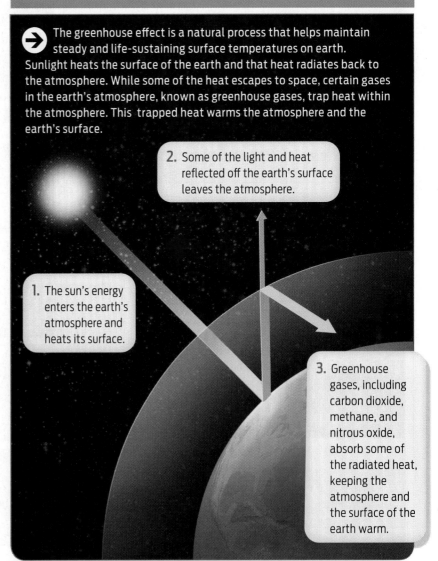

The greenhouse effect is a natural process that helps maintain steady and life-sustaining surface temperatures on earth. Sunlight heats the surface of the earth and that heat radiates back to the atmosphere. While some of the heat escapes to space, certain gases in the earth's atmosphere, known as greenhouse gases, trap heat within the atmosphere. This trapped heat warms the atmosphere and the earth's surface.

1. The sun's energy enters the earth's atmosphere and heats its surface.

2. Some of the light and heat reflected off the earth's surface leaves the atmosphere.

3. Greenhouse gases, including carbon dioxide, methane, and nitrous oxide, absorb some of the radiated heat, keeping the atmosphere and the surface of the earth warm.

INFOGRAPHIC 23.5

The Earth's Surface Temperature Is Rising with Carbon Dioxide Levels

As measured directly by thermometers, and as documented by historical records and other biological indicators (including tree rings, corals, and ice cores) the temperature on earth has increased rapidly in the past 140 years, along with increasing levels of carbon dioxide.

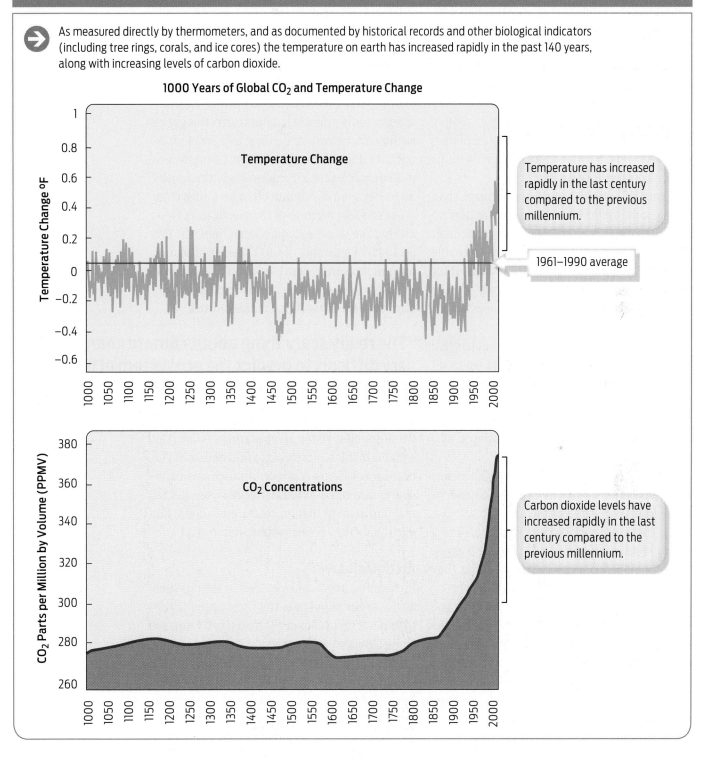

1000 Years of Global CO$_2$ and Temperature Change

Temperature Change

Temperature has increased rapidly in the last century compared to the previous millennium.

1961–1990 average

CO$_2$ Concentrations

Carbon dioxide levels have increased rapidly in the last century compared to the previous millennium.

increase in the earth's average temperature (Infographic 23.5).

For ecologist Hector Galbraith, director of the Climate Change Initiative at the Manomet Center for Conservation Sciences, in Massachusetts, one of the most worrying things about climate change is how quickly it is happening, and how sensitive species are to the changes. "Most people think of climate change as something that's 30 years out," says Galbraith. But that's simply not true, he notes. "We began seeing responses in ecosystems 20 years ago. The ecosystems knew about it before we did."

Plants, of course, are slower to adapt than animals; they cannot simply get up and move (although they may change their range over time by dispersing seeds into more favorable climes). But some animals can change their ranges quite quickly. "A bird can simply open its wings, and within two hours it's 50 miles farther north," says Galbraith.

What will be the outcome of all these changes? The answer, says Galbraith, is that we don't really know. "We're seeing changes to systems that have been relatively stable for thousands of years . . . . The really scary thing about climate change is it's very difficult to predict the ecosystem effects of these changes."

Nevertheless, there are disturbing scenarios. Take the relationship between birds and insects. Many forests are susceptible to insect attacks. Given their insect-rich diet, flycatchers are a natural form of pest control. If the birds move north, as evidence suggests they are doing, they leave behind a forest susceptible to predation by insects that might be less vulnerable to the changed climate or more adaptable. The maple-tree-loving pear thrip and the forest tent caterpillar are just two examples of insects that might be happy to see the flycatchers go. More insects means more dead trees, which in turn means more fuel for forest fires (Infographic 23.6).

Not all species will be negatively affected by climate change–some may actually benefit. But one species' success in coping with climate change may contribute to another's demise. For example, the adaptable red fox (*Vulpes vulpes*), found throughout the northern hemisphere, is venturing into the range of the endangered Arctic fox (*Vulpes lagopus*), whose habitat–the Arctic tundra–has gotten warmer. When the two species share a range, the Arctic fox inevitably suffers because the red fox out-competes it for food and also preys on Arctic fox pups.

While some species can adapt to a changing climate by shifting range, future global warming will likely exceed the ability of many species to adapt, as hospitable habitats can no longer be found or accessed. According to a 2004 study published in the journal *Nature*, as many as a million species could be driven to extinction by 2050 because of climate change. Using computer models, a group of 15 investigators from around the world estimated that between 15% and 37% of a sample of 1,103 species of plants, mammals, birds, reptiles, amphibians, and invertebrates would be "committed to extinction" because of warming temperatures.

"The really scary thing about climate change is it's very difficult to predict the ecosystem effects of these changes." –Hector Galbraith

---

The study's authors found that for many of these species, rising temperatures will make suitable habitat impossible to find or reach. The natural residents of mountaintops are especially vulnerable: as temperatures rise, species may move up to higher, colder elevations, but eventually they have nowhere left to go.

Arctic Meltdown

Predictably, snow- and ice-covered regions such as the Arctic stand to suffer most immediately from a warming climate, as frozen habitats start to melt. But the situation is worse than one might imagine. As Mark Serreze, director of the National Snow and Ice Data Center at the University of Colorado, notes, the Arctic has warmed, on average, twice as much as the rest of the world. This is what is known among climate scientists as Arctic amplification, and it has to do with how sea ice affects temperature. As Serreze explains, sea ice both reflects solar radiation and insulates the ocean. As global temperatures rise, ice begins to melt. With less sea ice, more solar radiation is

Rising Temperatures Mean Widespread Ecosystem Change

 Climate change is having dramatic impacts on entire ecosystems. With warming temperatures, songbirds are expanding their habitat into more northern territories. As birds move northward, they leave behind their insect prey that are free to devastate the now unprotected forests. Dead trees, in turn, lead to more forest fires, which further alter the landscape.

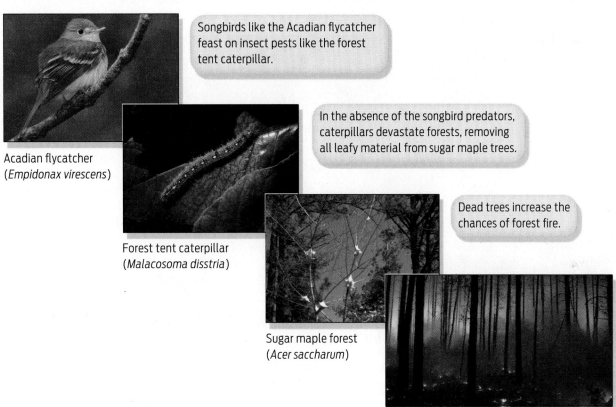

Songbirds like the Acadian flycatcher feast on insect pests like the forest tent caterpillar.

Acadian flycatcher
(*Empidonax virescens*)

In the absence of the songbird predators, caterpillars devastate forests, removing all leafy material from sugar maple trees.

Forest tent caterpillar
(*Malacosoma disstria*)

Dead trees increase the chances of forest fire.

Sugar maple forest
(*Acer saccharum*)

absorbed by the ocean and more of the relatively warm ocean is exposed to air, raising the air temperature even more. It's a positive feedback loop, which means that as additional ice is lost, temperatures will rise at an accelerated pace.

According to the extensive Arctic Climate Impact Assessment, the result of 4 years' work by more than 300 scientists around the world published in 2004, Arctic temperatures are projected to rise by an additional 4°-7°C (7°-13°F) over the next 100 years **(Infographic 23.7)**.

Warming temperatures could spell disaster for species that call the Arctic their home. Polar bears, for example, spend most of the year roaming the Arctic on large swaths of floating sea ice that blanket a good portion of the Arctic Ocean from September through March. The massive mammals use the sea ice to hunt for seals, which periodically pop up through

"whack-a-mole"-like breathing holes in the ice and are nabbed by bears. Yet the size of this frozen habitat has been shrinking, greatly reducing the bears' ability to obtain food.

Moreover, over the past few decades the ice has been breaking up earlier and earlier in spring. The sea ice in Hudson Bay, Canada, for example, now breaks up nearly 3 weeks earlier than it did in the 1970s. In the absence of unbroken summer sea ice, the polar bears are stuck on land (where there are no seals), or are forced to swim long distances to reach sea ice. Some, exhausted by the journey, drown. Those that do survive have fewer opportunities to hunt. Canadian polar bears now weigh on average 55 pounds less than they did 30 years ago, seriously compromising their reproductive ability.

Scientists have monitored sea ice on a daily basis by satellite since 1979. Over the past three

Arctic Temperatures Are Rising Fast

 Current measurements suggest that the Arctic is warming faster than other parts of the earth. 2008 was the ninth warmest year on record (since measurements began in 1880). Much of the earth was warmer in 2008 than in the period between 1951 and 1980 (regions in yellow, orange, red, and brown). The Arctic, Antarctic, and Eurasia warmed more than the rest of the planet.

2008 Surface Temperature Change Compared to 1951–1980 Average (°C)

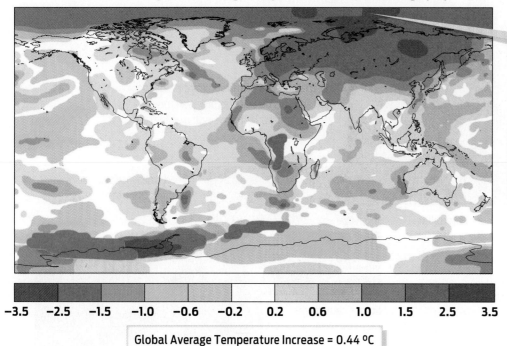

Temperature is increasing faster at the earth's poles than at its equator.

At the current rate of temperature increase, climate experts predict an additional 4°–7° C increase over the next 100 years.

−3.5　−2.5　−1.5　−1.0　−0.6　−0.2　0.2　0.6　1.0　1.5　2.5　3.5

Global Average Temperature Increase = 0.44 °C

decades, the area of Arctic sea ice has shrunk by more than 1 million square miles, an area roughly four times the size of Texas, according to Walt Meier, a research scientist with the National Snow and Ice Data Center in Boulder, Colorado. Arctic sea ice hit a record low in September 2007, at the end of the summer melt season, shrinking to a level that climate change models had predicted wouldn't happen until at least 2050. Scientists now fear that nearly all of the polar bear's summer sea ice could vanish by 2040—possibly sooner (Infographic 23.8).

Warming temperatures are also causing glaciers and ice caps on land to melt. Unlike sea ice, which, like an ice cube in a glass of water, doesn't raise the water level as it melts, melting glaciers and ice caps do. How much will seas rise? "By 2100, you're looking at probably about a meter," says Serreze. "Here in Boulder we're at 5,400 feet, [so] we're not worried about that.

Scientists now fear that nearly all of the polar bear's summer sea ice could vanish by 2040.

But if you're living in Miami, this is something that should concern you."

It's important to note that much of the data we have on climate change relates to global, long-term trends. From year to year, there may be slight variations—slightly warmer summers and less sea ice one year, slightly cooler summers and more sea ice the next. And indeed, from its all-time low in 2007, sea ice did indeed bounce back a bit in 2008 and 2009. But the trend is still unmistakably downward—toward less sea ice. By 2030 or 2040, says Serreze, there could be no summer ice to speak of. "You could take a ship across the north pole."

The evidence we have for global warming is clear, unmistakable, and alarming. Yet despite such evidence, recent surveys of public opinion show that large percentages of the American and British public do not believe in the reality of global warming. In response to such views, and media controversy, a group of roughly 250 scientists–all members of the U.S. National Academy of Science–signed a letter testifying to the legitimacy of climate data and climate science. Published in the May 7, 2010, issue of the journal *Science,* the letter concludes that the data we currently have establish with a 90% degree of confidence that the planet is warming.

Follow the Carbon

What's behind this planetary warming? The immediate cause is a fired-up greenhouse effect. And that, scientists argue, is the result of human activity. As they testified in their letter, "There is compelling, comprehensive, and consistent objective evidence that humans are changing the climate in ways that threaten our societies and the ecosystems on which we depend." How did we get to be the culprits in this situation? In short, by pumping more carbon dioxide into the atmosphere.

Carbon dioxide is the most notorious player in the greenhouse effect, and scientists believe it is responsible for most of the warming. In fact, atmospheric carbon dioxide concentrations are higher now than they have been in more than 700,000 years.

As we saw in Chapter 2, carbon is a natural ingredient in every living organism, part of the

INFOGRAPHIC 23.8

Arctic Sea Ice Is Melting

 Rising temperatures have caused the polar ice cap to melt and break apart earlier in the season. The reduction in the extent of summer sea ice is threatening the survival of polar bears, which require the sea ice to hunt for seals.

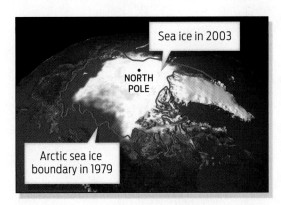

Sea ice in 2003

NORTH POLE

Arctic sea ice boundary in 1979

Since 1979, more than 20% of the polar ice cap has melted away.

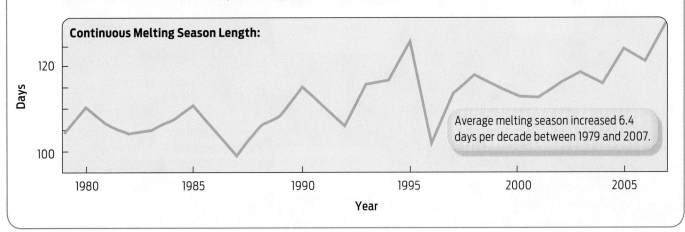

Continuous Melting Season Length:

Average melting season increased 6.4 days per decade between 1979 and 2007.

Days

120

100

1980   1985   1990   1995   2000   2005

Year

backbone of all organic molecules. Carbon also exists in inorganic forms: as carbon dioxide in the atmosphere, as carbonic acid dissolved in water, as calcium carbonate in limestone rocks. If dead organisms are fossilized before being digested by decomposers, the organic molecules contained within their bodies become trapped below the earth's surface or under the seas. Over time, these compressed organic molecules turn into **fossil fuels**–coal, oil, and natural gas.

Like other chemical elements, the total amount of carbon on earth remains essentially constant. In contrast to the way energy flows through an ecosystem in one direction (from the sun to producers to consumers and out to the universe as heat; Chapter 22), elements such as carbon move in cycles. Individual carbon atoms are recycled as carbon-based organisms die and new life is born, and as geological processes slowly reshuffle carbon in the nonliving environment. The movement of carbon through the environment follows a predictable pattern known as the **carbon cycle.**

At it cycles through the environment, carbon moves between organic and inorganic forms. For

FOSSIL FUEL
A carbon-rich energy source, such as coal, petroleum, or natural gas, formed from the compressed, fossilized remains of once-living organisms.

CARBON CYCLE
The movement of carbon atoms between organic and inorganic molecules in the environment.

UP CLOSE   Chemical Cycles: Nitrogen and Phosphorus

Nitrogen atoms cycle between different chemical and biochemical compounds as they move from organisms to the soil, water, and air and back to organisms. A variety of natural processes as well as some human activities contribute to the transformation and movement of nitrogen through the ecosystem.

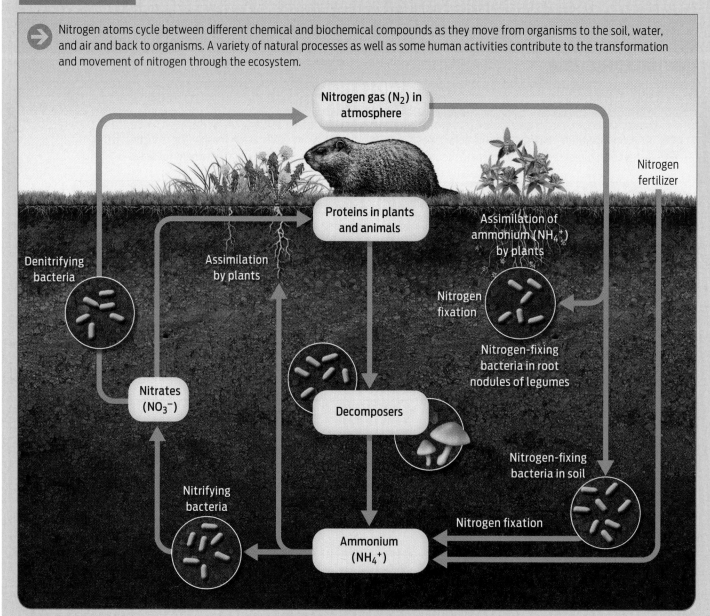

example, animals take in organic carbon when they eat other organisms and release inorganic gaseous CO_2 into the atmosphere as a by-product of cellular respiration. Similarly, when organisms die, decomposers in the soil use the dead organic material for food and energy, releasing some of the carbon during respiration as CO_2.

In turn, plants, photosynthetic bacteria, and algae take up CO_2 during photosynthesis and fix it into organic sugar molecules, thus reducing atmospheric CO_2 levels. Photosynthesis, respiration, and decomposition form a cycle that keeps carbon dioxide at a relatively stable level in the atmosphere. But human actions, such as deforestation and the burning of fossil fuels, can inject additional carbon dioxide into the cycle (**Infographic 23.9**).

Note that carbon isn't the only element that cycles through ecosystems. Other elements, such as nitrogen, phosphorus, and sulfur, as well as water (Chapter 24), also follow natural cycles (see **Up Close: Chemical Cycles**). But it's the carbon cycle that is most relevant to the phenomenon of global warming.

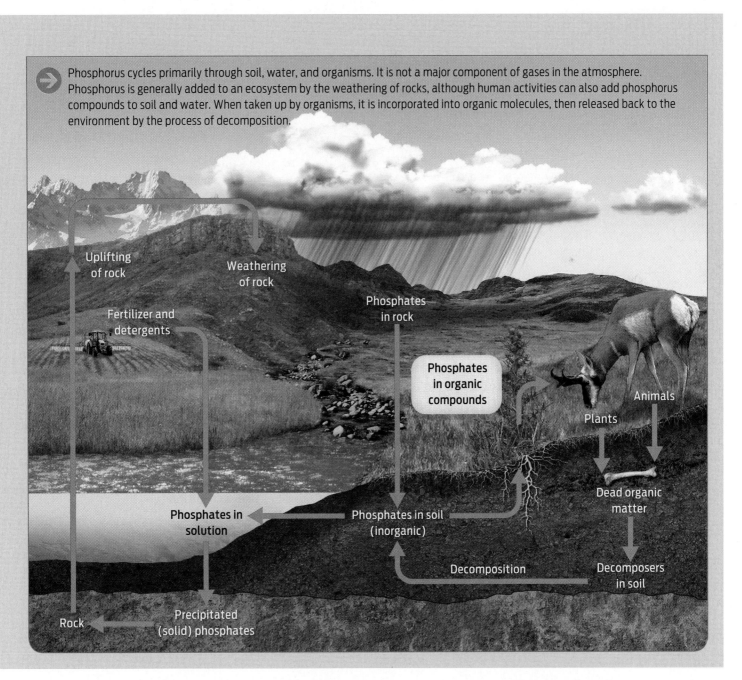

Phosphorus cycles primarily through soil, water, and organisms. It is not a major component of gases in the atmosphere. Phosphorus is generally added to an ecosystem by the weathering of rocks, although human activities can also add phosphorus compounds to soil and water. When taken up by organisms, it is incorporated into organic molecules, then released back to the environment by the process of decomposition.

The Carbon Cycle

The carbon cycle describes the movement of carbon atoms between organic molecules and inorganic CO_2 via natural processes such as photosynthesis, respiration, and decomposition. Since the 1700s, human activities, including burning fossil fuels and deforestation, have made significant contributions to the cycle.

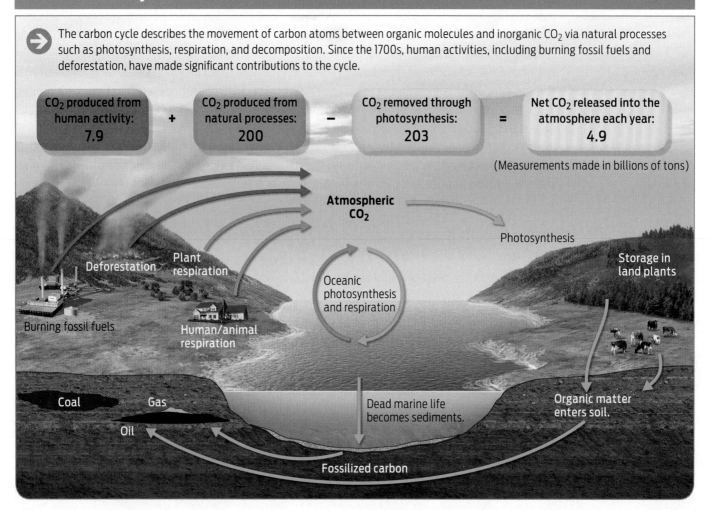

| CO_2 produced from human activity: 7.9 | + | CO_2 produced from natural processes: 200 | − | CO_2 removed through photosynthesis: 203 | = | Net CO_2 released into the atmosphere each year: 4.9 |

(Measurements made in billions of tons)

Atmospheric CO_2

Photosynthesis

Deforestation

Plant respiration

Oceanic photosynthesis and respiration

Storage in land plants

Burning fossil fuels

Human/animal respiration

Coal

Gas

Oil

Dead marine life becomes sediments.

Organic matter enters soil.

Fossilized carbon

For the most part, the amount of carbon present in the atmosphere as carbon dioxide has remained fairly constant. But since the late 1700s, with the rise of industry and the internal combustion engine, people have begun to alter the carbon cycle, adding increasing amounts of CO_2 to the atmosphere.

Before the industrial revolution, the carbon trapped in fossil fuels was not easily accessible, and therefore it wasn't cycling as part of the carbon cycle. But modern drilling and mining methods have unlocked the deep reserves of this ancient planetary energy. The CO_2 released when humans burn fossil fuels is the largest source of the carbon being added to the atmosphere by humans and is a major contributor to the enhanced greenhouse effect.

Air bubbles trapped in ice cores from the glacial ice of Greenland—an indirect measure of carbon dioxide in the atmosphere—show relatively constant amounts of CO_2 until 300 years ago. Since direct measurement of atmospheric CO_2 began late in the 19th century, its concentration has increased about 35% (Infographic 23.10).

Virtually all climate scientists agree that greenhouse gases emitted by human activities—primarily driving gasoline-powered cars and burning coal to generate electricity—have caused most of the global rise in temperature observed over the past 50 years. In 2001, an international group of scientists and policymakers known as the Intergovernmental Panel on Climate Change concluded that the global

Measuring Atmospheric Carbon Dioxide Levels

 Ice cores provide a way to measure biological and atmospheric conditions from the distant past. Cylinders of ice representing a time frame covering thousands of years can be extracted from glaciers. Gas bubbles present in the ice reveal the atmospheric composition thousands of years ago.

Present day carbon dioxide levels are measured directly from the air:
Direct measurements of carbon dioxide are currently taken from the Mauna Loa Research Station in Hawaii.

Mauna Loa Research Station

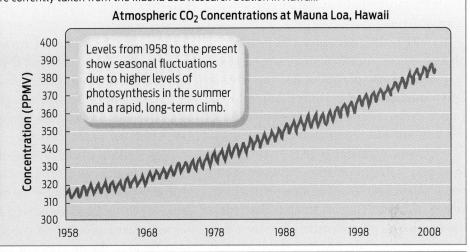

Atmospheric CO$_2$ Concentrations at Mauna Loa, Hawaii

Levels from 1958 to the present show seasonal fluctuations due to higher levels of photosynthesis in the summer and a rapid, long-term climb.

Historic carbon dioxide levels are measured in glacial ice cores:
Ice cores are long tubes of ice removed from a glacier. As each annual layer of ice was deposited on a glacier, it trapped the gas present in the earth's atmosphere at that time.

Ice cores contain layers of ice harboring gas bubbles that reveal the composition of the ancient atmosphere.

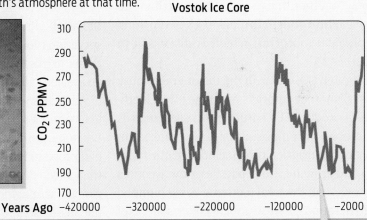

Vostok Ice Core

Together, these data provide a complete picture of atmospheric CO$_2$ levels over time:

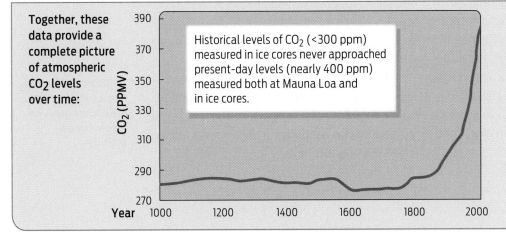

Historical levels of CO$_2$ (<300 ppm) measured in ice cores never approached present-day levels (nearly 400 ppm) measured both at Mauna Loa and in ice cores.

Some ice cores contain layers of ice from hundreds of thousands of years ago, and can be used to measure historical levels of CO$_2$. These measurements show that CO$_2$ levels cycle in patterns that correlate with major ice ages.

rise in average yearly temperature over the past 50 years was primarily anthropogenic—that is, caused by humans.

"Fossil fuels are incredibly efficient sources of energy," says Serreze, from the University of Colorado. "We've built our whole infrastructure around that. But what we didn't realize is that it's a trap, and that's what we're coming to grips with now."

Activities that decrease the number of photosynthetic organisms also increase global CO_2 levels. Since photosynthesizers are the only consumers of carbon dioxide in the carbon cycle, removing them not only reduces the amount of carbon dioxide they might have consumed, but also—in the case of large trees and stable populations of algae—eliminates what are in essence long-term storage vessels of carbon. Human activities that reduce the number of photosynthetic organisms on the planet include large-scale slash-and-burn agriculture, development that leads to deforestation, and various forms of pollution. Together, these activities contribute to our **carbon footprint,** a subset of our total ecological footprint, which is discussed in Chapter 24.

Though CO_2 is one of the major greenhouse gases, another culprit is methane (CH_4). Methane is produced by natural processes, such as microbes decomposing organic material in swamps, but agriculture, including cattle farming and growing rice in paddies, now accounts for over half the total methane being pumped into the atmosphere. One of the main sources of methane is the digestive gas produced by archaea living in the digestive systems of cattle. Emitted as flatulence, it adds an estimated 100 million tons of methane a year to the atmosphere. Although the atmospheric concentration of methane is far less than that of CO_2, atmospheric methane is more worrisome because it is 30 times more potent as a greenhouse gas (**Infographic 23.11**).

No Time for Fatalism

The United States is among the world's biggest emitters of greenhouse gases, yet it has been, for political reasons, reluctant to make significant reductions. It is one of the few countries that refused to ratify the Kyoto Protocol, a United Nations agreement adopted in 1997 that obligates endorsing countries to reduce carbon dioxide emissions. In 2010, the Obama administration signaled support for the Copenhagen Accord, which would commit the United States to a 17% reduction in greenhouse gases from 2005 levels by 2020, but carrying out those goals depends on Congress's passing a climate bill, which is very problematic.

Even if all the world's greenhouse gas emissions were turned off today like a faucet, a daunting problem remains: we would still face years of warming and its consequences because of past emissions—what climate scientists refer to as "heat in the pipeline." It's a grim reality that could lead some to take a fatalistic attitude. That would be a dangerous mistake, says Hector Galbraith. "We've got to get beyond the deer in the headlights stage and begin to think as conservation biologists about what we're going to do about this to help to mitigate the impact." It's an area he calls "adaptation."

Adaptation will not be easy. For many species, like Vermont's maples, it may already be too late. But doing nothing, say scientists, risks turning a bad problem into a catastrophic one. There are things each of us can do to mitigate the effects of global warming—for example, living a more sustainable lifestyle, one that uses fewer fossil fuels—and voting for government officials who support sustainable practices. Chapter 24 discusses the topic of sustainability, and how you can live more sustainably, in more detail.

"The real problem is not so much change," says Serreze. "Change has always happened; change always will happen . . . . The real key is we've got to get a handle on the problem before it gets out of hand." ∎

CARBON FOOTPRINT
A measure of the total greenhouse gases we produce by our activities.

Anthropogenic Production of Greenhouse Gases

 Human activities are increasing the levels of greenhouse gases in the atmosphere.

Burning Fossil Fuels
Burning fossil fuels (coal, natural gas, petroleum) liberates the carbon that was once stored as organic molecules in the earth and releases it into the atmosphere as carbon dioxide gas.

Methane from Cattle
The cattle that we raise have methane-producing microbes in their guts which help them digest the plant matter they eat. Cattle release large amounts of methane gas as flatulence.

Deforestation
Destroying and burning forests liberates carbon that was stored as organic molecules in trees and releases it into the atmosphere as carbon dioxide. In addition, it diminishes the capacity for carbon dioxide–capturing photosynthesis.

Rice Agriculture
Methane is released from rice paddies because of the methane-producing bacteria that live in the flooded, and therefore anoxic (oxygen-free), soil. Preparing land for growing rice may also destroy forests that formerly stored carbon in organic form.

Concrete Production
The production of concrete results in the release of large amounts of carbon dioxide. As we continue to develop cities on the earth, the impact of this process grows.

▶ Summary

- Ecosystems are made up of the living and nonliving components of an environment, including the communities of organisms present and the physical and chemical environment with which they interact.

- Temperature is an important physical feature of any ecosystem and serves as a clock to time many biological events, such as breeding and hibernation.

- Biomes are large, geographically cohesive ecosystems, defined by their characteristic plant life, which in turn is determined by temperature and levels of moisture.

- Global climate change is a persistent pattern of change in the climate of the earth. Global warming is an increase in earth's average temperature over time.

- Global climate change, and especially global warming, is having widespread effects on plant and animal life on the planet—altering seasonal life cycles, shifting ranges, and contributing to species loss by extinction.

- The greenhouse effect is a natural process by which heat from the earth's surface is radiated to heat-trapping gases in the atmosphere, maintaining a global temperature that can support life. Rising levels of greenhouse gases have led to the enhanced greenhouse effect.

- Elements cycle through ecosystems. The carbon cycle is the movement of carbon atoms through living and nonliving components of the environment by the biotic processes of photosynthesis, cellular respiration, and decomposition, as well as by long-term geological processes.

- Global warming is the result of an increase in the amount of carbon dioxide and other greenhouse gases in the atmosphere, due primarily to human activities, such as burning fossil fuels and deforestation.

- Global warming is leading to melting sea ice in the Arctic, which is diminishing habitat for the organisms that rely on it. Melting of glaciers and ice caps is leading to rising sea levels.

- Methane is a significant greenhouse gas whose levels have increased because of human activities, including raising cattle and farming rice in paddies.

Data References

Infographic 23.5: (*Top*): IPCC Fourth Assessment Report; (*bottom*): http://cdiac.ornl.gov/trends/co2/lawdome.html; http://www.esrl.noaa.gov/gmd/ccgg/trends/

Infographic 23.7: http://data.giss.nasa.gov/gistemp/2008/

Infographic 23.8: (*Top left*): Arctic Climate Impact Assessment, 2004; 2 Intergovernmental Panel on Climate Change 4th Assessment Report, 2007; (*bottom*): http://earthobservatory.nasa.gov/IOTD

Infographic 23.10: (*Top*): http://www.esrl.noaa.gov/gmd/ccgg/trends/; (*middle*): Petit J.R., et al. 2001. *Nature*, 399, pp.429-436; (*bottom*): http://cdiac.ornl.gov/trends/co2/lawdome.html

ECOSYSTEMS AND CLIMATE CHANGE

Species are adapted to the ecosystems of which they are part. Climate change can alter their natural patterns and therefore change the dynamics of entire ecosystems.

HINT See Infographics 23.1–23.3 and 23.6.

KNOW IT

1. Which of the following are parts of an ecosystem:
   a. the plant life present in an area
   b. the animals living there
   c. the amount of annual rainfall
   d. soil chemistry
   e. none of the above
   f. all of the above

2. From what you've read in this chapter, list several examples of species that changed their geographic distributions or the timing of events in their life cycle as a result of global climate change.

3. If you were asked to identify a biome, which of the characteristics below would be most important to have data on (select all that apply):
   a. monthly rainfall
   b. temperatures throughout the year
   c. plant life
   d. animal life
   e. human population size in the area

4. Which biome is characterized principally by evergreen trees?

5. Looking at Up Close: Biomes, where in North and South America do you find temperate deciduous forest? Tropical forest?

6. If global warming causes Arctic sea ice to melt, what will be the effect on sea levels in a low-lying region like Miami? What about if large parts of the Antarctic polar ice cap melted—what would be the effect on sea level?

USE IT

7. Although trees may not be able to walk away from increasingly warm regions, evolutionary adaptations may allow trees to survive in warmer regions. Discuss each of the adaptations listed below and decide if it is likely to be helpful or harmful in a warming environment. (Think about water—water is taken up by the roots of plants, and lost through pores in the leaves; CO_2 levels—CO_2 is taken up by plants through pores in leaves, then used by leaves for photosynthesis; and the movement of other species, for example insects, in response to global warming.)
   a. having smaller leaves
   b. having a larger number of pores on each leaf
   c. having thicker and waxier bark

8. What is a possible risk for humans if insects that carry pathogenic bacteria or viruses expand their range northward?

GREENHOUSE EFFECT

Certain gases in the atmosphere act to trap heat. This heat trapping is essential for life on earth, but it can be altered by human activities in ways that harm life.

HINT See Infographics 23.4, 23.5, 23.7, 23.8 and 23.11.

KNOW IT

9. Which greenhouse gas is emitted every time you breathe out?
   a. oxygen
   b. carbon dioxide
   c. methane
   d. nitrogen
   e. water vapor

10. Which of the following organisms will contribute to reducing atmospheric CO_2 levels?
   a. maple trees
   b. most algae
   c. polar bears
   d. pear thrips
   e. a & b
   f. a, b and d

11. Fossil fuels are most immediately derived from
   a. organic molecules.
   b. CO_2.
   c. methane.
   d. melting ice caps.
   e. photosynthesis.

12. Could we live in the absence of the greenhouse effect? Explain your answer.

USE IT

13. Describe the evidence that increasing levels of greenhouse gases are responsible for global climate change. What if someone suggested to you that global climate change was due to increased intensity of solar radiation? What kind of evidence would you ask them to provide to support their hypothesis?

CARBON CYCLE AND GREENHOUSE EFFECTS

Carbon cycles through the environment, moving between organic molecules and inorganic carbon dioxide gas. Human activities can change the dynamics of the carbon cycle.

HINT See Infographics 23.9–23.11.

➔ KNOW IT

14. Fill in the blanks in the image below.

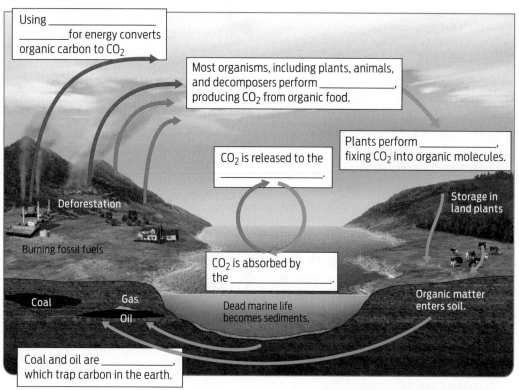

Using _____ _____ for energy converts organic carbon to CO_2

Most organisms, including plants, animals, and decomposers perform _____, producing CO_2 from organic food.

CO_2 is released to the _____.

Plants perform _____, fixing CO_2 into organic molecules.

Deforestation

Storage in land plants

Burning fossil fuels

CO_2 is absorbed by the _____.

Coal Gas Oil

Dead marine life becomes sediments.

Organic matter enters soil.

Coal and oil are _____, which trap carbon in the earth.

15. Decomposers _____ CO_2 by the process of _____.

   a. emit; photosynthesis
   b. take up; photosynthesis
   c. emit; cellular respiration
   d. take up; cellular respiration
   e. store; cellular respiration

➔ USE IT

16. How is ice useful in measuring atmospheric levels of CO_2?

17. Explain how each of the following contributes to an elevation of levels of greenhouse gases:
   a. large-scale slash-and-burn agriculture
   b. driving gasoline-fueled cars
   c. producing cattle for beef and dairy products
   d. rice production

18. Which of the following data would you use to determine the levels of atmospheric CO_2 in 1750? Justify your choice, including an explanation of why the other alternatives would not be as effective.
   a. historical weather records of daily temperatures

   b. archives of the Manua Loa observatory (to examine 1750 records)
   c. tree-ring analysis (to look for evidence of extreme fires)
   d. ice cores from ice formed in 1750

SCIENCE AND ETHICS

19. Visit an online carbon footprint or carbon emissions calculator (for example, http://www.epa.gov/climatechange/emissions/ind_calculator.html) and calculate your total carbon emissions.
   a. What is your largest source of emissions?
   b. What steps can you take to decrease your carbon emissions?
   c. Explain how line-drying (that is, air-drying) your laundry rather than drying it in the dryer can decrease your carbon emissions.

20. Using the carbon footprint calculator, design a low-carbon footprint menu for 1 day. Explain the basis for your food choices. Do you think that a low-carbon menu would be different in different parts of the country? Why or why not?

Eco-Metropolis

Eco-Metropolis

Designing the city of the future

Bumper-to-bumper traffic, a noxious cloud of gray smog, towering skyscrapers that seem to be straight out of the futuristic movie *Blade Runner*: welcome to Shanghai, China's largest city. With a population of 19 million and growing, it's not exactly a place you'd call environmentally friendly. But just 15 miles from this concrete jungle, on the island of Chongming at the mouth of the Yangtze River, something unprecedentedly green is in the works: the world's first eco-metropolis built completely from scratch. About three-fourths the geographic size of Manhattan, the eco-city known as Dongtan will be an urban oasis of green-roofed buildings, tree-lined streets, and pedestrian-friendly neighborhoods—the polar opposite of its dystopian neighbor.

More than just a nice place to live, Dongtan is being designed to incorporate lessons of cutting-edge ecological science. According to its designers, Dongtan will be entirely self-sufficient in food, water, and energy. It will produce no net carbon emissions and zero pollution.

Agriculture will be entirely organic and local. All trash will be recycled, composted, or used to generate electricity. Vehicles will be powered entirely by renewable energy. In short, Dongtan will be a model of **sustainability** for the rest of the world to emulate.

Cities occupy just 2% of the terrestrial surface area of the earth and have only half the world's population, yet they consume more than 75% of its natural resources.

Many urban planners and environmentalists would agree that it's a model the planet badly needs. According to the United Nations, cities occupy just 2% of the terrestrial surface area of the earth and have only half the world's population, yet they consume more than 75% of its natural resources. The reason for this imbalance? Our cities are flawed in their very design, say urban planners—built as if natural resources like land and water were unlimited, and waste was something that would magically disappear.

SUSTAINABILITY
The use of the earth's resources in a way that will not permanently destroy or deplete them; living within the limits of the earth's biocapacity.

Will Dongtan ever be built?

Consider London, which imports more than 80% of its food from other countries. That's a population of 7.5 million people unable to feed itself. Or the mega-metropolis of New York City, which produces some 16,000 tons of garbage every day, sending it by truck, rail, and barge to landfills as far away as Virginia and South Carolina. Cities, it seems, are bursting at the seams.

China has roughly 20% of the world's population but only 7% of the world's land area.

Yet they continue to grow. The United Nations estimates that the ratio of people living in cities versus the country is the highest it's ever been. As of 2008, for the first time more people lived in cities than in rural areas. And the mass migration from countryside to urban center shows no signs of abating: by 2050, 70% of the world's population will live in a city. Compare this figure to 1900, when only 10% of the world's population lived an urban life. That's a huge shift, and one that poses significant environmental challenges, which urban planners are increasingly

being asked to address. In China, where a population of 1.3 billion people represents an ecological force to be reckoned with, the task is both daunting and urgent.

China has roughly 20% of the world's population but only 7% of the world's land area, and its population is increasing by more than 10 million people each year. Every year the resources that each person uses increase as well, thanks to a rising standard of living. "China is one of the first countries in the world to realize this is an unsustainable direction and therefore is desperately trying to improve its energy efficiency and reduce its carbon intensity," says Peter Head, Director of Arup, the London-based engineering firm hired to design Dongtan.

But for Head and his colleagues, Dongtan was a chance to demonstrate to the world that urban growth can happen in a sustainable way, and that ecological challenges can be met with creative design solutions. While the plan does indeed look good on paper, at this point the eco-city is closer to fantasy than reality. Building has not yet begun on Dongtan, and it remains to be seen whether it will ever get off the drawing board. Arup is no longer working on the project

An aerial view of Dongtan.

(now that its design is complete), and the city's future prospects now rest with the Chinese government. But even if it remains just a twinkle in an architect's eye, Dongtan will have already achieved something important, shining a city-size spotlight on one of the most pressing issues facing humanity today.

Our Expanding Footprint

Judging by our numbers, humans are an extraordinarily successful species. Two thousand years ago, we numbered just 300 million globally–less than the current population of the United States. In 2010, there were 6.8 billion of us on the planet. Much of that growth occurred since 1950, thanks in large part to antibiotics and other advances in public health that have allowed people to live longer. And each hour more than 10,000 new people are added to the planet–roughly 3 per second, or 90 million per year. By 2050, demographers estimate, we'll hit the 9 billion mark. The human population is growing exponentially (Infographic 24.1).

As the human population grows, so does our environmental impact. Ecologists measure that impact with a tool known as the **ecological footprint**, which calculates the amount of land and water area a population requires to supply the resources it consumes and to absorb the wastes it generates. Humans require a vari-

INFOGRAPHIC 24.1

Human Population Growth

→ Since the advent of agriculture, the human population has been following an exponential growth pattern, and is approaching 7 billion people. Some estimates predict that the human population will number 9 billion by 2050.

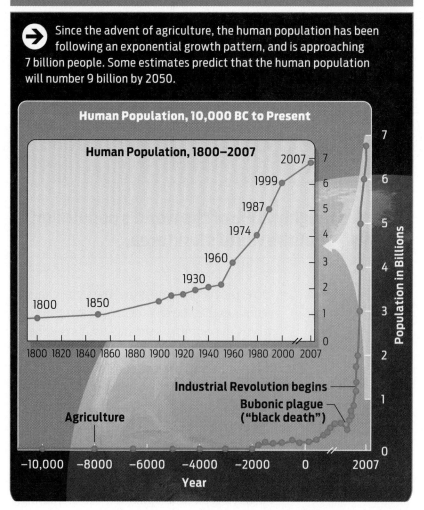

The Human Ecological Footprint

→ How much of the earth's resources does your lifestyle require? The ecological footprint is a measure of people's demand on nature. It uses 5400 different measures gathered from government agencies and scientific publications to calculate a footprint in global hectares, a measure of how much biologically productive land and water area (cropland, forests, grazing lands, fishing area, and built-up land) a human population requires to produce the resources it consumes and to absorb the waste it produces.

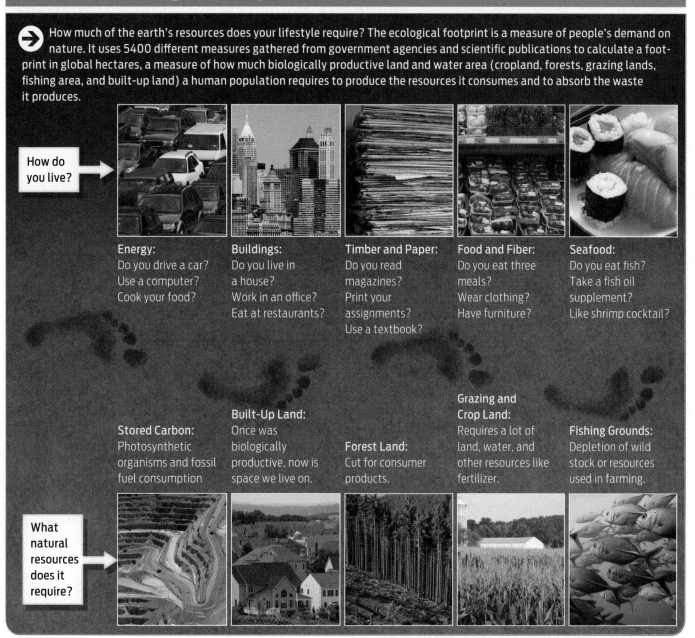

How do you live?

Energy:
Do you drive a car?
Use a computer?
Cook your food?

Buildings:
Do you live in a house?
Work in an office?
Eat at restaurants?

Timber and Paper:
Do you read magazines?
Print your assignments?
Use a textbook?

Food and Fiber:
Do you eat three meals?
Wear clothing?
Have furniture?

Seafood:
Do you eat fish?
Take a fish oil supplement?
Like shrimp cocktail?

What natural resources does it require?

Stored Carbon:
Photosynthetic organisms and fossil fuel consumption

Built-Up Land:
Once was biologically productive, now is space we live on.

Forest Land:
Cut for consumer products.

Grazing and Crop Land:
Requires a lot of land, water, and other resources like fertilizer.

Fishing Grounds:
Depletion of wild stock or resources used in farming.

GLOBAL HECTARE
The unit of measurement of the ecological footprint, representing the biological productivity of an average hectare of land.

ety of **natural resources** to live: farmland to grow crops or raise cattle, gasoline to power cars, oxygen to fill our lungs, to name just a few. All these resources come, directly or indirectly, from the earth.

In addition to providing us with natural resources, the earth also acts like a sponge, absorbing our wastes: the carbon dioxide we emit, for example, and the garbage we produce. By quantifying the amount of biologically productive earth area it takes to sustain our life-

styles, the ecological footprint puts a number on our environmental impact (**Infographic 24.2**).

Ecological footprints are expressed in units called **global hectares,** with 1 global hectare representing the biological productivity (both the resource-providing and waste-absorbing capacity) of an average hectare of land. A hectare is 10,000 square meters–about the size of a soccer field. As of 2006, the global average ecological footprint was 2.6 global hectares per person per year. In other words, it takes that much land and water area to support one average human for 1 year.

But, of course, not everyone uses resources to the same extent. Patterns of consumption vary greatly from region to region and country to country. An average American, for instance, has an ecological footprint of about 9 global hectares, while the average Haitian uses just 0.48 global hectares. These are per capita figures, averages for one resident in each of those countries. It's also possible to calculate the ecological footprint of a whole country. For example, China has a per capita footprint of 1.85 global hectares, much smaller than that of the United States, but because China's population is so large, its *total* footprint is only slightly smaller than that of the United States, which has many fewer people (Infographic 24.3).

Moreover, China's footprint is expanding rapidly—about 3% a year. "Three percent a year doesn't sound like very much," says Head, "but it means that China ... needs to find about 90 million hectares of new land every year for all the resources needed to support the growth and footprint of urbanization."

The human ecological footprint is often compared with the earth's **biocapacity**—its ability to sustain human demand given its available natural resources and its ability to absorb waste. If we think of the footprint as our demand on the earth, the biocapacity is the amount of supplies that the earth can produce to meet that demand.

The earth's biocapacity can't always keep up with our demand. Currently, there are 6.8 billion people living on the earth. As of 2006, the earth had approximately 11.9 billion hectares of biologically productive land and sea (which doesn't include areas like deserts, glaciers, and open ocean), which works out to about 1.8 hectares available per person. Since our current average ecological footprint is 2.6 global hectares per person, we are clearly exceeding the earth's biocapacity, using resources faster than the earth can rejuvenate them. In other words, our current lifestyles are unsustainable (Infographic 24.4).

"The human appetite for resources may be unlimited, but the planet's ability to sustain these needs is finite," says Mathis Wackernagel, co-creator of the ecological footprint concept and executive director of the nonprofit Global Footprint Network. "As our rising demand on ecological services pushes our natural systems to the breaking point, we are not only putting other species at risk, we are jeopardizing our own livelihoods and well-being."

According to Wackernagel, if everyone on the planet were to live like the average resident of the

INFOGRAPHIC 24.3

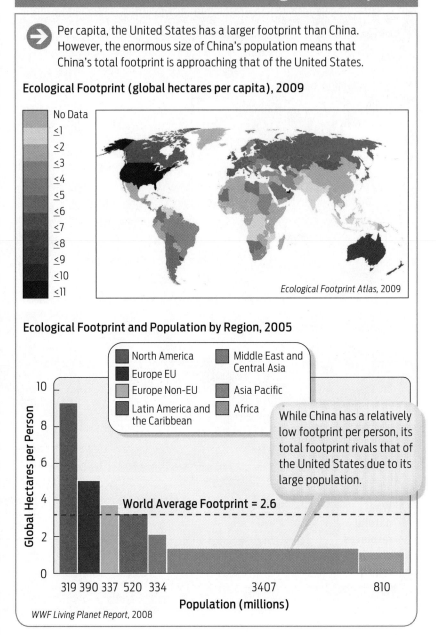

Countries Differ in Their Ecological Footprint

→ Per capita, the United States has a larger footprint than China. However, the enormous size of China's population means that China's total footprint is approaching that of the United States.

Ecological Footprint (global hectares per capita), 2009

Legend:
No Data
≤1
≤2
≤3
≤4
≤5
≤6
≤7
≤8
≤9
≤10
≤11

Ecological Footprint Atlas, 2009

Ecological Footprint and Population by Region, 2005

Legend:
- North America
- Europe EU
- Europe Non-EU
- Latin America and the Caribbean
- Middle East and Central Asia
- Asia Pacific
- Africa

Y-axis: Global Hectares per Person (0–10)
X-axis: Population (millions): 319 390 337 520 334 3407 810

World Average Footprint = 2.6

> While China has a relatively low footprint per person, its total footprint rivals that of the United States due to its large population.

WWF Living Planet Report, 2008

BIOCAPACITY
The amount of the earth's biologically productive area—cropland, pasture, forest, and fisheries—that is available to provide resources to support life.

United States, it would take about five earths to support us. By contrast, if everyone in the world lived like the average person in India, we would need less than half an earth to satisfy our demands.

> **"The human appetite for resources may be unlimited, but the planet's ability to sustain these needs is finite."** –Mathis Wackernagel

What is it about the U.S. lifestyle that leaves such a heavy footprint? Energy consumption, by far, is the largest culprit. The cars and SUVs we drive, the computers we work (and play) with and televisions that entertain us, the washers and dryers that clean our clothes, the air conditioners that cool our homes, the food we truck across the country or fly around the world–all these require energy. Globally, the energy component of our ecological footprint increased roughly 700% between 1961 and 2006, accounting for roughly half our total ecological footprint by 2006.

In the United States, as in most parts of the world, most of this energy comes from fossil fuels–oil, coal, and natural gas. As Chapter 23 discussed, burning fossil fuels releases carbon dioxide to the atmosphere, contributing to global warming. Therefore, the ecological footprint takes into account the amount of land and water area needed to absorb CO_2. This, combined with increased consumption, is what makes our energy footprint so large.

Because they take millions of years to form naturally, fossil fuels are considered **nonrenewable resources**, meaning that once depleted, they are essentially gone for good.

INFOGRAPHIC 24.4

The Human Ecological Footprint Is Greater Than Earth's Biocapacity

When comparing our biological demand, or ecological footprint, with the earth's biocapacity, it is clear that our footprint has been exceeding biocapacity since the mid-1970s. Our greatest demand relates to energy use, indicated by our large carbon footprint.

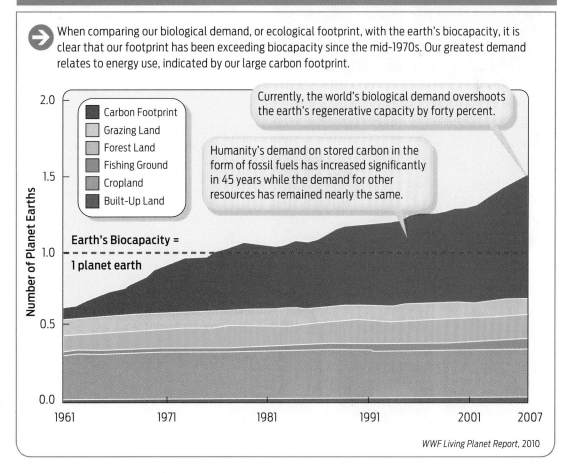

Carbon Footprint
Grazing Land
Forest Land
Fishing Ground
Cropland
Built-Up Land

Currently, the world's biological demand overshoots the earth's regenerative capacity by forty percent.

Humanity's demand on stored carbon in the form of fossil fuels has increased significantly in 45 years while the demand for other resources has remained nearly the same.

Earth's Biocapacity = 1 planet earth

WWF Living Planet Report, 2010

NONRENEWABLE RESOURCES
Natural resources that cannot be replaced.

Besides contributing to our carbon footprint, burning these nonrenewable resources releases pollutants such as sulfur dioxide and nitrogen dioxide. Both of these emissions can combine with water in the atmosphere to form acid rain, which damages both terrestrial and aquatic life. Coal also contains toxic elements such as arsenic and fluorine, which can cause bone and lung disease when inhaled or consumed (Infographic 24.5).

Can anything be done to reduce our ecological footprint? The most significant changes may ultimately have to come from government energy policy, but there are things that individuals can do as well.

Already, individuals and communities around the globe are devising creative ways to live more sustainably. At Carleton College in Minnesota, a wind turbine supplies 40% of the school's electricity. Energy-conscious residents of Calgary,

INFOGRAPHIC 24.5

Fossil Fuels Are Non-Renewable

 Most of the natural resources we use to supply our energy needs are non-renewable. Coal, oil, and natural gas are fossil fuels that take millions of years to form as organic material is compressed by layers of sedimentary rock. While plentiful today, and relatively cheap to obtain, fossil fuels come with significant environmental and human costs.

Coal: Supplies 23% of the World's Energy

Natural Gas: Supplies 24% of the World's Energy

Petroleum Oil: Supplies 37% of the World's Energy

Why Do We Use It?	Environmental Impact
Coal is burned in power plants to produce steam to turn turbines that generate electricity. Coal is relatively cheap to mine, and there is currently an abundance of it in the earth.	Mining coal from the earth often damages the habitat on large tracts of land. Greenhouse gases and pollutants like arsenic, nitrogen dioxide, and sulfur dioxide are released when coal is burned to make electricity. Coal miners have increased risk of respiratory illness.
Natural gas is burned to heat buildings and water. It is relatively cheap to extract, and there are currently large reservoirs of it deep in the earth.	Natural gas is extracted from underground and off-shore reservoirs. Drilling platforms can disrupt ocean habitat. Burning natural gas releases greenhouse gases to the atmosphere.
Oil is used to produce gasoline, petroleum products, and plastics. It is relatively cheap to extract, and there are currently large reservoirs of it deep in the earth.	Oil is drilled from undergound and off-shore reservoirs. Drilling platforms can disrupt ocean habitat. Oil spills can devastate ocean ecology and the seafood economy. Burning products made from oil produces pollution and emits greenhouse gases. Plastics do not biodegrade and therefore create a huge amount of landfill waste.

Unsustainable Mega-City Practices:

Dongtan-Style Alternatives:

Solar-powered water taxis

Living needs are within biking distance.

Taxis are electric or hydrogen fuel-cell powered.

Fossil Fuel-Based Transportation
Sprawl means a daily commute, burning fossil fuels, and creating greenhouse gas emissions.

Canada, can ride on a light rail transit system that obtains all its power from wind turbines. Outside London, England, the community of 100 residences known as BedZED satisfies all its energy needs from renewable sources such as solar panels and locally grown firewood. Apartments in Stockholm, Sweden, come equipped with stoves that burn gas extracted from organic waste generated in the community. In San Diego, garbage trucks run on methane gas captured from decaying garbage in landfills, while residents of Vermont can purchase "cow power"—energy obtained from cow manure—from their local utility company. The residents of Vienna and Paris bike freely around the city on municipally owned bicycles, greatly lowering their footprint.

The plan is for Dongtan to support 80,000 people by 2020, and 500,000 people by 2050.

But a whole city that is entirely self-sufficient in terms of energy and environmentally neutral in terms of carbon emissions and pollution? It sounds too good to be true.

Sustainable by Design

The plan is for Dongtan to support 80,000 people by 2020, and 500,000 people by 2050—the latter being about the population of central Atlanta. And yet, Dongtan's per capita ecological footprint will be a fraction of Atlanta's: 2.6 global hectares for a Dongtan resident versus 13 for an Atlantan.

How will Dongtan achieve a lower footprint? For one, the designers are not at all focused on superficial aesthetics, such as the decorative ornamentation of buildings, and are instead concentrating on what architects call performance-based design.

Focusing on performance and efficiency means rethinking the way cities work from the ground up—starting with transportation. "[T]the essential character of a city's land use comes down to how it manages its transportation," write Peter Newman and Jeff Kenworthy in their book *Sustainability and Cities: Overcoming Automobile Dependence*. When it comes to sustainability, car-based transportation is just about the worst thing that can happen to a city. Yet in many modern cities, such as Los Angeles and Houston, people have few alternatives to driving.

Not so in the eco-city. To eliminate the need for cars in Dongtan, all residential neighborhoods will be within 7 minutes' walking distance of public transportation, which will provide easy access to schools, hospitals, and businesses. Solar-powered water taxis and hydrogen fuel cell buses will provide the primary means of public transport. Bicycle paths and pedestrian walkways will crisscross the city. Cars will not be banned in the city, but car parks placed outside Dongtan will encourage leaving them behind. Curbing car culture will greatly reduce Dongtan's energy consumption.

The city will also be built in such a way that less energy is required to heat and cool it. Conventional cities are essentially "heat islands"—on average, 1°C (1.8°F) warmer than the countryside during the day, and up to 6°C (10.8°F) warmer at night. That's because concrete and asphalt absorb solar radiation. On a hot summer day, air

Unsustainable Mega-City Practices:

Concrete Heat Islands
Heat is generated, raising energy consumption required to cool buildings. Fresh rainwater is polluted as it runs off into drains.

Dongtan-Style Alternatives:

Plant trees along streets to cool cement neighborhoods.

The green roof at Chicago City Hall cuts air-conditioning costs.

conditioning can consume more energy in a city than any other single activity.

One very simple way to beat the heat is to plant more trees. Trees cool cities by providing shade. They also intercept solar radiation that would otherwise generate heat if it were absorbed by concrete or asphalt. In the eco-city, tree-lined streets, rooftop gardens, and green roofs will all temper the heat-island effect.

The buildings themselves will be constructed differently in Dongtan, with walls and windows designed to provide natural insulation and ventilation. To discourage the overuse of electricity, easy-to-read meters placed in obvious locations inside homes and offices will allow residents to see how much they use. Cost will be commensurate with usage.

Most important, Dongtan will generate all of its electricity and heat entirely from **renew-able resources**—those that can be naturally replenished as long as the rate of consumption is not greater than the rate of replacement. Wind turbines and solar panels, for example, will provide the bulk of electricity. In addition, a combined heat and power plant will turn biomass such as leftover rice husks—the region has plenty—into valuable energy for human use. Heat given off during the process will in turn be piped into homes and businesses. Even human waste won't go to waste: treated sewage will be composted to fertilize crops. With such measures, designers estimate that Dongtan will use 65% less energy than a conventional city of the same size **(Infographic 24.6)**.

While cities get a bad rap for being resource hogs, they do have a key advantage over more-spread-out ways of living—they operate as economies of scale. In other words, the density of

RENEWABLE RESOURCES
Natural resources that are replenished after use as long as the rate of consumption does not exceed the rate of replacement.

Unsustainable Mega-City Practices:

Waste
Cities produce tons of solid waste per day, which ends up in landfills. Sewers work to capacity to remove human waste in high-density populations.

Dongtan-Style Alternatives:

Dump trucks fill their tanks with methane waste collected from landfils.

Batch reactors digest human waste to use for fertilizer.

Eco-Cities: Sustainable by Design

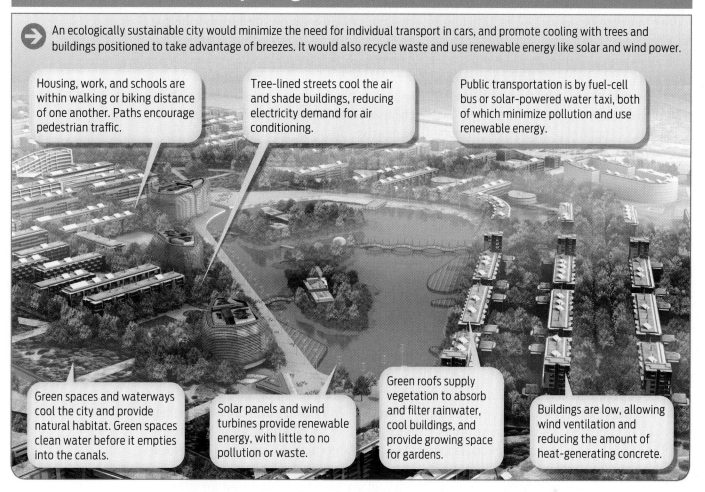

An ecologically sustainable city would minimize the need for individual transport in cars, and promote cooling with trees and buildings positioned to take advantage of breezes. It would also recycle waste and use renewable energy like solar and wind power.

Housing, work, and schools are within walking or biking distance of one another. Paths encourage pedestrian traffic.

Tree-lined streets cool the air and shade buildings, reducing electricity demand for air conditioning.

Public transportation is by fuel-cell bus or solar-powered water taxi, both of which minimize pollution and use renewable energy.

Green spaces and waterways cool the city and provide natural habitat. Green spaces clean water before it empties into the canals.

Solar panels and wind turbines provide renewable energy, with little to no pollution or waste.

Green roofs supply vegetation to absorb and filter rainwater, cool buildings, and provide growing space for gardens.

Buildings are low, allowing wind ventilation and reducing the amount of heat-generating concrete.

people makes possible a more efficient and affordable utilization of resources, which is ultimately more sustainable. For instance, a public transportation system that transports tens of thousands of people who might otherwise be driving gas-guzzling cars can reduce the carbon footprint significantly.

New York is a good example. The average New Yorker who takes the subway 1 mile to work uses much less fossil fuel energy than a suburbanite who commutes 10 miles to work in a car. In fact, according to David Owen, author of *Green Metropolis: Why Living Smaller, Living Closer, and Driving Less are the Keys to Sustainability*, New Yorkers have the smallest per capita carbon footprint in the United States, a statistic that has even led some commentators to refer to New York as the Big Green Apple.

The solution to the problems of urbanism, then, is not to de-urbanize, but to make cities part of the solution rather than part of the problem.

Moving Forward

Construction of Dongtan was supposed to begin in 2007, with the first phase of development—including arrival of the city's first 10,000 residents—completed by 2010, in time for the Shanghai Expo. That didn't happen. Development has stalled, and it's unclear whether the ambitious plans will ever be fully realized. Critics of Dongtan have long held that the city was a utopian fantasy, more useful as a public relations ploy than a place to live.

Peter Head of Arup emphatically challenges that view. While the future of Dongtan itself is uncertain, he says China has plans to make the whole of Chongming an eco-island, using specifications developed for Dongtan. "In many ways, all of the ideas and thinking [are] alive and well," says Head. He acknowledges, though, that there is much to be done and that Dongtan is only a start. The eco-city's hypothetical footprint of 2.6, for instance, is still more than a truly sustainable one of 1.8. Reducing resource use even further will depend partly on the will of future residents.

Even if Dongtan is never built, there are hopeful signs that urban sustainability is catching on around the world. Eco-cities are currently being planned or built in countries as diverse as Argentina, Australia, Finland, Vietnam, and the United States. And key elements of sustainability—such as finding alternatives to fossil fuels—are increasingly being recognized as an issue of national and global importance. For example, as of 2009, the United States obtained 17% of its energy from renewable energy sources (including nuclear energy). While wind and solar constitute a small fraction of these renewable sources, their contribution is growing. In 2010, the U.S. Department of the Interior approved the first offshore wind farm, to be operated off the coast of Cape Cod, Massachusetts. And in 2008, the United States became the world leader in wind power investments—though in terms of the proportion of energy it obtains from wind, it is still greatly outperformed by many European countries. Highest marks for use of wind power go to Denmark, which in 2009 obtained 20% of its electricity from wind.

The appeal of renewable sources of energy such as wind and solar is undeniable: they are plentiful, powerful, and environmentally neutral in terms of their carbon emissions. Solar power alone could theoretically provide more than enough clean energy to supply the needs of everyone on the planet many times over—assuming we could adequately and inexpensively harvest it.

The technologies to harness wind and solar power are currently much more expensive to build and operate than coal-fired power plants, for example. What makes fossil fuels such convenient and inexpensive sources of energy is the fact that the difficult work of harvesting the energy of sunlight has already been done, by the fossilized photosynthetic organisms that have been compressed over millions of years into oil, coal, and gas. (In a way, we are already using the energy of sunlight to power our lifestyles, but indirectly.) Unless the price of the new technologies comes down, or governments decide to subsidize these alternatives to make them cheaper or tax fossil fuels to make them more expensive, free markets will tend to favor cheaper options.

Of course, when you consider the environmental and human costs of obtaining and burning fossil fuels—from coal-mine explosions, to air pollution, to oil spills—they aren't actually that cheap; think of the 2010 Gulf oil disaster, which is estimated to cost $40 billion and counting. These downstream costs of fossil fuels, which are not reflected in their market price, are known among economists as externalities. If externalities were included in the price, as some economists and environmentalists suggest they should be, then the playing field with other forms of energy would be more level.

Then there are issues of space: solar panels and windmills can take up lots of it. The Mojave Desert, in California, for example, is home to an increasing number of solar power plants. As of 2010, plans have been approved for projects that, when completed, will cover some 39 square miles of land. To some environmentalists, this represents a threat to local wildlife.

And while many people support the idea of renewable energy in theory, many would also prefer not to have the technology located in their backyards. Wind turbines, especially, are seen by many as a kind of "sight pollution," cluttering the landscape. (In fact, this was a controversial aspect of Cape Wind, the Cape Cod wind farm: residents didn't want to look at it.)

And of course, wind does not always blow and sun does not always shine, so they are less reliable than other forms of energy. Given the limitations of our current technologies, it is not yet possible to satisfy our energy demands with only the existing infrastructure of wind turbines, solar panels, biofuels, and the other renewable energy sources so far developed. At least for the next decade, we cannot take fossil fuels out of our energy mix (Infographic 24.7).

Renewable Resources Reduce Our Ecological Footprint

 While many renewable resources are available to us, economic, technological, and environmental considerations currently limit their use as alternatives to fossil fuels.

	Why Don't We Use It More?	**Environmental Impact:**

Solar: 0.07% of World's Energy

Solar power is currently much more expensive to produce than non-renewable options. Producing solar panels involves using toxic chemicals, and the resulting waste must be properly disposed of.

Solar energy traps energy from the sun and converts it into electricity and heat with little impact on the environment. As nothing is burned to make the electricity there are zero polluting emissions from this process.

Wind: 0.49% of World's Energy

Wind power is currently much more expensive to produce than non-renewable options. Wind generators take up space, either on land or in the water, and must be located in windy areas. Some people don't want a visible wind farm near their homes.

Wind energy is used to turn wind turbines, producing electricity with little impact on the environment. In the absence of combustion, no pollutants are released to the environment. Bird species may be affected as turbines encroach on their air space.

Nuclear: 9% of World's Energy

Nuclear reactors are expensive to design and build, and a reactor has a limited life span. Extracting uranium from mines has an environmental impact, and the waste from uranium mines is radioactive. Most important, the waste from nuclear reactors is highly radioactive, making storage complicated. As well, weapons-grade plutonium can be made from reactor waste, which poses a security threat.

Nuclear energy uses radioactive elements harvested from the earth and concentrated. As these elements decay, they give off tremendous heat, which is used to produce electricity. As nothing is burned in the process, there are no polluting emissions.

Biofuels: 3.7% of World's Energy

While being intensively researched, biofuels have not yet become feasible replacements for fossil fuels. In some cases, significant emissions are associated with their production. In other cases, more research and investment is required to optimize the production process.

Biofuels are made from plant material. When burned, the only CO_2 released to the atmosphere is what the plants and algae took in through photosynthesis, so fossil deposits of carbon are not used. Biofuels can be made directly from plant material, or from energy-rich oils that algae make. In some cases, growing plants for biofuels competes with growing crops.

Hydroelectric: 2.4% of World's Energy

Hydroelectric power uses dams to block rivers, creating lakes with immense amounts of potential energy. Building dams destroys habitat, impacts local fish populations, and can force human populations to relocate.

Hydroelectric power relies on the conversion of potential energy (stored in the position of accumulated water behind a dam) to kinetic energy, which can turn a generator. There are no emissions associated with hydroelectric power. Hydro plants have long life spans, and hydro power can potentially power half the projected energy demands of the planet.

Geothermal: 0.35% of World's Energy

Geothermal energy is used extensively in Iceland and in some areas of California. However, it has yet to be fully developed in other areas, primarily because optimal technologies require further development.

Geothermal energy relies on naturally occurring heat from the magma layer beneath the earth's crust. This is a sizeable and sustainable resource that can be tapped to drive generators or directly heat homes and businesses. In some cases, noxious pollutants are released with the steam from geothermal resources.

Not a Drop to Drink

Fossil fuels aren't the only natural resources being overdrawn by a growing human population—water is, too. Although 70% of the globe's surface is covered with water, only 2.5% of it is freshwater, and most of that is locked up in ice caps and glaciers. A measly 1% of the total water on earth is available for human consumption. Nevertheless, freshwater is considered a renewable resource because the supply in lakes, rivers, reservoirs, and underground **aquifers** is continually being replenished by the water cycle. As long as the rate of water withdrawal from these sources is less than the rate of replacement, the supply of freshwater remains relatively constant **(Infographic 24.8)**.

Although, as a renewable resource, water is not consumed in the same way as coal or oil, our supply of freshwater is being divided among more and more people, meaning there is less available for everyone. According to the Food and Agriculture Organization of the United Nations, water use increased sixfold during the 20th century, more than twice the rate of population increase. Today, more than half of all the accessible freshwater contained in rivers, lakes, and aquifers is appropriated by humans, most of it for irrigation in agriculture.

Food and lifestyle choices also affect water availability. According to environmental scientist Arjen Y. Hoekstra, who developed the concept of the water footprint, it takes 900 liters of

AQUIFER
Underground layers of porous rock from which water can be drawn for use.

INFOGRAPHIC 24.8

Water Is a Renewable Resource

Fresh water is a valuable resource. In addition to its role in keeping us hydrated, it irrigates crops, sustains fisheries, and provides recreational opportunities. Although water is "used," it is not "used up": it is ultimately returned to the global ecosystem as it evaporates to the atmosphere, flows into rivers or streams, or enters underground aquifers.

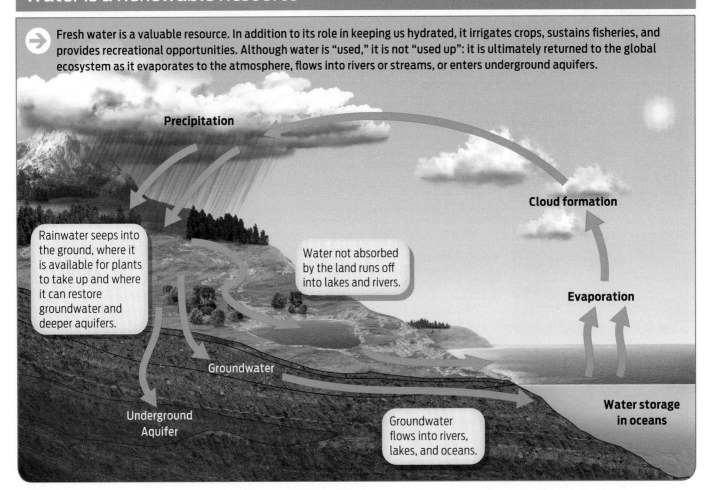

Precipitation

Cloud formation

Rainwater seeps into the ground, where it is available for plants to take up and where it can restore groundwater and deeper aquifers.

Water not absorbed by the land runs off into lakes and rivers.

Evaporation

Groundwater

Underground Aquifer

Groundwater flows into rivers, lakes, and oceans.

Water storage in oceans

Depletion of Fresh Water by a Growing Population

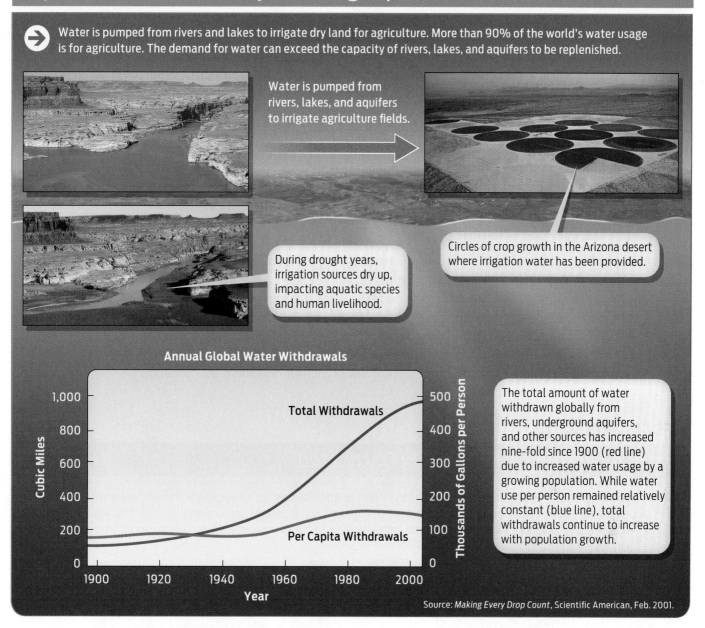

→ Water is pumped from rivers and lakes to irrigate dry land for agriculture. More than 90% of the world's water usage is for agriculture. The demand for water can exceed the capacity of rivers, lakes, and aquifers to be replenished.

Water is pumped from rivers, lakes, and aquifers to irrigate agriculture fields.

During drought years, irrigation sources dry up, impacting aquatic species and human livelihood.

Circles of crop growth in the Arizona desert where irrigation water has been provided.

Annual Global Water Withdrawals

Total Withdrawals

Per Capita Withdrawals

The total amount of water withdrawn globally from rivers, underground aquifers, and other sources has increased nine-fold since 1900 (red line) due to increased water usage by a growing population. While water use per person remained relatively constant (blue line), total withdrawals continue to increase with population growth.

Source: *Making Every Drop Count*, Scientific American, Feb. 2001.

water to produce a kilogram of corn but more than 15 times that much to produce a kilogram of beef.

One striking example of the consequences of increased water use can be seen in the Colorado River, which is often so depleted that in dry periods it fails to reach the Gulf of California, a sign that water is being withdrawn from this resource faster than it is being replenished (Infographic 24.9).

Though water is renewable, pollution shrinks the total amount of available clean freshwater on the planet. Agriculture, industry, and cities all play a role here. Runoff from streets carries pollutants such as motor oil and sewage; fertilizers, pesticides, and toxic chemicals leach from fields and factories. These substances can eventually reach aquifers, rivers, and oceans, contaminating the water that both humans and wildlife depend on.

Some countries experience water scarcity more acutely than others. That's because the geographic distribution of freshwater does not match the distribution of the world's population. Canada, for example, hosts just 0.5% of the world's population, but 20% of the global freshwater supply is within Canada's borders. China, on the other hand, has 20% of the world's people but only 7% of the world's water. The United Nations estimates that at least a billion people in the world currently lack access to clean and safe drinking water, and by 2025, two-thirds of the world's population will live in areas of moderate to severe water stress. And climate change, if it changes precipitation patterns, may also affect the global availability of water in unpredictable ways (Infographic 24.10).

Finding the Limits

When the designers at Arup were first approached about designing an eco-city from scratch, they were intrigued but skeptical. Surely there must be a catch, they thought. And indeed, there was a big one: the city, developers stipulated, must not disturb the migration path

of the rare birds that use the wetlands on Chongming Island as a stopping point along their way between Siberia and Australia.

To protect the birds and their flyway, the Dongtan master plan calls for a buffer zone between the city and the bird resting area. The zone will be more than 2 miles wide and will help prevent pollutants—including light, sound, air, and water pollution—from reaching the surrounding wetlands.

These efforts to protect bird species illustrate a final point about our expanding footprint: it takes a toll not only on humans, but on other species as well. Every few years, the World Wildlife Fund's Living Planet report documents the health of nearly 1,700 species of vertebrates around the world. Between 1970 and 2005 the number of individuals in the populations studied declined 30% overall, reflecting not only increased hunting and fishing by humans but also the degradation of habitat as humans

INFOGRAPHIC 24.10

Water Availability Is Not Equally Distributed

Fresh water is not evenly distributed across the globe, and its availability does not always follow international borders. In addition, access to even a sufficient water supply may be limited by economic, social, and political circumstances, such as war and ethnic conflict. As the human population continues to grow, and access to clean fresh water continues to decline, these problems are likely to intensify, particularly in areas with existing scarcities of water.

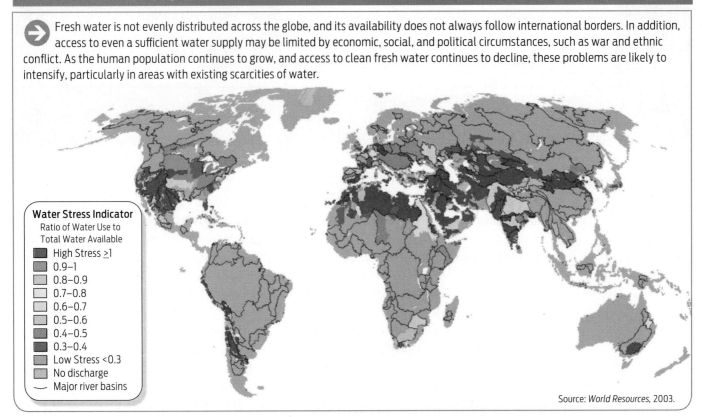

Water Stress Indicator
Ratio of Water Use to Total Water Available
- High Stress ≥1
- 0.9–1
- 0.8–0.9
- 0.7–0.8
- 0.6–0.7
- 0.5–0.6
- 0.4–0.5
- 0.3–0.4
- Low Stress <0.3
- No discharge
- Major river basins

Source: *World Resources*, 2003.

Species Loss Increases as Human Population Grows

→ As the human population increases, so does the number of species lost to extinction. There are two major contributors: habitat destruction related to human development and agriculture; and animals becoming food for a growing human population.

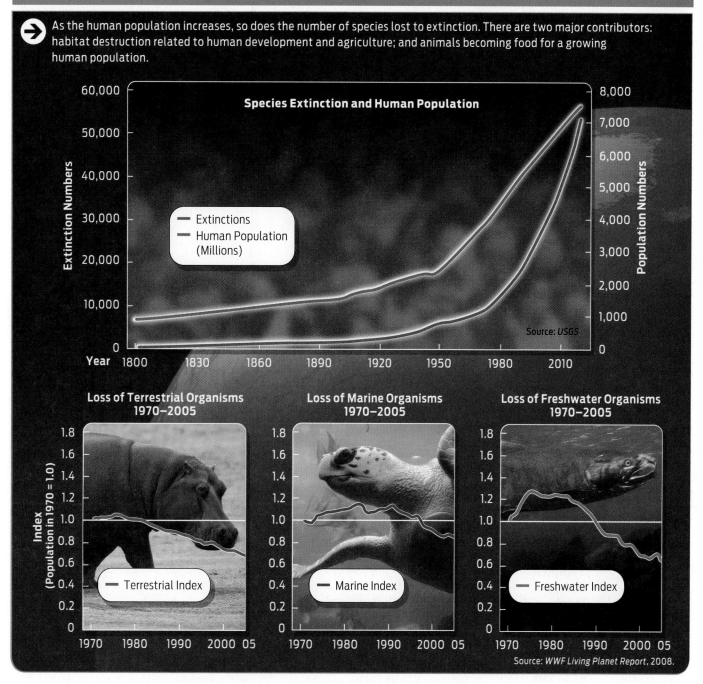

Species Extinction and Human Population

Legend: Extinctions; Human Population (Millions)

Source: *USGS*

Loss of Terrestrial Organisms 1970–2005 — Terrestrial Index

Loss of Marine Organisms 1970–2005 — Marine Index

Loss of Freshwater Organisms 1970–2005 — Freshwater Index

Index (Population in 1970 = 1.0)

Source: *WWF Living Planet Report*, 2008.

expanded into areas once occupied only by wild creatures **(Infographic 24.11)**.

Common sense tells us that the human population cannot continue to grow unchecked indefinitely–otherwise, within a few hundred years people would cover every square foot of the globe and many other species would be long gone. Experience with other species also tells us that the size of the human population will even-

tually reach an upper limit, at which point lack of resources and available space will curb growth.

What that precise limit is remains up for debate. The United Nations has estimated that the earth's carrying capacity (Chapter 21) is between 7 and 13 billion people; other researchers put the number much higher. Why is it difficult to pin down an exact figure? In part because the carrying capacity of the earth can change. The use of fertilizer and genetically modified crops, for example, has greatly increased the amount of food that can be produced from a given size plot of land. Likewise, modern sewer systems and advances in public health, which have helped prevent communicable diseases such as cholera, have allowed people to live in closer proximity, at much higher densities, than ever before. Quality of life concerns also muddy the calculation of carrying capacity. The earth may theoretically be able to support more than 13 billion people, but the question is, would anyone want to live in such a place?

Demographers tell us that there is little we can do in the short term to stem the human

INFOGRAPHIC 24.12

What You Can Do to Live More Sustainably

Take Action	Why?	Your Impact!
Reduce Home Water Use	The average U.S. household uses over 22,000 gallons of water per year for showers and baths. Water is almost always heated, resulting in increased fossil fuel consumption and greenhouse emissions.	If only 1,000 of us install faucet aerators ($2 – $5) and efficient showerheads (<$20), we can save nearly 8 million gallons of water and prevent over 450,000 pounds of carbon dioxide emissions each year!
Install Compact Fluorescent Lightbulbs (CFLs)	Electricity production is the largest source of greenhouse gas emissions in the United States, and lighting accounts for about 25 percent of American electricity consumption.	By replacing just four standard bulbs with CFLs, you can prevent the emission of 5000 pounds of carbon dioxide and reduce your electricity bill by more than $100 over the lives of those bulbs.
Eat Less Feedlot Beef	Feedlot beef is particularly wasteful. Producing one pound of feedlot beef in California, for example, requires five pounds of grain and over 2,400 gallons of irrigation water.	Eat more veggies. More of the energy in plants will go directly to you if it doesn't have to pass through a cow first. Plant food goes farther to support you than it does to raise feedlot beef.
Reduce Vampire Energy Waste	Electronics use energy even when they are turned off! This standby "vampire energy" accounts for 5 to 8 percent of a single family's home electricity use per year.	When you plug your electronics into a power cord that you turn off each night, you will save the equivalent of one month's electric bill each year.
Drive Less and Invest in Fuel Economy	With less than 5% of the world's population, America consumes a quarter of the world's oil and emits a quarter of the greenhouse gases, largely from automobiles.	Driving smaller vehicles and those with more fuel efficiency cuts carbon dioxide emissions and reduces dependence on non-renewable fossil fuels.
Recycle	75% of our trash can be recycled. The U.S. population discards each year enough glass bottles and jars to fill 12 giant skyscrapers. Recycling materials uses fewer non-renewable resources, saves energy, results in less air and water pollution, and creates more jobs than making new materials.	Recycling one aluminum can saves enough energy to run a TV for 3 hours. To produce each week's Sunday newspapers, 500,000 trees are cut down. Recycling a single run of the Sunday *New York Times* would save 75,000 trees. Taking reusable bags to do your weekly grocery trip reduces our demand for petroleum for plastic bags.

population climb. World population will continue to grow for years because of past growth: so many young people are alive today that even if they have only two children to replace themselves, the population will continue to expand for at least the next 50 years. This is true even in China where, since 1979 official government policy has limited couples to having one child.

For the longer term, demographers say that the best way to limit population growth is to raise the education level and quality of life for the world's women. According to Robert Engelman, Vice President for Programs at the Worldwatch Institute, there is a direct correlation between education level and the number of children women have. When women have more opportunities available to them, they tend to limit the size of their families.

"[T]he evidence suggests," Engelman writes in *Scientific American*, "that what women want–and have always wanted–is not so much to have *more children* as to have *more for* a smaller number of children they can reliably raise to healthy adulthood. Women left to their own devices, contraceptive or otherwise, would collectively 'control' population while acting on their own intentions."

Cities will also play a role. "In global terms, population growth is actually slowed by the growth of cities," writes Peter Newman. Far from being the harbingers of catastrophe, "[c]ities could indeed be helping to save the planet."

In fact, in many countries today, the birthrate is actually declining and may eventually stabilize as standards of living and educational levels of women rise around the world and as the world becomes more urbanized. Before it does, however, we will have to learn to adjust to a world occupied by at least a billion more people. And that means using all our resources, both renewable and nonrenewable, more wisely, more responsibly, and–indeed–more sustainably **(Infographic 24.12)**. ∎

▶ Summary

◾ As of 2010, the human population totaled 6.8 billion people. Some demographers say the number could hit 9 billion by 2050.

◾ As the human population grows, so does our ecological footprint, a measure of our demand on nature. Ecological footprint is measured in units called global hectares, the number of average hectares of land it takes to supply us with resources and to absorb our wastes.

◾ The ecological footprint of the current human population is greater than the earth's biocapacity, its total natural resources and ability to absorb our wastes.

◾ Natural resources include renewable resources, such as sunlight, wind, and water, and nonrenewable resources, such as fossil fuels (oil, coal, and gas).

◾ Burning fossil fuels generates harmful wastes (for example, greenhouse gases and pollutants) and increases our ecological footprint.

◾ Sustainability refers to the ability of humans to live within earth's biocapacity, without depleting nonrenewable resources. Sustainable practices minimize the consumption of nonrenewable resources by using renewable resources like wind and solar power instead of fossil fuels to generate electricity and heat.

◾ At their current level of development, technologies to harvest renewable energy cannot meet our total energy demands. Fossil fuels cannot yet be taken out of our energy mix.

◾ Although freshwater is a renewable resource, the world's supply is not distributed equally, and many people around the world suffer from water scarcity, a problem exacerbated by a rising population and the demands of agriculture.

◾ City dwellers have a high per capita ecological footprint compared to people living in rural areas, in large part because of more intensive fossil fuel energy use linked to driving cars and rising consumption.

◾ Cities can be more efficient than nonurban areas and can reduce their ecological footprint by limiting car use and incorporating sustainable technologies such as green roofs, public transportation, and renewable energy sources.

◾ Individually, we can decrease our ecological footprint by driving less, reducing water use, eating less meat, and recycling.

HUMAN POPULATION GROWTH AND ECOLOGICAL FOOTPRINT

The human population has grown exponentially, and this growth has substantial impacts on the planet. Ecologists measure our environmental impact by calculating ecological footprints.

HINT See Infographics 24.1–24.4.

⊙ KNOW IT

1. From what you've read in this chapter, explain some of the advances that have permitted the human population to grow exponentially.

2. Describe an ecological footprint.

3. From your understanding of an ecological footprint and what you read in this chapter, which of the following places likely has a population with the greatest ecological footprint?
  a. Dongtan, China
  b. a rural village in China
  c. Calgary, Canada
  d. Houston, Texas
  e. New York, New York

⊙ USE IT

4. For each place listed in Question 3, characterize its footprint as relatively high or relatively low. Justify your characterization by describing some of the factors that contribute to its footprint. (Refer to Infographics 24.4, 24.6, and 24.12 and the Global Footprint Network, http://www.footprintnetwork.org/en/index.php/GFN/.)

5. On the outskirts of a small town, a farmer has just sold his 5 acres of cropland to a developer who is planning to build 20 single-family condominium units on that land. Discuss the ways that this transaction will affect the size of the nearby town's population and the ecological footprint of the residents of the town.

6. What building considerations could the developer in Question 5 take into account to minimize the impact of this development on the ecological footprint of the town?

NATURAL RESOURCE USE

Natural resources occur naturally and cannot be produced by industrial processes. Some natural resources are renewable, while others are essentially irreplaceable.

HINT See Infographics 24.5 and 24.7–24.11.

⊙ KNOW IT

7. Which of the following waste products is/are associated with the burning of fossil fuels?
  a. water
  b. carbon dioxide
  c. nitrogen dioxide
  d. all of the above
  e. b and c

8. Mark each of the following natural resources as renewable (R) or nonrenewable (N):
  Freshwater _____
  Coal _____
  Codfish populations in the North Atlantic _____
  Wind _____
  Sunlight _____

9. If oil is formed from fossilized remains of once-living organisms, and if organisms keep dying, why is oil considered to be a nonrenewable resource?

⊙ USE IT

10. The renewability of some resources can depend on human choices and activities. List some such resources, and explain how human activities may lead a renewable resource to become essentially nonrenewable. (Look at Question 8 for some ideas.)

11. Think about your local region—for example, do you live in the desert southwest or on the northeast ocean shore? Describe the nonrenewable and renewable energy resources that are available in your region or that your region can harvest. What are some of the challenges that must be overcome in order to tap into the renewable energy resources in your region?

SUSTAINABILITY

Sustainability means living within the biocapacity of the earth. This includes using resources at a sustainable rate and not generating wastes faster than they can be decomposed or absorbed by the earth.

HINT See Infographics 24.6 and 24.12.

⊙ KNOW IT

12. The plans for Dongtan include many ideas that will contribute to sustainability. For each of the plans listed below, describe its impact on resource consumption and/or waste production.

a. schools and shops to be located near residences

b. buildings to have green roofs

c. solar panels to be mounted throughout the city

13. What are some ways in which waste can be used as a productive resource?

⊘ USE IT

14. Dongtan would be an entire city with sustainable practices pre-engineered into its design. If you live in a traditional city, what practices can you adopt to reduce your ecological footprint and embrace the philosophy of sustainable living? For each practice that you think of, explain how it would contribute to sustainability and the reduction of your ecological footprint.

15. Many cities have been developed in the hot and dry southwestern states of the United States. What are some of the sustainability implications of living in the desert?

SCIENCE AND ETHICS

16. Infographics 24.9 and 24.10 provides a dramatic illustration of some of the choices we face in resource management. Water from the Colorado River is being used to irrigate crops in Arizona. How might this use affect the use of the water for recreation (for example, swimming, fishing, boating) in the downstream regions? Water currently destined for agriculture could instead be retained in the river to help preserve the endangered silvery minnow. How would you balance the competing agricultural, recreational, and ecological concerns involved in this choice?

17. In general, how do you think the ecological footprint of the United States compares to that of Bangladesh? If footprint expansion accompanies economic development, what will happen to the carrying capacity of the global human population as developing countries continue to develop? Do you think that developed countries such as the United States have an obligation to reduce their ecological footprint in order to make room for the development of other countries?

Man versus Mountain

Man versus Mountain

Physiology explains a 1996 disaster on Everest

At 1:17 P.M., on May 10, 1996, Jon Krakauer planted one foot in China, the other in Nepal, and stood on the roof of the world. He was at the highest point above earth's sea level that any human has ever reached—short of standing on the moon. Yet he didn't feel like celebrating. It had taken him 6 long weeks to climb to the top of Mount Everest, and now that he was here his toes ached in the sub-zero cold, his breath came in short, painful bursts, and his head pounded from the altitude. It was a struggle just to stay upright. "I cleared the ice from my oxygen mask, hunched a shoulder against the wind, and stared absently at the vast sweep of earth below," wrote Krakauer in an account of his climb for *Outside* magazine later that year.

Krakauer's journey to Everest was a lifetime in the making. While other kids were idolizing astronaut John Glenn and baseball player Sandy Koufax, Krakauer's childhood heroes

> **"I cleared the ice from my oxygen mask, hunched a shoulder against the wind, and stared absently at the vast sweep of earth below."**
> **—Jon Krakauer**

were Tom Hornbein and Willi Unsoeld—two men from his hometown in Oregon who, in 1963, became the first climbers to scale the daunting western ridge of Everest. As a teenager, Krakauer became a skilled climber, vanquishing many of the world's most difficult peaks, and he dreamed of one day climbing Everest himself. By his mid-twenties, though, he had largely abandoned the idea as a boyhood fantasy. But old dreams die hard.

In 1995, Krakauer was working as a journal-ist when a call came to shadow an Everest climb and report on it for *Outside* magazine. The 42-year-old writer-adventurer jumped at the chance. He would join a team headed by the cele-brated climbing guide Rob Hall, whose company, Adventure Consultants, had successfully put 39 amateur climbers on top of Everest. Reaching the summit himself would mean enduring a weeks-long ascent from Base Camp, giving his body time to adjust to the high altitude.

It would also mean risking his life on a daily basis.

The icy tip of Mount Everest sits at 29,035 feet above sea level; cruising altitude of most commercial jetliners is 30,000 feet. A human plucked from sea level and deposited at this altitude would quickly lose consciousness and die. A climber who has spent weeks adjusting to the altitude can function better at the summit, but not very well, and not for very long. Everest is not only the highest place on earth, it is also one of the coldest. At the summit, where windchill temperatures average −53°C (−63°F), freezing to death is a real possibility.

Despite these dangers—or perhaps because of them—a handful of fearless men and women try to climb Everest every season. And every season, some of them don't come back. There are many reasons for these disasters—poor training,

At the summit, where windchill temperatures average −53°C (−63°F), freezing to death is a real possibility.

unforeseen accidents, raw egotism—but among the most important is basic biology: the human body is not equipped to survive at such extreme altitude, and such extreme temperature, for long.

The Body as Machine

Like a car or a computer, a human body is made up of many parts working together in a coordinated fashion. The parts are organized hierarchically, so that smaller components are organized into increasingly larger units, which are themselves organized into more complex systems. The study of all this intricate hardware is known as **anatomy.**

The result of millions of years of evolutionary tinkering, human bodies have an anatomical structure that is impressively well adapted to living in certain environments and performing

ANATOMY
The study of the physical structures that make up an organism.

certain functions. Our species evolved in the hot, flat savannahs of Africa where environmental conditions favored big brains, opposable thumbs, and bipedal posture–as well as the ability to keep cool (Chapter 20). As a result, modern humans excel at grasping a pencil or looking through a microscope; we do less well swimming at the bottom of the ocean or living on mountaintops. Fundamentally, that's because of how we're put together.

For all living things, the smallest anatomical unit is the cell. Human bodies are made up of trillions of cells, each of which can be classified as one of a few hundred different types. Cells, in turn, are organized into **tissues**–layers of specialized cells working together to execute a particular function. Humans and other animals have four different kinds of specialized tissue– epithelial, connective, muscle, and nervous– which carry out specific tasks in the **organs** of which they are a part. The stomach, for example, is an organ composed of the four types of tissue organized into a compartment for churning and digesting food. At the highest level of organization, organs interact chemically and physically as part of **organ systems.** The digestive system, for instance, consists not only of the stomach, but also of other organs, including the esophagus, small intestine, and liver, which all work together to digest and absorb food **(Infographic 25.1)**.

TISSUE
An organized collection of a single cell type working to carry out a specific function.

ORGAN
A structure made up of different tissue types working together to carry out a common function.

ORGAN SYSTEM
A set of cooperating organs within the body.

INFOGRAPHIC 25.1

How the Human Body Is Organized

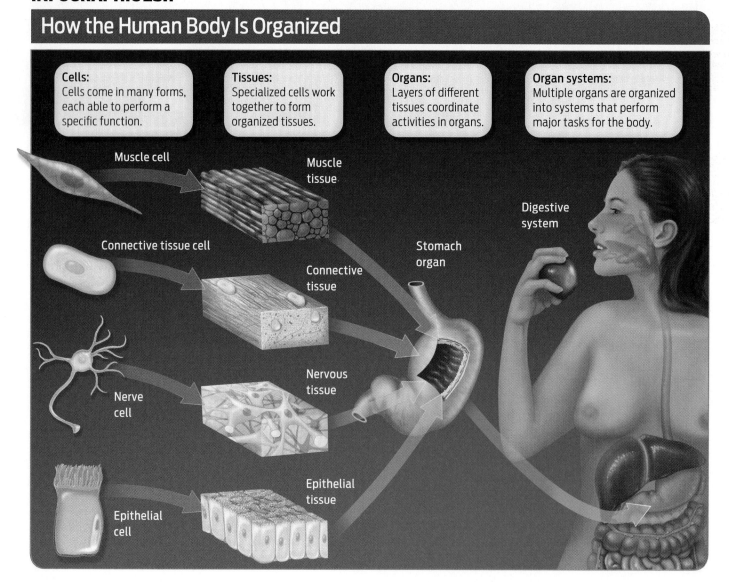

Cells:
Cells come in many forms, each able to perform a specific function.

Tissues:
Specialized cells work together to form organized tissues.

Organs:
Layers of different tissues coordinate activities in organs.

Organ systems:
Multiple organs are organized into systems that perform major tasks for the body.

Muscle cell

Muscle tissue

Connective tissue cell

Connective tissue

Stomach organ

Digestive system

Nerve cell

Nervous tissue

Epithelial cell

Epithelial tissue

If the body is like a machine, then physiologists–the scientists who specialize in **physiology,** the study of how a living organism's physical parts function–are interested in how this machine keeps running smoothly. Physiologists want to understand how organ systems cooperate to accomplish basic tasks, such as obtaining energy from food, taking in nutrients to build new molecules during growth and repair, and ridding the body of wastes. To the physiologist, the body is an integrated system for processing inputs and outputs and maintaining **homeostasis,** the maintenance of a relatively stable internal environment.

Thermoregulation: The Physiology of Staying Warm

Like many other animals, humans have an optimal operating temperature and are exquisitely sensitive to temperature changes. Although we

> **Although we can tolerate a wide range of external temperatures in our daily lives, we cannot tolerate even minute changes in our internal temperature.**

can tolerate a wide range of external temperatures in our daily lives, we cannot tolerate even minute changes in our internal temperature. That's because the enzymes that catalyze the chemical reactions in our body function only within a very narrow temperature range (for more on enzymes, see Chapter 4). The body thus works hard to maintain a relatively constant internal temperature compatible with life. Through **thermoregulation,** our body temperature is kept at a consistent–and toasty–37°C (98.6°F).

Keeping a consistent body temperature is just one example of how the body tries to maintain homeostasis. "What we're really thinking about with homeostasis are certain set points that your body needs to maintain," explains Robert Kenefick, a research physiologist with the U.S. Army Research Institute of Environmental Medicine in Natick, Massachusetts. The body has a number of such set points, he says, for things like temperature, blood pH, and blood pressure, and works hard to keep these factors balanced within a very narrow range, even in the face of a changing external environment **(Infographic 25.2).**

Kenefick is an exercise physiologist, and he has spent his career trying to understand how the human body maintains homeostasis during strenuous activities like hiking and running marathons. He works in the Army Research Institute's Thermal and Mountain Medicine Division, where a main focus of his research is understanding how the body performs in extreme cold.

Staying warm is hard to do when ambient temperatures drop below −50°C (−58°F), as they routinely do on Everest. To seal in heat, mountain climbers wear multiple layers of protective gear designed to trap in heat, wick away moisture, and insulate their bodies from the wind and cold. When not hiking, climbers consume copious amounts of hot tea or coffee to warm their insides. But insulated clothes and hot beverages would be of little help if it

INFOGRAPHIC 25.2

The Body Works to Maintain Homeostasis

➡ The body expends a great deal of energy to maintain a constant internal environment. Only small fluctuations are tolerated, even in the presence of extreme external conditions.

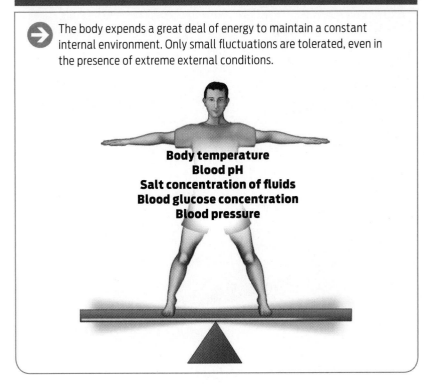

Body temperature
Blood pH
Salt concentration of fluids
Blood glucose concentration
Blood pressure

weren't for the body's natural way of keeping warm.

As Kenefick explains, human bodies respond to cold in two main ways: by conserving the heat they already have and by generating more. To conserve heat, the body performs what is called peripheral **vasoconstriction**–the decrease in diameter of blood vessels just below the surface of the skin. This clamping down of blood flow near the skin surface is why hands, feet, and noses are the first to feel cold on a cold day, and it's a sign that your body is trying to retain heat.

By constricting blood vessels in the skin, peripheral vasoconstriction pushes blood from the skin to the body core, where the internal organs are. "A lot of people believe this is done to increase the amount of blood that goes to your core to help protect those organs," says Kenefick. That's true to a degree, he notes, but the more important reason for peripheral vasoconstriction is "to decrease the amount of heat loss from your skin to the environment."

Like most things in the universe, heat moves along a gradient from higher to lower. "If the temperature is higher in your skin and lower in the air, then you're going to lose heat to the air. By bringing [these temperatures] closer together, you lose much less heat to the environment."

The second way the body responds to cold is by trying to generate more heat. It does this by shivering, which is the involuntary contraction of normally voluntary muscles. "We know that the by-products of cellular respiration–any time cells work, and that includes your muscle cells–are CO_2, heat, and water," says Kenefick. "So when your muscles contract through shivering, they create heat, and that heat helps to warm up the core of your body."

Of course, to maintain a constant temperature, our bodies must not only keep from getting too cold–they must also keep from getting too hot. Two main physiological responses help prevent overheating: peripheral **vasodilation,** the expansion of the diameter of blood vessels, which increases blood flow to the skin; and

evaporative cooling (otherwise known as sweating), which cools the body by releasing heat to the air. In other words, you have a set point for body temperature: if you get too cold, you vasoconstrict and shiver; if you get too hot, you vasodilate and sweat. A precise balance between the two must be maintained to keep tissues healthy. If peripheral vasoconstriction goes on for too long, for example, the result is frostbite–the death of tissues caused by lack of blood flow **(Infographic 25.3)**.

VASOCONSTRICTION
The reduction in diameter of blood vessels, which helps to retain heat.

VASODILATION
The expansion in diameter of blood vessels, which helps to release heat.

INFOGRAPHIC 25.3

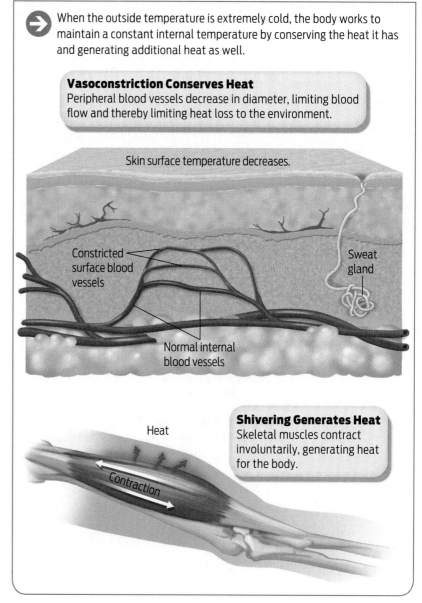

Thermoregulation in Response to Cold

When the outside temperature is extremely cold, the body works to maintain a constant internal temperature by conserving the heat it has and generating additional heat as well.

Vasoconstriction Conserves Heat
Peripheral blood vessels decrease in diameter, limiting blood flow and thereby limiting heat loss to the environment.

Skin surface temperature decreases.

Constricted surface blood vessels

Sweat gland

Normal internal blood vessels

Heat

Contraction

Shivering Generates Heat
Skeletal muscles contract involuntarily, generating heat for the body.

Krakauer and his teammates were no strangers to the cold. After weeks of slowly ascending from Base Camp to camps along the route, they reached the launching pad for the summit, the South Col, at 1 P.M. on May 9. "It is one of the coldest, most inhospitable places I have ever been," Krakauer wrote. A wind-swept saddle of rock and ice that sits between the peaks of Everest and neighboring Lhotse, the Col ("col" is a Welsh word meaning "saddle" or "pass") sits at 26,000 feet above sea level. Climbers pitch their tents on the relatively flat terrain and try not to think about the fact that they have entered what's known as the death zone.

Conditions were particularly bad on the Col that day. Gale-force winds blew through the camp, limiting visibility. As Krakauer's teammate, Beck Weathers, later recalled, "The weather was so crummy that when we first got in there, I didn't think there was any chance that we were going to climb that night."

But at 7 P.M., conditions improved markedly. It was still cold–minus 26°C (−15°F)–but the wind had abruptly ceased, and by 11 P.M., above their heads, the stars appeared, while a gibbous moon reflected off the mountain snow. It was the perfect night for a climb.

Into Thin Air

The 15-member team left camp shortly after 11 P.M. Night climbing is necessary in order to arrive at the most difficult parts of the climb during daylight hours and still have enough time to get back down to camp before nightfall. Krakauer led the pack that night, along with the team's head Sherpa, Ang Dorjee.

The pair reached the Southeast Ridge, the penultimate stop along the way to the summit, at 5:30 A.M., just as the sun was peering over the eastern peaks. By this time, Krakauer's hands and feet felt like unwieldy blocks of ice, nearly useless in performing the delicate work

of laying ropes and scaling ice. But it wasn't just the cold he had to deal with. His brain and body were also showing the effects of altitude: "Plodding slowly up the last few steps to the summit," wrote Krakauer in *Into Thin Air,* his 1997 book about the expedition, "I had the sensation of being underwater, of moving at quarter speed."

> **"Plodding slowly up the last few steps to the summit, I had the sensation of being underwater, of moving at quarter speed."**
> –Jon Krakauer

At high altitudes, the percentage of oxygen molecules in the air is the same as at sea level (about 20%), but the barometric pressure of O_2—the number of oxygen molecules banging around in a given volume of the atmosphere—is much lower, as it is with all molecules in air at high altitude. The lower pressure means that fewer oxygen molecules bind to the hemoglobin in blood, which means that blood is less saturated with oxygen, a condition called **hypoxia.** Since all cells require oxygen to function, hypoxia has many bodily consequences. The most serious and immediate occur in the brain. "I've been at 19- and 20,000 feet climbing myself," says physiologist Kenefick, "and I can tell you that doing simple tasks like tying your shoes—even though you've tied your shoes many times—is much more difficult." For the hikers on Everest, he says, each step would have been a struggle.

To help cope with conditions of low oxygen, climbers spend about 6 weeks **acclimatizing** their bodies to the conditions, spending a few nights at progressively higher elevations. Their bodies respond by increasing the output of red blood cells, the cells that contain hemoglobin and carry oxygen. The physiological adjustment of acclimatization allows them to carry more oxygen than could someone coming straight from sea level. But even well-acclimatized hikers usually need bottled oxygen to climb successfully.

At 1:17 P.M., after more than 12 hours of climbing, Krakauer finally reached the summit. It was smaller than he expected—a patch of ice the size of a picnic table, Buddhist prayer flags flapping on a string. He stood and took in the 360° panorama. The towering peaks of the

HYPOXIA
A state of low oxygen concentration in the blood.

ACCLIMATIZATION
The process of physiologically adjusting to an environmental change over a period of time. Acclimatization is generally reversible.

Climbers on Mount Everest, May 1996.

A camp below Mount Everest at night.

surrounding Himalayas were below him, draped in low-lying clouds, like distant swells in a choppy ocean. Beyond the mountain range, endless miles of continent stretched to the horizon, arching slightly with the curve of the earth.

Standing on top of the world, Krakauer was surprised by his own lack of elation. He had just cleared a huge personal hurdle, yet the victory felt hollow. Partly, he was too exhausted to truly care: he hadn't slept soundly in more than 50 hours, and the only food he had been able to choke down in 3 days was a bowl of ramen soup and some peanut M&Ms: sleep disturbances and digestive difficulties are additional side effects of elevation. But another thought lurked in his brain: the oxygen tank he had slung on his back to help him breathe was running low, and he still had to get down the mountain.

"With enough determination, any bloody idiot can get up this hill," guide Rob Hall had famously said. "The trick is to get back down alive." Keenly aware of the clock, Krakauer snapped a few perfunctory photos, and within 5 minutes was headed back down the mountain toward Camp IV.

Fifteen minutes later, after scaling the steep ice fin of the Southeast Ridge, he arrived at the pronounced notch in the mountain known as the South Summit, just below the main peak. As he prepared to rappel over the edge, he caught a glimpse of an alarming sight: a queue of 20 climbers, from three separate expeditions, waiting to come up. They were backed up at the notorious Hillary Step—a 40-foot wall of rock and ice named for Sir Edmund Hillary, who, with Tenzing Norgay, was the first to successfully scale it in 1954. Getting up the Step requires ropes, so climbers must go up one by one, and on this day there was a traffic jam.

While waiting for his turn to get down the Step, Krakauer peered into the distance and saw something he hadn't noticed before: on the horizon, dark clouds were sweeping in from the south, filling up a corner of what had been a clear blue sky. A storm was brewing.

By this point, it was well past the agreed-upon turnaround time of 1 P.M., set by Hall. The climbers who were still headed up the mountain

at this hour were willfully flouting safety rules. Not only that, the weather conditions were getting worse. Snow had started to fall, and it had become hard to see where mountain ended and sky began. The lower Krakauer got on the mountain, the worse the weather became.

Krakauer made it back to Camp IV on the Col just before 6 P.M. The bedraggled climber fell into his tent and quickly passed out. He was delirious, shivering uncontrollably, and exhausted. But he was alive.

Sensors Working Overtime

Even as he slept, Krakauer's body was working hard to thermoregulate. Like many physiological processes, thermoregulation is not something that requires conscious thought. It is more like the automated response of a home heating system, triggered when the thermostat is tripped.

The body's thermostat is the **hypothalamus,** a grape-size structure that sits at the base of the brain, right above the brain stem. The hypothalamus receives signals from many different **sensors,** specialized cells in the body that detect changes in both the internal and external environment. For cold, the major sensors are thermoreceptors in the skin and in the hypothalamus itself. Information from various sensors is fed to the hypothalamus, where it is integrated.

Acting as a thermostat, the hypothalamus has a specific temperature set point below which a warning message is triggered that body temperature is dropping. When that happens, the hypothalamus essentially tells the body to take corrective action. For example, it sends a signal to blood vessels in the skin, causing them to constrict in peripheral vasoconstriction. It can also send a signal to muscles to start shivering. Both signals are sent from the hypothalamus to their target tissues along nerve fibers. The cells, tissues, or organs that respond to such signals are known as **effectors:** they act to cause a change in the internal environment. Once the effectors have raised the body temperature, the sensors sense the changed conditions and the signals are turned off.

This circuit of sensing, processing, and responding is an example of a homeostatic

HYPOTHALAMUS
A master coordinator region of the brain responsible for a variety of physiological functions.

SENSOR
A specialized cell that detects specific sensory input like temperature, pressure, or solute concentration.

EFFECTOR
A cell or tissue that acts to exert a response on the basis of information relayed from a sensor.

Homeostasis Feedback Loops Require Sensors and Effectors

By means of sensors, the body constantly monitors factors like body temperature. The sensors relay temperature information to the hypothalamus. If the temperature is too hot or too cold, the hypothalamus sends signals to effector tissues and organs that work to return the temperature to homeostasis levels.

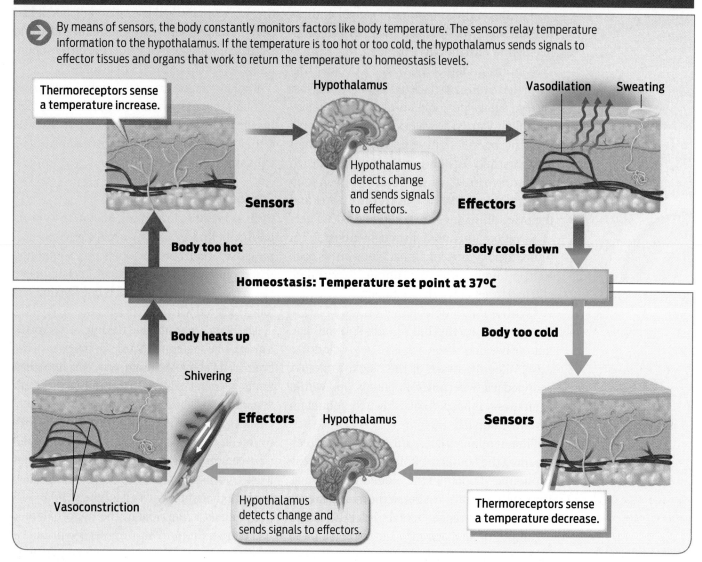

Thermoreceptors sense a temperature increase.

Hypothalamus

Vasodilation   Sweating

Hypothalamus detects change and sends signals to effectors.

Sensors

Effectors

Body too hot

Body cools down

Homeostasis: Temperature set point at 37°C

Body heats up

Body too cold

Shivering

Effectors

Hypothalamus

Sensors

Vasoconstriction

Hypothalamus detects change and sends signals to effectors.

Thermoreceptors sense a temperature decrease.

feedback loop (Infographic 25.4). In this case, the loop is a negative feedback loop because the output of the circuit (for example, rising body temperature) inhibits the input of the circuit (for example, peripheral vasoconstriction and shivering), thereby helping to bring the system back to its set point. Not all feedback loops act in a negative fashion. Positive feedback loops occur when the output of a system acts to further increase the input of the system. An example is the formation of a blood clot when you cut yourself, which is critical to preventing blood loss. Blood platelets stick to damaged blood vessels and release molecules that attract even more platelets to the area, which in turn attract even more platelets, and eventually a blood clot forms. Positive feedback loops are effective at rapidly amplifying a response, but negative feedback loops tend to be more common in physiology as they help return the body to its set point and ensure that homeostasis is maintained.

The hypothalamus does more than regulate body temperature. In fact, it is the body's main homeostasis control center, regulating many bodily states including hunger, thirst, and sleep.

FEEDBACK LOOP
A pathway that involves input from a sensor, a response via an effector, and detection of the response by the sensor.

The hypothalamus is part of what Kenefick calls our "lizard brain"– the evolutionarily ancient parts of the brain, which control our most basic physiological responses through unconscious reflexes.

The hypothalamus is able to play such an important role in homeostasis because it is so well connected to sensors and effectors. The hypothalamus is not only a key part of the **nervous system,** connected to parts of the body through nerves, it is also connected intimately to the **endocrine system,** through the pituitary gland. A pea-size structure that sits right below the hypothalamus, the pituitary gland releases **hormones,** which travel through the bloodstream and act on many tissues in the body, including other glands. The endocrine system, with its numerous hormone-secreting glands, is just one of many organ systems found in the human body that cooperate to maintain homeostasis (see **Up Close: Organ Systems** and subsequent chapters in this unit).

During the night, Krakauer was awakened by a teammate who gave him grave news: a number of his teammates, including Hall, had not yet returned to Camp IV. They were still out in the blistering subzero cold somewhere above 26,000 feet. Krakauer's heart sank. He knew the chances of surviving in the cold for that long were slim. By 5 P.M., everyone's oxygen tank would have been empty. It was now midnight. Krakauer feared for the others' lives. But he was also dumbfounded. Hall and the rest of his team were not far behind him on the mountain. What had gone wrong?

The storm began as a cyclone in the Bay of Bengal. It came in low from the valley and then rose up the mountain, gaining in ferocity and strength as it climbed. "One minute, we could look down and we could see the camp below. And the next minute, you couldn't see it," recalled Lou Kasischke, a member of Krakauer's team, who was one of 11 people trapped on the Col when the storm hit and who recounted his

> **"Within the space of five minutes, it changed from really a good day with a little bit of wind to desperate conditions, something I'd never experienced the ferocity of before."**
> —John Taske

experience in the PBS documentary *Storm over Everest.* "Within the space of five minutes, it changed from really a good day with a little bit of wind to desperate conditions, something I'd never experienced the ferocity of before," said John Taske, another member of Krakauer's team, on the same program.

According to Kent Moore, a physics professor at the University of Toronto, the storm that hit Everest that day also caused a particularly severe drop in barometric pressure, greatly reducing the availability of oxygen. "At these altitudes climbers are already at the limits of endurance," says Moore. "The sudden drop in pressure could have driven some of these climbers into severe physiological distress." In particular, they would have experienced the mental side effects of anoxic shock, which include confusion and disorientation.

Unable to tell in which direction they were going, and not wanting to take a wrong turn and step off a cliff, the climbers were forced to hunker down in the hurricane-force winds and wait for the storm to abate. Eventually, after 4 long hours, the clouds parted long enough for one of them to see where they were. Six climbers who were able to walk made it back to camp during this lull. An additional three were brought back safely by the efforts of Anatoli Boukreev, a Russian guide who, having descended to Camp IV, went back to search for them.

But others were not so lucky. Two climbers, too weak to make it back to camp, suffered severe frostbite before being rescued. One lost all his fingers and toes; the other, Beck Weathers, had to have his right hand amputated. Those climbers stuck higher on the mountain– including Hall–could not be rescued. Trapped without shelter in the subzero temperatures all night, their supplemental oxygen and food long gone, the hikers eventually lost their ability to cope with the cold and succumbed to **hypothermia,** a precipitous drop in body

NERVOUS SYSTEM
The collection of organs that sense and respond to information, including the brain, spinal cord, and nerves.

ENDOCRINE SYSTEM
The collection of hormone-secreting glands and organs with hormone-secreting cells.

HORMONE
A chemical signaling molecule that is released by a cell or gland and travels through the bloodstream to exert an effect on target cells.

HYPOTHERMIA
A drop of body temperature below 35ºC (95ºF), which causes enzyme malfunction and, eventually, death.

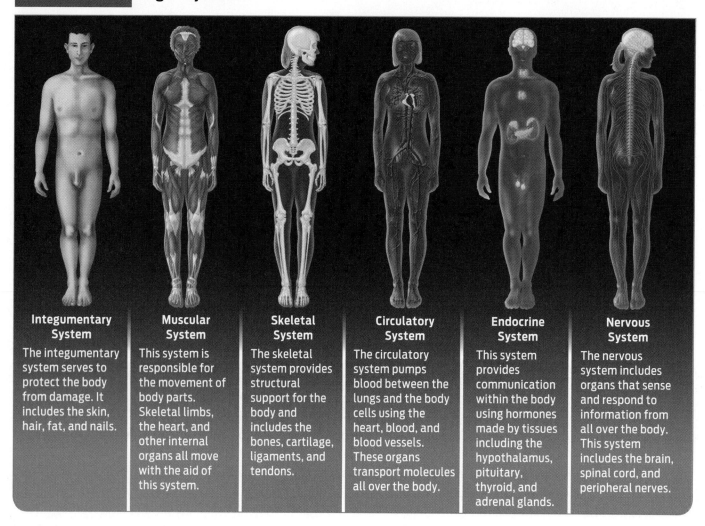

Integumentary System	Muscular System	Skeletal System	Circulatory System	Endocrine System	Nervous System
The integumentary system serves to protect the body from damage. It includes the skin, hair, fat, and nails.	This system is responsible for the movement of body parts. Skeletal limbs, the heart, and other internal organs all move with the aid of this system.	The skeletal system provides structural support for the body and includes the bones, cartilage, ligaments, and tendons.	The circulatory system pumps blood between the lungs and the body cells using the heart, blood, and blood vessels. These organs transport molecules all over the body.	This system provides communication within the body using hormones made by tissues including the hypothalamus, pituitary, thyroid, and adrenal glands.	The nervous system includes organs that sense and respond to information from all over the body. This system includes the brain, spinal cord, and peripheral nerves.

temperature. In all, eight climbers died on Everest that day.

This was not the first time that disaster had struck the summit. A 2008 study of all reported Everest deaths, from 1921 to 2006, led by researchers at Harvard's Massachusetts General Hospital and published in the *British Medical Journal,* found that more than 80% of deaths occurred above 26,000 feet, either during or the day after a summit attempt. While many of these deaths were attributable to traumatic injuries resulting from falls and avalanches, nearly as many were caused by hypoxia and hypothermia.

No Fuel Left to Burn

Although the body is able to cope with cold temperatures for some time through vasoconstriction and shivering, it cannot do so indefinitely. Thermoregulation is work, and work takes energy—roughly 150-300 Calories per hour for a 150-pound man. Eventually, if the body is not consuming food, it will run out of fuel.

The main fuel the body uses in times like this is the sugar glucose, a breakdown product of carbohydrate digestion. When we eat carbohydrates, sugars are released and absorbed into the circulation, and blood sugar increases (see Chapter 4). Some of this sugar may be used immediately as fuel for aerobic respiration in cells of the body (see Chapter 6). Whatever is not needed right away will be converted into **glycogen,** which is stored in muscles and the liver. By converting excess glucose to glycogen, the body maintains a relatively stable blood-glucose level—another example of homeostasis.

Blood-glucose levels are controlled by endocrine tissue in the **pancreas,** an organ that

GLYCOGEN
An energy-storing carbohydrate found in liver and muscle.

PANCREAS
An organ that secretes the hormones insulin and glucagon, as well as digestive enzymes.

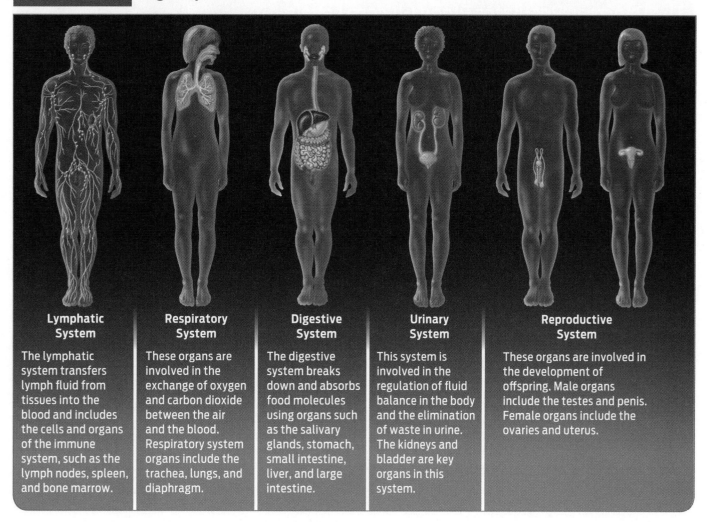

Lymphatic System

The lymphatic system transfers lymph fluid from tissues into the blood and includes the cells and organs of the immune system, such as the lymph nodes, spleen, and bone marrow.

Respiratory System

These organs are involved in the exchange of oxygen and carbon dioxide between the air and the blood. Respiratory system organs include the trachea, lungs, and diaphragm.

Digestive System

The digestive system breaks down and absorbs food molecules using organs such as the salivary glands, stomach, small intestine, liver, and large intestine.

Urinary System

This system is involved in the regulation of fluid balance in the body and the elimination of waste in urine. The kidneys and bladder are key organs in this system.

Reproductive System

These organs are involved in the development of offspring. Male organs include the testes and penis. Female organs include the ovaries and uterus.

functions in both the endocrine and digestive systems. In response to high blood sugar, the pancreas produces the hormone **insulin,** which binds to receptors on muscle and liver cells, signaling them to remove sugar from the blood. Insulin also signals these cells to make glycogen, using the sugars taken up from the blood.

When blood-sugar levels are low, the body first prompts us to eat by sending a signal to the hypothalamus. If eating isn't an option, the body begins to break down its stored glycogen. The key signal here is **glucagon,** another hormone released by the pancreas in response to low blood sugar, which triggers muscle and liver cells to convert their stored glycogen to glucose. Glucose from the liver is then released into the blood. More glucose in the blood means more energy available to shiver and stay active–and thus warm **(Infographic 25.5)**.

INSULIN
A hormone secreted by the pancreas that regulates blood sugar.

GLUCAGON
A hormone produced by the pancreas that causes an increase in blood sugar.

The trapped climbers hadn't eaten in hours, which meant they were operating on glycogen reserves. But the human body can store only so much glycogen. Eventually, after hiking and shivering for many hours, you will exhaust your fuel supply. Without fuel, your body can't continue to remain active and shiver. And if you can't remain active and shiver, then you can't generate heat and your body temperature will fall. That's when hypothermia can set in.

The average adult has enough stored glycogen to power about 12 to 14 hours of routine activity. When a person is exercising strenuously–say, running or hiking–glycogen stores can be depleted in as little as 2 hours. Marathon runners often refer to this point, which occurs at about mile 20, as "hitting the wall." Then, in order to continue exercising, you must eat something–preferably something with carbohydrates.

The Pancreas Regulates Blood Glucose Levels

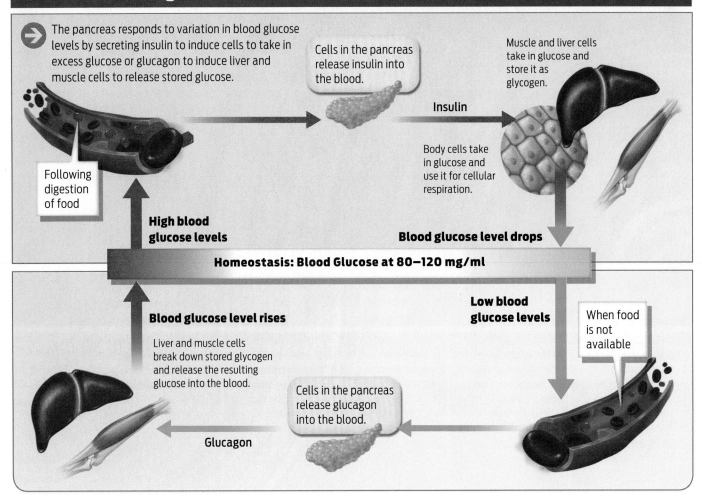

The pancreas responds to variation in blood glucose levels by secreting insulin to induce cells to take in excess glucose or glucagon to induce liver and muscle cells to release stored glucose.

Cells in the pancreas release insulin into the blood.

Muscle and liver cells take in glucose and store it as glycogen.

Insulin

Body cells take in glucose and use it for cellular respiration.

Following digestion of food

High blood glucose levels

Blood glucose level drops

Homeostasis: Blood Glucose at 80–120 mg/ml

Blood glucose level rises

Low blood glucose levels

When food is not available

Liver and muscle cells break down stored glycogen and release the resulting glucose into the blood.

Cells in the pancreas release glucagon into the blood.

Glucagon

"A lot of mountaineering communities think you need fat," says Kenefick. "And that's true–fat has more Calories per gram–9 kcals per gram compared to 4 kcals per gram of protein or carbohydrate. But when you're doing things like shivering, those types of contractions, especially, use a lot of glucose." Fats–though a good source of stored energy–are not as readily available for immediate use. And glucose is the primary fuel for the brain.

Fitness also likely played a role in how the Everest climbers fared. Being fit means having more muscle mass than fat. Having more muscle mass means you have more glycogen and can exercise longer and generate more heat through cellular respiration. Someone who is less fit, or who simply has less muscle mass, will tire sooner, need to sit down and rest, and continue to lose heat to the environment. This may be what happened to the climbers who were too weak to hike back to camp: they used up their glycogen stores faster than other climbers.

Another exacerbating factor, says Kenefick, would have been dehydration–a little-known cold-weather risk. In winter, our bodies are working harder under the weight of extra clothing, and sweat evaporates quickly in cold, dry air. We also lose a great deal of water as water vapor when we exhale. The body is about two-thirds water, and when the total water level drops by only a few percentage points, we become dehydrated–which can cause dangerous side effects like delirium, confusion, and convulsions. Kenefick points out that people do not feel as thirsty when it's cold, and thus become even more dehydrated. "We're really tropical animals," says Kenefick. "We came from the Sub-Sahara. We do much better in the heat."

The Kidneys Respond to Changes in Water Balance

The amount of water in the bloodstream controls the concentration of dissolved molecules in the blood and also determines blood volume and blood pressure. The kidneys control water availability by responding to a variety of signals.

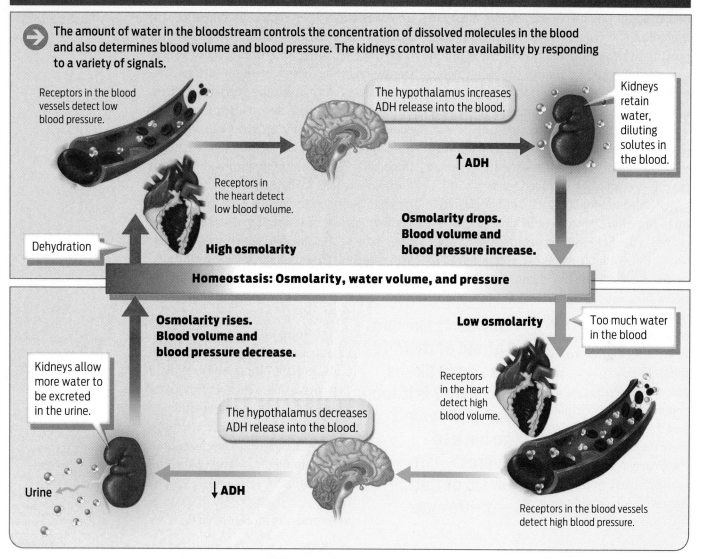

Receptors in the blood vessels detect low blood pressure.

The hypothalamus increases ADH release into the blood.

Kidneys retain water, diluting solutes in the blood.

Receptors in the heart detect low blood volume.

Dehydration

High osmolarity

↑ ADH

Osmolarity drops. Blood volume and blood pressure increase.

Homeostasis: Osmolarity, water volume, and pressure

Osmolarity rises. Blood volume and blood pressure decrease.

Low osmolarity

Too much water in the blood

Kidneys allow more water to be excreted in the urine.

Receptors in the heart detect high blood volume.

The hypothalamus decreases ADH release into the blood.

Urine

↓ ADH

Receptors in the blood vessels detect high blood pressure.

OSMOLARITY
The concentration of dissolved solutes in blood and other bodily fluids.

OSMOREGULATION
The maintenance of relatively stable volume, pressure, and solute concentration of bodily fluids, especially blood.

KIDNEY
An organ involved in osmoregulation, filtration of blood to remove wastes, and production of several important hormones.

As Kenefick explains, our body's sense of thirst relies on **osmolarity,** the concentration of dissolved solutes in the blood. Among the solutes dissolved in blood are electrolytes–ions such as sodium and potassium that are critical for nerve signaling and muscle contraction. Osmolarity is monitored by the hypothalamus as part of **osmoregulation.** When you are dehydrated–when you have less fluid in your blood–the concentration of dissolved solutes is higher. If the hypothalamus registers that the concentration of solutes in the blood is high, it will trigger a sense of thirst, encouraging you to drink. At the same time, it triggers the release of antidiuretic hormone (ADH) from the pituitary, which travels through the bloodstream and acts on the **kidneys.** ADH signals the kidneys to excrete less water in the urine. By reducing the amount of water lost in urine, ADH causes more water to be reabsorbed by the kidneys back into the bloodstream. Water in the bloodstream dilutes dissolved solutes and lowers the osmolarity. That's why people who are dehydrated have darker urine–it contains less water and is more highly concentrated.

Osmoregulation also depends on sensors that detect changes in blood volume and pressure, both of which depend on the amount of water in the blood. Sensors in the heart, for example, sense how full the heart's chambers are; sensors in blood vessels sense how stretched the vessels are. When low blood volume and pressure are detected, the hypothalamus responds by triggering the release of ADH from the pituitary into the blood, which acts on the kidneys to help retain water **(Infographic 25.6).**

With these multiple sensors for detecting dehydration, why do people feel less thirsty in the cold? The reason, says Kenefick, is that peripheral vasoconstriction pushes blood toward the core. All that extra blood being pushed centrally is sensed by the body as a normal amount of hydration. As a result, the sensation of thirst is reduced, despite the fact that you're dehydrated. This is why it's very important to drink adequate amounts of water in winter, even when you aren't thirsty.

"Because water plays such a large role in cellular function," says Kenefick, "being dehydrated is going to put a greater stress on your body." Dehydration can alter the concentration of electrolytes in the blood, and therefore alter nerve function and muscle contraction. Dehydration also lowers blood pressure and thus makes the heart work harder. Together, these effects can have dangerous consequences, impairing thinking and coordination—two things that matter a great deal when you're navigating the treacherous terrain of the world's tallest mountain during a blizzard, fighting against hypothermia.

> **"It wasn't like 'Am I afraid of this?' It was more like 'Is this right? Is it too selfish?' I won't go back to Everest—I'm afraid of that."**
> —Jon Krakauer

None of the climbers who died on Everest in 1996 was an inexperienced climber—three, in fact, were professional guides. Why didn't they heed these physiological warning signs? Part of the reason is that there was simply no time. The swift-moving storm made the decision for them. But the climbers had also made questionable choices earlier that affected their fate. Whether from hubris or brain-addled thinking, they continued climbing toward the summit even when the hour was late. In that sense, the clock was as much to blame as the weather. In the end, the climbers made a fatal wager with biology: in their race to the summit, they pushed themselves beyond the breaking point, overestimating, in Krakauer's words, "the thinness of the margin by which human life is sustained above 25,000 feet."

Not long after the disaster, Krakauer returned to climbing mountains. In a 1997 interview with *Bold Type* magazine, Krakauer was asked whether he was fearful of climbing again after the trauma he experienced on Everest. The chastened climber replied: "It wasn't like 'Am I afraid of this?' It was more like 'Is this right? Is it too selfish?' I won't go back to Everest—I'm afraid of that."

Warning Signs

Hypothermia isn't only a danger for death-defying mountain climbers: it's a leading cause of death during outdoor recreation like rafting and skiing, and is the number one way to lose your life while outdoors in cold weather. The Centers for Disease Control and Prevention estimate that hypothermia causes more than 1,000 deaths each year in the United States.

Wilderness medicine experts say the best way to prevent hypothermia—in addition to dressing appropriately and carrying plenty of food and water—is to be aware of its signs. In particular, watch for the "umbles": stumbles, mumbles, fumbles, and grumbles, which show changes in motor coordination and altered brain function. If you experience any of these signs in cold conditions, it's time to seek shelter.

Krakauer speaks to journalists on May 16, 1996, after he was flown to Kathmandu, Nepal, from Mount Everest.

How Do Other Organisms Thermoregulate?

In the face of extreme cold, humans strive to maintain a constant body temperature. Although we may pile on warm clothes or sit by a fire, most of our heat is coming from metabolic reactions occurring inside our bodies. Because we use internal metabolic heat to thermoregulate, humans are classified as **endotherms** ("endo," inside). We expend a great deal of energy maintaining a warm body temperature in a cold environment.

Like us, whales are mammals and endotherms. As you can imagine, whales face a great challenge in maintaining a sufficiently warm body temperature in the cold ocean depths. They are protected from this cold by a thick layer of fat tissue called blubber, a type of adipose tissue. Not only is blubber an excellent insulator that

helps prevent heat loss to the environment, blubber does not have many surface blood vessels—an adaptation that prevents heat loss from blood to the environment. Many endothermic animals rely on feathers, fur, or fat as insulation.

In contrast to endotherms, other animals must obtain their body heat from the environment and are therefore known as **ectotherms** ("ecto," outside). While these animals are commonly referred to as "cold-blooded," their body temperature actually mimics that of their environment. By using behavioral adaptations, many ectotherms also maintain a relatively stable body temperature (although by definition not as stable as endotherms'). For example, lizards bask in the sun to warm up and seek out shade or a

ENDOTHERM
An animal that can generate body heat internally to maintain its body temperature.

ECTOTHERM
An animal that relies on environmental sources of heat, such as sunlight, to maintain its body temperature.

Insulation Helps Keep Some Endotherms Warm

→ Whales insulate their bodies from extreme temperature fluctuations with thick layers of fat called blubber.

Blubber (white)
Fewer blood vessels

Muscle (pink)
More blood vessels

Skin

protected burrow to prevent overheating. Through these behaviors, lizards keep their bodies in a temperature range compatible with their metabolism.

While basking in the sun works well for terrestrial ectotherms, this is not an option for certain fish, such as marlins. The surrounding water absorbs most of the heat of the sun, so marlins cannot heat themselves by swimming in sunny waters. Their brains would be dangerously close to freezing if not for specialized "heater tissue" that generates heat to keep their brains warm. In this case, the heater tissue is modified eye muscle that acts to generate heat rather than force or movement. This form of heating is a type of "nonshivering thermogenesis" ("thermo," heat; "genesis," origin).

Another form of nonshivering thermogenesis, used by bats (and human babies), occurs in a tissue known as brown fat. Brown fat is located in the neck and shoulder areas and is several degrees Celsius warmer than the rest of the body. In brown fat, specialized mitochondria convert energy to heat rather than to ATP, and the many blood vessels in brown fat deliver that heat to other parts of the body.

In the sun, lizards increase their body temperature and become more active.

In the shade, they decrease their body temperature to safe levels.

Some Organisms Generate Heat from Nonshivering Thermogenesis

Ectothermic Marlins Heat Their Brains with Modified Eye Muscle Cells

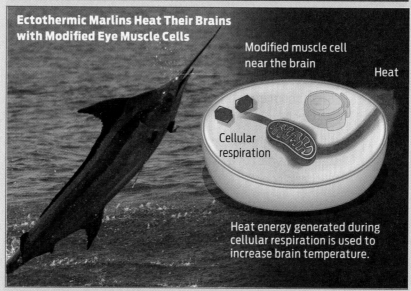

Modified muscle cell near the brain

Heat

Cellular respiration

Heat energy generated during cellular respiration is used to increase brain temperature.

Small Hibernating Endotherms Keep Warm with Brown Fat

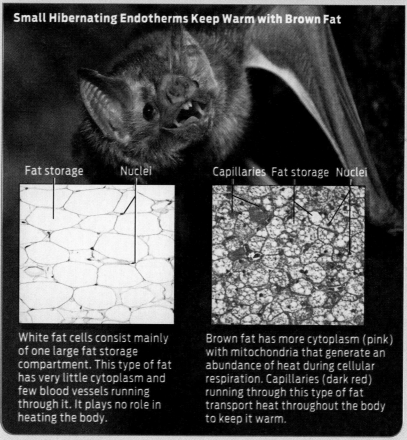

Fat storage   Nuclei

Capillaries Fat storage Nuclei

White fat cells consist mainly of one large fat storage compartment. This type of fat has very little cytoplasm and few blood vessels running through it. It plays no role in heating the body.

Brown fat has more cytoplasm (pink) with mitochondria that generate an abundance of heat during cellular respiration. Capillaries (dark red) running through this type of fat transport heat throughout the body to keep it warm.

▶ Summary

- Living organisms have an anatomical structure that is adapted to suit their physiological functions.

- Humans and other multicellular organisms are organized hierarchically: cells assemble to make up tissues; tissues congregate to form organs; organs work together as part of organ systems.

- Humans have many different organ systems that cooperate to accomplish basic physiological tasks, such as obtaining energy, taking in nutrients to build new molecules during growth and repair, and ridding themselves of wastes.

- Most organisms cannot tolerate wide fluctuations in their internal environment; their bodies work to maintain a stable internal environment, known as homeostasis.

- The process whereby organisms maintain a relatively constant internal temperature is called thermoregulation.

- The body responds to cold temperatures in two main ways: by conserving the heat it has, and by generating more. Vasoconstriction and shivering are key mechanisms of thermoregulation.

- Maintaining homeostasis requires both sensors and effectors. Sensors include nervous system receptors that detect changes in a variety of internal states (for example, temperature and blood pressure). Effectors include the glands or muscles that respond to an abnormal state in an effort to correct it.

- Sensors and effectors work together as part of a circuit or feedback loop.

- Hormones are chemical messengers that travel through the bloodstream, bind to receptors on a target cell, and effect a change in that cell. Insulin and glucagon are hormones that regulate blood-glucose levels.

- Osmoregulation is the control of water balance in the body. Sensors detect blood pressure, blood volume, and solute concentration. Kidneys are important effectors in maintaining water balance.

- Maintaining homeostasis is work and requires adequate energy and oxygen to power cellular respiration.

- Humans (and other mammals) are endotherms: we generate heat internally. Many other organisms, such as reptiles and fish, are ectotherms: they rely on behavior and the environment to maintain a temperature compatible with life.

BIOLOGICAL ORGANIZATION

Living things are organized hierarchically: cells are grouped into tissues, tissues into organs, and organs into organ systems.

HINT See Infographic 25.1.

➲ KNOW IT

1. Compare and contrast anatomy and physiology.

2. Organize the following terms on the basis of level of structure, from the simplest (1) to the most complex (4).
   ___ small intestine
   ___ mucus-secreting cell of the small intestine
   ___ digestive system
   ___ the layer of muscle that contributes to the function of the small intestine

➲ USE IT

3. An emergency room doctor setting a complex bone fracture is relying primarily on knowledge of
   a. anatomy.
   b. physiology.
   c. thermoregulation.
   d. homeostasis.
   e. osmoregulation.

4. Is a personal trainer who works with clients to help them lose weight through a combination of diet and exercise focusing primarily on anatomy or physiology? Explain your answer.

5. Why is the heart considered an organ and not a tissue?

HOMEOSTASIS AND THERMOREGULATION

Most living things work hard to maintain a relatively stable internal environment in the face of a changing external environment. Thermoregulation is the body's way of maintaining a stable internal temperature.

HINT See Infographics 25.2–25.6.

➲ KNOW IT

6. What is homeostasis?

7. Why is glucagon released as part of the response to a drop in body temperature?

8. Describe the feedback loop involved in thermoregulation in cold conditions. Use the following terms in your answer: hypothalamus, sensor, muscle, effector, low body temperature, normal body temperature.

9. People who are severely dehydrated produce _____ urine that is _____.
   a. a high volume of; highly concentrated and dark in color
   b. a high volume of; dilute and light in color
   c. a low volume of; highly concentrated and dark in color
   d. a low volume of; dilute and light in color
   e. a normal volume; a normal color (neither very light nor very dark)

➲ USE IT

10. How could damage to the hypothalamus prevent someone from shivering even if the core body temperature drops dramatically?

11. What conditions might cause high levels of insulin in the circulation? What events would follow?

12. In this chapter you read about homeostatic mechanisms for keeping warm. What responses do you think could help the body dissipate heat during exertion on a hot day? For each mechanism that you propose, give a brief explanation.

13. Tibetan Sherpas, many of whom serve as guides and rescuers on Everest, often do not require bottled oxygen to reach the summit. Why might Tibetans, who have lived at high elevations for many generations, have an easier time than others with hypoxia? (Think about both short-term and long-term changes.)

SCIENCE AND ETHICS

14. What arguments can you make to use tax dollars to pay for basic research into physiology? Refer to some specific examples from the chapter in your answer.

15. The U.S. National Park Service has to rescue stranded hikers, often at great expense. Do you think that hikers' level of preparation should be a factor in determining whether or not they should bear the cost of their rescue? What factors would you consider to determine whether or not a hiker was adequately prepared? Give a physiological reason for each factor that you propose.

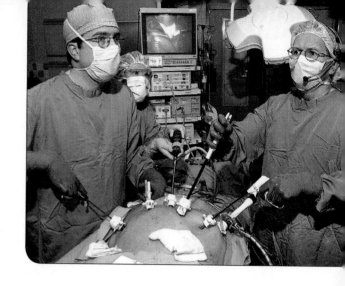

Drastic Measures

Drastic Measures

For the morbidly obese, stomach-shrinking surgery is a last resort

Amy Jo Smith hardly recalls a time growing up when her family wasn't dieting. Her parents were both obese, and they were always trying to lose weight.

Smith herself was relatively slender until her senior year in high school, when her weight began to creep up. She grew up on a horse farm in northeast Maryland. As a teenager she spent much of her spare time on the road, taking her horses to shows. She attributes her weight gain to a diet that consisted primarily of fast food. "I was always eating on the run," says Smith, now 36 years old and a computer literacy teacher. But, she says, her growing girth "never stopped me from doing the things I wanted to do." The extra weight did bother her, though, and she tried several diets and diet pills, only to see her weight yo-yo up and down.

In 2004, at a routine checkup, Smith's doctor noticed that she suffered a number of ills that were likely caused by Smith's 264-pound weight. For one thing, she had been suffering from migraines. She also had stress incontinence—a bladder that leaked when stressed by coughing or laughing, for example. "I thought that was just normal," she says. And she went months at a time without having a period—a telltale sign of a hormonal imbalance often associated with obesity. Smith's physician suggested that, to lose weight, Smith consider having a surgery that would shrink her stomach to the size of a golf ball.

Though we often use the term "fat" casually, obesity is actually a medical condition. A person is considered obese when he or she weighs 20% or more than his or her ideal body weight, based on body mass index. Body mass index (BMI), discussed in Chapter 6, is an estimate of body fat based on a person's height and weight. Morbid obesity—sometimes called clinically severe obesity—is defined as being 100 pounds or more over one's ideal body weight or having

a BMI of 40 or higher. Obesity becomes "morbid" when it significantly increases the risk of one or more obesity-related health conditions or serious diseases. At 5' 2" and with a BMI of 48, Smith was not only obese, she had become morbidly obese.

Smith's weight fluctuated from diet pills and yo-yo dieting.

Even then, Smith had a hard time accepting that she needed such a drastic method to lose weight. "At first I thought he was a quack," she says of her doctor. But she began to think more about the surgery after a friend underwent the stomach-shrinking surgery with stunning results.

There are several types of bariatric, or weight-loss, surgery. ("Bariatric" refers to the study and treatment of obesity; the word is

At 5´ 2″ and with a BMI of 48, Smith was not only obese, she had become morbidly obese.

from the Greek *baros,* meaning "heaviness.") The two most common procedures are adjustable gastric banding and gastric bypass ("gastric" means "of the stomach"). In adjustable gastric banding, a surgeon wraps an adjustable band around the stomach to make it smaller so that it holds less food. In gastric bypass, the stomach is surgically made smaller and the small intestine is rerouted. Gastric bypass surgery not only shrinks the size of the stomach but also alters the way food is digested so that the body absorbs fewer Calories. The type of surgery recommended depends on the individual patient's medical history and weight-loss goal. Because there are many associated risks, a National Institutes of Health panel of experts has recommended surgery only for people considered morbidly obese—people whose risk of death from diabetes or heart disease because of excess weight is five to seven times greater than for those of average weight.

But bariatric surgery is no miracle cure. Because it shrinks the stomach, patients must live the rest of their lives on a strict diet. If they overeat, they suffer nasty side effects such as vomiting and diarrhea.

"It's sort of barbaric," says Monica Skarulis, director of the Metabolic Clinical Research Unit at the National Institutes of Health. Because the surgery so drastically reduces the size of the stomach and restricts how much a person can eat, it amounts to "forced behavior control," she says. On top of that, some gastric bypass patients suffer mineral and vitamin deficiencies over the long term that cause bone loss and potentially other health impairments. And the surgery itself is risky: as many as 20% of patients suffer complications a year after the surgery that are severe enough to put them back in the hospital.

For some morbidly obese people, however, the risk of dying from obesity-related diseases is higher than the risk of surgical complications. And in terms of weight reduction, the surgery is more effective than lifestyle changes alone. Most patients lose 30% to 50% of their excess weight in the first 6 months and 77% after about a year. Studies also show that even 10 years after surgery, most patients still weigh 25% to 30% less than they did before the surgery. Consequently, demand for the surgery is soaring.

Digestion Basics

In Unit 1 we saw that all heterotrophic organisms require food as a source of nutrients and energy, and that to extract both nutrients and energy, our digestive systems break down foods into usable subunits. Here we explore the anatomy and function of the digestive system in more detail, an investigation that will help us understand how a surgery that changes the anatomy of the digestive system can help people lose weight.

Digestion begins immediately following **ingestion**—the act of putting food into our mouths—and consists of both mechanical and chemical processes. These processes occur in the central structure of the digestive system, which is known as the **digestive tract**—essentially a long tube lined with muscles that extends from the mouth to the anus. As the muscles relax and contract, the tube pushes food along. Along its length, this tube receives inputs from various other organs including the salivary glands, gallbladder, liver, and pancreas. The digestive tract's main function is to transform the food we eat into a form our bodies can use. It must also rid the body of the waste left over once usable nutrients and energy are removed from food we have taken in **(Infographic 26.1)**.

DIGESTION
The mechanical and chemical breakdown of food into subunits so that nutrients can be absorbed.

INGESTION
The act of taking food into the mouth.

DIGESTIVE TRACT
The central pathway of the digestive system; a long muscular tube that pushes food between the mouth and the anus.

SALIVARY GLANDS
Glands that secrete enzymes, including salivary amylase, which digests carbohydrates, into the mouth.

INFOGRAPHIC 26.1

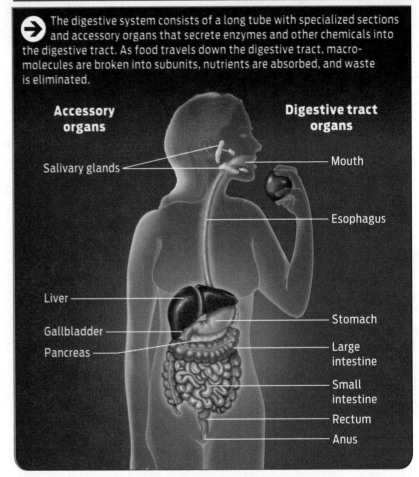

The Digestive System

➔ The digestive system consists of a long tube with specialized sections and accessory organs that secrete enzymes and other chemicals into the digestive tract. As food travels down the digestive tract, macromolecules are broken into subunits, nutrients are absorbed, and waste is eliminated.

Accessory organs
- Salivary glands
- Liver
- Gallbladder
- Pancreas

Digestive tract organs
- Mouth
- Esophagus
- Stomach
- Large intestine
- Small intestine
- Rectum
- Anus

When food enters the mouth, chewing mechanically breaks it down into smaller pieces. **Salivary glands** secrete enzymes into saliva, which chemically dismantle macromolecules into their subunits. Salivary amylase, for example, begins to break down carbohydrates into simpler sugars. Meanwhile, the **tongue** compresses the food into a ball and works it to the back of the mouth.

When we swallow, food is propelled along the **esophagus,** the section of the digestive tract between the mouth and the stomach, by rhythmic waves of contracting muscles in a process called **peristalsis.** Food then enters the **stomach.** The stomach contains acid that can inactivate potentially harmful bacteria ingested with our food. Stomach acid has a pH of close to 1, approximately the same as battery acid, and its action helps protect us against food-borne diseases. Stomach acid also denatures proteins in the food, unfolding their three-dimensional structures into linear strands. This makes it easier for the enzyme **pepsin,** which is produced in the stomach, to chemically break proteins apart into individual amino acids **(Infographic 26.2)**.

Like the esophagus, the stomach is also muscular, expanding and contracting as it accepts food and churns it. Each time it contracts, stomach acid mixes with food, creating a soupy

INFOGRAPHIC 26.2

The Upper Digestive Tract

→ The upper digestive tract includes the mouth, esophagus, and stomach as well as enzymes and other chemicals secreted by the salivary glands and the stomach.

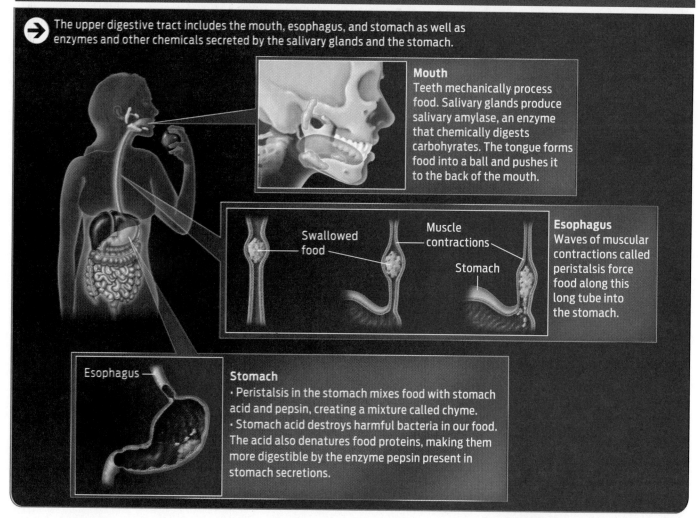

Mouth
Teeth mechanically process food. Salivary glands produce salivary amylase, an enzyme that chemically digests carbohyrates. The tongue forms food into a ball and pushes it to the back of the mouth.

Swallowed food

Muscle contractions

Stomach

Esophagus
Waves of muscular contractions called peristalsis force food along this long tube into the stomach.

Esophagus

Stomach
· Peristalsis in the stomach mixes food with stomach acid and pepsin, creating a mixture called chyme.
· Stomach acid destroys harmful bacteria in our food. The acid also denatures food proteins, making them more digestible by the enzyme pepsin present in stomach secretions.

mixture called **chyme.** Stomach acid is strong and can dissolve most foods. But despite so much acid churning inside, the stomach itself remains intact. This is possible because the stomach is lined with a thick layer of protective mucus. Occasionally, this mucus layer is damaged—as from a bacterial infection—and the stomach lining becomes more vulnerable to gastric juices; the result is a painful sore known as an ulcer.

While the stomach can absorb some substances, such as water, ethanol, and certain drugs, directly into the bloodstream, most of the chyme is pushed farther down in the digestive tract, primarily into the small intestine, where it is further processed.

Although the stomach is only a small part of the upper digestive tract, it plays a large part in weight gain. Evolutionarily speaking, the reason we have a stomach in the first place is to enable us to store food. Without a stomach, we would have to eat constantly to fuel our activities. When we eat a large meal, the stomach expands greatly to accommodate and store all that food. It's partly because of this elasticity that we can eat enough to sustain us for hours. But we can eat more than our bodies need.

Surgical Treatments for Obesity

Because of Smith's weight-related illnesses, her doctor recommended that she have bariatric surgery. The doctor explained that bariatric surgery changes the anatomy of the digestive system to limit the amount of food a person can eat and digest before feeling full. There are several surgical methods, all based on the fact that restricting the amount of food we put into our stomachs will cause weight loss. In gastric bypass, the most common type of bariatric surgery, a surgeon uses staples to seal the stomach except for a small pouch at the top. The pouch is about the size of a golf ball. The surgeon then cuts the small intestine and sews part of it directly onto the pouch. Consequently, food is redirected, bypassing most of the stomach and the first section of the small intestine. The rest of digestion continues normally **(Infographic 26.3)**.

Bypassing part of the small intestine contributes to weight loss because the **small intestine** is the primary organ that extracts nutrients from

CHYME
The acidic "soup" of partially digested food that leaves the stomach and enters the small intestine.

SMALL INTESTINE
The organ in which the bulk of chemical digestion and absorption of food occurs.

DUODENUM
The first portion of the small intestine; the duodenum receives chyme from the stomach and mixes it with digestive secretions from other organs.

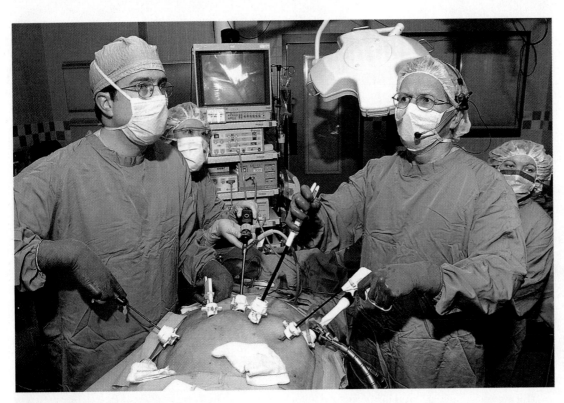

Bariatric surgery in progress.

Gastric Bypass Surgery

 In the most common type of gastric bypass surgery, most food is rerouted directly to the small intestine. The surgery reduces the amount of food that a person can ingest and also reduces the amount of nutrients and Calories that a person can absorb.

Before surgery

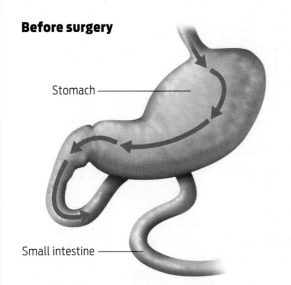

Stomach

Small intestine

Food passes into the stomach, which can stretch from the size of a large sausage (when empty) to hold 3–4 liters of food. Chyme made in the stomach enters the small intestine, where digestion releases food nutrients.

After surgery

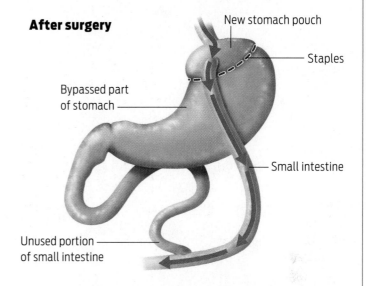

New stomach pouch

Staples

Bypassed part of stomach

Small intestine

Unused portion of small intestine

The stomach is reduced to the size of a golf ball, diminishing drastically the amount of food it can hold. Also, stomach contents bypass the upper part of the small intestine, reducing the amount of nutrients (and therefore of Calories) the person can absorb.

PANCREAS
An organ that helps digestion by producing enzymes (such as lipase) that act in the small intestine, and by secreting a juice that neutralizes acidic chyme.

LIVER
An organ that aids digestion by producing bile salts that emulsify fats.

BILE SALTS
Chemicals produced by the liver and stored by the gallbladder that emulsify fats so that they can be chemically digested by enzymes.

food. The "small" intestine isn't actually small. It winds through the lower part of the abdomen for about 20 feet. It's called "small" because its diameter is smaller than that of the large intestine (which is a later part of the digestive tract). The first part of the small intestine (about 10 inches or so) is the **duodenum,** and is the place where chyme from the stomach combines with digestive secretions from several other organs. The **pancreas,** for example, secretes pancreatic juice, a basic (pH greater than 7) fluid, into the small intestine to neutralize the acidic (pH less than 7) chyme, which would otherwise damage the small intestine. The pancreas also secretes enzymes that help chop organic mole-

> The "small" intestine isn't actually small. It winds through the lower part of the abdomen for about 20 feet.

cules such as carbohydrates, proteins, and fats into smaller pieces. Enzymes secreted by the small intestine itself further break down macromolecules into building blocks such as amino acids, sugars, fatty acids, and glycerol.

Whereas proteins, carbohydrates, and nucleic acids are easily digested by this powerful mixture of digestive enzymes, fats pose a special challenge. Because fats don't dissolve in water-based solutions (remember, fats are hydrophobic), they don't mix well with the watery solutions in the small intestine. This makes it difficult for fat-digesting enzymes to break them down. To help the process along, the **liver** secretes **bile salts,** which are chemically

Most Chemical Digestion Occurs in the Small Intestine

→ The small intestine is the major organ that digests food. Accessory organs secrete enzymes and other substances into the small intestine, which itself also produces digestive enzymes.

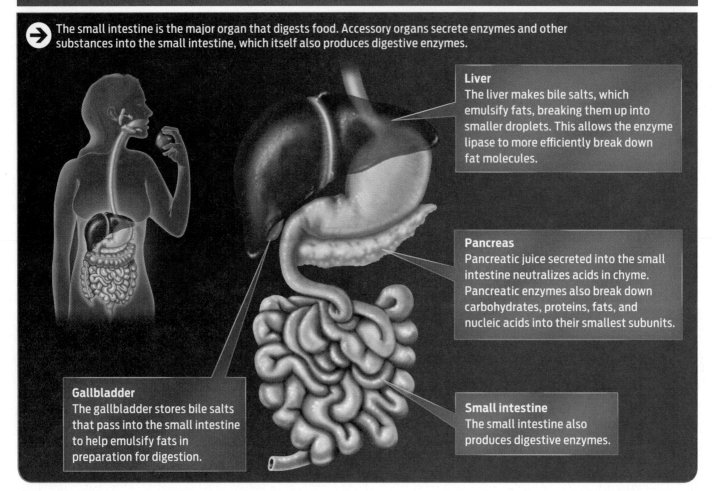

Liver
The liver makes bile salts, which emulsify fats, breaking them up into smaller droplets. This allows the enzyme lipase to more efficiently break down fat molecules.

Pancreas
Pancreatic juice secreted into the small intestine neutralizes acids in chyme. Pancreatic enzymes also break down carbohydrates, proteins, fats, and nucleic acids into their smallest subunits.

Gallbladder
The gallbladder stores bile salts that pass into the small intestine to help emulsify fats in preparation for digestion.

Small intestine
The small intestine also produces digestive enzymes.

suited to dividing large hydrophobic fat globules into smaller droplets—that is, **emulsifying** them. These bile salts pass from the liver into the **gallbladder,** which in turn stores them for future use. When we eat a high-fat meal, bile salts pass from the gallbladder into the small intestine, where they help emulsify the fats. Once the fats are emulsified, a lipid-digesting enzyme secreted by the pancreas called **lipase** chemically breaks them down to release their constituent fatty acids and glycerol **(Infographic 26.4).**

The liver and gallbladder are crucial organs of the digestive system. A person with no gallbladder or a dysfunctional one—for example, someone with gallstones who has had the gallbladder removed—cannot digest fats well. Although the liver might secrete bile salts, they can't be stored. If that person eats a high-fat meal, most of the fat taken in will be excreted in the feces.

Once digested into their smallest subunits, food molecules are absorbed by **epithelial cells** lining the small intestine in a stage of digestion known, not surprisingly, as **absorption.** The inner lining of the small intestine is folded into fingerlike projections called **villi** that are composed of many densely packed epithelial cells. The folds greatly increase the surface area through which the intestine can absorb nutrients. The food molecules then pass into blood vessels of the circulatory system, which transport them throughout the body, where they are used as a source of nutrients and energy to build and maintain cells **(Infographic 26.5).**

EMULSIFY
To break up large fat globules into small fat droplets that can be more efficiently chemically digested by enzymes.

GALLBLADDER
An organ that stores bile salts and releases them as needed into the small intestine.

LIPASE
A fat-digesting enzyme active in the small intestine.

EPITHELIAL CELLS
Cells that line organs and body cavities; in the digestive tract they sit in direct contact with food and its breakdown products.

The Small Intestine Absorbs Nutrients

→ The small intestine is the primary organ that absorbs nutrients from food. Nutrients enter the circulatory system via blood vessels connected to the small intestine.

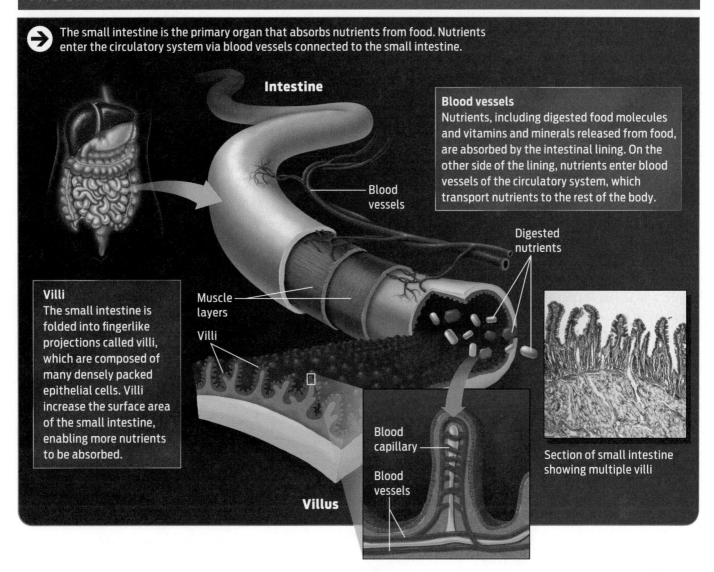

Intestine

Blood vessels
Nutrients, including digested food molecules and vitamins and minerals released from food, are absorbed by the intestinal lining. On the other side of the lining, nutrients enter blood vessels of the circulatory system, which transport nutrients to the rest of the body.

Blood vessels

Digested nutrients

Villi
The small intestine is folded into fingerlike projections called villi, which are composed of many densely packed epithelial cells. Villi increase the surface area of the small intestine, enabling more nutrients to be absorbed.

Muscle layers

Villi

Blood capillary

Blood vessels

Villus

Section of small intestine showing multiple villi

ABSORPTION
The uptake of digested food molecules by the epithelial cells lining the small intestine.

VILLI (SINGULAR: VILLUS)
Fingerlike projections of folds in the lining of the small intestine that are responsible for most nutrient and water absorption.

Because most digestion occurs in the small intestine, that structure is a common target for diet pills and other supplements. Alli, for example, which is an over-the-counter weight-loss pill, prevents the breakdown up to 25% of the fat consumed. The drug works by interfering with lipase in the small intestine; consequently, some fat remains undigested and passes out in feces.

While diet pills can help people lose weight, they are largely ineffective in people who need to lose more than a few pounds. Most people who are morbidly obese, Smith included, have already tried various diet pills

and lost weight, but not enough to improve their health. Even more troublesome, diet pills aren't a long-term solution. Over time, most people gain all of the lost weight back, and more, after they stop taking the pills. Smith's weight fluctuated over the years, and though she weighed 264 pounds when her doctor recommended surgery, this weight was likely not her heaviest, she says.

The second weight-loss surgery that Smith considered does not involve altering the small intestine. In adjustable gastric banding, a surgeon partitions the stomach into two parts by wrapping an inflatable plastic band around the upper part of the stomach to seal it off and create a small pouch. The principle is similar to that of gastric bypass; the difference is that in adjustable gastric banding food still passes through the entire stomach to the intact small intestine, although in much smaller quantities. The pouch holds approximately ½ cup of food, whereas the typical stomach holds about 6 cups of food. When the upper part of the stomach is filled, it sends a message to the brain that the entire stomach is full. This helps one feel full for longer periods of time, eat smaller portions, and lose weight over time. The band can also be adjusted or surgically removed if necessary. Banding is becoming more popular because it's simpler and has a lower complication rate than gastric bypass **(Infographic 26.6)**. In fact, the Food and Drug Administration recently approved it for more widespread use: now people with a BMI of at least 30 may qualify for the procedure.

Gastric bypass, however, is still the more commonly performed procedure—in part because it's more effective in reducing weight than adjustable gastric banding. That's because gastric bypass fundamentally alters the normal path of digestion through the stomach and small intestine; food bypasses not only most of the stomach but also more than half of the small intestine. Consequently, not as many nutrients and associated Calories are absorbed into the bloodstream—which is what makes the procedure so effective at reducing weight. Smith chose gastric bypass surgery over adjustable banding because, she says, she "wanted a permanent fix." She went on: "Bands can always be removed. This was a lifelong decision and commitment for me so the idea of being able to change your mind was not for me."

While both surgeries alter the first parts of the digestive system, they leave the final part,

INFOGRAPHIC 26.6

Gastric Banding Surgery

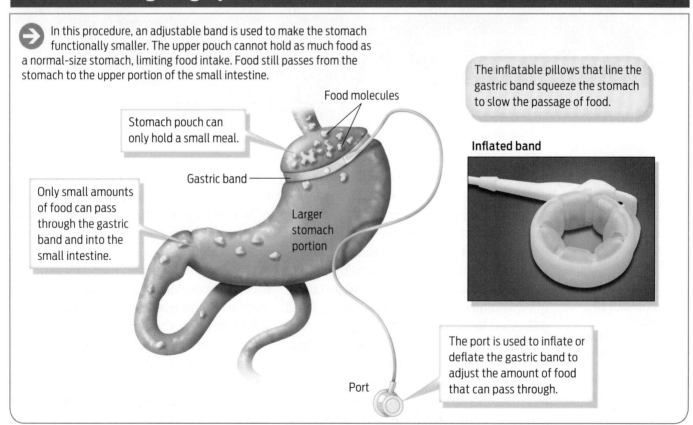

➔ In this procedure, an adjustable band is used to make the stomach functionally smaller. The upper pouch cannot hold as much food as a normal-size stomach, limiting food intake. Food still passes from the stomach to the upper portion of the small intestine.

The inflatable pillows that line the gastric band squeeze the stomach to slow the passage of food.

Inflated band

Food molecules

Stomach pouch can only hold a small meal.

Gastric band

Only small amounts of food can pass through the gastric band and into the small intestine.

Larger stomach portion

The port is used to inflate or deflate the gastric band to adjust the amount of food that can pass through.

Port

The Large Intestine

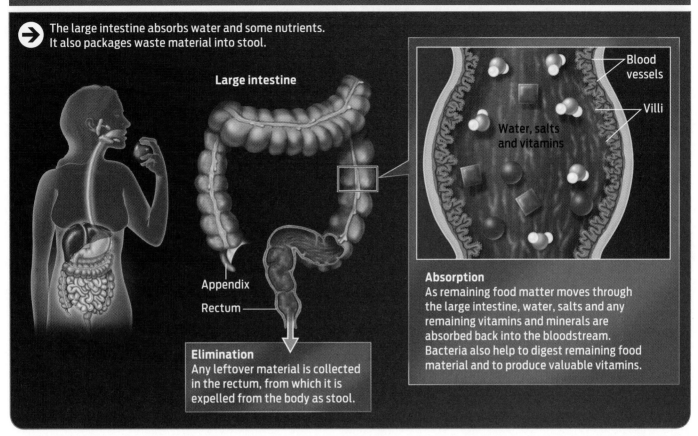

→ The large intestine absorbs water and some nutrients. It also packages waste material into stool.

Large intestine

Appendix

Rectum

Elimination
Any leftover material is collected in the rectum, from which it is expelled from the body as stool.

Blood vessels

Villi

Water, salts and vitamins

Absorption
As remaining food matter moves through the large intestine, water, salts and any remaining vitamins and minerals are absorbed back into the bloodstream. Bacteria also help to digest remaining food material and to produce valuable vitamins.

LARGE INTESTINE
The last organ of the digestive tract, in which remaining water is absorbed and solid stool is formed.

COLON
The first and longest portion of the large intestine; the colon plays an important role in water reabsorption.

STOOL
Solid waste material eliminated from the digestive tract.

ELIMINATION
The expulsion of undigested matter in the form of stool.

the **large intestine,** intact. After chyme passes through the small intestine, it moves on to the large intestine. The large intestine functions like a trash compactor—it holds and compacts material that the body can't use or digest, such as plant fiber. Within the **colon,** the first section of the large intestine, fiber, small amounts of water, vitamins, and other substances mix with mucus and bacteria that normally live in the large intestine. As this waste travels through the colon, most of the water and some vitamins and minerals are reabsorbed into the body through the colon lining. Bacteria chemically break down some of the fiber to produce nutrients for their own survival and also to nourish cells lining the colon (this is one reason fiber is an important dietary nutrient). As the large intestine expands and contracts, it pushes what ultimately becomes **stool** into the rectum, from

which it is **eliminated** through the anus as feces (Infographic 26.7).

Costs and Benefits of Surgery

After looking into various medical centers and attending informational sessions, Smith decided to have her surgery at Christiana Care Hospital in Wilmington, Delaware, in August 2009. She went through the hospital's presurgical screening program, which entailed a number of medical tests and consultation with a team of doctors that included a cardiologist, a pulmonologist, a psychologist, a dietician, and other specialists to ensure that she was physically and mentally fit for the surgery. Often there are undiagnosed medical problems that must be considered or treated before a patient can undergo surgery.

After the surgery Smith lost weight, but it wasn't a smooth ride. The stomach takes time to heal, and so doctors advise patients to ingest only liquids for the first few weeks, puréed foods for the next few weeks, and then gradually progress to solid foods. Because the stomach is now so small, it can carry only about an ounce of food at a time–a handful of crackers or a few broccoli florets. Eating too much at once can cause vomiting or intense stomach pain. But some people find it hard to control their eating habits and wind up in a lot of pain.

Moreover, patients must also stick to a special diet after the surgery or suffer other unpleasant consequences. For example, patients must introduce carbohydrates like breads and pasta into their diet very slowly, says Skarulis, of the National Institutes of Health. If they eat too many simple carbohydrates, the carbs enter the small intestine too quickly. This effect, called gastric dumping, causes nausea and massive diarrhea. But at the same time, patients lose weight.

And research suggests that this weight loss, despite the nasty side effects, does more good than harm: it saves lives. In 2007, Swedish researchers published results from a study in which they followed about 2,000 obese patients who had undergone weight-loss surgery–either gastric bypass or surgical banding–over 15 years and compared them to about 2,000 similarly obese people who didn't have surgery but who were counseled in diet and exercise. After 10 years, those who had gastric bypass surgery weighed 25% less than their presurgery weight; those who had stomach-banding surgery were down about 15%. Those who got traditional diet advice lost no more than 2% of their weight **(Infographic 26.8)**.

More significantly, there were 129 deaths in the diet-only group, mostly from weight-related heart disease and cancer, and 101 deaths in the surgery group–a large difference statistically. Deaths in the surgery

INFOGRAPHIC 26.8

Surgical Procedures and Long-Term Weight Loss

A 2007 study of obese people showed that those who underwent weight-loss surgery lost significantly more weight than those given only weight-loss counseling.

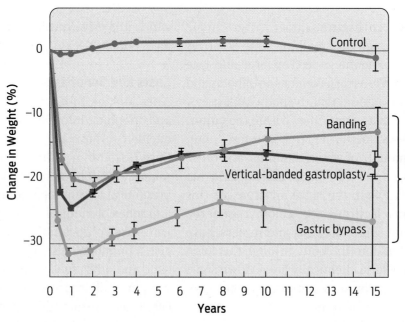

All forms of surgery resulted in significant weight loss compared with those who did not receive surgery (the control group). The people in the surgery groups maintained significant weight loss over 15 years.

Data represent the mean percent weight change during a 15-year period. Error bars indicate the range of weight loss expected if the experiment were to be repeated.

Source: **Effects of Bariatric Surgery on Mortality in Swedish Obese Subjects** Sjöström et al. *The New England Journal of Medicine*. 2007 August 23; 357 (8): 741–52.

Obesity Surgery Saves Lives

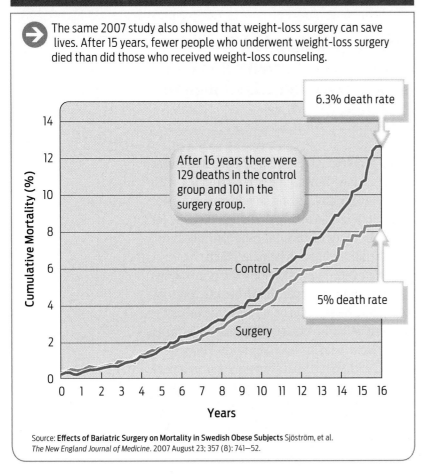

→ The same 2007 study also showed that weight-loss surgery can save lives. After 15 years, fewer people who underwent weight-loss surgery died than did those who received weight-loss counseling.

6.3% death rate

After 16 years there were 129 deaths in the control group and 101 in the surgery group.

Control

5% death rate

Surgery

Cumulative Mortality (%) — Years

Source: **Effects of Bariatric Surgery on Mortality in Swedish Obese Subjects** Sjöström, et al. *The New England Journal of Medicine.* 2007 August 23; 357 (8): 741–52.

group were also mainly from heart disease and cancer, although there were half the number of heart attack deaths in this group compared with the diet group **(Infographic 26.9)**.

Even more encouraging to scientists is the finding that both major types of weight-loss surgery can reverse type 2 diabetes, a condition in which the body can no longer regulate blood-sugar levels and that over time can damage organs and nerves. In 2004, a review in the *Journal of the American Medical Association* of 130 studies of more than 22,000 patients found that 77% of diabetics who have bariatric surgery are cured of diabetes and that 86% are either cured or have their symptoms improve.

No Cure for Obesity

The surgery isn't for everyone. As with any major surgery, there are risks. Although bariatric surgery is much safer today than it was 10 years ago,

1 in 200 patients still dies from the surgery, which can cause complications such as blood clots, hernias, or bowel obstructions. Patients can also end up back in the hospital to repair intestinal leaks that can lead to serious infection.

Smith has had a host of complications that have put her back in the hospital. A few months after her surgery, she felt terrible cramping in her side. Tests showed that scar tissue had formed at the site where her small intestine had been cut from her stomach. Surgery to remove the tissue revealed that part of her intestine and stomach had twisted and anchored onto this scar tissue, which was partly what was causing her pain. Soon after the scar tissue was removed and her stomach and intestines put back in place, she was still having stomach pains after eating. In addition to a feeding tube to her stomach, doctors decided to insert a catheter into a vein in her arm through which she could take nutrients directly into her bloodstream. Smith spent weeks in and out of the hospital between January and April of 2010. But she has had no additional complications that have landed her in the hospital since.

However, her health is still at risk. Since people who have gastric bypass surgery (as opposed to gastric banding) end up with part of the small intestine bypassed, they absorb fewer of the micronutrients they eat. Patients must take such

vitamin supplements as iron, folate, vitamin B_{12}, and calcium for the rest of their lives. There may be additional micronutrient deficiencies that scientists haven't yet recognized; only long-term follow-up of these patients will reveal how serious a problem this is. To monitor her micronutrient levels, Smith has a blood test every 3 months.

What's more, the surgery is not a permanent cure for obesity. As statistics show, most people who have the surgery regain various amounts of weight over time. This is because appetite is controlled by a complicated interaction between the digestive system and the brain. While surgery may reduce the size of the stomach, it doesn't alter the desire to eat, which is controlled by the brain. While scientists are still studying the dynamics of appetite control, they do know that if people do not exercise control over their diet and lifestyle, even those who have had surgery can regain significant amounts of weight.

Although the stomach pouch may stretch over time, it can never be as large as it was before gastric surgery. Most patients never weigh as much as before the surgery. More important, the Swedish study showed for the first time that long-term weight loss for the

morbidly obese, even when they remain overweight, is enough to save lives.

Long-term weight loss for the morbidly obese, even when they remain overweight, is enough to save lives.

The surgery, however, is a drastic measure, as Smith's case shows. She still struggles with nausea every day; strong smells can cause her to vomit. She also feels pain in her left side, for which she takes medication. She also takes anti-anxiety pills at night to help her sleep.

Despite these complications and return visits to the hospital, however, Smith has not "one day of regret," she says. On her first "surgiversary"—her surgical anniversary—she wrote a letter to her surgical team in which she said: "I have been blessed with 35 birthdays but none can compare to my surgiversary. I never imagined in a year that I would lose over 100 lbs, run a 5K the day of my surgiversary . . . sit sideways in a student desk, wear a size 12 pants from a 24-26 . . . be able to sit comfortably in a restaurant booth, and be able to stand on a table or chair without thinking, 'my gosh am I going to break this?'"

By March 2011, Smith had dropped down to 146 pounds. To maintain her weight loss, Smith wakes up at 3:30 every morning to get to the gym to exercise, and she adheres to a strict diet that is high in protein and sparse on simple sugars like sweets. "I don't recognize myself anymore," Smith says.

Of her strict regimen, Smith says, "The surgery is merely a tool. If you aren't willing to make a lifestyle change, it's not going to work for you."

"The surgery is merely a tool. If you aren't willing to make a lifestyle change, it's not going to work for you."
– Amy Jo Smith

How Do Other Organisms Process Food?

Humans and many other animals have what's known as a complete digestive tract–one with two openings, a mouth and an anus. Not all organisms have such a tube-like digestive tract. In fact, many organisms have no digestive tract at all and yet are still able to obtain and process food from their environment.

Take fungi, for example. Fungi are eukaryotic organisms. Some, such as yeasts, are unicellular, while the majority are multicellular. Multicellular fungi have "bodies" that are made up of microscopic filaments called hyphae. However, these "bodies" do not have distinct organs or organ systems, and fungi do not have digestive tracts. To obtain nutrients, fungi extend hyphae into soil (or a piece of bread), where they encounter food such as dead plants and animals. Indi-

vidual hypha cells then release digestive enzymes directly onto their food and absorb the released nutrients directly into their cells. Since digestion occurs outside their bodies, fungi do not need a stomach or a mouth, or even a digestive tract. Rather, each cell can absorb the products of external digestion–digestion that occurs in the environment–directly.

As another example, consider sea anemones, invertebrate animals that live in the oceans, attached to surfaces such as rocks. While sea anemones do digest their food internally, they don't have a digestive tract like humans. Rather, they have a single, multifunctional digestive cavity called a gastrovascular cavity. They can capture food and shove it into this multifunctional digestive cavity using the

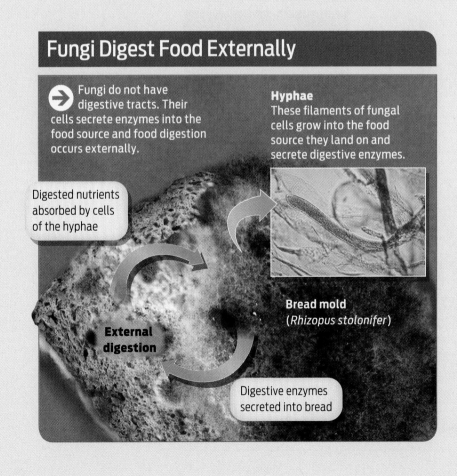

Fungi Digest Food Externally

→ Fungi do not have digestive tracts. Their cells secrete enzymes into the food source and food digestion occurs externally.

Hyphae
These filaments of fungal cells grow into the food source they land on and secrete digestive enzymes.

Digested nutrients absorbed by cells of the hyphae

External digestion

Bread mold
(*Rhizopus stolonifer*)

Digestive enzymes secreted into bread

tentacles that surround their mouths. From the mouth, food enters the gastrovascular cavity, where digestion and absorption of nutrients occur. The gastrovascular cavity has only one opening, the mouth. Food enters through the mouth, and wastes exit through it. This is an example of an incomplete digestive tract, in contrast to our own complete digestive tract in which food flows one way from the mouth to the anus.

Equipped with a stomach or not, all heterotrophic (that is, nonphotosynthetic) organisms need some way of digesting food and absorbing nutrients. As the above examples show, there are many ways to accomplish this task. Photosynthetic organisms such as plants, of course, do not need stomachs or a digestive system because they are autotrophs: they make their own food and therefore don't need to "eat" it (see Chapter 5). Instead, they take up carbon dioxide and use water and the energy of sunlight to convert the atmospheric carbon dioxide to carbohydrates in the plant body.

Sea Anemones Have an Incomplete Digestive Tract

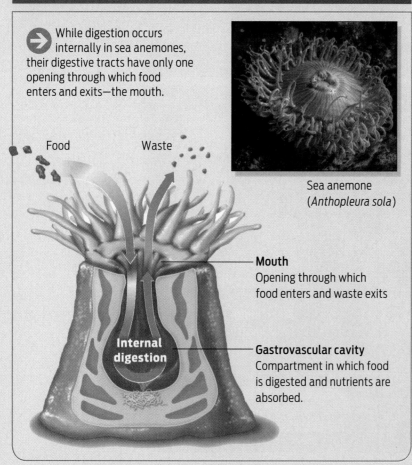

→ While digestion occurs internally in sea anemones, their digestive tracts have only one opening through which food enters and exits—the mouth.

Food

Waste

Sea anemone (*Anthopleura sola*)

Internal digestion

Mouth
Opening through which food enters and waste exits

Gastrovascular cavity
Compartment in which food is digested and nutrients are absorbed.

Plants Have No Digestive Tract

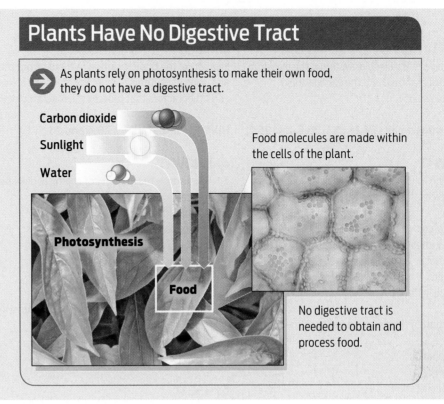

As plants rely on photosynthesis to make their own food, they do not have a digestive tract.

Carbon dioxide

Sunlight

Water

Photosynthesis

Food

Food molecules are made within the cells of the plant.

No digestive tract is needed to obtain and process food.

▷ Summary

■ The digestive tract is a long muscular tube that extends from the mouth to the anus and contains specialized organs situated along its length.

■ Digestion begins in the mouth, where teeth chew food (mechanical digestion) and salivary enzymes begin breaking down carbohydrates (chemical digestion).

■ Food passes from the mouth into the stomach through the esophagus, propelled by waves of muscular contractions called peristalsis.

■ The stomach is muscular and acidic and contains a protein-digesting enzyme called pepsin. It is elastic and can expand after a large meal to store food for a few hours.

■ Food processed in the stomach is called chyme. Chyme passes into the small intestine, where enzymes further digest it.

■ Bariatric surgeries reduce the size of the stomach so that it can no longer hold as much food. They may also shorten the small intestine.

■ Enzymes from the pancreas help to digest organic molecules in the small intestine.

■ Bile salts, produced in the liver and stored in the gallbladder, emulsify fats and help the body digest them. Epithelial cells that line the small intestine absorb the broken-down products of food; from there, food molecules enter the bloodstream and are transported throughout the body.

■ The large intestine absorbs water and forms solid stool from indigestible matter in food such as fiber.

■ Humans and many other animals have a complete digestive tract—one with a mouth and an anus. Not all organisms have a complete digestive tract; many have no digestive tract at all.

DIGESTIVE SYSTEM ANATOMY

The function of the digestive system is the digestion and absorption of food and the elimination of indigestible wastes.

HINT See Infographics 26.1–26.3, 26.6, and 26.7.

➔ KNOW IT

1. Place the following structures of the digestive system in order (from the entry of food to the exit of waste).

___ esophagus
___ large intestine
___ stomach
___ mouth
___ small intestine

2. Which part of the digestive tract has the most acidic pH?

a. esophagus
b. colon
c. small intestine
d. stomach
e. mouth

3. Why is it helpful to have an expandable stomach?

➔ USE IT

4. What do the gallbladder, liver, and pancreas have in common with respect to the digestive system? How do they differ from the mouth, stomach, and small intestine?

5. Gastric bypass surgery causes the _____ to become _____.

a. stomach; smaller
b. small intestine; larger
c. stomach; less acidic
d. small intestine; less acidic
e. stomach; larger

6. Muscle paralysis in the digestive tract would compromise which digestive function?

a. digestion in the stomach
b. digestion in the small intestine
c. absorption in the small intestine
d. digestion in the mouth
e. movement of food

DIGESTIVE PROCESSES

In order for the body to obtain nutrients from the diet, food encounters a variety of digestive enzymes and other factors that process the macromolecules.

HINT See Infographics 26.4 and 26.5.

➔ KNOW IT

7. Where does the majority of chemical digestion take place?

a. small intestine
b. esophagus
c. mouth
d. stomach
e. colon

8. What do pepsin and salivary amylase have in common?

9. Which organ produces lipase?

➔ USE IT

10. A person who has had his or her gallbladder surgically removed will have trouble processing

a. fats.
b. carbohydrates.
c. minerals.
d. vitamins.
e. proteins.

11. Compare and contrast the roles of bile salts and lipase.

12. Why would someone with a blocked duct between the pancreas and the small intestine experience pancreatic inflammation (pancreatitis)? Note that in this case inflammation is a response to tissue damage.

13. If you stand on your head, can processed food still pass from your small intestine into your large intestine? Explain your answer.

14. Why do both people who have had their gallbladders removed and people who take Alli experience "greasy" diarrhea if they eat a high-fat meal?

SCIENCE AND SOCIETY

15. What measures can a person with a very high BMI take to reduce the risk of health complications? If a person with a very high BMI chooses not to alter his or her diet and lifestyle, or is unsuccessful in the attempt to cut the risk of serious medical conditions, do you think that public or private health insurance should cover the cost of treating such a condition? Explain your answer. Consider societal, personal, economic, and genetic circumstances that can contribute to a high BMI.

16. From what you have read about gastric bypass surgery, what would you tell someone who is morbidly obese and who is considering this surgery about its known risks, benefits, and any "unknowns"? Would you say the same to someone considering the surgery who is simply overweight, not morbidly obese? Explain your answer.

Smoke on the Brain

Smoke on the Brain

Nicotine and other drugs of abuse alter the brain and are hard to kick

Jack Ward thought he could resist picking up a cigarette, but the temptation was too strong. He had happily quit smoking in 2006. But when he walked into a smoke-filled poker room 3 years later at a friend's house in Brooklyn, New York, the scene before him seemed to run in slow motion. He watched intently as smokers sucked languidly on their cigarettes, deeply drawing in each puff and then exhaling with sighs of contentment. He resisted that night. But poker night became a weekly event, and finally he gave in: he began smoking again.

Ward knew that smoking is risky. Cigarette smoke is associated with various health problems. He had had countless arguments with his wife, who had insisted he stop smoking to protect his own health and hers as well, which is partly why he quit in the first place. But during poker night, none of that seemed to matter.

Cigarette smoking is highly addictive—smokers can develop a physical and psychological need to smoke. And while anyone might be able to smoke one cigarette or even several and not become addicted, most people find it extremely difficult to stop if they have smoked for an extended length of time. Ward had started smoking casually in high school. By the time he went to graduate school, he was smoking a pack and a half a day. "I was surrounded by people who smoked," Ward recalls. "It never seemed unusual to smoke so much."

Brain scientists have long known that cigarettes, caffeine, and other drugs of abuse such as cocaine, heroin, alcohol, and marijuana stimulate the brain's reward system, a complex circuit of brain cells that evolved to make us feel good after eating or having sex—activities we must engage in if we are to survive and pass along our genes. Without a feeling of pleasure from these activities that ensure our survival, we might never seek out sex or food, and our species would die out. Drugs of abuse stimulate the very same pleasure pathways, which is why

they make people feel so good, and also why they can be so hard to resist.

Nonchemical addictions can do the same. Certain behaviors–gambling, shopping, sex–can start out as habits but slide into addictions. Even food can be addictive. Studies of obese people, for example, have shown that the brains of compulsive eaters are hyperactive in areas that respond to food. For

After prolonged use, drugs of abuse change the structure and function of the brain in ways that can wreak havoc on users' lives.

these people, the mere thought of eating floods the brain with pleasure. In other words, almost anything deeply enjoyable can become an addiction.

But the pleasure comes at a massive cost. After prolonged use, drugs of abuse change the structure and function of the brain in ways that can wreak havoc on users' lives. Severely addicted people may stop eating, stop working, in fact stop all activities because nothing matters except the drug–and they will do anything to get it. Without it, they have intense cravings, obsessive thoughts, and are deeply depressed.

Once established, addiction is difficult to treat. The stories of celebrities who have been in and out of rehab and the deaths of superstars such as Michael Jackson and Heath Ledger from drug overdoses are testament to just how difficult. But scientists today are gaining a better understanding of how physiological changes in the brain cause addiction, and that knowledge is leading to better treatments, which will help addicts reclaim their brains and their lives.

Addiction and the Brain

That cigarettes are addictive is likely no surprise to you. In fact, about one-third of Americans smoke some kind of tobacco, according to the National Institute on Drug Abuse (NIDA). Smoking is so common that you probably know at least one smoker.

The component in tobacco that makes cigarettes so addictive is the chemical nicotine. When a person smokes, nicotine floods the lungs, where cells absorb it and transport it to the bloodstream. The body absorbs nicotine very quickly–in fact, nicotine reaches the brain within 8 seconds after tobacco smoke is inhaled. Chewing tobacco also contains nicotine, which the body absorbs through the mucous membranes that line the mouth.

Nicotine is addictive because it stimulates feelings of pleasure in the **brain.** The brain is the master coordinator of the body, controlling virtually all of the body's activities, including sensation, movement, intellect, and just about all involuntary actions. The brain is also the seat of memory. Pleasurable experiences recorded in our memories serve as motivation; we tend to seek out the same pleasurable experiences over and over. Ultimately, the decision to smoke a cigarette, or to carry out any other behavior, is made in the brain (**Infographic 27.1**).

The brain doesn't act in isolation: it communicates with the rest of the body, sending out and receiving signals to coordinate our many activities. The brain accomplishes this via **nerves**–long fibers made up of specialized cells and supportive tissue that transmit signals. A major collection of nerves is the **spinal cord,** a bundle of nerve fibers about the

BRAIN
An organ of the central nervous system that integrates and coordinates virtually all functions of the body.

NERVE
A bundle of specialized cells that transmit information.

SPINAL CORD
A bundle of nerve fibers, contained within the bony spinal column, that transmits information between the brain and the rest of the body.

CENTRAL NERVOUS SYSTEM (CNS)
The brain and the spinal cord.

Left: Students gather outside the student center on the University of Kentucky campus in Lexington to protest the school's ban on smoking. Right: Smokers in a designated area outside Tokyo's Shibuya Station.

Nicotine Stimulates Pleasure in the Brain

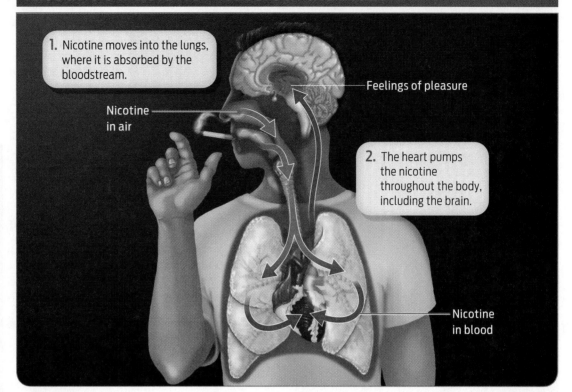

1. Nicotine moves into the lungs, where it is absorbed by the bloodstream.

Nicotine in air

Feelings of pleasure

2. The heart pumps the nicotine throughout the body, including the brain.

Nicotine in blood

PERIPHERAL NERVOUS SYSTEM (PNS)
All the nervous tissue outside the central nervous system. The PNS collects sensory information and transmits instructions from the CNS.

NEURONS
Specialized cells of the nervous system that generate electrical signals in the form of action potentials.

SENSORY NEURONS
Cells that convey information from both inside and outside the body to the CNS.

MOTOR NEURONS
Neurons that control the contraction of skeletal muscle.

CELL BODY
The part of a neuron that contains most of the cell's organelles, including the nucleus.

DENDRITES
Branched extensions from the cell body of a neuron, which receive incoming information.

AXON
The long extension of a neuron that conducts action potentials away from the cell body toward the axon terminal.

diameter of a finger that extends from the base of the brain down to the lower back, contained in and protected by our bony spinal column. Together, the brain and the spinal cord make up the **central nervous system (CNS).** Obviously, information from the brain must also travel to distant body sites, like fingers and toes. This is accomplished by the **peripheral nervous system (PNS).** The peripheral nervous system includes all nervous tissue outside the brain and spinal cord leading to and from the limbs and organs.

The CNS and the PNS work together to transmit messages between the brain and the rest of the body and to respond to stimuli in the environment. Sensory receptors of the PNS present in our ears, our eyes and in our skin, for example, enable us to sense our environment; muscles and other tissues with effector cells allow us to respond. Peripheral nervous tissue present in other organs, including the lungs, heart, and digestive organs, keeps our bodies operating without conscious thought. The nervous system

allows us to perceive and understand the world around us and to translate thought into action **(Infographic 27.2).**

Nerve Transmission

There is a mechanism—a step-by-step process—through which the CNS and PNS communicate. This process begins at the cellular level. The nervous system is equipped with specialized cells called **neurons,** which transmit information signals along nerves. There are different types of neurons. For example, **sensory neurons,** such as sensory receptors in the eyes or skin, receive information from the external world and transmit it to the CNS. **Motor neurons** transmit information from the PNS to muscle cells, signaling them to contract or relax.

Neurons are especially adept at transmitting signals because of their unique structure: a neuron consists of a large **cell body** together with branched extensions called **dendrites** that receive signals, and a single large **axon**

The Nervous System Has Two Main Parts

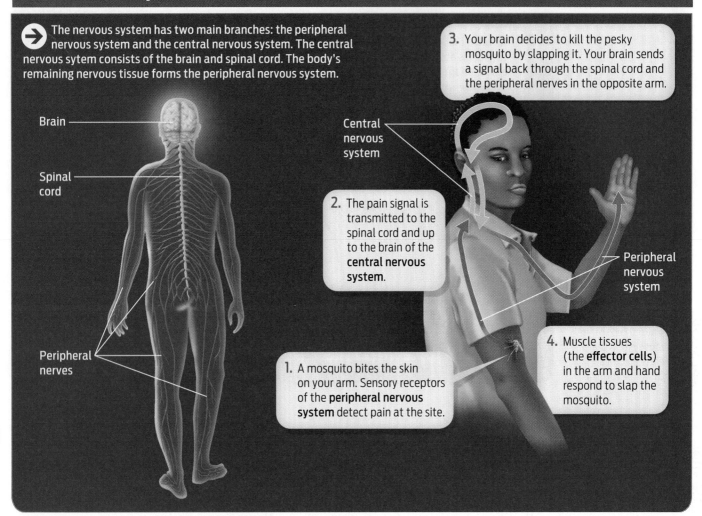

→ The nervous system has two main branches: the peripheral nervous system and the central nervous system. The central nervous sytem consists of the brain and spinal cord. The body's remaining nervous tissue forms the peripheral nervous system.

Brain

Spinal cord

Peripheral nerves

Central nervous system

3. Your brain decides to kill the pesky mosquito by slapping it. Your brain sends a signal back through the spinal cord and the peripheral nerves in the opposite arm.

2. The pain signal is transmitted to the spinal cord and up to the brain of the **central nervous system**.

Peripheral nervous system

1. A mosquito bites the skin on your arm. Sensory receptors of the **peripheral nervous system** detect pain at the site.

4. Muscle tissues (the **effector cells**) in the arm and hand respond to slap the mosquito.

that carries signals away from the neuronal cell body. The end of the axon, called the **axon terminal,** transmits signals to the next cell in the pathway **(Infographic 27.3).**

The signals sent along a neuron are actually electric currents. Dendrites and axons are like electrical cords that help collect and transmit information in the nervous system.

All animal cells have different charges inside and outside their cell membranes. And as in all animal cells, the charge on the inside of the neuron is negative relative to the charge outside the cell. But unlike the vast majority of cells, neurons can change the distribution of the charge. When axons of neurons allow positive ions to cross the cell membrane,

the charge distribution across the membrane changes. This initiates an **action potential,** a coordinated pattern of ion flow across the membrane, resulting in a characteristic pattern of changes in charge distribution across the membrane. Like the fuse leading to a stick of dynamite, an action potential firing at one point along an axon can "ignite" an action potential farther along the axon, and so on **(Infographic 27.4).**

Action potentials travel down the length of a neuron very quickly. Just think how quickly your arm moves in response to touching a hot stove—the movement seems almost instantaneous; an action potential can last for less than 1 millisecond.

AXON TERMINAL
The tip of an axon, which communicates with the next cell in the pathway.

ACTION POTENTIAL
An electrical signal within neurons caused by ions moving across the cell membrane.

MYELIN
A fatty substance that insulates the axons of neurons and facilitates rapid conduction of action potentials.

INFOGRAPHIC 27.3

Neuron Structure Is Highly Specialized

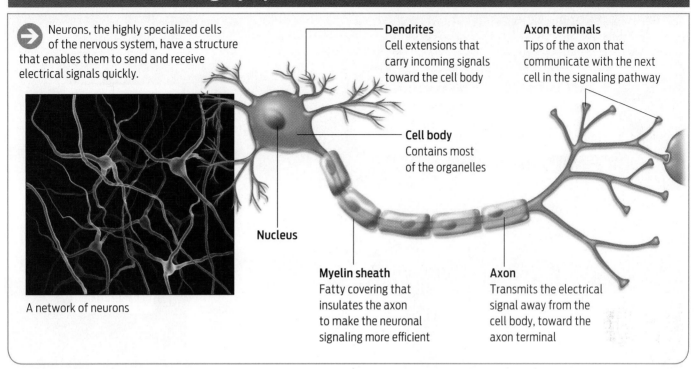

Neurons, the highly specialized cells of the nervous system, have a structure that enables them to send and receive electrical signals quickly.

Dendrites
Cell extensions that carry incoming signals toward the cell body

Axon terminals
Tips of the axon that communicate with the next cell in the signaling pathway

Cell body
Contains most of the organelles

Nucleus

Myelin sheath
Fatty covering that insulates the axon to make the neuronal signaling more efficient

Axon
Transmits the electrical signal away from the cell body, toward the axon terminal

A network of neurons

GLIAL CELLS
Supporting cells of the nervous system.

CEREBELLUM
The part of the brain that processes sensory information and is involved in movement, coordination, and balance.

BRAIN STEM
The part of the brain that is closest to the spinal cord and which controls vital functions such as heart rate, breathing, and blood pressure.

CEREBRUM
The region of the brain that controls intelligence, learning, perception, and emotion.

CEREBRAL CORTEX
The outer layer of the cerebrum, the cerebral cortex is involved in many advanced brain functions.

Signals can travel so quickly because neurons, much like an electrical cord, are insulated: they are coated with a sheath of **myelin** that surrounds the axon and prevents the electrical charge from leaking across the membrane. Myelin is critical to nerve transmission: without it, action potentials weaken, losing strength as they pass along the axon, and can peter out before reaching the axon terminal. Diseases in which myelin degenerates over time, such as multiple sclerosis, can cause progressive paralysis, primarily because neurons lacking myelin can't efficiently transmit signals to muscles. The myelin sheath is produced by a type of **glial** (or supporting) **cell.**

The Anatomy of Addiction

Neurons communicate not only to move muscles that control limbs, but also to solidify thoughts and lay down memories. This constant chatter within the nervous system is the way in which the brain grows and adapts to new environments and learns new tasks.

Millions of neurons make up each of the many regions of the brain. Scientists typically categorize any given brain region according to its function. The brain consists of more than 40 subregions, and there are four major regions: the cerebellum, the brain stem, the cerebrum, and the diencephalon. The **cerebellum** is part of the rearmost portion of the brain, the hindbrain, and is involved in processing sensory information and in movement, coordination, and balance. The **brain stem** consists of the hind- and midbrain and coordinates involuntary functions like reflexes, heart rate, and breathing. This part of the brain is often referred to as the "old brain," since it is the most primitive and oldest part of the brain. The **cerebrum** is the largest part of the forebrain. Its outer layer, the **cerebral cortex,** is the seat of our more advanced brain functions, including perception and thinking, and gives us our distinct personalities and most human characteristics. Its inner portion transmits signals (in the form of action potentials) from the cortex

Signals Move along Neurons

→ Neurons communicate through electrical signals called action potentials, which are generated as ions flow across the cell membrane. When a neuron "fires," Na⁺ and K⁺ ions enter and leave the axon in a characteristic pattern, creating an action potential. The firing of one action potential triggers the firing of another in the neighboring region of the axon. In this way, signals pass along the axon.

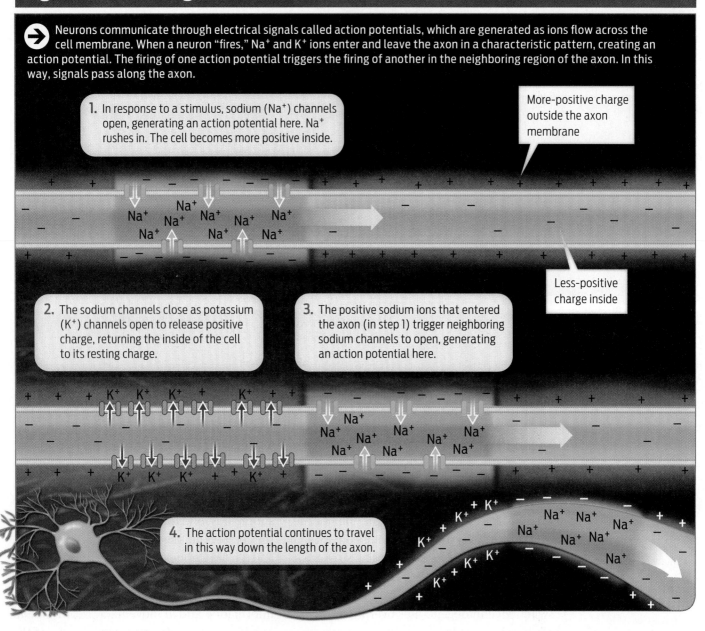

1. In response to a stimulus, sodium (Na⁺) channels open, generating an action potential here. Na⁺ rushes in. The cell becomes more positive inside.

More-positive charge outside the axon membrane

Less-positive charge inside

2. The sodium channels close as potassium (K⁺) channels open to release positive charge, returning the inside of the cell to its resting charge.

3. The positive sodium ions that entered the axon (in step 1) trigger neighboring sodium channels to open, generating an action potential here.

4. The action potential continues to travel in this way down the length of the axon.

to various brain regions and to other parts of the body. The cerebrum is made up of about 10 billion neurons, and is divided into left and right hemispheres; each hemisphere is divided into four lobes. Each lobe processes a variety of functions, including smell, hearing, speech, and vision. The **diencephalon,** located between the brain stem and the cerebrum,

> **The cerebrum is made up of about 10 billion neurons, and is divided into left and right hemispheres; each hemisphere is divided into four lobes.**

includes the thalamus and hypothalamus, among other structures, and regulates homeostatic functions like body temperature, hunger, thirst, and the sex drive.

The diencephalon, part of the limbic system, is a set of structures that lie on both sides of the thalamus, just under the cerebrum. The

DIENCEPHALON
A brain region located between the brain stem and the cerebrum that includes the thalamus and hypothalamus, among other structures, and regulates homeostatic functions like body temperature, hunger, thirst, and the sex drive.

Parts of the Brain

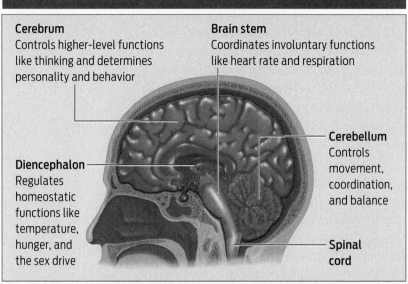

Cerebrum
Controls higher-level functions like thinking and determines personality and behavior

Brain stem
Coordinates involuntary functions like heart rate and respiration

Cerebellum
Controls movement, coordination, and balance

Diencephalon
Regulates homeostatic functions like temperature, hunger, and the sex drive

Spinal cord

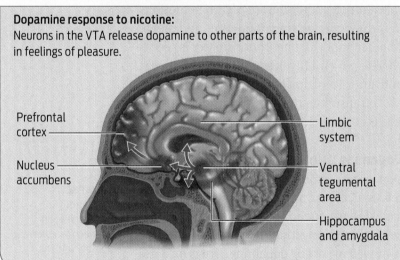

Dopamine response to nicotine:
Neurons in the VTA release dopamine to other parts of the brain, resulting in feelings of pleasure.

Prefrontal cortex

Nucleus accumbens

Limbic system

Ventral tegumental area

Hippocampus and amygdala

HIPPOCAMPUS
The region of the brain involved in learning and memory.

AMYGDALA
A subregion of the brain that processes emotions, especially fear and anxiety, and is the seat of emotional memories.

DOPAMINE
A neurotransmitter that is involved in conveying a sense of pleasure in the brain.

limbic system includes not only the hypothalamus and the thalamus, but also the **hippocampus,** the **amygdala,** and several other subregions of the brain. The limbic system is the primary seat of our emotions and memories and, along with two nearby regions, the nucleus accumbens and the ventral tegumental area, is a major region involved in addiction.

Although the brain has many regions and subregions, these areas are not discrete units. The brain is an integrative organ. Any action or thought often involves a coordinated effort by several brain regions at once. Indeed, neurons in one region may have axons and dendrites that extend to neighboring regions. Many brain regions contribute to forming addictive behavior. So although the so-called reward system is housed primarily in the nucleus accumbens and ventral tegumental area, drugs of abuse stimulate these regions as well as the prefrontal cortex of the cerebrum. This reward system also connects to the amygdala and to the hippocampus of the limbic system. When a smoker lights up, nicotine that reaches the brain stimulates the production of **dopamine**–one of the body's primary pleasure signals–in several brain regions, giving the smoker a pleasurable sensation. And while these regions of the brain are likely the most important in forming addictive behaviors, they are not the only parts of the brain involved in addiction, illustrating how complex interactions within the brain can be **(Infographic 27.5)**.

That drugs can affect the brain so broadly explains why they can be so hard to kick. It also explains why even addicts who successfully stop taking drugs for decades can relapse simply by being in an environment that conjures up memories of the drug. The tinkling of a whiskey glass, meeting an old drinking buddy, being in a bar or a room filled with smokers puffing away can be reminders powerful enough to lure former addicts back to their old ways.

How Neurons Communicate

For Ward, who turned 38 in 2011, it wasn't just the smoking at the weekly poker game that rekindled his desire to smoke; it was a combination of events. For one thing, he says he was "extremely stressed" about work. He had started working as a freelance writer, and he was having trouble juggling his deadlines. In the past, the radio journalist would often rely on smoking to calm himself. "I'm a nervous worker," he says. "Some people shake their legs, or pace, or exercise to work off stress. For me it was smoking." And then a difficult family situation arose. Up to that point, he had managed to avoid smoking, even in the presence of smokers, because, he says, he had learned to think of himself as a nonsmoker: holding to that self-concept was a way to persuade himself not to pick up a cigarette. But in the end, the combination of stressful events and easy access to cigarettes pushed him to smoke again.

Of course, many former abusers are able to resist temptation in circumstances like Ward's. This is true because resistance, like addiction, is a complex behavior. The decision to act on any impulse is influenced by several factors: other people present at the time, past experiences, and the ability to control one's behavior all play a role in determining the final action. In other words, many brain circuits are involved in both addiction and resistance, and they all communicate with one another to ultimately influence a decision either to remain abstinent or to reach for a fix.

This communication begins at the level of the individual cell, in the transmission of an action potential along the axon of a single neuron. But how does that neuron then communicate with the next cell in the pathway?

When the electrical signal reaches the axon terminal, the information must be communicated to the next cell in the pathway, which could be another neuron, or an effector cell in a muscle or gland. The site of communication between a neuron and the next cell in the pathway is called the **synapse,** and includes a small space, or gap, between the two cells.

When an action potential reaches a neuron's axon terminal, it stimulates the neuron to release **neurotransmitters**–chemical messengers–from its axon terminal into the space between the neuron and the cell with which it is communicating. Neurotransmitters diffuse across this space, known as the **synaptic cleft,** to carry messages from one cell to another. Each neurotransmitter fits into a receptor on the surface of the cell receiving the signal. The act of a neurotransmitter binding to its receptor initiates a signaling pathway in the receiving cell. If the receiving cell is another neuron, the binding of neurotransmitter may spark another action potential **(Infographic 27.6).**

Cells in the nervous system use several types of neurotransmitters to communicate different types of messages. For example, the neurotransmitter serotonin, which is active in the CNS and the gastrointestinal tract, plays myriad roles in the body, from regulating emotions to influencing anxiety, appetite, and sleep. Another important neurotransmitter is acetylcholine, which is involved in learning, memory, and muscle contraction. In addiction, the neurotransmitter that plays a particularly large role is dopamine.

Dopamine's Role in Addiction

Dopamine's main job is to convey information related to elation and pain. The joy we get from a meal, a promotion, a winning poker hand, or sex–or anything that gives us pleasure–is conveyed in part by dopamine. Drugs of abuse stimulate dopamine production, which is why they can become so addictive.

But these drugs cause such a high that over time they can alter the dopamine communication system so that the pleasures of everyday life become meaningless compared to the pleasure of the drug. Normally the brain produces dopamine at a relatively constant rate, and dopamine occupies only a portion of dopamine receptors at any given time. But when a person smokes, snorts cocaine, or takes heroin, for example, dopamine levels increase dramatically in the synaptic cleft. With so much dopamine available, practically all of the brain's dopamine receptors become activated simultaneously.

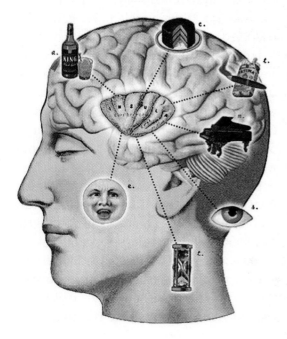

Neurons Use Chemical Signals to Communicate

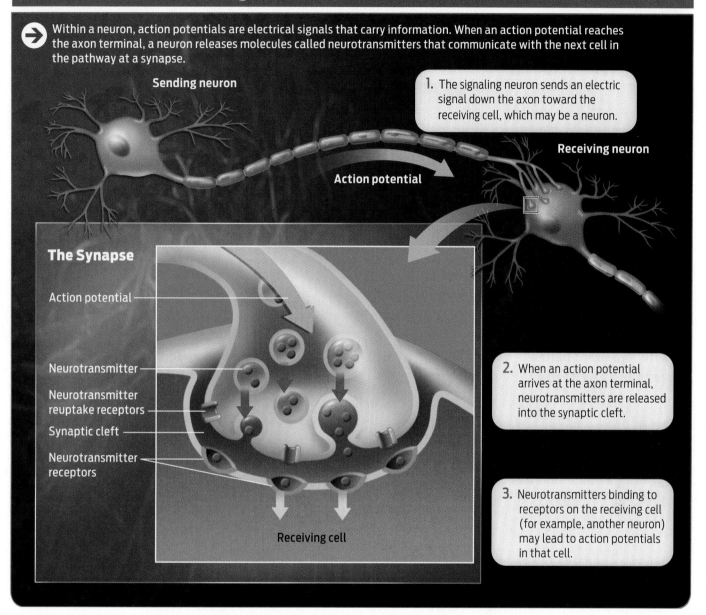

→ Within a neuron, action potentials are electrical signals that carry information. When an action potential reaches the axon terminal, a neuron releases molecules called neurotransmitters that communicate with the next cell in the pathway at a synapse.

Sending neuron

1. The signaling neuron sends an electric signal down the axon toward the receiving cell, which may be a neuron.

Receiving neuron

Action potential

The Synapse

Action potential

Neurotransmitter

Neurotransmitter reuptake receptors

Synaptic cleft

Neurotransmitter receptors

Receiving cell

2. When an action potential arrives at the axon terminal, neurotransmitters are released into the synaptic cleft.

3. Neurotransmitters binding to receptors on the receiving cell (for example, another neuron) may lead to action potentials in that cell.

The immediate effect is euphoria. But there is a downside. Because so much dopamine is produced, the brain becomes overwhelmed and tries to dampen the drug's effect by switching off some of its dopamine receptors. When the drug wears off, fewer receptors are functioning–bringing down mood. Shutting down receptors that would otherwise be available to receive information is called down-regulation.

The low can be so low that normal pleasures such as eating or socializing become dull and listless affairs. The user's mood may now be even lower than it was before taking the drug. As dopamine receptors shut down, ever-larger quantities of the drug are required to produce a high–and the high may never be as high as the user experienced the first time. As they come

As dopamine receptors shut down, ever-larger quantities of the drug are required to produce a high—and the high may never be as high as the user experienced the first time.

Addictive Drugs Alter Dopamine Signaling

 Elevated levels of dopamine cause initial feelings of pleasure but then lead to withdrawal symptoms if the drug is stopped or to the need for even more of the drug to achieve the same feelings of pleasure.

a. Normal Dopamine Signaling

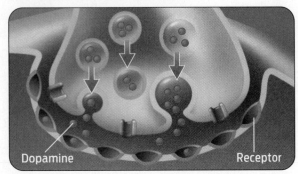

Dopamine
Receptor

Normally, cells release moderate amounts of dopamine into the synapse and not all dopamine receptors are occupied.

b. After Drug Use

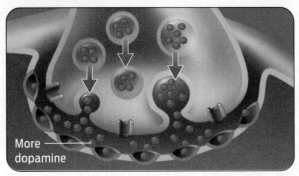

More dopamine

Certain drugs cause massive dopamine release into the synapse. Many more dopamine receptors become occupied. The result is an intense feeling of pleasure.

c. After Repeated Drug Use

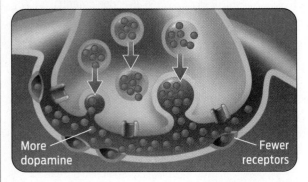

More dopamine
Fewer receptors

Dopamine overstimulation causes the receiving cell to down-regulate (that is, shut down) some dopamine receptors. Because the person now has fewer receptors, more of the drug is required to feel high, because more of the drug boosts the odds that more of the remaining dopamine receptors will be occupied.

d. After Drug Withdrawal

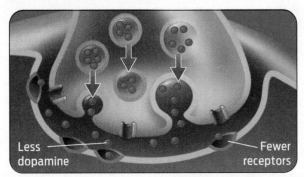

Less dopamine
Fewer receptors

In the absence of the drug and with fewer active dopamine receptors available, normal dopamine signaling is at low levels. As a result, the user may feel sick and depressed.

down from a high, addicts will likely feel even more unhappy and depressed as the dopamine response system is dampened. Eventually, many addicts need to take drugs simply to feel normal. Without drugs, they suffer the physical symptoms of withdrawal, which may include depression, anxiety, and intense cravings. The specific symptoms and their intensity vary depending on the drug (Infographic 27.7).

A down-regulated dopamine system isn't the only structural brain change that scientists have observed in drug addicts. Researchers have also shown that drugs such as cocaine can change the shape of neurons in specific parts of the brain and consequently may impair their ability to transmit messages. Brain-imaging studies have also shown that addicts consistently have lower than normal levels of blood flow in the frontal regions of the cerebrum, a region involved in decision making, during withdrawal from cocaine, and higher than normal levels while they are on the drug.

Even more troubling, adolescents who take drugs may be preventing their brains from developing normally. A 2004 study by researchers at University of California, Los Angeles, and the National Institute of Mental Health (NIMH) that imaged the brains of 13 children and young people age 4 to 21 over 10 years showed that that some parts of the brain, such as the prefrontal cortex, which among other tasks controls behavior, are not fully developed until young adulthood—the mid-twenties or so. Subsequent research confirms this finding and also suggests that the prefrontal cortex may not fully mature until even later. Taking drugs at an early age may hinder normal development of this region.

And that's not the end of the story. There are several other neurotransmitters, in addition to dopamine, involved in addiction, says Joe Frascella, director of the division of Clinical Neuroscience and Behavioral Research at the National Institute on Drug Abuse, and scientists are just starting to study how drugs of abuse affect them. While scientists have shown that almost every drug of abuse affects the dopamine system in varying degrees, Frascella says, "It's certainly more complex than just dopamine." Scientists have only just scratched the surface when it comes to learning exactly how long-term drug use affects the brain **(Table 27.1)**.

Born Addicts?

Of course, not everyone who takes drugs becomes addicted. Exactly why some people seem to be more at risk than others isn't clear. But researchers have a few hypotheses based on existing evidence. For some of us, it may be a matter of biology: a predisposition to addiction caused by an inherent shortage of dopamine or other types of receptors. Some people may just have been born with fewer receptors,

or their brains may have lost receptors over time because of difficult life experiences. Consequently, drugs provide these people a high that they can't get from any other stimulus. And the drug feels too good to stop.

And just as some of us may be biologically predisposed to addiction, others may be predisposed to avoid it: some people's brains may simply be better at overriding the pleasure-seeking impulse. As we've seen, drug addiction involves several brain circuits and we vary in our ability to control our behaviors.

A 2011 study by researchers at the Scripps Research Institute Florida, for example, identified a brain pathway involved in nicotine addiction. The researchers found that the number of a brain receptor called alpha-5 influences how susceptible mice are to nicotine addiction. When given the opportunity to self-administer nicotine, normal mice will stop after reaching a certain dose. Mice with no alpha-5 receptors, however, won't stop until they've taken a much higher dose. Humans also have alpha-5 receptors in varying numbers, and scientists hypothesize that people with fewer receptors are less sensitive to nicotine and may become more easily addicted.

Regardless of the reason people become addicted, addiction is a serious public health problem. Tobacco, for example, is responsible in some way for one out of every five deaths in the United States, according to NIDA. That's because nicotine not only affects the brain but also targets cells throughout the body. Nicotine acts directly on the heart, for example, to change heart rate and blood pressure, and also acts on the nerves that control respiration to change breathing patterns. Smokers have a higher incidence of both heart and lung disease. And smoking causes cancer.

Drug use also affects health in ways that diminish the quality of life. For example, deficits in the dopamine system caused by drug use weaken memory and motor skills. Nora Volkow, head of NIDA, has shown that methamphetamine

> **Tobacco, for example, is responsible in some way for one out of every five deaths in the United States, according to NIDA.**

TABLE 27.1

Potentially Addictive Drugs and Their Effects

	MODE OF ACTION	EFFECT
COCAINE	Causes a large release of dopamine into the synapse and inhibits its removal by reuptake receptors; causes an amplified signal between neurons.	Highly addictive; produces exaggerated feelings of well-being and confidence and loss of interest in life activities. High doses lead to paranoia, anxiety, and increased blood pressure.
HEROIN	Mimics endorphins that are normally sent into a synapse to relieve pain; binds opiate receptors in the limbic system, brain stem, and spinal cord, artificially affecting mood, respiration, and pain response.	Rush of euphoria followed by a foggy feeling. Powerful withdrawal symptoms make this drug extremely addictive. Overdose slows breathing to dangerous levels. Associated with transmission of HIV through contaminated needles.
CAFFEINE	Mimics adenosine and binds to adenosine receptors in the forebrain and blocks the natural sleep response. Stimulated cells also signal the production of adrenaline and dopamine, further supporting an alert and pleasurable response.	Withdrawal symptoms include headaches, jittery feelings, and increased anxiety; can interfere with deep-sleep cycles, leading to exhaustion and depression. The adrenaline produced constricts blood vessels, affecting heart rate.
NICOTINE	Increases levels of dopamine in the synapse of neurons primarily in the ventral tegmental area and prefrontal cortex.	Enhanced short-term feelings of pleasure; very addictive. Other chemicals in inhaled smoke are also toxic.
ECSTASY	Causes excessive release of serotonin in parts of the brain that control the senses, sleeping, and emotions; destroys nerve cells that produce serotonin.	Causes paranoia, anxiety, confusion, and difficulty concentrating. Long-term use causes difficulty differentiating reality and fantasy.
ALCOHOL	A general depressant of the central nervous system; changes communication patterns between neurons in the hypothalamus and frontal lobe.	Acts as an anesthetic, affecting involuntary responses like breathing, diminishing senses, and influencing motor and behavior control; kills cells in the brain and other organs.
MARIJUANA	Mimics the chemical anandamide; artificially stimulates the anandamide receptors in areas of the brain that affect memory, emotion, and sensory perception.	Weakens short-term memory and can block the production of long-term memory; weakens problem-solving ability and coordination.
INHALANTS (GLUE, HAIR SPRAY, PAINT THINNER, ETC.)	Vapors destroy the myelin sheath that insulates the axons of neurons and thus slow down or stop nerve signaling; damages cells in the brain, lungs, heart, liver, kidneys, and bones.	Causes headaches, nausea, and disorientation and can diminish the ability to learn, remember, and solve problems; may make the heart more sensitive to brain signaling.
METHAMPHE-TAMINE	Causes release of dopamine and norepinephrine into the synapse. Dopamine signaling creates feelings of pleasure; norepinephrine signaling increases blood pressure and heart rate.	Creates feelings of pleasure and euphoria, paranoia, and hallucinations; may alter dopamine-producing neurons connected with Parkinson disease.
RITALIN	Causes increased release of dopamine into the synapse of nerve cells in the prefrontal cortex; strengthens dominant signals in the brain and weakens background signals that are less organized.	At lower doses increases the ability to focus and concentrate; at higher doses can inhibit formation of new nerve pathways, interfering with cognition and brain development. Prescribed for attention deficit disorder, but has become a common street drug.

INFOGRAPHIC 27.8

Drugs Diminish Memory and Motor Skills

Methamphetamine has been shown to decrease the number of dopamine receptors in the brain, impairing motor skills and memory.

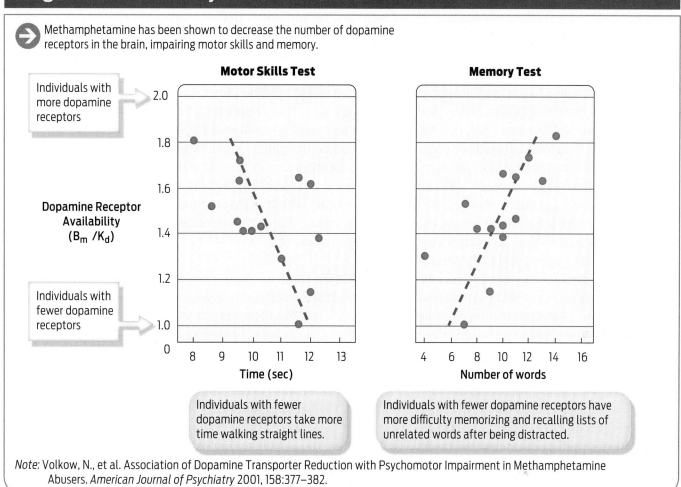

Individuals with more dopamine receptors

Dopamine Receptor Availability (B_m /K_d)

Individuals with fewer dopamine receptors

Motor Skills Test

Time (sec)

Individuals with fewer dopamine receptors take more time walking straight lines.

Memory Test

Number of words

Individuals with fewer dopamine receptors have more difficulty memorizing and recalling lists of unrelated words after being distracted.

Note: Volkow, N., et al. Association of Dopamine Transporter Reduction with Psychomotor Impairment in Methamphetamine Abusers. *American Journal of Psychiatry* 2001, 158:377–382.

(crystal meth) users not only have fewer dopamine receptors than nonusers, they also have poorer short-term memory and score much lower on tests of motor skills, such as quickly walking a straight line **(Infographic 27.8)**.

Kicking the Habit

Research on the neurobiology of drug addiction is informing new and better ways to treat addicts. Volkow and other researchers have shown that after a period of abstinence from drugs, the dopamine system can repair itself–but it generally takes longer than a year. And while studies have shown that functional skills such as short-term memory and motor control do come back, it's not clear whether or not they come back completely.

Given the nature of many of the drug-induced brain changes, experts now see addiction as a

chronic disease. Like heart disease or diabetes, diseases for which patients require long-term treatment, patients with addictions require long-term treatment plans. And the occasional relapse is only a predictable setback, not a failure of the treatment, says Volkow.

Experts also now know that the best treatments should target addictive behaviors in several ways. They should, for example, decrease the reward value of the drug, according to Volkow, perhaps by counseling addicts to seek out other pleasurable experiences and to repeat them to reinforce their value in the brain. Avoiding the drug and focusing on other pleasurable experiences will, over time, weaken conditioned memories of the drug and drug-related stimuli.

And a better understanding of the effects on the brain of substance abuse is informing efforts to develop better medications. Success rates for medications in use today are low (in some cases as few as 20% of addicts taking the medication have quit using the damage-causing drug after 6 months of treatment), and the medications have serious side effects. Two popular antismoking drugs for example, Chantix and Zyban, can cause serious mental disturbances and increase aggression and suicidal thoughts. Consequently, scientists are studying medications that work in different ways—medications that block the transport of dopamine, for example, or alleviate stress or interfere with stressful memories.

For example, Nabi Biopharmaceuticals, based in Rockville, Maryland, is testing a nicotine vaccine in late-stage clinical trials. According to Nabi, this inhalable vaccine stimulates the immune system to produce antibodies that bind to nicotine, preventing nicotine from reaching receptors in the brain. Smoking then no longer feels pleasurable and smokers should more easily be able to quit. The vaccine has side effects, however, and preliminary results suggest that it is no more effective than existing medications.

In the end, it's likely a combination of behavioral strategies and medications that target specific neurotransmitters or brain circuits that will work the best to help addicts kick the habit, says Frascella, of NIDA.

Before Ward quit smoking in 2006, he was up to a pack and a half a day. Smoking had increasingly become a point of tension between him and his nonsmoking wife. She desperately wanted him to kick the habit, both for the sake of his health and because of the risks to herself of heart disease and cancer from secondhand smoke. A medical examination showed that Ward already had risk factors for heart disease, including high blood pressure. His doctor told him to stop smoking. "She didn't ask me to stop," Ward says. "She simply said, 'You are going to stop smoking in two weeks.'" She prescribed a nicotine patch—a skin patch that delivers nicotine to the bloodstream and enables smokers to gradually kick smoking by easing nicotine withdrawal symptoms—and other medication. Within a few months, Ward had stopped smoking.

Old Habits Die Hard

But after that first poker night, Jack Ward went back to smoking. This time, instead of his old pack-and-a-half-a-day habit, he managed to cut down to only a few cigarettes a week, all of them smoked on his weekly poker night.

Two years after Ward started smoking again, he moved. The weekly poker nights were gone, and gone, too, were the social cues that had tempted him to pick up the habit again. There are more antismoking social influences in his life now than ever. Most of his friends do not smoke, he exercises more, and perhaps most important of all, he wants to be a good role model for his 4-year-old daughter: "I don't want her to see me smoking." Ward no longer smokes. ■

How Do Other Organisms Sense and Respond?

As we've seen, humans have a sophisticated central nervous system that consists of a brain and spinal cord. Organisms distantly related to humans have simpler nervous systems.

Invertebrate animals like jellyfish and sea anemones, for example, have neurons that sense and send information. But because they lack a defined master control organ such as a brain, they do not have a central nervous system. Instead, they have what are known as nerve nets, which consist of sensory neurons, motor neurons that control muscles, and neurons that connect the sensory and motor neurons; but they do not have the equivalent of a brain. Nerve nets enables these animals to respond to stimuli with behaviors such as swimming and eating—but the response is more like a reflex than a thought process.

Jellyfish Have a Nerve Net

➔ Some organisms, like jellyfish, have a system of nerves that allow them to make a general response to physical contact, but have no central brain.

Nerve net
Nerve nets are series of interconnected neurons that allow cnidarians like jellyfish to respond to physical contact. In response to stimuli, neurotransmitter impulses are sent between neurons across the nerve net rather than to a central brain.

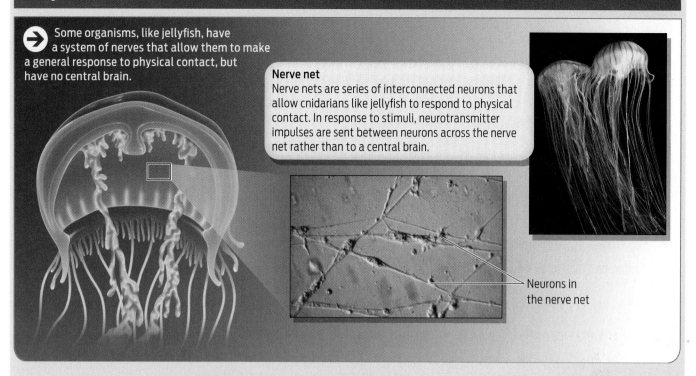

Neurons in the nerve net

Flatworms Have a Primitive Brain

➔ Flatworms were the first group of organisms to have both a central and a peripheral nervous system.

The flatworm, *Turbellaria*

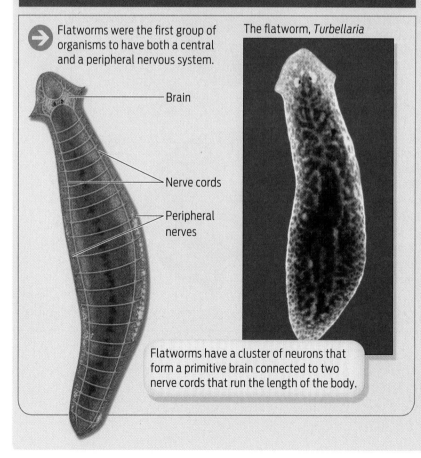

Brain

Nerve cords

Peripheral nerves

Flatworms have a cluster of neurons that form a primitive brain connected to two nerve cords that run the length of the body.

The first animals thought to have evolved primitive brains are flatworms, invertebrates that include several human parasites such as tapeworms and flukes. Their primitive brain consists of a collection of neurons at the head end of the flatworm. Flatworms also have two large nerves, called nerve cords, that extend along the body and sprout smaller nerves that reach the body tissues—a simple version of a peripheral nervous system.

As animals evolved over time, their brains became larger and more complex, allowing them to perform more-complex functions. In vertebrate animals such as fish, frogs, birds, and humans, brains and peripheral nervous systems are larger and more complex than the primitive brains of flatworms. But the brains of vertebrate organisms vary, too, which accounts for differences in computational and processing ability. The human cerebrum, for example, contains more folds than do the cerebrums of other vertebrates, for example birds. The folded structure enables more tissue to be packed into a given amount of space. In addition, the human brain is organized into hemispheres and lobes. This folded structure and complex organization account for the intellectual advantage that humans have over other organisms. In fact, the relative size and degree of folding of the cerebrum of any organism is associated with higher processing abilities. A fish cerebrum is small and has a relatively smooth surface, whereas mammalian cerebrums, such as those of humans, dolphins, and whales, are larger and contain many more folds.

Vertebrates Have Large and Complex Brains

All vertebrates have both a peripheral and a central nervous system. They have true brains that vary in size and complexity.

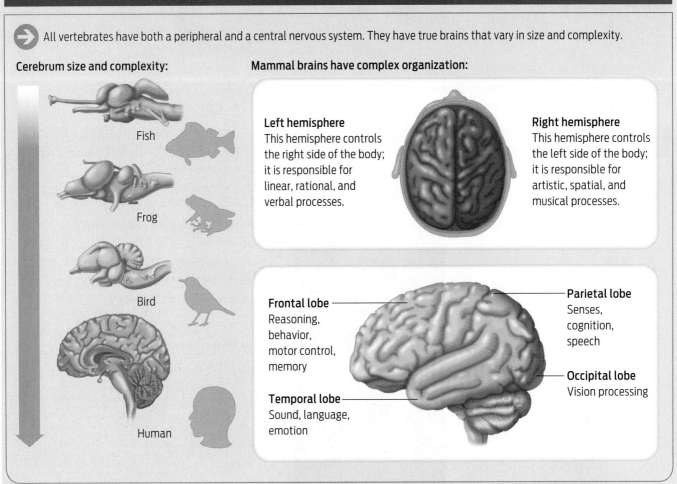

Cerebrum size and complexity:

Fish

Frog

Bird

Human

Mammal brains have complex organization:

Left hemisphere
This hemisphere controls the right side of the body; it is responsible for linear, rational, and verbal processes.

Right hemisphere
This hemisphere controls the left side of the body; it is responsible for artistic, spatial, and musical processes.

Frontal lobe
Reasoning, behavior, motor control, memory

Temporal lobe
Sound, language, emotion

Parietal lobe
Senses, cognition, speech

Occipital lobe
Vision processing

▶ Summary

■ The nervous system senses and responds to signals from the environment and coordinates bodily actions, including moving, breathing, and thinking.

■ The central nervous system (CNS) consists of the brain and the spinal cord, a thick bundle of nerves extending from the base of the brain to the lower back; the peripheral nervous system (PNS) consists of all the nerves extending from the spinal cord to the limbs and internal organs.

■ The PNS senses and responds to information both inside and outside our bodies. It includes the sensory receptors of our sensory organs, such as the eyes, ears, and skin, as well as the effectors, such as the muscles, that carry out instructions sent from the CNS.

■ Neurons are specialized cells that consist of a cell body, branched dendrites, and a long axon. Neurons conduct electrical signals known as action potentials.

■ The coordinated movement of positively charged ions causes action potentials across the neuron membrane. When an action potential reaches the end of a neuron, it causes the neuron to release neurotransmitters.

■ Neurotransmitters are chemical signaling molecules released by neuron axon terminals into the synaptic cleft of a synapse to communicate with other cells (for example, neurons, muscle cells, and endocrine gland cells). Important neurotransmitters include dopamine, serotonin, and acetylcholine.

■ Dopamine is a neurotransmitter that is involved in producing feelings of pleasure. It is one of the primary neurotransmitters involved in addiction.

■ Neurotransmitters bind to receptors on their target cells. The number of receptors can be down-regulated in response to persistently high levels of a neurotransmitter.

■ Different parts of the brain coordinate different functions. For example, decision making, learning, and memory formation; feelings of pleasure; and involuntary functions such as heart rate and breathing are all coordinated by different parts of the brain, which communicate with other brain regions to carry out most actions.

■ All organisms have ways of sensing and responding to their environment, but not all have a central nervous system with a brain and spinal cord.

NERVOUS SYSTEM AND NEURONS

The nervous system allows us to respond to stimuli and, among other functions, to learn, think, and feel emotions. Neurons in the nervous system send and receive signals.

HINT See Infographics 27.2, 27.3, and 27.5.

➡ KNOW IT

1. For each of the following, indicate whether it is a part of the CNS or of the PNS.

___ a light-detecting receptor in the eye
___ the amygdala
___ a pain receptor in the skin
___ the spinal cord
___ the thalamus

2. Neurons receive information through their
　a. axons.
　b. axon terminals.
　c. cell bodies.
　d. dendrites.
　e. nuclei.

3. Action potentials are a type of _____ signaling that relies on _____.
　a. electrical; neurotransmitters
　b. electrical; charged ions
　c. electrical; electrons
　d. chemical; neurotransmitters
　e. chemical; charged ions

4. Which part of the brain coordinates movement? Which part of the brain maintains body temperature?

➡ USE IT

5. Is information flow in the spinal cord one way or two way? Explain your answer.

6. How does multiple sclerosis cause muscle weakness? Does multiple sclerosis directly affect muscles?

7. Gatorade and other sports drinks contain replacement electrolytes (ions necessary to enable muscles to continue to contract, especially the ions lost during sweating). Which ions are likely to be in Gatorade?
　a. calcium
　b. hydrogen
　c. phosphorus and calcium
　d. potassium and sodium
　e. calcium and phosphorus

8. A blow to the brain that causes the loss of the ability to speak most likely affected the
　a. cerebellum.
　b. cerebrum.
　c. diencephalon.
　d. brain stem.
　e. hypothalamus.

COMMUNICATION BY NEURONS

Neurons transmit electrical signals along their axons and use chemical signals to communicate with other neurons and effector cells at synapses.

HINT See Infographics 27.6 and 27.7.

➡ KNOW IT

9. Neurons release neurotransmitter from their
　a. cell bodies.
　b. dendrites.
　c. axon terminals.
　d. all of the above
　e. b and c

10. What happens when a neurotransmitter is released into a synaptic cleft?

11. Compare and contrast electrical and chemical signaling by neurons.

➡ USE IT

12. Cocaine prevents dopamine from being removed from the synapse. Why does this cause feelings of pleasure?

13. Botox is a chemical treatment injected into skin to prevent wrinkling. It is primarily a bacterial toxin that prevents certain neurons from releasing the neurotransmitter acetylcholine. Acetylcholine is normally released by motor neurons to signal muscles to contract. Does Botox therefore paralyze muscles in a relaxed state or a contracted state?

14. Is more or less of the neurotransmitter acetylcholine released by the axon terminals of neurons in patients with multiple sclerosis compared to those in people who do not have multiple sclerosis? Explain your answer.

ADDICTION

Addiction is associated with alterations in brain chemistry and in the levels of certain neurotransmitters and receptors.

HINT See Infographic 27.7.

◉ KNOW IT

15. Addictive substances confer a sense of pleasure because they

    a. decrease the amount of dopamine in synaptic clefts.

    b. increase the amount of dopamine in synaptic clefts.

    c. increase the number of dopamine receptors on the axon terminals of cells that release dopamine.

    d. increase the number of dopamine receptors on dendrites of cells that release dopamine.

    e. c and d

16. Why do drug users need to take ever-increasing amounts of drugs to get the same high?

◉ USE IT

17. Would you expect a person born with a relatively low number of dopamine receptors to be happier or sadder than the average? Explain your answer.

18. Parkinson disease is caused primarily by a gradual loss of dopamine-producing neurons in the brain. Why are mood swings often among the debilitating symptoms of Parkinson disease?

SCIENCE AND SOCIETY

19. Given that many addictive substances and behaviors (for example, the nicotine in cigarettes; cocaine; gambling) act at least partially through the same biological mechanism—dopamine pathways in the brain—why do you think that each is regulated differently by the government?

20. If you were a researcher at the NIDA, how would you explain some of the broader impacts of your work to a group of taxpayers who, as one of them put it, don't understand "why my tax dollars are going to help a bunch of junkies"?

Chapter 28 Reproductive System

Too Many Multiples?

⊙ What You Will Be Learning

28.1 Female Reproductive System

28.2 Male Reproductive System

28.3 Fertilization Occurs in the Female Oviduct

28.4 In Vitro Fertilization (IVF)

28.5 Hormones Regulate the Menstrual Cycle

28.6 Hormones Support Pregnancy

28.7 How Sperm Form

28.8 Causes of Infertility

28.9 Assisted Reproductive Technologies Can Result in Multiple Births

UP CLOSE Prenatal Development

How Do Other Organisms Reproduce?

557

Too Many Multiples?

The birth of octuplets raises questions about the fertility business

The live birth of octuplets is an extremely rare occurrence, which has happened only twice in U.S. history. So the arrival of the second set in a California hospital in January 2009 was greeted with fanfare. Headlines read "Octuplets Stun Doctors" and "Eight Babies!"

But days after news of the miracle multiple birth spread worldwide, the public reception turned sour when it came to light that the 33-year-old mother, Nadya Suleman, already had six children all under the age of 7 who were, like the octuplets, born through a process of assisted reproduction called **in vitro fertilization (IVF).**

Even more disturbing to some, Suleman was an unemployed single mother on welfare. The public outcry was fierce. How could she support her children? Was she psychologically disturbed? And why had her doctor agreed to give her another round of IVF when she already had six children?

There were many more questions than answers. And far more important, the case cast a spotlight on the business of fertility treatment: in the United States fertility clinics are unregulated. Although the American Society of Reproductive Medicine (ASRM) issues guidelines on how doctors should administer fertility services, in most states there are no laws regulating what doctors can and cannot do in this regard. Though critics called Suleman's doctor irresponsible, he had not violated any laws.

Infertility treatment wouldn't be nearly as controversial if it didn't increase the odds that a woman would conceive more than one child during the treatment. "Multiples," as these babies are called, are often born prematurely. Consequently they are underweight and have underdeveloped organs, and so are at risk for birth defects such as cerebral palsy and for health problems later in life. Carrying multiples also increases the risk that the mother will develop dangerously high blood pressure, diabetes,

IN VITRO FERTILIZATION (IVF)
A form of assisted reproduction in which eggs and sperm are brought together outside the body and the resulting embryos are inserted into a woman's uterus.

Thanks to in vitro fertilization, Suleman has 14 children.

vitamin deficiencies, or one or another of several other medical conditions during her pregnancy that may affect the health of her unborn children. Even women who have twins have a higher risk of medical complications during pregnancy, and their babies are at a higher risk of premature birth than singletons. Suleman's eight children are the only surviving set of octuplets ever.

Multiple births without fertility treatment are extremely rare. Scientists estimate that the incidence of natural triplets is 1 in 6,000 to 8,000 births—of quadruplets, 1 in 500,000 births. The incidence of higher-order multiple births is even rarer. But because assisted reproduction procedures have skyrocketed, so, too, have the number of multiple births. In 2006, for example, the number of triplet births was 12 times higher than normal, according to statistics compiled by the U.S. Centers for Disease Control and Prevention. The medical complications associated with such pregnancies place a large burden on the health care system.

Although there have been efforts to improve assisted reproductive technology and thus reduce the likelihood of multiple births, extreme cases such as Suleman's have drawn lawmakers to the scene. Many states are considering legislation that would

> ## Scientists estimate that the incidence of natural triplets is 1 in 6,000 to 8,000 births—of quadruplets, 1 in 500,000 births.

place restrictions on fertility doctors. Working from another perspective, some groups are fighting for legislation that would require health insurance providers to pay for fertility treatments. Because many patients pay out of pocket, patients with limited financial resources can, because of their circumstances, pressure doctors to be aggressive with treatment, despite the health risks associated with a multiples pregnancy. The goal of this legislation is to make assisted reproduction safer for everyone, not just for those who can afford it.

The Anatomy of Sex

Up to 15% of couples have trouble becoming pregnant–perhaps a surprising statistic, considering that many women spend years trying *not* to conceive. And most couples are unaware that

> **Successful pregnancies require properly functioning sexual anatomy, balanced and precise hormone secretion, and some lucky timing.**

they have a fertility problem until they begin trying to have a child and nothing happens.

While having children is a natural part of life, the actual process of reproduction is a highly complicated affair. Successful pregnancies require properly functioning sexual anatomy, balanced and precise hormone secretion, and some lucky timing.

Suleman claimed she suffered from a medical condition that prevented her from conceiving a child naturally. If she did, she wasn't alone. In the United States, an estimated one out of eight couples experiences infertility, which is defined as the inability to conceive within a year or to bring a pregnancy to term. Many things can cause infertility: advanced age, infections, hormonal imbalances, chromosomal abnormalities, and physical blockage of reproductive passages.

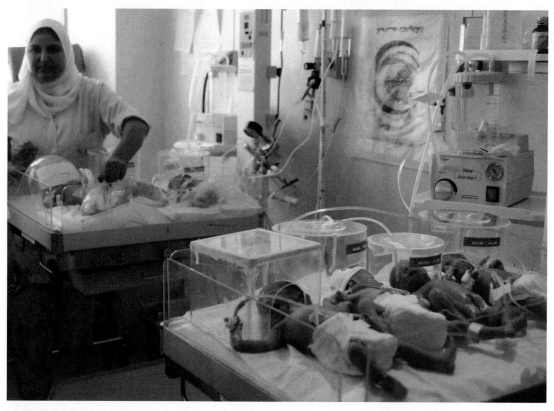

A nurse monitors the newborn septuplets of a woman who gave birth to four males and three females at the hospital of the University of Alexandria, Egypt, in 2008.

OVARIES
Paired female reproductive organs; the ovaries contain eggs and produce sex hormones.

ESTROGEN
A female sex hormone produced by the ovaries.

PROGESTERONE
A female sex hormone produced by endocrine cells in the ovaries (particularly in the cells of the corpus luteum) that prepares and maintains the uterus for pregnancy.

OVIDUCT
The tube connecting an ovary and the uterus in females. Eggs are ovulated into and fertilized within the oviducts.

UTERUS
The muscular organ in females in which a fetus develops.

ENDOMETRIUM
The lining of the uterus.

TESTES (SINGULAR: TESTIS)
Paired male reproductive organs, which contain sperm and produce androgens (primarily testosterone).

SCROTUM
The sac in which the testes are located.

TESTOSTERONE
The primary male sex hormone, which stimulates the development of masculine features and plays a key role in sperm development.

ANDROGEN
A class of sex hormones, including testosterone, that is present in higher levels in men and causes male-associated traits like deep voice, growth of facial hair, and defined musculature.

SEMINIFEROUS TUBULES
Coiled structures that constitute the bulk of the testes and in which sperm develop.

EPIDIDYMIS
Tubes in which sperm mature and are stored before ejaculation.

VAS DEFERENS
Paired tubes that carry sperm from the testes to the urethra.

URETHRA
The passageway through the penis, shared by the reproductive and urinary tracts.

SEMEN
The mixture of fluid and sperm that is ejaculated from the penis.

Men can suffer from a low sperm count or have abnormal sperm. In many cases, the reason for a couple's infertility remains unknown.

According to the ASRM, modern medicine can offer treatment to 90% of infertile couples. But fertility isn't an exact science, and treatment isn't always effective. There are many organs and hormones involved in human reproduction, and communication among them is a highly orchestrated process.

The female reproductive system consists of the **ovaries**—paired structures that hold eggs in various stages of development and that produce the sex hormones **estrogen** and **progesterone,** the major female sex hormones. Each month, one ovary typically releases one egg into an adjacent organ called the **oviduct** (also known as the fallopian tube). Eggs travel via the oviducts to the **uterus,** an elastic muscular compartment that can support a growing fetus. The uterus is lined with tissue called the **endometrium (Infographic 28.1).**

In males, paired glands called **testes** (or testicles) are contained in a sac of skin called the **scrotum.** The testes produce **testosterone,** the primary male sex hormone. Testosterone is one of the **androgens**—a class of hormones that are present in higher amounts in males than in females. Each testis contains tightly coiled **seminiferous tubules** within which sperm develop. Remarkably, each testis contains approximately 250 meters of seminiferous tubules. Sperm travel through the seminiferous tubules and enter the **epididymis,** where they mature and are stored until ejaculated. From the epididymis, sperm travel through paired tubes called the **vas deferens** and exit the body through the **urethra.** Along the way, the male reproductive system adds additional fluid to sperm to make the sperm hearty enough to survive in the female reproductive tract. Ejaculated sperm and the accompanying fluid are called **semen (Infographic 28.2).**

INFOGRAPHIC 28.1

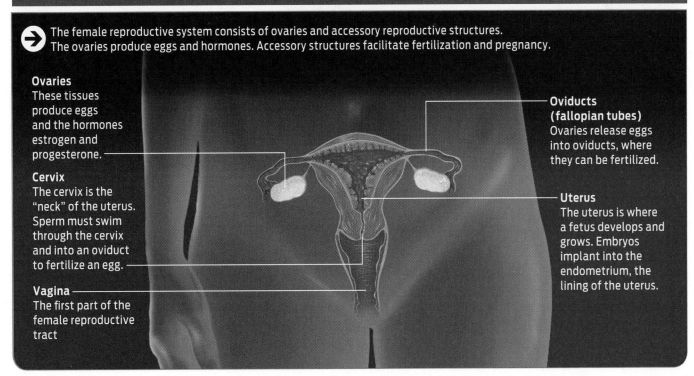

Female Reproductive System

→ The female reproductive system consists of ovaries and accessory reproductive structures. The ovaries produce eggs and hormones. Accessory structures facilitate fertilization and pregnancy.

Ovaries
These tissues produce eggs and the hormones estrogen and progesterone.

Cervix
The cervix is the "neck" of the uterus. Sperm must swim through the cervix and into an oviduct to fertilize an egg.

Vagina
The first part of the female reproductive tract

Oviducts (fallopian tubes)
Ovaries release eggs into oviducts, where they can be fertilized.

Uterus
The uterus is where a fetus develops and grows. Embryos implant into the endometrium, the lining of the uterus.

Male Reproductive System

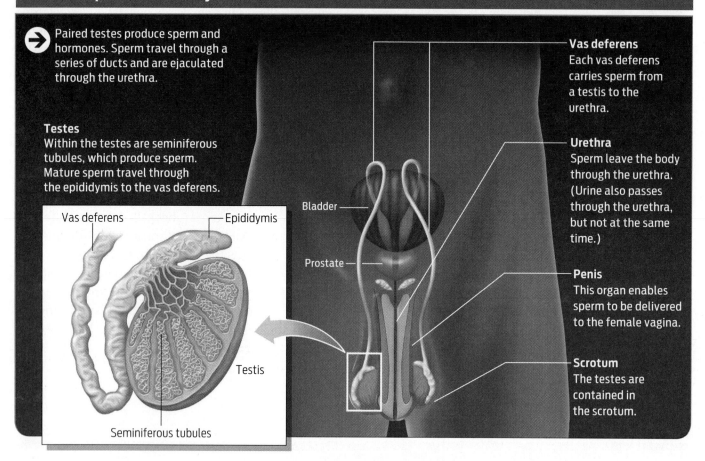

→ Paired testes produce sperm and hormones. Sperm travel through a series of ducts and are ejaculated through the urethra.

Testes
Within the testes are seminiferous tubules, which produce sperm. Mature sperm travel through the epididymis to the vas deferens.

Vas deferens

Epididymis

Testis

Seminiferous tubules

Bladder

Prostate

Vas deferens
Each vas deferens carries sperm from a testis to the urethra.

Urethra
Sperm leave the body through the urethra. (Urine also passes through the urethra, but not at the same time.)

Penis
This organ enables sperm to be delivered to the female vagina.

Scrotum
The testes are contained in the scrotum.

Conception and Infertility

During sex, when a man ejaculates into a woman's **vagina,** sperm swim up the reproductive tract, through the opening of the uterus, called the **cervix** (so called from the Latin word for "neck"), and into the oviducts. If a single sperm fuses with an egg, the egg is **fertilized (Infographic 28.3).** The fertilized egg, called a **zygote,** then travels into the uterus, where—if a successful pregnancy is to occur—it must implant into the uterine wall and eventually develop into a fetus.

That's the way normal conception happens—but several things can go wrong along the way. Physical damage to the reproductive organs can prevent fertilization. In some cases, a woman's oviducts can be blocked or damaged, preventing eggs from entering the uterus or sperm from getting into the oviduct, where fertilization normally takes place. The most common cause of blocked tubes is pelvic inflammatory disease, which can be caused by sexually transmitted diseases such as chlamydia, a bacterial infection. An untreated infection can cause scar tissue to build up in the oviducts and block them.

In men, any obstructions in the vas deferens or epididymis will block sperm transport. Varicose veins in the testicles and sexually transmitted bacterial infections such as chlamydia or gonorrhea can also cause blocked tubes.

Fertility clinics can test for physical blockages and, in some cases, surgically correct them. When surgery isn't an option, IVF is often recommended. In IVF, hormones are

VAGINA
The first part of the female reproductive tract, extending up to the cervix; also known as the birth canal.

CERVIX
The opening or "neck" of the uterus, where sperm enter and babies exit.

FERTILIZATION
The fusion of an egg and a sperm to form a zygote.

ZYGOTE
A fertilized egg.

Fertilization Occurs in the Female Oviduct

During intercourse, sperm ejaculated from the penis enters the female reproductive tract. Sperm must swim through the tract into the oviducts to fertilize an ovulated egg. While many sperm may make it to the oviduct, only one will actually fertilize the egg. Blockages in the female reproductive tract can impede sperm passage to an egg, and consequently compromise fertilization.

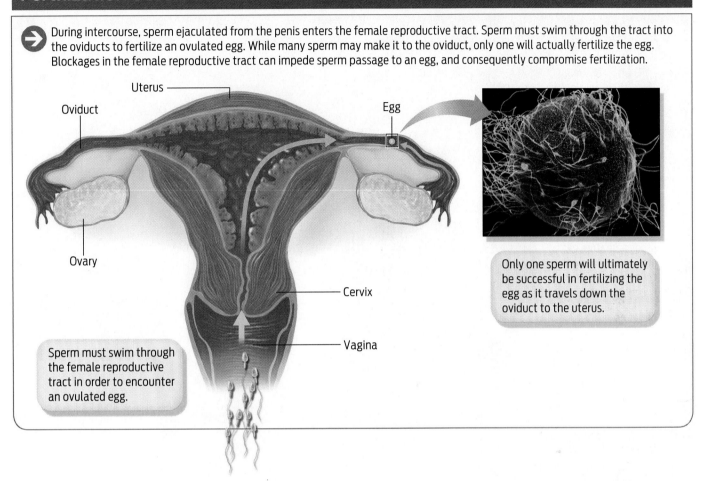

Uterus

Oviduct

Egg

Ovary

Cervix

Vagina

Only one sperm will ultimately be successful in fertilizing the egg as it travels down the oviduct to the uterus.

Sperm must swim through the female reproductive tract in order to encounter an ovulated egg.

EMBRYO
An early stage of development, an embryo forms when a zygote undergoes cell division.

administered to a woman to promote egg development, then eggs are extracted from her ovaries through a long, thin needle inserted through the vagina. Sperm are extracted from ejaculate or, in cases of physical blockage, from the epididymis. The sperm and eggs are combined outside the body in a petri dish to allow the sperm to fertilize the eggs. The resulting **embryos** are then inserted into the woman's uterus in hope that at least one will develop into a fetus **(Infographic 28.4)**. IVF has been used to help infertile couples conceive children since 1978, when Louise Brown, the first IVF baby, was born.

IVF may also be recommended in cases of low sperm number, abnormal sperm, and even in cases in which the cause of infertility can't be determined. In such cases the term "infertility" can actually be a misnomer. Many couples with defective sperm or unexplained infertility can still conceive a child naturally—it just may take longer. But because no one can predict how long it might take to achieve a successful pregnancy and because fertility decreases with age, IVF makes conception more likely by bringing sperm and egg together artificially.

Hormones and Fertility

Physical blockages and sperm quality account for only about 30% to 40% of all cases of infertility; another 20% to 30% are caused by hormonal

In Vitro Fertilization (IVF)

→ In vitro fertilization involves extracting eggs and sperm and mixing them together outside the body to allow fertilization. The resulting embryos are then inserted back into a woman's uterus in hope that at least one of them will implant into the uterus and grow into a fetus.

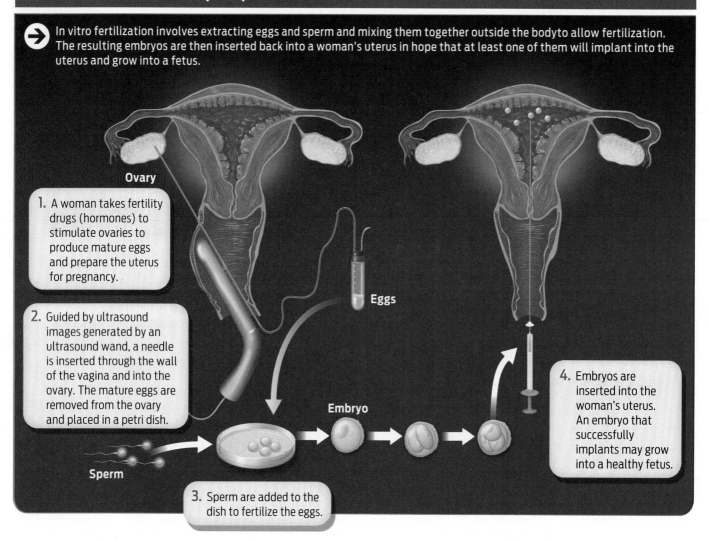

Ovary

1. A woman takes fertility drugs (hormones) to stimulate ovaries to produce mature eggs and prepare the uterus for pregnancy.

2. Guided by ultrasound images generated by an ultrasound wand, a needle is inserted through the wall of the vagina and into the ovary. The mature eggs are removed from the ovary and placed in a petri dish.

Eggs

Embryo

Sperm

3. Sperm are added to the dish to fertilize the eggs.

4. Embryos are inserted into the woman's uterus. An embryo that successfully implants may grow into a healthy fetus.

imbalances. Hormones are responsible for regulating the production of gametes, both sperm and egg. In females, estrogen and progesterone are key reproductive hormones without which a woman would be infertile. In males, testosterone is the primary hormone that stimulates sperm to develop.

In females, estrogen and progesterone are responsible for the menstrual cycle, a reproductive cycle that repeats roughly once every 28 days after the onset of puberty. During each cycle, estrogen and progesterone levels rise and fall to trigger the ovaries to release an egg and

prepare a woman's uterus for pregnancy should an egg be fertilized.

The hypothalamus controls levels of estrogen and progesterone in the body. The hypothalamus is the master coordinator of a number of important physiological functions, including hunger, thirst, and temperature regulation–in addition to reproduction. To regulate fertility, the hypothalamus secretes hormones that signal the **anterior pituitary gland** to produce two other key hormones. These two hormones, follicle-stimulating hormone and luteinizing hormone, travel through

ANTERIOR PITUITARY GLAND
The gland in the brain that secretes luteinizing hormone (LH) and follicle-stimulating hormone (FSH).

the bloodstream and directly stimulate the ovaries.

Follicle-stimulating hormone (FSH) acts on structures in the ovaries called **follicles,** each of which contains an immature egg. FSH signals follicles in the ovary to enlarge and to produce estrogen. Estrogen has several effects, one of which is to cause the endometrium to start to thicken. Estrogen also stimulates eggs within the ovaries to mature.

As hormone-secreting structures, the hypothalamus, the anterior pituitary, and the ovaries are all part of the endocrine system. All endocrine glands secrete hormones into the circulation. Hormones then travel though the bloodstream to reach their target cells. Hormones interact with their target cells by binding to specific receptors present on the target cells. For example, FSH binds to specific receptors on certain ovarian cells, triggering them to enlarge and release estrogen.

At about the midpoint of the cycle–roughly 10 to 14 days after a woman begins menstrual bleeding–rising estrogen levels trigger the brain to release a large amount of **luteinizing hormone (LH).** This LH surge triggers **ovulation**–the release of an egg from the follicle and the ovary itself into the oviduct. After the egg has been ovulated, the remaining follicle becomes a structure called the **corpus luteum,** which secretes progesterone. One of the most important roles of progesterone is to encourage the endometrium to continue to thicken. The thickened endometrium contains blood vessels and nutrients and is prepared to receive an embryo if the egg is fertilized.

Although both ovaries can release eggs during the same cycle, they typically take turns and only one egg is released during each cycle (identical twins occur when a single egg is released and fertilized and then splits into two embryos early in embryonic development). In about 1% of cycles, however, there are multiple ovulations, in which case fraternal twins, triplets–or as many embryos as there are ovulations–can develop.

Ovulation presents a crucial time window during which a woman can become pregnant. Sperm must swim through the cervix and uterus and into the oviduct containing the released egg in order to fertilize it. Because sperm can survive in the female reproductive tract anywhere from 3 to 7 days, a woman can become pregnant even if she has sex before she ovulates. Sperm can in effect "wait" in the oviduct for ovulation to occur. Once the egg leaves the oviduct, however, the odds that it will be fertilized are extremely small.

If an egg is not fertilized within 24 hours of being ovulated, it is no longer viable. The corpus luteum degenerates at about day 26 of the cycle, progesterone levels drop, and the uterine lining sloughs off, leading to **menstruation (Infographic 28.5).**

If the egg is fertilized it is called a zygote. As the zygote travels to the uterus it begins dividing and developing into an embryo. The embryo implants itself in the endometrium of the uterus about 1 week after the egg is fertilized. Once implanted in the uterus, the embryo secretes a hormone called **human chorionic gonadotropin (hCG),** which signals the corpus luteum to continue producing progesterone to support the thickening endometrium. This is the hormone that, in effect, tells the reproductive system that pregnancy has begun (hCG is the hormone that most pregnancy tests use as a marker to detect a pregnancy). Once the embryo implants, embryonic and maternal endometrial tissues interact to form the **placenta,** a disc-shaped structure that provides nourishment and support (like waste removal) to the developing fetus. In addition to its role in delivering oxygen, nutrients, and other key molecules (like antibodies from the mother that help protect the embryo against infections), the placenta eventually takes over the task of producing estrogen and progesterone from the corpus luteum. The ovaries stop responding to follicle-stimulating hormone and luteinizing hormone at the time of menopause, which in American women occurs at about age 51; at this time women stop

Hormones Regulate the Menstrual Cycle

→ A complex interplay of hormones from the hypothalamus, anterior pituitary gland, and ovaries drives the monthly female reproductive cycle.

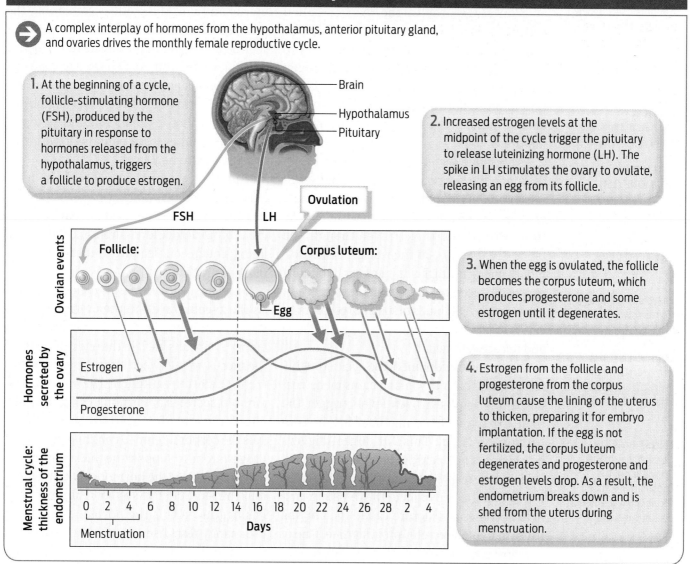

1. At the beginning of a cycle, follicle-stimulating hormone (FSH), produced by the pituitary in response to hormones released from the hypothalamus, triggers a follicle to produce estrogen.

Brain
Hypothalamus
Pituitary

2. Increased estrogen levels at the midpoint of the cycle trigger the pituitary to release luteinizing hormone (LH). The spike in LH stimulates the ovary to ovulate, releasing an egg from its follicle.

Ovulation

FSH
LH

Follicle:
Corpus luteum:
Egg

Ovarian events

3. When the egg is ovulated, the follicle becomes the corpus luteum, which produces progesterone and some estrogen until it degenerates.

Hormones secreted by the ovary

Estrogen

Progesterone

4. Estrogen from the follicle and progesterone from the corpus luteum cause the lining of the uterus to thicken, preparing it for embryo implantation. If the egg is not fertilized, the corpus luteum degenerates and progesterone and estrogen levels drop. As a result, the endometrium breaks down and is shed from the uterus during menstruation.

Menstrual cycle: thickness of the endometrium

0 2 4 6 8 10 12 14 16 18 20 22 24 26 28 2 4
Menstruation
Days

---

ovulating and stop having monthly reproductive cycles, and are therefore no longer able to conceive (**Infographic 28.6**).

Since hormones play such a crucial role in pregnancy and reproduction, many types of **contraception** make use of this phenomenon and interfere with the normal female hormone cycle (**Table 28.1**). In the combination birth control pill that contains both estrogen and progesterone, for example, these hormones are at levels that prevent the anterior pituitary gland from releasing follicle-stimulating and luteinizing hormones. This prevents ovulation

and consequently a woman taking this pill does not release eggs to be fertilized. Progesterone in birth control pills prevents successful pregnancy in other ways, too: it thickens the cervical mucus, blocking sperm from entering the uterus and oviducts, and also reduces endometrial thickening—a process that is necessary to support an embryo.

Men do not have a monthly hormone cycle, but beginning at puberty, sperm go through recognizable stages of development. The seminiferous tubules house precursor sperm cells that go through cell division—that is,

CONTRACEPTION
The prevention of pregnancy through physical, surgical, or hormonal methods.

Hormones Support Pregnancy

Estrogen and progesterone support the implanted embryo as it develops. Early in pregnancy, the embryo secretes human chorionic gonadatropin (hCG), which signals the corpus luteum to continue to produce these hormones. The placenta, once it has formed, takes over estrogen and progesterone production.

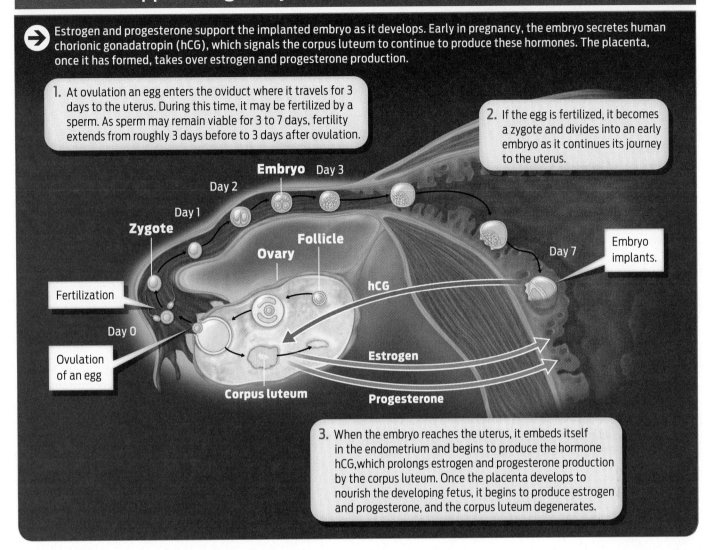

1. At ovulation an egg enters the oviduct where it travels for 3 days to the uterus. During this time, it may be fertilized by a sperm. As sperm may remain viable for 3 to 7 days, fertility extends from roughly 3 days before to 3 days after ovulation.

2. If the egg is fertilized, it becomes a zygote and divides into an early embryo as it continues its journey to the uterus.

Embryo Day 3
Day 2
Day 1
Zygote
Follicle
Ovary
hCG
Day 7
Embryo implants.
Fertilization
Day 0
Ovulation of an egg
Estrogen
Corpus luteum
Progesterone

3. When the embryo reaches the uterus, it embeds itself in the endometrium and begins to produce the hormone hCG, which prolongs estrogen and progesterone production by the corpus luteum. Once the placenta develops to nourish the developing fetus, it begins to produce estrogen and progesterone, and the corpus luteum degenerates.

meiosis—and specialization to produce mature sperm cells. It takes approximately 6 weeks for sperm to mature. Maturation is stimulated by testosterone, and although men produce testosterone continuously throughout their adult lives, they produce less of the hormone as they age and consequently sperm quality declines over time (Infographic 28.7).

The Many Causes of Infertility

Given how important hormones are to egg maturation and ovulation, it's not surprising that hormonal imbalances are a common cause of female infertility. Disruption in the part of the brain that regulates ovulation can cause low levels of luteinizing hormone and follicle-stimulating hormone, which in turn may prevent ovulation or make it erratic. Even slight irregularities in the hormone system can prevent the ovaries from releasing eggs. Specific causes of such hormonal imbalances include injury, tumors, excessive exercise, and starvation. Some medications are associated with ovulation disorders, and some studies have shown that stress can negatively affect fertility, as can poor nutrition.

TABLE 28.1

Contraception

 There are many ways to prevent conception. Currently available contraceptives include behavioral methods, physical and chemical barriers, hormones, and surgery.

METHOD	FAILURE RATE (% OF PREGNANCIES/ YEAR WITH EACH METHOD)	DESCRIPTION
No Contraception	85%	Sexual intercourse without any method of contraception.
Behaviors	3–27% 0% (abstinence)	The rhythm method (avoiding intercourse around the time a woman ovulates); withdrawal (the male withdraws his penis before ejaculating); sexual abstinence.
Barriers	2–21%	The male and female condom, the diaphragm, and the cervical cap. All of these prevent sperm from entering the uterus and are typically used with spermicidal foams or jellies that contain sperm-inactivating chemicals.
Hormones	0.3–8%	Female hormonal contraceptives contain a combination of synthetic estrogen and progesterone or progesterone only. Women can take these hormones in the form of combination pills, a skin patch, a cervical ring, a minipill, a regular injection, or an implant. All hormonal methods prevent pregnancy in three major ways: they thicken the cervical mucus, making it less likely that sperm will be able to swim through it and get into the oviducts; they prevent ovulation; and they thin the endometrium, so it cannot support the implantation of an embryo.
Surgery	0.5–0.15%	Surgical options include a vasectomy for men and tubal ligation for women. In a vasectomy, the vas deferens is cut, so sperm can no longer be ejaculated. In tubal ligation, the oviducts are cut and tied off, so sperm can no longer reach an ovulated egg. Both surgeries are essentially permanent, although they can be reversed with limited success. Those who do not want to have children, or any more children, typically choose surgical sterilization.
Intrauterine device	0.2–0.8%	An intrauterine device (IUD) is a small T-shaped device that typically contains copper or another metal. It is inserted into the base of the uterus through the cervix. The IUD is a long-term contraceptive option. It prevents pregnancy by thickening cervical mucus and consequently impeding sperm from entering the uterus. It also weakens the endometrium, making it less able to support an embryo.

Source: Trussell J. Contraceptive efficacy. In Hatcher RA, Trussell J, Nelson AL, Cates W, Stewart FH, Kowal D. *Contraceptive Technology: Nineteenth Revised Edition.* New York NY: Ardent Media, 2007.

How Sperm Form

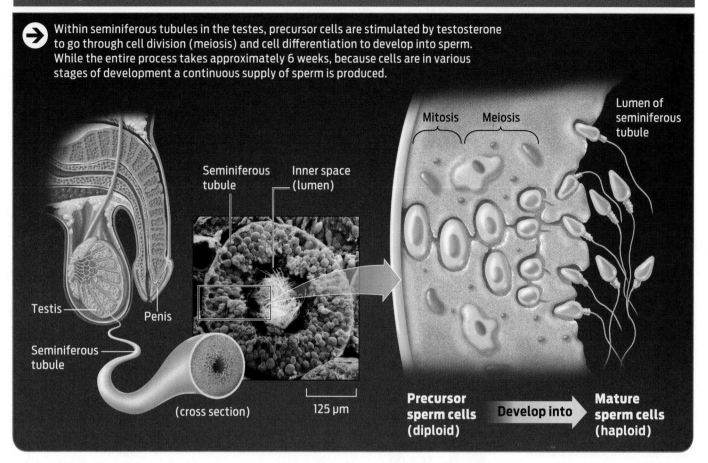

Within seminiferous tubules in the testes, precursor cells are stimulated by testosterone to go through cell division (meiosis) and cell differentiation to develop into sperm. While the entire process takes approximately 6 weeks, because cells are in various stages of development a continuous supply of sperm is produced.

Testis

Penis

Seminiferous tubule

(cross section)

Seminiferous tubule

Inner space (lumen)

125 μm

Mitosis   Meiosis

Lumen of seminiferous tubule

Precursor sperm cells (diploid) → **Develop into** → **Mature sperm cells (haploid)**

Some women suffer from polycystic ovary syndrome, a condition associated with the production of excessive amounts of androgens (the "male" sex hormones, typically made in only small amounts in females). This syndrome is one of the most common hormonal disorders, affecting an estimated 10% of women of reproductive age. In addition to affecting ovulation, the condition is associated with ovarian cysts, irregular menstrual cycles, diabetes, and obesity.

Although men, too, can suffer hormonal imbalances, these are much less common than in women. Males may have low testosterone levels if they have a pituitary tumor that causes a reduction in the levels of FSH and LH (which in men are required to stimulate sperm and testosterone production by the testes), or some kind of damage to the testes. Injury, che-

motherapy, and radiation can also interfere with testosterone production. Low testosterone levels reduce both sperm production and libido.

Overall, there is no single factor that is the most common cause of infertility, but rather a combination of physical and hormonal abnormalities that together contribute to the 10% to 15% infertility rate we see in the U.S. population **(Infographic 28.8)**. And there remains a small percentage of cases, approximately 10% of all infertile couples, in which the cause of infertility remains unknown.

The Problem of Multiples

Women with polycystic ovary syndrome or other hormonal disorders who wish to become pregnant are typically prescribed a course of

Causes of Infertility

In women:

Nonfunctional ovaries
Ovaries may fail to ovulate because of a variety of causes, including hormonal imbalances, genetic abnormalities, undeveloped ovarian tissue, endometriosis, and cancer.

Cysts, fibroids, polyps
Each of these is a type of abnormal growth that may block passages or interfere with normal function of the tissue.

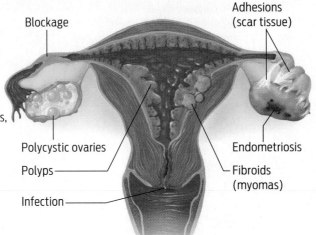

Blockage

Adhesions (scar tissue)

Polycystic ovaries

Polyps

Infection

Endometriosis

Fibroids (myomas)

Endometriosis
The tissue that lines the uterus grows abnormally and invades other tissues in the pelvis. The wayward tissue irritates the nerve endings of these organs and interferes with their function.

Nonreceptive fluids
Cervical mucus may be nonreceptive to sperm. Antisperm chemicals may be secreted.

Blockages
Passages may become blocked or disabled in both male and female reproductive systems because of tissue scarring, infection, cancer, or abnormal tissue growth.

Prostatitis
Sperm pass through and receive fluid from the prostate on their way into the urethra. An enlarged prostate can block the passage of semen.

Sperm abnormalities
Men may have low numbers of healthy sperm and/or physically abnormal sperm.

In men:

Erectile dysfunction
Genetic abnormalities, neurological problems, hormonal imbalances, and physical blockages inhibit the ability of blood to flood the penis tissue to support an erection.

Testicular varicose veins
Valves in the veins that keep blood flowing in one direction deteriorate, causing blood to back up and pool. These enlarged veins can interfere with sperm production and transport.

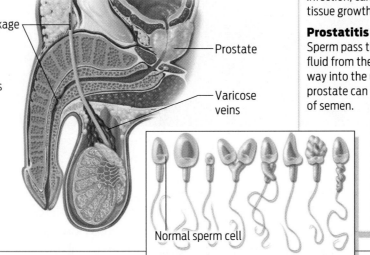

Blockage

Prostate

Varicose veins

Normal sperm cell

---

fertility drugs. These are usually anterior pituitary hormones to stimulate multiple ovarian follicles to develop and multiple eggs to ovulate. Then, in a form of artificial insemination called **interuterine insemination,** sperm are injected into the uterus. Ultrasound technology is used to monitor ovulation, so the insemination can be performed when chances are highest that one or more eggs have been released into the oviduct.

Fertility drugs almost always cause two or more ovulations. This is desirable during IVF,

in which the number of eggs that are fertilized outside the body and implanted into the uterus can be controlled. But this advantage turns into a liability with intrauterine insemination. Since sperm are inserted into the uterus during insemination, it's difficult to control the number of eggs that are fertilized and, consequently, the number of conceptions. Multiple births result when more than one egg is fertilized (each by a different sperm), leading to more than one embryo implanting in the uterus and developing into a fetus (**Infographic 28.9**).

INTRAUTERINE INSEMINATION (IUI)
A form of assisted reproduction in which sperm are injected directly into a woman's uterus.

Assisted Reproductive Technologies Can Result in Multiple Births

One hazard of assisted reproduction is a high probability of multiple births. Babies born as multiples are more likely to be born underweight and premature, putting them at risk for a variety of serious health conditions.

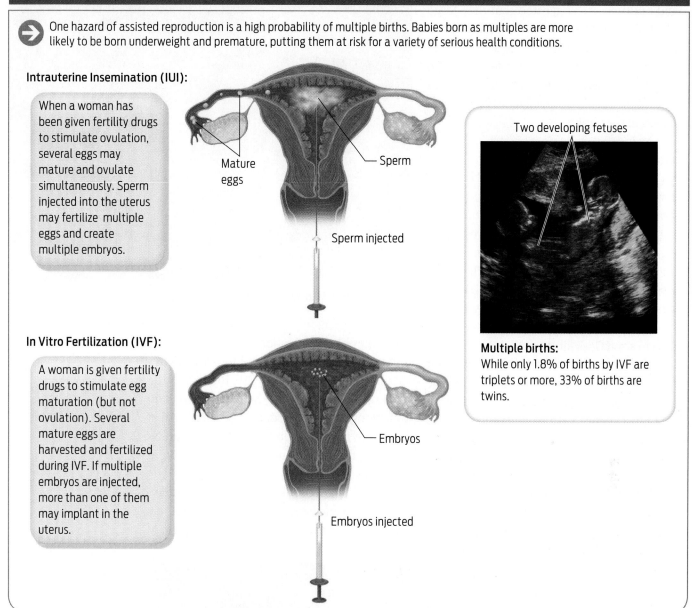

Intrauterine Insemination (IUI):

When a woman has been given fertility drugs to stimulate ovulation, several eggs may mature and ovulate simultaneously. Sperm injected into the uterus may fertilize multiple eggs and create multiple embryos.

Mature eggs

Sperm

Sperm injected

In Vitro Fertilization (IVF):

A woman is given fertility drugs to stimulate egg maturation (but not ovulation). Several mature eggs are harvested and fertilized during IVF. If multiple embryos are injected, more than one of them may implant in the uterus.

Embryos

Embryos injected

Two developing fetuses

Multiple births:
While only 1.8% of births by IVF are triplets or more, 33% of births are twins.

In Suleman's case IVF treatment was too effective. Fertility treatment was also too effective for Jon and Kate Gosselin, the famous couple who starred in the television reality series *Jon & Kate Plus 8*. The couple first had twins and then later sextuplets, all of whom were conceived through intrauterine insemination.

The book *Multiple Blessings,* by the Gosselins and Beth Carson, tells the story. Kate's first treatment consisted of "painful injections" of fertility drugs to stimulate her ovaries. She ended up having twins. A year later, Kate longed for another baby. This time, Kate saw a different specialist. Kate and her husband told their doctor that they did not want multiples but would not selectively "reduce," or abort a fetus. She received another round of fertility drugs.

When Kate had an ultrasound examination during the treatment, the Gosselins saw three, possibly four, developing eggs. Their doctor told them that they were unlikely to end up with three or four babies, but to be sure to avoid multiples he suggested that they forgo intrauterine insemination then and try again later. They decided to go ahead with the insemination anyway. When the sonogram showed seven developing embryos (one would later disappear), Jon dropped to his knees, Kate recalled in the book.

Costly Care

Most insurance policies don't cover the cost of assisted reproduction, and many patients pay out of pocket. Because the costs of treatment are so high, many doctors find themselves under pressure to be aggressive with treatment with patients who have limited resources. Therefore, financial considerations can dictate a course of treatment, even when the treatment is unlikely to be successful or to be too successful, as in Suleman's and Gosselin's cases.

As of 2010, a single insemination treatment cost an average of $6,000–almost half the cost of an average in vitro fertilization cycle. Since most health insurance companies do not cover assisted reproduction, couples are often left to decide on a procedure based on how much they can afford. Many couples may choose insemination over in vitro fertilization simply because it is less expensive, even though the procedure has a low success rate. And when it is successful, it carries a high risk of multiples.

The average cost of a single treatment is more than $10,000. Since each round of IVF carries only about a 40% success rate according to ASRM, some couples must undergo multiple rounds to achieve a successful pregnancy.

The success rate of a single round of intrauterine insemination is much lower. On average, only 10% to 20% of all insemination treatments result in a live birth. But younger women and women who take fertility drugs and, as a result, have multiple ovulations, tend to have higher success rates. This last point is key: multiple ovulations double the odds of a successful

Kate Gosselin and her eight children.

pregnancy by insemination. Similarly, the more embryos that are transferred into a woman's uterus after each round of IVF, the higher the odds of a pregnancy. In the end, doctor and patient together should decide which course of treatment to pursue on the basis of a couple's specific fertility condition and financial resources.

Although most of Suleman's health and financial records have remained confidential, it's unlikely that the unemployed 33-year-old could

Suleman (right) with some of her 14 children.

have afforded to pay for multiple rounds of IVF. The Associated Press and various newspapers have reported that Suleman received at least $165,000 in disability benefits from the state of California. But considering she already had six children to care for and had been unemployed for many years, it's unclear how her second round of IVF (which resulted in eight babies) was paid for.

Medical details revealed during a court case brought by the California Medical Board against Suleman's doctor, Michael Kamrava, show that Kamrava created 14 embryos and implanted a dozen of them. Only eight survived, and the babies were born 9 weeks prematurely. The Associated Press reported that during the hearing Kamrava said he regretted implanting the 12 embryos and "would never do it again."

Kamrava further stated that Suleman was adamant about using all 12 embryos, even though he suggested implanting only 4. "She just wouldn't accept doing anything else with those embryos. She did not want them frozen, she did not want them transferred to another patient in the future," he said, according to the AP story. Kamrava said that he consented only after Suleman agreed to have a fetal abortion if necessary. After the implantation, however, he only heard from Suleman after the birth of her octuplets, despite his numerous attempts to contact her.

From a medical perspective, the birth of multiples from any assisted reproduction procedure is problematic. According to the March of Dimes, 50% of twins and 90% of triplets are born prematurely, as are virtually all quads and quintuplets. Their lungs are often immature and so the babies must be hooked up to mechanical breathing ventilators, which sometimes scar the lungs so that these children will for the rest of their lives be prone to asthma, pneumonia, chronic lung disease, and other respiratory disorders. And because their brains aren't fully developed, premature babies are susceptible to brain hemorrhages and to developmental difficulties, including learning disabilities (see **Up Close: Prenatal Development**). These medical conditions pose a huge burden on the health care system.

To reduce the birth of multiples and prevent the associated health problems, the American Society for Reproductive Medicine recommends transferring no more than two embryos in women under 35 years of age (and only one if possible) and no more than five in women over 40, because as a woman ages the success rate tends to drop (Suleman had six times the number of recommended embryos implanted). These guidelines appear to have been effective: in 2007, only 1.8% of live births to patients under 35 were triplets or more, as compared to 6.4% in 2003. But assisted reproduction still results in a high number of twins—in 2007, 33% percent of births from IVF alone were twins.

Despite the drop in multiple births, some lawmakers want more control. The risk that doctors may act against the ASRM guidelines, with negative consequences, remains. The California Medical Board continues to seek to revoke Kamrava's medical license, claiming that he was negligent not only in Suleman's case, but also in the cases of two other women who suffered serious medical complications because of aggressive fertility treatments. Lawmakers are concerned that there is nothing to prevent cases like these from happening again. Several states are considering legislation that would regulate doctors. A Missouri bill, for example, would require doctors not to exceed the ASRM's embryo-transfer guidelines. But experts fear that regulation would hinder good care. Fertility doctors need flexibility to tailor treatment to a couple's individual condition of fertility, they argue. Such legislation "seeks to substitute the judgment of politicians for that of physicians and their patients," ASRM president R. Dale McClure has said in a statement evaluating a similar bill.

Instead of putting controls on doctors, some groups favor regulation that would broaden access to treatment—legislation that would require health insurers to cover infertility diagnostics and treatment, for example. As long as fertility treatment continues to be a financial burden on couples, fertility doctors will face pressure from patients to give them the most for their money, says Barbara Collura, executive director of RESOLVE, an infertility advocacy

The embryonic stage begins 1 week after the embryo implants and continues until the 8th week after fertilization. After that, the developing baby is known as a fetus. Pregnancy is divided into three trimesters.

1st Trimester: development of tissue layers and vital organs

Embryonic stage

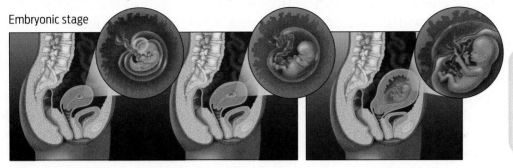

| Month 1 | Month 2 | Month 3 |

The first trimester includes the embryonic stage of development and the early fetal stage. The embryo and fetus grow rapidly and critical organs develop.

2nd Trimester: growth and gender determination

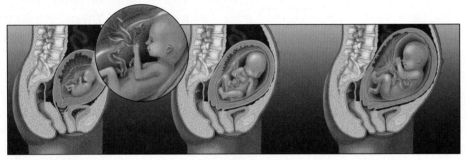

| Month 4 | Month 5 | Month 6 |

During the second trimester, the fetus continues to grow and develop features, including external genitalia.

3rd Trimester: weight gain and organ system development

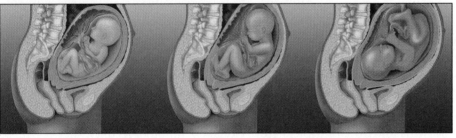

| Month 7 | Month 8 | Month 9 |

During the third trimester the fetus becomes fully developed and gains sufficient weight to be born at full term about 35 weeks after fertilization.

---

group. Broad health insurance coverage would eliminate the cost factor. The result might be that couples would be able to forgo treatments such as insemination that have high risks with low success rates and skip directly to IVF when appropriate. Cost would not be a factor when considering treatment options.

"If nothing else, these high profile cases have served as a wake-up call," says Collura. The community of health care workers is examining its own procedures and methods because it would rather self-regulate than have regulation imposed on it from the outside, she adds. "After Suleman, the community is really looking into how it [the birth of octuplets] happened and how it can prevent it from happening again." ∎

> **"If nothing else, these high profile cases have served as a wake-up call."**
> — Barbara Collura

How Do Other Organisms Reproduce?

All organisms reproduce, but not all organisms have sex. Reproduction is a fundamental facet of all life–without the ability to reproduce any species would die out. Humans reproduce sexually, which means that reproduction involves two parents: one provides eggs, and the other provides sperm. Not all organisms reproduce in the same way, nor do they have the same sexual anatomy as humans.

In humans and other mammals, fertilization of an egg by sperm occurs internally, in a female oviduct. This system allows sperm and egg to be protected but limits the number of eggs that can be fertilized at any one time and requires a "sperm delivery system" (that is, a penis) to place the sperm in a location where they can easily swim to the egg.

Many nonmammalian species also rely on internal fertilization, but via a different sexual anatomy. Internal fertilization in birds and reptiles, for example, occurs in an organ called the cloaca (plural: cloacae). The cloaca is the shared opening for solid waste, urine, and the reproductive system. Birds briefly touch their cloacae together (in the so-called cloacal kiss) and this speedy kiss-and-run is sufficient for sperm to be transferred from the male to the female.

Birds Reproduce Sexually and Fertilize Eggs Internally

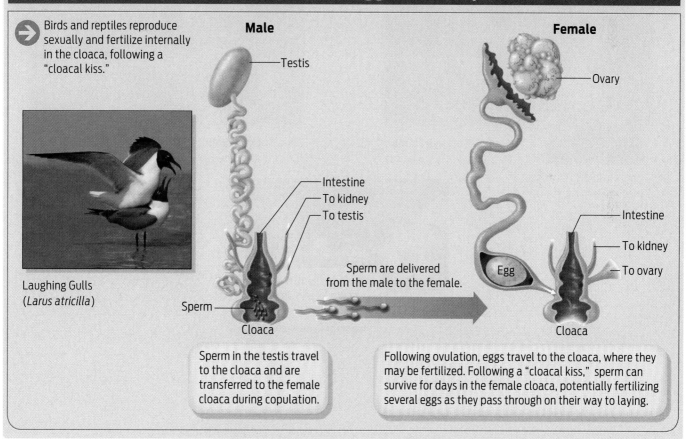

Birds and reptiles reproduce sexually and fertilize internally in the cloaca, following a "cloacal kiss."

Male

Testis

Intestine
To kidney
To testis

Sperm

Cloaca

Laughing Gulls
(*Larus atricilla*)

Sperm are delivered from the male to the female.

Female

Ovary

Intestine

To kidney

To ovary

Egg

Cloaca

Sperm in the testis travel to the cloaca and are transferred to the female cloaca during copulation.

Following ovulation, eggs travel to the cloaca, where they may be fertilized. Following a "cloacal kiss," sperm can survive for days in the female cloaca, potentially fertilizing several eggs as they pass through on their way to laying.

Interestingly, as embryos human females have a cloaca, but the single cloacal opening divides into separate openings during embryonic development. In very rare cases, baby girls are born with an intact cloaca–having only one opening, rather than the usual three. Surgical correction to form distinct anal, vaginal, and urethral openings is possible in these cases.

In contrast, many organisms rely on external fertilization–their gametes fuse in the outside environment. Female salmon, for example, deposit their eggs in gravel nests in streambeds. Male salmon then swim over the eggs and release sperm, fertilizing thousands of eggs simultaneously. Other aquatic species such as coral and hydra also reproduce this way. Many plant species also practice external fertilization.

Fish Reproduce Sexually and Fertilize Eggs Externally

Organisms like fish reproduce sexually, with two parents each contributing a gamete. The gametes combine externally during fertilization.

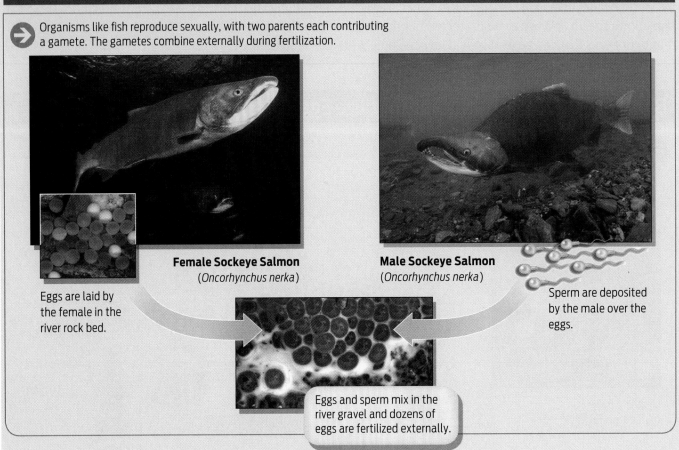

Female Sockeye Salmon
(*Oncorhynchus nerka*)

Male Sockeye Salmon
(*Oncorhynchus nerka*)

Eggs are laid by the female in the river rock bed.

Sperm are deposited by the male over the eggs.

Eggs and sperm mix in the river gravel and dozens of eggs are fertilized externally.

Still other species use an entirely different form of reproduction altogether: they reproduce asexually. In asexual reproduction, a single parent produces offspring without additional genetic input from another individual. All bacteria reproduce asexually. Some organisms, such as certain types of fungi, can reproduce both sexually and asexually. Baker's yeast (*Saccharomyces cerevisiae*) is a fungus that is commonly used to make bread rise. *S. cerevisiae* can produce gametes called spores, and these gametes can fuse to generate zygotes that develop into unicellular yeast. But both the spores and the yeast can also make exact copies of themselves by mitotic cell division, resulting in identical populations of cells.

Why is there so much variation in the way that organisms reproduce? The short answer, as always, is evolution. Natural selection has favored adaptations that allow organisms to thrive in their particular environments. For example, while asexual reproduction produces offspring that are genetically identical to one another and to the parent, this type of reproduction is generally rapid and allows populations to grow quickly, which is an advantage if you are a bacterium or have a short life span or are easily killed.

Bacteria Reproduce Asexually

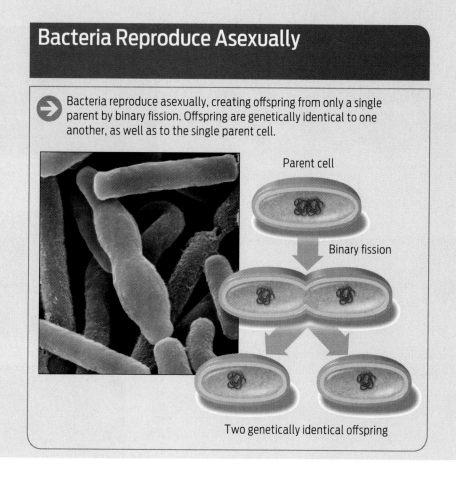

→ Bacteria reproduce asexually, creating offspring from only a single parent by binary fission. Offspring are genetically identical to one another, as well as to the single parent cell.

Parent cell

Binary fission

Two genetically identical offspring

Relative to external fertilization, internal fertilization places a large demand on the mother because she not only supplies all nutrients to a growing fetus, but also provides a protective environment within her body during the long time it takes for a fetus to grow into a baby. However, this maternal investment, while leaving the mother vulnerable during the pregnancy, often results in the successful live birth of a baby from each egg that is fertilized, compared to the exposed embryos that result from external fertilization. Over time, different organisms have adapted different reproductive strategies to ensure that their offspring survive in the particular environment each organism occupies.

Yeast Reproduce Both Sexually and Asexually

Yeast can reproduce asexually by going through mitosis. They can also reproduce sexually by producing spores.

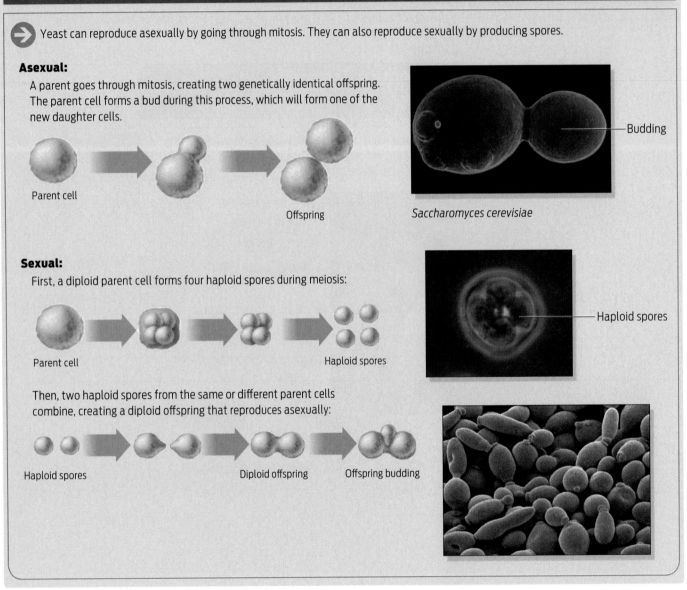

Asexual:

A parent goes through mitosis, creating two genetically identical offspring. The parent cell forms a bud during this process, which will form one of the new daughter cells.

Parent cell

Offspring

Budding

Saccharomyces cerevisiae

Sexual:

First, a diploid parent cell forms four haploid spores during meiosis:

Parent cell

Haploid spores

Haploid spores

Then, two haploid spores from the same or different parent cells combine, creating a diploid offspring that reproduces asexually:

Haploid spores   Diploid offspring   Offspring budding

▶ Summary

■ The female reproductive system consists of paired ovaries, which produce eggs and secrete the hormones estrogen and progesterone, as well as accessory structures that enable fertilization and support pregnancy.

■ The male reproductive system consists of paired testes, which produce sperm, the hormone testosterone, as well as accessory structures that enable fertilization.

■ Fertilization of an egg by a sperm occurs in a female oviduct. Only one sperm can fertilize an egg at one time.

■ The female reproductive cycle is coordinated by a complex balance of hormones that is controlled by the brain.

■ Hormones from the hypothalamus trigger the anterior pituitary gland to release follicle-stimulating hormone and luteinizing hormone, which in turn stimulate eggs to develop and the ovary to secrete estrogen and progesterone.

■ Ovulation, the monthly release of an egg from an ovarian follicle, is caused by a spike in LH at the midpoint of the cycle.

■ Estrogen and progesterone stimulate eggs to develop and the endometrium to thicken and prepare for a possible pregnancy.

■ Upon fertilization, the zygote divides and travels to the uterus, where it implants in the nutrient-rich endometrium.

■ The implanted embryo secretes the hormone hCG, which acts on the corpus luteum to maintain progesterone secretion.

■ If an egg is not fertilized, progesterone levels fall and the endometrium sloughs off during menstruation.

■ Sperm develop in the seminiferous tubules of the testes. The process is stimulated by testosterone.

■ Infertility has many causes, including blocked passageways caused by infection and scar tissue, genetic abnormalities, and hormonal deficiencies.

■ Assisted reproduction involves artificially bringing sperm and egg together, either inside the body (in IUI) or outside the body (in IVF).

■ Fertility drugs increase the number of eggs that mature and are ovulated by a female at one time. Multiple pregnancies result when sperm fertilize more than one available egg.

■ Humans and other mammals, as well as birds and reptiles, reproduce sexually using internal fertilization. Other sexually reproducing animals, such as fish, use external fertilization. Some organisms are asexual and have no sex at all.

Chapter 28 Test Your Knowledge

REPRODUCTIVE ANATOMY

Human males and females have distinct reproductive tracts that enable successful fertilization and pregnancy.

HINT **See Infographics 28.1–28.3 and 28.7.**

KNOW IT

1. Sperm develop in
   a. the epididymis.
   b. the vas deferens.
   c. the seminiferous tubules.
   d. the urethra.
   e. the penis.

2. Why can untreated pelvic inflammatory disease lead to infertility?
   a. because it prevents ovulation
   b. because it scars and blocks the oviducts
   c. because it scars and blocks the cervix
   d. because it interferes with estrogen production by the ovaries
   e. because it interferes with FSH and LH production by the anterior pituitary gland

3. Describe the relationship between (a) the uterus and the cervix and (b) the uterus and the endometrium.

USE IT

4. List the structures that sperm must pass through to reach and fertilize an egg. Begin with the seminiferous tubules.

5. A friend tells you that her boyfriend has been diagnosed with gonorrhea, a sexually transmitted disease. She isn't worried for herself because she doesn't have any symptoms of infection. What can you tell her about the invisible risks of an untreated sexually transmitted bacterial infection?

HORMONES AND REPRODUCTION

A complex interplay of hormones regulates gamete development and supports a successful pregnancy.

HINT **See Infographics 28.5 and 28.6.**

KNOW IT

6. What is the source—testes, ovaries, anterior pituitary gland, or embryo—of each of the following hormones?
   Luteinizing hormone (LH)
   Follicle-stimulating hormone (FSH)
   Testosterone
   Estrogen
   Progesterone
   hCG

7. The hormone hCG is an indicator of pregnancy; it also
   a. signals the corpus luteum to keep producing progesterone.
   b. triggers ovulation.
   c. acts on the anterior pituitary gland, causing it to release a surge of LH.
   d. acts on the endometrium, causing it to thicken.
   e. attracts sperm.

USE IT

8. Which of the following would most directly cause reduced levels of estrogen production?
   a. an anterior pituitary tumor that increases secretion of LH
   b. an increase in hypothalamus hormones that target the anterior pituitary
   c. an anterior pituitary tumor that increases secretion of FSH
   d. a decrease in hypothalamus hormones that target the anterior pituitary
   e. anterior pituitary damage that prevents synthesis and release of FSH

9. In an episode of *Law & Order: Special Victims Unit,* a blood sample from a crime scene was found to have extremely low levels of FSH and LH. Detectives used this information to determine that the blood came from a prepubescent girl, not a woman of reproductive age. Explain how they reached this conclusion.

10. As discussed in this chapter, oral contraceptives (such as the combination birth control pill) are designed to block ovulation in women. As males do not ovulate, a male hormonal contraceptive would

have to target sperm development. Which hormone would have to be blocked to prevent sperm development in males? What would be a likely undesired consequence of this type of male contraception?

INFERTILITY AND ASSISTED REPRODUCTION

There are many possible causes of infertility. Successful pregnancy can be achieved through a variety of assisted reproductive techniques, each of which has its own challenges, risks, and benefits.

HINT See Infographics 28.4, 28.8, and 28.9.

➔ KNOW IT

11. Which of the following could interfere with ovulation?
   a. blocked oviducts
   b. chronically low levels of LH
   c. excessive production of cervical mucus that blocks the cervix
   d. presence of sperm in the oviduct
   e. low levels of hCG

12. Compare and contrast in vitro fertilization (IVF) and intrauterine insemination (IUI).

➔ USE IT

13. Assume that there is an array of diagnostic methods available to you, including blood tests to determine hormone levels and ultrasound to visualize internal structures. What results might confirm each of the following infertility-associated conditions? Be as specific as possible.
   a. a blocked epididymis
   b. polycystic ovary syndrome
   c. menopause
   d. oviduct scarring

14. Why does IUI have a higher risk of multiple births than IVF?

SCIENCE AND SOCIETY

15. What do you think about Suleman's decision to have 14 children? Should local authorities or the federal government be empowered to regulate the number of children someone is allowed to have?

16. From the perspective of a fertility specialist, how would you respond to a congressional representative about a proposed increased regulation of fertility clinics? In order to make a convincing argument, include both pros and cons, medical and scientific considerations, and a description of the patient population that this specialist serves.

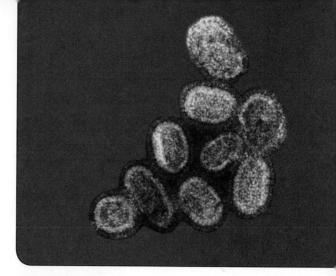

Viral Mysteries

Viral Mysteries

After nearly a century, scientists learn what made the 1918 influenza pandemic so deadly

In the fall of 1918, World War I was coming to an end–but another insidious threat was surfacing. In communities around the world, a deadly illness was gaining momentum, spreading like wildfire from person to person.

At first the disease seemed much like a bad cold. But many people soon became severely ill with high fevers and body aches, symptoms typical of influenza, or "flu." Some developed even more-severe symptoms–such as coughing so intense that they spat up blood. Some died so quickly after falling ill that doctors wondered whether the pestilence was in fact something entirely different from flu.

In many cases there wasn't even enough time to seek medical help. Stories were told of victims who were completely healthy in the evening but were dead by morning. In a study of the outbreak published in the *British Medical Journal* in 1979, Norman Roy Grist, an infectious disease specialist at the University of Glasgow, in Scotland, wrote that patients with seemingly ordinary influenza would rapidly "develop the most vicious type of pneumonia that has ever been seen" and later

would "struggle for air until they suffocated." Influenza patients "died struggling to clear their airways of a blood-tinged froth that sometimes gushed from their nose and mouth," recalled Isaac Starr, a physician who was a third-year medical student in Philadelphia during the war and who published his recollections of the illness in the *Annals of Internal Medicine* in 1976.

Between 1918 and 1920, the disease swept around the globe. Had the world not been at war, the pandemic–an epidemic that spreads globally–might never have happened. Troops carried the illness with them across Europe and back to the United States, infecting relatives, friends, and strangers along the way.

Influenza killed more people in a year than the Black Death of the Middle Ages killed in a century; it killed more people in 24 weeks than

Influenza killed more people in a year than the Black Death of the Middle Ages killed in a century; it killed more people in 24 weeks than AIDS has killed in 24 years.

Police in Seattle, Washington, wearing protective gauze face masks during the 1918 flu pandemic.

AIDS has killed in 24 years. This was perhaps the greatest pandemic in history. By the time it was over, at least 50 million people around the world had died.

The causative agent was no Ebola virus or other exotic germ. This was indeed influenza—caused by the same **virus** that today causes flu in people every year, the majority of whom recover completely within a week's time. When it first appeared the disease was called the Spanish flu because of the large number of early cases documented in Spain. Why this particular flu virus inflicted so much more devastation than any pandemic that came before or after, scientists of the time couldn't say. For decades after the pandemic had subsided, it seemed that it would never be possible to study the 1918 flu virus—samples of it had never been collected and stored. The reason for the severity of the pandemic remained a mystery.

It would be more than 75 years before scientists would understand what made the 1918 flu so deadly. That understanding may help us prepare for potential future pandemics.

A Deadly Virus

The annual flu virus does kill people. Each year in the United States about 36,000 people die from flu-related complications and more than 200,000 people are hospitalized from flu-related illnesses.

But most of these deaths strike primarily young children and the elderly, whose bodies are less able to fight off the infection. The 1918 flu, by contrast, struck down people in the prime of life.

The influenza virus, like all viruses, is an infectious particle that consists of nucleic acid surrounded by a protein shell. Viruses are considered to be nonliving because even though all viruses have genes, they are not made of cells. Because they don't have cells, viruses can reproduce only by infecting host cells. They then use the host cell's machinery to replicate and make more viruses. Eventually they destroy the host cell (Infographic 29.1).

Unlike cells, which rely on DNA as their hereditary material, viruses can store their genetic information in the form of either DNA or RNA. RNA viruses include the influenza virus and the viruses that cause colds, measles, and mumps. RNA viruses also cause more serious diseases, such as AIDS and polio. Diseases caused by DNA viruses include hepatitis B, chicken pox, and herpes. Each type of virus has a characteristic shape, with a distinctive protein shell.

Some viruses cause more damage than others. The severity of any illness typically depends on how quickly the **immune system**

IMMUNE SYSTEM
A system of cells and tissues that acts to defend the body against foreign cells and infectious agents.

INFOGRAPHIC 29.1

Viruses Infect and Replicate in Host Cells

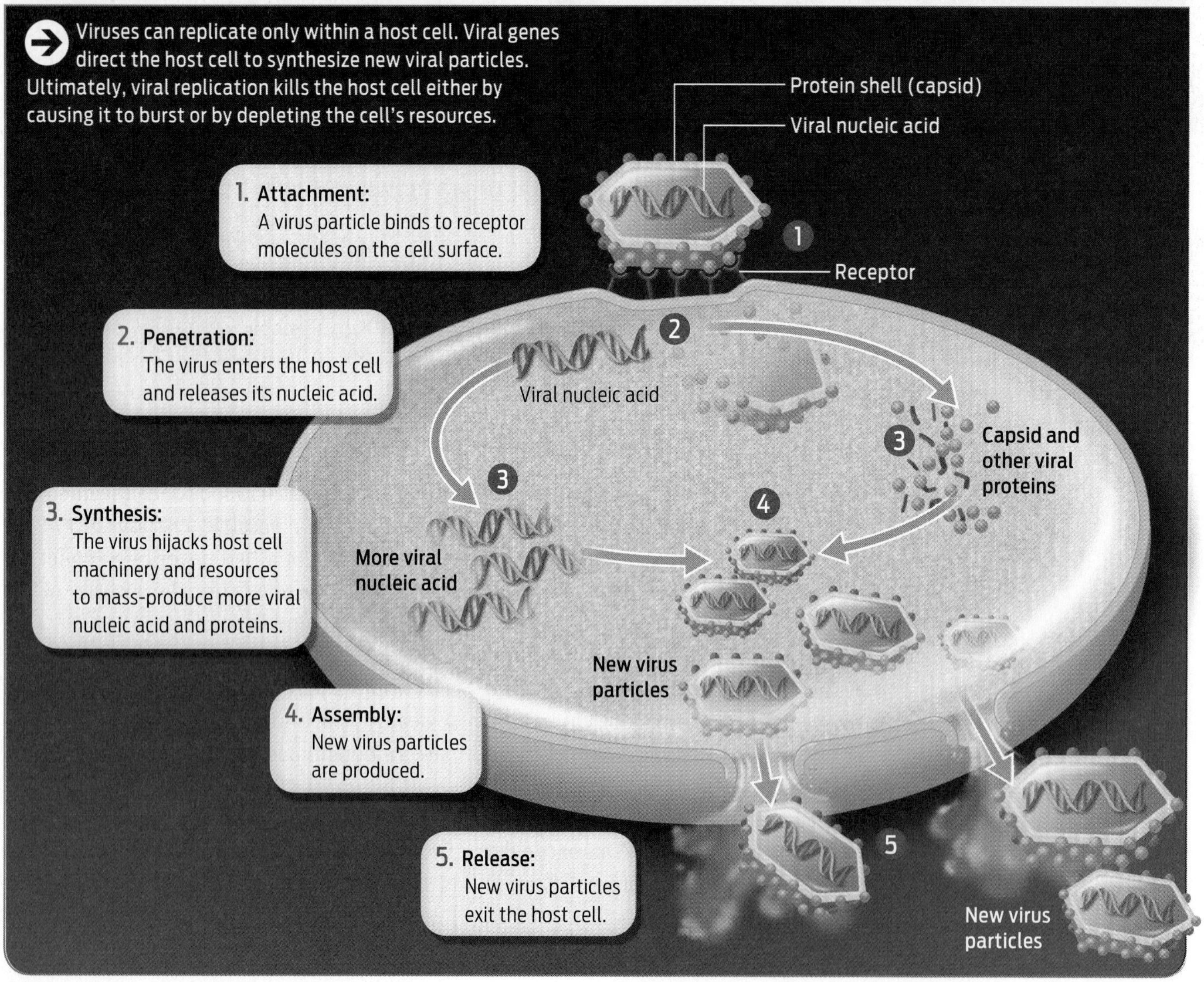

→ Viruses can replicate only within a host cell. Viral genes direct the host cell to synthesize new viral particles. Ultimately, viral replication kills the host cell either by causing it to burst or by depleting the cell's resources.

Protein shell (capsid)
Viral nucleic acid

1. **Attachment:**
   A virus particle binds to receptor molecules on the cell surface.

① Receptor

2. **Penetration:**
   The virus enters the host cell and releases its nucleic acid.

Viral nucleic acid

Capsid and other viral proteins

3. **Synthesis:**
   The virus hijacks host cell machinery and resources to mass-produce more viral nucleic acid and proteins.

More viral nucleic acid

New virus particles

4. **Assembly:**
   New virus particles are produced.

5. **Release:**
   New virus particles exit the host cell.

New virus particles

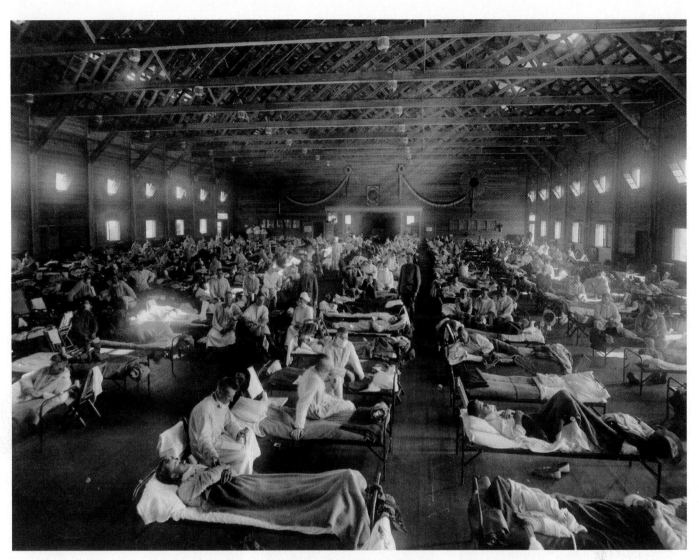

An emergency hospital in Camp Funston, Kansas, during the outbreak of Spanish influenza.

responds and how well the infected tissue can repair itself. Most people recover completely from colds because the respiratory tract contains rapidly dividing cells that quickly replace damaged ones. Polio virus, however, attacks nerve cells that almost never divide, which is why polio infection may cause permanent damage to motor skills. People who survive polio, however, will likely never contract the disease again–their bodies have become immune to the infection. Today children are vaccinated against polio, and the disease is practically nonexistent among the vaccinated in the Western world.

In severe cases, the lung looked torn apart, as if ravaged by shrapnel.

A Microscopic Battleground

Some clues as to what made the 1918 flu infection so deadly came from examining how the virus affected the body. Autopsies performed at the time showed that the 1918 infection wasn't typical of any influenza previously seen. Almost no internal organ was left untouched. Brain tissues looked flat and uncommonly dry; heart muscle was flabby, in contrast to its normally firm condition; and kidney damage had occurred in almost every case. By far the most damaged organ was the lung. In severe cases, the lung looked torn apart, as if ravaged by shrapnel.

The Immune System Defends against a Variety of Pathogens

Viruses

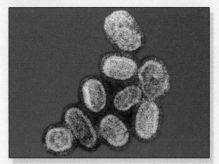

Influenza virus
Flu

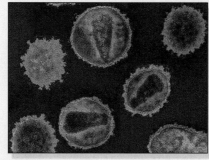

Human immunodeficiency virus
HIV

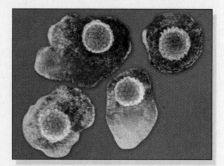

Varicella zoster virus
Chicken pox

Bacteria

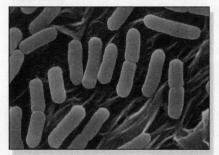

Escherichia coli
Food poisoning

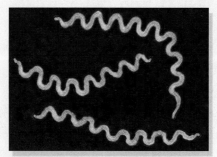

Treponema pallidum
Syphilis

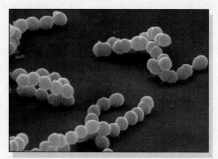

Streptococcus pyogenes
Strep throat, scarlet fever

Parasites

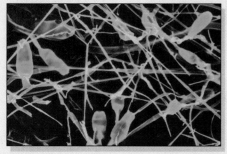

Trychophyton mentagrophytes (fungus)
Athlete's foot

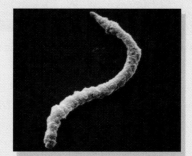

Plasmodium falciparum (worm)
Malaria

Giardia lamblia (protist)
Gastroenteritis

Today we know that victims' lungs were severely damaged not by the virus itself, but by all the defenses the body unleashed to fight the virus. The victims' immune systems reacted vehemently against the infection and in the process destroyed so many cells that some of the body's organs could no longer function. The immune system worked hardest in the lungs to fight the infection–the lungs were the battle-ground between the influenza virus and the immune system.

Immunity is the resistance to infections conferred by the immune system. Like an army, the body's immune system defends the body from different kinds of **pathogens**– foreign particles such as viruses, bacteria, and parasites that cause an immune response **(Infographic 29.2)**.

IMMUNITY
The resistance to a given pathogen conferred by the activity of the immune system.

PATHOGEN
Infectious agents including certain viruses, bacteria, fungi, and parasites. Many pathogens trigger an immune response.

Innate and Adaptive Immunity

Innate defenses protect us from a variety of invaders, but do not distinguish between specific pathogens. They are always present, and require little or no time to become active. Adaptive immunity relies on the action of white blood cells called lymphocytes to mount a unique defense against each specific invader.

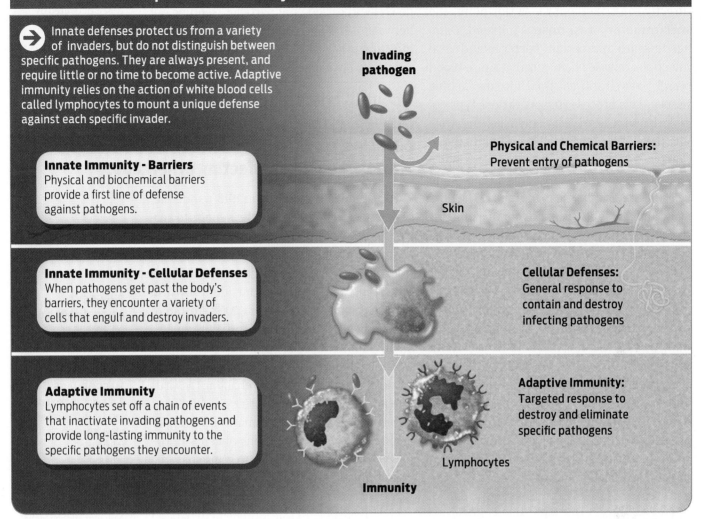

Innate Immunity - Barriers
Physical and biochemical barriers provide a first line of defense against pathogens.

Innate Immunity - Cellular Defenses
When pathogens get past the body's barriers, they encounter a variety of cells that engulf and destroy invaders.

Adaptive Immunity
Lymphocytes set off a chain of events that inactivate invading pathogens and provide long-lasting immunity to the specific pathogens they encounter.

Invading pathogen

Physical and Chemical Barriers: Prevent entry of pathogens

Skin

Cellular Defenses: General response to contain and destroy infecting pathogens

Adaptive Immunity: Targeted response to destroy and eliminate specific pathogens

Lymphocytes

Immunity

INNATE IMMUNITY
Nonspecific defenses, such as physical and chemical barriers and phagocytic cells, that are present from birth and are always active.

ADAPTIVE IMMUNITY
A protective response, mediated by lymphocytes, that confers long-lasting immunity against specific pathogens.

LYMPHOCYTE
A specialized white blood cell of the immune system.

HISTAMINE
A molecule released by damaged tissue and during allergic reactions.

The human immune system has two primary lines of defense that coordinate to protect us from pathogens and other harmful substances. The first includes physical and chemical barriers such as the skin and mucous membranes, which block invaders from entering the body. It also includes defensive proteins and white blood cells that engulf invaders. Because we are born with these defense mechanisms and they are always active, they are referred to as **innate immunity.** Innate defenses are nonspecific, in that they do not specifically recognize foreign invaders and substances. The second line of defense, called **adaptive immunity,** includes the coordinated actions of specialized white blood cells called **lymphocytes.** The adaptive response is highly diverse: it learns to respond to specific pathogens and substances. This is important because there

are many types of pathogens and they evolve quickly **(Infographic 29.3).**

Inborn Defenses

The innate immune system starts defending at sites where the body is exposed to the outside world, and it is always present and active. Enzymes in saliva destroy some pathogens; nasal hair and the mucus that lines the throat trap others. Underneath the layer of mucus is a layer of cells that carry hairlike projections called cilia, which sweep away pathogens and thus prevent them from taking hold in the throat. When pathogens do successfully breach these physical barriers, the body tries to flush them out with more fluid–chemicals, like **histamine,** are released by damaged tissues and trigger runny noses, watery eyes, coughs, and sneezes to expel the invaders.

When pathogens manage to overcome these chemical defenses, they begin to replicate. In some cases, replication of the pathogens leads directly to tissue damage. At this point, the **inflammatory response** is initiated. Damaged tissues and certain bacterial and viral infections release chemicals that cause blood vessels to swell and leak fluid into surrounding tissues. This process attracts various types of white blood cells to the inflamed site. The swelling, pain, and redness—all signs of inflammation—we sometimes see in response to an injury come from leaky blood vessels and immune cells that rush to the site of the injury. The fluid at inflamed sites also contains clotting proteins that stop the bleeding and prevent pathogens from spreading to neighboring tissues.

Several different types of white blood cells contribute to our innate defenses. Most important are **phagocytes,** which include **macrophages,** found in tissues, and **neutrophils,** cells that "patrol" the bloodstream: when phagocytes come into contact with pathogens, they release killing enzymes, then bind to and engulf the invaders. Some phagocytes enhance the inflammatory response, while others help initiate an adaptive response against the specific invader. Other white blood cells, called **natural killer cells,** attack virus-infected cells—and even cancer cells—and cause them to die.

The body has other nonspecific defenses. These include proteins that either attack microbes directly or prevent them from reproducing. Virus-infected cells can send out "SOS" signals in the form of **interferon.** Interferon proteins impede viral replication in neighboring cells and also signal other immune cells to become active. Other defensive proteins, called **complement proteins,** coat the surface of microbes, flagging them for destruction by phagocytes, or directly destroy some pathogens by punching holes in the cell membrane (**Infographic 29.4**).

Inflammation Overdrive

The influenza virus easily evades the body's physical barriers and takes up residence in the upper respiratory tract: the mouth, the nose, and the throat. Within minutes, virus particles attach to epithelial cells that line the entire respiratory tract and slip inside the body. Viruses then hijack the host cell's cellular machinery to replicate their own genetic material. Generally, about 10 hours after the virus invades a cell the cell begins to release newly synthesized viral particles, sending out between 1,000 and 10,000 viruses capable of infecting other cells. Eventually the infected cells die, weakening the respiratory tract.

About 10 hours after the virus invades a cell, the cell begins to release newly synthesized viral particles, sending out between 1,000 and 10,000 viruses capable of infecting other cells.

For decades following the 1918 pandemic, researchers were stymied in their efforts to analyze the original virus. Then, in 1997, pathologist and physician Johan Hultin extracted the virus from the body of a 1918 flu victim who had been buried in the permafrost outside a town in Alaska. About 10 years later, in 2008, Yoshihiro Kawaoka, at the University of Wisconsin–Madison, and his colleagues discovered what made the 1918 virus so deadly: four alleles that encoded versions of proteins enabling the virus to penetrate the lungs more effectively. While less lethal flu viruses replicate primarily in the upper respiratory tract, the versions of genes in the 1918 virus allowed it not only to cruise past the mouth, nose, and throat and infect the lungs, but also to trigger a massive inflammatory response.

Our innate defenses are sufficient to fight off many types of infection. The inflammatory response, in particular, is good at destroying invaders. The balance, however, between defense and destruction is delicate. The inflammatory response can go into overdrive and destroy the very organ it is trying to save. It was the upsetting

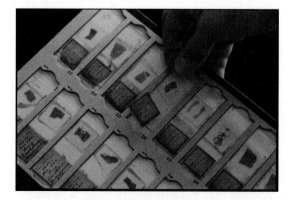

Samples of lung tissue from the 1918 pandemic are studied to reconstruct the gene sequence of the virus that caused Spanish flu.

INFLAMMATORY RESPONSE
An innate defense that is activated by local tissue damage.

PHAGOCYTE
A type of white blood cell that engulfs and ingests damaged cells and pathogens.

MACROPHAGE
A phagocytic cell that resides in tissues and plays an important role in the inflammatory response.

NEUTROPHIL
A phagocytic cell in the circulation that plays an important role in the inflammatory response.

NATURAL KILLER CELL
A type of white blood cell that acts during the innate immune response to find and destroy virally infected cells and tumor cells.

INTERFERON
Antiviral proteins produced by virally infected cells to help protect adjacent cells from becoming infected.

COMPLEMENT PROTEINS
Proteins in blood that help destroy pathogens by coating or puncturing them.

Some Important Features of Innate Immunity

a. Physical barriers

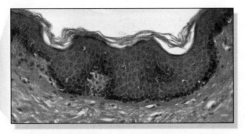

Skin (on the outside of the body) and mucous membranes (lining the inside of the body) have layers of tightly packed cells that prevent pathogens from entering the body. The mucus that coats mucous membranes traps foreign substances.

b. Inflammation

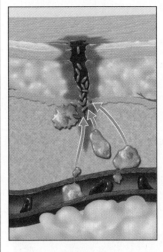

1. Microbes get past physical barriers.
2. Damaged cells and microbes release molecules that increase blood flow and attract white blood cells to infected areas.
3. Blood vessels leak, causing surrounding tissue to swell with fluid that contains clotting factors and white blood cells.
4. White blood cells ingest pathogens and also trigger an adaptive response. Clotting reactions contain the infection.

c. Phagocytes

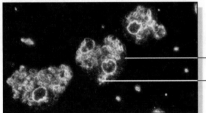

— Macrophage
— Yeast cell

Macrophage engulfs yeast cells

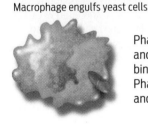

Phagocytes (including macrophages and neutrophils) can recognize, bind to, and ingest pathogens. Phagocytes also trigger inflammation and adaptive immune responses.

d. Antimicrobial chemicals

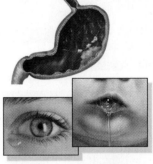

Acid in the stomach kills many of the microorganisms that we ingest.

Tears and **saliva** contain an enzyme that breaks down the cell walls of bacteria, causing them to burst.

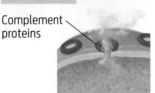

Complement proteins

Complement proteins in blood puncture holes in bacterial cells or coat the cell's surface, making them more easily destroyed by phagocytes.

---

of this balance that played a crucial role in the 1918 pandemic. In the face of such an aggressive flu virus, the inflammatory response was so massive that it damaged the very tissue it was supposed to protect. As fluids leaked out of blood vessels, they impaired oxygen uptake in the lungs; and as phagocytes produced chemicals to kill pathogens, those same chemicals destroyed host cells and tissues and destroyed lung tissue. What killed many people was not the virus itself but the overly aggressive inflammatory response, according to research described by John Barry in his 2004 book, *The Great Influenza* (Infographic 29.5).

Others died not from the virus itself or its respiratory consequences but from secondary infections. Because influenza weakened the defenses of the respiratory tract, other organisms, such as bacteria, had unimpeded entry into the lungs. Bacterial pneumonia—an infection of the lungs—was responsible for at least half of all deaths during the 1918 pandemic, according to research by Jeffery Taubenberger, a virologist at the U.S. Centers for Disease Control and Prevention.

The inflammatory response was so massive that what killed many people was not the virus itself, but the overly aggressive inflammatory response.

Lung Inflammation Can Kill

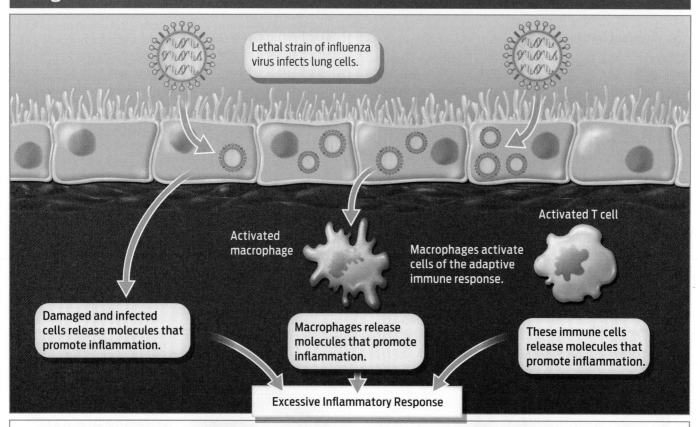

Lethal strain of influenza virus infects lung cells.

Activated macrophage

Macrophages activate cells of the adaptive immune response.

Activated T cell

Damaged and infected cells release molecules that promote inflammation.

Macrophages release molecules that promote inflammation.

These immune cells release molecules that promote inflammation.

Excessive Inflammatory Response

Acute respiratory distress:

- Cell death and debris
- Dilation of blood vessels
- Massive fluid influx
- Influx of white blood cells
- Loss of gas exchange
- Pneumonia

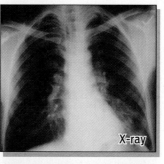

X-ray

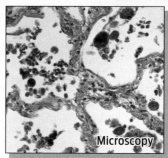

Microscopy

Infected lung is cloudy on x-ray and filled with immune cells, fluid, and debris, preventing normal function.

Normal lung:

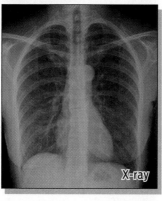

X-ray

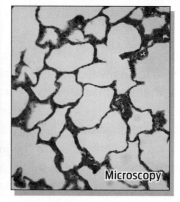

Microscopy

Normal lung is clear on x-ray and has space between cells that can fill with air.

Immunological Memory

Not everyone who became infected with the 1918 flu died, despite its virulence. In fact, while there were approximately 50 million deaths, experts estimate that some 525 million people were infected. How so many people were able to fight the infection while others died remains mostly a mystery. But scientists do have some clues.

Of those who became infected and then recovered, some may have had partial immunity from

B CELLS
White blood cells that mature in the bone marrow and produce antibodies during the adaptive immune response.

The Lymphatic System: Where B and T Lymphocytes Develop

➔ The immune system is made up of specialized cells called lymphocytes that develop, mature, and act in a variety of tissues, including the bone marrow, thymus, spleen, and lymph nodes, which are part of the lymphatic system.

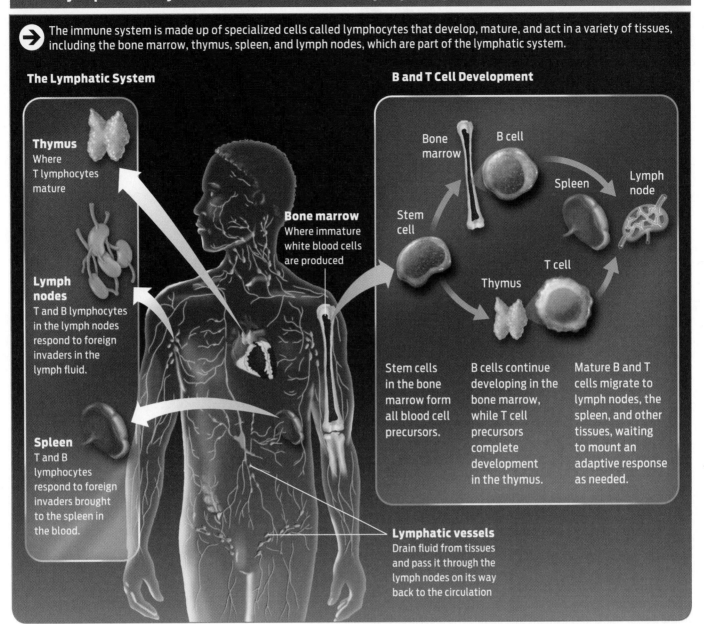

The Lymphatic System

Thymus
Where T lymphocytes mature

Lymph nodes
T and B lymphocytes in the lymph nodes respond to foreign invaders in the lymph fluid.

Spleen
T and B lymphocytes respond to foreign invaders brought to the spleen in the blood.

Bone marrow
Where immature white blood cells are produced

Lymphatic vessels
Drain fluid from tissues and pass it through the lymph nodes on its way back to the circulation

B and T Cell Development

Bone marrow

B cell

Stem cell

Spleen

Lymph node

Thymus

T cell

Stem cells in the bone marrow form all blood cell precursors.

B cells continue developing in the bone marrow, while T cell precursors complete development in the thymus.

Mature B and T cells migrate to lymph nodes, the spleen, and other tissues, waiting to mount an adaptive response as needed.

THYMUS
The organ in which T cells mature.

T CELLS
White blood cells that mature in the thymus and can destroy infected cells or stimulate B cells to produce antibodies, depending on the type of T cell.

LYMPH NODES
Small organs in the lymphatic system where B and T cells may encounter pathogens.

an earlier infection. Such long-lasting immunity is conferred by the adaptive immune system.

Whereas the innate immune system is always ready to fight, the adaptive immune system must be primed over time. From birth into adulthood, our bodies are continuously assaulted by a barrage of infectious agents. With repeated exposure, our bodies develop a memory of every infectious agent we encounter that gets past our innate defenses. Should we confront the same pathogen twice, immunological memory helps our bodies fight off infection before it can take hold.

The cells of the adaptive immune system are the B and T lymphocytes. These cells are produced in the bone marrow. Some immature lymphocytes in bone marrow become **B cells.** Other lymphocytes migrate from the bone marrow to the **thymus,** a gland in the chest, where they become **T cells.** Both B cells and T cells eventually make their way to the **lymph nodes** and other organs of the **lymphatic system,** where they lie in wait for pathogens. With these two types of immune cells, the adaptive immune system mounts a dual defense **(Infographic 29.6).**

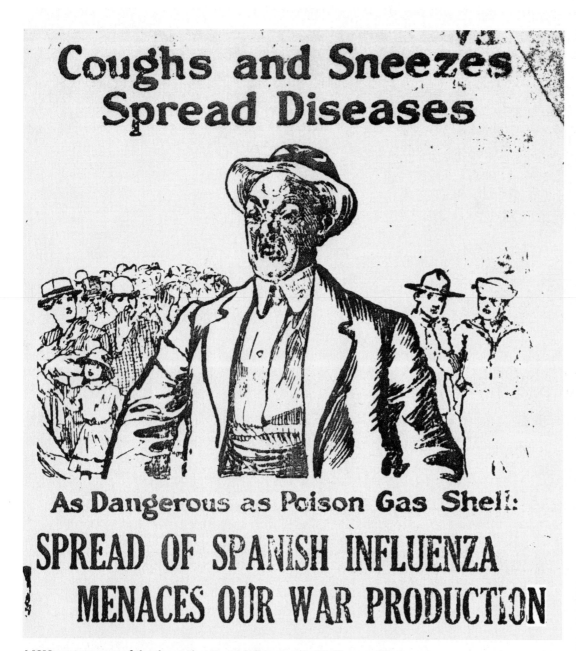

Coughs and Sneezes
Spread Diseases

As Dangerous as Poison Gas Shell:

SPREAD OF SPANISH INFLUENZA
MENACES OUR WAR PRODUCTION

A 1918 poster warns of the threat that Spanish flu presented to the war effort.

Part of adaptive immunity is called **humoral immunity.** In this immune response, a type of adaptive immunity, specialized T cells called **helper T cells** and B cells work together to recognize bacterial and viral **antigens.** When a helper T cell recognizes a particular antigen, it can activate a corresponding B cell. That B cell will divide repeatedly to create an army of **plasma cells**–cells that secrete many copies of an **antibody** specific to that particular antigen.

Antibodies are secreted into the bloodstream, where they bind to specific antigens and neutralize the antigen-carrying pathogen. Antibodies fight infections in several ways. First, by binding to antigens, they may physically block the infectious organism from infecting other cells, rendering it harmless. Second, they can flag a pathogen to be digested by phagocytes. Finally, they can mark the pathogen so that complement proteins will bind to it and destroy it. In other words, antibodies that bind to antigens are like giant red flags, alerting the body's immune system to take action **(Infographic 29.7).** Antibodies produced against one

Humoral Immunity: B Cells and Antibody Production

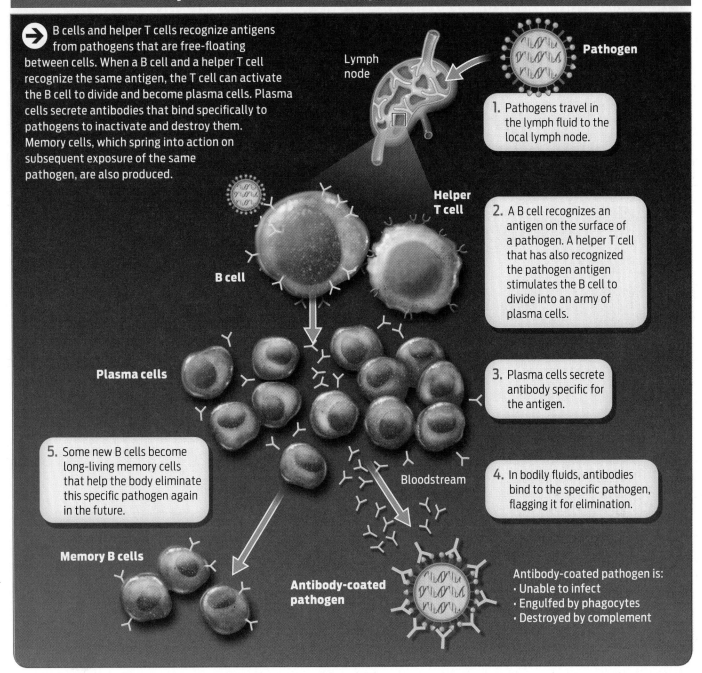

B cells and helper T cells recognize antigens from pathogens that are free-floating between cells. When a B cell and a helper T cell recognize the same antigen, the T cell can activate the B cell to divide and become plasma cells. Plasma cells secrete antibodies that bind specifically to pathogens to inactivate and destroy them. Memory cells, which spring into action on subsequent exposure of the same pathogen, are also produced.

Lymph node

Pathogen

Helper T cell

B cell

Plasma cells

1. Pathogens travel in the lymph fluid to the local lymph node.

2. A B cell recognizes an antigen on the surface of a pathogen. A helper T cell that has also recognized the pathogen antigen stimulates the B cell to divide into an army of plasma cells.

3. Plasma cells secrete antibody specific for the antigen.

5. Some new B cells become long-living memory cells that help the body eliminate this specific pathogen again in the future.

Bloodstream

4. In bodily fluids, antibodies bind to the specific pathogen, flagging it for elimination.

Memory B cells

Antibody-coated pathogen

Antibody-coated pathogen is:
· Unable to infect
· Engulfed by phagocytes
· Destroyed by complement

CELL-MEDIATED IMMUNITY
The type of adaptive immunity that rids the body of altered (that is, infected or foreign) cells.

CYTOTOXIC T CELL
A type of T cell that destroys altered cells, including virally infected cells.

antigen are usually ineffective against any other antigen—that is, they are highly specific.

Another part of the adaptive immune response is **cell-mediated immunity.** In this response **cytotoxic T cells** recognize infected or foreign cells (from a transplanted organ or tissue, for example) because these cells display foreign antigens on their surfaces. These antigens are either proteins

derived from the infectious agent or proteins from foreign cells. Once activated, cytotoxic T cells bind to antigens on the altered cells and release cytotoxic chemicals that cause the altered cells to self-destruct (Infographic 29.8).

Note that both humoral and cell-mediated immune responses destroy invading pathogens, but there is a key difference between them: humoral immunity acts by releasing antibodies that bind to antigens on free-floating pathogens

Cell-Mediated Immunity: Cytotoxic T Cells Kill Infected Cells

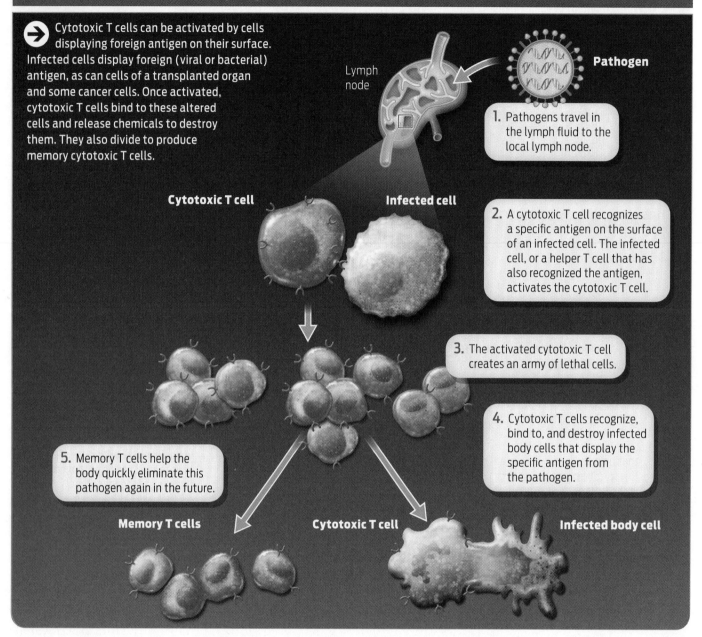

➡ Cytotoxic T cells can be activated by cells displaying foreign antigen on their surface. Infected cells display foreign (viral or bacterial) antigen, as can cells of a transplanted organ and some cancer cells. Once activated, cytotoxic T cells bind to these altered cells and release chemicals to destroy them. They also divide to produce memory cytotoxic T cells.

Lymph node

Pathogen

1. Pathogens travel in the lymph fluid to the local lymph node.

Cytotoxic T cell

Infected cell

2. A cytotoxic T cell recognizes a specific antigen on the surface of an infected cell. The infected cell, or a helper T cell that has also recognized the antigen, activates the cytotoxic T cell.

3. The activated cytotoxic T cell creates an army of lethal cells.

4. Cytotoxic T cells recognize, bind to, and destroy infected body cells that display the specific antigen from the pathogen.

5. Memory T cells help the body quickly eliminate this pathogen again in the future.

Memory T cells

Cytotoxic T cell

Infected body cell

in lymph and blood; in a cell-mediated immune response cytotoxic T cells bind to and destroy infected or foreign cells in body tissues.

The diversity of adaptive immunity enables our bodies to fight off countless numbers of pathogens. Occasionally, however, for reasons that are not entirely clear, this specificity runs amok and immune cells become active even against antigens that aren't harmful to us, or worse, they begin attacking healthy cells in the body.

When the immune system attacks antigens from outside the body, such as those in the environment, like dust or certain types of food, an **allergy** is the result. Allergies are quite common, often accompanied by a runny nose, watery eyes, and sneezing. A malfunctioning immune system can cause more severe disease when it attacks the body's own healthy cells. Multiple sclerosis, lupus, and rheumatoid arthritis are all **autoimmune diseases,** diseases in which the body's immune system attacks specific body

ALLERGY
A misdirected immune response against environmental substances such as dust, pollen, and foods that causes discomfort in the form of physical symptoms.

AUTOIMMUNE DISEASE
A misdirected immune response in which the immune system mistakenly attacks healthy cells.

Memory Cells Mount an Aggressive Secondary Response

The adaptive immune system's primary humoral response is slow and produces low levels of antibodies. Upon subsequent exposures, memory cells produced during the primary response respond quickly and trigger B cells to produce high levels of antibodies. This secondary response is so rapid and effective that illness does not occur. The secondary response of cell-mediated immunity and cytotoxic T cells is also quick and effective.

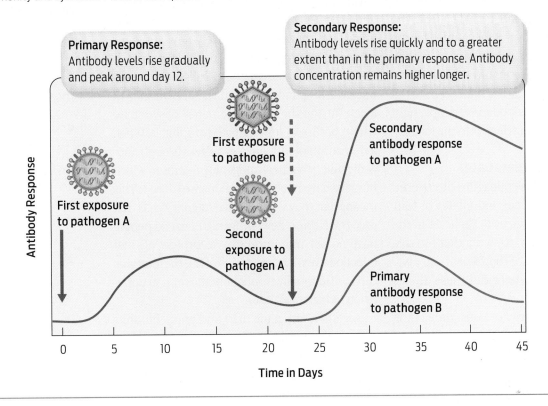

Primary Response:
Antibody levels rise gradually and peak around day 12.

Secondary Response:
Antibody levels rise quickly and to a greater extent than in the primary response. Antibody concentration remains higher longer.

First exposure to pathogen B

First exposure to pathogen A

Second exposure to pathogen A

Secondary antibody response to pathogen A

Primary antibody response to pathogen B

Antibody Response

Time in Days

PRIMARY RESPONSE
The adaptive response mounted the first time a particular antigen is encountered by the immune system.

MEMORY CELL
A long-lived B or T cell that is produced during the primary response and that is rapidly activated in the secondary response.

SECONDARY RESPONSE
The rapid and strong response mounted when a particular antigen is encountered by the immune system subsequent to the first encounter.

tissues. An autoimmune disease is the body's inappropriate immune response against itself.

Building a Line of Defense

How do the innate and adaptive immune systems work together? First-time exposure to a pathogen will almost certainly cause illness, because the adaptive response takes 7 to 10 days to develop. Over time an exposed individual will recover as T and B cells are activated and antibody levels increase. This initial slow response is the **primary response.** As B and T cells are churned out, some of them become **memory cells.** These memory cells remain in the bloodstream and "remember" the infection. The next time the same pathogen is encountered, memory B and T cells become active, dividing rapidly and producing very high levels of antibodies. They

fight the specific pathogen so quickly that the illness usually doesn't occur a second time. This rapid reaction is called the **secondary response** (Infographic 29.9).

The secondary response is what makes us immune to a particular infection. It's also how **vaccines** work. The source of all vaccines is the pathogen itself–some vaccines are made of only the antigens that cause an immune response, while others are a weakened or essentially dead version of the entire pathogen. However the vaccine is made, the goal is the same: to create a primary response in the body that's strong enough to create memory cells, yet weak enough not to cause disease symptoms. Thus, if the pathogen is subsequently encountered naturally, the secondary response is prepared. In this way, vaccination is like being infected with a pathogen without having the disease.

Even if they haven't been vaccinated, people exposed to a pathogen that is similar to a pathogen with which they were previously infected may be partly protected from the disease caused by the new pathogen. Memory B and T cells may still respond, although only partially–in which case the illness may occur, but mildly.

Evidence suggests that such partial immunity may have helped some of those infected survive the 1918 flu pandemic. Statistical data from the time show that people over 65 accounted for the fewest influenza cases, suggesting that they might over the years have acquired immunity or partial immunity from earlier infection. Partial immunity might have helped these people fight off the virus before it dug deep into the lungs.

Lessons Learned 75 Years Later

When Kawaoka and his colleagues discovered that the 1918 flu virus carried alleles of four genes that enabled the virus to replicate in the lower respiratory tract, thus making the disease so deadly, a piece of the 75-year-old mystery was solved. But where did these alleles come from?

New influenza viruses are constantly being produced by two mechanisms: by mutation and by grabbing genes from other viruses. Because influenza viruses replicate their genetic material so rapidly and don't "proofread" the replicated copies, mistakes often occur in the genomes of newly replicated viruses. **Antigenic drift** is the gradual accumulation of mutations that causes small changes in the antigens on the virus surface. Antigenic drift explains why there can be different types, or strains, of influenza circulating at the same time.

Two important antigens on the influenza virus are hemagglutinin (which binds to receptors on host cells and enables the virus to enter host cells) and neuraminidase (which helps new viruses exit host cells). The host immune system mounts an adaptive response specifically to these two antigens. When there is a change in hemagglutinin or neuraminidase or both, the memory cells no longer recognize them. These new antigens prompt a new and slow primary immune response. Anti-genic drift causes the seasonal variation in circulating flu viruses.

An influenza virus strain can also swap genes with other strains of influenza. While every influenza strain contains the same set of genes, the particular alleles of these genes that are present differ from strain to strain. The exchange of alleles between two strains that have infected the same cell does not simply create a small change in viral gene sequence: it introduces an entirely new allele, and therefore an entirely new antigenic protein. This process, called **antigenic shift,** is responsible for pandemic outbreaks of flu, including the 1918 pandemic, and the appearance of avian flu and the 2009 emergence of swine flu (H1N1), both of which originated in animals. When a new strain of influenza emerges through antigenic shift, it has new antigenic proteins to which humans have not been previously exposed. This means that they will not have memory cells that recognize these new antigens. Because people have no existing immunity to protect against infection by emerging strains of flu, these strains can spread rapidly throughout the human population (**Infographic 29.10**).

Together, antigenic drift and antigenic shift create an increasing variety of strains over time until one of the variants is able to infect human cells so efficiently that it sweeps through the population and causes a pandemic. These strains are named by the type of H (hemagglutinin) and N (neuraminidase) proteins they carry, with some antigenic combinations more deadly than others. Tracking the strains as they move through the population helps public health officials predict how severe a coming flu season will be. But since viruses can mutate every time they reproduce, they continue to evolve during epidemics. This unpredictability is what makes influenza so frightening: an apparently mild outbreak can suddenly become deadly.

This unpredictability is what makes influenza so frightening: an apparently mild outbreak can suddenly become deadly.

Preventing Pandemics

Because influenza viruses change so quickly, public health officials closely monitor the types of influenza viruses that infect animals. Viruses that infect animals may accumulate mutations or pick up genes that enable them to infect human cells.

VACCINE
A preparation of killed or weakened microorganisms or viruses that is given to people or animals to generate a memory immune response.

ANTIGENIC DRIFT
Changes in viral antigens caused by genetic mutation during normal viral replication.

ANTIGENIC SHIFT
Changes in antigens that occur when viruses exchange genetic material with other strains.

Antigenic Drift and Shift Create New Influenza Strains

→ Mutation and gene exchange are two mechanisms by which influenza viruses can change. Mutations that accumulate gradually can cause variations in surface antigen. This process is called antigen drift. Different strains can also swap genes and cause surface antigens to change more dramatically. This is called antigenic shift. Drift is responsible for annual seasonal variation in influenza; shift is responsible for dramatic pandemics.

Antigenic Drift:
· Gradual change
· Caused by point mutations that occur when the virus replicates.

Antigenic Shift:
· Rapid change
· Caused by gene exchange between two different viruses that simultaneously infect the same cell.

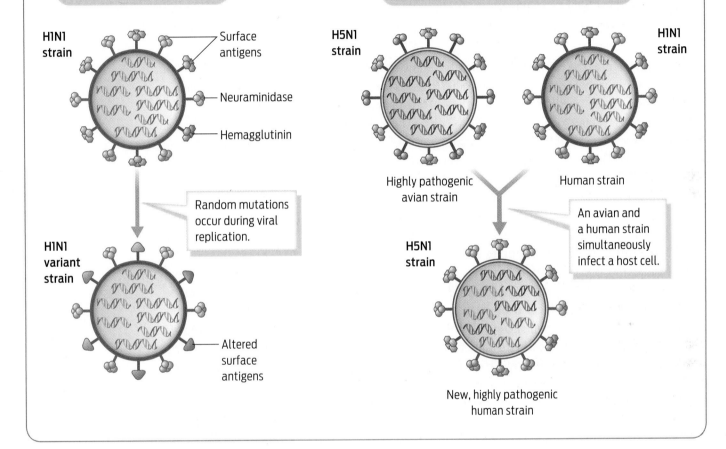

Severe acute respiratory syndrome (SARS), for example, is a respiratory illness caused by a virus. In 2003, more than 8,000 people worldwide became sick with SARS, according to the World Health Organization (WHO), and 774 died. Experts think the virus jumped from animals to people, although they have not yet determined the identity of the animal host.

They do, however, suspect that the source was a species of bird. Birds are a primary source of potentially pandemic viruses. Wild birds are routinely infected with viruses that live in their gastrointestinal tracts but cause little or no dis-ease in the bird. Sometimes, however, bird flu viruses can jump from wild birds to poultry or other farm animals and become lethal. And since humans maintain such close contact with poultry, pigs, and other domestic animals, the risk of an animal flu infecting humans is high.

In the 1990s, a lethal strain of the avian flu virus (H5N1) crossed over into humans from domesticated birds, but it was not easily transferred from person to person. However, scientists continue to monitor this bird strain, not only because it can be deadly to the commercial poultry business, but also because the 1918 flu

TABLE 29.1

Known Flu Pandemics

NAME	DATE	VIRUS ANTIGEN SUBTYPE	NUMBER OF DEATHS
Asiatic (Russian) flu	1889–1890	H2N2	1 million
1918 (Spanish) flu	1918–1919	H1N1	20–100 million
Asian flu	1857–1958	H2N2	1–1.5 million
Hong Kong flu	1968–1969	H3N2	0.75–1 million
Swine flu	2009–present	H1N1	10,000 in 2009

carried a variation of the H5N1 bird flu genes. In fact, most human influenza pandemics have been caused by flu viruses that carry variations of bird flu genes (**Table 29.1**).

To help people ward off flu infection, public health officials offer flu shots every year. A flu shot is a vaccine: a weakened version or part of a flu virus is injected into the body in hope of generating a primary immune response and memory cells against that strain of virus. However, since influenza viruses mutate so frequently, a yearly flu shot may not protect us from getting the flu the following year. In fact, a flu shot may not confer protection for the duration of a season. Scientists create a new vaccine each year by tracking which strains of influenza are circulating worldwide and studying the antigens on their surfaces. But their predictions can be wrong. If public health officials decide to vaccinate against a strain of influenza with one variant of hemagglutinin antigen, but a strain with a different variant of the antigen is the one that strikes, even those who are vaccinated may become ill anyway. The specific memory cells they carry against flu won't protect them from a different strain of the virus.

This is one reason why public health officials are so concerned about the 2009 H1N1 influenza virus, or swine flu. H1N1 is a new virus strain that was first detected in people in the United States in April 2009 and became a worldwide pandemic in 2010. It was called swine flu because many of its genes were similar to those in influenza viruses that normally occur in North American pigs. Further study has shown,

however, that this new virus is actually a mix of genes from flu viruses that normally circulate in European and Asian pigs, birds, and humans.

There are other reasons for concern. Whereas 90% of deaths from seasonal flu occur in people over 65 years old, H1N1 causes most severe disease in people under 25. This is the age group primarily affected by the 1918 virus, suggesting that swine flu, like Spanish flu, has the potential to become a more virulent infection that can kill within hours.

In 2009, Kawaoka reported in the journal *Nature* that the H1N1 virus could replicate in the lungs of mice, ferrets, and monkeys much better than a seasonal flu virus could. That finding "suggests the virus can cause serious respiratory illness in many people," Kawaoka stated in the paper. He also reported that some people born before 1918 have antibodies to the H1N1 virus, suggesting it shares some antigens in common with the 1918 virus.

Today, scientists have better surveillance, better communication, and can make more informed decisions about whether and how to quarantine groups to prevent a dangerous virus from spreading—even when extensive global travel might enable a new pandemic to spread rapidly.

In early 2010, the Bill and Melinda Gates Foundation awarded a $9.5 million 5-year grant to University of Wisconsin–Madison research scientists, who will be led by Kawaoka, to study viral mutations that could be early warning signs of pandemic flu viruses.

The 1918 flu pandemic taught scientists a lot about how viruses can become lethal in a matter of days, and demonstrated that developing vaccines can be like taking aim at a moving target. But by finding influenza genes common to strains that mutate less frequently than others, researchers may be able to develop a universal flu vaccine that would be effective over a number of years.

There are now at least two universal flu vaccines in clinical trials. Some companies are scaling up their vaccine manufacturing efforts, while others are boosting research and development of new universal vaccines. Some companies are also working on different antiviral drugs to help reduce the impact and spread of the influenza virus. The hope is that such measures will suppress the next big pandemic, whenever it happens, before it begins to rage. ■

How Do Other Organisms Defend Themselves?

Although not all organisms have as highly developed an immune system as do humans and other mammals, virtually all organisms have evolved ways to defend themselves from threats large and small.

Some organisms are able to repel invaders and predators by physical and chemical means. Sea sponges, for example, which represent early branches at the base of the evolutionary tree, are animals with different cell types but no distinct and differentiated tissues. They have no immune system, but they combine physical defenses with poisonous secretions to ward off danger. An internal support system made of collagen fibers and stiffened with hard spikes called crystalline spicules acts as a physical defense against predators like fish who may want to take a bite. In order to deter such predators from getting close enough to nibble, sponges also secrete chemicals that are toxic to many potential predators.

Many bacteria are able to protect themselves from a class of viruses known as bacteriophage–"phage" for short. Phage infection causes a bacterial cell to burst, or lyse–a life-ending event for that bacterium. But bacteria have defenses. They have restriction enzymes that act like scissors, cutting up any DNA that a phage injects into the bacterial cell. By destroying the phage DNA, bacteria prevent phages from replicating. At the same time, bacteria's own DNA is protected from these powerful molecular scissors.

Sea Sponges Employ Physical and Chemical Defenses

Crystalline spicules are a painful defense against fish that try to take a bite.

Toxins released by sponges keep predators from getting too close.

Bacterial cells that try to infect a sponge are lysed by molecules that punch holes in their cell walls.

They have a mechanism to chemically modify their DNA, which ensures their DNA won't be cut up by their own restriction enzymes.

Not all organisms that are evolutionarily ancient lack an immune system. Sea stars, for example, have a primitive immune system that consists of specialized cells that can attack or neutralize invaders. It has long been known that sea stars very rarely develop bacterial infection. It turns out that their ability to resist bacterial pathogens comes from a class of phagocytic cells called amoebocytes, which act much like the macrophages of the mammalian innate immune system. Amoebocytes circulate in the body cavity fluid of sea stars. When bacteria are injected into the body cavity of sea stars, the vast majority of amoebocytes actively

Bacteria Use Enzymes to Defend against Infection

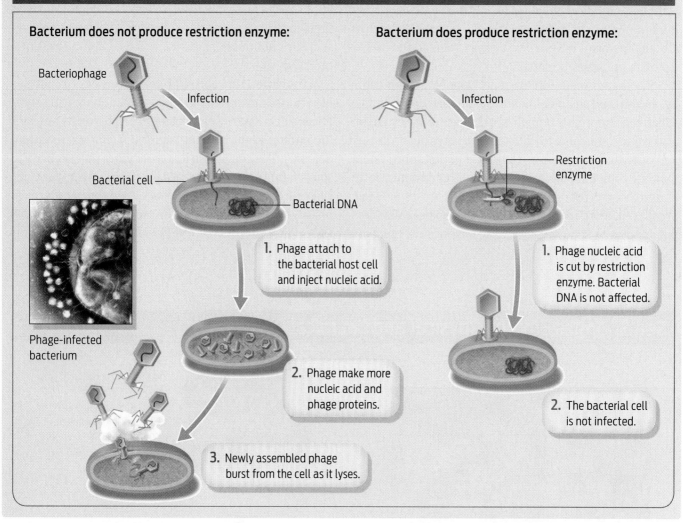

Bacterium does not produce restriction enzyme:

Bacteriophage

Infection

Bacterial cell

Bacterial DNA

Phage-infected bacterium

1. Phage attach to the bacterial host cell and inject nucleic acid.

2. Phage make more nucleic acid and phage proteins.

3. Newly assembled phage burst from the cell as it lyses.

Bacterium does produce restriction enzyme:

Infection

Restriction enzyme

1. Phage nucleic acid is cut by restriction enzyme. Bacterial DNA is not affected.

2. The bacterial cell is not infected.

engulfs and destroys the bacteria within 10 minutes. This rapid and efficient phagocytic response likely helps sea stars remain infection free.

Fruit flies also have several defense mechanisms that resemble the mammalian innate immune system. Phagocytic cells known as plasmatocytes act like human macrophages, ingesting and destroying foreign cells. Fruit flies also have cells called lamellocytes, which can coat and essentially wall off foreign objects, such as the eggs of parasitic wasps, that are too big to be phagocytosed. This walling off, or encapsulation, helps destroy the foreign object. Humans carry out a similar process in response to certain kinds of lung infections (including tuberculosis). In humans, immune cells surround and wall off the pathogen in a structure known as a granuloma. Thus, even some organisms that branched early on the evolutionary tree have defense mechanisms that are very much like ours.

Sea Stars Have Innate Cellular Defenses

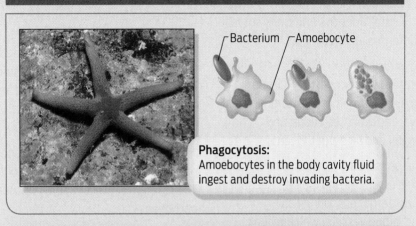

Bacterium Amoebocyte

Phagocytosis:
Amoebocytes in the body cavity fluid ingest and destroy invading bacteria.

Fruit Flies Have Multiple Physical, Chemical, and Cellular Defenses

External Barriers:
Fruit flies have a hard exoskeleton and specialized cells that line the airways and digestive system.

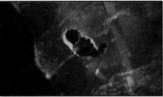

Encapsulation:
Lamellocytes (green) surround and wall off infecting parasites (black) inside living *Drosophila* larvae.

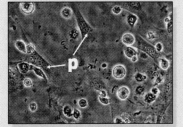

Phagocytosis:
Phagocytic cells known as plasmatocytes (labeled **p**) ingest and destroy invaders.

▶ Summary

■ The immune system defends the body against infection by pathogens.

■ Pathogens are particles that cause an immune response; they include certain bacteria, viruses, fungi, and parasites.

■ Viruses are noncellular; they consist of a genome of nucleic acid (DNA or RNA) contained within a protein shell.

■ The immune system has two components: innate immunity and adaptive immunity. Innate immunity is the first defense against invaders; adaptive immunity comes into play once innate defenses are breached.

■ The innate immune system includes defenses with which we are born, and which are always active: they include barriers such as skin and mucous membranes; antimicrobial chemicals in tears, saliva, and other secretions; and phagocytic cells that engulf and digest pathogens.

■ The inflammatory response, part of the innate immune system, is triggered by tissue injury and is a key mechanism for dealing with invading pathogens. During an inflammatory response, blood vessels swell and leak, marshaling phagocytic cells and protective molecules to the area to contain the infection.

■ Adaptive immunity is conferred by specialized lymphocytes called B and T cells.

■ B cells produce antibodies that specifically recognize antigens on pathogens and mark the pathogens for destruction.

■ T cells destroy infected cells and also stimulate B cells to produce antibodies.

■ The immune response can go awry, causing allergies, which are responses to intrinsically harmless substances (for example, pollen), and autoimmune conditions that result when the immune system mounts a response against the body's own cells and tissues.

■ At the first exposure to a particular pathogen, a primary response is generated that takes time to become fully effective; the primary response also produces memory cells. Memory cells remain in the body and become active at the time of a subsequent exposure to the same pathogen, producing a more rapid and vigorous secondary response that fights the pathogen and usually prevents the associated illness.

■ Vaccination elicits a primary response. Memory cells produced during the primary response protect against illness following subsequent exposure to the actual pathogen.

■ The adaptive response is highly specific for particular pathogens. If the pathogen changes, by antigenic shift or drift, the body will need to mount a separate response for each strain of the pathogen.

■ All organisms have ways of defending themselves against infection. Many organisms, including evolutionarily ancient ones, protect themselves by mechanisms similar to those of the human innate immune system.

VIRUSES AND OTHER PATHOGENS

Many viruses and bacteria can cause disease and are therefore pathogens.

HINT See Infographics 29.1, 29.2, 29.10, and 29.11.

◉ KNOW IT

1. Which of the following is found in all viruses?
    a. DNA
    b. RNA
    c. a membranous envelope
    d. a protein shell
    e. a cell membrane

2. Explain how viruses replicate within humans.

3. Why does poliovirus cause long-lasting damage, whereas those infected by influenza virus typically make a full recovery?

4. What is the difference between antigenic shift and antigenic drift?

◉ USE IT

5. Both viruses and bacteria can be human pathogens. Describe some key differences between them.

6. Why do poliovirus, influenza virus, and HIV infections cause different symptoms?

INNATE IMMUNITY

Innate immunity includes physical and chemical barriers and the inflammatory response, the first lines of defense of the body against pathogens.

HINT See Infographics 29.3 and 29.4.

◉ KNOW IT

7. Name three components of the innate immune system. For each, provide a brief description of how it offers protection.

8. What do macrophages and neutrophils have in common?

◉ USE IT

9. From what you know about innate immunity, would you predict different or identical innate responses to infections from *E. coli* (a bacterium) and *S. aureus* (another bacterium)? Explain your answer.

10. Neutropenia is a deficiency of neutrophils. Would you expect someone with neutropenia to be able to mount an effective inflammatory response? Explain your answer.

11. Why are those with influenza infections susceptible to bacterial pneumonia?

12. Why might those taking anti-inflammatory drugs be more susceptible than others to bacterial infections?

ADAPTIVE IMMUNITY

Adaptive immunity is a specific response to pathogens and other foreign antigens.

HINT See Infographics 29.3 and 29.6–29.9.

◉ KNOW IT

13. Compare and contrast the features of innate and adaptive immunity.

14. B cells, plasma cells, and antibodies are all related. Describe this relationship, using words, a diagram, or both.

◉ USE IT

15. Anti–hepatitis C antibodies in a patient's circulation indicate
    a. that the patient is mounting an innate response.
    b. that the patient has been exposed to HIV.
    c. that the patient has been exposed to hepatitis C within the last 24 hours.
    d. that the patient has been exposed to hepatitis C at least 2 weeks ago.
    e. that the patient has hepatitis.

16. Vaccination against a particular pathogen stimulates what type of response?
- **a.** innate
- **b.** primary
- **c.** secondary
- **d.** autoimmune
- **e.** b and c

17. Will someone who has been exposed to seasonal influenza in the past
- **a.** have memory B cells?
- **b.** still be at risk for seasonal influenza next year? Why?
- **c.** still be at risk for H1N1 (swine flu)? Why?

18. What processes are responsible for the emergence of pandemic influenza strains, such as H1N1 swine flu? Explain how these strains can spread so successfully through the human population.

19. *Staphylococcus aureus* causes a bacterial skin infection that can become very serious.
- **a.** Why does the body exhibit innate and adaptive responses to *Staphylococcus aureus* but not to its own skin cells?
- **b.** Will the innate response to *Staphylococcus aureus* be equally effective against *Streptococcus pyogenes*, another bacterium that can cause skin infections? Explain your answer.
- **c.** Will the adaptive response to *Staphylococcus aureus* be equally effective against *Streptococcus pyogenes*? Explain your answer.

20. HIV is a virus that infects and eventually destroys helper T cells. Why do people with AIDS (advanced HIV infections) often die from infections by other pathogens?

SCIENCE AND SOCIETY

21. Vaccination against a potentially pandemic disease, such as measles or influenza, lessens the probability of developing disease. Would it be a good idea for an agency such as a local school board to require that everyone in its jurisdiction be vaccinated?

22. What are the relative merits of investing in disease prevention (for example, by vaccines) or in disease treatment (for example, by antibiotics and antivirals)?

Q & A: Plants

Bugs are drawn to what look like dewdrops on this carnivorous plant, the Australian sundew. The "dewdrops" turn out to be sticky mucus bubbles that trap the insects so the plant can digest them.

Q & A: Plants

Plants have evolved a unique set of solutions to nature's challenges

Like other organisms, plants are products of evolution. They are equipped with adaptations that help them survive and flourish in their particular environments. Plants face many of the same life challenges as do other organisms—obtaining

nutrients, reproducing, protecting themselves—but their solutions to these challenges are often unique and surprising. We encountered the evolution of plant diversity in Chapter 19; here we focus on plants that have evolved to succeed on dry land.

ROOT SYSTEM
The belowground parts of a plant, which anchor it and absorb water and nutrients.

SHOOT SYSTEM
The aboveground parts of a plant, including the stem and photosynthetic leaves.

PLANT STRUCTURE

◨ Plants don't have bones—so how do they stand up?

A From tiny 3-inch-tall crocus flowers to massive redwood trees standing 300 feet, nearly all land plants share the same basic design: a below-ground **root system** for absorbing water and nutrients and an aboveground **shoot system** made up of stems and leaves. Bones not included. In the absence of a bony skeleton, what keeps a plant from flopping over?

Like other eukaryotes, plants are made of cells packed with organelles, including a nucleus, endoplasmic reticulum, and mitochondria. Plant cells contain a few plant-specific parts as well, including a supportive **cell wall; chloroplasts,** the sites of photosynthesis; and a **central vacuole,** essentially a large water balloon occupying the center of the cell. When filled with water, vacuoles create **turgor pressure** against the cell wall, keeping a plant body rigid and upright. When the vacuoles are less than full, turgor pressure is reduced, and the plant wilts.

Cell walls contribute to a plant's stiffness in another way. Plant cells are packed tightly together, much like bricks in a wall; carbohydrates that make up the cell walls act like glue, helping adjacent cells stick together. With many cells held together in this way, plant tissues are exceptionally strong, which is why they make such durable ropes and fabrics.

Some plant tissues, such as those that make up the stem, are made of cells with two cell walls: an outer cell wall containing cellulose, and a second, inner cell wall containing cellulose and **lignin.** Lignin is a hard, durable material, which lends added support and strength to plant tissues; it is what makes them "woody." A plant with a thick, woody stem (or trunk, in the case of a tree) can grow tall and still not topple over. The tallest of all plants is the

> **The tallest of all plants is the California redwood (*Sequoia sempervirens*), which can reach heights of more than 350 feet.**

California redwood (*Sequoia sempervirens*), which can reach heights of more than 350 feet.

Plants would not be able to grow so tall were it not for the extensive system of roots that anchors them in the ground. Some plants have root systems that are much deeper than the plants are tall, and some plants' roots extend out horizontally to about three times the branch spread. A vast number of tiny **root hairs** increase the surface area of the root, enhancing its ability to absorb water and nutrients. When root hairs are included, a plant's total root length can be hundreds of miles long. In addition, some plants produce very long **taproots**—deep-reaching roots that help the plant reach the underground water table. Taproots also store water and carbohydrates for the plant, making them appealing and nutritious to animals. Carrots and turnips are actually taproots **(Infographic 30.1).**

◨ Why don't plants bleed?

A Like animals, plants have a **vascular system** for transporting valuable fluids. Instead of blood, however, a plant's vascular system transports nutrients like water and sugar throughout the plant body. Plants need water to live and grow (you may have learned this the hard way if you've ever killed a plant with neglect). Among other things, water is crucial for photosynthesis in leaves. But you don't water a plant's leaves; you water the roots by pouring water in the soil. How does water get from the roots to the rest of the plant?

A plant's water-carrying tissue is known as **xylem** (pronounced ZYE-lum). Xylem tissue is made of cells arranged into long, stiff tubes; the cells have holes in each end and are stacked one on top of the next. Water moves up from the roots through these tubes to the aboveground stems, and eventually into the leaves, where it is used during photosynthesis to make sugar (for more on photosynthesis, see Chapter 5). Plant

All Plants Share the Same Basic Design

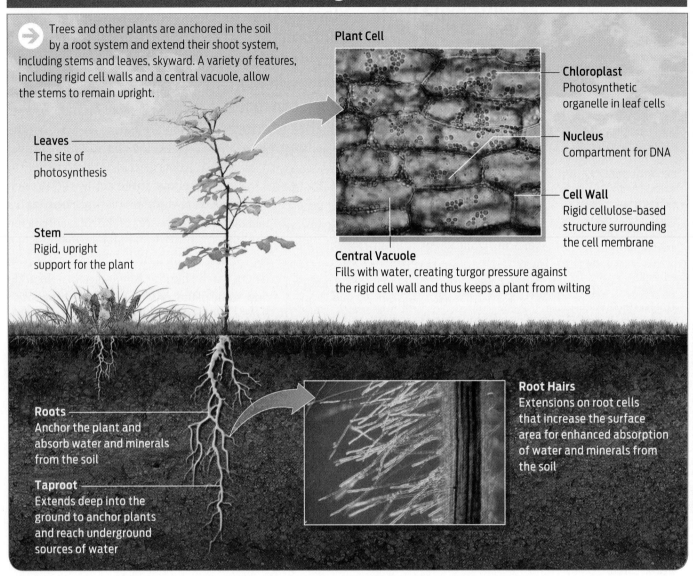

Trees and other plants are anchored in the soil by a root system and extend their shoot system, including stems and leaves, skyward. A variety of features, including rigid cell walls and a central vacuole, allow the stems to remain upright.

Plant Cell

Chloroplast
Photosynthetic organelle in leaf cells

Nucleus
Compartment for DNA

Cell Wall
Rigid cellulose-based structure surrounding the cell membrane

Central Vacuole
Fills with water, creating turgor pressure against the rigid cell wall and thus keeps a plant from wilting

Leaves
The site of photosynthesis

Stem
Rigid, upright support for the plant

Roots
Anchor the plant and absorb water and minerals from the soil

Taproot
Extends deep into the ground to anchor plants and reach underground sources of water

Root Hairs
Extensions on root cells that increase the surface area for enhanced absorption of water and minerals from the soil

cells use the newly synthesized sugar as food, so the sugar must be transported out of the leaves and back down through the plant. Another series of tubes, known as **phloem** (pronounced FLO-um; think "f" for "food"), transports sugar through the plant. Phloem supports two-way transport: sugar moves down to the roots, where some of it is stored; later, sugar moves up to the shoot system, where it provides nutrition and energy for growing fruit, buds, and leaves.

> **Xylem and phloem, like the vessels of the human circulatory system, transport all the essential nutrients the plant body needs to live and grow.**

Xylem and phloem, like the vessels of the human circulatory system, transport all the essential nutrients the plant body needs to live and grow.

Moving water and other nutrients up to the top of a 300-foot tree against the force of gravity is no easy task—hundreds of pounds of water must be lifted up through what is essentially a long water pipe. And unlike animals, which have a heart to pump blood along, plants have no such mechanical help. Instead, plants rely on evaporation of water from the

PHLOEM
Plant vascular tissue that transports sugars throughout the plant.

TRANSPIRATION
The loss of water from plants by evaporation, which powers the transport of water and nutrients through a plant's vascular system.

leaves to siphon up water in a process called **transpiration.** Because water is a polar molecule that can form hydrogen bonds with other water molecules, it has great cohesive strength (see Chapter 2). As water evaporates from leaves into the air, water in the xylem is pulled up to replace it. The cohesive strength of water is enough to counteract the force of gravity and pull water up to astonishing heights.

For a plant, transpiration is life sustaining: it is the force that carries water through the plant. But losing too much water can be dangerous. On a hot, dry, or windy day, a large tree can lose hundreds of gallons of water vapor from its leaves. To control the amount of water lost by transpiration, a plant's leaves are coated with a waxy **cuticle** that functions somewhat like a rubber suit, sealing in moisture. At regular intervals, the cuticle is punctuated by pores, called **stomata,** which open and close. When stomata are open, water vapor leaves freely and other gases enter and exit–specifically, carbon dioxide and oxygen. When stomata are closed, water and gases are sealed in. Many plants keep their stomata open during the day to let in carbon dioxide for photosynthesis. At night, they close the stomata to conserve water **(Infographic 30.2)**.

INFOGRAPHIC 30.2

Plants Transport Water and Sugar through Their Vascular System

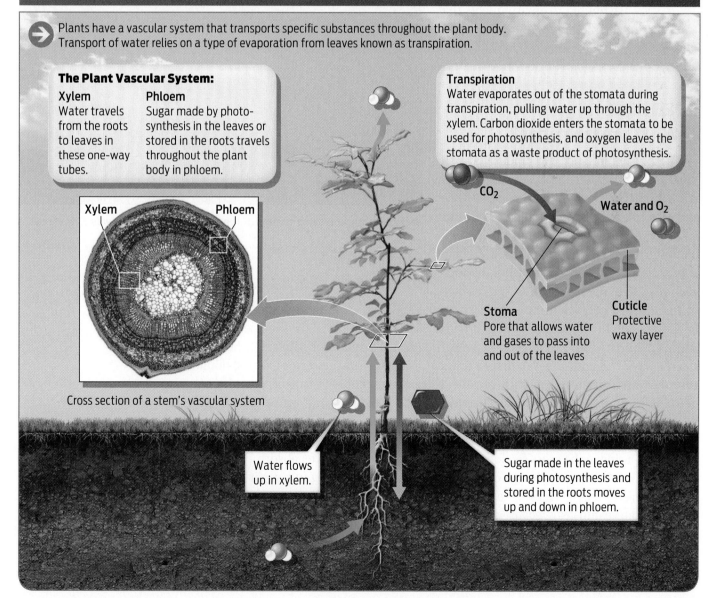

Plants have a vascular system that transports specific substances throughout the plant body. Transport of water relies on a type of evaporation from leaves known as transpiration.

The Plant Vascular System:

Xylem
Water travels from the roots to leaves in these one-way tubes.

Phloem
Sugar made by photosynthesis in the leaves or stored in the roots travels throughout the plant body in phloem.

Transpiration
Water evaporates out of the stomata during transpiration, pulling water up through the xylem. Carbon dioxide enters the stomata to be used for photosynthesis, and oxygen leaves the stomata as a waste product of photosynthesis.

CO_2

Water and O_2

Xylem

Phloem

Cross section of a stem's vascular system

Stoma
Pore that allows water and gases to pass into and out of the leaves

Cuticle
Protective waxy layer

Water flows up in xylem.

Sugar made in the leaves during photosynthesis and stored in the roots moves up and down in phloem.

Eventually, if a plant loses too much water, it will die. During an extreme drought, if you put a stethoscope to a tree and listen closely, you may hear clicks. This is the sound of the rigid column of water in the xylem breaking because there is no water in the soil to replace the water being lost by transpiration.

The presence of specialized tissues for transporting water is what distinguishes plants from their water-dwelling ancestors, the algae. The evolution of this vascular tissue is what allowed plants to colonize nearly every part of the land, from valley to mountaintop. A few primitive nonvascular land plants, such as mosses and liverworts, also lack true roots and shoots containing vascular tissue. Without these specialized tubes for transporting water, they are limited to environments that are saturated with water, where they grow close to the ground in squat, spongy mats.

Q What are tree rings?

A Plants experience two kinds of growth: primary and secondary. Primary growth is growth in length–a tree getting taller. Trees and other plants also grow wider as well–this is secondary growth. Tree trunks increase in diameter because they add a new layer of xylem tissue–otherwise known as **wood**–each year. In temperate regions, such as most of the United States, xylem growth is dormant in the winter. In the spring, xylem growth starts again. The first xylem cells to grow in the spring are usually larger in diameter and thinner-walled than xylem cells produced later in the summer. The boundary between the smaller xylem cells from one summer and the larger xylem cells from the next spring appears as a ring in the cross section of a tree trunk. You can determine the age of a tree by counting the number of rings. Tree rings vary in width from year to year because differences in temperature and rainfall affect the amount of xylem tissue that is produced in any one season.

Each dark ring represents the end of 1 year of growth. This tree was 37 years old.

Q What do plants eat?

A This is a a trick question, because plants don't "eat" in the same sense as animals. Plants are autotrophs, meaning they make their own food from inorganic materials found in the environment. Give a plant some sunshine, carbon dioxide, and water–plus a few soil nutrients–and you will have a happy plant, capable of feeding and nourishing itself. From the air, plant leaves absorb carbon dioxide, which they convert into sugar through photosynthesis. Sugar is a source of energy and building blocks for the plant. From the soil, plant roots absorb nutrients–primarily nitrogen– that are required to make plant proteins and other molecules.

WOOD
Secondary xylem tissue found in the stem of a plant.

Although nitrogen gas is plentiful in air spaces in the soil, it exists in a form that is essentially unusable by plants. Luckily, bacteria within the soil are able to convert nitrogen into a form that can be used by plants, a process known as **nitrogen fixation** (for more, see Chapter 23). Some of these bacteria, like members of the genus *Rhizobium*, can even live symbiotically in a plant's roots, forming lumpy structures known as root nodules on the roots of legumes. By making plant growth possible, nitrogen-fixing bacteria play a critical role in supporting nearly all life on the planet.

Fertile soil naturally contains nitrogen-fixing bacteria and therefore provides adequate supplies of usable nitrogen for plants to absorb and use to grow. And soil can be supplemented with fertilizer, giving plants a boost of artificial nitrogen and other nutrients. But in certain natural environments, such as bogs or rock outcroppings, it's hard for plants to obtain the nitrogen they need. The acidity of a bog, for example, prevents organic matter from breaking down, so nutrients are recycled more slowly. In these environments, plants have evolved novel ways to obtain scarce nitrogen—some of which would put animal carnivores to shame.

Trumpet pitchers (*Sarracenia*), for example, lure insects with brightly colored flowers and nectar "bribes." But the rim of the plant's trumpet-shaped flower is slippery. Unsuspecting trespassers climb onto the rim, lose their grip, and tumble into a deep cavity filled with digestive juices. Prevented from escape by downward-pointing spikes, the tiny prisoners drown and are slowly dissolved. The resulting insect soup—a rich source of nitrogen—is absorbed by the plant.

The Venus flytrap (*Dionaea muscipula*) takes an even more dramatic approach. The plant's "flower" is actually a spring-loaded trap that snaps shut around unsuspecting prey. Tiny hairs inside the flower act as sensors; when the sensors are tripped by a moving insect, the trap slams shut and the feasting begins (**Infographic 30.3**).

Even when plants obtain nitrogen in this "carnivorous" way, they must still perform photosynthesis to make sugar. The plant body is composed of complex carbohydrates, such as cellulose, which the plant makes by stringing sugar molecules together. And the starting material for photosynthesis is carbon dioxide gas. Thus the air, rather than the soil, is where a plant obtains the material to put on weight.

Can plants photosynthesize at night?

A By definition, there is a crucial part of photosynthesis that can occur only during daylight hours—the light-absorbing "photo" part. So technically, the answer is, no—plants can't photosynthesize at night. However, some plants can collect carbon dioxide in the dark, a useful capability in hot, dry climates.

Because daylight hours tend to be the time of day when it's hottest and driest, plants lose a lot of water during the day. To conserve water, plants can close their stomata. But closing stomata also prevents carbon dioxide—another crucial ingredient for photosynthesis—from entering the leaf. In many plants, wheat and rice, for example, the result of this trade-off is reduced output of plant food: sugar.

Some plants have adapted to sun-scorched surroundings. Corn and sugar cane, for example, are able to thrive in hot, sunny climates by keeping their stomata closed as much as possible. They use a molecule known as PEP carboxylase to capture CO_2 even when it is present only at low concentrations in air pockets in the leaf when the stomata are closed. Once the CO_2 has been captured, it can be "fed" into the synthesis part of photosynthesis to generate the sugar that the plants rely on.

Still other plants have adapted to hot, dry conditions by splitting up two parts of photosynthesis that usually occur simultaneously: taking in CO_2 and making sugar. Pineapples as well as cacti and other succulent plants—juicy aloe and jade plants, for instance—conserve water by keeping their stomata closed during the day when it's hot. But at night, when it's cooler, they open their stomata to allow carbon dioxide in. The CO_2 that is captured at night isn't

NITROGEN FIXATION
The process of converting atmospheric nitrogen into a form that plants can use to grow.

Plants Obtain Nutrients in a Variety of Ways

 Plants obtain carbon from atmospheric CO_2 during photosynthesis. They obtain minerals and nutrients such as nitrogen from the soil, often with a little "supplementation."

Plants are autotrophs: they make their own food through photosynthesis, with atmospheric CO_2 providing the carbon to build molecules such as sugars.

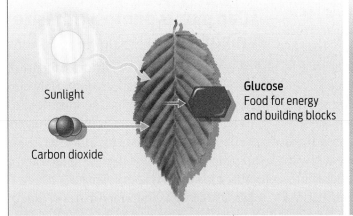

Sunlight

Carbon dioxide

Glucose
Food for energy
and building blocks

Plants get other nutrients necessary for plant growth from the soil.

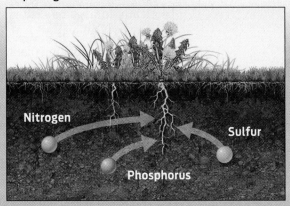

Nitrogen

Sulfur

Phosphorus

Carnivorous plants in nutrient-poor soil obtain nutrients such as nitrogen by digesting insects.

Trumpet pitcher plants attract insects with bright colors and sweet nectar.

Fly trapped inside a trumpet pitcher plant

Some plants get their nitrogen by "partnering" with nitrogen-fixing bacteria that live in association with their roots.

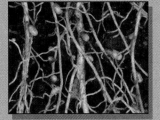

Rhizobium
A nitrogen-fixing
bacterium

Clover plant roots with nodules containing nitrogen-fixing bacteria

used right away but instead is stored. During the day, these plants can leave their stomata closed and use the stored CO_2 to complete photosynthesis and make sugar. By segregating in time the steps of photosynthesis, these well-adapted plants have found a way to thrive in conditions that would wither their less physiologically adapted cousins (**Infographic 30.4**).

Like other eukaryotic organisms, plants use sugar to perform aerobic respiration to make ATP (Chapter 6). All plants respire both during the day and at night. Because the total amount of carbon dioxide given off by plants during cellular respiration is less than the total amount taken in for photosynthesis, plants are carbon dioxide sinks that ultimately lower the amount of carbon dioxide in the atmosphere.

Ⓠ Why do leaves change color in the fall?

Ⓐ For most of the year, leaves are photosynthesis factories, using sunlight, water, and carbon dioxide to make sugar. A key component of the photosynthetic machinery is a pigment called **chlorophyll**, which absorbs red and

CHLOROPHYLL
The dominant pigment in photosynthesis, which makes plants appear green.

Beating the Heat: Some Plants Conserve Water

→ All plants must perform the "photo" reactions of photosynthesis during the day. But many plants experience reduced levels of photosynthesis in hot and dry conditions. Other plants are adapted to live in hot and dry climates, and have different strategies to take in CO_2 while minimizing water loss.

Stomata open all day: too much water loss

CO_2 in H_2O out

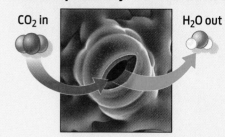

When stomata are open, carbon dioxide can enter the leaves for photosynthesis. However, this increases transpiration and therefore water loss.

Stomata closed all day: too little sugar produced

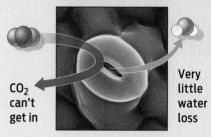

CO_2 can't get in

Very little water loss

In dry climates, some plants keep stomata closed during the hot hours of the day. This conserves water, but inhibits the uptake of carbon dioxide for photosynthesis. These plants make less sugar food.

Aloe ferox

To combat this problem, some plants in dry climates:

1. Conserve water by having thick, waxy succulent leaves.
2. Open stomata at night to capture CO_2, while reducing water loss by transpiration.
3. Extract CO_2 from airspaces within the plant even when stomata are mostly closed to conserve H_2O.

blue wavelengths of light and reflects green wavelengths. Chlorophyll is the reason plants appear green. To keep photosynthesis running, plants make abundant chlorophyll in spring and summer. But they also produce a smaller amount of other pigments—yellow-reflecting xanthophyll and orange-reflecting carotene. In the leaves, these pigments capture additional wavelengths of light and therefore expand the range of light that is effective for photosynthesis. You can't see these other pigments in leaves during spring and summer because leaves are chock full of green chlorophyll, camouflaging the other hues (although you can see them elsewhere in some plants: in the flesh of a pineapple, for example, or the root that is a carrot).

After a summer of intense sugar stockpiling, trees, bushes, and other deciduous plants that seasonally drop their leaves start to settle in for the winter and begin to shut down their photosynthesis machinery. During the winter months in temperate regions, there isn't enough sunlight or water to drive photosynthesis; days are shorter and the water in the ground is frozen and can't be absorbed. As temperatures cool and daylight wanes, plants turn off the production of their light-absorbing pigments. Of all the pigments, chlorophyll is the most chemically unstable and therefore the most short lived: levels fall quickly once production stops. By contrast, xanthophyll and carotene linger longer. As green fades, the other colors peak through—mostly yellow and orange, which have been there all along, but hidden.

The really intense colors—the fiery reds and deep purples—that some trees and bushes turn in autumn are the result of a fourth pigment, called anthocyanin. This distinctive chemical, which is

Plants Produce Multiple Light-Capturing Pigments

 Leaves contain chlorophyll and other pigments. These pigments help capture a wide range of wavelengths of light to maximize the efficiency of photosynthesis.

More Chlorophyll:

Green leaves produce an abundance of the green pigment chlorophyll during the warm, sunny months. They also produce smaller amounts of red and yellow pigments.

Less Chlorophyll:

When the hours of daylight are shorter and the temperature is cooler, the trees perform less photosynthesis and do not produce chlorophyll. The red, purple, and yellow pigments produced by the leaves are revealed.

---

the same one that gives apples and acai berries their color, is produced in leaves in response to high sugar concentration. Why does sugar concentration rise in the fall? As trees prepare to lose their leaves in winter, a corklike membrane develops between the leaf stem and the branch, sealing off the leaf and preventing phloem from transporting sugar out of the leaf. Unable to move out of the leaf, sugar begins to collect. The more sunny days there are in fall, the more photosynthesis occurs, the more sugar is created and trapped in the leaves. More sugar means more anthocyanin is produced, yielding the bright colors we associate with fall.

Some trees naturally produce lots of anthocyanin—maple and oak, for example—which explains why these trees produce such brilliant colors. Others, such as aspen and poplar, produce less anthocyanin and turn yellow or orange in fall **(Infographic 30.5)**.

Eventually the corky membrane that seals off the leaf from the tree causes the leaf to dry out. With a slight gust of wind, the leaf flutters to the ground. By shedding its leaves, a tree protects itself from water loss by transpiration in cold, dry air. The tree is now prepared for winter.

PLANT REPRODUCTION

Q Do plants have sex?

A They may not go on blind dates or place personal ads, but plants are among the most "flirtatious" and sexually active organisms on earth. The many bright colors, provocative shapes, and alluring fragrances of flowers are the plant equivalents of lipstick, muscle shirts, and perfume—evolutionary novelties designed to attract pollinating suitors.

Like other sexually reproducing creatures, plants have distinct male and female reproductive

STAMEN
The male sexual organ of a flower.

POLLEN
Small, thick-walled structures that contain cells that will develop into sperm.

PISTIL
The female reproductive organ of a flower.

POLLINATION
The transfer of pollen from a male stamen to a female pistil.

organs that produce haploid gametes, sperm and egg. In flowering plants, male and female sexual organs are housed within a plant's flowers. The male sexual organ, called a **stamen,** consists of a stalklike filament that supports a structure, the anther, that contains **pollen.** Pollen grains contain cells that will develop into sperm. The female structure, called a **pistil,** consists of a tubelike style topped with a sticky stigma, a "landing pad" for pollen. At the base of the pistil is a plump ovary stuffed with egg-containing ovules. Transfer of pollen from a male stamen to a female pistil results in **pollination,** which may then lead to fertilization. To fertilize eggs, sperm from the pollen must travel down the pistil to the ovule, where the eggs are located. When a haploid sperm fuses with a haploid egg, the result is a diploid plant embryo.

> **Plants are among the most "flirtatious" and sexually active organisms on earth.**

In some flowering plant species, male and female organs are found on separate male and female plants. Ginkgo trees and holly bushes are examples of such unisexual plants. More commonly, however, flowering plants–from dandelions to lilies–are hermaphrodites: their flowers contain both male and female reproductive parts. Fittingly, a few such hermaphroditic plants can self-pollinate–transferring pollen from stamen to pistil within the same flower or plant (Mendel's pea plants could do this). But most hermaphroditic plants have evolved ways to prevent self-pollination and to encourage outcrossing instead. They may have pistils and stamens of different heights, for example, or their pollen may be chemically rejected by eggs of the same plant.

Plants Reproduce Sexually

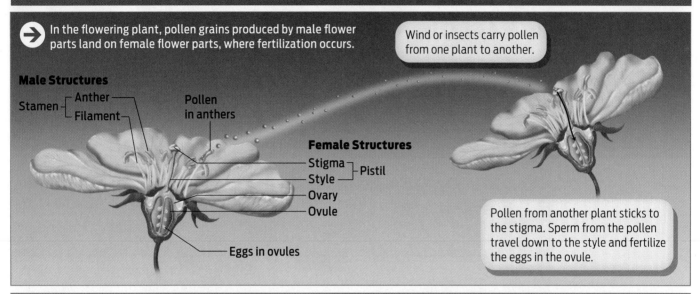

→ In the flowering plant, pollen grains produced by male flower parts land on female flower parts, where fertilization occurs.

Wind or insects carry pollen from one plant to another.

Male Structures

Stamen
- Anther
- Filament

Pollen in anthers

Female Structures

Stigma
Style
} Pistil

Ovary

Ovule

Eggs in ovules

Pollen from another plant sticks to the stigma. Sperm from the pollen travel down to the style and fertilize the eggs in the ovule.

Birds and bats are also essential pollinators for many flowering plants.

Plants have evolved many elaborate ways to spread their pollen from plant to plant. For many flowering plants, pollinators such as bees and hummingbirds transfer pollen grains from the anther of one flower to the stigma of another. Other plants rely on wind to transfer pollen (Infographic 30.6).

Not all plants are so sexually adventurous. Some are asexual, or have asexual phases, and can create new individuals through underground root runners or bulbs that develop into whole new plants. Asexual reproduction, in the form of cuttings and grafting, is especially important in agriculture. A cutting taken from one plant can be planted in soil to form a new plant; sugar cane and pineapples are often reproduced this way. A cutting taken from a plant can also be grafted to the root system of a different plant–a procedure commonly used to perpetuate vineyard grapes.

❓ What spins like a helicopter, shoots like a rocket, and contains its own parachute?

A The answer to this biological riddle: a seed. Unlike animals, land plants cannot move to seek out more hospitable living conditions for their offspring when resources are scarce or the neighborhood gets too crowded. Forest or desert, valley or mountaintop, plants are stuck where nature put them. Their solution to this enforced sedentary existence is to disperse their offspring far and wide through seeds. A **seed** is a small embryonic plant contained within a sac of stored nutrients—essentially, a plant starter kit. It develops from a fertilized egg and is a perfect package for delivering an immature plant to its new home. It's also a tremendously successful evolutionary adaptation, which is why two types of seed-bearing plants dominate the plant world: **gymnosperms,** which include seed-cone-producing conifers, and **angiosperms,** the flowering plants. "Gymnos" is Greek for "naked," so the name literally means "naked seeds"; in a pinecone, for example, seeds sit nestled under the cone's scales. In angiosperms, seeds are located inside a fruit or nut: "angio" is derived from the Greek for "vessel" or "container." Angiosperms make up nearly 90% of all plants—including most of our agricultural crops (see Chapter 22).

There are many shapes and sizes of seeds, and they are dispersed in myriad ways. Some are small and hitch a ride on fur or clothing by means of tiny hooks or burrs. Some, like coconuts, are large and buoyant and can float across oceans to reach distant beaches. The delicate parachutes of dandelions, the spinning helicopters of maples and pine trees, and other lightweight seeds sail on the wind. Cottonseeds are little more than hairy specks of dirt, but on a steady breeze they can windsurf for miles. Other seeds are packaged inside fruit: tempted by its bright color and sugary content, animals eat the fruit and then deposit the seeds in feces some distance away from the original plant. Some seeds are dispersed through a ballistic mechanism. The seedpod of the squirting cucumber (*Ecballium elaterium*), for example, fills with slimy juice as it ripens. Eventually, the mounting pressure of

> **Land plants cannot move to seek out more hospitable living conditions for their offspring when resources are scarce.**

the increased volume of juice causes the cucumber to shoot off the plant like a rocket, trailing a plume of seeds and slime in its wake. At the slightest touch, the bulging seedpods of the aptly named touch-me-not plant explode, spraying seeds like bullets. When a seed lands in favorable conditions, with enough water, it will germinate and grow into a young plant, or seedling **(Infographic 30.7)**.

PLANT HORMONES

Q Can plants see?

A Observe an old building and you'll likely see ivy scaling up its walls. Keep houseplants next to a window and you'll find them bending toward the sunlight. How does a plant know where it's going? While plants don't have eyes and therefore cannot see, they are quite adept at sensing and responding to their environment, which they do through various kinds of tropism (from the Greek "tropos," meaning "turn").

The growth of a plant shoot toward light is called **phototropism.** This is how leaves get the sunlight they need for photosynthesis, and it's why ivy climbs up walls and houseplants lean into the window. Shoots grow towards the light because of **auxin,** a plant hormone that promotes cell elongation as one of its effects. When light hits one side of a plant shoot, auxin moves to the shaded side, creating a gradient of the hormone in the stem. The side receiving the most direct sunlight contains the least auxin, while the shaded side contain the most. Auxin promotes elongation of cells on the shady side of the stem. This causes the shaded side to elongate faster than the sunny side, pushing the stem toward the sun. The whole stem doesn't sense light and produce auxin, though—just the tip of the stem. Cover the tip of a young plant and it won't turn toward the light, but will instead grow straight up because, in the absence of

SEED
An embryonic plant contained in a protective structure.

GYMNOSPERMS
Cone-bearing seed plants.

ANGIOSPERMS
Flowering plants.

PHOTOTROPISM
The growth of the stem of a plant towards light.

AUXIN
A plant hormone that causes elongation of cells as one of its effects.

INFOGRAPHIC 30.7

Seeds Carry a Young Plant to a New Destination

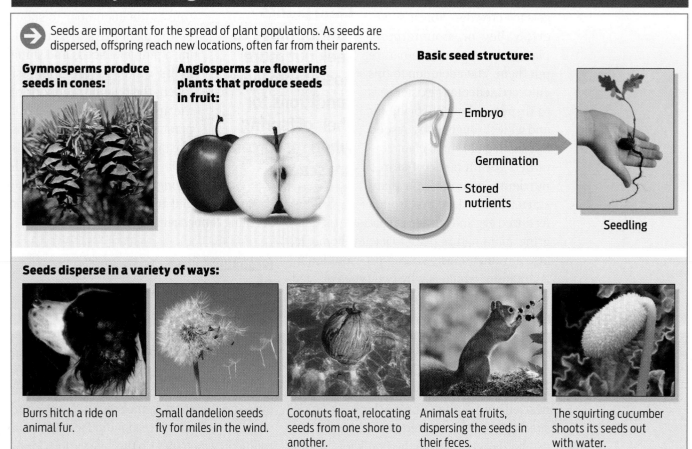

→ Seeds are important for the spread of plant populations. As seeds are dispersed, offspring reach new locations, often far from their parents.

Gymnosperms produce seeds in cones:

Angiosperms are flowering plants that produce seeds in fruit:

Basic seed structure:

— Embryo

Germination

— Stored nutrients

Seedling

Seeds disperse in a variety of ways:

Burrs hitch a ride on animal fur.

Small dandelion seeds fly for miles in the wind.

Coconuts float, relocating seeds from one shore to another.

Animals eat fruits, dispersing the seeds in their feces.

The squirting cucumber shoots its seeds out with water.

---

sunlight, auxin is not differentially localized to one side of the stem.

Other mechanisms help a plant sense where it is and orient itself in space. **Gravitropism** is the growth of plants in response to gravity– roots grow downward, with the force of gravity, and shoots grow upward, against it. Auxin is again the main player in this mechanism. When a plant is placed on its side, more auxin is sent to the down side of the stem, in the direction of gravity. This causes the cells on the down side to elongate more. Stems begin to curve away from gravity. Root cells, however, respond the opposite way to auxin: more auxin on the gravity side of roots inhibits root cell elongation on the down side, so roots bend toward gravity. Gravitropism allows a planted seed to send its shoots toward the light and its roots toward the soil. The same chemical signal–auxin–is able to produce these opposite effects because different tissues respond differently to the hormone.

A plant's sense of touch is called **thigmotropism;** it's how vines sense their way around poles or trellises and carnivorous

An example of gravitropism in the remains of a cellar in a Roman villa in the Archeologic Park in Baia, Italy. The leaf-bearing stems of this upside-down tree bend upward away from gravity.

GRAVITROPISM
The growth of plants in response to gravity. Roots grow downward, with gravity; shoots grow upward, against gravity.

THIGMOTROPISM
The response of plants to touch and wind.

ETHYLENE
A gaseous plant hormone that promotes fruit ripening as one of its effects.

plants sense their prey. Touch-sensitive growth is also controlled by auxin. Vine cells touching a pole, for example, get less auxin and consequently elongate less, while cells on the other side elongate more. The result is another kind of lopsided growth in the shoot, which eventually causes the vine to coil around whatever it's touching. A plant's sense of touch can be exquisitely sensitive–more sensitive than a human's. A human can detect the presence of a thread weighing 0.002 mg laid across the arm. By contrast, the feeding tentacle of the insectivorous sundew plant can sense a thread of less than half that weight. The legs of a single gnat are enough to trigger the tentacle into swift action **(Infographic 30.8)**.

Q Does one bad apple really spoil the bunch?

A This old adage is true, and its truth can be shown empirically: put a ripe apple in a bowl of unripe ones, and the unripe neighbors will quickly ripen. That's because ripe fruit–bananas and apples, especially–produce **ethylene,** a gaseous plant hormone, one effect of which is to promote ripening. In a confined space, the ethylene gas collects and causes nearby fruit to ripen through the loss of chlorophyll and the breakdown of cell walls. The result is the conversion of a hard, green fruit to a soft, ripe one. Commercial fruit growers take advantage of the action of ethylene when they ship fruit

INFOGRAPHIC 30.8

Plants Sense and Respond to Their Environment

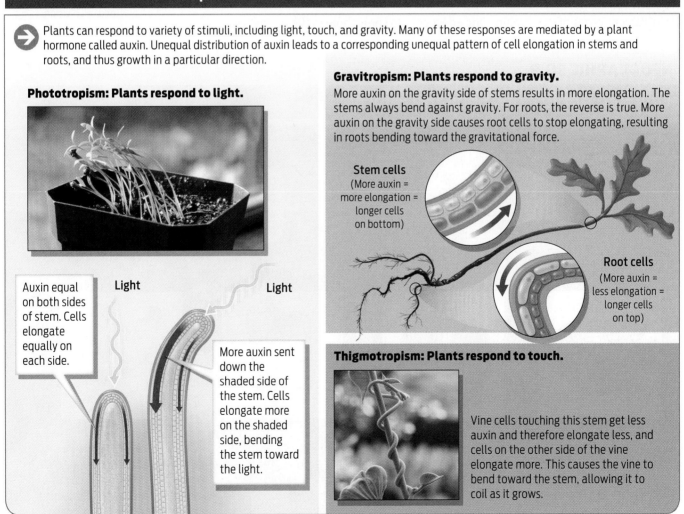

Plants can respond to variety of stimuli, including light, touch, and gravity. Many of these responses are mediated by a plant hormone called auxin. Unequal distribution of auxin leads to a corresponding unequal pattern of cell elongation in stems and roots, and thus growth in a particular direction.

Phototropism: Plants respond to light.

Auxin equal on both sides of stem. Cells elongate equally on each side.

Light

Light

More auxin sent down the shaded side of the stem. Cells elongate more on the shaded side, bending the stem toward the light.

Gravitropism: Plants respond to gravity.

More auxin on the gravity side of stems results in more elongation. The stems always bend against gravity. For roots, the reverse is true. More auxin on the gravity side causes root cells to stop elongating, resulting in roots bending toward the gravitational force.

Stem cells
(More auxin = more elongation = longer cells on bottom)

Root cells
(More auxin = less elongation = longer cells on top)

Thigmotropism: Plants respond to touch.

Vine cells touching this stem get less auxin and therefore elongate less, and cells on the other side of the vine elongate more. This causes the vine to bend toward the stem, allowing it to coil as it grows.

to distributors. Often, fruit is picked while still green and then exposed to natural or synthetic ethylene just before arrival in grocery stores to hasten ripening. This method works especially well with tomatoes, avocados, bananas, and cantaloupe. In some cases, fruit growers remove ethylene from storage containers in order to prevent ripening so that fruits can be stored for a long time—apples that are picked in fall and sold in summer, for example (**Infographic 30.9**).

As a plant hormone, ethylene does more than ripen fruit. It is responsible for a number of different aging effects in plants—including leaf dropping in autumn.

Q Can plants take hormones to improve their performance?

A In a manner of speaking, yes. **Gibberellins** are plant hormones that promote growth. Scientists have identified more than 100 different types of gibberellin. One effect of gibberellins is to provide the chemical cue for seeds to germinate and grow. When conditions are right for a seed to germinate—when rising temperatures begin to melt frost, for example—these growth-promoting hormones give the green light for a seedling to grow.

Applying gibberellins to a young plant can increase the length of its stem, which also indirectly increases the size of its fruits—a fact that makes them very useful in agriculture. Gibberellins are commonly used to increase the size of seedless grapes, for example. On an untreated seedless grape plant, the stem remains relatively short, so the bunches of grapes growing on the stem are clustered densely together, resulting in small grapes. When sprayed with gibberellins, the stems grow longer, giving the grapes more room to grow. Because seeds are the normal source of gibberellins, seedless grapes have no source of gibberellins to help them grow naturally, which is why farmers need to spray them in the first place.

GIBBERELLINS
Plant hormones that cause stem elongation and cell division.

INFOGRAPHIC 30.9

One Bad Apple Does Spoil the Bunch!

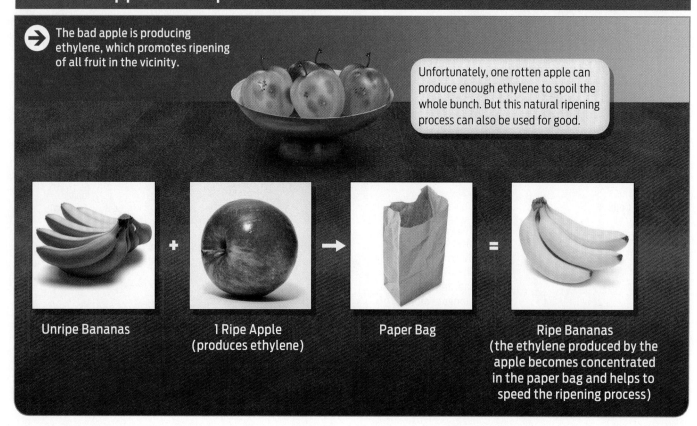

→ The bad apple is producing ethylene, which promotes ripening of all fruit in the vicinity.

Unfortunately, one rotten apple can produce enough ethylene to spoil the whole bunch. But this natural ripening process can also be used for good.

Unripe Bananas + 1 Ripe Apple (produces ethylene) → Paper Bag = Ripe Bananas (the ethylene produced by the apple becomes concentrated in the paper bag and helps to speed the ripening process)

ABSCISIC ACID (ABA)
A plant hormone that helps seeds remain dormant.

Gibberellins play many roles in modern agriculture. Some genetically modified crop plants are *less* sensitive to gibberellins and as a result produce dwarf varieties. Dwarf wheat plants, for example, produce shorter stems and remain small. Yet when treated with fertilizer to encourage growth by providing important nutrients, these dwarf plants devote more of their energy and resources to making wheat grains rather than growing tall stems, greatly improving the yield of wheat grains relative to the less valuable stems (**Infographic 30.10**).

In many species of plant, seed germination is controlled by the balance between the growth-promoting properties of gibberellins and the growth-inhibiting effects of another hormone, **abscisic acid (ABA).** You can think of ABA as the brakes and gibberellins as the gas pedal. Heavy spring rainfalls dilute the levels of ABA in a seed, allowing the gibberellins to step on the gas, ushering in a new season's blooms.

INFOGRAPHIC 30.10

Hormones Trigger Plant Growth and Development

 Certain plant hormones regulate growth and development of plants. When humans apply these hormones to crops, plant growth can be dramatically enhanced.

Gibberellin hormones enhance plant growth:

This pumpkin and the larger bunch of grapes on the right were treated with gibberellin hormones. Spraying plants with gibberellins enhances the number and size of the fruit produced.

The relative amounts of ABA and gibberellin trigger germination:

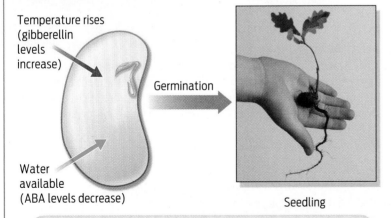

Temperature rises (gibberellin levels increase)

Germination

Water available (ABA levels decrease)

Seedling

Gibberellins are produced when temperatures rise. ABA is diluted by spring rains. The presence of more growth hormone and the absence of inhibitory ABA result in germination and seedling growth.

PLANT DEFENSES

🔍 Why are some plants poisonous?

A From juicy peaches and succulent strawberries to zesty basil and peppery arugula, many plants are incontestably delicious. Humans aren't the only ones who think so. Many herbivores such as insects, birds, and rodents find plant parts tasty and irresistible. This is both helpful and hurtful to a plant. On the one hand, plants rely on animals to eat their fruits and disperse their seeds. On the other hand, plants must ensure that only noncrucial parts of the plants are eaten by other organisms. Eating a plant's fruit is one thing; eating all of its leaves is quite another.

Over the course of their evolution, plants have adapted in many ways to protect their important parts from an herbivore's chomping. Some defenses are mechanical: the stems of a raspberry plant are covered in prickly spines to prevent unwanted chewing; holly leaves are waxy and difficult for insect jaws to grasp. Other defenses are chemical: leaves of the tobacco plant produce nicotine, which is toxic to insects; the bark of the South American evergreen cinchona tree produces quinine, an extremely bitter substance that many animals find distasteful (except certain humans, who use it in their gin and tonics). Such antiherbivory chemicals are a highly effective way to deter pests from eating a plant's leaves.

While a plant's fruits are often tasty and meant to be digested, seeds generally are not. The seed contains a new plant, and therefore it must be protected. Many seeds are encased within an indigestible shell that prevents them from being destroyed by an animal's stomach juices. The unlucky animal that succeeds in breaking open the shell and eating the seed is in for an unwelcome surprise. Seeds are sources of some of the most potent poisons on earth, including ricin, cyanide, and strychnine. Ricin, found in castor beans, can be lethal to animals in quantities as little as two beans. Cyanide, which is found in small doses in the seeds of peaches, apricots, and apples, kills by interrupting cellular respiration in mitochondria; unable to make ample amounts of the short-term energy-storage molecule ATP, nerve and muscle tissues quickly shut down (**Infographic 30.11**).

Seeds are sources of some of the most potent poisons on earth, including ricin, cyanide, and strychnine.

Plants also use chemicals to combat other plants. Blue gum eucalyptus trees (*Eucalyptus globulus*), for example, secrete a sticky gumlike substance that acts as a deterrent to the germination and growth of noneucalyptus plants in the nearby vicinity.

Perhaps the most interesting method of deterrence that plants use is a kind of unwitting alliance. Some plants, such as wild tobacco, emit potent vapors when they are eaten by insect pests. The vapors, in turn, attract natural predators of the insects, which are thus enlisted in the plant's defense. This "enemy of my enemy" approach is an example of mutualism (see Chapter 22).

Individual plants can even communicate with other individuals of their species and unite against a common enemy. At the first sign of grazing by a hungry giraffe or elephant, for example, African acacia trees release a bad-tasting poison into their leaves. At the same time, they release a gaseous chemical–the versatile hormone ethylene–which drifts out of the stomata of their leaves. Other acacias in a 50-yard radius detect the gas and are prompted to start releasing poison in their leaves, too.

Although antiherbivory chemicals complicate an herbivore's life, they are often quite useful to

INFOGRAPHIC 30.11

Plants Defend Themselves

Physical Defenses:

Raspberry thorns can impale hungry insects.

The jaws of insects that try to feed on holly leaves may not be strong enough to penetrate their waxy coat.

Chemical Defenses:

The leaves of the tobacco plant contain nicotine which is toxic to insects.

The bark of the cinchona tree contains the bitter chemical quinine, which deters insects from feasting on it.

Peach pits have a tough exterior that protects the embryo inside. If an animal is successful in cracking the seed open, it is met with lethal toxins like cyanide.

When a hornworm feeds on a wild tobacco plant, the plant releases chemicals that both repel the worm and attract the worm's predators.

humans. Some of our most important medicines, in fact, are extracts of plant chemical defenses, including aspirin, morphine, digitalis, and the anticancer drug paclitaxel (its proprietary name is Taxol), which was originally obtained from the bark of the Pacific yew tree. ■

◉ Summary

■ Nearly all plants share the same basic structure: an aboveground shoot system that includes the stem and photosynthetic leaves, and an underground root system that anchors the plant in the soil and absorbs water and nutrients.

■ All plant cells are surrounded by a cell wall made of the complex carbohydrate cellulose. A central water-filled vacuole creates turgor pressure against the cell wall and helps a plant stand up.

■ Some plant tissues have cells with an additional inner cell wall made of cellulose and lignin. Lignin is extremely durable and is what makes plant tissues woody.

■ Plants have a vascular system. Xylem transports water and dissolved nutrients from the roots to the shoots. Phloem transports dissolved sugars throughout the plant body.

■ Water transport is powered by evaporation through stomata in leaves, a process known as transpiration. Plants have mechanisms to control the amount of water lost by transpiration.

■ Plants undergo primary (vertical) and secondary (horizontal) growth. Secondary growth is mostly growth of xylem tissue. Tree rings reflect the growth of xylem tissue from season to season.

■ Plants are autotrophs, able to make their food from sunlight and carbon dioxide (by photosynthesis). They also require a few additional nutrients, primarily nitrogen. Plants obtain nitrogen from the soil, where bacteria convert it into a usable form. In nutrient-poor soils, some plants obtain nitrogen by trapping and digesting insects.

■ Some plants have adapted to hot, dry climates keeping their stomata closed during the day and by collecting carbon dioxide for photosynthesis primarily at night.

■ A variety of pigments assist with photosynthesis. Chlorophyll is the predominant of these pigments and is responsible for the green color of plant leaves. Other pigments contribute to photosynthesis and are visible in the fall, after chlorophyll has broken down.

■ Many plants reproduce sexually and have male and female sexual organs that produce haploid gametes. Plants can be either unisexual or hermaphroditic.

■ Some plants rely on pollinators such as insects to deliver male pollen to female reproductive structures. Other plants use wind to deliver pollen to the female structures.

■ Seed-bearing plants (gymnosperms and angiosperms) produce embryonic plants encased within a protective seed that can disperse a great distance from the parent plant.

■ Plants respond to their environment by various tropisms. Phototropism is growth toward light; gravitropism is growth in response to gravity; thigmotropism is growth in response to touch or wind.

■ Plants produce hormones that contribute to growth and development: ethylene contributes to fruit ripening; gibberellins and ABA regulate germination and stem growth; auxin controls cell elongation.

■ Plants have physical and chemical defenses to fend off herbivores. Humans make use of these defenses in developing pharmaceuticals.

Chapter 30 Test Your Knowledge

Plants have unique structures, as well as specific nutritional and reproductive strategies.

⊛ KNOW IT

1. Draw and label a diagram of two adjacent plant cells. Include key intracellular structures.

2. Which of the following statements represents a true distinction between xylem and phloem?
a. Xylem provides support only; phloem provides transport.
b. Xylem provides water and nutrient transport; phloem provides sugar transport.
c. Xylem transports materials from shoots to roots; phloem transports materials in either direction.
d. Xylem transports sugars in either direction; phloem transports water from roots to shoots
e. all of the above

3. What is the function of the cuticle?
a. It enables neighboring cells to stick together.
b. It provides rigidity to the cell wall.
c. It is toxic to many herbivorous insects.
d. It prevents water loss.
e. It is sticky and helps pollen to stick to a plant during pollination.

4. Plants are autotrophs and can make sugar from CO_2. How do they obtain their CO_2?
a. through stomata
b. by absorption through the root system
c. by digesting insects
d. by breaking down carbon-rich carbohydrates stored in roots
e. a and b

5. What characterizes a dormant seed?
a. the presence of sperm
b. the presence of eggs
c. the presence of an embryonic plant
d. the absence of ABA
e. the presence of several pollen grains

6. Describe how fertilization follows pollination in an angiosperm. What has to happen, and what plant structures are involved?

7. Which of the following pigments are present in green leaves late in the summer?
a. chlorophyll
b. carotene
c. xanthophyll
d. a and b
e. all of the above

8. Which hormone helps fruit to ripen?
a. auxin
b. ethylene
c. gibberellins
d. estrogen
e. anthocyanin

9. If you wanted a plant to grow really tall, which hormone should you apply?
a. auxin
b. ethylene
c. gibberellins
d. anthocyanin
e. ABA

⊛ USE IT

10. Paper is made from wood that is broken down to pulp. Why are lignin-digesting enzymes included in the pulping process? Would these enzymes have to be included in the pulping process if paper were made from green leaves? Explain your answer.

11. Describe the "conflict" that plants face with respect to opening and closing their stomata.

12. Are trumpet pitchers strict autotrophs? Explain your answer.

13. If you applied to the soil around your plants a chemical that kills bacteria (but not plants), why might your plants die?

14. Why may the bright coloration of a trumpet pitcher have a different function than the bright colors of yellow or orange squash blossoms? (Think about this: squash are pollinated by bees.)

15. If a plant could not make chlorophyll, would you expect it to survive? Why or why not?

16. Why do seedless grapes need hormone treatment to develop big clusters of big grapes, while seeded varieties can develop large fruits without exogenous (that is, externally applied) hormones?

17. Nopales are cactus pads (the large, thick "leaves" of the prickly pear cactus) and make a delicious salad. What antiherbivory mechanism fails when we succeed in making ensalada de nopales—prickly pear salad?

SCIENCE AND ETHICS

18. Paclitaxel (Taxol) is a cancer chemotherapy drug that was initially discovered in the bark of the Pacific yew tree (*Taxus brevifolia*). This species is very slow growing, and occurs only in a small region of the Pacific Northwest. Discuss the conflicts between human needs (for example, anticancer drugs) and the impact of those needs on sensitive plant populations. What are possible solutions to these tensions?

Answers

Chapter 1

1. c

2. e

3. Peer review means that the study has been reviewed by other scientists who are considered experts in that particular field. Ideally, both the results of the study and the methods used to conduct the study are reviewed and the article is rejected or revised accordingly. Peer review is important to ensure that sloppy studies are not reported and to ensure that scientists draw only appropriate conclusions from their results.

4. b

5. c

6. Ideally, the characteristics of both the control and experimental groups should be as close to identical as possible, meaning that there should be no intentional significant differences between the two groups with regard to age or breast cancer status. Because the study is intending to look at risk of developing breast cancer, both groups should be composed of women who do not currently have breast cancer.

7. The group of participants should be large and include frequent, infrequent, and non-coffee drinkers. The group should be composed of members of both sexes and multiple nationalities, ethnicities, ages, and socioeconomic groups. Randomly divide participants into two groups and give one group (the experimental group) caffeine in a drink and the other group (the control group) a placebo (no caffeine in the drink). Participants should not know whether they are part of the experimental or the control groups (the study is "blind"). Additionally, the scientist conducting the study should not know who is receiving which treatment until all the data are collected (the study is "double blind"). Multiple tests for brain function should be given. A larger study could include multiple experimental groups which receive differing amounts of caffeine to test dosage.

8. b

9. a

10. No. The study shows a correlation between abstaining from caffeine consumption and developing Parkinson disease but does not show that caffeine consumption can prevent the development of the disease.

11. a: No. The opinion expressed is anecdotal at best, meaning that it is reporting the experience of only a single person. Further, there is little reason to believe that the testimonial is even truthful; it is likely a paid endorsement, and there are no regulations regarding the truth of such endorsements. b: Results of either an experimental or epidemiological study reported in a peer-reviewed scientific journal.

12. Answers may vary. News organizations, and the corporations that own these organizations, are ultimately responsible for reports in the media. Scientists currently have little control over such reports.

13. Answers may vary. Items for the checklist may include the following:
    Was there a study done?
    Who conducted it?
    Was it peer reviewed?
    Was it reported in a respectable scientific journal?
    Who were the subjects in the study?
    Was it an experimental study or an epidemiological study?
    How many subjects?
    If the study was experimental, was it randomized?
    Do the authors of the study think that the results merit a behavioral change?

Chapter 2

1. c

2. Homeostasis is the ability of a cell or an organism to maintain a stable internal environment, usually in terms of pH, temperature, and chemical makeup, even when the external environment changes. The processes and molecules of life are delicate, and therefore homeostasis is generally important because small changes can destroy these molecules or disrupt the processes of life.

3. b

4. e

5. a

6. A polymer is a molecule is composed of smaller, sometimes repeating, subunits called monomers. Examples include complex carbohydrates (polysaccharides), which are composed of smaller sugar molecules called monosaccharides; proteins, which are composed of subunits called amino acids; and nucleic acids (DNA and RNA), which are composed of subunits called nucleotides.

7. a

8. Answers may vary. The characteristics of life are described in the text. Additional ideas include looking for common molecules representative of life on earth, including complex organic molecules, macromolecules, and water.

9. Answers may vary. However, dead organisms will generally still have a cellular structure and will not be reproducing, sensing and responding to the environment, or using energy.

10. Arguments for: Viruses reproduce. Viruses are generally able to maintain a stable internal chemical environment and make up for a lack of complex homeostatic mechanisms by being more resistant to harsh environments. Viruses can sense and respond to certain stimuli, such as detecting the presence of living host cell to invade. When inside living cells, viruses utilize the resources inside the cell for energy. Arguments against: Viruses require other living organisms in order to reproduce. Viruses generally do not grow, that is, they do

not increase in size, but only replicate. Viruses do not obtain or use energy on their own. Unlike living organisms, viruses are not made up of cells.

11. Life would be likely, but there is not sufficient evidence to conclude that life is present. There are nonliving (abiotic) means of producing glucose from carbon dioxide and water.

12. a: Sterols and triglycerides differ both in structure and in function. Sterols are composed of four carbon rings and function as color-producing pigments, hormones, or components of membranes (such as the cell membrane). Triglycerides are composed of three fatty acids covalently linked to the molecule glycerol. Triglycerides commonly function as energy storage molecules which in animals also serve to thermally insulate the organism from the environment. b: Phospholipids and triglycerides both have a glycerol backbone to which two fatty acids are attached. The difference is that in the third position either a third fatty acid is attached (triglycerides) or a phosphoryl-containing group (containing oxygen and the element phosphate) is attached (phospholipids). Because of this difference in structure, these molecules have different chemical properties and biological functions.

13. Olive oil is made up of triglycerides, which are nonpolar molecules that are hydrophobic ("water-fearing"). Salt is made up of Na^+ and Cl^- ions that are charged and interact with the polar ends of water molecules. Therefore salt is considered hydrophilic.

14. e

15. aqueous. a: The solvent is water. b: The solute is sugar. Additionally, both coffee and tea contain many other organic molecules that become dissolved in the water when tea or coffee is brewed. These molecules are produced by the tea leaves or coffee bean, and are extracted and dissolved into the hot water when brewed.

16. c

17. a

18. Hydrogen bonds and ionic bonds are both electrostatic attractions between charged atoms of two different molecules. Both are noncovalent interactions. Hydrogen bonds are between partially charged atoms (usually atoms of a water molecule), whereas ionic bonds are between ions, which have fully positive or negative charges.

19. Oil does not dissolve in the aqueous vinegar because the oil is hydrophobic. Water molecules are strongly attracted to one another (via hydrogen bonds) but not to the oil. Water excludes the oil in favor of interacting with itself or other hydrophilic molecules. Because salt is hydrophilic, it will dissolve in the vinegar but not in the oil. The ions in salt will remain together in oil because they are attracted more strongly to each other (by ionic interactions) than to the nonpolar triglycerides that make up the oil. In the vinegar, the ions that make up salt will become dissolved because they are attracted to the polar water molecules.

20. Liquid water can absorb and store large amounts of heat without evaporating and lose large amounts of heat before freezing. Therefore, seaside towns are buffered against changes in temperature because heat is absorbed by the ocean during the day and is transferred to the air from the

relatively warm ocean water at night. In contrast, the desert sand is made primarily of silicon dioxide, which cannot absorb nearly as much heat during the day as can water, and rapidly cools at night.

21. b

22. Answers will vary. Some considerations if Martian dirt samples are brought to earth are disease; competition with native species (the invasive species effect); ecological effects, public fear. If an earth life form is released on Mars, effects may be destruction of the Martian environment and competition with Martian life forms if any are present, and possible extinction of those Martian life forms. Ethical implications include effects on religious views, cultural effects, the possibility of creating a disease-causing organism, and possible medical breakthroughs.

Chapter 3

1. Cell theory posits that the fundamental units of life are cells. All living organisms are made up of one or more cells. Cells arise only from other living cells.

2. d

3. prokaryotic cells and eukaryotic cells

4. a: No. According to cell theory, neither viruses nor prions (the protein aggregates that cause mad-cow disease) are considered to be living organisms. b: No. Nonliving agents can cause disease.

5. All living organisms contain genetic instructions in the form of DNA. All living organisms also synthesize the four classes of biological molecules (proteins, carbohydrates, lipids, and nucleic acids). Additionally, all known living organisms have a cell membrane and ribosomes.

6. e

7. c

8. a: Both involve the movement of a solute moving down a concentration gradient (that is, moving from a higher concentration to lower concentration). In both cases, additional energy is not required. b: Both facilitated diffusion and active transport require the function of a protein embedded in a membrane. However, the solute is moving in the opposite direction (relative to its concentration gradient) in these two cases.

9. Facilitated diffusion is necessary for molecules that cannot cross the phospholipid bilayer easily by themselves. Generally this is true of larger molecules and charged molecules. Small, nonpolar molecules, such as molecular oxygen (O_2) can pass freely through membranes, whereas polar molecules (like water) cannot.

10. c

11. Although the bacterial cell wall protects bacteria cells from lysis by keeping the cell from swelling, water can still escape from a cell placed in an environment that is high in salt. Because the concentration of salt outside the cell is high, the concentration of water is low. Water rushes out of the bacteria cell toward the lower concentration. As a result, bacterial cells shrink and die through desiccation. Some bacteria are more tolerant to high salt concentrations than others.

12. b

13. d

14. e

15. a: A mitochondrion (plural: mitochondria) is a rod-shaped organelle, approximately the size of some bacteria. Mitochondria are thought to have arisen by endosymbiosis, in which a bacterium is engulfed within another, larger cell. They are bound by a double membrane of two phospholipid bilayers. Mitochondria are considered the powerhouse of the cell; many of the reactions that extract energy from nutrients are housed within the mitochondria. b: The nucleus is a spherical membrane-bound organelle, bound by a double membrane composed of two lipid bilayers called the nuclear envelope. The nucleus houses the DNA of the cell. c: The endoplasmic reticulum (ER) is a network of single-membrane-bound tubes, actually an outgrowth of the outer layer of the nuclear envelope. The ER acts as a transport system within the cell. Proteins synthesized by ribosomes that coat part of the ER (the rough ER) are transported to other parts of the cell, often by first passing to the Golgi apparatus. d: The chloroplast is a membrane-bound organelle found in plants and other photosynthetic eukaryotes; also believed to have arisen from bacteria (again, by endosymbiosis). Reactions that capture energy from light are housed within the chloroplast, including the protein chlorophyll, which is responsible for the green color of plant tissues.

16. No. If you took an antibiotic that stopped bacterial reproduction, there would be no need for the bacteria to synthesize more peptidoglycan. So penicillin would not be able to interfere with new peptidoglycan production, as none would be synthesized in this situation.

17. Assuming that the concentration of solutes in the solution is the same as that inside the cell, then the cells will not burst, despite having weakened cell walls, because the osmotic pressure will not change–there will be no net movement of water into or out of the bacterial cells.

18. It is more challenging because both fungi and humans are eukaryotic organisms. Therefore the cell components of a fungal cell are much more similar to those of human than are those of bacteria. It is much harder to find a chemical that will selectively kill fungal cells without causing harm to human cells.

19. Muscles and nervous tissues are much more metabolically active than other tissues. Thinking and moving take a lot of energy. These tissues expend more energy and thus require more mitochondrial activity to produce energy. Tissues such as skin can rely more on alternative, but less effective, mechanisms for cellular energy production (which will be discussed later in the text).

20. Answers may vary. Physicians might explain that antibiotics will not have any effect on the viruses that cause flu or the common cold. Additionally, a physician might point out that using antibiotics decreases their effectiveness because of the development of antibiotic-resistant bacterial strains, which can be fatal. Because of the use of antibiotics over the past 60 years, there are now strains of the bacterium *Staphylococcus aureus* (and of other bacteria) that are resistant to all available antibiotics. For this reason, antibiotics should be used only when they are necessary to treat a bacterially caused disease.

Chapter 4

1. c

2. catabolic

3. c

4. e

5. b

6. c

7. a

8. A meal heavy in starch and fiber would cause less of a spike in blood sugar than a meal of only starch.

9. A diet rich in fiber will result in lower blood-glucose levels because the body cannot break the covalent chemical bonds holding the glucose monomers in fiber together. Therefore, it cannot be digested and only serves to slow the breakdown and absorption of other sugars. Fiber also has additional health benefits beyond controlling blood-sugar levels.

10. Because phospholipids are part of the cell membranes of both plants and animals, they can be directly acquired by consuming any food that contains cells, which includes all meats, vegetables, and fruits. However, phospholipids are not typically acquired directly from the diet in this way, but are "constructed" by anabolic reactions from fatty acids and other molecules. Fats and oils are the dietary sources of fatty acids.

11. Insulin causes cells of the body to take up glucose from the blood. Therefore, for a type I diabetic, who does not produce insulin naturally, insulin will be most effective if taken with, immediately before, or immediately after a meal, when blood-glucose levels are the highest or are rising. Blood-glucose levels steadily decline between meals.

12. a: calcium; b: calcium, vitamin D; c: vitamin C, very small amounts of calcium (not significant as a dietary source); d: vitamin C, calcium; e: vitamin D, very small amounts of calcium (not significant as a dietary source). None of the foods listed contains all three of these nutrients

13. e

14. Both cofactors and coenzymes are accessory molecules that enzymes use to accomplish their function. All coenzymes are cofactors, but not all cofactors are coenzymes. A coenzyme is a cofactor that is an organic molecule. Most vitamins are organic molecules that are consumed and are either coenzymes themselves or are converted into coenzymes.

15. b

16. c

17. The shape of the active site is very important in both the function of an enzyme and in specifying the substrates upon which the enzyme can act. Depending on how the shape of the active site is altered, the enzyme could act on different substrates, the enzyme could be completely nonfunctional, or (less likely) there could be no effect at all. Generally, however, most enzymes are highly evolved and delicately tuned molecular machines, and so most changes will result in a complete loss of enzyme activity.

18. High fever can cause the enzymes of the body to malfunction. This is the main danger of hyperthermia, in which the body becomes overheated. However, this same mechanism is believed to be the reason why fever is a common response by the body to infection: the increased body temperature causes the enzymes in bacteria or other infectious organisms to malfunction, slowing their growth and aiding their removal from the body.

19. Osteoporosis is a disorder characterized by the thinning of the bones through loss of bone density. It is common with increasing age and more common in women than in men. Diet and exercise can both help to reduce the effects of osteoporosis.
Exercise reduces risk by stimulating bone growth. Dietary calcium reduces risk by preventing calcium from being reabsorbed from bone when it is needed elsewhere by the body. Dietary vitamin D regulates calcium metabolism and prevents bone loss due to calcium reabsorption. Dietary vitamin C is required for new bone deposition.

20. Vitamin D is important for regulating calcium levels in the body, and vitamin D deficiency can lead to loss of bone mass. Vitamin D is not found in many foods and is mainly synthesized in the skin. A critical reaction in this synthesis requires UV light (from sunlight). Typically only a few minutes in intense summer sunlight are necessary, but during the winter, or at latitudes far from the equator, the sunlight is not intense enough to adequately produce vitamin D. Vitamin D deficiency is a growing concern in the United States and around the world.

21. a: Perhaps the primary advantage to a nutritionally engineered diet is that it requires little time to prepare, and may involve consuming foods that are less expensive than whole-food alternatives. Additionally, with such a diet it can be easier to keep track of both macro- and micronutrient consumption because the exact amounts of each of these components in the food is known. Finally, such a diet could allow people with certain dietary restrictions such as food allergies to consume foods rich in nutrients that would otherwise be rare in their restricted diet. Disadvantages include the possibility that people do not understand enough about nutrition to adequately construct diets in this way. Certain micronutrients that are found only in unprocessed food may provide health benefits that are not yet appreciated, and therefore would not be included in the engineered diet. b: Engineered diets provide a particular advantage to those individuals for whom it is especially important to control diet.

22. Answers may vary. Responses should consider what evidence is necessary to make such a claim, and whether it is worth delaying the release of a potentially helpful food alternative to wait for extensive scientific studies to be done. Another important factor to consider might be whether the studies that demonstrated the ability of this particular fiber to reduce blood-sugar levels have been repeated with type I diabetics.

Chapter 5

1. b

2. a

3. Algae appear green because the chlorophyll within algae absorbs red and blue wavelengths and reflects the green wavelengths of sunlight. Our eyes perceive the reflected green wavelengths.

4. c

5. Photosynthetic algae obtain energy by using sunlight to create sugars by photosynthesis. The energy stored in these sugars can be used by the algae to carry out essential life processes. Animals are not able to use the sun's energy directly. Instead, to obtain energy animals eat plants or other organisms that have eaten plants. The energy the animals use ultimately comes from the sun.

6. oxygen (O); carbon dioxide (I); photons (I); glucose (O); water (I)

7. Increasing carbon dioxide levels should increase photosynthesis because carbon dioxide is an input for the reaction. If forests become immersed in water the plants will not be healthy because they are not adapted to living in water; thus this aspect of climate change will produce a negative effect.

8. Lipids harvested from algae are more "useful" because there are fewer steps required before the lipids can be used. To produce fuel from carbohydrates found in plants the plants need to be broken down and fermented to produce ethanol. A lot of energy is lost in these processing steps.

9. c

10. c

11. You should eat the algae directly to gain the most energy. If you feed the algae to a cow first, the cow will use some of the energy, and thus the burger made from the cow will contain less of the energy from the algae.

12. a

13. g

14. Biodiesel from algae requires the transfer of energy from sunlight to organic molecules (including oils). There is one major energy transfer (sun to organic molecule). Lipids from animals are only produced by the animal after the animal has eaten a plant or another animal that ate a plant. There are more steps between sunlight and animal lipids than between sunlight and algal lipids, and some energy is lost at each step.

15. One major advantage of growing algae in enclosed tubes is that the conditions can be controlled to maximize oil production. Disadvantages are the need for providing sufficient CO_2 for photosynthesis and the cost of maintaining an elaborate growing system.

16. Algae may be the perfect solution to the food versus fuel debate. Algae can be grown on land that is not suitable for crops, like the desert, and algae is not a major human food source and therefore algae production does not compete with food production. A conflicting factor is that algae still require water for growth, but much less than a crop would require.

Chapter 6

1. d

2. d

3. Her BMI would be ~27, which would make her overweight according to the CDC.

4. e

5. a

6. Your lifestyle would need to be modified to compensate for the additional Calories, regardless of the type of food eaten. Ultimately, the only ways to avoid weight gain are either to reduce Calories eaten from other food sources or to increase the number of Calories burned, preferably by additional exercise.

7. a: 884g; b: 3,536 Calories (assuming perfect aerobic metabolism); c: 442g; d: She will run for 2.91 hours at this pace and burn an estimated 2,575 Calories in that amount of time. Again assuming perfect aerobic metabolism, it will take her 4 hours at this pace to burn through her glycogen stores. She can run 36 miles in this time at this pace, which is well beyond the 26.2 required to complete a marathon. e: Her body will need to extract energy from another fuel source. Fat (from adipose tissue) and protein (from muscle) are the next sources, in that order. However, it is important to note that in cases of complete carbohydrate starvation (such as this one), protein from muscle will still be broken down to provide glucose (blood sugar) to the brain. This happens because humans are incapable of making glucose from fats. This process of protein wasting continues until the brain adjusts to another energy source (ketone bodies), which can be produced from fats.

8. Many possible reasons have been suggested, among them smaller portions, less energy-rich but nutrient-poor foods, longer meal times, less snacking, and greater self-control at meal times.

9. c

10. d

11. e

12. b

13. See Infographic 6.10. The carbon atom (in CO_2) will be taken up into the spinach leaf, into the plant tissue, into the plant cell cytoplasm, and eventually into the chloroplast. There it will be converted, along with water (H_2O), into carbohydrate (glucose). This glucose may be used to construct the plant cell wall (fiber) or stored for later use by the plant (starch). If it is stored as starch, then when it is consumed by a human, the glucose will be broken down again into CO_2 by glycolysis and the citric acid cycle. Thus, the carbon atom will have come full circle. (If the glucose had been converted to fiber, other organisms, particularly fungi and bacteria, will also break down the fiber to glucose and then to CO_2, so the atom of carbon can always return to CO_2.).

14. a

15. In the presence of oxygen, aerobic respiration is the preferred method by which glucose is utilized by most cells. The process involves catabolism of the glucose molecule in a process called glycolysis. Following glycolysis, each carbon atom originally found in the glucose molecule is converted to CO_2 in the citric acid cycle. Electrons removed from glucose during this process are transferred to the electron transport chain, by which they are eventually transferred to oxygen to form water (along with two hydrogen atoms). The energy from the electrons transported in this way is captured by the cell as ATP. The entire process produces ~36 ATP from each glucose molecule. In the absence of oxygen, aerobic respiration is not possible. Cells utilize fermentation to acquire a small amount of ATP, but only two ATP per glucose molecule are produced by fermentation.

16. c

17. A common hypothesis is that by eating longer meals, the French eat more slowly, and therefore allow time for stomach to signal the brain to quit eating. Often we do not realize that we are full until several minutes, or dozens of minutes, after eating. Therefore, by eating quickly, we may consume far more food than is necessary to feel full. Because of this difference, longer meal times may result in the French eating less food (and fewer Calories) at every meal.

18. Answers may vary. See the discussion in "The culture of eating" section of the text.

19. Answers may vary. They should be accompanied by a justification of how a particular intervention might result in reduced obesity. Examples: (1) Taxing high-fat foods might encourage people to purchase foods that contain higher amounts of carbohydrates and proteins. Theoretically these foods might be lower in Calories, given the lower energy density of these other nutrients. (2) Financial incentives might encourage people to exercise more, thereby increasing total Calories burned. (3) Fruits and vegetables, particularly vegetables, are high in nutrients but relatively low in Calories. Encouraging their consumption might decrease overall Calorie consumption in the American public. (4) Providing incentives to teachers to encourage teaching about nutrition might increase awareness of nutrition and its effect. The idea is that those who understand the dangers of poor nutrition and/or obesity will be motivated to change eating and exercise habits.

Chapter 7

1. e

2. c

3. b

4. (1) The two original strands of the DNA molecules are separated by means of *heat*. (2) The enzyme *DNA polymerase* "reads" each template strand and adds complementary nucleotides to make a new strand.

5. Step 1. The strands separate:
ATCGGCTAGCTACGGCTATTTACGGCATAT

TAGCCGATCGATGCCGATAAATGCCGTATA

Step 2. DNA polymerase adds complementary nucleotides, forming two new double helices:.

ATCGGCTAGCTACGGCTATTTACGGCATAT
TAGCCGATCGATGCCGATAAATGCCGTATA

ATCGGCTAGCTACGGCTATTTACGGCATAT
TAGCCGATCGATGCCGATAAATGCCGTATA

6. a: F; b: T; c: F; d: F; e: T

7. Statement b in Question 6 is true because DNA is the molecule of heredity that is passed from parents to offspring. Statement e in Question 6 is true because all body fluids, including saliva, contain DNA that can be used for DNA evidence. Unfortunately, in the case of Roy Brown, the technicians were not able to extract enough DNA from the bite marks for PCR.

8. Statement a in Question 6 should be, "G pairs with C and A pairs with T." Statement c in Question 6 should be, "DNA consists of coding sequences, which encode information to produce proteins, and noncoding sequences, which regulate when coding regions are turned on or off." Statement d in Question 6 should be, "The number of STR repeats on your maternal chromosome can be different from the number of STR repeats on your paternal chromosome."

9. a

10. d

11. d

12. Lane B

13. a: Suspect B's profile matches the profile from the blood collected at the crime scene for all of the markers tested. b: Suspect A is most likely unrelated to the victim since they share very few STR bands. Suspect B is likely either the victim's parent or child since they share at least one band at each of the markers tested.

14. d

15. c

16. AMELY and AMELX

17. For AMELY: If the perpetrator is female, you would expect no bands, and if the perpetrator is male, you would expect one band. For AMELX: If the perpetrator is female, you would expect two bands or one thick band if the repeat is the same length on both X chromosomes. If the perpetrator is male, you would expect only one band.

18. a.

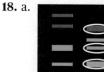

b: Lane M1 identifies the father of the child.

19. The advantages of having a DNA bank include clearing individuals who have been wrongly convicted and helping to identify suspects in unsolved cases. Banking DNA could be problematic, depending on who has access to the information and how it is used. For example, people's DNA markers can indicate their susceptibility to certain diseases, enabling insurance companies with access to this information to discriminate against those individuals. This is particularly troubling because the markers indicate a correlation and not a definitive likelihood of developing the disease.

Chapter 8

1. A protein's function is determined by the shape, which is determined by the interaction of the amino acids that make up the protein–specifically, the order of the amino acids and how their side chains interact.

2. c

3. The protein will not be able to function because the heat will cause the protein shape to change and shape is critical for function.

4. The insulin gene is made up of DNA nucleotides that provide information about how to make the insulin protein. The insulin gene is transcribed into mRNA in the nucleus; then the mRNA is transported into the cytoplasm and translated into insulin protein.

5. d

6. e

7. The problem is likely in the regulatory sequence because the patient has reduced levels of normal antithrombin. The regulatory sequence controls how much mRNA is made and therefore how much protein is made. Alternatively, the patient could carry one allele with changes in the coding sequence that make a nonfunctional protein. The patient would still have low levels of functional protein made from the allele with the normal coding sequence.

8. To increase the level of antithrombin, the regulatory sequence should be modified. The regulatory sequence controls the timing and location of transcription of antithrombin. The amount of mRNA made is directly related to the protein produced.

9. To express a gene in skin cells, combine the regulatory sequence from the melanin gene with the coding sequence of the gene of interest. The regulatory sequence of melanin is specific to skin cells, so the gene of interest will only be produced there. To express melanin in yeast cells, use the regulatory sequence from a yeast gene and the coding sequence from melanin. The coding sequence is necessary to produce the correct melanin protein.

10. The beta casein regulatory sequence was used to express antithrombin in milk because the beta casein gene is expressed only in the mammary glands. This was important to ensure that the goats were not harmed by the production of antithrombin.

11. RNA polymerase (N); ribosome (C); tRNA (C); mRNA (C) (mRNA is transcribed in the nucleus and transported to the cytoplasm for transcription, so it is active and carries out its main function in the cytoplasm.)

12. a: The complementary DNA strand is TCTATGCTTTGT. b: The complementary mRNA strand is UCUAUGCUUUGU. c: The mRNA sequence contains four amino acids–Ser, Met, Leu, Cys–but only three will be translated into protein. Met is the start codon where translation will occur, so Ser will not be translated. The final protein will include three amino acids: Met, Leu, Cys.

13. a: If RNA polymerase cannot bind to the regulatory region, the gene will not be transcribed; thus neither mRNA nor protein will be produced. b: A change in the coding sequence will not have an effect on transcription of the mRNA. Depending on the change in the sequence, the protein structure and levels could be unchanged if the change were to an amino acid with similar properties; or the protein could become nonfunctional because the new amino acid causes the protein to be shaped differently. A third possibility is that the change will create a stop codon in the middle of the protein, creating a truncated, nonfunctional protein. c: A change in the regulatory region that increases transcription will increase the amount of mRNA and protein that is produced. The function and structure of the protein will be normal. d: The change to the regulatory region will increase the level of transcription creating higher levels of mRNA. The change to the coding region may result in a nonfunctional protein resulting from changes in the shape of the protein. The combination of these changes will lead to an increase in nonfunctional protein which the cell will degrade; and the phenotype will be similar to that observed if the protein is not produced.

14. a: proline (Pro); b: proline (Pro); c: leucine (Leu)

15. The benefit of producing insulin in either pigs or bacteria is that more diabetics are able to live a long and healthy life because of the greater availability of insulin. The ethical question is whether or not we should modify organisms to produce unnatural proteins. One might argue that producing insulin in pig pancreas or bacteria is invasive and will harm the pig or bacteria. On the other hand, these human-insulin-producing animals were produced for the purpose, so harming the pig or bacteria for production of insulin is serving the greater good. In some ways it is easier to accept using bacteria to produce insulin: there are millions of bacteria, they have a short life cycle and rapid regeneration, and we can't see a single bacterium with the naked eye.

Chapter 9

1. e

2. c

3. Embryonic development, wound healing, and replacement of blood cells all require mitosis to create more cells.

4. c

5. Pregnant women should not take drugs that interfere with cell division because the cells of the developing embryo are rapidly undergoing cell division. If a woman were to take these drugs, the developing embryo would cease to grow and would die or have major defects.

6. b

7. Chemotherapy interferes with cell division to kill cancer cells. However, the drugs target any actively dividing cells such as intestinal cells, blood cells, and hair follicles. The side effects of killing these cells may include nausea and diarrhea (by interfering with normal cell division in the intestinal tract), and hair loss (by interfering with cell division in hair follicles).

8. a

9. c

10. Chemotherapy targets actively dividing cells like the lining of the digestive tract. When these cells are killed, side effects such as nausea may occur. Cognitive symptoms are not a side effect because neurons rarely (if ever) divide and so are not affected.

11. c

12. a: Irinotecan slows the growth of the tumor—the rate of cell division—by interfering with cell division. Thus the tumor grows more slowly in the presence of irinotecan than in the absence of chemotherapy. b: PHY906 plus irinotecan slows the growth of the tumor more than irinotecan alone, as can be seen in the graph by the very limited growth of the tumor in the presence of both irinotecan and PHY906. PHY906 enhances the effect of irinotecan.

13. Eating whole foods that are rich in beta-carotene may have more benefits than a beta-carotene supplement because there may be interactions between beta-carotene and other molecules in the whole food that will make the beta-carotene more potent.

14. anaphase

15. interphase (S phase)

16. If a cell does not complete cytokinesis there will be one cell with twice the number of chromosomes relative to the parent.

17. A drug interfering with spindle fiber shortening would be an effective cancer drug because anaphase cannot occur if the spindle fibers cannot shorten. Disrupting anaphase will lead to apoptosis and therefore slow the growth of the tumor.

18. Over-the-counter herbal supplements are not regulated by the FDA and may have varying amounts of the effective compound. Additionally, these supplements may contain other, potentially harmful, compounds.

19. Establishing the efficacy of PHY906 in humans will require a clinical study. The study should include cancer patients that are given chemotherapy alone or chemotherapy with PHY906. The patients should be closely monitored and allowed to stop taking PHY906 if negative effects occur at any point in the study. The patients should be informed of all possible risks before being asked to agree to participate in the study.

Chapter 10

1. Mutations in both tumor suppressor genes and in oncogenes increase the risk of developing cancer. Tumor suppressor genes cause cancer when the proteins the genes code for become nonfunctional; oncogenes cause cancer when the proteins become permanently activated, or "turned on." Both types of genes play important roles in cell division and its regulation. Tumor suppressors typically signal the cell to pause cell division in order to fix errors; oncogenes tend to promote cell division.

2. *BRCA1* is a tumor suppressor gene that produces a DNA repair protein that helps detect and repair mutations.

3. b

4. e

5. b

6. c

7. At birth, all of Lorene Ahern's cells–including her breast cells and her liver cells–were genetically identical and carried a mutation in one of her two *BRCA1* alleles. For cancer to develop, some of her breast cells must have accumulated additional genetic mutations, which would make these cells genetically different from her normal breast cells.

8. If there is no family history of breast cancer it is unlikely that the niece has a mutation in *BRCA1*. Therefore, she should be able to reduce her risk of developing cancer by adopting lifestyle changes like not smoking, using sunscreen, and minimizing exposure to carcinogens, which will decrease her chance of accumulating cancer-causing mutations.

9. e

10. a

11. The normal BRCA1 protein acts as a tumor suppressor to halt cell division and promote DNA repair. This means that it will take only one additional mutation in *BRCA1* (in the other allele) for them to lose all BRCA1 function. Nonfunctional alleles of *BRCA1* encode proteins that do not act properly to detect and repair damaged DNA.

12. There are many possible answers, as well are concerns about privacy. Nellie's doctor might advise Nellie to tell both her sister, Anne, and her brother that she carries the *BRCA2* mutation, but ultimately it is up to the sister to decide whether or not she wants to be tested for the mutation. Anne may not want to live with the burden of knowing that she has a higher risk but not a guarantee of developing breast cancer. A counselor might suggest that Anne be tested because there is evidence to suggest that there are treatment options available to carriers of the mutation, including prophylactic surgery. The brother should also be advised since men with mutations in *BRCA2* are also at higher risk for developing breast and prostate cancers. Another consideration is that their children's risk is affected if a parent carries the mutation.

Chapter 11

1. 46 (23 pairs)

2. 23

3. A person with CF is homozygous recessive at the *CFTR* gene and carries two of the CF-associated alleles in all of his or her lung cells. A heterozygous carrier for CF has one CF-associated allele and one normal allele at the *CFTR* gene. Someone who is homozygous dominant carries two of the normal alleles at the *CFTR* gene.

4. a: A heterozygous genotype will have a normal phenotype (like Emily's parents). b: A homozygous dominant genotype will have a normal phenotype. c: A homozygous recessive genotype will have cystic fibrosis.

5. Two individuals with different phenotypes may have different mutations at the *CFTR* gene or different alleles in other modifier genes that may affect the severity of the disease.

6. c

7. f

8. Maternal   Paternal

Each haploid gamete could contain one of the following chromosomes:

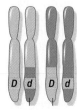

9. Since meiosis halves the total number of chromosomes, 41 unpaired chromosomes would be present in one of the alien's gametes.

10. Mitosis starts with a diploid cell and produces two genetically identical diploid cells. Meiosis also starts with a diploid cell but, because of recombination, results in the formation of four haploid cells containing new genetic combinations of alleles.

11. If meiosis I is skipped, there would be no recombination. The 46 replicated chromosomes would each separate their chromatids during meiosis II, resulting in gametes with 46 chromosomes (instead of 23).

12. b

13. recessive inheritance pattern

14. c

15. a: All of Emily's gametes will carry the allele that is associated with CF (*a*). The man's gametes will all carry the allele that is wildtype (*A*).

b:

	A	A
a	Aa	Aa
a	Aa	Aa

c: 0%

d: 100%

16. Since Huntington disease is a dominant disorder, the friend has a 50% probability of developing it, as shown in the Punnett square below:

	t	t
T	Tt	Tt
t	tt	tt

17. If you take a genetic test for a disease for which there is no cure, you do so knowing you would not be able to undergo treatment to improve your prognosis; thus, even if you are presently asymptomatic, you may become anxious about developing the disease. Knowing if you will develop a disease may affect your choices about having children; it may help ongoing research; and may be helpful knowledge for your doctor in monitoring your overall health.

Chapter 12

1. c

2. Males have only one X chromosome, whereas females have two X chromosomes. Recessive traits occur when only recessive alleles are present. Males have only one allele of each X-linked gene because they have one X chromosome; therefore, if they have the recessive allele they will develop the recessive genetic disease. Females are less likely to be affected because their recessive allele can be masked by a dominant allele.

3. If a male has an X-linked recessive disease he cannot pass it to his son because the father must pass the Y chromosome to his son. The X chromosome in males will always come from the mother.

4. c

5. a: XX, female; b: XXY, male; c: XY, male; d: X, female

6. a: The brother and son of a female will not have identical Y chromosomes because the Y chromosome is inherited from the father. The exception is if the brother is the son's father. b: The brother and son of a male will have essentially identical Y chromosomes because the Y chromosome is inherited from the father. The two brothers will have received essentially the same Y chromosome from their father and then pass that Y chromosome to their sons.

7. a: DMD is X-linked recessive. 50% of the sons will have DMD and 0% of the daughters will have DMD. b: Rickets is X-linked dominant. 50% of the sons will have rickets and 100% of the daughters will have rickets.

8. There are many genes that contribute to the phenotype of height, so it is a polygenic trait.

9. d

10. Polygenic inheritance is primarily due to the influence of effects from multiple genes. Multifactorial inheritance includes an interaction with the environment.

11. Incomplete dominance describes traits in which heterozygous individuals have an intermediate phenotype between that of the homozygous dominant and homozygous recessive phenotypes. Traits that are codominant produce heterozygotes that display both the dominant and recessive phenotypes.

12. The only possible recipient for an A+ donor is type A+. The possible donors to an A+ recipient are type A+, type A-, type O+, and type O-.

13. Environment influences overall height. The two women may have had different diets while they were growing.

14. The hypothesis that genes and environment influence phenotype is supported by the significantly higher number of people diagnosed with depression who have two copies of the short allele and have had four or more stressful life experiences. If depression were controlled only by the number of short alleles, then all people with short alleles would have the same propensity for depression. Similarly, if depression were based entirely on environment, then all people who experience four or more stressful events should have a high propensity for depression.

15. From these data, the probability of people with two short alleles becoming depressed increases after three or more stressful life events.

16. Phenotype is the result of both genotype and environment. Even if two people have the same genotype for a predisposing allele, their environments may be different and thus change their probability of developing the disease.

17. a: 23 chromosomes (human egg); b: 23 chromosomes (human sperm); c: 46 chromosomes (zygote)

18. Genotypes cannot be deduced from karyotype analysis. Karyotype analysis is used to determine the number of chromosomes present.

19. c and d

20. Research supports a correlation between some of the most obvious birth defects and the age of a woman's eggs, but there are findings that the age of the male can also influence the frequency of cognitive disorders. It is more likely that the egg cells will include chromosomal abnormalities.

21. Factors for considering genetic counseling include age, family history, and medical history. The value of having this information is to be better prepared to support a child regardless of his or her abilities.

Chapter 13

1. A five-year-old child does have adult stem cells. The "adult" stem cells are somatic cells that are still able to divide to regenerate specific cell types.

2. a

3. d

4. Tissues are made up of different specialized cell types that work together. Neurons and glial cells are different cell types that work together to allow the firing of electrical impulses.

5. It would not be sufficient to replace only the neurons as the nervous tissue is made up of neurons and glial cells, and glial cells help the neurons with sending rapid signals.

6.

	Photoreceptor cells of the retina	Heart muscle fibers	Helper T cells
Myosin gene present?	X	X	X
Myosin mRNA present?		X	
Myosin protein present?		X	
Retinal gene present?	X	X	X
Retinal mRNA present?	X		
Retinal protein present?	X		
CD4 gene present?	X	X	X
CD4 mRNA present?			X
CD4 protein present			X

7. c

8. Advantages of using one's own cells include the following: There is no need to wait for a donor match because the cells will come from the recipient. Cells will not be rejected by the recipient's immune system: the cells come from the recipient so the immune system will recognize them as self. The recipient will not have to be on immune-suppressant medication, which can lead to other illness: the cells come from the recipient so the immune system will recognize them as self.

9. Embryonic stem cells can differentiate into almost any cell type and are found in early embryos. Adult stem cells are more limited as to the types of cells they can differentiate into; they are found in tissues.

10. c

11. It is more challenging to engineer a bladder because there are several types of cells (including muscle cells and nerve cells) that are required to make the organ, whereas skin is made up only of skin cells.

12. a: heart muscle; b: none; c: pancreas; d: neurons

13. Embryonic stem cells have a wider utility than adult stem cells in that embryonic stem cells can differentiate into almost any other cell type. These cells do not have an identity, so there is the promise that they could be put into a patient's body to stimulate regeneration in the damaged tissue. Adult stem cells are useful for a narrow range of cell types that are similar to the adult stem cell. For example, blood stem cells can make other blood cells, but could not make a liver cell. Embryonic stem cells are derived from early embryo cells and thus there are ethical concerns with using this cell type. Adult stem cells come from tissues and so there is less controversy about using them.

14. The genes that were inserted functioned to de-differentiate the cells back to an embryonic state. The genes had to be added into the cell because the endogenous genes had been shut off during development.

15. There are many possible opinions: one is to choose to allocate funds to all types of stem cells because each type will serve a unique purpose. Additionally, this field is relatively new and it is important to continue exploring all research avenues for the best solutions. Funding should also be allocated to those researching the ethical questions related to this research so that we don't cross a line that can't be undone.
The technologies to create cloned embryos for "reproductive" or "therapeutic" cloning do not differ except that in reproductive cloning the embryo is implanted into a woman's uterus. There are many opinions and also regulations regarding reproductive cloning of humans. Many people think that reproductive cloning of humans should be illegal because they are concerned that the humans born from this process would not have the same rights as humans born from traditional means.

Chapter 14

1. a

2. Colonization means the bacteria are carried on or in the body without causing disease; infections are associated with disease.

3. c

4. MRSA can be passed from person to person by direct skin contact or touching contaminated surfaces. If the bacteria can find their way into a wound, they can get into the body and cause an infection. Athletes with cuts or scrapes can get MRSA from contact with other people or contaminated objects (for example, towels).

5. Once I confirmed that the infection is really caused by a MRSA strain (methicillin-resistant–also resistant to other beta-lactam antibiotics such as penicillin), I would try non-beta-lactam antibiotics. If these were not successful, then I would consider prescribing vancomycin as this antibiotic is reserved for severe MRSA infections that don't respond to other types of antibiotics. I would recommend that the teammates increase their hand washing, decrease contact when possible, and that the locker room be thoroughly cleaned to cut down the frequency that the other players will come in contact with MRSA.

6. Beta-lactam antibiotics work by interfering with the bacteria's ability to synthesize cell walls. Our eukaryotic cells are not affected because there is no cell wall in animal cells.

7. c

8. d

9. "Fitness" describes the ability of an individual to survive and reproduce in a given environment. An organism that has a higher fitness will be able to reproduce and pass its genes to the next generation at a higher frequency than a less fit individual.

10. a

11. c

12. Asexual reproduction occurs when an organism replicates its own genome and divides into two daughter cells. The daughter cells are a copy of the mother cell. The daughter cells could be different from the mother or the other daughter cell if mutations occur during replication.

13. Evolution is the result of change in allele frequency over generations (time). Bacteria evolve at a high rate because the generation time is minutes or hours compared to years for other organisms.

14. c

15. Some of the cells from the population will grow because during replication their DNA has accumulated mutations that make them resistant to vancomycin.

16. The genotype determines the phenotype, so if the genotype frequency is changed then the phenotype frequency will change.

17. a: The snails will be greenish in color to blend in with the grass. The snails that do not blend into their surroundings have a higher probability of being eaten by birds and therefore cannot pass their genes to the next generation. b: Individual snails will not be able to mutate to change their color in response to the environmental selective pressure. Snails that are brown will be selected for and will reproduce, resulting in more brown snails. The green snails will have a low fitness in this environment because they will stand out and have a higher probability of being eaten before they reproduce. Over time the color phenotypes will shift to brown because these snails have a higher fitness in this environment and thus will reproduce more, leaving their brown alleles in the next generation of snails. The new color phenotype is the result of random mutation and recombination, leading to changes in the genome that are passed to the gametes and resulting in the brown phenotype.

18. a: It is troublesome to hear this story because antibiotics will not kill the virus and the increase in antibiotics in the environment will increase the chance that bacteria in the environment will become resistant. b: No. c: The risk to the friend is that bacteria in the friend's body are being exposed to the antibiotic and a few bacteria in his body may be resistant to that antibiotic. Those resistant bacteria will continue to replicate, leaving him with a population of bacteria that are resistant to that antibiotic. If those bacteria should infect his bloodstream (for example, through a break in the skin), that infection will be hard to treat (as the bacteria are already resistant to at least one class of antibiotic). The risk to the community is that if those bacteria should be transferred to others, they could cause an antibiotic-resistant infection in those affected.

Chapter 15

1. 0.34

2. b

3. Yes, evolution has occurred. The genetic definition of evolution is a change in the allele frequencies within a population.

4. Population B would be the most likely to survive a sudden environmental change, because it has the greatest allele and genotype diversity. This diversity increases the chances that some individuals will be better adapted to survive changes in their environment.

5. Populations 1 and 4 are the most threatened. Population 1 has both a low total size and a single gene for which there is only one allele in the population. However, population 1 has good genetic diversity with regard to the other two genes being studied. Population 4 has the least genetic diversity of all the populations but has a 20-fold higher total population. Either of these populations could be of great concern to conservationists.

6. If PKU occurs in 1 in 15,000 people, then q^2 is 1/15,000, and q is the square root of 1/15,000 (= 0.008). Therefore $p = 1 - q$ (= 0.992), therefore the carrier frequency = $2 \times p \times q$ (= 0.016, or 1.6% of the population).

7. $p = 0.45$, $q = 0.55$; predicted frequency of homozygous dominant (*AA*) is 20.25%, actual percentage is 5%; the population is not in Hardy-Weinberg equilibrium.

8. d

9. c

10. a: The frequency of alleles *A* through *L* has changed drastically. *A-L* but not *B* frequencies are reduced to zero. *B* frequency is now 1.0. b: This is an example of a bottleneck effect.

11. Genetic drift is an example of evolution because genetic drift changes the frequency of alleles in a population (the definition of evolution). Genetic drift differs from natural selection in that genetic drift does not necessarily lead to adaptation, since the changes in allele frequencies are not due to selection for a particular beneficial trait but rather are due to random events.

12. In this example, the descendant population's allele frequencies might remain similar to the founder's because the descendants are more likely to mate with other members of this same population rather than with members of the population around them.

13. d

14. Geographic isolation prevents gene flow. Due to natural selection or genetic drift, allele frequencies of the two separated populations will diverge over time. Without being able to exchange genetic material, the two populations may eventually experience changes in allele frequencies and evolve traits which prevent successful mating, and thus will have speciated.

15. According to the biological species concept, these populations are still the same species if they can mate and produce fertile offspring. You would need to observe mating between members of the previously separated populations and then follow their progeny to see if they are fertile–that is, if they can mate and produce offspring.

16. Inbreeding is detrimental because it decreases the number of heterozygotes in a population, increasing the proportion of individuals which are homozygous for recessive alleles. Many recessive alleles are mutations that are detrimental but which do not confer a phenotype in heterozygous individuals because the nonmutated gene is a dominant allele; however, in a homozygous recessive individual, these traits are expressed, usually with extremely negative consequences including decreased fitness, fertility, or viability (that is, the trait is lethal). This phenomenon is called inbreeding depression.

17. Over time the gene pools of these groups would converge, becoming more similar to one another.

18. *A* and *E* frequencies will increase; all other frequencies will decrease.

19. a: Answers will vary depending on date and sources used. In general, endangered species are at risk for becoming extinct, whereas threatened species are at risk for becoming endangered in the near future. b: Because the genetic diversity of the population has already been reduced, and further habit preservation will not restore genetic diversity in the short term. Other interventions are necessary to create gene flow and restore genetic diversity, such as the introduction of pumas from another area as described in the text. c: There are a number of possible answers, which include but are not limited to: genetic testing to identify any individuals with new alleles followed by intentional breeding of those individuals to increase rare allele frequencies; attempts to breed cheetahs with other closely related cats in the hope that some combinations may produce fertile offspring; genetic engineering to introduce new traits artificially; separation of the species into multiple distinct environments in the hope that each new founder population will evolve new traits distinct to that environment.

Chapter 16

1. the shallowest layers (those closest to the surface)

2. c

3. This fossilized skeleton appears to be most similar to extant (that is, currently living) bony fish. Specific characteristics noted could include the presence of dorsal, pectoral, pelvic, and tail fins, which are all clearly visible and are useful for swimming, indicating the organism likely lived in the water. Additionally, the connected skull and shoulder bones indicate protected gills, again indicating that this is a marine organism related to bony fish. The presence of teeth and large mouth indicate that this organism was likely predatory.

4. No. The hypothesis that these sea cucumbers existed does not predict that their fossils should be present in the fossil record, as organisms which had only soft-tissues (such as a sea cucumber) rarely produce fossils.

5. The barnacles are at least as old as the oysters, which are at least as old as the surrounding rock, which is dated at 100 million years. Fossils must be at least as old as the surrounding rock which encased them. Likewise, organisms found to coexist in the fossil record must have lived at approximately the same time.

6. c

7. elongated bony ribs and weight-bearing pectoral fins that include ankles

8. Transitional fossils represent midpoints between two groups of organisms. Often they are extinct organisms that represent a transitional form between the ancestors of two groups of extant (that is, currently living) organisms. In this case, *Tiktaalik* represents a transitional form between bony fish and amphibians (or all tetrapods). Often transitional fossils help scientists understand how organisms changed morphologically over time.

9. See Infographic 16.4. The first real tetrapod would likely have more distinguishable hindlimbs that would be capable of supporting weight, rather than pelvic fins. This would likely be the defining characteristic of the first tetrapod and would be the primary difference distinguishing it from *Tiktaalik*. Given the trends observed in the *Tiktaalik* fossil, you might also expect the tetrapod fossil to also have longer and thicker ribs, a less defined or smaller gill slit, longer neck, more developed fore-limbs with more defined digits ("fingers").

10. The land represented a new ecological niche into which life could expand, but this does not mean that the oceans and freshwater environments are not places in which life could thrive. Fish were already well adapted to surviving and proliferating in these marine environments, and therefore descendents of those ancient fish exist today.

11. The skeletal anatomy of a chicken wing and a human arm are very similar. All major bones are present and in the same locations relative to other bones. The primary difference is found in the fine bones that make up the digits. In the human hand, these bones are longer, more numerous, and arranged in a way that allows independent movement, an important feature in human evolution required to elegantly manipulate objects.

12. Middle ear bones in humans; gills in fish

13. The presence of five digits indicates that having five digits may have provided an evolutionary advantage to the ancestors of both otters and humans. Both humans and otters have evolved to utilize fine motor movements of their hands or paws. It is likely that having five digits may have improved motor skills, providing a reproductive advantage that is reflected in the complexity that is the human hand. Because otters are known to use their paws for grasping and, in simple tool use, using stones to crack open shellfish, we might predict that otters will continue to evolve paws that are more and more functionally complex. Because humans and otters share an ancient ancestor, both humans and otters use the homologous bones in their hands and paws.

14. It depends on the stage of development at which the embryos are observed. At early stages of development, both human and chicken embryos have post-anal tails, so the presence of a tail at that stage cannot be used to distinguish the two. Later, the post-anal tail disappears in the human embryo.

15. By comparing the three sequences to one another in a pairwise fashion. By counting the number of differences between each pair of sequences, a quantitative measure of similarity (% similarity) can be established. The more similar the sequences (the higher the % similarity between them), the more likely it is that they are closely related. Sequence evidence from a single gene is often combined with evidence from other genes to establish relationships between organisms. DNA sequence data is just one means by which to determine such relationships. Comparisons of morphological traits, such as the arm bones, are another way to gather evidence to establish such relationships.

16. With only a few rare exceptions, the genetic code is universal for all living organisms known on this planet. Therefore, the same piece of DNA encodes the exact same amino acids in bacteria as it does in humans.

17. The two proteins differ in the specific amino acid sequence that makes them up. This is encoded by the specific nucleotide sequence of the gene. Although the code is the same, the specific sequence of nucleotides differs between distantly related organisms. Because the more closely related two organisms are, the more similar their gene sequences will be, it is possible to determine relationships from the similarities between these genes.

18. Answers will vary. Considerations may include the importance of protecting important scientific finds, the role of government in such protection, and the extent to which the government may act to protect such finds at the cost of limiting personal freedom. There are many things which people do not have a right to collect because the act of collecting, trading or selling such items causes harm to others or the environment, and therefore there are bans on the collection and trade of pelts, tusks, teeth, horns, feathers, or other animal parts of protected species. Likewise, visitors to many protected geological sites are prohibited from taking rocks or other items.

Chapter 17

1. They are all radioisotopes that decay at steady and predictable rates, changing into other elements.

2. You would use uranium-238 because it has the longest half-life (4.5 billion years). Other isotopes with shorter half-lives will be barely detectable in a sample that is extremely old, having long ago decayed to levels that are below detectable limits.

3. (1) the first prokaryotes (~3.0 billion years ago), (2) an increase in oxygen in the atmosphere (~2.5 billion years ago), (3) the first multicellular eukaryotes (~1.2 billion years ago), (4) the Cambrian explosion (545 million years ago) (5) the first animals (~540 million years ago), (6) the Permian extinction (248 million years ago), (7) the extinction of dinosaurs (~65 million years ago)

4. a: approximately 4.5 billion years old; b: approximately half

5. ~9 billion years old

6. Many of the ancestors of these organisms may have been evolving for a long time without appearing in the fossil record because not all organisms leave fossils behind. The sudden appearance of numerous organisms in the fossil record of the Cambrian explosion may be largely due to the development of shells and other hard-body parts, which are more likely to leave behind a fossil.

7. Amphibians were certainly present. Early reptiles and sharks might also be found in these layers.

8. No. They may indeed be closely related, but similar morphology does not necessarily indicate homology. The two organisms may share common characteristics because of convergent evolution rather than homology.

9. See Infographic 17.5. The continents were then generally closer together than they are now. Since that time, because of plate tectonics–the movements of independent continental plates in the earth's mantle or crust–the continents have largely drifted apart. One major exception is the Indian continent, which has since collided with the Asian continent, forming the Himalayan mountain range. As landmasses moved and separated, so did the organisms that lived on those landmasses.

10. The two species may look alike because they have evolved similar traits independently because they are adapting to similar environments (the desert climate). This is an example of convergent evolution.

11. They would have migrated to both the north and south polar regions, in which case both penguins and polar bears would be found at both regions today.

12. Bats are mammals and hence share a common ancestor with all mammals. This common ancestor is not thought to have possessed any structures homologous to insect wings that could have been inherited. Therefore, it is more likely that bat wings evolved separately. This is another example of convergent evolution.

13. f

14. domain, kingdom, phylum, genus, species

15. d

16. Monera was divided because DNA evidence showed that it was made up of two distinct groups of organisms. These were later separated into the current domains of Bacteria and Archaea.

17. c

18. Answers will vary. For instance, because organisms are related, we can study possible cancer treatments in yeast or mice to determine if these treatments interfere with, for example, cell division in these organisms. If they do, they may also stop the division of cancerous cells in humans. If we understand the evolutionary adaptations of organisms to their environments, we may be able to identify organisms that will do well in habitats that have been degraded by human activities, or that will do well in environments that are changing as the result of global climate change.

Chapter 18

1. c

2. The fundamental difference between the two groups is that prokaryotes lack internal membrane-bound organelles. The lack of a nucleus enclosing the chromosomal DNA is the defining characteristic of a prokaryotic organism.

3. d

4. They were originally grouped together because of their similar size and morphology. Both are prokaryotic organisms.

5. "Archaebacteria" literally means "ancient bacteria." The name was originally used because archaea were then seen as a particular sort of bacteria, one that might be very old. The strength of this term is that it emphasizes the structural similarity of archaea to bacteria; the great weakness of this term is that it implies that archaea are a subset of bacteria. We now understand this to not be the case–bacteria and archaea are separate but related groups of organisms.

6. a

7. e

8. c

9. No. Many archaea live in environments that are difficult to replicate in the lab. Therefore, there are many archaea that scientists are unable to culture, but they are nonetheless present in the environment.

10. No. Bacteria and archaea generally look similar, and therefore DNA sequence evidence is usually used to distinguish between these two groups of prokaryotes.

11. These processes are important because they convert CO_2 and N_2 gases into forms that humans can use.

12. No. *N. gonorrhoeae* use pili to attach to human cells and evade host defenses. Without the pili, *N. gonorrhoeae* would not be a very effective pathogen.

13. High temperatures: Most organisms cannot survive outside a narrow temperature range. Temperatures outside this range lead to protein denaturation and membrane instability, resulting in cell death. High pressure: Most organisms are evolved to live within a

specific pressure range. The high pressures found at the bottom of the ocean would crush many other organisms. Alkalinity: Most organisms are evolved to live within a specific pH range. The high pH (basic) conditions at Lost City would kill most other organisms. Toxic gases: Many other organisms have not evolved the ability to tolerate the high concentrations of certain toxic gases found at Lost City.

14. c

15. Their presence supports the idea that these compounds can be produced abiotically (that is, without life). This supports the idea that the molecules that make up living organisms could have been produced by the harsh conditions present on the early earth. The organisms that live inside the vents may be more similar to the first organisms on earth than any other extant organisms on the planet.

16. It might modify the hypothesis that life may have started at specifically this type of thermal vent; however, if no methane was produced abiotically at this particular vent, that does not establish that methane cannot be produced at thermal vents or abiotically in other ways.

17. No. You would not expect to find photosynthetic organisms at Lost City because there is no appreciable sunlight at that depth.

18. Because of the harsh conditions present at the vents, it seems unlikely that the scientists working there might contaminate the area with surface organisms that would be able to survive and compete with the natural microbes. However, microbial life is so diverse and unpredictable that such contamination is certainly possible. Should such microbes survive, they might out-compete natural organisms (as an invasive species) and could disrupt or totally destroy the current ecosystem. This could lead to the extinction of a species, and so this threat should be taken seriously by the scientists studying the vents.

19. Answers will vary. There are countless examples of how breakthroughs in basic science (that is, science for the sake of understanding the world, rather than for a specific purpose, such as treating a disease) have led to breakthroughs which benefit human society. By studying the thermal vents, scientists hope to better understand the diversity of life on earth and its origins. The vents have already led to changes in our understanding of evolution and the mechanisms by which organisms harness energy on earth. For example, this ecosystem is one of the only known ecosystems that does not ultimately rely on the energy of the sun, as photosynthetic organisms do not make up the base of this ecosystem's food chain.

Chapter 19

1. Olympic National Park contains many species not found anywhere else, like the Olympic torrent salamander and the Olympic gopher, because of the park's isolation and topography. The park is an ecological island with saltwater on three sides and during the last ice age was separated from the rest of the United States. A large diversity of animals is able to survive in the various habitats found in the park, which include glacier-topped mountains, temperate rain forest, lakes, rivers, and the Pacific coastline.

2. The domain Eukarya encompasses all eukaryotic organisms, including plants, animals, fungi, and protists, which contain membrane-bound organelles.

3. Both the fisher and the Douglas fir are eukaryotes and inhabit the low-elevation rain forests of the Olympic National Park.

4. a: One would expect less diversity in Lake Michigan because there is less variation in habitat and Lake Michigan has been subjected to repeated glaciations, decreasing the time for evolution of new species. b: One would expect less diversity in the Sonoran Desert in Arizona because the climate limits the diversity of organisms to those that can survive in a dry habitat. There are fewer protists and bryophytes because these organisms must avoid drying out. c: One would expect less diversity in the prairies of Kansas because there are fewer trees, so there are fewer potential habitats.

5. A fungicide would kill the fungi, which are critical for decomposition. The lack of fungi would lead to an accumulation of nonliving organic matter and there would be fewer nutrients available for other eukaryotes.

6. The first group of plants to live on land was the bryophytes, which lack roots and tissue for transporting nutrients and water. The lack of a vascular system limits these plants to damp environments. So even though they were first, they don't have all the adaptations necessary to live in a variety of terrestrial habitats.

7. d

8. The evolution of seeds allowed plants to survive harsh conditions and spread to new locations. Seeds are protected within cones or fruit, and can be spread relatively easily.

9. A hungry animal is more likely to disperse the seeds of an angiosperm than a gymnosperm since the seeds are enclosed in fruit, a tasty treat for a hungry animal.

10. Ferns were the first true vascular plants. They were able to grow taller and overran the landscape during the Carboniferous period. Since then, vascular plants such as trees have dominated many landscapes.

11. b

12. c

13. Both the fisher and the human are predators and have a backbone, mammary glands, and hair on their bodies.

14. d

15. a: Using flight as a criterion, woodpecker and wasp would be grouped together and the nonflying group would include human, ant, and fisher. The "two-legged" group includes human and woodpecker; the "more-than-two-legged" group includes wasp, ant, and fisher. The only animal having feathers is the woodpecker; the "nonfeather" group includes human, wasp, ant, and fisher. b: These groupings do not accurately reflect the taxonomic relationship since, for example, wasps and ants are arthropods and woodpecker, human, and fisher are chordates. Thus, it is necessary to use molecular biology and multiple characteristics to characterize the

relationship between organisms correctly. c: Wasps and ants share characteristics of arthropods, including segmented bodies with jointed appendages and a hard exoskeleton, and characteristics of insects, including three pairs of jointed legs and a three-part body consisting of head, thorax, and abdomen. The human and fisher are grouped together on the basis of the presence of an endoskeleton, the production of milk, and the presence of hair.

16. All arthropods have an exoskeleton for protection from predators, to prevent them from drying out, and to support movement. Some arthropods, like spiders, have evolved the ability to produce venom for protection and predation. Beetles have been extraordinarily successful because of the development of wings and specialized mouthparts. Wings allow beetles to escape predators and to access habitats and remote food sources. Ants have evolved complex social behavior that allows them to coordinate the behavior of the group.

17. a: Fungi are heterotrophs and cannot carry out photosynthesis. b: Fungi do not ingest their food–they perform external digestion instead. c: Fungi digest food by secreting enzymes onto their food, which break down molecules into smaller organic compounds that can be absorbed by the fungi.

18. c

19. Both fungi and plants are eukaryotes containing cell walls. Neither are mobile. The key difference between the two is how they obtain their nutrients. Plants are autotrophs, producing their own food through photosynthesis; fungi are heterotrophs, obtaining their nutrients by decomposing organic matter.

20. b

21. f

22. Protists are no longer considered a separate kingdom because protists do not form a cohesive evolutionary group: some members undergo photosynthesis like plants, other members eat other organisms like animals, and still other members are decomposers like fungi. Genetic information will be the most useful basis for creating new taxonomic "homes" for the protists.

23. If the protist is living in a freshwater environment, water will enter it by osmosis, decreasing the concentration of solutes in the protist and potentially causing the protist to swell and burst. The contractile vacuole prevents this from happening by removing some of the excess water.

24. a: The Mexican gray wolf is an endangered species, so its reintroduction may prevent them from becoming extinct. In addition, there may be benefits for the habitat by reintroducing wolves. For example, the wolves may perform a necessary evolutionary function by removing unfit individuals from the prey population. b: The reintroduction of species may negatively affect the habitat and other human activities. Negative effects on the environment include changes in other species that may increase the likelihood of certain diseases in the area or alter the biodiversity of the area. With the Mexican gray wolf, some of the considerations relating to humans include concern that the wolves may attack pets, livestock, or humans, may affect military activity in the area, and may decrease prey of interest to hunters.

25. Answers will vary depending on locality.

Chapter 20

1. c

2. b

3. Folate is normally destroyed by UV light, and darker skin evolved as a mechanism to maintain folate. a: If this were no longer true, and folate levels were unaffected by UV light, levels of melanin would likely decrease in populations living at the equator because the role of UV light in producing vitamin D would be beneficial. b: There would be no change expected in this population. The levels of UV are low in this environment, so there is no pressure for dark skin to protect folate, and light skin allows vitamin D production.

4. d

5. c

6. Mitochondrial DNA is inherited solely from the mother. Mitochondrial DNA mutates at a fairly regular rate. A mother with a DNA mutation in her mitochondrial DNA will pass it down to all her children, and her daughters will pass it to all their children. Because these mutations are passed on intact, without recombination, mtDNA is a useful tool to track human ancestry back through generations.

7. a

8. e

9. See Infographic 20.7. Africans have the highest levels of genetic diversity, as they are the descendants of the most ancient populations. The next highest levels would be expected in Asian populations (the migration from Africa to Asia took place ~67,000 years ago), followed by Europeans (~40,000 years ago) and then South Americans (13,000 years ago).

10. Tools would have aided in hunting and food preparation, allowing Australopithecines to have better nutrition and less risk of starving to death.

11. If there were another benefit to having an opposable big toe (faster movement, better walking stance) it would have been maintained by selection, and the number of individuals with this feature would increase. Since humans no longer have an opposable big toe, there must not have been a benefit to this feature and it was lost from the population.

12. a: If there was better hunting or mate selection on the ground, an early hominid in a forested environment might have moved to the upright walking position and lost an opposable big toe. In grassland, where there is nothing to climb, the arboreal traits would confer no selective advantage. b: Other traits that might be favored in a forested environment include the ability to see well in low-light conditions (for example, under a dark canopy of leaves) and a good sense of balance (necessary to walk on branches).

13 & 14. Answers will vary. Students should address race as a construct that is often produced by looking at genetics, physical traits, geography, history, and cultural traditions.

Chapter 21

1. A population is a group of organisms living in a particular geographic region. A community is all the organisms (species) in a geographic area. Communities differ from populations in that they include interactions between populations of different species.

2. e

3. Scat (feces) reveals information about the organism that produced it. Studying scat samples can elucidate what an organism has been eating, as well as providing a source of the organism's DNA. DNA analysis can be used to identify individuals, as well as to look for genetic diseases and the effects of inbreeding.

4. c.

5. Ecology is the study of organisms, interactions between organisms, and between organisms and the nonliving components of the environment. Ecologists look at (1) the way individual organisms respond to the environment (temperature, pH, light, etc.); (2) how a population of organisms grows, breeds, and changes genetically over time; (3) how populations interact with each other—such as the effect of predators or symbiosis between populations; (4) how elements like rain, disease, fire, etc., affect communities and populations.

6. You could travel and track the squirrels by their tracks and count them at their nests, but this would be very time consuming. Alternatively, you could set up a series of square plots that represent 20%-25% of the nature preserve. You could then count the number of squirrels in that area, and extrapolate to find an estimate of how many squirrels are likely to inhabit the whole nature preserve. These approaches could also be used to determine the population size of maple trees, but since the trees do not move, they could be counted more individually more easily than squirrels.

7. The populations of moose and wolves are linked—the moose provide virtually all of the wolf's food. By knowing the changes to the size of the moose herd, researchers would be able to predict changes to the size of the wolf pack. It is important to know the moose's cause of death because if other (nonwolf) factors are causing the moose to die, researchers would have to take that additional factor into consideration when assessing the future moose population size as well the wolf population size. For example, if researchers find that many moose are dying because of tick infestation or starvation, then the moose population will be affected independently of wolf predation, and this in turn will affect the wolf population.

8. Scat (feces) reveals information about the organism that produced it. Studying scat samples can reveal what an organism has been eating. If the scat from a particular type of herbivore has a variety of vegetation in it in approximately the same proportion as vegetation in the local environment, we could infer that those herbivores tend to eat what is available, without a strong preference.

On the other hand, if a herbivore has a strong preference for a particular type of vegetation, then that vegetation would be present in the scat at a higher proportion than the proportion of the plant in the local environment, suggesting that the herbivore is eating that plant preferentially.

9. c and d

10. Carrying capacity is the maximum number of individuals that an environment can support given its space and resources. At carrying capacity the population will level off, fluctuating slightly but maintaining a relatively constant size around the carrying capacity.

11. Because the parasite passes from fish to fish through the water, but can't survive for long periods of time alone in the water, the parasite will have the biggest impact at a high fish density. This is because at high density, the fish will be crowded, and so the parasite will likely find new host in a short amount of time. At low population density, it might take a parasite a long time to encounter a new host, and it may die before reaching a new host. A hot summer and drought may cause a lake to start to dry up (or at least lose a lot of its volume). This would leave less volume for the same number of fish, thereby increasing their population density, and permitting greater opportunities for density-dependent factors to affect the population.

12. a: Abiotic. Hot summer temperatures increase the ticks on moose and also may weaken the health of the moose, eventually causing death, decreasing the size of the moose population, and affecting the size of the wolf pack in subsequent years. b: Biotic. Ticks weaken the health of the moose, eventually causing death, decreasing the size of the moose population, and affecting the size of the wolf pack in subsequent years. c: Biotic. The lack of trees would mean a lack of food for the moose, which would weaken the health of the moose, eventually causing death, decreasing the size of the moose population, and affecting the size of the wolf pack in subsequent years. d: Biotic. The illness in the wolves would make them less able to hunt the moose or result in fewer wolves in the population. This will allow the moose population to grow. e: Abiotic. Deep snowfall is likely to trap the moose, making it easier for the wolves to catch and kill. This will cause death, decrease the size of the moose population, and affect the size of the wolf pack in subsequent years.

13. a: The introduction of a new herbivore will cause a decrease in the population of trees on Isle Royale. This is because the new herbivore and the moose will both be eating the trees. b: The moose population will decline with the introduction of a new herbivore that is not preyed on by wolves. This is because there will now be competition for food between the organisms, and some moose will die from starvation. c: The wolf population will decrease as the moose population decreases after the introduction of the herbivore. However, if the wolves can prey on the new herbivore, the wolf population may increase, because of this new food source.

14. If the population of moose remains stable on the island, the wolf population could be influenced by factors such as weather, disease, and genetic inbreeding.

15. a: Population R (10,000) would add more individuals at the end of the first year. This is because 5% of 10,000 is 500 individuals added to the population, whereas 5% of 100 (the size of population Q) is an increase of only 5 individuals.

b: After 5 years the size of each population would be:

Year	Population Q	Population R
0	100	10,000
1	105	10,500
2	110	11,025
3	116	11,576
4	121	12,155
5	127	12,763

c: If the populations reached carrying capacity at year 3, the level of resources would stop them from growing any larger. The populations would then remain very close to the carrying capacity (with minor fluctuations above and below the carrying capacity). The larger population would level off at approximately 11,576 individuals, which would remain its size in subsequent years.

16. a: Answers will vary. Some people may value the wolves when they are at remote locations, but less so when they are near (and potential predators of) pets and herds of cattle or flocks of chickens. b: A number of strategies could be considered. (1) Introduction of wolves from another population would not only increase the number of wolves but also provide new alleles to increase the genetic diversity in the inbred population of wolves. (2) Wolf pups could be protected in a refuge where they would be fed and have adequate shelter in a cold winter to increase their survival rate. Once the wolf pups were judged to be healthy and strong, they could be released to the pack. If more wolf pups survived each year, that would increase the size of the population (although it wouldn't help with the problem of inbreeding). (3) Instead of introducing both male and female wolves from another population, only females could be introduced. This would increase the number of mates for males and contribute to increasing the genetic diversity of the population.

17. If the carp are eating huge amounts of algae, there will be fewer algae for other organisms that rely on algae as their major food source. Many of these organisms will be other fish, whose populations will suffer as a result. If these fish are commercial or recreational sport fish, the fishing industry will be negatively affected. Furthermore, if the carp are seriously devastating the algae population (much as the moose can do to the tree population on Isle Royale), there may be local impacts on CO_2 levels. Photosynthetic algae take up CO_2 as they photosynthesize, and this helps mitigate climate change caused by elevated levels of CO_2. Management strategies are certainly challenging. One is to essentially try to overfish the Asian carp, rewarding the capture and removal of the carp. Another (and riskier) strategy might be to introduce a parasite that will attack the carp but not native fish.

Unfortunately, this strategy tends to backfire, in that the introduced parasite has unintended negative consequences on other species.

Chapter 22

1. A population is a group of organisms of the same species that live in the same area and can mate with one another and produce fertile offspring. A community consists of populations that interact and are connected by their actions. Communities contain more than one species.

2. Keystone species are those that are very important to a community because of their central role in supporting all the species in the community.

3. d

4. a

5. There are many possible answers. An example: Phytoplankton are a keystone species in the ocean, providing food for many organisms in the ocean and producing a large amount of oxygen (via photosynthesis) that is essential for life on earth.

6. Those suffering from pollen allergies are most likely allergic to pollen that is wind-carried because that pollen is in the air and can be inhaled.

7. b

8. f

9. The answer to Question 8 provides an example. A bear is part of a terrestrial food chain. The bear eats blueberries, which are pollinated by bees. Humans and birds also eat blueberries. Those birds in turn can be eaten by predatory birds, such as hawks. The bear can also "cross over" to an aquatic food chain and eat salmon. Those salmon eat organisms lower on the food chain, for instance algae.

10. The energy stored in the grain is released as the cow digests the grain. Energy is used in digestion and some energy is lost to heat. Energy is also used to sustain the life of the cow. When the meat of a cow is eaten it contains less energy than the grain because much of the energy stored in the grain has been lost or used.

11. The herbivore eats only producers (plants). Herbivores receive ~10% of the energy that is stored in the producer; the rest is burned as fuel or given off as heat. The top carnivore would eat only meat, and depending on the size of the food chain would have access to 1% or less than 1% of the energy. This is because at each level up the food chain an organism can extract only 10% of the energy in the level below.

Producer (100%) → herbivore (10%) → carnivore 1 (1%) → carnivore 2 (> 1%)

12. Bees are attracted to flowers with yellow, blue, or purple petals; other nectar-seeking organisms are attracted to other colors of flowers. Bees and other pollinators have also coevolved with some plants in such a way that the shape of the flower and the shape of the bee fit together to maximize pollen pick-up and release and provide the bee with abundant nectar and pollen. Although bees and other pollinators both feed on nectar, they do not try to feed on the same flowers and thus are not in direct competition.

13. d

14. Although mussels and barnacles are both filter feeders, they might be able to coexist because the sizes of their filters are different. The mussel may be able to eat larger organisms than the barnacles, so there is not competition for the same food.

15. The number and diversity of bees in the area would likely decrease drastically because corn is wind pollinated and does not make nectar or have flowers flowers to the bees. The bees would have to leave the area in search of a food source.

16. b

17. The relationship between bees and the bacteria that live within them can be characterized as mutualistic symbiosis. The bacteria get nutrients and a safe place to live from the bees, and the bees benefit from the bacteria because the bacteria help the bees to defend themselves from disease.

18. Researchers studying colony collapse disorder (CCD) noticed that the bees were very sick and seemed to have weakened immune systems. The virus IAPV was a good candidate for causing weakened immune response, and it was found in 96% of the hives with CCD. However, further research showed that not all colonies that are infected with IAPV have CCD, so there must be another factor causing CCD. A second hypothesis was that the parasitic varroa mite was feeding on the bees' blood, causing a weakened immune system and making them more susceptible to disease. However, research has shown that the levels of mite infection in colonies with CCD are no higher than the levels seen in previous years when CCD was not a problem.

19. The relationship between *E. coli* and humans is a symbiosis. Most of the time *E. coli* is beneficial to humans because these bacteria can prevent other pathogenic bacteria from colonizing (competition), and some types of *E. coli* can produce vitamin K_{12}. *E. coli* benefits from living in the human intestine because of the available nutrients and environment that is conducive to growth.
This type of relationship is called mutualism. Some strains of *E. coli* are parasitic: the bacteria get nutrients and a place to live and the humans get sick.

20. Point out the importance of bees in helping plants to reproduce. These plants are used as food sources by humans or by other animals which humans eat. Without the bees the plants would not be fertilized and no seeds would form.

21. Planting a single crop over a wide area decreases the number of pollinators that can be supported because there will be competition for the common resource. Also, there might not be enough variety in food sources for the pollinators to maintain a healthy diet. Similarly, a single crop will flower (and produce pollen and nectar) all at once, creating a situation of feast at the time of flowering and famine for the rest of the year. These periods of famine can cause the loss of pollinator species, affecting many other crops.

Chapter 23

1. f

2. Species discussed in this chapter that might be affected by global climate change include maple trees, seahorses, turtles, fish, yellow-bellied marmots, the Arctic fox, the red fox, and polar bears.

3. a, b, and c.

4. The coniferous forest biome is characterized principally by evergreen trees.

5. Temperate deciduous forest–eastern North America; tropical forest–Central America and northern South America; tundra–northern North America

6. Melting sea ice does not cause sea levels to rise. However, when ice caps melt, they cause sea levels to rise, thus putting low-lying areas at risk of flooding.

7. a: Smaller leaves would decrease the surface area that can lose water to the environment, and that means that less water would be lost by evaporation. Less surface area also means a decreased ability to take up CO_2 from the environment. However, given that increased global temperatures are caused by increased CO_2 levels, smaller leaves might not be a negative factor. Overall this could be a *useful adaptation* for the increased temperatures. b: More pores on each leaf would increase the amount of water lost because there would be more exposed area from which to lose water. Having more pores would also increase the amount of CO_2 that could be taken into the plant and used for photosynthesis. Overall the loss of water would be more detrimental to the plant than the increased CO_2, so this would *not be a useful adaptation to global climate change*. c: Thicker, waxier bark would serve to maintain water in the trunk of the plant, since water won't evaporate out of these surfaces. This waxy layer would not affect CO_2 uptake since this tissue is not photosynthetic. Similarly, thicker and waxier bark would be harder for insects to munch on, so this feature would be protective in the face of migrating populations of plant-eating insects. Overall this would be an *effective adaptation* for global climate change.

8. People in northern climates could contract new and different diseases if insects that carry diseases expand their ranges northward because of global climate change.

9. b

10. b

11. a

12. In the greenhouse effect, the heat trapped by greenhouse gases raises the temperature of the atmosphere and in turn raises the temperature of the surface of the earth. Without the greenhouse effect, the temperature on earth would be -18°C; we could not survive at that temperature.

13. The evidence that increasing levels of greenhouse gases (particularly CO_2) are responsible for global climate change includes these points: (1) Carbon dioxide concentrations are higher than they have been in 700,000 years. (2) Since direct measurement of atmospheric carbon dioxide began in the late 19th century, its concentration has increased ~35%. Figure 23.5 shows how CO_2 levels increased with temperatures over the last 1,000 years. Both temperature and CO_2 stayed relatively level until the start of the industrial revolution, when CO_2 levels increased.

Data from the past 50+ years showing that the level of solar radiation had increased over the entire planet would support the hypothesis. Also supportive would be a graph and data analysis that show a correlation between increased levels of solar radiation and the increasing temperatures of the planet.

14. Using *fossil fuels* for energy converts organic carbon to CO_2. Most organisms, including plants, animals and decomposers, perform *respiration,* producing CO_2 from organic food. CO_2 is released to the *atmosphere.* CO_2 is absorbed by the *photosynthetic organisms in the ocean.* Plants perform *photosynthesis,* fixing CO_2 into organic molecules. Coal and oil are *fossil fuels* that trap carbon in the earth.

15. c

16. Surface ice is not useful, but ice cores provide a way to measure atmospheric conditions from the distant past. Cylinders of ice can be extracted from glaciers, and the composition of gas bubbles within them can be analyzed. This analysis reveals the atmospheric conditions of thousands of years ago, when those bubbles were initially trapped in the ice.

17. a: Large-scale slash and burn agriculture releases carbon that was stored in trees directly to the atmosphere as carbon dioxide. Additionally, as the land does not generally support long-term crop production, the overall levels of photosynthesis (which removes CO_2 from the atmosphere) are reduced relative to those of the original forest. b: Driving gasoline-fueled cars (in fact, burning any fossil fuel) releases carbon (as CO_2) that had been stored for a very long time in the earth. This carbon is essentially "new" carbon being introduced into the atmosphere, adding to the amount of carbon already cycling. c: Cattle raised for beef and dairy products have methane-producing microbes in their guts that help them digest food, but the cattle release large amounts of methane gas as flatulence. Methane is a potent greenhouse gas. d: Rice production also releases methane gas into the environment. Methane is a potent greenhouse gas.

18. d. Ice cores that contain ice formed in 1750 would be the most effective way to determine the CO_2 concentration at that time. The Mauna Loa observatory has been continuously monitoring and collecting data related to atmospheric change only since the 1950s–it would not have data from 1750. Neither tree rings nor historical records would give accurate values from which to determine CO_2 levels but could provide clues to the weather in 1750.

19. Answers will vary. Considerations include: emissions from automobiles, which could be mitigated by using public transportation, carpooling, or cycling or walking on a regular basis; the use of household electrical products (which often rely on coal-fired electrical plants to generate the electricity), which can be mitigated by replacing worn-out appliances with energy-efficient versions, line-drying laundry rather than using a dryer, and turning off and unplugging electrical devices when they are not in use.

20. Answers will vary. These actions would contribute to a low-carbon footprint menu: (1) purchasing locally grown food, thus reducing the consumption of fossil fuels required to transport food long distances from the site of production to the site of consumption (different parts of the country would rely on different foods, depending on what is produced locally; (2) eating foods that do not require cooking, to reduce use of electricity or gas); (3) using a solar oven rather than a gas or electric oven to cook food

Chapter 24

1. Some advances discussed in the chapter are the development of agriculture and the use of antibiotics and other advances in public health.

2. An ecological footprint is a tool used to calculate how much of the earth's resources a population's lifestyle requires. It calculates how much biologically productive land and water area a human population needs to produce the resources it consumes and to absorb the waste it produces.

3. d

4. The factors contributing to the ecological footprint of each area are healthy food, energy for mobility and heat, fiber for paper, clothing and shelter, fresh air and clean water.

City	Footprint	Factors Contributing to Footprint
Dongtan, China	Relatively low	Designed to be sustainable: no cars are necessary because everything is within walking distance; energy is derived from solar and wind sources; buildings are designed to remain cool without reliance on air conditioning; extensive recycling to minimize waste and pollution.
Rural village in China	Relatively low	Less developed, therefore less use of fossil fuels for cars; food is produced locally, reducing transportation costs.
Calgary, Canada	Relatively high	Calgary is considering its ecological footprint in development decisions, and residents can use a light rail system that is powered through the use of wind turbines. Thus its footprint is probably smaller than would be expected for a developed city in North America.
Houston, Texas	Relatively high	Large area and highways necessitate travel by automobile, increasing fossil fuel use. Hot summers require high use of electricity for air conditioning; city acts as a heat trap.
New York	Relativity low	As noted in the chapter, the high population density and relatively small area in New York allows New Yorkers to use public transportation, rather than driving cars for their daily commutes.

5. The addition of 20 families will increase the population of the town. If the children remain in the town and raise families, the town's population will likely continue to increase in the future. Generally speaking, urban populations have a higher footprint than rural populations. As the new condominiums are on the outskirts of the town, the families will need to drive into town for school, work, and shopping. If each family has two cars, this represents 40 new vehicles, more than what was likely used on the farm. Similarly, 20 families will produce a variety of waste that will likely be in excess of that generated by farming a crop. There are now 20 households to be heated and cooled rather than the original farmhouse. It is thus likely that this population will use more energy and generate more waste than the farm, thereby enlarging the ecological footprint.

6. The developer could use solar or wind power to replace traditional fossil fuels for energy sources. The developer could also take into consideration public transportation and provide alternatives to driving, as well as making housing, workplaces, and school within walking and biking distance.

7. e

8. freshwater (R); coal (NR); codfish populations in the North Atlantic (R); wind (R); sunlight (R)

9. Coal, oil, and natural gas are fossil fuels that take millions of years to form from organic material in dead organisms and therefore these resources are not renewable on a useful timescale.

10. Renewable resources like food and water may not have the time to renew themselves when the population is growing so quickly and demand for them exceeds the space available to grow and harvest them. For example, at one point the North Atlantic codfish industry was in danger of collapsing because of high demand and overfishing–taking cod faster than they could reproduce was making the cod a renewable resource. Governments chose to limit the catch of cod, and the numbers of cod are gradually increasing. Similarly, while water is technically renewable, we can choose how to use water, and how to maintain clean water for drinking, even at the expense of recreation. And while we are unlikely to run out of wind or sunlight, we can make choices to make these resources more accessible. For example, we can agree to look at perhaps unattractive wind farms for the sake of taking advantage of this resource.

11. Answers will vary, depending on locale. Challenges might be space (solar panels and wind mills) and the expense of new technology.

12. a: Schools and shops located near residences will increase walking, biking, or the use of public transportation. These measures would decrease use of fossil fuels (because they reduce the use of fossil-fuel-powered cars), save energy, decrease CO_2 production, and reduce air pollution. b: If the buildings have green roofs, they have vegetation to absorb and filter rainwater, and provide growing space for food. This measure would save energy needed to bring food into the city from farms and recycles water and help keep buildings cool. b: Solar panels are mounted

throughout a city would provide renewable energy with little pollution or waste.

13. Waste in the form of methane collected from landfills can be used to power dump trucks; batch reactors can digest human waste to use as fertilizer.

14. Examples: reducing home water use would help maintain water availability; installing compact fluorescent lightbulbs would help save electricity and decrease the amount of CO_2 released into the atmosphere; eating less meat–meat takes a great deal of grain and water to produce, affecting energy consumption; unplug electronics not in use would reduce vampire energy flow–energy flow from devices that are turned off; driving less would reduce fossil fuel burning and CO_2 emissions; recycling would reduce the amount of waste produced.

15. Cities in the desert must draw water from underground aquifers or rely on water from rivers. As the desert is typically hot (especially in the summer), the energetic costs of home cooling in the summer are substantial: a great deal of electricity is required to cool homes in the summer. Many southwestern cities are spread out, creating a demand for fossil fuels.

16. When water is taken out of the river to irrigate crops, water levels beyond this point will be much lower. This means that there is less water in the river (or reservoirs) for fishing, boating, and swimming. On the other hand, the water is being used to irrigate crops that represent food for the population. In addition to negative impacts on recreation downstream, the withdrawal of water for irrigation means that there is not enough water in the river to support an endangered species of minnow. This puts human needs (irrigation of crops) in direct conflict with the needs of another species. Different people will have different opinions about this issue.

17. In general, the United States has a much higher footprint than Bangladesh, primarily because the United States is a highly developed country. As countries develop, their footprints increase. This will lead to an overall increase in the global footprint. As the global footprint increases, the number of people that the earth can support will decrease (that is, carrying capacity will decrease). Different people will have different opinions about whether or not this is a shared responsibility–whether or not developed countries should reduce their footprints to accommodate the development of other countries.

Chapter 25

1. Anatomy is the study of the structure of organisms, for example the structure of the human body and its "parts." Physiology is the study of how organisms function and maintain homeostasis.

2. (1) mucus-secreting cell of the small intestine (cell); (2) the layer of muscle that contributes to the function of the small intestine (tissue) (3) small intestine (organ); (4) digestive system (organ system)

3. a

4. Physiology. Diet influences a number of physiological processes (for example, levels of blood sugar, glycogen

storage, energy balance), as does exercise (for example, thermoregulation, energy expenditure).

5. Because the heart is composed of a number of tissue types, including nervous tissue and muscle.

6. Homeostasis is the ability to maintain a relatively constant internal environment, even when the external environment changes.

7. When body temperature drops, one response is to generate heat through shivering. This requires that glucose be available to the muscle, so that the muscle has an energy source for its contraction during shivering. Glucagon is a hormone that promotes the breakdown of glycogen to glucose, providing fuel for muscles.

8. Cold conditions produce a *low body temperature*. *Sensors* can detect this drop in body temperature and send a signal to the *hypothalamus*. The hypothalamus in turn sends a signal to *muscles*, which act as *effectors* in this case—contracting by shivering to produce heat and bring about a *normal body temperature*.

9. c

10. The hypothalamus receives input about body temperature from sensors and sends out instructions to effectors. If the hypothalamus is damaged, then it may not receive the information that the core temperature is dropping, or it may not be able to send information to muscles to instruct them to shiver.

11. High insulin is the result of high blood sugar. High blood sugar triggers the pancreas to release insulin. In the presence of insulin, many cells are signaled to take up glucose from the blood. This causes a drop in blood-sugar levels. Now that blood sugar is lower, there is no longer a signal to the pancreas to release insulin, so insulin levels drop as well.

12. Sweating: as sweat evaporates, it cools the surface (in this case, the skin) from which it is evaporating. Vasodilation: by increasing blood flow to the skin, more heat can be released to the environment.

13. In the short term, individual Tibetan Sherpas have acclimatized to the relatively hypoxic conditions on Mount Everest: for example, they likely have relatively large numbers of oxygen-transporting red blood cells in their circulation. In the long term, the Sherpa population has evolved to have high fitness in the low-oxygen environment at high altitude: for example, some alleles encoding hemoglobin may be more efficient at binding oxygen when it is present only at low levels, and these alleles would have been selected for over generations.

14. Research into how the body functions can contribute to understanding of what happens during the course of a variety of diseases, as well as suggesting possible treatments. For example, the study of endocrinology can lead to a better understanding of diabetes (characterized by high blood sugar), stress (many of the hormones produced by the adrenal glands are associated with stress), and cardiovascular disease (heart and blood vessel function).

15. Answers regarding financial responsibility may vary. Preparation for unexpected weather (in a winter climate) should include sufficient clothing to stay warm (prevention of hypothermia), sufficient food to provide energy for the hike and more (in case of being lost or stranded, or needing to fuel shivering muscles), adequate water (to prevent dehydration). If the hike is in an extreme environment, adequate preparation would include sufficient acclimatization to that environment (for example, time to acclimatize to low oxygen pressure).

In a climate that may become extremely hot, adequate water to stay hydrated is critical, especially as water will be lost by evaporation (sweating) as the body tries to remain cool. Adequate food will also be important. In this environment, severe sunburn can be as dangerous as severe frostbite in cold climates, so adequate sun protection is essential.

Chapter 26

1. (1) mouth; (2) esophagus; (3) stomach; (4) small intestine; (5) large intestine

2. d

3. An expandable stomach allows you to eat more than you can immediately process—the stomach can expand to accommodate the extra food and hold on to it until it can be processed through the small intestine. Without the expandable stomach, you would have to continuously eat small amounts of food.

4. The gallbladder, liver, and pancreas are all accessory organs that secrete enzymes and other substances into the digestive tract. They are not part of the main tube of the digestive tract, which includes the mouth, stomach, and small intestine.

5. a

6. e

7. a

8. They are both digestive enzymes.

9. pancreas

10. a

11. Both are involved in processing fats, but they do so in distinct ways. Bile salts help emulsify large globules of fat to small droplets. These droplets can then be acted on by lipase, which chemically digests triglycerides.

12. The duct between the pancreas and the small intestine allows digestive enzymes produced in the pancreas to enter the small intestine, where they will digest macromolecules in food. When the duct is blocked, the digestive enzymes remain in the pancreas, where they will digest (and damage) the pancreatic tissue.

13. Yes. Food will still pass from your small intestine into your large intestine because food is propelled through your digestive tract by muscular contractions, not gravity.

14. Gallbladder removal leads to the inability to store bile salts. This in turn means that fats cannot be fully processed. Alli is a lipase inhibitor, so taking Alli means that fats cannot be digested. Because undigested fats cannot be absorbed, they are eliminated as fats from the digestive tract, contributing to greasy and loose stools.

15. Someone with a high BMI can reduce the number of Calories consumed and work to burn more Calories

through exercise. Some considerations about financing weight-loss surgery and treating obesity-related diseases include the cost of healthful food compared to the cost of highly processed junk food and compulsive behavior around food as well as other behavioral factors that influence food consumption.

16. The surgery is not without risks and has a long recuperation time. Major changes to eating habits after the surgery are necessary to avoid serious discomfort and to lose weight. Not everyone loses the same amount of weight, and some people regain lost weight. Someone at high risk for obesity-related diseases should considering this surgery in consultation with a physician and surgeon. Someone not yet in the very high risk zone in terms of BMI should discuss with a physician the relative risks and benefits of the surgery and the options for weight-loss interventions that do not require surgery.

Chapter 27

1. a light-detecting receptor in the eye–PNS; the amygdala–CNS; a pain receptor in the skin–PNS; the spinal cord–CNS; the thalamus–CNS

2. d

3. b

4. movement: cerebellum; body temperature: diencephalon (specifically the hypothalamus)

5. It is two way. Information from the environment comes in toward the brain, and instructions from the brain go out to the appropriate effectors.

6. Muscles are effectors and receive signals to contract from effector neurons. If these effector neurons lose their myelin sheath, the signal doesn't reach the axon terminal, and therefore the information is not communicated to the muscle. The muscle is capable of contracting but is not receiving the appropriate signals to do so.

7. d

8. b

9. c

10. It diffuses across the synapse and binds to receptors on the membrane of the next cell in the signaling pathway (the postsynaptic cell), which could be another neuron or an effector such as a muscle or gland.

11. Electrical signaling occurs along the length of the axon as characteristic patterns of ion flow, that is, action potentials. Chemical signaling occurs by the release of neurotransmitters (chemical signaling molecules) from the axon terminals of neurons. The neurotransmitters cross the synapse and bind to receptors on the receiving cell.

12. If dopamine remains in the synapse, there will effectively be more dopamine present, which will trigger feelings of pleasure.

13. As acetylcholine normally signals muscles to contract, and Botox prevents the release of acetylcholine into synapses, the muscles will not receive signals to contract, and will be paralyzed in a relaxed state.

14. Because multiple sclerosis is a demyelinating disease, the action potentials in the motor neurons will not reach the axon terminal. Therefore, there will be less acetylcholine

released. This explains why those with multiple sclerosis experience muscle weakness: their muscles are not getting the neurotransmitter signal to contract.

15. b

16. In the presence of the drug, dopamine levels are elevated, and cells are bombarded with dopamine. The response of these cells is to down-regulate their dopamine receptors (so that there are fewer active receptors to respond to dopamine). With fewer receptors, there has to be a stronger stimulus (that is, more drug) to achieve equivalent feelings of pleasure.

17. Probably sadder. Even if they produced the same amount of dopamine as others, with fewer dopamine receptors they have an overall reduced response to dopamine. As dopamine is involved in feelings of pleasure, those who respond less well to dopamine will likely have lower levels of pleasure, and hence may feel sadder than others.

18. The loss of dopamine-producing neurons means that there will be less dopamine present. This will result in lower levels of happiness or pleasure, seriously affecting someone with Parkinson disease.

19. This difference is likely to result from politics and social conventions. Cigarette smoking has long been a socially accepted norm (even if not a healthy one), while cocaine is not.

20. There are many possible responses. Some considerations are: not all people with addictions are "junkies" or morally inferior; addictions can have serious health and behavioral consequences, involving treatments that can be costly to the taxpaying public; by understanding the causes of addiction, we may be able to contribute to reducing the need for these treatments; much of the research is basic research that contributes to our fundamental understanding of the nervous system, which can lead to treatments for many neurological diseases, such as Parkinson disease and Alzheimer disease; research can also benefit people who suffer from neurological traumas, include posttraumatic stress disorder and severe nervous system injury.

Chapter 28

1. c

2. b

3. a: The uterus is the organ in which embryos and fetuses develop. The cervix is the opening between the cervix and the vagina. b: The endometrium is the lining of the uterus. The embryo implants into the endometrium.

4. seminiferous tubules→epididymis→vas deferens→urethra

5. One of the risks is inflammation that can lead to permanent scarring of the oviducts. The inflammation can be silent, in that there are no symptoms. Similarly, scarred oviducts present no problems until a woman tries to become pregnant. Then she may encounter infertility, as the scarring prevents egg and sperm from meeting.

6. luteinizing hormone (LH) (anterior pituitary); follicle-stimulating hormone (FSH) (anterior pituitary); testosterone (testis); estrogen (ovary); progesterone (ovary); hCG (embryo)

7. a

8. e

9. FSH and LH are not produced until puberty (when females begin to experience menstrual cycles). Thus, women of reproductive age have substantially higher levels of FSH and LH than girls who have not reached sexual maturity.

10. Testosterone is the primary hormone responsible for sperm development in males and would have to be targeted in order to prevent sperm development. However, blocking testosterone production or activity would also cause a loss of libido and loss of many masculine characteristics, side effects that would not be tolerable.

11. b

12. Both IVF and IUI generally rely on hormonal stimulation of women, so that they will ovulate several mature eggs at the same time. In IVF, those eggs are surgically removed, mixed with sperm in a petri dish, and the resulting embryos placed in the uterus, where it is hoped that they will implant into the endometrium. In IUI, at about the time of ovulation (of possibly multiple eggs), sperm are placed directly into the uterus, in hope that fertilization will occur in the uterus, and that any embryos produced will implant in the endometrium.

13. a: blocked epididymis–this could be detected by imaging; b: polycystic ovary syndrome–this could be diagnosed by a combination of symptoms (for example, irregular menstrual cycles) and and assay of androgen levels (elevated levels of androgens will be found in an affected woman); c: menopause–as menopause is marked by the cessation of menstrual cycles, a woman probably near the age of 50 who has stopped menstruating and whose levels of reproductive hormones are low is probably in menopause; d: oviduct scarring–imaging (for example, ultrasound to visualize an internal blockage) would confirm this diagnosis.

14. In IVF, a fertility specialist can choose how many (or few) embryos to implant. In IUI, there is no way to control the number of eggs that are ovulated in response to the hormonal treatment. Thus, a very large number of eggs could be ovulated, and if all were to be fertilized by the introduced sperm, a large number of embryos could result.

15. Answers will vary. Some of the factors to consider include the health risks to premature babies (and the fact that multiples are often premature); the health care costs (and the question of who bears the burden of those costs); the ability of parents to raise many children of the same age; and the strong desire of infertile couples to be able to have their own children.

16. There are many considerations. In favor of regulation: (1) Clinics wouldn't compete on the basis of success rates, thus reducing the pressure to implant many embryos. All clinics would be operating on a level field. (2) If the number of embryos is regulated, that would reduce the number of multiple pregnancies in which parents may have to consider selective reduction. (3) If regulation reduces the number of multiple births, then the number of premature babies should decline, reducing the burden of illness. Against regulation: (1) Many regulations would reduce the number and rate of successful pregnancies. This places a burden on parents who desperately want to have their own children, as they may have to endure repeated disappointment. (2) If the success rate is reduced, insurance may be less likely to pay for a procedure that has a small chance of success, increasing the financial burden on the patients. (3) Doctors will be unable to pursue all possible options for their patients. (4) Some unscrupulous doctors may set up illegal clinics, and desperate patients may seek treatment in clinics that do not meet general safety standards.

Chapter 29

1. d

2. They must infect the appropriate cell type. Once they enter that cell, they take over the host cell, directing it to replicate the virus.

3. Poliovirus infects and damages nerve cells, which are unable to replicate (and therefore to repair themselves). Influenza infects cells lining the respiratory tract, which are frequently replaced.

4. Antigenic drift is the result of small changes (mutations) in the viral genome. Antigenic shift is the result of segmented viruses exchanging genome segments. This means that a given virus can end up with genes from a completely different virus.

5. Viruses are not made of cells–they must infect host cells in order to replicate. Bacteria are prokaryotic organisms. The vast majority are capable of replicating without entering into and infecting host cells.

6. Each of these viruses has a different host cell. Poliovirus ultimately infects cells of the nervous system, ultimately causing paralysis. Influenza virus infects cells lining the respiratory tract, causing respiratory tract symptoms. HIV infects cells of the immune system, causing an immune deficiency.

7. skin, which provides protection against pathogen entry; enzymes in tears and saliva that digest components of pathogens; phagocytes that ingest and digest pathogens

8. They are both phagocytic cells.

9. As innate immunity is nonspecific, the innate responses to *E. coli* and to *S. aureus* would be the same. The innate response does not recognize specific pathogens.

10. Neutrophils are critical players in the inflammatory response. Someone with a deficiency of neutrophils would therefore not be able to mount an effective inflammatory response.

11. As influenza can directly and indirectly (through the inflammatory response) damage lung tissue, it destroys the first line of defense against the entry of pathogenic bacteria.

12. Anti-inflammatory drugs suppress the inflammatory response, leaving a person more vulnerable to infections, as bacteria may not be contained or destroyed at the site of entry.

13. Innate immunity is always active and nonspecific. It is present since birth and does not strengthen over time. In contrast, adaptive immunity is specific to a particular pathogen. It is not always "on"–it is induced by the presence of the specific pathogen and strengthens with repeated exposure (because of memory B cells).

14. B cells are lymphocytes that can be activated during an adaptive immune response. Upon activation, B cells specialize into antibody-producing plasma cells, which produce large numbers of antibody molecules.

15. d

16. b

17. a: Yes. b: The memory B cells from a previous exposure will be specific for the particular strain of influenza that was circulating at the time of that exposure. Because of antigenic drift, the influenza virus changes seasonally, so next year's virus will not be identical to the previous virus and this person will need to raise a new primary response against the new flu. c: Because H1Ni arose from antigenic shift, it is substantially different from the seasonally circulating influenza strains. This person's memory B cells will not be specific for this strain of influenza.

18. Antigenic shift results in "mixing and matching" of viral genes from different viruses. This results in a virus that has genes from viruses that typically infect pigs or birds and genes from viruses that typically infect humans. Because the proteins encoded by the pig or bird virus have not been in viruses that infect humans, there is no existing immunity to these essentially completely new proteins, so these "hybrid" viruses can run rampant through the global population.

19. a: Because our own skin cells are recognized as self and (except in cases of autoimmunity) the immune responses are not directed at self. b: Yes. The innate responses are nonspecific. They do not recognize individual pathogens, and are equally effective against all pathogens. c: No. The adaptive response to *Staphylococcus aureus* is specific to *Staphylococcus aureus* and will not be effective against a different bacterium.

20. As helper T cells are critical in mounting an adaptive immune response, the destruction of helper T cells by HIV renders someone who is infected unable to produce an effective adaptive response to pathogens and thus vulnerable to a variety of infections.

21. A person who has been vaccinated has a much lower risk of becoming infected with that pathogen. If everyone is vaccinated, there will be no reservoir of susceptible individuals in whom the pathogen can survive. This can reduce the spread of the pathogen. Also, it will reduce the number of sick people, reducing the burden on health care resources.

22. There are many considerations. One approach is to think about the impact of other disease-preventive conditions (for example, clean water) and what they have done to reduce the burden of disease. Another is to consider a recently developed vaccine (for example, the Heamophilus influenzae type B vaccine for children) and look at the mortality before and after the introduction of this vaccine, keeping in mind that this vaccine was introduced in the era of modern medicine, when there were effective antibiotics and hospital treatments available.

Chapter 30

1. See Infographic 30.1.

2. b

3. d

4. a

5. c

6. Pollination is the delivery of pollen to the female pistil. The pollen grains land on the sticky stigma, and then sperm travel down the style to reach the egg contained in the ovule. Fertilization will generate an embryo that will be contained in a seed.

7. e

8. b

9. c

10. Only wood contains lignified cell walls. Wood pulping requires lignin-digesting enzymes to break down the lignin. Green leaf cells do not have lignified cell walls, so would not require lignin-digesting enzymes to break down.

11. Plants need to open their stomata to let in CO_2 so that they can carry out photosynthesis. Most plants have to do this during the day (when photosynthesis, which is dependent on light, occurs). However, when plants open their stomata to let in CO_2, they lose water by evaporation. In dry climates, this water loss can be substantial and detrimental to the plant.

12. They are not, as they supplement their nutrition with nitrogen from other organisms, specifically insects.

13. Because you might have killed the nitrogen-fixing bacteria in the soil, depriving your plant of fixed nitrogen in the soil.

14. The trumpet pitcher's bright coloration attracts insects that are a source of nitrogen–that is, of "food." Squash blossoms also attract insects (for example, bees), but in this case insects provide "transportation," transferring pollen from the anthers to the insect's body so that the pollen can be carried to the pistil of another plant as the insect makes a tour of blossoms.

15. It would not be able to survive. Chlorophyll is essential for photosynthesis, and without chlorophyll the plant would not be able to make its food.

16. Gibberellins are responsible for the stem elongation that allows large bunches of grapes to develop. These hormones are naturally produced by seeds. Seedless grapes therefore do not produce gibberellins on their own, so must be treated with gibberellins.

17. The prickly spines on the prickly pear cactus are meant to protect the plant from herbivores. By removing the spines for the salad, humans are circumventing this antiherbivory mechanism.

18. Humans have identified many valuable compounds from plants (see Chapter 9). If the plants are fast growing and

easy to cultivate, we can grow enough plants to meet our needs. However, if the plants are very slow growing or can't be grown in fields or orchards, then we run the risk of driving the plants to extinction in order to meet our needs. In the long run, this result is not beneficial to either humans or plants. Possible solutions include legislation to protect the plants and their environment, finding alternative sources of the valuable compounds, using alternative strategies to cultivate the plant of interest (for example, grafting), or using chemical synthesis to produce the valuable compound independently of the plant.

Glossary

abiotic Refers to nonliving components of the environment such as temperature and precipitation.

abscisic acid (ABA) A plant hormone that helps seeds remain dormant.

absorption The uptake of digested food molecules by the epithelial cells lining the small intestine.

acclimatization The process of physiologically adjusting to an environmental change over a period of time. Acclimatization is generally reversible.

acid A substance that increases the hydrogen ion concentration of solutions, making them more acidic.

action potential An electrical signal within neurons caused by ions moving across the cell membrane.

activation energy The energy required for a chemical reaction to proceed. Enzymes accelerate reactions by reducing their activation energy.

active site The part of the enzyme that binds to substrates.

active transport The energy-requiring process by which solutes are pumped from an area of lower concentration to an area of higher concentration with the help of transport proteins.

adaptation The response of a population to environmental pressure, so that advantageous traits become more common in the population over time.

adaptive immunity A protective response, mediated by lymphocytes, that confers long-lasting immunity against specific pathogens.

adaptive radiation The spreading and diversification of organisms that occur when they colonize a new habitat.

adenosine triphosphate (ATP) The molecule that cells use to power energy-requiring functions; the cell's energy "currency."

adult stem cells (somatic stem cells) Stem cells located in tissues that help maintain and regenerate those tissues.

aerobic respiration A series of reactions that occurs in the presence of oxygen and converts energy stored in food into ATP.

alga (plural: algae) A uni- or multicellular photosynthetic protist.

allele frequency The relative proportion of an allele in a population.

alleles Alternative versions of the same gene that have different nucleotide sequences.

allergy A misdirected immune response against environmental substances such as dust, pollen, and foods that causes discomfort in the form of physical symptoms.

allopatry Speciation that occurs because of geographic or climatic barriers to gene flow.

amino acids The building blocks of proteins. There are 20 different amino acids found in proteins.

amniocentesis A procedure that removes fluid surrounding a fetus to obtain and analyze fetal cells to diagnose genetic disorders.

amygdala A subregion of the brain that processes emotions, especially fear and anxiety, and is the seat of emotional memories.

anabolic reaction Any chemical reaction that combines simple molecules to build more-complex molecules.

anatomy The study of the physical structures that make up an organism.

androgen A class of sex hormones, including testosterone, that is present in higher levels in men and causes male-associated traits like deep voice, growth of facial hair, and defined musculature.

anecdotal evidence An informal observation that has not been systematically tested.

aneuploidy An abnormal number of one or more chromosomes (either extra or missing copies).

angiosperm A seed-bearing flowering plant with seeds typically contained within a fruit.

animal A eukaryotic, usually multicellular, organism that obtains nutrients by ingesting other organisms or molecules produced by other organisms.

annelid A segmented worm, such as an earthworm.

anterior pituitary gland The gland in the brain that secretes luteinizing hormone (LH) and follicle-stimulating hormone (FSH).

antibiotic A chemical that can slow or stop the growth of bacteria; many antibiotics are produced by living organisms.

antibody A protein produced by B cells that binds to antigens and either neutralizes them or flags other cells to destroy pathogens.

anticodon The part of a tRNA molecule that binds to a complementary mRNA codon.

antigen A specific molecule (or part of a molecule) to which specific antibodies can bind, and against which an adaptive response is mounted.

antigenic drift Changes in viral antigens caused by genetic mutation during normal viral replication.

antigenic shift Changes in antigens that occur when viruses exchange genetic material with other strains.

apoptosis Programmed cell death; often referred to as cellular suicide.

aquifer Underground layers of porous rock from which water can be drawn.

archaea One of the two domains of prokaryotic life, the other is Bacteria.

arthropod An invertebrate having a segmented body, a hard exoskeleton, and jointed appendages.

atom The smallest unit of an element that cannot be chemically broken down into smaller units.

autoimmune disease A misdirected immune response in which the immune system mistakenly attacks healthy cells.

autosomes Paired chromosomes present in both males and females; all chromosomes except the X and Y chromosomes.

autotrophs Organisms such as plants, algae, and certain bacteria that capture the energy of sunlight by photosynthesis.

auxin A plant hormone that causes elongation of cells as one of its effects.

axon The long extension of a neuron that conducts action potentials away from the cell body toward the axon terminal.

axon terminal The tip of an axon, which communicates with the next cell in the pathway.

B cells White blood cells that mature in the bone marrow and produce antibodies during the adaptive immune response.

bacteria One of the two domains of prokaryotic life; the other is Archaea.

base A substance that reduces the hydrogen ion concentration of solutions, making them more basic.

bilateral symmetry The pattern exhibited by a body plan with clear right and left halves that are mirror images of each other

bile salts Chemicals produced by the liver and stored by the gallbladder that emulsify fats so that they can be chemically digested by enzymes.

binary fission A type of asexual reproduction in which one parental cell divides into two.

biocapacity The amount of the earth's biologically productive area–cropland, pasture, forest, and fisheries–that is available to provide resources to support life.

biogeography The study of how organisms are distributed in geographical space.

biological species concept The definition of a species as a population whose members can interbreed to produce fertile offspring.

biome A large geographic area defined by its characteristic plant life, which in turn is determined by temperature and levels of moisture.

biotic Refers to living components of the environment.

blastocyst The stage of embryonic development in which the embryo is a hollow ball of cells. Researchers can derive embryonic stem cell lines from cells of a blastocyst stage embryo.

body mass index (BMI) An estimate of body fat based on height and weight.

bottleneck effect A type of genetic drift that occurs when a population is suddenly reduced to a small number of individuals, and alleles are lost from the population as a result.

brain An organ of the central nervous system that integrates and coordinates virtually all functions of the body.

brain stem The part of the brain that is closest to the spinal cord and which controls vital functions such as heart rate, breathing, and blood pressure.

bryophyte A nonvascular plant that does not produce seeds.

calorie The amount of energy required to raise the temperature of 1 gram of water by 1° Celsius.

cancer A disease of unregulated cell division: cells divide inappropriately and accumulate, in some instances forming a tumor.

capsule A sticky coating surrounding some bacterial cells used to adhere to surfaces.

carbohydrate An organic molecule made up of one or more sugars. A one-sugar carbohydrate is called a monosaccharide; a carbohydrate with multiple linked sugars is called a polysaccharide.

carbon cycle The movement of carbon atoms between organic and inorganic molecules in the environment.

carbon fixation The conversion of inorganic carbon (for example, CO_2) into organic forms (for example, sugars).

carbon footprint A measure of the total greenhouse gases we produce by our activities.

carcinogen Any chemical agent that causes cancer. Many carcinogens are mutagens.

carrier An individual who is heterozygous for a particular gene of interest, and therefore can pass on the recessive allele without showing any of its effects.

carrying capacity The maximum population size that a given environment or habitat can support, given its food supply and other natural resources.

catabolic reaction Any chemical reaction that breaks down complex molecules into simpler molecules.

cell The basic structural unit of living organisms.

cell body The part of a neuron that contains most of the cell's organelles, including the nucleus.

cell cycle An ordered sequence of stages that a cell progresses through in order to divide during its life; the stages includes preparatory phases (G_1, S, G_2) and division phases (mitosis and cytokinesis).

cell cycle checkpoint A cellular mechanism that ensures that each stage of the cell cycle is completed accurately.

cell division The process by which a cell reproduces itself; cell division is important for normal growth, development, and repair of an organism.

cell membrane A phospholipid bilayer with embedded proteins that forms the boundary of all cells.

cell theory The concept that all living organisms are made of cells and that cells are formed by the reproduction of existing cells.

cell wall A rigid layer surrounding the cell membrane of some cells, providing shape and structure. In plant cells, the cell wall is made of cellulose, a complex carbohydrate.

cell-mediated immunity The type of adaptive immunity that rids the body of altered (that is, infected or foreign) cells.

cellular differentiation The process by which a cell specializes to carry out a specific role.

central nervous system (CNS) The brain and the spinal cord.

central vacuole A fluid-filled compartment in plant cells that contributes to cell rigidity by exerting turgor pressure against the cell wall.

centromere The specialized region of a chromosome where the sister chromatids are joined. This site is critical for proper alignment and separation of sister chromatids during mitosis.

cerebellum The part of the brain that processes sensory information and is involved in movement, coordination, and balance.

cerebral cortex The outer layer of the cerebrum, the cerebral cortex is involved in many advanced brain functions.

cerebrum The region of the brain that controls intelligence, learning, perception, and emotion.

cervix The opening or "neck" of the uterus, where sperm enter and babies exit.

chemotherapy The treatment of disease, specifically cancer, by the use of chemicals.

chlorophyll The pigment present in the green parts of plants that absorbs photons of light energy during the light reactions of photosynthesis.

chloroplast The organelle in plant and algal cells that is the site of photosynthesis.

chromosome A single, large DNA molecule wrapped around proteins. Chromosomes are located in the nuclei of most eukaryotic cells.

chyme The acidic "soup" of partially digested food that leaves the stomach and enters the small intestine.

citric acid cycle A set of reactions that takes place in mitochondria and helps extract energy (in the form of high-energy electrons) from food; the second step of aerobic respiration.

coding regions Sequences of DNA that serve as instructions for making proteins.

coding sequence The part of a gene that specifies the amino acid sequence of a protein. Coding sequences determine the identity, shape, and function of proteins.

codominance A form of inheritance in which both alleles contribute equally to the phenotype.

codon A sequence of three mRNA nucleotides that specifies a particular amino acid.

coenzyme A small organic molecule, such as a vitamin, required for enzyme activity.

cofactor An inorganic substance, such as a metal ion, required for enzyme activity.

colon The first and longest portion of the large intestine; the colon plays an important role in water reabsorption.

commensalism A type of symbiotic relationship in which one member benefits and the other is unharmed.

community A group of interacting populations of different species living together in the same area.

competitive exclusion principle The concept that when two species compete for resources in an identical niche, one is inevitably driven to extinction.

complement proteins Proteins in blood that help destroy pathogens by coating or puncturing them.

complementary Two strands of DNA are said to be complementary in that A always pairs with T, and G always pairs with C.

complex carbohydrate (polysaccharide) A carbohydrate made of many simple sugars linked together, that is, a polymer of monosaccharides; examples are starch and glycogen.

consumers Heterotrophs that eat other organisms or the organic molecules produced by organisms to obtain energy.

contraception The prevention of pregnancy through physical, surgical, or hormonal methods.

control group The group in an experiment that experiences no experimental intervention or manipulation.

convergent evolution The process by which organisms that are not closely related evolve similar adaptations as a result of independent episodes of natural selection.

corpus luteum The structure in the ovary that remains after ovulation and secretes progesterone.

correlation A consistent relationship between two variables.

covalent bond A strong chemical bond resulting from the sharing of a pair of electrons between two atoms.

cuticle The waxy coating on leaves and stems that prevents water loss.

cytokinesis The physical division of a cell into two daughter cells.

cytoplasm The gelatinous, aqueous interior of all cells.

cytoskeleton A network of protein fibers in eukaryotic cells that provides structure and facilitates cell movement.

cytotoxic T cell A type of T cell that destroys altered cells, including virally infected cells.

decomposer An organism such as a fungus or bacterium that digests and uses the organic molecules in dead organisms as sources of nutrients and energy.

dendrites Branched extensions from the cell body of a neuron, which receive incoming information.

density-dependent factor A factor whose influence on population size and growth depends on the number and crowding of individuals in the population (for example, predation).

density-independent factor A factor that can influence population size and growth, regardless of the numbers and crowding within a population (for example, weather).

deoxyribonucleic acid (DNA) The molecule of heredity, common to all life forms, that is passed from parents to offspring.

dependent variable The measured result of an experiment, analyzed in both the experimental and control groups.

descent with modification Darwin's term for evolution, combining the ideas that all living things are related and that organisms have changed over time.

diabetes A disease characterized by abnormally high blood-sugar levels.

diencephalon A brain region located between the brain stem and the cerebrum that includes the thalamus and hypothalamus, among other structures, and regulates homeostatic functions like body temperature, hunger, thirst, and the sex drive.

differential gene expression The process by which different genes are "turned on" (that is, expressed) in different cell types.

digestion The mechanical and chemical breakdown of food into subunits so that nutrients can be absorbed.

digestive tract The central pathway of the digestive system; a long muscular tube that pushes food between the mouth and the anus.

diploid Having two copies of every chromosome.

directional selection A type of natural selection in which organisms with phenotypes at one end of a spectrum are favored by the environment.

distribution pattern The way that organisms are distributed in geographic space, depending on resources and interactions with other members of the population.

diversifying selection A type of natural selection in which organisms with phenotypes at both extremes of the phenotypic range are favored by the environment.

DNA polymerase An enzyme that "reads" the sequence of a DNA strand and helps to add complementary nucleotides to form a new strand during DNA replication.

DNA profile A visual representation of a person's unique DNA sequence.

DNA replication The natural process by which cells make an identical copy of a DNA molecule.

domain The highest category in the modern system of classification; there are three domains–Bacteria, Archaea, and Eukarya.

dominant allele An allele that can mask the presence of a recessive allele.

dopamine A neurotransmitter that is involved in conveying a sense of pleasure in the brain.

double helix The spiral structure formed by two strands of DNA nucleotides bound together.

duodenum The first portion of the small intestine; the duodenum receives chyme from the stomach and mixes it with digestive secretions from other organs.

ecology The study of the interactions between organisms, and between organisms and their environment.

ecological footprint A measure of how much land and water area is required to supply the resources a person or population consumes and to absorb the wastes they produce.

ecosystem The living and nonliving components of an environment, including the communities of organisms present and the physical environment with which they interact.

ectotherm An animal that relies on environmental sources of heat, such as sunlight, to maintain its body temperature.

effector A cell or tissue that acts to exert a response based on information relayed from a sensor.

electron A negatively charged subatomic particle with negligible mass.

electron transport chain A process that takes place in mitochondria and produces the bulk of ATP during aerobic respiration; the third step of aerobic respiration.

element A chemically pure substance that cannot be chemically broken down; each element is made up of and defined by a single type of atom.

elimination The expulsion of undigested matter in the form of stool.

embryo An early stage of development reached when a zygote undergoes cell division to form a multicellular structure.

embryonic stem cells Stem cells that make up an early embryo, which can differentiate into nearly every cell in the body.

emulsify To break up large fat globules into small fat droplets that can be more efficiently chemically digested by enzymes.

endocrine system The collection of hormone-secreting glands and organs with hormone-secreting cells.

endometrium The lining of the uterus.

endoplasmic reticulum A membrane-enclosed series of passages in eukaryotic cells in which proteins and lipids are synthesized.

endoskeleton A solid internal skeleton found in many animals, including humans.

endosymbiosis The theory that free-living prokaryotic cells engulfed other free-living prokaryotic cells billions of years ago, forming eukaryotic organelles such as mitochondria and chloroplasts.

endotherm An animal that can generate body heat internally in order to maintain its body temperature.

energy The ability to do work. Cellular work includes processes such as building complex molecules and moving substances in and out of the cell.

enzyme A protein that speeds up the rate of a chemical reaction.

epidemiology The study of patterns of disease in populations, including risk factors.

epididymis Tubes in which sperm mature and are stored before ejaculation.

epithelial cells Cells that line organs and body cavities; in the digestive tract they sit in direct contact with food and its breakdown products.

esophagus The section of the digestive tract between the mouth and the stomach.

essential amino acids Eight amino acids the human body cannot synthesize and must obtain from food.

essential nutrient A substance that cannot be synthesized by the body and must be obtained preassembled from the diet; certain amino acids and fatty acids, vitamins, and minerals are essential nutrients.

estrogen A female sex hormone produced by the ovaries.

ethylene A gaseous plant hormone that promotes fruit ripening as one of its effects.

eukaryote Any organism of the domain Eukarya; eukaryotic cells are characterized by the presence of a membrane-enclosed nucleus and organelles.

evolution Change in allele frequencies in a population over time.

exoskeleton A hard external skeleton covering the body of many animals, such as arthropods.

experiment A carefully designed test, the results of which will either support or rule out a hypothesis.

experimental group The group in an experiment that experiences the experimental intervention or manipulation.

exponential growth The unrestricted growth of a population growing at a constant growth rate.

extinction The elimination of all individuals in a species; extinction may occur over time or in a sudden mass die-off.

facilitated diffusion The process by which large or hydrophilic solutes move across a membrane from an area of higher concentration to an area of lower concentration with the help of transport proteins.

falsifiable Describes a hypothesis that can be ruled out by data that show that the hypothesis does not explain the observation.

feedback loop A pathway that involves input from a sensor, a response via an effector, and detection of the response by the sensor.

fermentation A series of chemical reactions that takes place in the absence of oxygen and converts some of the energy stored in food into ATP. Fermentation produces far less ATP than does aerobic respiration.

fern The first true vascular plants; ferns do not produce seeds.

fertilization The fusion of an egg and a sperm to form a zygote.

fiber A complex plant carbohydrate that is not digestible by humans.

fitness The relative ability of an organism to survive and reproduce in a particular environment.

flagella (singular: flagellum) Whiplike appendages extending from the surface of some bacteria, used in movement of the cell.

folate A B vitamin also known as folic acid. Folate is an essential nutrient, necessary for basic bodily processes such as DNA replication and cell division.

follicle The part of the ovary where eggs mature.

follicle-stimulating hormone (FSH) A hormone secreted by the anterior pituitary gland. In females, FSH triggers eggs to mature at the start of each monthly cycle.

food chain A linked series of feeding relationships in a community in which organisms further up the chain feed on ones below.

food web A complex interconnection of feeding relationships in a community.

fossils The preserved remains or impressions of once-living organisms.

fossil fuel Carbon-rich energy source, such as coal, petroleum, or natural gas, formed from the compressed, fossilized remains of once-living organisms.

fossil record An assemblage of fossils arranged in order of age, providing evidence of changes in species over time.

fungus (plural: fungi) A single-cell or multicellular eukaryotic organism that obtains nutrients by secreting digestive enzymes onto organic matter and absorbing the digested product.

gallbladder An organ that stores bile salts and releases them as needed into the small intestine.

gametes Specialized reproductive cells that carry one copy of each chromosome (that is, they are haploid). Sperm are male gametes; eggs are female gametes.

gel electrophoresis A laboratory technique that separates fragments of DNA by size.

gene A sequence of DNA that contains the information to make at least one protein.

gene expression The process of using DNA instructions to make proteins.

gene flow The movement of alleles from one population to another, which may increase the genetic diversity of a population.

gene pool The total collection of alleles in a population.

gene therapy A type of treatment that aims to cure disease by replacing defective genes with functional ones.

genetically modified organism (GMO) An organism that has been genetically altered by humans.

genetic code The particular amino acids specified by particular mRNA codons.

genetic drift Random changes in the allele frequency of a population between generations; genetic drift tends to have more dramatic effects in smaller populations than in larger ones.

genome One complete set of genetic instructions encoded in the DNA of an organism.

genotype The genetic makeup of an organism.

gibberellins Plant hormones that cause stem elongation and cell division.

glial cells Supporting cells of the nervous system.

global hectare The unit of measurement of the ecological footprint, representing the biological productivity of an average hectare of land.

global warming An increase in the earth's average temperature.

glucagon A hormone produced by the pancreas that causes an increase in blood sugar.

glycogen A complex animal carbohydrate made of linked chains of glucose molecules; a source of stored energy.

glycolysis A series of reactions that breaks down sugar into smaller units; glycolysis takes place in the cytoplasm and is the first step of both aerobic respiration and fermentation.

Golgi apparatus An organelle made up of stacked membrane-enclosed discs that packages proteins and prepares them for transport.

Gram-negative Refers to bacteria with a cell wall that includes a thin layer of peptidoglycan surrounded by an outer lipid membrane that does not retain the Gram stain.

Gram-positive Refers to bacteria with a cell wall that includes a thick layer of peptidoglycan that retains the Gram stain.

gravitropism The growth of plants in response to gravity. Roots grow downward, with gravity; shoots grow upward, against gravity.

greenhouse effect The normal process by which heat is radiated from earth's surface and trapped by gases in the atmosphere, helping to maintain the earth at a temperature that can support life.

greenhouse gas Any of the gases in earth's atmosphere that absorb heat radiated from the earth's surface and contribute to the greenhouse effect; for instance, carbon dioxide and methane.

growth rate The difference between the birth rate and the death rate of a given population; also known as the rate of natural increase

gymnosperm A seed-bearing plant with "naked" seeds typically held in cones.

habitat The physical environment where an organism lives and to which it is adapted.

half-life The time it takes for one-half of a substance to decay.

haploid Having only one copy of every chromosome.

Hardy-Weinberg equilibrium The principle that, in a nonevolving population, both allele and genotype frequencies remain constant from one generation to the next.

heat The kinetic energy generated by random movements of molecules or atoms.

Helper T cell A type of T cell that helps activate B cells during humoral responses.

herbivory Predation on plants, which may or may not kill the plant.

heterotrophs Organisms, such as humans and other animals, that obtain energy by eating other organisms or molecules produced by other organisms.

heterozygous Having two different alleles of a given gene.

hippocampus The region of the brain involved in learning and memory.

histamine A molecule released by damaged tissue and during allergic reactions.

homeostasis The maintenance of a relatively stable internal environment, even when the external environment changes.

homologous chromosomes The two copies of each chromosome in a diploid cell. One chromosome in the pair is inherited from the mother, the other is inherited from the father.

homology Anatomical, genetic, or developmental similarity among organisms due to common ancestry.

hominid Any living or extinct member of the family Hominidae, the great apes–humans, orangutans, chimpanzees, and gorillas.

homozygous Having two identical alleles of a given gene.

hormone A chemical signaling molecule that is released by a cell or gland and travels through the bloodstream to exert an effect on target cells.

human chorionic gonadotropin (hCG) A hormone produced by an early embryo that helps maintain the corpus luteum until the placenta develops.

humoral immunity The type of adaptive immunity that fights infections and other foreign substances in the circulation and lymph fluid.

hydrogen bond A weak electrical attraction between a partially positive hydrogen atom and another atom with a partial negative charge.

hydrophobic "Water-fearing"; hydrophobic molecules will not dissolve in water.

hydrophilic "Water-loving"; hydrophilic molecules dissolve in water.

hypha (plural: hyphae) A long, threadlike structure through which fungi absorb nutrients.

hypothalamus A master coordinator region of the brain responsible for a number of physiological functions.

hypothermia A drop of body temperature below 35°C (95°F), which causes enzyme malfunction and, eventually, death.

hypothesis A testable and falsifiable explanation for a scientific observation or question.

hypoxia A state of low oxygen concentration in the blood.

immune system A system of cells and tissues that acts to defend the body against foreign cells and infectious agents.

immunity The resistance to a given pathogen conferred by the activity of the immune system.

in vitro fertilization (IVF) A form of assisted reproduction in which eggs and sperm are brought together outside the body and the resulting embryos are inserted into a woman's uterus.

inbreeding Mating between closely related individuals. Inbreeding does not change the allele frequency within a population, but it does increase the proportion of homozygous individuals to heterozygotes.

inbreeding depression The negative reproductive consequences for a population associated with having a high frequency of homozygous individuals possessing harmful recessive alleles.

incomplete dominance A form of inheritance in which heterozygotes have a phenotype that is intermediate between homozygous dominant and homozygous recessive.

independent assortment The principle that alleles of different genes are distributed independently of one another during meiosis.

independent variable The variable, or factor, being deliberately changed in the experimental group.

induced pluripotent stem cell A pluripotent stem cell that was generated by manipulation of a differentiated somatic cell

inflammatory response An innate defense that is activated by local tissue damage and that acts to recruit phagocytes to the site of infection to destroy and contain invading pathogens.

ingestion The act of taking food into the mouth.

innate immunity Nonspecific defenses, such as physical and chemical barriers and phagocytic cells, that are present from birth and are always active.

inorganic molecule A molecule that lacks a carbon-based backbone and C-H bonds.

insect A six-legged arthropod with three body segments: head, thorax, and abdomen.

insulin A hormone secreted by the pancreas that regulates blood sugar.

insulin A hormone secreted by the pancreas that regulates blood sugar.

interferon Antiviral proteins produced by virally infected cells to help protect adjacent cells from becoming infected.

interphase The stage of the cell cycle in which cells spend most of their time, preparing for cell division. There are three distinct phases within interphase (G_1, S, and G_2).

intrauterine insemination (IUI) A form of assisted reproduction in which sperm are injected directly into a woman's uterus.

invertebrate An animal lacking a backbone.

ion An electrically charged atom, the charge resulting from the loss or gain of electrons.

ionic bond A strong electrical attraction between oppositely charged ions.

karyotype The chromosomal makeup of cells. Karyotype analysis can be used to detect trisomy 21 prenatally.

keystone species A species on which other species depend, and whose removal has a dramatic impact on the community.

kidney An organ involved in osmoregulation, filtration of blood to remove wastes, and production of several important hormones.

kinetic energy The energy of motion or movement.

large intestine The last organ of the digestive tract, in which remaining water is absorbed and solid stool is formed.

light energy The energy of the electromagnetic spectrum of radiation.

lignin A stiff strengthening agent found in secondary cell walls of plants.

lipase A fat-digesting enzyme active in the small intestine.

lipids Organic molecules that generally repel water.

liver An organ that aids digestion by producing bile salts that emulsify fats.

logistic growth A pattern of growth that starts off fast and then levels off as the population reaches the carrying capacity of the environment.

luteinizing hormone (LH) A hormone secreted by the anterior pituitary gland. In females, a surge of LH triggers ovulation.

lymph nodes Small organs in the lymphatic system where B and T cells may encounter pathogens.

lymphatic system The organ system that works with the immune system to defend the body by removing toxins and pathogens from the blood, allowing B and T cells to respond to pathogens.

lymphocyte A specialized white blood cell of the immune system.

lysosome An organelle in eukaryotic cells filled with enzymes that can degrade worn-out cellular structures.

macromolecules Large organic molecules that make up living organisms; they include carbohydrates, proteins, and nucleic acids.

macronutrients Nutrients, including proteins, carbohydrates, and fats, that organisms must ingest in large amounts to maintain health.

macrophage A phagocytic cell that resides in tissues and plays an important role in the inflammatory response.

mammals Members of the class Mammalia; all members of this class have mammary glands and a fur-covered body.

matter Anything that takes up space and has mass.

meiosis A specialized type of cell division that generates genetically unique haploid gametes.

melanin A pigment, produced by a specific type of skin cell, that gives skin color.

memory cell A long-lived B or T cell that is produced during the primary response and that is rapidly activated in the secondary response.

menstruation The shedding of the uterine lining (the endometrium) that occurs when an embryo does not implant.

messenger RNA (mRNA) The RNA copy of an original DNA sequence formed during transcription.

metabolism All the chemical reactions taking place in the cells of a living organism that allow it to obtain and use energy, including breaking down food molecules and building new molecules.

metastasis The spread of cancer cells from one location in the body to another.

micronutrients Nutrients, including vitamins and minerals, that organisms must ingest in small amounts to maintain health.

mineral An inorganic chemical element required by organisms for normal growth, reproduction, and tissue maintenance; examples are calcium, iron, potassium, and zinc.

mitochondria Membrane-bound organelles responsible for important energy-conversion reactions in eukaryotes.

mitochondrial DNA (mtDNA) The DNA in mitochondria that is inherited solely from the mother.

mitosis The segregation and separation of duplicated chromosomes during cell division.

molecule Atoms linked by covalent bonds.

mollusk A soft-bodied invertebrate, generally with a hard shell (which may be tiny, internal, or absent in some mollusks).

monomer One chemical subunit of a polymer.

monosaccharide The building block, or monomer, of a carbohydrate.

motor neurons Neurons that control the contraction of skeletal muscle.

multifactorial inheritance An interaction between genes and the environment that contributes to a phenotype or trait.

multipotent Describes a cell with the ability to differentiate into a limited number of cell types in the body.

mutagen Any chemical or physical agent that can damage DNA by changing its nucleotide sequence.

mutation A change in the nucleotide sequence of DNA.

mutualism A type of symbiotic relationship in which both members benefit; a "win-win" relationship.

mycelium (plural: mycelia) A spreading mass of interwoven hyphae that forms the often subterranean body of multicellular fungi.

myelin A fatty substance that insulates the axons of neurons and facilitates rapid conduction of action potentials.

natural killer cell A type of white blood cell that acts during the innate immune response to find and destroy virally infected cells and tumor cells.

natural resources Raw materials that are obtained from the earth and are considered valuable even in their relatively unmodified, natural form.

natural selection Differential survival and reproduction of individuals in response to environmental pressure that leads to change in allele frequencies in a population over time.

nerve A bundle of specialized cells that transmit information.

nervous system The collection of organs that sense and respond to information, including the brain, spinal cord, and nerves.

neurons Specialized cells of the nervous system that generate electrical signals in the form of action potentials.

neurotransmitter A chemical signaling molecule released by neurons to communicate with neighboring cells.

neutron An electrically uncharged subatomic particle found in the nucleus of an atom.

neutrophil A phagocytic cell in the circulation that plays an important role in the inflammatory response.

niche The space, environmental conditions, and resources that a species needs in order to survive and reproduce.

nitrogen fixation The process of converting atmospheric nitrogen into a form that plants can use to grow.

nonadaptive evolution Any change in allele frequency that does not by itself lead a population to become more adapted to its environment; the causes of nonadaptive evolution are mutation, genetic drift, and gene flow.

noncoding regions DNA sequences that do not hold instructions to make proteins.

nondisjunction Failure of chromosomes to separate accurately during cell division; nondisjunction in meiosis leads to aneuploid gametes.

nonrenewable resources Natural resources that cannot be replaced.

nuclear envelope The double membrane surrounding the nucleus of a eukaryotic cell.

nucleic acids Organic molecules made up of linked nucleotide subunits; DNA and RNA are examples of nucleic acids.

nucleotides The building blocks of DNA. Each nucleotide consists of a sugar, a phosphate, and a base.

nucleus (atomic) The dense core of an atom.

nucleus (eukaryotic) The organelle in eukaryotic cells that contains the genetic material.

nutrients Components in food that the body needs to grow, develop, and repair itself.

obese Having 20% more body fat than is recommended for one's height, as measured by a body mass index greater than 30.

oncogene A mutated and overactive form of a proto-oncogene. Oncogenes drive cells to divide continually.

organ A structure made up of different tissue types working together to carry out a common function.

organ system A set of cooperating organs within the body.

organelles The membrane-bound compartments of eukaryotic cells that carry out specific functions.

organic molecule A molecule with a carbon-based backbone and at least one C-H bond.

osmolarity The concentration of dissolved solutes in blood and other bodily fluids.

osmoregulation Maintenance of a relatively stable volume, pressure, and solute concentration of bodily fluids, especially blood.

osmosis The diffusion of water across a semipermeable membrane from an area of lower solute concentration to an area of higher solute concentration.

osteoporosis A disease characterized by thinning bones.

ovaries Paired female reproductive organs; the ovaries contain eggs and produce sex hormones.

oviduct The tube connecting an ovary and the uterus in females. Eggs are ovulated into and fertilized within the oviducts.

ovulation The release of an egg from an ovary into the oviduct.

paleontologist A scientist who studies ancient life by means of the fossil record.

pancreas An organ that secretes the hormones insulin and glucagon, as well as digestive enzymes.

parasitism A type of symbiotic relationship in which one member benefits at the expense of the other.

pathogen A disease-causing agent, usually an organism.

peer review A process in which independent scientific experts read scientific studies before their publication to ensure that the authors have appropriately designed and interpreted their study.

pepsin A protein-digesting enzyme that is active in the stomach.

peptidoglycan A macromolecule that forms all bacterial cell walls and provides rigidity to the cell wall.

peripheral nervous system (PNS) All the nervous tissue outside the central nervous system. The PNS collects sensory information and transmits instructions from the CNS.

peristalsis Coordinated muscular contractions that force food down the digestive tract.

pH A measure of the concentration of H^+ in a solution.

phagocyte A type of white blood cell that engulfs and ingests damaged cells and pathogens.

phenotype The visible or measurable traits of an individual.

pheromones Chemical signaling molecules released into the environment in order to signal to other members of the same species.

phloem Plant vascular tissue that transports sugars throughout the plant.

phospholipid A type of lipid that forms biological membranes.

phospholipid bilayer A double layer of phospholipid molecules that characterizes all biological membranes.

photons Packets of light energy, each with a specific wavelength and quantity of energy.

photosynthesis The process by which plants and other autotrophs use the energy of sunlight to make energy-rich molecules using carbon dioxide and water.

phototropism The growth of the stem of a plant toward light.

phylogenetic tree A branching tree of relationships showing common ancestry.

phylogeny The evolutionary history of a group of organisms.

physiology The study of the way a living organism's physical parts function.

pili (singular: pilus) Short, hairlike appendages extending from the surface of some bacteria, used to adhere to surfaces

pistil The female reproductive structure of a flower, made up of a stigma, style, and ovary.

placebo A fake treatment given to control groups to mimic the experience of the experimental groups.

placebo effect The effect observed when members of a control group display a measurable response to a placebo because they think that they are receiving a "real" treatment.

placenta A structure made of fetal and maternal tissues that helps sustain and support the embryo and fetus.

plant A multicellular eukaryote that has cell walls, carries out photosynthesis, and is adapted to living on land.

plasma cell An activated B cell that divides rapidly and secretes an abundance of antibodies.

plate tectonics The movement of the earth's upper mantle and crust, which influences the geographical distribution of landmasses and organisms.

pluripotent Describes a cell with the ability to differentiate into nearly any cell type in the body.

polar molecule A molecule in which electrons are not shared equally between atoms, causing a partial negative charge at one end and a partial positive charge at the other; for example, water.

pollen Small, thick-walled structures that contain cells that will develop into sperm.

pollination The transfer of pollen from male to female plant structures so that fertilization can occur.

polygenic trait A trait whose phenotype is determined by the interaction between alleles of more than one gene.

polymer A molecule made up of individual subunits, called monomers, linked together in a chain.

polymerase chain reaction (PCR) A laboratory technique used to replicate, and thus amplify, a specific DNA segment.

population A group of organisms of the same species living and interacting in a particular geographic area.

population density The number of organisms per given area.

potential energy Stored energy.

predation An interaction between two organisms in which one organism (the predator) feeds on the other (the prey).

primary response The adaptive response mounted the first time a particular antigen is encountered by the immune system.

prion A protein-only infectious agent.

producers Autotrophs (photosynthetic organisms) that form the base of every food chain.

progesterone A female sex hormone produced by endocrine cells in the ovaries (particularly in the cells of the corpus luteum) that prepares and maintains the uterus for pregnancy.

prokaryote A usually single-cell organism whose cell lacks internal membrane-bound organelles and whose DNA is not contained within a nucleus.

prokaryotic cells Cells that lack internal membrane-bound organelles.

protein An organic molecule made up of linked amino acid subunits. Proteins play many critical roles in living organisms.

protist A eukaryote that cannot be classified as a plant, animal, or fungus; usually unicellular.

proton A positively charged subatomic particle found in the nucleus of an atom.

proto-oncogene A gene that codes for a protein that helps cells divide normally.

punctuated equilibrium The theory that most species change occurs in periodic bursts as a result of sudden environmental change.

Punnett square A diagram used to determine probabilities of offspring having particular genotypes, given the genotypes of the parents.

radial symmetry The pattern exhibited by a body plan that is circular, with no clear left and right sides.

radiation therapy The use of ionizing (high-energy) radiation to treat cancer.

radioactive isotope An unstable form of an element that decays into another element by radiation, that is, by emitting energetic particles.

radiometric dating The use of radioactive isotopes as a measure for determining the age of a rock or fossil.

randomized clinical trial A controlled medical experiment in which subjects are randomly chosen to receive either an experimental treatment or a standard treatment (or placebo).

recessive allele An allele that reveals itself in the phenotype only if the organism has two copies of that allele.

recombination The stage of meiosis in which maternal and paternal chromosomes pair and physically exchange DNA segments.

regulatory sequence The part of a gene that determines the timing, amount, and location of protein produced.

relative dating Determining the age of a fossil on the basis of its position relative to layers of rock or fossils of known age.

renewable resources Natural resources that are replenished after use as long as the rate of consumption does not exceed the rate of replacement.

reproductive isolation Mechanisms that prevent mating (and therefore gene flow) between members of different species.

Rhizobium A genus of nitrogen-fixing bacteria that form symbiotic associations with the roots of legumes.

ribosome The cellular machinery that assembles proteins during the process of translation.

RNA polymerase The enzyme that accomplishes transcription. RNA polymerase copies a strand of DNA into a complementary strand of mRNA.

root hairs Tiny extensions of root cells that increase the surface area of roots to enhance their ability to absorb water and nutrients.

root system The belowground parts of a plant, which anchor it and absorb water and nutrients.

salivary glands Glands that secrete enzymes, including salivary amylase, which digests carbohydrates, into the mouth.

sample size The number of experimental subjects or the number of times an experiment is repeated. In human studies, sample size is the number of subjects.

saturated fat An animal fat, such as that found in butter; saturated fats are solid at room temperature.

science The process of using observations and experiments to draw evidence-based conclusions.

scientific theory A hypothesis that is supported by many years of rigorous testing and thousands of experiments.

scrotum The sac in which the testes are located.

secondary response The rapid and strong response mounted when a particular antigen is encountered by the immune system subsequent to the first encounter.

seed The embryo of a plant, along with a starting supply of food, encased in a protective covering.

semen The mixture of fluid and sperm that is ejaculated from the penis.

semi-conservative DNA replication is said to be semi-conservative because each newly made DNA molecule has one original and one new strand of DNA.

seminiferous tubules Coiled structures that constitute the bulk of the testes and in which sperm develop.

sensor A specialized cell that detects specific sensory input like temperature, pressure, or solute concentration.

sensory neurons Cells that convey information from both inside and outside the body to the CNS.

sex chromosomes Paired chromosomes that differ between males and females, XX in females, XY in males

shoot system The aboveground parts of a plant, including the stem and photosynthetic leaves.

short tandem repeats (STRs) Sections of a chromosome in which DNA sequences are repeated.

simple diffusion The movement of small, hydrophobic molecules across a membrane from an area of higher concentration to an area of lower concentration; simple diffusion does not require an input of energy.

simple sugar (monosaccharide) A carbohydrate made up of a single sugar subunit; an example is glucose.

sister chromatid One of the two identical DNA molecules that make up a duplicated chromosome following DNA replication.

small intestine The organ in which most chemical digestion and absorption of food occurs.

solute A dissolved substance.

solution A mixture of solutes dissolved in a solvent.

solvent A substance in which other substances can dissolve; for example, water.

speciation The genetic divergence of populations owing to a barrier to gene flow between them, leading over time to reproductive isolation and the formation of new species.

spinal cord A bundle of nerve fibers, contained within the bony spinal column, that transmits information between the brain and the rest of the body.

stabilizing selection A type of natural selection in which organisms near the middle of the phenotypic range of variation are favored.

stamen The male reproductive structure of a flower, made up of a filament and an anther.

starch A complex plant carbohydrate made of linked chains of glucose molecules; a source of stored energy.

statistical significance A measure of confidence that the results obtained are "real," rather than due to random chance.

stem cells Immature cells that can divide and differentiate into specialized cell types.

stomach An expandable muscular organ that stores, mechanically breaks down, and digests proteins in food.

stomata (singular: stoma) Pores on leaves that permit the exchange of oxygen and carbon dioxide with the air and also allow water loss.

stool Solid waste material eliminated from the digestive tract.

substrate A compound or molecule that an enzyme binds to and on which it acts.

sustainability Using the earth's resources in a way that will not permanently destroy or deplete them; living within the limits of earth's biocapacity.

symbiosis A situation in which two different organisms live together, often interdependently

synapse The site of communication between a neuron and another cell; the synapse includes the axon terminal of the communicating neuron, the space between the cells, and the site of reception on the receiving cell.

synaptic cleft The physical space between a neuron and the cell with which it is communicating.

T cells White blood cells that mature in the thymus and can destroy infected cells or stimulate B cells to produce antibodies, depending on the type of T cell.

taproot A long, straight root produced by some plants to store water and carbohydrates.

taxonomy The process of identifying, naming, and classifying organisms on the basis of shared traits.

testable A hypothesis is testable if it can be supported or rejected by carefully designed experiments or nonexperimental studies.

testes (singular: testis) Paired male reproductive organs, which contain sperm and produce androgens (primarily testosterone).

testosterone The primary male sex hormone, which stimulates the development of masculine features and plays a key role in sperm development.

tetrapod An organism with four true limbs, that is, bony appendages with jointed wrists, ankles, and digits; mammals, amphibians, birds, and reptiles are tetrapods.

thermoregulation The maintenance of a relatively stable internal body temperature.

thigmotropism The response of plants to touch and wind.

thymus The organ in which T cells mature.

tissue An organized collection of a single cell type working to carry out a specific function.

tongue A muscular organ in the mouth that aids swallowing.

totipotent Describes a cell with the ability to differentiate into any cell type in the body.

trans fat A type of vegetable fat which has been hydrogenated, that is, hydrogen atoms have been added, making it solid at room temperature.

transcription The first stage of gene expression, during which cells produce molecules of messenger RNA (mRNA) from the instructions encoded within genes.

transfer RNA (tRNA) A type of RNA that helps ribosomes assemble chains of amino acids during translation.

transgenic Refers to an organism that carries one or more genes from a different species.

translation The second stage of gene expression. Translation "reads" mRNA sequences and assembles the corresponding amino acids to make a protein.

transpiration The loss of water from plants by evaporation, which powers the transport of water and nutrients through a plant's vascular system.

transport proteins Proteins involved in the movement of molecules across the cell membrane.

triglyceride A type of lipid found in fat cells that stores excess energy for long-term use.

trisomy 21 Carrying an extra copy of chromosome 21; also known as Down syndrome.

trophic levels Feeding levels, based on positions in a food chain.

tumor suppressor genes Genes that code for proteins that monitor and check cell cycle progression. When these genes mutate, tumor suppressor proteins lose normal function.

turgor pressure The pressure exerted by the water-filled central vacuole against the plant cell wall, giving a stem its rigidity.

unsaturated fat A plant fat, such as olive oil; unsaturated fats are liquid at room temperature.

urethra The passageway through the penis, shared by the reproductive and urinary tracts.

uterus The muscular organ in females in which a fetus develops.

vaccine A preparation of killed or weakened microorganisms or viruses that is given to people or animals to generate a memory immune response.

vagina The first part of the female reproductive tract, extending up to the cervix; also known as the birth canal.

vas deferens Paired tubes that carry sperm from the testes to the urethra.

vascular plant A plant with tissues that transport water and nutrients through the plant body.

vascular system Tube-shaped vessels and tissues that transport nutrients throughout an organism's body.

vasoconstriction The reduction in diameter of blood vessels, which helps to retain heat.

vasodilation The expansion in diameter of blood vessels, which helps to release heat.

vertebrate An animal with a bony or cartilaginous backbone.

vestigial structure A structure inherited from an ancestor that no longer serves a clear function in the organism that possesses it.

villi (singular: villus) Fingerlike projections of folds in the lining of the small intestine that are responsible for most nutrient and water absorption.

virus An infectious agent made up of a protein shell that encloses genetic information.

vitamin An organic molecule required in small amounts for normal growth, reproduction, and tissue maintenance.

vitamin D A fat-soluble vitamin necessary to maintain a healthy immune system and build healthy bones and teeth. The human body produces vitamin D when skin is exposed to UV light.

wood Secondary xylem tissue found in the stem of a plant.

X chromosome One of the two sex chromosomes in humans.

X-linked trait A phenotype determined by an allele on an X chromosome.

xylem Plant vascular tissue that transports water from the roots to the shoots.

Y chromosome One of two sex chromosomes in humans. The presence of a Y chromosome signals the male developmental pathway during fetal development.

zygote A cell that is capable of developing into an adult organism. The zygote is formed when an egg is fertilized by a sperm.

Photo Credits

Chapter 1

p. 1: Courtesy of Ann Warren. **p. 3:** Steve Bronstein/Getty Images. **p. 4:** Courtesy of Ann Warren. **p. 5:** *Infographic 1-1* (TL) Used by permission of *North Carolina Medical Journal* (Morrisville, NC: North Carolina Institute of Medicine), November/December 2003, www.ncmedicaljournal.org, (TR) Courtesy of *British Medical Journal*, 28 May 2009, Vol. 338, Issue 7706, (B) Aleksej Vasic/iStockphoto. **p. 6:** *Infographic 1.2* (TL) *Science*, Vol. 324, no. 5935, 26 June 2009. Reprinted with permission from AAAS, (TR) PLoS Med 2 (2), cover image, Krista Steinke. **p. 7:** Micha Pawlitzki/Zefa/Corbis: (TL) *Cell Metabolism*, March 4, 2009, cover illustration by Chris Lange, © Elsevier, 2009, (TC) © Elsevier, (TR) *Science*, Vol. 320, no. 5882, 13 June 2008. Reprinted with permission from AAAS; (B) Rick Wilson/The Florida Times-Union. **p. 10:** *Infographic 1.5* (L) Jessica Peterson/Photolibrary, (TR) Nancy Nehring/iStockphoto, (CR) Interfoto/Alamy, (BL) Mark Moffett/Getty Images, (BR) G. Lasley/VIREO. **p. 12:** Vicki Wagner/Alamy. **p. 15:** *Infographic 1.8* Tony West/Alamy.

Chapter 2

p. 19: NASA/JPL. **p. 21:** Calvin J. Hamilton. **p. 22:** (T) NASA/JP; (B) NASA/JSC. **p. 23:** *Infographic 2.1* (T) Bildarchiv/AgeFotostock, (C) globestock/iStockphoto, (B) Kazuo Ogawa/AgeFotostock. **p. 27:** NASA/JSC. **p. 33:** *Infographic 2.6* (T) Mixa/Superstock, (C) B. Runk/S. Schoenberger/Grant Heilman, (B) AgeFotostock/Superstock. **p. 34:** NASA/Time Life Pictures/Getty Images. **p. 35:** GSFC/NASA.

Chapter 3

p. 39: Joe Raedle/Getty Images. **p. 41:** Pictorial Press Ltd/Alamy. **p. 42:** Fleming, Alexander. 1929. On the Antibacterial Action of Cultures of a Penicillium, with Special Reference to Their Use in the Isolation of B. Influenzae. *British Journal of Experimental Pathology*, Vol. 10, pp. 226–236, Fig. 2; *Infographic 3.1* (L) The British Library/Photolibrary, (C and R) Biophoto Associates/Photo Researchers. **p. 43:** *Infographic 3.2* (TL) Scenics & Science/Alamy, (TC) Roland Birke/Photolibrary, (TR) Dennis Kunkel/Visuals Unlimited, (BL) Ed Reschke/Photolibrary, (inset) David Toase/Photolibrary, (BC) Ed Reschke/Photolibrary, (inset) Christian Fischer/WIKI, Creative Commons, (BR) Michael Abbey/Photo Researchers, (inset) A. & F. Michler/Photolibrary. **p. 45:** SPL/Photo Researchers. **p. 47:** (T) Daily Herald Archive/SSPL/Getty Images; (B) Research and Development Division, Schenley Laboratories, Inc., Lawrenceburg, Indiana. **p. 54:** Joe Raedle/Getty Images. **p. 55:** NIH.

Chapter 4

p. 59: ma-k/iStockphoto. **p. 61:** ma-k/iStockphoto. **p. 62:** Courtesy of Nestlé SA. **p. 63:** *Infographic 4.1* (TL) Denis Pepin/Featurpics, (TC) Shadow216/Dreamstime, (TR) Fresh Food Images/Photolibrary, (CL) Jeffrey Coolidge/Getty Images, (CC) AgeFotostock/Superstock, (CR) Juanmonino/iStockphoto, (BL) AgeFotostock/Superstock, (BC) adlifemarketing/iStockphoto, (BR) Creative Commons: <http://www.flickr.com/photos/tellumo/232317103/sizes/o/in/photostream/>. **p. 65:** (T) BostjanT/iStockphoto; (BL and BR) The Photo Works. **p. 66:** The Photo Works. **p. 67:** Boissonnet/AgeFotostock. **p. 71:** *Infographic 4.6* (TL) Rob Owen/Whal/StockXchng, (TR) ktphotog/iStockphoto, (C) Dr. Michael Klein/Peter Arnold, (B) shironosov/iStockphoto. **p. 73:** *Infographic 4.7* (L) Reprinted with permission © 2008 Southwest Research Institute. All rights reserved. (R) David M. Phillips/Photo Researchers. **p. 75:** *Infographic 4.8* Copyright © 2008. For more information about The Healthy Eating Pyramid, please see The Nutrition Source, Department of Nutrition, Harvard School of Public Health, http://www.thenutritionsource.org, and *Eat, Drink, and Be Healthy* by Walter C. Willett, M.D. **p. 76:** (L) Bengt-Göran Carlsson/TIOFOTO/Nordic Photos; (R) Alison Wright/Photo Researchers

Chapter 5

p. 81: Philip Hart. **p. 83:** (T) Courtesy of Sapphire Energy, Inc., (B) Philip Hart. **p. 84:** egdigital/iStockphoto. **p. 85:** Courtesy of Sapphire Energy, Inc. **p. 86:** *Infographic 5.2* (T, from left) Kimberly Deprey/iStockphoto, mihtiander/FeaturePics, Bios/Photolibrary, (B) AP Photo/Arthur Max, (CL and CR) Visuals Unlimited/Corbis, (R, from top) Gudella/FeaturePics, moori/FeaturePics, Yobro10/Dreamstime.com, Photo168/Dreamstime.com. **p. 88:** *Infographic 5.3* Corbis/SuperStock. **p. 89:** *Infographic 5.4* (from left) Pixelgnome/Dreamstime.com, Vasily Smirnov/Dreamstime.com, Rocky Reston/Dreamstime.com, Shevelartur/Dreamstime.com, Aleksandr Lazarev/iStockphoto. **p. 91:** Ashley Cooper/Alamy; *Infographic 5.5* (top panel) (L) Visuals Unlimited/Corbis, (TR) Mark Hamblin/AgeFotostock, (BR) Visuals Unlimited/Corbis; (bottom panel) (L) 2ndLookGraphics/iStockphoto, (C) Jfybel/Dreamstime, (R) NNehring/iStockphoto. **p. 93:** Michael Macor/San Francisco Chronicle/Corbis. **p. 94:** *Infographic 5.7* (L) Brand X Pictures, (R) Maxrale/iStockphoto. **p. 97:** Courtesy of Jim Sears/A2BE Carbon Capture LLC.

Chapter 6

p. 101: Leaf/Dreamstime. **p. 103:** (L) Jose Luis Pelaez Inc./Getty Images; (TR) Tim Platt/Getty Images; (BR) Sian Kennedy/Getty Images. **p. 105:** (B) Courtesy of Paul Rozin; *Infographic 6.2* (L) Christian Handl/Photolibrary, (C, inset) Michael Gray/Dreamstime.com, (CT) iperl/Featurepics, (CB) Stephen Bonk/Dreamstime.com, (R) Photodisc. **p. 106:** (L) Steve Stock/Alamy; (R) Hartmann Christian/SIPA; *Infographic 6.3* (T) Royalty-Free/Corbis, (B) Robert Fried/Alamy. **p. 107:** *Infographic 6.4* Ilena Eisseva/Featurepics. **p. 110:** Tips Italia/Photolibrary. **p. 111:** *Infographic 6.6* (T) apcuk/iStockphoto, (C) EricGerrard/iStockphoto, (B) bluestocking/iStockphoto, (R) DNY59/iStockphoto. **p. 115:** (L) Leaf/Dreamstime; (R) Konstik/Dreamstime. **p. 116:** AP Photo/Harry Cabluck.

Chapter 7

p. 121: The Innocence Project. **p. 122:** (logo) The Innocence Project. **p. 123:** JUPITERIMAGES/Brand X/Alamy. **p. 124:** *Infographic 7.1* ISM/Phototake. **p. 125:** Uli Holz, Yeshiva

University. **p. 127:** Courtesy of Kary Mullis. **p. 128:** The Innocence Project. **p. 132:** *Infographic 7.6* (L) Biophoto Associates/Photo Researchers, (C) Dr. Gopal Murti/ Photo Researchers, (R) Biophoto Associates/Photo Researchers. **p. 133:** Kevin Rivoli/The New York Times/Redux.

Milestones in Biology: The Model Makers

p. 137: SSPL/Getty Images. **pp. 138–139:** James D. Watson Collection, Cold Spring Harbor Laboratory Archives. **p. 140:** *Rosalind Franklin and the Shape of DNA* (L) National Portrait Gallery, London, (R) Omikron/Photo Researchers. **p. 141:** *Erwin Chargaff's Work Provided a Clue to Base Pairing* National Library of Medicine. **p. 142:** SSPL/Getty Images.

Chapter 8

p. 143: Edwin Remsberg/Alamy. **p. 145:** Sebastian Knight/ Dreamstime.com. **p. 146:** Nigel Cattlin/Visuals Unlimited. **p. 148:** *Infographic 8.3* Eye of Science/Photo Researchers. **p. 151:** *Infographic 8.6* (from left) MedicalRF.com/ AgeFotostock, Lauritzasoare/Dreamstime.com, Courtesy of GTC Biotherapeutics, Courtesy of Lundbeck Inc. **p. 152:** Edwin Remsberg/Alamy. **p. 156:** Courtesy of GTC Biotherapeutics, Inc. **p. 159:** Dan Reynolds/CartoonStock.

Milestones in Biology: Sequence Sprint

p. 161: Mario Tama/Getty Images. **pp. 162–163:** Alex Wong/ Newsmakers/Getty Images. **p. 166:** Mario Tama/Getty Images. **p. 167:** Sinclair Stammers/Photo Researchers.

Chapter 9

p. 169: Stefano Lunardi/AgeFotostock. **p. 171:** Emilio Ereza/ Alamy. **p. 173:** *Infographic 9.1* (TL) NIH, (TR) Courtesy of Bristol-Myers Squibb Company, (BL) Steffen Hauser/ Botanikfoto/Alamy, (BR) Stefano Lunardi/AgeFotostock. **p. 174:** *Infographic 9.2 Targeting Connexin43 Expression Accelerates the Rate of Wound Repair.* (2003). Cindy Qiu, Petula Coutinho, Stefanie Frank, Susanne Franke, Lee-yong Law, Paul Martin, Colin R. Green and David L. Becker. *Current Biology,* 13(19), 1697-1703. **p. 177:** Michael Abbey/Photo Researchers. **p. 180:** *Infographic 9.7* (L) Jim West/ AgeFotostock, (R) Peter Arnold, Inc./Alamy. **p. 181:** Steve Gschmeissner/Photo Researchers. **p. 182:** Alex Segre/ Photographers Direct.

Chapter 10

p. 187: Du Cane Medical Imaging Ltd./Photo Researchers. **p. 189:** UPI/BIll Greenblatt/Newscom. **p. 190:** *Infographic 10.1* (L) Stockbyte, (R) Ingram Publishing/Photolibrary. **p. 191:** Du Cane Medical Imaging Ltd./Photo Researchers. **p. 194:** *Infographic 10.4* (TL) Lepas/Dreamstime, (TC) Larry Jordan/ FeaturePics, (TR) Photodisc, (BL and BC) Royalty-Free/Corbis, (BR) Mary Lane/FeaturePics. **p. 197:** Glow Wellness/ SuperStock. **p. 199:** Courtesy of Lorene Ahern.

Chapter 11

p. 203: AP Photo/Carlos Osorio. **p. 205:** Courtesy of Emily Schaller. **p. 207:** *Infographic 11.2* ISM/Phototake. **p. 211:** Jeffrey Sauger. **p. 212:** *Infographic 11.6* Simon Fraser/Photo Researchers. **p. 217:** AP Photo/Carlos Osorio.

Milestones in Biology: Mendel's Garden

p. 221: Authenticated News/Getty Images. **pp. 222–223:** Malcolm Gutter/Visuals Unlimited. **p. 224:** *Concepts of Inheritance before Mendel* (L) Preformation, drawn by N. Hartsoecker, 1695, (R) Wellcome Library, London. **p. 227:** (T) Authenticated News/Getty Images, (B) Garden World Images/ AgeFotostock.

Chapter 12

p. 229: Courtesy of Peter Morenus. **p. 230:** Sebastian Kaulitzki/Alamy. **p. 232:** *Infographic 12.1* (L and R)) ISM/ Phototake. **p. 234:** Biophoto Associates/Photo Researchers. **p. 237:** AP Photo/Leslie Close. **p. 240:** (T, from left) Ludo Kuipers Photography; photographersdirect.com; Yuri Arcurs/ Fotolia; ImageSource/AgeFotostock; Tatiana Morozova/ FeaturePics; (C, from left) Denis Pepin/FeaturePics; PhotosIndia/Alamy; Jacob Langvad/Getty Images; Thomas Cockrem/Alamy; (BL) leaf/FeaturePics; (BR) Dmitriy Shironosov/Dreamstime.com. **p. 241:** Susumu Nishinaga/ Photo Researchers; *Infographic 12.6* (L) ImageSource/ AgeFotostock, (C) leaf/FeaturePics, (R) Ryan McVay/Getty Images. **p. 243:** Courtesy of Peter Morenus. **p. 245:** Jason Sitt/ Fotolia. **p. 246:** *Infographic 12.10* National Institute of Mental Health. **p. 247:** Markus Moellenberg/Corbis. **p. 249:** *Infographic 12.13* ISM/Phototake.

Chapter 13

pp. 253 and 254: AP Photo/PA. **p. 255:** Deco Images II/Alamy. **p. 256:** *Infographic 13.1* (T) Ed Reschke/Photolibrary, (C) Robert Knauft/Biology Pics/Photo Researchers, (B) Phototake/ Alamy. **p. 257:** Courtesy of Robert Langer, photo by Stu Rosner. **p. 258:** *Infographic 13.3* AP Photo/Brian Walker. **p. 259:** Wake Forest University Health Sciences/Center for Regenerative Medicine/Urology. **p. 264:** *Infographic 13.7* (T) James King-Holmes/Science Photo Library, (B) AP Photo/PA/ Files. **p. 266:** *Infographic 13.8* NIH.

Chapter 14

p. 271: ISM/Phototake. **p. 273:** CDC/ Janice Carr; Jeff Hageman. **p. 274:** *Infographic 14.1* (L) USDA/ARS, (TR) Paulo Cruz/Dreamstime.com, (CR, inset) DNY59/iStockphoto, (CR) Offscreen/Dreamstime.com, (BR) Genevieve Astrelli/ iStockphoto. **p. 281:** *Infographic 14.7* (T) CDC/Janice Haney Carr/Jeff Hageman, M.H.S., (C) Binh Tran/iStockphoto, (B) W. Lane/Minden Pictures. **p. 282:** ISM/Phototake. **p. 284:** *Infographic 14.8* (TL) Gallo Images/Getty Images, (TR) alandj/ iStockphoto, (CL) walik/iStockphoto, (CR) Steve Shepard/ iStockphoto, (BL) Ewa Walicka/Dreamstime.com, (BR) Christine Schuhbeck/AgeFotostock. **p. 287:** Nick D. Kim/ CartoonStock.

Milestones in Biology: Adventures in Evolution

p. 289: London Stereoscopic Company/Getty Images. **pp. 290–291:** HMS *Beagle* in the Galápagos by John Chancellor (1925-984). Courtesy of Gordon Chancellor. **p. 292:** *Lamarckianism* Bettmann/Corbis. **p. 293** *The Evolution of Darwin's Thought* (TC) public domain, (TR) Reproduced with permission from John van Wyhe, ed., The Complete Work of Charles Darwin Online (http://darwin-online.org.uk/), (BL) Classic Image/Alamy, (BC) Reproduced with permission from

Life in Extreme Environments, Portland State University. **p. 374:** Courtesy of the University of Washington, Lost City Science Team, IFE, URI-IAO, and NOAA.

Chapter 19

p. 377: Courtesy of Northwest Trek. **p. 378:** (L and R) National Park Service, Olympic National Park, photo by Janis Burger. **p. 379:** Georgette Douwma/Getty Images. **p. 381** *Infographic 19.2* (TL) Patrick Robbins/Dreamstime.com, (TR) Courtesy of Bob Wightman, (BL) National Park Service, (BC) Marcopolo/ FeaturePics, (BR) Courtesy of Peter Wigmore. **p. 382:** *Infographic 19.3* (L) Fotogal/FeaturePics, (inset) Mariya Bibikova/iStockphoto, (CL) Ferns at Muir Woods, CA, Sanjay ach/http://en.wikipedia.org/wiki/Fern, (inset) George Bailey/ Dreamstime.com, (CR) Michael P. Gadomski/Photo Researchers, (inset) Ray Roper/iStockphoto, (R) Mark Turner/ Photolibrary, (inset) Courtesy of Greg Rabourn. **p. 383** Courtesy of Northwest Trek. **p. 384:** *Infographic 19.4* (from left) Courtesy of Brooke et al., NOAA-OE, HBOI; Anky10/ Dreamstime.com; Ed Reschke/Photolibrary; Photolibrary/ Alamy; Manipulateur/Fotolia; London Scientific Films/ Photolibrary; Stock.xchng; U.S. National Park Service; Karen Arnold/Dreamstime.com. **p. 385:** David Gomez/iStockphoto. **p. 387:** (TL) Jan Gottwald/iStockphoto, (TC) Outdoorsman/ Dreamtime.com, (TR) Mark Conlin/Alamy, (BL) © Gary Nafis, (BC) Lon E. Lauber/Photolibrary, (BR) Chris Mattison/Alamy. **p. 389:** *Infographic 19.5* (C) Ed Reschke/Photolibrary, (TL) London Scientific Films/Photolibrary, (CL) Eye of Science/ Photo Researchers, (BL) Steve Gschmeissner/Photo Researchers, (TR) Mike Norton/Dreamstime.com, (BR) Alexander Makarov/iStockphoto. **p. 390:** *Infographic 19.6* (L) Gary Retherford/Photo Researchers, (CL) Roland Birke/ Photolibrary, (CR) http://en.wikipedia.org/wiki/File:Dog_ vomit_slime_mold.jpg, (R) Oxford Scientific /Photolibrary. **p. 391:** *Infographic 19.7* (L) Wim van Egmond/Visuals Unlimited, (CL) Stock.xchng, (C) Zefiryn/Fotolia, (CR) Ximinez/Fotolia, (R) Wim van Egmond/Visuals Unlimited.

Chapter 20

p. 395: AP Photo/Rick Bowmer. **p. 397:** Janine Wiedel Photolibrary. **p. 398:** Mark Wilson/Getty Images; *Infographic 20.1* (TL) Library of Congress Prints and Photographs Division [LC-USW3-037939-E], (TR) Bachmann/AgeFotostock, (BL) AP Photo/Rick Bowmer, (BR) Courtesy of www.worldmap.com. **p. 399:** *Infographic 20.2* Carolina Biological Supply Company/ Phototake. **p. 400:** *Infographic 20.3* (L) Elena Rostunova/ Alamy, (R) Living Art Enterprises, LLC/Photo Researchers. **p. 401:** Nina Jablonski. **p. 402:** *Infographic 20.4* Adapted from Chaplin G., Geographic Distribution of Environmental Factors Influencing Human Skin Coloration, *American Journal of Physical Anthropology* 125:292-302, 2004; map updated in 2007. Designer: Emmanuelle Bournay, UNEP/GRID-Arendal. **p. 404:** *Infographic 20.6* (T) Dennis Kunkel/Visuals Unlimited, (B) CNRI/Photo Researchers. **p. 405:** *Infographic 20.7* (L) Anthropological Skull Model-KNM-ER 406, Omo L.7a 125, 3B Scientific®, (R) © 2001 David L. Brill/Brill Atlanta. **p. 410:** *Infographic 20.10* Adapted from Chaplin G., Geographic Distribution of Environmental Factors Influencing Human Skin Coloration, *American Journal of Physical Anthropology* 125:292-302, 2004; map updated in 2007. Designer: Emmanuelle Bournay, UNEP/GRID-Arendal.

Chapter 21

p. 413: Tom Ulrich/Visuals Unlimited. **p. 415:** John Vucetich. **p. 416:** *Infographic 21.1* (T) U.S. Fish and Wildlife Service, (TC) AP Photo/Michigan Technological University, John Vucetich, (BC) John Vucetich, (B) sherwoodimagery/iStockphoto. **p. 418:** *Infographic 21.2* (TL) Flirt/SuperStock, (CL) Tom Hansch/ Dreamstime.com, (BL and BC) John Vucetich, (TR and CR) Ecological Studies of Wolves on Isle Royale Annual Report 2009-10 by John A. Vucetich and Rolf O. Peterson, School of Forest Resources and Environmental Science, Michigan Technological University, Figs. 6 and 11, courtesy of John Vucetich. **p. 419:** Russell Burden/Photolibrary; *Infographic 21.3* (L) Les Cunliffe/Dreamstime.com, (C) Melvinlee/ Dreamstime.com, (R) Marcel Krol/Dreamstime.com. **p. 420:** *Infographic 21.4* Oksana Churakova/Dreamstime.com. **p. 421:** John Vucetich. **p. 422:** *Infographic 21.6* (T) Tom Ulrich/Visuals Unlimited, (C) Steve Kazlowski/DanitaDelimont.com, (B) Andrey Rozov/Dreamstime.com. **p. 423:** *Infographic 21.7* (T) Imagebroker/Alamy, (TC) Mark Duffy/Alamy, (BC) Cliff Keeler/ Alamy, (B) John Vucetich. **p. 425:** *Infographic 21.8* (TL) Photobac/Dreamstime.com, (TR)   Terry Morris/iStockphoto. com, (C) James Mattil/AgeFotostock, (BL) Oksana Churakova/ Dreamstime.com, (BC) Jim Kruger/iStockphoto, (BR) John Vucetich. **p. 426:** *Infographic 21.9* (L) John Vucetich, (R) Courtesy of Sandy Updyke. **p. 427:** Ann & John Mahan.

Chapter 22

p. 431: OJO Images Ltd/Alamy. **p. 433:** Danish Ismail/Reuters/ Landov. **p. 434:** OJO Images Ltd/Alamy. **p. 435:** *Infographic 22.2* (T) Dennis MacDonald/Photolibrary, (B) Ragnar/ FeaturePics. **p. 436:** *Infographic 22.3* Olga Demchishina/ iStockphoto. **p. 437:** Accent Alaska.com/Alamy. **p. 438:** *Infographic 22.4* (T) David Kay/Dreamstime.com, (C) Rui Miguel da Costa Neves Saraiva/iStockphoto, (B) Kyu Oh/ iStockphoto. **p. 439:** *Infographic 22.5* Row 1: (L) Blend Images/ Superstock, (C) Michael Sewell/Photolibrary, (R) Benny Rytter/iStockphoto; Row 2: (L) James Phelps Jr/Dreamstime. com, (CL) Abdolhamid Ebrahim/iStockphoto, (C) Lunamarina/Dreamstime.com, (CR) Rui Miguel da Costa Neves Saraiva/iStockphoto, (R) Image by Larry D. Moore, used under a Creative Commons ShareAlike License, http:// en.wikipedia.org/wiki/File:Rabbit_in_montana.jpg; Row 3: (L) James Urbach/Photolibrary, (C) ElementalImaging/ iStockphoto, (R) Lukrecja/FeaturePics; Row 4: (L) Kyu Oh/ iStockphoto, (C) brytta/iStockphoto, (R) Roy T. Free/ AgeFotostock. **p. 440:** *Infographic 22.6* (TL) Jon Yuschock/ Fotolia, (TR) Courtesy of Donald Stahly, (CL) D. Harms/ Photolibrary, (CR) Crown Copyright courtesy of Central Science Laboratory/Photo Researchers, (B) Harry Rogers/ Photo Researchers. **p. 442:** *Infographic 22.7* (TL) Willi Schmitz/iStockphoto, (TR) Jim McKinley/Getty Images, (BL) Tim Martin/Dreamstime.com, (BC) Joanne Green/iStockphoto (BR) Steve Byland/iStockphoto. **p. 443:** *Infographic 22.8* (T) P-59 Photos/Alamy, (C) ElementalImaging/iStockphoto, (B) Nic Bothma/epa/Corbis. **p. 444:** (TL) Maigi/Dreamstime.com, (TC) Curt Pickens/iStockphoto, (TR) mrolands/Featurepics. com, (BL) Valentyn75/Dreamstime.com, (BR) Courtesy of Häagen-Dazs. **p. 445:** *Infographic 22.9* (TL) Custom Life Science Images/photographersdirect.com, BL) Custom Life Science Images/photographersdirect.com, (R, from top) Ann

Johansson Photography, Du an Kosti /iStockphoto, Courtesy of Mariano Higes and Raquel Martin-Hernandez, Courtesy of Beeologics and Professor Ilan Sela (Maori, et al., IAPV, a bee-affecting virus associated with Colony Collapse Disorder can be silenced by dsRNA ingestion, *Insect Molecular Biology* (2009) 18(1), 55-60).

Chapter 23

p. 449: WorldFoto/Alamy. **p. 450:** Lauri Patterson/iStockphoto. **p. 451:** Jeff Lepore/Panoramic Images. **p. 454:** Ned Therrien/Visuals Unlimited. **p. 455:** (L, from top) Richard Walters/iStockphoto, malerapaso/iStockphoto, Art33art/Dreamstime.com, (R, from top) Andoni Canel/Photolibrary, gsk/FeaturePics, Tom Bean/Alamy, Krzysztof Odziomek/iStockphoto, Jacka/Dreamstime.com. **p. 459:** *Infographic 23.6* (from left) Jim Zipp/Photo Researchers; Jack Thomas/Alamy; Robert L. Anderson, USDA Forest Service, Bugwood.org.; Evgeny Dubinchuk/Dreamstime.com. **p. 461:** *Infographic 23.8* (L) Goddard/NASA, (R) WorldFoto/Alamy. **p. 465:** *Infographic 23.10* (T) U.S. Dept. of Commerce, NOAA, Earth System Research Laboratory, (BL) Vin Morgan/AFP/Getty Images, (BR) Courtesy of Karin Kirk/Science Education Resource Center at Carleton College. **p. 467:** *Infographic 23.11* (TL) Anna Lubovedskaya/iStockphoto, (TC) Matt Meadows/Photolibrary, (TR) Stringer/epa/Corbis, (BL) olyniteowl/iStockphoto, (BR) Justin Kase Ztwoz/Alamy.

Chapter 24

p. 471: AP Photo/Chicago Department of Environment, Mark Farina. **p. 473:** Artist's impression of Dongtan, designed by Arup, © Arup. **p. 474:** Nir Elias/Reuters/Corbis. **p. 475:** *Infographic 24.2* (T, from left) Capricornis/Dreamstime.com, tank_bmb/FeaturePics, Photong/Dreamstime.com, Dimaberkut/Dreamstime.com, hfng/FeaturePics, (B, from left) Gary Whitton/Dreamstime.com, surpasspro/FeaturePics, Borut Trdina/iStockphotos, stu99/FeaturePics, Tommy Schultz/Dreamstime.com. **p. 476:** *Infographic 24.3* (map) http://commons.wikimedia.org/wiki/File:World_map_of_countries_by_ecological_footprint.png, (graph) © WWF, 2006. Living Plant Report 2006. WWF, Gland, Switzerland. **p. 478:** *Infographic 24.5* (TL) Frank Roeder/Dreamstime.com, (TR) Airwolf01/Dreamstime.com, (CL) Dennis Macdonald/Photolibrary, (CR) Lukasz Koszyk/iStockphoto, (BL) Hanhanpeggy/Dreamstime.com, (BC) Shaun Lowe/iStockphoto, (BR) Dmitro Tolokonov/iStockphoto. **p. 479:** (from left) Enrique Garcia Medina/Archivolatino, © 2009 Tyler Rush Photography and Water Taxi, LLC., Don Nichols/iStockphotos, Chine Nouvelle/Sipa/Newscom. **p. 480:** (T, from left) Rigucci/FeaturePics, (C) Jim West/Alamy, (R) AP Photo/Chicago Department of Environment, Mark Farina, (B, from left) photoneer/FeaturePics, Ashley Cooper/Visuals Unlimited, Frances Roberts/Alamy. **p. 481:** *Infographic 24.6* Artist's impression of Dongtan, designed by Arup, © Arup. **p. 483:** *Infographic 24.7* (from top) Aerial Archives/Alamy, Ron and Patty Thomas Photography/iStockphoto, Petr Nad/iStockphoto, LIU XIN/Xinhua/Landov Pavle Marjanovic/Dreamstime.com, Rob Broek/iStockphoto. **p. 485:** *Infographic 24.9* (TL and BL) Courtesy of John C. Dohrenwend/USGS, (R) Jim Wark/Photolibrary. **p. 486:** *Infographic 24.10* (map) © 2003 World Resources Institute. **p. 487:** *Infographic 24.11* (T) AP Photo/Rick Bowmer, (BL) Chris

Fourie/Dreamstime.com, (BC) http://en.wikipedia.org/wiki/Loggerhead_sea_turtle, (BR) Mark Conlin/Alamy. **p. 488:** *Infographic 24.12* (from top) stuartbur/iStockphoto, rockphoto/FeaturePics, okea /FeaturePics, Don Nichols/iStockphoto, Dimitri Vervitsiotis/Digital Vision/Getty Images, (inset) AP Photo/Damian Dovarganes, Yuliyan Velchev/Dreamstime.com, (inset) Dmitro Tolokonov/iStockphoto.

Chapter 25

p. 493: Galen Rowell/Corbis. **p. 495:** Painted Sky Images/SuperStock. **p. 499:** Courtesy of www.alanarnette.com. **p. 500:** (L) Ken Kamler; (R) Galen Rowell/Corbis. **p. 508:** Devendra M. Singh/AFP/Getty Images. **p. 509:** *Insulation Helps Keep Some Endotherms Warm* (T) Gerrit Vyn, (B) Dennis Scott/Corbis. **p. 510:** (T) jimbo/FeaturePics; (B) Clement Philippe/AgeFotostock. **p. 511:** *Some Organisms Generate Heat from Nonshivering Thermogenesis* (T) AfriPics.com/Alamy, (C) Michael Lynch/Alamy, (BL and BR) Dr. Gladden Willis/Visuals Unlimited/Getty Images.

Chapter 26

p. 515: AP Photo/The Daily Reflector, Rhett Butler. **p. 517:** (T) Pete Saloutos/Corbis; (B) Courtesy of Amy Jo Smith. **p. 520:** AP Photo/The Daily Reflector, Rhett Butler. **p. 523:** *Infographic 26.5* Biophoto Associates/Photo Researchers. **p. 524:** *Infographic 26.6* Courtesy of Allergan Medical. **p. 527:** Science Photo Library/Alamy. **p. 528:** Courtesy of Amy Jo Smith. **p. 529:** *Fungi Digest Food Externally* (L) Jack Bostrack/Visuals Unlimited, (R) Biophoto Associates/Photo Researchers. **p. 530:** *Sea Anemones Have an Incomplete Digestive Tract* http://simple.wikipedia.org/wiki/File: Anemone_monterey_madrabbit.jpg [public domain], via Wikimedia Commons from Wikimedia Commons. **p. 531:** *Plants Have No Digestive Tract* (L) Lessadar/FeaturePics, (R) Nancy Nehring/iStockphoto.

Chapter 27

p. 535: Tyler Sipe/The New York Times/Redux. **p. 535:** Dontcut/Dreamstime.com. **p. 538:** (L) AP Photo/The Lexington Herald-Leader, David Perry; (R) Tyler Sipe/The New York Times/Redux. **p. 541:** *Infographic 27.3* Gary Carlson/Photo Researchers. **p. 544:** Image by Lou Beach. **p. 551:** *Jellyfish Have a Nerve Net* (L) Courtesy of Peter A. V. Anderson, The Whitney Laboratory for Marine Bioscience, (R) Lesya Castillo/Featurepics; *Flatworms Have a Primitive Brain* M. I. (Spike) Walker/Alamy.

Chapter 28

p. 557: Jason Winslow/Splash News/Newscom. **p. 559:** Eyevine/Polaris. **p. 560:** Adel Al-Masry/AFP/Getty Images/Newscom. **p. 563:** *Infographic 28.3* David M. Phillips/Photo Researchers. **p. 568:** (from top) Photomac/FeaturePics.com; (L) Superstock, (R) Jenny Swanson/iStockphoto; (L) Tina Sbrigato/iStockphoto, (R) Moodboard/SuperStock; Daniel Garcia/Dreamstime.com; Imagebroker/Alamy. **p. 569:** *Infographic 28.7* Richard Kessel/Visuals Unlimited. **p. 571:** *Infographic 28.9* UHB Trust/Getty Images. **p. 572:** (T) Jason Winslow/Splash News/Newscom, (B) Jeff Steinberg/Matt Smith, PacificCoastNews/Newscom. **p. 575:** *Birds Reproduce Sexually and Fertilize Eggs Internally* James Urbach/Alamy. **p. 576:** *Fish Reproduce Sexually and Fertilize Eggs Externally* (TL) Mark Conlin/Alamy, (inset) Natural

Visions/Alamy, (TR) Mark Conlin/Alamy, (B) Genevieve Anderson. **p. 577:** *Bacteria Reproduce Asexually* Dennis Kunkel/Visuals Unlimited. **p. 578:** *Yeast Reproduce Both Sexually and Asexually* (T) J. Forsdyke/Gene Cox/Photo Researchers, (C) Dr. George J. Wong, University of Hawaii at Manoa, (B) David Scharf/Photolibrary.

Chapter 29

p. 583: Science Vu/CDC/Visuals Unlimited. **p. 585:** Time Life Pictures/Getty Images. **p. 587:** Courtesy of the National Museum of Health and Medicine, Armed Forces Institute of Pathology, Washington, D.C. **p. 588:** *Infographic 29.2* Row 1: (L), Science Vu/CDC/Visuals Unlimited, (C) Eye of Science/Photo Researchers, (R) George Musil/Getty Images; Row 2: (L) Dennis Kunkel Microscopy/Visuals Unlimited, (C) James Cavallini/ Photo Researchers, (R) Eye of Science/Photo Researchers; Row 3 (L) Dennis Kunkel Microscopy/Visuals Unlimited/Corbis, (C) Masamichi Aikawa, M.D./Phototake/Alamy, (R) Original image by Arturo Gonzalez, CINVESTAV, Mexico. **p. 590:** Karen Kasmauski/National Geographic Stock. **p. 591:** *Infographic 29.4* (T) Carolina Biological Supply Company/Phototake, (BL) Cecil H. Fox/Photo Researchers, (BC) Ocean/Corbis, (BR) Evan Kafka/ Getty Images. **p. 592:** *Infographic 29.5* (TL) CDC/ Dr. Thomas Hooten, (TR) Carolina Biological Supply Company/Visuals Unlimited, (BL) temet/iStockphoto, (BR) Ed Reschke/ Photolibrary. **p. 594:** Michigan Tech Archives. **p. 601:** *Sea Sponges Employ Physical and Chemical Defenses* (L) Dennis Sabo/iStockphoto, (R) Courtesy of Kate Hendry. **p. 602:** *Bacteria Use Enzymes to Defend against Infection* http:// en.wikipedia.org/wiki/File:Phage.jpg. **p. 603:** *Sea Stars Have Innate Cellular Defenses* Paul Kay/Photolibrary; *Fruit Flies Have Multiple Physical, Chemical, and Cellular Defenses* (from top) playTOME, FeaturePics.com, Carl-Johan Zettervall and Dan Hultmark, Umeå University, Willott, E., Tran, H. Q. 2002. Zinc and Manduca sexta hemocyte functions. 9 pp. *Journal of Insect Science,* 2.6. Available online: insectscience.org/2.6. Photo by H. Q. Tran.

Chapter 30

p. 607: Pixtal Images/Photolibrary. **p. 608:** Helene Schmitz. **p. 610:** *Infographic 30.1* (T) Nuridsany et Perennou/Photo Researchers, (BL) Perennou Nuridsany/Photo Researchers, (BR) Krilt/Dreamstime.com. **p. 611:** *Infographic 30.2* Steve Gschmeissner/Photolibrary. **p. 612:** GYRO Photography/ amanaimagesRF/Getty Images. **p. 614:** *Infographic 30.3* (from left) Pixtal Images/Photolibrary, Courtesy of Chris Moody; Courtesy of Frank Dazzo, Center for Microbial Ecology, Michigan State University; Dr. Jeremy Burgess/Photo Researchers. **p. 615:** *Infographic 30.4* (TL and BL) Dr. Jeremy Burgess/SPL/Photo Researchers, (R) EastEggImages/Alamy. **p. 616:** *Infographic 30.5* (L) Jens Stolt/iStockphoto, (inset) sovlanik/iStockphoto, (R) Don Johnston/Alamy, (inset) Natallia Khlapushyna/Dreamstime.com. **p. 617:** Image Plan/Corbis. **p. 618:** *Infographic 30.6* (L) Chas53/FeaturePics, (R) Rolf Nussbaumer Photography/Alamy. **p. 619:** *Infographic 30.7* (TL) Ray Roper/iStockphoto, (TC) cross section, Olga Demchishina/ iStockphoto, (TR) Shaun Pimlott/iStockphoto, (B, from left) Stephen Dalton/Getty Images, Suijo/Dreamstime.com, Mark A. Johnson/Alamy, Juniors Bildarchiv/Alamy, Petra Gurtner/AgeFotostock. **p. 620:** http://en.wikipedia.org/ wiki/File:Upsidedown-tree.JPG. **p. 621:** *Infographic 30.8* (L) Maryann Frazier/Photo Researchers, (TR) Martin Shields/ Alamy, (BR) Tonnywu76/Dreamstime.com. **p. 622:** *Infographic 30.9* (from left) pixhook/iStockphoto, http://en.wikipedia.org/ wiki/File:Red_Apple.jpg, devon/FeaturePics, saiko3p/ iStockphoto. **p. 623:** *Infographic 30.10* (TL) Linda Kloosterhof/ iStockphoto, (TR) Sylvan Wittwer/Visuals Unlimited, (B) Shaun Pimlott/iStockphoto. **p. 624:** *Infographic 30.11* (TL) Cathleen Abers-Kimball/iStockphoto, (TR) hisforhome/iStockphoto, (CL) Photo by Derek Ramsey. http://en.wikipedia.org/wiki/ File:Nicotiana_Tobacco_Plants_1909px.jpg, (CC) Michel Viard/ Photolibrary, (CR) The Photo Works, (BL) Serghei Piletchi/ Dreamstime.com, (BR) Stefanie Timmermann/iStockphoto.

Index

Note: page numbers followed by f indicate figures; those followed by t indicate tables.

immune compromise, staph infection and, 274
immune system, 583-604
    definition of, 586
    development of, 597-598, 597f
    immunological memory, 592-597, 593f, 595f, 596f
    inborn defenses, 587-590, 588f-591f
    influenza, 585-587, 586f, 590-591, 592f
    of other organisms, 601-603
    pathogens and, 588f
    and preventing pandemics, 598-600, 600t
immune-suppressing drugs, for preventing tissue rejection, 259
immunity, 588-589
    adaptive, 589, 589f, 593, 597f
    cell-mediated, 595, 596f
    humoral, 594, 595f
    innate, 589-589f
    partial, 598
immunological memory, 592-597, 593f, 595f, 596f
imprints, fossil, 324
In Darwin's Shadow (Shermer), 297
in vitro fertilization (IVF), 558-559, 562-563, 564f, 570-574, 571f
inborn defenses, 589-590, 588f, 590f, 591f
inbreeding depression, 307, 310
inbreeding, 307, 310-311, 311f
incomplete dominance by, 240, 241f
independent assortment, 209-210, 210f
    Mendel's law of, 227, 228f
independent variable, 7, 8f
individual organism, 416, 416f
induced pluripotent stem cells (IPS cells), 265-267, 266f
infection
    antibiotic revolution and, 275-277, 276f
    MRSA, 272-276, 274f, 281-284
    pathogens as source of, 369, 370f
    staph, 273-275, 274f, 277-278, 278f, 369
    by superbugs, 282-284, 284f
    treatment and prevention of, 282-284, 284f
infertility, 560-561
    causes of, 567-569, 570f
    conception and, 562-563, 562f, 564f
    hormones and, 563-567, 566f, 568t, 569f
infertility, hybrid, 313f
inflammation, 591f
inflammatory response, 590-591
    excessive, 592f
    in flu pandemic of 1918, 591
influenza, 584-585
    action of, 590-591, 592f
    cause of, 585-587
    flu shot for, 600
    gene exchange in, 598, 599f
    mutations in, 598, 599f

ingestion, 518
inhalants, 548t
inheritance. See genetic inheritance
inherited antithrombin deficiency, 148-149
innate immunity, 589-590, 589f
    features of, 591f
    of sea stars, 602-603
Innocence Project, 123-125, 127, 132, 134
inorganic molecule, 25f, 26
    creating organic molecules from, 343-344
insects, 389
insemination. See intrauterine insemination
insulin, 66-67, 70f, 504-505, 506f
integumentary system, 504
interferon, 590
intermediate fossils, 330-331, 331f
internal fertilization, 575-578
interphase, 175, 175f, 179
intersexual person, 231-232
Into Thin Air (Krakauer), 500
intrauterine insemination (IUI), 570, 571f, 572-574
invertebrate iridescent virus (IIV), 441
invertebrates, 329, 385-386
inviability, hybrid, 313f
ion, 31
ionic bond, 31, 32f
ipecac, 172t
IPS cells. See induced pluripotent stem cells
irinotecan, 178, 180f-181f, 181
iron, 74t
Isle Royale
    ecology of, 416, 416f
    monitoring animal health, 421-423, 422f
    moose arrival on, 417-418, 420-421
    population density, 424
    population growth on, 417-421, 420f
    population patterns on, 423
    wolf and moose study, 414-427
    wolves arrival on, 421
isolation, reproductive, 312-313, 313f
isotopes, radioactive, 341-343, 342f-343f
Israeli acute paralysis virus (IAPV), 441, 445f
IUI. See intrauterine insemination
IVF. See in vitro fertilization

Jablonski, Nina, 398-402, 402f, 407-409
Jackson, Michael, 538
Jason, 363-365
Jefferson Westerinen, Julia, 239
Jefferson, Field, 237, 239f
Jefferson, John Weeks, 239f
Jefferson, Randolph, 238
Jefferson, Thomas, 234-240, 238f-239f
jellyfish, 550-551

journalism, health in, 14-15, 15f
junk science, DNA v., 124-126, 126f

Kamrava, Michael, 573
karyotype, 248, 249f
Kasischke, Lou, 503
Kawaoka, Yoshihiro, 590, 598, 600
Kenefick, Robert, 497-498, 500, 503, 505-506, 508
Kenworthy, Jeff, 479
keystone species, 434
Khan, Genghis, 237
kidneys, 507, 507f
"killer bee," 443
kinetic energy, 88-89, 88f-89f
Klebsiella, resistance in, 283
Klevens, Monica, 272
Knoll, Andrew, 22
Krakauer, Jon, 494, 499-501, 503, 508
Kulakowski, Sabina, 122, 124, 126-127, 129, 133-134
Kyoto Protocol, 466

labeling, of food, 65-66
lactic acid, 113-114, 114f
Lactobacillus bulgaricus, 368
Lamarck, Jean-Baptiste, 292, 292f
Lamarckianism, 292-293, 292f
lamellocytes, 603
land, first vertebrates on, 322-326, 328-331, 331f
Langer, Robert, 257
Lannetti, Rick, 275
Lannetti, Ricky, 272, 275, 281, 283
large intestine, 525, 525f
law of independent assortment, 227, 228f
law of segregation, 226f, 227
leaves, changing colors of, 614-616, 616f
Ledger, Heath, 538
Lemba, 237
leukemia, transplantation for, 261
leukocytes, transplantation of, 261
Lewis, Jeffrey, 378, 383, 392
LH. See luteinizing hormone
life, 19-35
    beginning of, 343-344, 372-373, 372f
    characteristics of, 22-24, 23f
    classification of, 348-354, 350f-352f, 354f
    cycle of, 388, 389f
    elements of, 24f-25f, 25-26, 28-29
    geologic timeline of, 345-347, 346f
    Martian bacteria, 27, 30-31, 30f
    molecules of, 24f-25f, 25-26, 28-29
    search for alien, 22-24, 23f
    traces of alien, 26-27
    water's role in, 31-34, 32f-34f
    "weird," 34-35
life history timeline, 344-345
light energy, 91f, 94, 94f
light reactions, in photosynthesis, 96f
light-capturing pigments, 614-616, 616f

lignin, 609
limbic system, 542-543
lipase, 522, 522f
lipid membrane, in beginning of life, 344
lipid, 26, 29
liver, 521-522, 522f
livestock, antibiotic overuse in, 283
lizard, phylogenetic tree of, 352-353, 352f
lizards, 509-510
lobe-finned fish, 329, 332f
logistic growth, 420, 420f
Lohsen, Rebecca, 272, 275-276, 281, 283
London, 473
Loomis, W. Farnsworth, 400-401
Lost City, 359-374
    as clue to beginning of life, 372-373, 372f
    exploration of, 363-364, 364f
    extremophiles in, 369-372, 371f
    methane production and consumption in, 369-372, 372f
    types of life in, 364-367, 365f-367f
    unique microbial life in, 360-362, 363f
lung damage, in CF patients, 211, 212f
luteinizing hormone (LH), 565, 566f
Lyell, Charles, 293, 293f, 297
lymph nodes, 593, 593f
lymphatic system, 505, 593, 594, 593f
lymphocytes, 589, 589f, 593
Lynfield, Ruth, 276, 282-283
lysosome, 51, 53

MacMahon, Brian, 2
macromolecule, 26-27
macronutrient, 61-62, 63f-64f
macrophages, 590, 591f, 592f
magnesium, 74t
Malacosoma disstria, 458, 459f
male
    determination of, 231-234, 232f, 233t
    sex-linked inheritance in, 234, 235f-236f
male reproductive system, 561, 562f
Malthus, Thomas, 293f, 295, 296f
mammals, 350, 386-388
manatees
    adaptation by, 315-316
    Amazonian, 312-314, 314f
    analysis of evolution of, 306-310
    boat collisions with, 300-302
    Florida, 300-316, 303f-305f, 311f, 313f-316f
    gene flow in, 311-312, 311f
    gene pools of, 302-303, 303f
    genetic drift in, 303-306, 304f-305f
    geographic ranges of, 303f
    inbreeding in, 307, 310-311
    natural selection in, 313
    speciation in, 312-315, 313f-316f
    West African, 312-314, 314f
    West Indian, 312-314, 314f
maple syrup production, 451-455, 452f, 454f

Margulis, Lynn, 54, 390
marijuana, 548t
marlins, 510-511
Mars, life on. *See* alien life
Martian bacteria, 27, 30-31, 30f
maternal age, Down syndrome risk and, 246-248, 248f-249f
matter, 24-25, 24f
McClure, R. Dale, 573
McKay, Chris, 31, 35
McKay, David, 20-21
Meade, Harry, 149-150
mechanical isolation, 313f
media, health in, 14-15, 15f
Meier, Walt, 459-460
meiosis, 208-210, 208f-209f, 213, 247, 248f
melanin, skin color and, 399-401, 399f
melanocytes, 399, 399f
Melatonin, 171
membrane. *See* cell membrane
memory cells, 595f-597f, 596-598
memory, drug abuse and, 549f
Mendel, Johann Gregor, 221-228
    concepts of inheritance before, 223, 224f
    pea plant experiments of, 224-227, 225f-226f, 228f
Mendel's law of independent assortment, 227, 228f
Mendel's law of segregation, 226f, 227
menstrual cycle, 566f
menstruation, 565
messenger RNA (mRNA), 152-155, 153f-155f
metabolism, 23, 66-67, 67f
metaphase, 179
metastasis, 177
meteorite ALH84001, 20-22, 26-27, 31-32, 34
methamphetamine, 547, 548t, 549, 549f
methane
    from cattle, 466, 467f
    in hydrothermal vents, 369-372, 372f
methanogenesis, 369-372, 372f
methanogens, 369-370, 371f
Methanopyrus kandleri, 370, 371f
methicillin-resistant *Staphylococcus aureus* (MRSA), 272-275, 274f, 284
    in communities, 281-283
microbial life
    collection and processing of, 363-364, 364f
    in hydrothermal vents, 360-362, 363f
    types of, 364-367, 365f-367f
micronutrient, 72, 73f, 74t-75t
milk, transgenic, 144-145, 155-157
Miller, Stanley, 343-344
mineral, 70-72, 71f, 74t
mineralization, 324
mitochondria, 50-51, 53-54

mitochondrial DNA (mtDNA), 402-404, 404f
    for assessing genetic diversity, 314-315, 315f
mitochondrial Eve, 402-404, 403f
mitosis, 175f-176f, 176, 179
modifier gene, 206, 215-216
Moffitt, Terrie, 245-246, 247f
molds, 389f
    fossil, 324
molecules, of life, 24f-25f, 25-26, 28-29
mollusks, 384f, 385
Monera, 367
Money, John, 231
monomer, 26-27
monosaccharide, 26-28, 68-69
Moore, Kent, 503
moose, 414-427
    arrival on Isle Royale, 417-418, 420-421
    distribution patterns of, 418f
    health monitoring of, 423
    population cycles of, 421, 421f
    population growth of, 417-423, 420f, 422f
    population sampling of, 417
    warming climate and, 425, 426f
morbid obesity, 516-518, 523
morphine, 624
Morrison, Nina, 134
Morse, Burr, 450
moths, niche of, 442, 442f
motor neurons, 539
motor skills, drug abuse and, 549f
Mount Everest 1996 disaster, 494-495, 497, 499-501, 503-506, 508
mouth, 519f
mRNA. *See* messenger RNA
MRSA. *See* methicillin-resistant *Staphylococcus aureus*
mtDNA. *See* mitochondrial DNA
mucus, in CF patients, 211, 212f
Mullis, Kary, 127-128
multifactorial inheritance, 242-246, 246f-247f
multiple births, 558-559, 569-572, 571f
multiple sclerosis, 541
multipotent cells, 262, 263f
muscle cells, differential gene expression in, 261
muscular system, 504
mushrooms, 388, 389f
mutagen, 193-194, 194f
mutation, of viruses, 597-598, 599f
mutation. *See* genetic mutation
mutualism, 439, 440f
mycelium, 388
myelin, 540-541, 541f
Myers, Richard, 404

Nabi Biopharmaceuticals, 550
national park. *See* Olympic National Park
natural gas, 478f